We Will Show Them!

Essays in Honour of Dov Gabbay on his 60th Birthday

Volume 2

We Will Show Them!

Essays in Honour of Dov Gabbay on his 60th Birthday

Volume 2

edited by

Sergei Artemov, Howard Barringer,

Artur d'Avila Garcez, Luis C. Lamb and John Woods

ISBN 1-904987-12-5
Published by College Publications
Scientific Directors: Dov Gabbay, Vincent F. Hendricks and John Symons
Managing Director: Jane Spurr
Department of Computer Science
King's College London
Strand, London WC2R 2LS, UK

http://www.collegepublications.co.uk

Cover design by Lydia Rivlin
Printed by Lightning Source, Milton Keynes, UK

Photograph Gallery

Dov Gabbay, today — actually 21st July 2005!

This picture was taken during a Logic Conference at Bar Ilan University in 1979 or 1980. You can see Dov, who is 5th from the right.

Dov, in his office at Imperial College in 1991.

This is a picture of Dov, in his office with his "muse", 7of9, and colleagues Odinaldo Rodrigues, Artur Garcez, Nicola Olivetti and Luis Lamb.

"Consistency is futile!"

CONTENTS

Preface

Visitors to Dov Gabbay's suite of offices at King's can hardly miss noticing the upper right quadrant of the white-board, directly opposite the room's sofa. In the board's other sectors, one sees the usual jottings of a logician at work, the detritus of definitions and proofs about modal and temporal logic, or labelled deduction and fibred semantics, or agenda relevance and the *GW*-schema for abduction, or neural networks and logic for learning. But the entries in the upper right aren't logic at all. In the centre is a single short sentence in English. In the rest of the space is a sprawl of different translations of it in German, Portuguese, Greek, Russian, Swedish, Hebrew, Dutch, Turkish, Latin, Italian and Hungarian and many more. In time, this babel grew to overflowing. To relieve the pressure, it was transferred by Lydia Rivlin, to an attractively backgrounded, large wooden panel affixed to the wall facing the entrance to Dov's rooms. Everyone knows what a rabbits' warren King's is, and it will come as no surprise that while the multi-lingual panel can hardly be seen by the entering visitor, it is squarely in the exitor's field of vision. What does the departing visitor see? In no fewer than 25 of the world's languages, one sees the sentence, "We will show them", having been contributed, one by one, by Dov's visitors streaming into King's from the world over.

For the most part, visitors react to "We will show them" with amusement. It is, they think, a funny thing for someone as well-known and esteemed as Dov Gabbay to be saying. For it is, is it not, a defiant saying, expressing a determination to persist until the outer world takes some notice and accords one a measure of respect? Clearly, then, an ironic joke? It is a joke of sorts, but not that one. It light-heartedly exploits an ambiguity, but it is not ironic.

When Dov was a first year student at the Hebrew University in 1963, he enrolled in the course in Mathematics and Physics. Before the year was out, he had lost his affection for physics, and his heart to pure mathematics and logic. As Dov would later say, "Physics was a speculative mess, scarcely different from romantic fiction! Only logic could deliver the certitude demanded by genuine knowledge." When he made that decision Dov was 18 years old. How could he know that logic was in process of the largest transformation since taking the mathematical turn in the middle and late nineteenth century? How could he know that he himself would be one of the driving forces of this change, and of how far it would take him from the austere dogmatism of his youth? Dov has called this transition "the practical turn in logic". Whereas the earlier, mathematical, turn had concentrated on the analysis of logically interesting properties in a highly decontextual way, the practical turn would reverse that abstraction. It would acknowledge the influence of time and place, of belief and belief change, of relevance and analogy, of consequence relations weaker than strict deduction, and the central place of agents. It would be a transformation that sought to recover something of logic's ancient and original remit, which was the exposure of the structure of human reasoning as transacted under the conditions of real-life. It is forty-two years since Dov repudiated physics for its speculativeness and its messiness. In the intervening years, he has learned a marked tolerance for these traits, and has only very recently observed,

> I believe that the computational point of view of logic, arising from computer science needs, which includes resource considerations and various other mechanisms such as abduction, revisions, actions, change, etc., will become increasingly predominant in logic. This point of view will eventually be extensively applied in traditional areas like analytic philosophy, law, economics, theology, language, and political theory among others. Logic will continue to evolve as a result of these applications to become an independent discipline akin to mathematics and physics, chemistry and biology. Mathematical logic will no longer be an important part of logic outside mathematics departments.

Dov is no enemy of mainstream mathematical logic, which is where he himself has done some of his most important work. No one admires more than he the foundational achievements of set theory, model theory, proof theory and recursion theory, not to overlook work of like importance in theoretical computer science and logic programming. Still, not every mainstream logician is happy with the turn toward the practical. It is a turn toward a messiness and a speculativeness that makes physics look rather neat and tidy. Some see the new logic as a return to the discredited Laws of Thought tradition in logic, and to a widely detested psychologism. Perhaps the shortest way of capturing the present cut of Dov Gabbay's jib as a logician is this. When Dov was a boy, he thought that logic was a purely abstract science. He now thinks, as did Aristotle, that logic is a humanities subject.

One sees, then, that there is in this view something indeed to be defiant about. Notwithstanding logic's inexorable turning toward the contextual, there are legions of practitioners of the mainstream, including some of its most eminent, for whom the last thing that a proper logic could be is one of the humanities. Dov has acquired more of his large reputation as an insider rather than an outsider. His insider reputation is wholly secure. His outsider reputation is a work in progress. Accordingly, a defiant "We will show them."

"We will show them" has another meaning that captures Dov's intellectual personality, and is not in the least defiant. People who know him will be aware that one good way to send Dov Gabbay into a quite undisguised coma is to discuss a problem rather than solve it. Colleagues who want to hold Dov's attention had better show him something, and had better not talk about showing it. Dov's hankering for demonstration wins him the status of honorary Missourian. Perhaps "hankering" is not the right word for it; it is better to think of it as a drive. Like most academics of high importance, Dov maintains a large network of contacts around the world. It would be interesting to know how many of these contacts arose from a "cold-call" from Dov, asking to be shown something. One sees this same drive at conferences. Dov is incapable of small talk. He isn't interested in tanking up at the bar or visiting baroque churches. Dov wants people to show him things. And, of course, the favour is returned as often as it is asked for. "Here, let me show you" could well be Dov's epitaph. Dov never lectures with notes. He has no affection for powerpoint presentations. These constrain spontaneity. But more importantly, they are not the best ways of showing things. The best way of showing things is with markers on a whiteboard or on overhead transparencies. The best way of showing something to others is to show it to oneself in their presence. So, again, "We will show them."

His evolving fidelity to logic's humanistic character is but one of the respects in which Dov Gabbay stands apart from the mainstream. Another is his extraordinary range. The chapters of this large book run to 60 and reflect several times that number of issues and problems. It is very nearly the literal truth to say that there isn't a single matter covered here that lies beyond Dov's own demonstrated reach. There is an amusing (and true) story about Dov as a youthful associate professor at Stanford in the 1970s. Georg Kreisel tapped on Dov's door. Would he please come to Kreisel's office to take a call from Gödel. Yes, of course. As it happens, Gödel wanted to say his piece in German; so Kreisel stayed on the phone and translated for Dov. What, Gödel demanded to know, did Dr. Gabbay think he was doing? Was it modal logic, was it temporal logic, was it intuitionistic logic, was it tense logic, or was it the formal semantics of natural languages, or was it statistics, or what? Dov allowed that it was all these, and more. Gödel was fit to be tied, fearful that an accursed dilettantism would soon strike at the brilliant young man's emerging profile. Dov did what he always does when under fire. He *negotiated.* After a good deal of to-ing and fro-ing, mediated by a growingly bemused Kreisel, Dov promised Gödel that he would cut back his involvement with statistics.

The mainstream's distrust of range is one of its defining features. Crossing the line of a sub-discipline invites rejection from both sides. Crossing over into a wholly different discipline risks opprobrium from both sides. This hostility to cross-disciplinary work is deeply entrenched. There are exceptions, of course. (One thinks of Weiner and von Neumann, to name just two.) But there are costs. To keep the mocking judgement of amateurism at bay, the discipline-crosser is expected to meet two targets. He must produce in accordance with each set of local standards. And he should develop results that couldn't (or most likely wouldn't) have been produced in either of the single disciplines alone. Part of what enables Dov to meet these expectations is that he is endlessly learning new things. Again, he never stops getting people to show him things. The things he most wants to be shown are the most important, the most central things — the essence of them. Dov is also a veritable pan-logician. Dov thinks that anything that interests him has a logic. Since his interests range all over the place, this helps explain why he is so receptive to a large pluralism in logic. It also helps explain why his multi-disciplinary efforts have been met with such success. Dov is an inveterate builder of formal models. Modelling comes to him as naturally as breathing. Dov will model anything that strikes his fancy. He will do so at the drop of a hat, especially if he has an audience. ("Here, let me show you"). What makes this so is that Dov is dominantly a visual thinker. He sees structure where others see nothing. It is also this architectural sense that mitigates Dov's drift toward the psychologically real. For all its messiness, Dov does not think that a psychologically real account of reasoning is unmodellable. This is important. While Dov wholly embraces the research targets of a practical logic of cognitive systems, his *methods* are still largely those of the logical mainstream. Any fool can make a bad model of just about anything. It takes something special to model the real in ways that subdue imprecision yet preserve complexity. Doubtless, it is Dov's successes in this regard that maintains him as a fully paid-up member of the mathematical mainstream. This is exactly as it should be for someone who thinks like Dov. Not only does the turn towards the practical re-orient logic's subject matter, it also re-sets the limits of a mathematical methodology. Thus logic's recovery of its humanistic focus is, in Dov Gabbay's hands, no cause

to abandon the applicable insights of mathematical re-expression.

It is less a fact than an expository convenience that logic's turn toward the practical is a discrete and recent event. It is, in fact, not a single event and it is not all that recent. During the period in which first order classical logic was achieving its hegemony, non-classical rivals were also gaining strength. Not least of these were intuitionistic logic and the modal logics of strict implication, each of which arose in the early years of the century just past, and each of which had an applicational cast. Intuitionistic logic sought better to serve a philosophically correct appreciation of mathematical reasoning as actually practised, just as modal logic aimed for a more realistic account of entailment. In their turn, relevant, epistemic, deontic, temporal, dynamic and linear logics made for a closer tie between logic's formalisms and the structure of facts in the real world. Similar engagements with similar results arose from developments in situation and belief-change logics, and, from computer science, the logics of default, defeasible and non-monotonic reasoning, as well as logic programming. The pace of change was abated by the research programmes of cognitive science, argumentation theory and informal logic. There is these days more logic than you can shake a stick at. It is clearly an *embarras de richesse.* It is not an altogether good state for logic to be in. It threatens with prospects that anything goes. Anything-goes logic is Cole Porter logic. It embodies the conceit of pluralism without bounds. One of the attractions of the conservative mainstream is its insistence that *hardly anything* goes. It should be easy for all to see that Cole Porter logic is not a live option, that unconstrained pluralism will test the mettle of intellectual integrity as sorely as the paradoxical proofs of Presocratic conjurers. Somewhere lines have to be drawn. Limits must be placed on the theorists' freedom to say that human reasoning is anything he sees fit to stick into one or other of his models. Models require mature data as inputs, and the distortions to which they inevitably give rise must be reined in. The task of pulling together the best of logic's present pluralism into comprehensive and stable theories of the empirically real and normatively sound still lies before us, as unperformed as it is urgent. Logicism was the last great synthesis in logic (and a failed one). Logic is long overdue for another, and we can only hope that it will not fail. Perhaps, in the end, it is the new synthesis that best animates Dov's motto.

"We will show them": How likely is this to be true? Let us say simply that its prospects are good with Dov Gabbay at the helm, and significantly better than were he not.

The Editors, August 2005

Acknowledgements

We are grateful to Jane Spurr for her very effective assistance during all the phases of this project. Without her support, we would not have been able to meet the tight schedule — starting with invitations being sent out to authors in March 2005 — for the edition, production, and publication of this book by October 2005.

We would also like to thank the authors for their overwhelming response to our invitation, which has forced us to have a two-volume book, and for producing such high-quality, interesting contributions in such a short period of time.

We also thank Lydia Rivlin for producing an eye-catching, unique cover for the book, and the International Federation for Computational Logic (IF-CoLog) for their financial support, which will promote the free distribution of the book to key libraries and universities across the world.

The Editors, October 2005

List of Contributors

Samson Abramsky. *Oxford University Computing Laboratory, UK*
Email: samson.abramsky@comlab.ox.ac.uk

Name. *City University of New York, USA.*
Email: sartemov@gc.cuny.edu

David Ahn. *University of Amsterdam, The Netherlands.*
Email: ahn@science.uva.nl

Sisay Fissaha Adafre. *University of Amsterdam, The Netherlands.*
Email: sfissaha@science.uva.nl

Atocha Aliseda. *Universidad Nacional Autónoma de México*
Email: atocha@filosoficas.unam.mx

Amilhood Amir. *Bar-Ilan University, Israel, and Georgia Instiute of Technology, USA.*
Email: amir@cs.biu.ac.il

Name. Carlos Areces *INRIA Lorraine, France.*
Email: Carlos.Areces@loria.fr

Wouter van Atteveldt. *Vrije Universtiteit Amsterdam, The Netherlands.*
Email: wh.van.attevedlt@fs2.vu.nl

Arnon Avron. *Tel Aviv University, Israel.*
Email: aa@math.tau.ac.il

Matthias Baaz. *Technische Universität, Austria.*
Email: baaz@logic.at

Sebastian Bader. *Dresden University of Technology, Germany.*
Email: sebastian.bader@inf.tu-dresden.de

Howard Barringer. *Univeristy of Manchester, UK.*
Email: Howard.Barringer@manchester.ac.uk

Johan van Benthem. *Amsterdam, The Netherlands, and Stanford, USA.*
Email: johan@science.uva.nl

Patrick Blackburn. *INRIA Lorraine, France.*
Email: Patrick.Blackburn@loria.fr

Alexander Bochman. *Holon Academic Institute of Technology, Israel.*
Email: bochmana@hait.ac.il

Krysia Broda. *Imperial College London, UK*
Email: kb@doc.ic.ac.uk

Peter D. Bruza. *University of Queensland, Australia.*
Email: bruza@dstc.edu.au

Richard J. Cole. *University of Queensland, Australia.*
Email: rcole@itee.uq.edu.au

Carlos Caleiro. *Technical University of Lisbon, Portugal*
Email: ccal@math.ist.utl.pt

Walter Carnielli. *State University of Campinas, Brazil.*
Email: carniell@cle.unicamp.br

Marcelo E. Coniglio. *State University of Campinas, Brazil*
Email: coniglio@cle.unicamp.br

Ariel Cohen. *Technion - Israel Institute of Technology, Israel.*
Email: arikc@bgumail.bgu.ac.il

Marcello D'Agostino. *Dipartimento di Scienze Umane, Università di Ferrara, Italy.*
Email: dgm@unife.it

Artur S. d'Avila Garcez. *City University, London, UK.*
Email: aag@soi.city.ac.uk

Anuj Dawar. *University of Cambridge Computer Laboratory, UK.*
Email: Anuj.Dawar@cl.cam.ac.uk

Jürgen Dix. *Clausthal University of Technology, Germany.*
Email: dix@informatik.tu-clausthal.de

Kosta Došen. *Mathematical Institute, SANU, Belgrade, Serbia.*
Email: kosta@turing.mi.sanu.ac.yu

Luis Fariñas del Cerro. *Université Paul Sabatier, Toulouse, France.*
Email: farinas@irit.fr

Paolo Ferraris. *University of Texas, USA.*
Email: otto@cs.utexas.edu

Melvin Fitting. *City University of New York, USA.*
Email: melvin.fitting@lehman.cuny.edu

Marcelo Finger. *Universidade de São Paulo, Brazil.*
Email: mfinger@ime.usp.br

Chris Fox. *University of Essex, UK.*
Email: foxcj@essex.ac.uk

Michael Gabbay. *King's College London, UK.*
Email: michael.gabbay@kcl.ac.uk

Murdoch Gabbay. *King's College London, UK.*
Email: jamie@dcs.kcl.ac.uk

Olivier Gasquet. *Université Paul Sabatier, Toulouse, France.*
Email: gasquet@irit.fr

Joseph Goguen. *University of California at San Diego, USA.*
Email: goguen@cs.ucsd.edu

John Grant. *Towson University, USA.*
Email: jgrant@towson.edu

Andreas Herzig. *Université Paul Sabatier, Toulouse, France.*
Email: herzig@irit.fr

Pascal Hitzler. *University of Karlsruhe, Germany.*
Email: phitzler@aifb.uni-karlsruhe.de

Wilfrid Hodges. *Queen Mary, University of London, UK.*
Email: hodges@qml.ac.uk

Ian Hodkinson. *Imperial College London, UK.*
Email: imh@doc.ic.ac.uk

Rosalie Iemhoff. *Technische Universität, Austria.*
Email: iemhoff@logic.at

Dale Jacquette. *The Pennsylvania State University, USA.*
Email: dlj4@email.psu.edu

Michael Kaminski. *Technion - Israel Institute of Technology, Israel.*
Email: kaminski@cs.technion.ac.il

Ruth Kempson. *King's College, London, UK.*
Email: ruth.kempson@kcl.ac.uk

Sarit Kraus. *University of Maryland, USA and Bar Ilan University, Israel.*
Email: sarit@macs.biu.ac.il

Jan Willem Klop. *Vrije Universiteit, The Netherlands*
Email: jwk@cs.vu.nl

Agi Kurucz. *King's College London, UK.*
Email: kuag@dcs.kcl.ac.uk

Ugur Kuter. *University of Maryland, USA.*
Email: ukuter@cs.umd.edu

Luís C. Lamb. *Federal University of Rio Grande do Sul (UFRGS), Brazil.*
Email: luislamb@acm.org

Shalom Lappin. *King's College London, UK.*
Email: shalom.lappin@kcl.ac.uk

Daniel Leivant. *Indiana University, USA.*
Email: leivant@cs.indiana.edu

Vladimir Lifschitz. *University of Texas, USA.*
Email: vl@cs.utexas.edu

Kai Lin. *University of California at San Diego, USA.*
Email: klin@cs.ucsd.edu

Lorenzo Magnani. *University of Pavia, Italy and Sun Yat-sen University, China.*
Email: lmagnani@unipv.it

David Makinson. *King's College London, UK.*
Email: david.makinson@kcl.ac.uk

Larisa Maksimova. *Siberian Branch of Russian Acad. Sci., Russia*
Email: lmaksi@math.nsc.ru

Erica Melis. *German Research Institute for Artificial Intelligence (DFKI), Germany.*
Email: melis@dfki.de

George Metcalfe. *Technische Universität, Austria.*
Email: metcalfe@logic.at

Alice G. B. ter Meulen. *Univesity of Groningen, The Netherlands.*
Email: atm@let.rug.nl

Wilfried Meyer-Viol. *King's College, London, UK.*
Email: meyervio@dcs.kcl.ac.uk

Johann A. Makowsky. *Technion - Israel Institute of Technology, Israel.*
Email: janos@CS.Technion.AC.IL

Ben Moszkowski. *De Montfort University, UK.*
Email: benm@dmu.ac.uk

Dana Nau. *University of Maryland, USA.*
Email: nau@cs.umd.edu

Rolf Nossum. *University of Kristiansand, Norway.*
Email: Rolf.Nossum@hia.no

Hans-Jürgen Ohlbach. *Universität München, Germany*
Email: ohlbach@lmu.de

Anjolina G. de Oliveira. *Universidade Federal de Pernambuco (UFPE), Brazil.*
Email: ago@di.ufpe.br

Nicola Olivetti. *Unversity of Torino, Italy.*
Email: olivetti@di.unito.it

Don Perlis. *University of Maryland, USA.*
Email: perlis@cs.umd.edu

Zoran Petrić. *Mathematical Institute, SANU, Belgrade, Serbia.*
Email: zpetric@mi.sanu.ac.yu

Gabriella Pigozzi. *King's College London, UK.*
Email: pigozzi@kcl.ac.uk

Amir Pnueli. *Weizmann Institute of Science, Israel and Courant Institute of Mathematical Sciences New York University, USA.*
Email: amir@cs.nyu.edu; amir@wisdom.weizmann.ac.il

Ruy J. G. B. de Queiroz. *Universidade Federal de Pernambuco (UFPE), Brazil.*
Email: ruy@di.ufpe.br

Mark Reynolds. *University of Western Australia, Australia.*
Email: mark@csse.uwa.edu.au

Maarten de Rijke. *University of Amsterdam, The Netherlands.*
Email: mdr@wins.uva.nl

Odinaldo Rodrigues. *King's College London, UK.*
Email: odinaldo.rodrigues@kcl.ac.uk

Alessandra Russo. *Imperial College London, UK*
Email: ar3@doc.ic.ac.uk

Vladimir V. Rybakov. *Manchester Mentropolitan University, UK.*
Email: v.rybakov@mmu.ac.uk

Mohamad Sahade. *Université Paul Sabatier, Toulouse, France.*
Email: sahade@irit.fr

Stefan Schlobach. *Vrije Universtiteit Amsterdam, The Netherlands.*
Email: schlobac@few.vu.nl

David E. Rydeheard. *University of Manchester, UK.*
Email: david@cs.man.ac.uk

Amílcar Sernadas. *Technical University of Lisbon, Portugal*
Email: acs@math.ist.utl.pt

Cristina Sernadas. *Technical University of Lisbon, Portugal*
Email: css@math.ist.utl.pt

Valentin Shehtman. *Institute for Information Transmission Problems, Russia*
Email: shehtman@lpcs.math.msu.su

Jörg Siekmann. *German Research Institute for Artificial Intelligence (DFKI), Germany.*
Email: siekmann@dfki.uni-sb.de

Patrick Suppes. *Stanford University, USA.*
Email: suppes@csli.stanford.edu

Roel de Vrijer. *Vrije Universiteit, The Netherlands*
Email: rdv@cs.vu.nl

Jon Williamson. *University of Kent, UK.*
Email: j.williamson@kent.ac.uk

Frank Wolter. *University of Liverpool, UK.*
Email: frank@csc.liv.ac.uk

John Woods. *University of Britsh Columbia, Canada, and King's College London, UK.*
Email: jhwoods@interchange.ubc.ca
Michael Zakharyaschev. *King's College London, UK.*
Email: mz@dcs.kcl.ac.uk

Specifying, Programming and Verifying with Equational Logic

JOSEPH GOGUEN AND KAI LIN

Dedication
This paper is dedicated, with respect and affection, to Prof. Dov Gabbay, whose vision of what logic could and should be has been an inspiration for many of us to pursue our dreams.

1 Introduction

Programming is difficult, as shown by the fact that debugging a program usually takes more time than creating it; moreover, the difficulty of debugging increases non-linearly with program size. One reason for such phenomena is the astonishing complexity and subtlety of the semantics of most widely used programming languages, due mainly to the desire for high efficiency on conventional processors. But rapid increases in the power and flexibility of hardware, and in the need for greater reliability and security in applications, suggest that it may be valuable to consider alternative approaches, based on higher level languages with much simpler semantics, despite the undoubted inertia of tradition, and the difficulty of learning new languages and new paradigms.

This paper focuses on the OBJ family of languages, which have semantics based on various extensions of (first order) equational logic. The OBJ languages are **logical programming languages**, in which programs are theories, and computation is deduction, which makes it possible to do specification, programming and verification in a unified framework. This paper is mainly intended to introduce and motivate the material that it covers, rather than to provide a thorough mathematical exposition. Consequently, there are many references and several examples, but all proofs and many technical details are omitted.

Equational logic of course cannot do everything, but when it is applicable, it has some significant advantages resulting from its simplicity, including ease of learning and use, and the decidability of problems that are intractable in more complex logics; moreover, algorithms for these problems are often quite efficient, e.g., term rewriting, unification, narrowing, and

Knuth-Bendix completion. Unsorted unconditional equational logic, which goes back to Whitehead [1998], and later Birkhoff [1935], is not expressive enough for most computer science applications, and has therefore been extended in many ways. The simplest extensions are to many sorted and conditional equational logic [Goguen and Meseguer, 1985a], which provide additional expressive power and support strong type checking. Two further extensions, to overloaded operations and subtypes, yield **order sorted equational logic** [Goguen and Meseguer, 1985b], which supports partial operations, exception handling, type conversion, multiple representations, and more [Meseguer and Goguen, 1993]; the core members of the OBJ family are all based on this, including OBJ2, OBJ3 [Goguen *et al.*, 2000], CafeOBJ [Diaconescu and Futatsugi, 1998a], and most recently, BOBJ [Goguen *et al.*, 2003; Goguen and Lin, 2003] which extends OBJ3 with hidden equational logic and higher order modules.

Two other ways to extend equational logic are partial algebra [Cerioli *et al.*, 1998], which is used in CASL [Mosses, 2004], and membership equational logic [Meseguer, 1997], which is used in Maude [Clavel *et al.*, 1996; Clavel *et al.*, 2001]. A more radical extension introduces behavioral abstraction, which permits specifying systems with states, infinite data structures (such as streams), non-determinism, and concurrency, and which also allows verifying much more sophisticated properties of systems; our version of this is called **hidden algebra** [Goguen, 1989; Goguen and Malcolm, 1997; Goguen and Roşu, 1999; Roşu, 2000]. Behavioral logic is a diverse research area, including not just hidden algebra, but also coherent hidden algebra od Diaconescu [Diaconescu and Futatsugi, 1998b; Diaconescu and Futatsugi, 1998a], the observational logic of Bidoit and Hennicker [1999; 2003], and coalgebra, e.g., see the survey [Jacobs and Rutten, 1997], and for recent results, [Fiadeiro *et al.*, 2005]. These approaches fall into two broad categories, depending on whether or not a fixed data algebra is assumed for all models. [Goguen *et al.*, 2000a; Roşu and Goguen, 2001]; further details are given later.

Perhaps the two most important innovations are the module system which appears in all current OBJ family members, and the C4RW coinduction algorithm of BOBJ [Goguen *et al.*, 2000a; Goguen *et al.*, 2000b; Goguen *et al.*, 2003; Goguen and Lin, 2003]. OBJ3 and BOBJ have higher order module systems, though BOBJ goes further, and has also extended its coinduction algorithm to handle mutual coinduction; therefore this paper is somewhat focused on these two recent contributions, for which it provides an introduction and motivation in the next two subsections. It is noteworthy that together they give a useful platform for specifying and verifying complex systems, such as communication protocols. Section 2.5 discusses logical programming.

1.1 Modularization

Modularization controls the complexity of large systems by composing them from parts; this eases both initial construction and later modification by making large grain structure explicit, and it also considerably facilitates reuse. Early designs for OBJ [Goguen, 1977] called for a module system like that of the Clear specification language [Burstall and Goguen, 1977]. This approach was later improved and formalized as **parameterized programming**, which provides parameterized modules and (so called) views among its "first class citizens," where the latter say how to fit the syntax of a formal parameter to an actual parameter, including defaults when there is only one obvious choice; moreover, views can be parameterized, and **module expressions** compose modules, and in particular, can describe software architectures. Parameterized programming was first fully implemented in OBJ3 [Goguen *et al.*, 2000], following partial implementations in earlier versions of OBJ, and it appears in all current members of the OBJ family, including CafeOBJ [Diaconescu and Futatsugi, 1998a], Maude [Clavel *et al.*, 2001], the European languages CASL [Mosses, 2004] and ACTTWO [Ehrig and Mahr, 1990], and of course BOBJ. Other languages that have been influenced by parameterized programming include Ada, ML, C++, and Modula, none of which has views, so that the syntax of an actual parameter module must contain the syntax of its formal parameter. Ada does not even allow instantiated parameterized modules to be used as actual parameters. ML [Milner *et al.*, 1997] comes closest to fully implementing parameterized programming, but it lacks views, and in particular, it lacks the convenience of default views.

The Clear module system [Burstall and Goguen, 1977] has semantics based on the category of theories, where views are given by theory morphisms and module composition is given by colimit, inspired by an earlier category theoretic approach to general systems [Goguen, 1971]. This semantics also applies to the OBJ family, and is here extended to higher order parameterized programming in Section 3.3, and illustrated with an inductive proof scheme.

It should not be thought that parameterized programming is limited to functional languages, let alone to algebraic specification languages; it can be implemented for almost any language, e.g., using techniques suggested for the LIL [Goguen, 1986] extension of Ada, and implemented in Lileanna [Tracz, 1993], which involve translating to intermediate compiled code (Dianna in the case of Ada) and then applying the compiler's backend optimization. Similar techniques can be used for the higher order case.

1.2 Behavioral Specification and Verification

The basic idea of **behavioral abstraction** is that an equation (or other axiom) need not actually be satisfied by all interpretations into a model of its free (or universally quantified) variables, but need only *appear* to do so with respect to a given set of "experiments," which consist of applying a sequence of state changing operations, and then one state observing operation. This idea, first suggested by Horst Reichel [1981; 1995], underlies the hidden algebra approach to behavioral specification and verification that is implemented in BOBJ and CafeOBJ. The weaker notion of satisfaction is important because many clever implementations used in practice only satisfy their specifications in this sense.

A **behavioral specification** consists of a signature Σ in which some operations are declared behavioral (as defined in Section 2.4), a set of equations, some of which are behavioral, and a subsignature of Σ, called a **cobasis**, the operations in which can be used in experiments. Then **behavioral verification** attempts to determine if a given equation is behaviorally satisfied by all models that behaviorally satisfy the given specification. In contrast to the corresponding problem for ordinary satisfaction, there cannot exist any finite complete set of rules of deduction for behavioral satisfaction [Buss and Roşu, 2000]. Nevertheless, there is an algorithm that is surprisingly useful in practice, as we will see. Moving from ordinary to behavioral logics allows much more natural treatment of systems with internal states, and also supports concurrency and non-determinism, which are essential for modern network based systems. Moreover, since the BOBJ implementation builds on order sorted algebra, classes, subclasses (inheritance), overloading, exceptions, and abstract data types, are also supported, making this approach suitable for non-trivial software engineering applications.

Although simulated computation with behavioral specifications is not feasible, specifying and verifying high level designs seems more useful in practice, because the bugs that are most difficult to find and correct typically arise at the design level; in fact, we argue that verification is a proper generalization of programming for the specification level. CafeOBJ [Diaconescu and Futatsugi, 1998a] and Spike [Berregeb *et al.*, 1998] also implement behavioral specification and verification, but circular coinductive rewriting is only implemented in BOBJ. Behavioral equivalence generalizes the notion of bisimilarity used in process algebra, for which there is a very large literature, including proof methods that are special cases of coinduction; we just mention Milner's CCS [Milner, 1980] and [Park, 1980], where the notion of bisimilarity seems to have originated. Hidden algebra generalizes process algebra and transition system to include non-monadic parameterized methods and attributes, which can sometimes dramatically simplify verification.

2 Basic OBJ Features

OBJ began around 1974 as a notation for algebraic specification, and soon after was implemented in the OBJ0 system based on term rewriting; the first publication on OBJ [Goguen, 1977] included early versions of parameterized programming and order sorted algebra. Subsequent implementations include OBJT [Goguen and Tardo, 1979], OBJ1 [Goguen and Meseguer, 1982], OBJ2 [Futatsugi *et al.*, 1985], OBJ3 [Goguen *et al.*, 2000], and most recently, BOBJ [Goguen *et al.*, 2003; Goguen and Lin, 2003], which is highly portable since implemented in Java, and is used for the examples in this paper. CafeOBJ [Diaconescu and Futatsugi, 1998a] and Maude [Clavel *et al.*, 2001] also share many basic design features. Readers already familiar with OBJ or equational programming may wish to skip Sections 2.1 to 2.3, but Section 2.4 introduces behavioral semantics, and Section 2.5 contains a new theory of logical programming.

2.1 Loose Semantics

The simplest semantics for used for OBJ is loose semantics, in which a theory specifies the variety of all algebras that satisfy its axioms. We begin with some basic concepts, notation and terminology from. Given a set S, an S-**sorted set** A is a family of sets A_s, one for each $s \in S$. The elements of S are called **sorts** and the notation $\{A_s \mid s \in S\}$ is used. A **signature** Σ is an $(S^* \times S)$-sorted set $\{\Sigma_{w,s} \mid \langle w,s\rangle \in S\}$. The elements of $\Sigma_{w,s}$ are called operation (or function) symbols of **arity** w, **sort** s, and **type** $\langle w,s\rangle$; in particular, $\sigma \in \Sigma_{[],s}$ is a constant symbol ($[]$ denotes the empty string). If σ has the type $\langle w,s\rangle$, we write $\sigma\colon\ w \to s$, and constants are written $c\colon \to s$ when $c \in \Sigma_{[],s}$.

Signatures are given in BOBJ by giving sorts after the keywords `sort` or `sorts`, and operations after the keywords `op` or `ops`. The **form** of an operation follows the `op` keyword, then a colon followed by a list of the sorts for arguments to that operation, followed by an arrow, followed by the value sort of the operation. Underbar characters serve as place holders within the form, to indicate where the arguments should go; the number of underbars and argument sorts should be the same. If there are no underbars but the argument sort list is non-empty, as with the `insert` operation below, the operation is assumed to have syntax that requires opening and closing parentheses, with commas between arguments, as in `insert(2, S)`.

```
sorts Elt Set .
op empty : -> Set .
op _in_ : Elt Set -> Bool   .
op insert : Elt Set -> Set .
```

Overloading is possible (and very helpful for readability) in this framework, since the same form can have more than one type. E.g., the form `_in_` could also be an operation on lists, with type $\langle$`List List`, `Bool`$\rangle$, where `Bool` is the sort of Booleans, from the builtin module `BOOL`, which is imported by default into every other module. Σ is a **ground signature** iff $\Sigma_{[],s} \cap \Sigma_{[],s'} = \emptyset$ whenever $s \neq s'$ and $\Sigma_{w,s} = \emptyset$ unless $w = [\,]$. Union is defined componentwise, by $(\Sigma \cup \Sigma')_{w,s} = \Sigma_{w,s} \cup \Sigma'_{w,s}$. A common case is union with a ground signature X, where we use the notation $\Sigma(X)$ for $\Sigma \cup X$.

A Σ**-algebra** A consists of an S-sorted set also denoted A, plus an **interpretation** of Σ in A, which is a family of arrows $i_{s_1...s_n,s} : \Sigma_{s_1...s_n,s} \rightarrow [A_{s_1} \times ... \times A_{s_n} \rightarrow A_s]$ for each type $\langle s_1...s_n, s\rangle \in S^* \times S$, which interpret the operation symbols in Σ as actual operations on A. For constant symbols, the interpretation is given by $i_{[],s} : \Sigma_{[],s} \rightarrow A_s$. Usually we write just σ for $i_{w,s}(\sigma)$, but if we need to make the dependence on A explicit, we may write σ_A. A_s is called the **carrier** of A of sort s. Given Σ-algebras A and B, a Σ**-homomorphism** $h\colon A \rightarrow B$ is an S-sorted arrow $h\colon A \rightarrow B$ such that $h_s(\sigma_A(m_1, ..., m_n) = \sigma_B(h_{s_1}(m_1), ..., h_{s_n}(m_n))$ for each $\sigma \in \Sigma_{s_1...s_n,s}$ and all $m_i \in A_{s_i}$ for $i = 1, ..., n$, and such that $h_s(c_A) = c_B$ for each constant symbol $c \in \Sigma_{[],s}$.

A Σ**-congruence relation** $\sim$ on a Σ-algebra A is a S-indexed equivalence relation such that if $\sigma\colon\ s_1...s_n \rightarrow s$ and $a_i, b_i \in A_{s_i}$ with $a_i \sim b_i$ for $1 \leq i \leq n$, then $\sigma(a_1, ..., a_n) \sim \sigma(b_1, ..., b_n)$. Given a Σ-congruence relation $\sim$ on A, the **quotient** Σ**-algebra** $A/{\sim}$ is a Σ-algebra $A/{\sim}$ such that $(A/{\sim})_s$ is $A_s/{\sim_s}$ for any sort s and $\sigma([a_1], ..., [a_n]) = [\sigma(a_1, ..., a_n)]$ for any $a_i \in A_{s_i}$ and $1 \leq i \leq n$ and $\sigma\colon\ s_1...s_n \rightarrow s \in \Sigma$.

Given an S-sorted signature Σ, the S-sorted set T_Σ of Σ**-terms** is the smallest S-sorted set such that $\Sigma_{[],s} \subseteq T_{\Sigma,s}$ and given $\sigma \in \Sigma_{s_1...s_n,s}$ and $t_i \in T_{\Sigma,s_i}$ then $\sigma(t_1...t_n) \in T_{\Sigma,s}$. Notice that T_Σ is a Σ-algebra by interpreting $\sigma \in \Sigma_{[],s}$ as just σ, and $\sigma \in \Sigma_{s_1...s_n,s}$ as the operation sending $t_1, ..., t_n$ to the list $\sigma(t_1...t_n)$. Thus, T_Σ is called the Σ**-term algebra** . Note that because of overloading, terms do not always have a unique parse. Below is the key property of this algebra; proofs are generally omitted, but can be found in the literature.

THEOREM 1. *Given a signature Σ with no overloaded constants and a Σ-algebra A, there is a unique Σ-homomorphism $T_\Sigma \rightarrow A$.* □

Given a signature Σ and a ground signature X disjoint from Σ, we can form the $\Sigma(X)$-algebra $T_{\Sigma(X)}$ and then view it as a Σ-algebra by forgetting the names of the new constants in X; let us denote this Σ-algebra by $T_{\Sigma(X)}$. It has the following universal **freeness** property:

THEOREM 2. *Given a Σ-algebra A and $a\colon X \to A$, there is a unique Σ-homomorphism $a\colon T_{\Sigma(X)} \to A$ extending a, in the sense that $\overline{a}_s(x) = a_s(x)$ for each $x \in X_s$ and $s \in S$; sometimes we will write just a instead of $\overline{a}$.*

A Σ-**equation** consists of a ground signature X of variable symbols (disjoint from Σ) plus two $\Sigma(X)$-terms of the same sort $s \in S$; we may write such an equation abstractly in the form $(\forall X)\ t = t'$ and concretely in the form $(\forall x, y, z)\ t = t'$ when $|X| = \{x, y, z\}$ and the sorts of x, y, z can be inferred from their uses in t and in t'. Similarly, a Σ-conditional equation consists of a ground signature X of variable symbols plus a set of pairs of $\Sigma(X)$-terms u_i, u'_i and t, t', each pair of the same sort and $1 \leq i \leq n$, written in the form $(\forall X)\ t = t'$ `if` $u_1 = u'_1, ..., u_n = u'_n$. Hereafter we use the word "equation" for both the conditional and unconditional cases. A **specification** or **theory** P is a pair (Σ, E), consisting of a signature Σ and a set E of Σ-equations.

If a Σ-equation e is $(\forall X)\ t = t'$ `if` $u_1 = u'_1, ..., u_n = u'_n$ and A is a Σ-algebra, we say A **satisfies** this equation, written $A \models_\Sigma e$, iff for any map $\theta\colon X \to A$, if $\theta(u_i) = \theta(u'_i)$ for $1 \leq i \leq n$, then $\theta(t) = \theta(t')$. Given a specification $P = (\Sigma, E)$, $A \models P$ iff $A \models_\Sigma e$ for every $e \in E$. Given a set of Σ-equations E, we define **provability** $\vdash_\Sigma$ for Σ-equations by the following:

1. $E \vdash_\Sigma (\forall X)\, t = t$
2. If $E \vdash_\Sigma (\forall X)\, t = t'$, then $E \vdash_\Sigma (\forall X)\, t' = t$
3. If $E \vdash_\Sigma (\forall X)\, t = t'$ and $E \vdash_\Sigma (\forall X)\, t' = t''$, then $E \vdash_\Sigma (\forall X)\, t = t''$
4. If $(\forall Y)\, t = t'$ `if` $u_1 = u'_1, ..., u_n = u'_n \in E$ and $\theta\colon Y \to T_{\Sigma(X)}$ and $E \vdash_\Sigma (\forall X)\, \theta(u_i) = \theta(u'_i)$ for $1 \leq i \leq n$, then $E \vdash_\Sigma (\forall X)\, \theta(t) = \theta(t')$.
5. If $E \vdash_\Sigma (\forall Y)\, t_i = t'_i$ and $t_i \in T_{\Sigma(X), s_i}$ for $1 \leq i \leq n$ and $\sigma\colon s_1 ... s_n \to s$, then $E \vdash_\Sigma (\forall X)\, \sigma(t_1, ..., t_n) = \sigma(t'_1, ..., t'_n)$.

THEOREM 3 (Soundness and Completeness). *If e is a Σ-equation, $E \models_\Sigma e$ iff $E \vdash_\Sigma e$.* □

EXAMPLE 4. A simple example of loose semantics is the theory of groups:

```
th GROUP is sort Elt .
  op e : -> Elt .
  op _-1 : Elt -> Elt [prec 5].
  op _*_ : Elt Elt -> Elt .
  vars X Y Z : Elt .
  eq X * e = X .
  eq X * (X -1) = e .
  eq (X * Y)* Z = X *(Y * Z).
end
```

The keyword `th` introduces theories with loose semantics; it is followed by the name of the module, `GROUP`, and closed by the keyword `end`. Declarations for variables and equations have the obvious keywords. The precedence "attribute" `[prec 5]` gives the operation `-1` a tight binding (in OBJ, lower precedence numbers indicate tighter binding).

Order sorted signatures can also include subsort declarations. We do not develop the theory of order sorted algebra, but do note that all major results generalize, sometimes with minor modifications.

2.2 Term Rewriting

Many simple equations can be proved by **reduction**, also called **term rewriting**, which applies the available equations to a given term, until no equation can be applied. This subsection gives some basic theory.

Given a signature Σ and ground signatures X, Y of **variable symbols** (disjoint from Σ), a **substitution** θ is a S-sorted set $\{\, \theta_s : X_s \to T_{\Sigma,s}(Y) \,\}$. By Theorem 2, every such θ extends uniquely to a Σ-homomorphism $\overline{\theta}\colon T_\Sigma(X) \to T_\Sigma(Y)$. For any term $t \in T_{\Sigma,s}(X)$, let $\theta(t) = \overline{\theta}_s(t)$. Given a term $p \in T_{\Sigma,s}(X)$ and a term $t \in T_{\Sigma,s}(Y)$, we say p **matches** t if there exists a substitution θ such that $\theta(p)$ is syntactically the same as t.

Given a signature Σ and a ground signature X of variable symbols (disjoint from Σ), a **Σ-rewrite rule** is a pair of terms, written $l \to r$, such that l and r have the same sort and all variables in r also appear in l. A **Σ-rewrite system**, or **term rewriting system**, abbreviated **TRS**, R is a set of Σ-rewrite rules. A term t **rewrites to** a term t' using R, written $t \to_R t'$ or just $t \to t'$, iff there exists a rewrite rule $l \to r$ in R and a substitution θ such that t has a subterm $\theta(l)$ and t' can be obtained from t by replacing $\theta(l)$ with $\theta(r)$; the term $\theta(l)$ is called the **redex** of the rewrite. Let $\to_R^*$ be the reflexive and transitive closure of $\to_R$. R is **confluent**, also called **Church-Rosser**, iff whenever $t \to_R^* t_1$ and $t \to_R^* t_2$, then there exists a term t' such that $t_1 \to_R^* t'$ and $t_2 \to_R^* t'$. R is **terminating** iff there is no infinite rewriting $t_1 \to_R^* t_2 \to_R^* \ldots$. A **normal form** of t under R is a term t' such that t' cannot be written and $t \to_R^* t'$; we may write $[[t]]_R$ for the normal form of t under R. A TRS is **canonical** iff it is confluent and terminating. It can be shown that in a canonical TRS, every Σ-term has a unique normal form, called its **canonical form**. Note that reduction applies to ground terms only, which means that any variables desired should be introduced as new constants[1]. See [Baadera nd Nipkow, 1998] for a basic survey of one sorted term rewriting.

---

[1]This is justified by the so called Theorem of Constants, which says $P \models_\Sigma (\forall X)\ \varphi$ iff $P \models_{\Sigma \cup X} \varphi$, where φ is a Σ-sentence

The OBJ languages use term rewriting to provide an operational semantics, by viewing equations as rewrite rules, i.e., by applying equations in the forward direction. Term rewriting for initial and loose theories is invoked with the command `red`, followed by a term (and a period). For example, given a module `INTSET` for sets of integers,

```
select INTSET .
red 3 in insert(1,insert(2,insert(3,empty))).
```

constructs the set $\{1, 2, 3\}$ and then tests whether 3 is in it, in the context of the module `INTSET`, which is made the module currently in focus by the `select` command. Here is the output:

```
reduce in INTSET: 3 in insert(1,insert(2,insert(3,empty)))
result Bool: true
rewrite time: 165ms        parse time: 4ms
```

For operations with attributes for associativity, commutativity, or identity, rewriting is done modulo those equations; details can be found in [Goguen *et al.*, 2000; Baadera nd Nipkow, 1998] and many other places. The built in module `TRUTH`, which is included in `BOOL` and is by default imported into every other module, provides a polymorphic binary operation `==` which compares the normal forms of its two arguments. For example,

```
red insert(3,insert(3,empty)) == insert(3,empty) .
```

returns `true`, since the two canonical forms are identical; otherwise it returns `false`. If the TRS is canonical, then `true` is returned iff the two terms are provably equal, but if the TRS is non-terminating, reduction may go into an infinite loop, and if the TRS is not confluent, reduction could return `false` when the terms are nonetheless provably equal.

2.3 Initial Semantics

Given a specification (Σ, E), a natural congruence relation $\equiv_E$ can be defined directly from $\vdash_\Sigma$ by $t \equiv_E t'$ iff $E \vdash_\Sigma (\forall\emptyset)\, t = t'$, and we have the following important *initiality* results:

THEOREM 5. *Given a specification $S = (\Sigma, E)$, for any Σ-algebra A with $A \models S$, there exists a unique Σ-homomorphism from $T_\Sigma/\equiv_E$ to A. Given a set of E of Σ-equations, a Σ-algebra A is initial iff it has no junk (the Σ-homomorphism $T_\Sigma \to A$ is surjective) and no confusion (it satisfies* only *the equations that can be deduced from E).*

The **initial semantics** of a specification (Σ, E) is the class of its initial algebras. It can be shown that all the initial algebras of a specification are Σ-isomorphic. By Theorem 5, $T_\Sigma/\equiv_E$ is an initial algebra of (Σ, E). Because any element in $T_\Sigma/\equiv_E$ can be generated by operations, induction is valid for proving properties of initial algebras. Generally, more than one induction scheme is valid for a given specification.

EXAMPLE 6. The module below defines natural numbers in Peano notation with five operations: the constant `0`, the successor function `s`, infix operations `+` and `*`, and `sum`(n) which computes $1+2+...+n$; the keyword `obj` indicates that it has initial semantics:

```
obj NATS is sort Nat .
  op 0 : -> Nat .
  op s_ : Nat -> Nat .
  op _+_ : Nat Nat -> Nat [ assoc comm prec 40 ] .
  op _*_ : Nat Nat -> Nat [ assoc comm prec 20 ] .
  op sum : Nat -> Nat .
  vars M N : Nat .
  eq 0 + M = M .
  eq s M + N = s(M + N) .
  eq 0 * M = 0 .
  eq s M * N = M * N + N .
  eq sum(0) = 0 .
  eq sum(s M) = s M + sum(M) .
end
```

The operations `+` and `*` are declared associative and commutative and given precedence. The equations define the non-constructor operations recursively over the constructors[2], `0` and `s`, and rewriting proceeds by matching modulo the equations for the attributes. (The `assoc` and `comm` attributes actually do more than the corresponding equations: they enable parsing and pattern matching modulo those equations.) These numbers differ from those of BOBJ's built in module `NAT`, which use Java integers and provides many additional operations.

Induction is an essential aspect of theorem proving, and is valid for modules with initial semantics (but not loose semantics). Although not directly supported by the OBJ languages, inductive proofs can be still be done by the method of proof scores [Futatsugi *et al.*, 2005], as illustrated by the following:

EXAMPLE 7. We prove the formula $1+2+...+n = n(n+1)/2$, in the form

$$(\forall\ \texttt{N}:\ \texttt{Nat})\ \texttt{sum(N)} + \texttt{sum(N)} = \texttt{N} * (\texttt{s N})\ .$$

The first `red` command below checks the base case, the constant `n` stands for the universally quantified variable `N` in the formula, the equation introduces

[2]A subsignature $\Pi \subseteq \Sigma$ is a signature of **constructors** for a theory P iff for every (ground) Σ-term t, there is a Π-term t' such that $P \vdash_\Sigma t = t'$.

the inductive hypothesis, and the second `red` checks the inductive step. The operation `==` uses rewriting modulo attributes, and returns `true` iff its two arguments reduce to the same thing modulo the given attributes.

```
open NATS .
   red sum(0) + sum(0) == 0 * (s 0) .
   op n : -> Nat .
   eq sum(n) + sum(n) = n * (s n) .
   red sum(s n) + sum(s n) == (s n) * (s s n) .
close
```

Since both reductions return `true`, this proof succeeds.

Because this same pattern is followed in many other proofs, encapsulating it in a reusable module would be useful. But induction is second order, so this cannot be done with a first order module system; Section 3.3 will show how to do it using BOBJ's higher order modules and views.

2.4 Hidden Algebra

Behavioral specifications characterize systems by how they behave in response to relevant experiments, rather than how they are implemented. Our hidden algebra formalization of this intuition distinguishes visible from hidden sorts, with equality being strict on visible sorts and behavioral on hidden sorts, in the sense of indistinguishability under experiments; thus hidden sorts are treated as black boxes, the state of which can only be observed and updated by certain specific operations. Therefore behavioral specifications impose fewer constraints on the semantics of modules, as a result of which some inference rules of ordinary equational reasoning are unsound, although a small modification restores soundness; another result of this extra freedom is that no finite set of inference rules can be complete for behavioral satisfaction [Buss and Roşu, 2000]. Context induction [Hennicker, 1990; Berregeb *et al.*, 1998] and general coinduction [Goguen and Malcolm, 1999; Goguen and Malcolm, 1997] are established proof techniques for behavioral properties, but both need creative human intervention. Circular coinduction [Roşu and Goguen, 2001] is a powerful circular coinductive rewriting algorithm implemented in BOBJ by C4RW, and used to automatically proved many behavioral properties [Goguen *et al.*, 2000a; Goguen *et al.*, 2000b].

A **hidden signature** Σ is a signature with its sorts partitioned into visible sorts V and hidden sorts H. Operations in Σ with one hidden argument and a visible result may be called **attributes**, and those with one hidden argument and a hidden result called **methods**. A **hidden Σ-algebra** is just a Σ-algebra; elements of visible sort in a hidden Σ-algebra represent *data*, and those of hidden sorts represent *states*; the subalgebra of visible sorts and operations of visible sort is called the **data algebra**. A **behav-**

ioral specification or **theory** is a triple (Σ, Γ, E) where Σ is a hidden signature, and Γ is a hidden subsignature of Σ, and E is a (finite) set of Σ-equations. The operations in Γ are called **behavioral.**

The definition of hidden algebra given above allows a loose interpretation for the data algebra, following the general approach of [Goguen *et al.*, 2000a; Roşu and Goguen, 2001]. However, this is not appropriate for some problems, for example if `true` and `false` become identified. This can be remedied by requiring every hidden algebra over a given signature to have a *fixed* data algebra, as in the original version of hidden algebra, or alternatively, by allowing so called data constraints, in the sense of [Goguen and Burstall, 1992], as additional sentences. Note that general results proved for the loose data approach will also apply to fixed data algebras, so there is no loss of generality in proceeding in this way.

Given a hidden signature Γ, a **Γ-context**, denoted $C[\,\square\,]$, for sort s is a Γ-term in $T_\Gamma(\{\square\} \cup Z)$ having exactly one special variable $\square$ of the sort s, where Z is an infinite set of special variables different from $\square$. If $C[\,\square\,]$ is a Γ-context of sort s and $t \in \Sigma_s$, let $C[\,t\,]$ denote the result of substituting t for $\square$. A Γ-context $C[\,\square\,]$ for hidden sort s is called **Γ-experiment** if its sort is visible.

If Γ is a subsignature of a hidden signature Σ and A is Σ-algebra and $\sim$ is an equivalence on A, then an operation σ in $\Sigma_{s_1 \ldots s_n, s}$ is **congruent** for $\sim$ iff $\sigma_A(a_1, ..., a_n) \sim \sigma_A(a'_1, ..., a'_n)$ whenever $a_i \sim a'_i$ for $1 \leq i \leq n$. A **hidden Γ-congruence** on A is an equivalence relation on A that is congruent for each operation in Γ and is the identity on visible sorts. The Γ-congruence $\equiv^\Gamma_\Sigma$, called **behavioral equivalence**, on A is defined as follows: two data elements are equivalent iff they are equal, and two states are equivalent iff they cannot be distinguished by Γ-experiments,i.e., iff any experiment produces the same value when applied to them. The following is a basic result:

THEOREM 8. *Given a hidden subsignature Γ of Σ and a Σ-algebra A, $\equiv^\Gamma_\Sigma$ is the largest hidden Γ-congruence on A.*

An operation σ is **Σ-behaviorally congruent** for A (or simply **congruent**) iff it is congruent for $\equiv^\Gamma_\Sigma$. A hidden Σ-algebra A **Γ-behaviorally satisfies** a Σ-equation $e = (\forall X)\ t = t'$ `if` $u_1 = u'_1, ..., u_n = u'_n$, written $A \models^\Gamma_\Sigma e$, iff for any mapping $\theta\colon X \to A$, if $\theta(u_i) \equiv^\Gamma_\Sigma \theta(u'_i)$ for $1 \leq i \leq n$, then $\theta(t) \equiv^\Gamma_\Sigma \theta(t')$. If E is a set of Σ-equations, then $A \models^\Gamma_\Sigma E$ iff $A \models^\Gamma_\Sigma e$ for any $e \in E$. We say A **behaviorally satisfies** a behavioral specification $\mathcal{B} = (\Sigma, \Gamma, E)$ iff $A \models^\Gamma_\Sigma E$; we write $A \models^\Gamma_\Sigma \mathcal{B}$. Define $E \models^\Gamma_\Sigma e$ iff $A \models^\Gamma_\Sigma E$ implies $A \models^\Gamma_\Sigma e$ for every algebra A. Define $\mathcal{B} \models^\Gamma_\Sigma e$ iff $E \models^\Gamma_\Sigma e$.

Given a behavioral specification $\mathcal{B} = (\Sigma, \Gamma, E)$, the **provability relation**

$\Vdash_{\Sigma}^{\Gamma}$ for Σ-equations is defined by the following rules:

1. Reflexivity: $E \Vdash_{\Sigma}^{\Gamma} (\forall X)\ t = t$.
2. Symmetry: If $E \Vdash_{\Sigma}^{\Gamma} (\forall X)\ t_1 = t_2$, then $E \Vdash_{\Sigma}^{\Gamma} (\forall X)\ t_2 = t_1$.
3. Transitivity: If $E \Vdash_{\Sigma}^{\Gamma} (\forall X)\ t_1 = t_2$ and $E \Vdash_{\Sigma}^{\Gamma} (\forall X)\ t_2 = t_3$, then $E \Vdash_{\Sigma}^{\Gamma} (\forall X)\ t_1 = t_3$.
4. Substitution: If $(\forall Y)\, t = t'$ `if` $u_1 = u'_1, ..., u_n = u'_n$ in E and $\theta \colon Y \to T_{\Sigma}(X)$ and $E \Vdash_{\Sigma}^{\Gamma} (\forall X)\ \theta(u_i) = \theta(u'_i)$ for $1 \leq i \leq n$, then $E \Vdash_{\Sigma}^{\Gamma} (\forall X)\ \theta(t) = \theta(t')$.
5. Congruence:

   (a) If $E \Vdash_{\Sigma}^{\Gamma} (\forall X)\ t = t'$ where $t, t' \in T_{\Sigma \cup X, v}$ and $v \in V$, and $t_1, ..., t_{i-1}, t_{i+1}, ..., t_n \in T_{\Sigma \cup X}$, then $E \Vdash_{\Sigma}^{\Gamma} (\forall X)\ \sigma(t_1, ..., t_{i-1}, t, t_{i+1}, ..., t_n) = \sigma(t_1, ..., t_{i-1}, t', t_{i+1}, ..., t_n)$.

   (b) If $E \Vdash_{\Sigma}^{\Gamma} (\forall X)\ t_i = t'_i$ for $1 \leq i \leq n$ and σ is congruent operation in Σ, then $E \Vdash_{\Sigma}^{\Gamma} (\forall X)\ \sigma(t_1, ..., t_n) = \sigma(t'_1, ..., t'_n)$.

Define $\mathcal{B} \Vdash (\forall X)\, t = t'$ iff $E \Vdash_{\Sigma}^{\Gamma} (\forall X)\, t = t'$. These rules specialize those of ordinary equational deduction by considering all sorts visible. Note that (5b) only applies to congruent operations. If all operations are congruent, then ordinary equational deduction is sound for behavioral satisfaction. The following expresses soundness with respect to both equational and behavioral satisfaction, generalizing a result in [Diaconescu and Futatsugi, 1998a] that equational deduction is sound when all operations are congruent.

THEOREM 9. *If* $\mathcal{B} \Vdash (\forall X)\, t = t'$, *then* $\mathcal{B} \models\!\equiv_{\Sigma}^{\Gamma} (\forall X)\, t = t'$ *and also* $E \models (\forall X)\, t = t'$.

General coinduction [Goguen and Malcolm, 1999; Goguen and Malcolm, 1997; Roşu, 2000] can be used to prove that a Σ-equation $(\forall X)\, t = t'$ is behaviorally satisfied by a behavioral specification $\mathcal{B}$ by the following steps:

- Define a binary relation R on terms (R is called the **candidate relation**).
- Show that R is a hidden Σ-congruence.
- Prove that $t\ R\ t'$.

Soundness of general coinduction follows directly from Theorem 8. Its major problem is that it requires human creativity to define an appropriate candidate relation. **Context induction** [Hennicker, 1990] can also be used to prove behavioral properties, using well-founded induction on the context structure to show that it is valid for all experiments. But in many real examples, context induction is not trivial and requires extensive human input, for example, in the form of inductive lemmas that can be difficult to discover and difficult to prove [Gaudel and Privara, 1991].

It often happens that some experiments are unnecessary in a context induction, because the equations imply that some experiments are equivalent to others. A similar but dual situation occurs in abstract data type theory when all the elements can be generated from a subset of operations, called the constructors, generators, or basis (when induction is involved). A general definition of cobasis is introduced in [Roşu and Goguen, 1998], and a simplified version can be given as follows: a **cobasis** Δ is a subset of operations in Γ that generates enough experiments, in the sense that no other experiment can distinguish any two states that cannot be distinguished by these experiments.

The denotational semantics of a behavioral module is the class of all algebras (i.e., implementations) that behaviorally satisfy specifications, and their operational semantics is given by behavioral rewriting. Behavioral modules in BOBJ are defined between the keywords `bth` and `end`. Sorts in behavioral modules are considered hidden unless declared with the keyword `dsort`, for visible sorts in behavioral modules. Similarly, operations in behavioral modules are considered congruent unless given the attribute `ncong`.

EXAMPLE 10 (A Behavioral Theory of Sets).

```
bth BSETNAT is sort Set .
  pr NATS .
  op empty : -> Set .
  op _in_ : Nat Set -> Bool   .
  op insert : Nat Set -> Set .
  vars N1 N2 : Nat . var S : Set .
  eq N1 in empty = false .
  eq N1 in insert(N2, S) = N1 == N2 or N1 in S   .
end
```

The first equation gives the result of observing `empty` with `_in_`, and the next equation gives the results of observing `insert` with `_in_`.

The most important difference between this behavioral theory and the initial theory for sets in Example 12 is that this theory does not have the equation `insert(E1, S) = S if E1 in S .` Although the other equations look the same, they are methodologically different. Data theories are usually designed with respect to constructors, but behavioral theories are designed with respect to observors. For example, `empty` and `insert` are constructors of the data theory `SET`, i.e., all ground sets can be created with them; and then all other operations can be defined based on the terms generated by these constructors.

We recommend designing a behavioral theory by selecting some operations as a cobasis to generate the behavioral equivalence relation, and then defining other operations with respect to these basic observers. For example, in `BSETNAT` above, the operation `_in_` is the unique observer in the cobasis, so that two sets are behaviorally equivalent iff they always return the same visible results under the observation of `_in_`, i.e., iff they have the same elements. Then for example, the traditional implementation of sets as lists with possible repetitions is behaviorally correct.

2.5 Logical Programming

Our claim that the OBJ languages are rigorously based on versions of equational logic is best demonstrated by defining the notion of logical programming language, and then showing how the various OBJ computations fit that definition. To be fully formal would require formalizing the notion of "a logic," including both deductive and model theoretic aspects. Such a formalization was sketched in the main paper on institutions [Goguen and Burstall, 1992], was carried further in a somewhat different way by Meseguer in [Meseguer, 1989], and was recently more fully realized in [Mossakowski *et al.*, 2005]. Here we leave that notion informal, assuming that a logic $\mathcal{L}$ has notions of signature Σ, Σ-sentence (with $Sen(\Sigma)$ the set of all these), Σ-model (with $Mod(\Sigma)$ the class of all these), Σ-satisfaction $\models_\Sigma$ and Σ-deduction $\vdash_\Sigma$ such that deduction is **sound**, i.e., such that $P \vdash_\Sigma e$ implies $P \models_\Sigma e$ where P is a set of Σ-sentences, e is a Σ-sentence, and $P \models_\Sigma e$ means that $M \models_\Sigma P$ implies $M \models_\Sigma e$ for all Σ-models M. Readers familiar with institutions with proofs [Mossakowski *et al.*, 2005] will see how to use that notion to fully formalize the above. Although it is less clear how to formalize the meta-logic $\mathcal{L}'$ introduced below, it suffices to let it be just the ordinary language of mathematics, applied to $\mathcal{L}$; in particular, it allows talk about proofs in $\mathcal{L}$.

A **program** P of a logical programming language over a logic $\mathcal{L}$ is a theory over $\mathcal{L}$, i.e., a set of Σ-sentences; a **query** is a sentence in $\mathcal{L}'$ of the form $(\exists X)\ q(X)$ where X is a set of variable symbols; and an **answer** to such a query is an assignment a from X to terms in $\mathcal{L}'$ such that $q(a)$ is in $Sen(\Sigma)$ and $P \vdash_\Sigma q(a)$, where $q(a)$ denotes the result of substituting $a(x)$ into q for each x in X; this is sound with respect to the intended models by assumption.

We now consider some examples. The first is **loose semantics** for (say) many sorted (or order sorted) equational logic, where signatures declare sorts (with subsorts) and operations, where sentences are equations, models are algebras, and satisfaction and deduction are as usual. Queries are of the form $(\exists p)\ \overline{p} = e$ where p is a variable for proofs over P, e is an equation, and $\overline{p}$ denotes the equation that p proves. Here we have **query completeness** in the sense that a query $(\exists X)\ q(X)$ has an answer a with $P \vdash_\Sigma q(a)$ iff $P \models_\Sigma q(a)$ (this generalizes the query completeness notion of [Meseguer, 1989] to our setting).

Our second example is **initial semantics** for many sorted (or order sorted) equational logic, where the relation $P \models_\Sigma e$ is restricted to initial models for P, and where queries have the form $(\exists p)\ \overline{p} = e$, and where deduction allows induction as well as equational reasoning. Query completeness holds here, although there is no algorithm that can realize it.

Our third example is **pure logic programming**, in the sense used in the Prolog community (e.g., [Lloyd, 1987]). Here signatures Σ declare relation symbols, Σ-sentences are Horn clauses, and Σ-queries have the form[3] $(\exists X)\ R(X)$ where R is a conjunction of relations applied to variables, and deduction is resolution. Then a suitable Herbrand theorem (e.g., see [Goguen and Meseguer, 1986]) implies that we can use either loose or initial semantics, and that query completeness holds. All this extends to many sorted Horn clause logic with equality [Goguen and Meseguer, 1986].

Our fourth example is **term rewriting** over equational theories P that are terminating. Here sentences are rewrite rules, queries have the form $(\exists t')\ t = t'$ where t' is reduced with respect to P, and deduction is term rewriting with P. This is not query complete over loose semantics, unless P is also Church-Rosser and therefore canonical, in which case initial semantics also applies.

Our fifth example is **behavioral semantics** as implemented by BOBJ's C4RW algorithm [Goguen *et al.*, 2003; Goguen and Lin, 2003]. Here programs are signatures (that include congruence declarations for some operations) with sets of rewrite rules and a cobasis declaration; satisfaction is behavioral; deduction is C4RW, shown sound in [Goguen *et al.*, 2003]; and queries have the form $(\exists p)\ \overline{p} = e$ where e is a behavioral equation and p is a C4RW proof. This is neither query complete nor reducible to initial semantics.

The first and fifth examples would not usually be called programming, because instead of computing a value, they try to prove an assertion. However, we claim that verification is the proper analog of programming for the level of specifications. Moreover, C4RW is surprisingly efficient when it terminates, e.g., [Goguen *et al.*, 2003] – which it may not, just as with ordi-

[3] It would be more consistent with our other examples to ask for proofs of $R(X)$ instead of just the substitution. Note that asking for a proof as output is just asking for a trace of a computation.

nary programming. Note that when C4RW fails to return `true`, the equation tested could still be valid, due to the necessary incompleteness of behavioral deduction. On the other hand, the first example is very far from being efficient, so that its computations, which essentially are blind searches, should not properly be called programming; in fact, the OBJ languages do not implement this, but rather allow users to construct proofs by hand. But if the program is canonical, then the validity of equations can be decided by checking whether or not the two terms have the same reduced forms, using the built in `==` operation; this can be seen as based on initial semantics. We conclude that Meseguer's initiality requirement in [Meseguer, 1989] is reasonable for those computations that are ordinarily called programming, because fixed data structures such as integers are likely involved, though this is not necessarily the case, e.g., for term rewriting proofs of equational identities in the theory of groups. However, query completeness is less reasonable, and neither requirement is appropriate if we wish to capture all of the semantics of OBJ family languages. Nevertheless, these two notions usefully enrich our understanding of the nature of computation.

We finally remark that the logical "existential" or "query" semantics sketched here is not limited to first order languages. For example, it also applies to pure functional programming languages, such as Haskell [Hudak *et al.*, 1992], though we omit the details, some of which can be found in [Meseguer, 1989]. Moreover, it applies to databases and to brokers in service oriented architectures.

3 Modularization

The module systems of parameterized programming go well beyond those of standard programming languages. We believe that views are not just a syntactic convenience, but are necessary for realizing the full potential of module parameterization. For example, we speculate that the lack of views explains the confusing multiplicity of semantics that have been given for ML functors ("functor" is ML terminology for parameterized module, see [Ullman, 1998]), as well as its awkward treatment of sharing.

3.1 Parameterization and Views

Given signatures Σ, Σ' with sorts S, S', then a **signature morphism** $\Sigma \to \Sigma'$ is a pair (f, g) where $f\colon S \to S'$, and g is an $(S^* \times S)$-indexed function $g_{w,s}\colon \Sigma_{w,s} \to \Sigma'_{f(w),f(s)}$. A **view**, or **theory morphism** , from a theory $T = (\Sigma, E)$ to a theory $T' = (\Sigma', E')$ is a signature morphism $v\colon \Sigma \to \Sigma'$ such that if $(\forall X)\ t = t'$ is an equation in E, then $E' \vdash (\forall \overline{X})\ \overline{v}(t) = \overline{v}(t')$ where $\overline{X}_{f(s)} = X_s$ for any sort $s \in \Sigma$ and $\overline{v}\colon T_{\Sigma(X)} \to T_{\Sigma'(\overline{X})}$ is the Σ-homomorphism induced by v; we may write $v\colon T \to T'$. The OBJ languages do not check semantic correctness of views, but only their syntax; therefore users should check the semantics.

EXAMPLE 11 (A Simple View).

```
  view V from GROUP to INT is
    sort Elt to Int .
    op (_ -1) to (-_) .
    op (_*_) to (_+_) .
end
```

View syntax is straightforward, except that when items are omitted, the system tries to figure out those missing items; the resulting views are called **default views**, see [Goguen *et al.*, 2000] for details.

A **parameterized specification** or **parameterized theory** is a pair (T_1, T_2) of specifications such that T_1 is included in T_2; we call T_1 the **parameter** or **interface theory** and T_2 the **body**. In Example 12 below, T_1 is ELT and T_2 is SET. Instantiation of (T_1, T_2) with an actual parameter P requires a view $T_1 \rightarrow P$ to describe the binding of actual to formal parameters; often a default view can be used. Following ideas developed for the Clear specification language [Burstall and Goguen, 1981; Goguen and Burstall, 1992], the instantiation is given by a colimit construction.

EXAMPLE 12 (A Parameterized Initial Theory of Sets). The initial theory SET below allows us to form sets of elements from any collection with an equality relation defined on it satisfying the law of identity, given in its interface theory ELT. Parameterization of a module M by an interface I is indicated with the notation M[X :: I], where X is the formal parameter of the parameterized module.

```
th ELT is sort Elt .
   op eq : Elt Elt -> Bool .
   var E : Elt .
   eq eq(E, E) = true .
end

dth SET[X :: ELT] is sort Set .
   op empty : -> Set .
   op _in_ : Elt Set -> Bool  .
   op insert : Elt Set -> Set .
   vars E1 E2 : Elt . var S : Set .
   eq E1 in empty = false .
   eq E1 in insert(E2, S) =
      eq(E1, E2) or E1 in S  .
```

```
    eq insert(E1, S) = S if E1 in S .
  end
```

The following tells BOBJ to instantiate SET with the builtin module INT of integers, and call the result INTSET:

```
dth INTSET is
  pr SET[INT] .
end
```

This uses a default view from ELT to INT and pr (for "protecting") indicates a importation.

Two additional features from parameterized programming are renaming and sums of modules. The first allows selected sorts and operations to be renamed within a module; this can be very helpful when reusing modules in new contexts. The sum just combines two or more modules, taking proper account of any shared submodules that may have arisen through importation. The syntax of these features is illustrated in the following:

```
dth NATS+INT is
  pr NATS *(sort Nat to Peano, op 0 to zero) + INT .
end
```

Here a sort and constant of NATS are renamed to avoid conflict with those of INT, and the two modules are then combined; the parser can determine whether the sort of any given term is Peano or Int, even though the operations _+_ and _*_ are overloaded.

EXAMPLE 13 (Behavioral Theory of Streams). The behavioral specification STREAM declares infinite streams parameterized by the "trivial" interface theory TRIV, which only requires that some sort be designated.

```
th TRIV is sort Elt . end

bth STREAM[X :: TRIV] is sort Stream .
  op head_ : Stream -> Elt .
  op tail_ : Stream -> Stream .
  op _&_ : Elt Stream -> Stream .
  var E : Elt . var S : Stream .
  eq head(E & S) = E .
  eq tail(E & S) = S .
end
```

The operation _&_ inserts an element into the head of a stream, and head and tail respectively return the first element, and the stream after removing its first element. The next specification adds an operation which "zips" two streams together by taking elements from them alternately:

```
  bth ZIP[X :: TRIV] is pr STREAM[X] .
    op zip : Stream Stream -> Stream .
    vars S S' : Stream .
    eq head zip(S,S') = head S .
    eq tail zip(S,S') = zip(S', tail S) .
  end
```

The picture below shows the application of `zip` to two input streams:

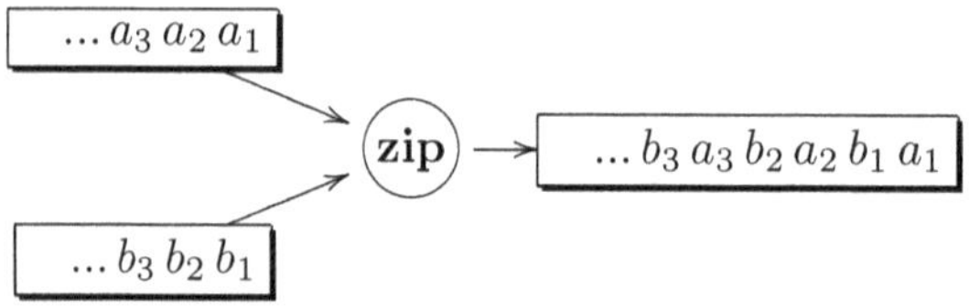

The command `red` does behavioral rewriting in the context of a behavioral theory. For example,

```
  open ZIP[NAT] .
    ops ones twos : -> Stream .
    vars S S' : Stream .
    vars N M  : Nat .
    eq head ones = 1 .
    eq tail ones = ones .
    eq head twos = 2 .
    eq tail twos = twos .
    red head tail tail zip(ones, twos).
  close
```

We will use these behavioral theories in later examples.

3.2 Behavioral Views

Behavioral parameterized theories can use any kind of theory as their interfaces, but the interfaces of non-behavioral theories must not be behavioral theories, i.e., behavioral theories are only allowed as interfaces for other (parameterized) behavioral theories. Given behavioral theories $\mathcal{B}_i = (\Sigma_i, \Gamma_i, E_i)$ for $i = 1, 2$, let the set of visible sorts and the set of hidden sorts in $\mathcal{B}_i$ be V_i and H_i, respectively. Then a **behavioral view** from $\mathcal{B}_1$ to $\mathcal{B}_2$ is a signature morphism $v\colon \Sigma_1 \to \Sigma_2$ such that: (1) $v(s) \in V_2$ for any sort $s \in V_1$; and (2) for any equation $(\forall X)\ t = t'$, if $\mathcal{B}_1 \models\!\!\equiv (\forall X)\, t = t'$, then $\mathcal{B}_2 \models\!\!\equiv (\forall \overline{X})\, \overline{v}(t) = \overline{v}(t')$ where $\overline{X}_{v(s)} = X_s$ for any sort $s \in \Sigma_1$ and $\overline{v}\colon T_{\Sigma_1(X)} \to T_{\Sigma_2(\overline{X})}$ is the homomorphism induced by v.

Notice that this definition of behavior views requires verifying all behavioral properties of the source module, which is impossible in practice. It is sufficient to define a signature morphism v from $\mathcal{B}_1$ to $\mathcal{B}_2$ such that (1)

all translated equations of $\mathcal{B}_1$ are behaviorally satisfied by $\mathcal{B}_2$; and (2) the image of a cobasis of $\mathcal{B}_1$ under v is a cobasis of $\mathcal{B}_2$. This is because it then follows that any behavioral property of $\mathcal{B}_1$ is also behaviorally satisfied by $\mathcal{B}_2$. In practice, the condition (2) above can be satisfied by making some operations non-congruent.

3.3 Higher Order Parameterized Programming

Since [Gpgiem, 1988] shows that first order parameterized modules give essentially all the programming power of higher order functional languages but with a first order logic, one may ask what higher order parameterized modules can add to this. The answer is that they add an architectural level of structural description and reuse that goes well beyond that of first order modules, as shown by the example below.

As already mentioned, all current OBJ family languages have first order parameterized modules and views, including Maude and CafeOBJ, as well as CASL. OBJ3 for some time has had formal parameters that are parameterized by previously introduced formal parameters [Goguen and Malcolm, 2000], but BOBJ is the only language that provides higher order views. A very different approach to higher order modularization based on the λ-calculus and type theory, appears in Extended ML [Sannella and Tarlecki, 1986], in ASL [Sannella *et al.*, 1992], and in its extension ASL+FPC [Aspinall, 2003]. Recent SML/NJ releases of ML [Milner *et al.*, 1997] include higher order parameterized modules, based on higher order parameterized signatures, but without views. C++ allows higher order parameterized templates, but these amount to little more than macro expansions with type checking. Larch [Garland, 1999] has a parameter passing mechanism that can simulate some uses of first order views, but it does not support views as reusable first class citizens. A new semantics for higher order parameterized programming is given in Section 3.3. We first illustrate the main ideas with an inductive proof scheme written in BOBJ.

A Reusable Induction Scheme

This subsection illustrates higher order parameterized programming by defining a reusable induction scheme[4] that builds on one in [Goguen and Malcolm, 2000], which in turn built on [Yatsu and Futatsugi, 1995]. The interface module `NIND` below requires constructors for basic Peano induction for natural numbers; it is the interface for the induction scheme, which will be instantiated with actual modules which specify the inductive problem to be solved. Because `NIND` has initial semantics, allowable actuals must also

[4]The modules in this example are generated by hand, but the Kumo system generates (first order) inductive proof schemes automatically [Goguen and Lin, 2001].

have initial semantics for their two operations that correspond to those in this module.

```
obj NIND is sort Term .
  op c : -> Term .
  op f : Term -> Term .
end
```

We now define terms over `NIND` by introducing a new constant symbol `x` of sort `Term`; this will be the induction variable.

```
obj TERM [X :: NIND] is
  op x : -> Term .
end
```

Because `TERM` has `NIND` as its interface, instanting it with an actual module `A` gives a module that defines terms over the operations of `A`. For example, if the formal parameter of `TERM` is instantiated with `NATS`, then `sum(x) + sum(x)` is one of the resulting terms. This module has initial semantics because we want its models to contain all and only terms in the variable `x`.

The interface theory below calls for two terms, for the left and right sides of an equational goal:

```
th GOAL [X :: NIND] [T :: TERM[X]] is
  ops l r : -> Term .
end
```

Because its first keyword is `th` (for theory), this module has loose semantics, which allows its two constants to be instantiated arbitrarily. It has two formal parameters, the first with interface `NIND`, with the second, `TERM[X]`, dependent[5] on the first. The two constants, `l` and `r`, represent the left and right sides of a goal.

A module with its interfaces separated into groups with brackets can be partially instantiated by providing actual modules for the parameters in the first group, with result a module parameterized by the remaining parameters, and having the partial instantiation as its body. Now we define the induction scheme:

```
th SCHEME [P :: NIND, G :: GOAL[P]] is
  us B is G[(c).(TERM[P])] .
  let base = l.B == r.B .
  us H is G[(x).(TERM[P])] .
  eq l.H = r.H .
  us C is G[(f(x)).(TERM[P])] .
  let step = l.C == r.C .
```

[5]This gives the effect of what is called a **dependent type** in type theory.

```
    let proof = base and step .
  end
```

The second interface of SCHEME is a parameterized module having GOAL[P] as its interface. The first line of the body of SCHEME instantiates G with TERM[P] by mapping the symbol x in its formal parameter TERM[P] to c in the actual parameter TERM[P], which is denoted (c).(TERM[P]) using the "dot" qualification convention; the result is then renamed B (for base) and is imported, where the keyword "us" indicates importation without requiring initiality or any other constraints to be satisfied. The equation base = l.B == r.B is well defined because l and r are constants of the sort Term.G[(c).(TERM[P])] in B. Similarly, the H and C importations are for the induction hypothesis and the inductive step. Lines 2,3,5 of the body of SCHEME correspond to lines 1,3,4 of the body of the above "open", except that SCHEME must be instantiated before the proof can be executed, and the two cases to be checked are conjoined into one by "and".

In more detail, to do an inductive proof using SCHEME, we first instantiate its first formal parameter P with an actual module containing appropriate functions over its constructors, then we instantiate its second formal parameter G with two terms over that, say defined by an actual module M. Then B is calculated as M[(c).(TERM[P])], and all the operations in G are replaced by operations from M. More precisely, a view from G[(c).(TERM[P])]) to M[(c).(TERM[P])] is created for replacing the operations of B in SCHEME. So to apply SCHEME to NATS, we first define a view from NIND to NATS,

```
  view NINDV from NIND to NATS is
    op f to s .
  end
```

which will instantiate the first parameter of SCHEME to NATS. Notice that the mappings sort Term to Nat and op c to 0 need not be stated here, since they are inferred by the default view mechanism. On the other hand, op f to s is needed because there are two unary operations in NATS. Once SCHEME is instantiated with NATS, its second parameter becomes G[NINDV], which we instantiate with

```
  view SUMV from GOAL[NINDV] to GOAL[NINDV] is
    op l to (sum(x) + sum(x)) .
    op r to (x * s x) .
  end
```

so that the complete instantiation is accomplished with the command

```
  make SUM-PROOF is SCHEME[NINDV, SUMV] end
```

This evaluates the module expression in its body and gives it the name SUM-PROOF. In the evaluation, B is calculated as GOAL[NINDV][(0).TERM

[NINDV]], using the view from G[P][(c).TERM[P]] to GOAL[NINDV][(0). TERM[NINDV]] that BOBJ automatically generates, with the body

```
op l to sum(0) + sum(0) .
op r to 0 * s 0 .
```

Since SUMV maps l to sum(x) + sum(x), when the (still parameterized) module GOAL[NINDV] is instantiated with TERM[NINDV], then x is mapped to 0, so that l is mapped to sum(0) + sum(0). Under the above view, the equation base = l.B == r.B becomes base = sum(0)+ sum(0) == 0 * s 0. Similar work is done for the modules H and C. Thus SUM-PROOF contains

```
eq base = sum(0) + sum(0) == 0 * s 0 .
eq sum(n) + sum(n) = n * s n .
eq step = sum(s n) + sum(s n) == s n * s s n .
```

and then the whole proof can be checked with just one command,

```
red proof .
```

for which BOBJ returns true after execution. However, users often want to see more detail, which can be accomplished by first giving the command

```
set trace on .
```

Actually, there is a simpler way to instantiate using a so-called *in-line view*:

```
make SUM-PROOF is SCHEME[NINDV, view to GOAL[NINDV] is
                        op l to sum(x) + sum(x). end] end
```

and NINDV could also be replaced by an in-line view.

Of course, higher order modules can do much more than this.....

Semantics for Higher Order Modules

This section sketches a semantics for higher order modules, based on the categorical general systems theory of [Goguen, 1971; Goguen and Ginali, 1978], particularly its higher order capability, the importance of which was emphasized (to Goguen) by Gregory Bateson in the early 1970s. This section assumes familiarity with category theory (for which see [Fiadeiro, 2004; Pierce, 1991] among many other sources), and necessarily begins rather abstractly. The intention is to develop an approach that is independent of any particular linguistic basis, and that in particular transcends the *ad hoc* peculiarities of the many architecture description languages that have been proposed. The approach also applies to mainstream imperative programming languages, e.g., by using underlying concrete institutions like those proposed in [Goguen and Tracz, 1999]. A semantics for the Maude module system in [Duran and Meseguer, 2003] is rather similar, but applies only to the first order case, and is formulated as an institution the signatures of which are diagrams of theories.

The most basic construction is the category $\mathbb{D}(\mathbb{C})$ of diagrams over a category[6] $\mathbb{C}$, which has objects functors $\mathbb{B} \to \mathbb{C}$ from a variable base category $\mathbb{B}$ to the fixed target $\mathbb{C}$, with morphisms from $a\colon \mathbb{B}_1 \to \mathbb{C}$ to $b\colon \mathbb{B}_2 \to \mathbb{C}$ being a functor $f\colon \mathbb{A} \to \mathbb{B}$ plus a natural transformation $\alpha\colon f;b \Rightarrow a$ (where ";" denotes composition), with the evident identities, and with composition $(f,\alpha);(g,\beta) = (f;g,\ (f{*}\beta);\alpha)$ where (g,β) is a morphism from b to $c\colon \mathbb{B}_3 \to \mathbb{C}$. Then $\mathbb{D}(\mathbb{C})$ is cocomplete if $\mathbb{C}$ is[7]; it will be convenient to write $\downarrow D$ for the colimit of D in $\mathbb{D}(\mathbb{C})$. Also, note that there is a natural injection $\mathbb{C} \to \mathbb{D}(\mathbb{C})$, for which we will use the notation $\lceil \_ \rceil$, and that $\downarrow \_$ is right adjoint to $\lceil \_ \rceil$.

Since $\mathbb{D}$ can be applied to any category, we can form $\mathbb{D}(\mathbb{D}(\mathbb{C}))$, which we denote by $\mathbb{D}2(\mathbb{C})$ or just $\mathbb{D}2$; now we can iterate to form $\mathbb{D}^n$, with $\mathbb{D}0 = \mathbb{C}$ by convention; moreover, we can form the colimit in $\mathbb{C}at$ of the sequence of natural injections

$$\mathbb{D}0 \to \mathbb{D}1 \to \mathbb{D}2 \to \mathbb{D}3 \to \ldots\ldots$$

since $\mathbb{C}at$ is cocomplete; denote this colimit $\mathbb{D}^\infty$. Also, there is a functor $Colim\colon \mathbb{D}2 \to \mathbb{D}$ which computes the diagram of colimits of a diagram of diagrams[8]; substituting $\mathbb{D}^{n-1}(\mathbb{C})$ for $\mathbb{C}$gives also $Colim\colon \mathbb{D}^{n+1} \to \mathbb{D}^n$. Similarly, let $\lceil \_ \rceil$ denote any injection $\mathbb{D}^n \to \mathbb{D}^{n+1}$, let $\downarrow \_$ denote any colimit functor $\mathbb{D}^{n+1} \to \mathbb{D}^n$, let $\downarrow 2$ denote $\downarrow\downarrow$ and more generally, let $\downarrow^k\colon \mathbb{D}^{n+k} \to \mathbb{D}^n$. Finally, let $\downarrow^\infty$ denote the map $\mathbb{D}^\infty \to \mathbb{C}$ induced by all the maps $\downarrow^n\colon \mathbb{D}^n \to \mathbb{C}$. Then it is not hard to see that $Colim\lceil D \rceil = D$, that $\downarrow \lceil D \rceil = D$, and that $\downarrow Colim\ D = \downarrow 2D$, among other such identities[9].

To apply this machinery to higher order modules, we substitute for $\mathbb{C}$ a category $\mathbb{T}$ of theories (which for BOBJ would involve constraints in the sense of [Goguen and Burstall, 1992] for initial semantics); $\mathbb{T}$ contains the basic, or zeroth order, modules. First order modules lie in $\mathbb{D}(\mathbb{T})$, second order modules in $\mathbb{D}2(\mathbb{T})$, an so on; from now on, we write just $\mathbb{D}, \mathbb{D}2$, etc. The functor $\downarrow^\infty$ computes the zeroth order specification of a system built by composing higher order modules. Note that the sum operation $(+)$ on theories is just coproduct, and that the same applies to diagrams of any order. We do not give a categorical semantics for renaming (the $*$ operation) because it is easier (almost trivial) to define it operationally, and in any case, colimits do not need to care much for names, since they keep careful track of where names come from.

---

[6]It helps to handle submodule sharing if $\mathbb{C}$ is a category with inclusions in the sense of [Căzănescu and Roşu, 1997; Căzănescu and Roşu, 2000].

[7]This follows from its being the indexed category $[\_, \mathbb{C}]$ using general results about indexed categories.

[8]It can be obtained by extending the colimit functors on each $[\mathbb{B}, \mathbb{C}]$ to all of $\mathbb{D}(\mathbb{C})$.

[9]However, these only hold up to isomorphism.

Parameterization and instantiation are more interesting. Parameterized programming [Goguen, 1989; Goguen *et al.*, 2000] defines a parameterized module to be a theory inclusion $i\colon P \to B$ where P is the parameter theory and B is the body (see Section 3.1). This works well, but it does not capture the idea that the inclusion itself is a module. However, we can do this with above machinery by encapsulating i, i.e., by viewing it as a diagram M, i.e., as an object in $\mathbb{D}$. This shift of level is part of a much richer, software architecture oriented point of view, in which module instantiation appears as a kind of module interconnection, instead of a perhaps *ad hoc* seeming pushout: let A be an actual parameter for M, i.e., let there be given a "fitting morphism" $f\colon P \to A$; then the instantiation of M by A is indicated by the module interconnection diagram $n\colon \lceil A \rceil \to M$ in $\mathbb{D}2$, where the functor component of n maps the one object of the category that underlies $\lceil A \rceil$ to the object underlying P in M, and the natural transformation of n is f.

Of course, much more can be done, by making use of more complex diagrams of higher orders. Sockets, pipes, connectors, ports, adaptors, channels – the entire zoo of contemporary software architecture is naturally modeled in this formalism, without needing to bring in any additional *ad hoc* features. The final chapters of a recent book [Fiadeiro, 2004] by José Fiadeiro contain much that is relevant to this topic, though with a different semantics; in particular, it provides excellent motivation for higher order parameterized modules, with many examples from software engineering. It seems likely that an alternative approach can be developed based on John Gray's Cartesian closed category of sketches [Gray, 1989].

We now return briefly to the induction scheme of Section 3.3. The module `TERM` is a first order parameter theory for the second order module `GOAL`, which in turn is a parameter theory for the third order module `SCHEME`, which has further structure arising from its internally defined modules `B`, `H`, and `C`. The modules `TERM`, `GOAL` and `SCHEME` are also all parameterized by `NIND`, but instantiating `NIND` with `NAT` using the view `NINDV` and taking the colimit still yields a third order module, because `SCHEME` is still parameterized by the second order (but now partially instantiated) module `GOAL`. The next step instantiates `GOAL` with itself, using a tricky view `SUMV` that introduces the terms to be proved equal. Now taking the colimit collapses to a zeroth order theory in which the computations can take place, triggered by a single command. (It is possible to draw some helpful diagrams for all this, but there isn't sufficient room in this paper.)

It is interesting to notice that techniques like those used for higher order modules can also be used to define higher order data types. We illustrate this with a simple example instead of giving a general construction. Let

$L\colon \mathbb{T} \to \mathbb{T}$ be the functor which sends a theory T to the theory $T+$LIST$[T]$, and construct the sequence

$$\mathbb{L}0 \to \mathbb{L}1 \to \mathbb{L}2 \to \mathbb{L}3 \to ....$$

where $\mathbb{L}0 = T$, $\mathbb{L}1 = L(T)$, $\mathbb{L}2 = L(L(T))$, etc., with the evident inclusions. Then its colimit includes elements, lists, lists of lists, lists of lists of lists, etc. It seems there may be an amusing analogy here with continued fractions that is worth further exploration.

4 Circular Coinductive Rewriting

Behavioral rewriting [Diaconescu and Futatsugi, 1998a] is to behavioral deduction what standard rewriting is to standard equational deduction, a simple but useful proof method. Based on the notion of cobasis, a more powerful proof method called **circular coinduction** is introduced in [Roşu and Goguen, 2001]. A enriched behavioral deduction system can be got by adding the following rule: Suppose Δ is a cobasis of a behavioral specification $\mathcal{B} = (\Sigma, \Gamma, E)$ and $<$ is a well founded partial order on Γ-contexts which is preserved by the operations in Γ. For any terms t_1 and t_2 in $T_\Sigma(X)$, if for any $\delta \in \Delta$ and for appropriate variables W, $\mathcal{B} \Vdash\!\vdash^{\Gamma}_{\Sigma} (\forall X)(\forall W)\, \delta(t_1, W) = c[\theta(t_1)]$ and $\mathcal{B} \Vdash\!\vdash^{\Gamma}_{\Sigma} (\forall X)(\forall W)\, \delta(t_2, W) = c[\theta(t_2)]$ and $c < \delta$, or else $\mathcal{B} \Vdash\!\vdash^{\Gamma}_{\Sigma} (\forall X)(\forall W)\, \delta(t_1, W) = u$ and $\mathcal{B} \Vdash\!\vdash^{\Gamma}_{\Sigma} (\forall X)(\forall W)\, \delta(t_2, W) = u$ for some Γ-term u, then $\mathcal{B} \Vdash\!\vdash^{\Gamma}_{\Sigma} (\forall X)\, t_1 = t_2$. Circular coinductive rewriting proves behavioral equalities by combining behavioral rewriting with circular coinduction [Goguen *et al.*, 2000a]; it also strengthens the duality with induction by allowing coinductive hypotheses to be used in proofs.

BOBJ provides a limited operational semantics for behavioral modules, by applying equations as behavioral rewrite rules. Because of non-congruent operations, ordinary rewriting is not in general sound, as illustrated by the following behavioral theory with a non-congruent operation:

EXAMPLE 14 (Nondeterministic Stacks). The following behavioral variant of a stack theory illustrates one way that nondeterminism can arise in hidden algebra specifications:

```
bth NDSTACK is sort Stack .
  protecting NAT .
  op push _ : Stack -> Stack [ncong] .
  op top _ : Stack -> Nat .
  op pop _ : Stack -> Stack .
  var S : Stack .
  eq pop push S = S .
end
```

The operation `push` places a nondeterministically chosen natural number on the stack's top. Even for behaviorally equivalent stacks `S1` and `S2`, `push(S1)` and `push(S2)` may insert different natural numbers onto `S1` and `S2`; therefore `push(S1)` and `push(S2)` may be distinguishable by the attribute `top`, so that `push` should be declared non-congruent. The equation in this specification says that a stack is not behaviorally changed by pushing a new element and then popping it. Notice that `push(pop(push(S))) == push(S)` is not behaviorally satisfied, although `pop(push(S))` and `S` are behaviorally equivalent. However, ordinary rewriting will reduce `push(pop(push(S)))` to `push(S)`.

Behavioral rewriting is invoked with the command `red`, which handles non-congruent operations properly. A term $C[\,\theta(l)\,]$ **behaviorally rewrites** to $C[\,\theta(r)\,]$, where $C[\,\Box\,]$ is a context and $l \rightarrow r$ is a rewrite rule, iff one of the following is satisfied:

1. The redex does not have a non-trivial context.
2. All operations from the top of C down to $\Box$ are congruent.
3. The context of the redex has a subcontext D such that all the operations from the top of D to $\Box$ are congruent and D has a visible sort.

For example, `push(pop(push(S)))` cannot be reduced to `push(S)`, because the context `push(`$\Box$`)` doesn't satisfy the conditions above.

Behavioral rewriting can prove simple behavioral properties, but more powerful methods are needed to verify more difficult behavioral properties. Unlike general coinduction [Goguen and Malcolm, 1997] and context induction [Berregeb *et al.*, 1998], conditional circular coinductive rewriting with provides a powerful way to prove behavioral properties, without intensive human intervention. The C4RW algorithm also includes very useful capabilities for automatic cobasis discovery and for case analysis; the algorithm is described in detail and proved correct in [Roşu and Goguen, 2001], and is also described and then illustrated with a correctness proof for a non-trivial version of the alternating bit protocol in [Goguen and Lin, 2003].

4.1 Mutual Coinduction

Behavioral operations can mutually depend on each other. In this case, behavioral properties $\{P_1, ..., P_n\}$ may also depend on each other, in such a way that no P_i can be proved by itself, but they can all be proved together. The BOBJ syntax for this is:

```
cred  (<goal>) ... (<goal>) .
```

where the goals may be conditional . Mutual circular coinductive rewriting is in the C4RW algorithm. The example below uses operations `odd` and **even** which take a stream and return streams respectively formed by the elements in the odd and even positions of the argument stream. Thus odd($e_1\, e_2\, e_3\, e_4\, e_5\, e_6\, e_7\, e_8 \ldots$) is $e_1\, e_3\, e_5\, e_7 \ldots$, while even($e_1\, e_2\, e_3\, e_4\, e_5\, e_6\, e_7, e_8 \ldots$) is $e_2\, e_4\, e_6\, e_8 \ldots$.

```
  bth ODD-EVEN[ X :: TRIV ] is pr ZIP[X] .
    var S : Stream .
    ops (odd_) (even_) : Stream -> Stream .
    eq head odd S  = head S .
    eq tail odd S  = even tail S .
    eq head even S = head tail S .
    eq tail even S = even tail tail S .
  end
```

This module imports ZIP, and all its operations are behavioral since they all preserve the intended behavioral equivalence, which is 'two streams are equivalent iff they have the same elements in the same order.' The property zip(odd S, even S) = S is proved by circular coinduction with:

```
  cred zip(odd S, even S) == S .
```

We next introduce a behavioral module for infinite binary trees:

```
  bth TREE[ X :: TRIV ] is sort Tree .
    op root_ : Tree -> Elt .
    ops left_ right_ : Tree -> Tree .
    op make : Elt Tree Tree -> Tree .
    var E : Elt .  vars T1 T2 : Tree .
    eq root  make(E, T1, T2) = E .
    eq left  make(E, T1, T2) = T1 .
    eq right make(E, T1, T2) = T2 .
  end
```

Given a tree, the operations **root**, **left** and **right** respectively return its root, its left subtree, and its right subtree; these three operations are a cobasis of **Tree**. The operation **make** takes an element and two trees to create a new tree. Next, define an operation **t2s**, which transforms trees into streams, and an operation **s2t** which does the inverse:

```
  bth TREE-STREAM [ X :: TRIV ] is
    pr TREE[X] + ODD-EVEN[X] .
    op t2s_ : Tree -> Stream .
    op s2t_ : Stream -> Tree .
    var T : Tree . var S : Stream .
    eq head t2s T  = root T .
    eq tail t2s T  = zip(t2s left T, t2s right T) .
    eq root s2t S  = head S .
```

```
    eq right s2t S = s2t tail odd S .
    eq left s2t S  = s2t even S .
  end
```

In converting a tree to a stream, `t2s` first outputs its root, and then interleaves streams from its left and right subtrees, as illustrated in the following:

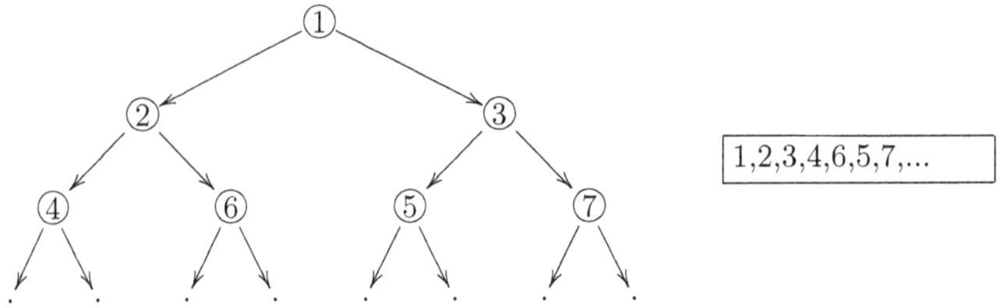

To build a tree from a stream, `s2t` uses the head of the stream as root, and uses the stream got by selecting all elements at odd positions except the first to build the right subtree, and all elements at even positions to create the left subtree. In the following, two behavioral properties are first proved and then introduced as lemmas for proving that `s2t` and `t2s` are inverse operations:

```
  open .
    vars S S1 S2 : Stream .
    var T : Tree .
    cred even zip(S1, S2) == S2 .
    eq even zip(S1, S2) = S2 .                *** lemma
    cred zip(even S, even tail S) == tail S .
    eq zip(even S, even tail S) = tail S . *** lemma
    cred even t2s T == t2s left T .
    eq even t2s T = t2s left T .              *** lemma
    cred s2t t2s T == T .
    cred t2s s2t S == S .
  close
```

The following defines functions `f` and `g` on trees and streams respectively, which will turn out to be identity functions. For any tree `T`, `left(f(T))` is defined by first taking the left subtree of `T`, and then converting it to a stream, and then applying `g` to the stream (since `g` is an identity function, this will be just the original stream), and then transforming the stream back to the tree, and finally applying `f` to this tree. Other cases are similar.

```
  bth SETUP [ X :: TRIV ] is pr TREE-STREAM [X] .
    op f_ : Tree -> Tree .
```

```
    op g_ : Stream -> Stream .
    var T : Tree . var S : Stream .
    eq root f T  = root T .
    eq left f T  = f s2t g t2s left T .
    eq right f T = f s2t g t2s right T .
    eq head g S  = head S .
    eq tail g S  = g t2s f s2t tail S .
  end
```

This way of defining `f` and `g` are not so unusual in practice. Sometimes it is hard to define an operation on a given sort directly, but we can transform it to an element on another sort, then use the operations on that sort, and finally transform the result back to the original sort. Now we prove `f` and `g` are identity functions in the following:

```
  set cobasis of TREE-STREAM .
  open .
    eq s2t t2s T = T .
    eq t2s s2t S = S .
    cred ( f T == T ) ( g S == S ) .
  close
```

This produces the following BOBJ output:

```
c-reduce in SETUP :
   f T == T
   g S == S
using cobasis of SETUP:
   op head _ : Stream -> Elt [prec 15]
   op root _ : Tree -> Elt
   op left _ : Tree -> Tree
   op right _ : Tree -> Tree
   op tail _ : Stream -> Stream [prec 15]
---------------------------------------
handled: f T == T
reduced to: f T == T
add rule (C1) : f T = T
-----------------------------------------
handled: g S == S
reduced to: g S == S
add rule (C2) : g S = S
-----------------------------------------
target is: f T == T
expand by: op root _ : Tree -> Elt
reduced to: true
     nf: root T
-----------------------------------------
target is: f T == T
expand by: op left _ : Tree -> Tree
deduced using (C1, C2) : true
```

```
     nf: left T
-----------------------------------------
target is: f T == T
expand by: op right _ : Tree -> Tree
deduced using (C1, C2) : true
     nf: right T
-----------------------------------------
target is: g S == S
expand by: op head _ : Stream -> Elt [prec 15]
reduced to: true
     nf: head S
-----------------------------------------
target is: g S == S
expand by: op tail _ : Stream -> Stream [prec 15]
deduced using (C2, C1) : true
     nf: tail S
-----------------------------------------
result: true
c-rewrite time: 327ms      parse time: 3ms
```

In this proof, two new "circularity" rules, C1 and C2, are added in the first two steps. The third step expands the goal `f(T) == T` by using `root`, and then the new goal is proved by behavioral rewriting directly. The next step gets a new goal by expanding the same goal using `left`, and the output above shows that both sides of this new goal reduce to `left(T)` by behavioral rewriting. The following shows the steps of this proof, using both C1 and C2:

	`left f T`	(using `left f T = f s2t g t2s left T`)
⟶	`f s2t g t2s left T`	(using C1, `f T = T`)
⟶	`s2t g t2s left T`	(using C2, `g S = S`)
⟶	`s2t t2s left T`	(using `s2t t2s T = T`)
⟶	`left T`	

The circularity `g S = S` which is used in the proof could not be proved if `f T == T` were the only coinductive goal.

EXAMPLE 15. **Fibonacci and Other Streams** We can define a generalized Fibonacci function by

$$fib(n+2) \quad = \quad fib(n+1) + fib(n)$$

where $fib(0)$ and $fib(1)$ may be given any values. Using `STREAM` as defined above, the following defines a stream `fib` of generalized Fibonacci numbers:

```
bth FIBO-STREAM is pr STREAM[NAT] .
  vars M N : Nat .
  op fib : Nat Nat -> Stream .
  eq head fib(M, N) = M .
```

```
    eq tail fib(M, N) = fib(N, M + N) .
  end
```

Note that for any natural numbers M and N, the first and second elements of the stream `fib(M, N)` are the first and second arguments of `fib`. All other elements in the stream `fib(M, N)` are the sums of the two prior elements. Now suppose the function g is defined on natural numbers by:

$$\begin{array}{rcll} g(n+2) & = & g(n+1)+g(n) & \text{if } n \text{ is even} \\ g(n+2) & = & g(n-1) & \text{if } n \text{ is odd} \end{array}$$

where $g(0)$ and $g(1)$ can again be given any values. The following defines streams of these numbers:

```
  bth G-STREAM is pr ODD-EVEN[NAT] .
    vars M N : Nat .
    op g : Nat Nat -> Stream .
    eq head g(M, N) = M .
    eq head tail g(M, N) = N .
    eq head tail tail g(M, N) = M + N .
    eq head tail tail tail g(M, N) = M .
    eq tail tail tail tail g(M, N) = g(M + M + N, M + N) .
  end
```

If $fib(0) = g(0) = M$ and $fib(1) = g(1) = N$, then the property $fib(n+2) = g(2n+1)$ can be expressed in the query

```
  tail tail fib(M,N) == tail odd g(N, M) .
```

However, this cannot be proved directly, because it generates infinitely many new proof tasks. Moreover, the property only covers the situation where g is applied to odd natural numbers, so we need a property covers even natural numbers, namely $fib(n) = g(2n)$. These two properties each need the other, and though neither can be proved by itself, they can be proved together by the following:

```
  open FIBO-STREAM + G-STREAM .
    vars M N : Nat .
    cred ( tail tail fib(M, N) == tail odd g(N, M) )
         ( fib(M, N) == even g(N, M) ) .
  close
```

5 Conclusions

Many theoretical and practical innovations have developed with the OBJ family of languages, some of which are listed below. We feel that this strongly supports the view that theory and practice should be pursued together, since each raises new ideas for the other, and each can test and support the validity of the other. Sometimes we implemented things that we did not yet have theory for, sometimes we puzzled over how to implement an existing theory, and often we struggled to extend both theory and implementation to cover some phenomenon of practical interest. Certainly both higher order parameterization and mutual coinduction fall into this area. Some other innovations, most of which have already been discussed somewhere in this paper, are: overloaded many sorted algebra, order sorted algebra, retracts, membership equational logic (in Maude [Meseguer, 1997]), hidden algebra, circular coinductive rewriting, parameterized programming, higher order parameterized modules, institutions, efficient term rewriting modulo equations (in Maude [Clavel *et al.*, 1996]), Grothendieck institutions (for the semantics of CafeOBJ [Diaconescu, 2002]), initial algebra (and model) semantics, e.g. for constraint logic programming [Goguen and Meseguer, 1986], and reflective equational programming (in Maude [Duran and Meseguer, 2003]). Many of these seem underexploited, and since the intellectual mines of OBJ are probably not yet exhausted, we may perhaps expect to see more interesting or unusual logical contributions from this area of research in the future.

Acknowledgements

The work surveyed in this paper is the result of dedicated effort by many people over a long period of time, as citations in this paper make clear, and the new work reported here would have been impossible without their important contributions. Although it is infeasible to thank everyone, we would like to particularly thank Rod Burstall, Răzvan Diaconescu, Grant Malcolm, José Meseguer, and Grigore Roşu for their very valuable collaborations, and Till Mossakowski and Andrzej Tarlecki for their valuable comments on the new ideas in this paper.

BIBLIOGRAPHY

[Aspinall, 2003] David. Aspinall. Type checking parameterised programs and specifications in ASL+FPC. In M. Wirsing, D. Pattinson, and R. Hennicker, eds., *Recent Trends in Algebraic Development Techniques, 16th International Workshop, WADT'02*, pp. 129–144. Springer, Lecture Notes in Computer Science, volume 2755, 2003.

[Baadera nd Nipkow, 1998] Franz Baader and Tobias Nipkow. *Term Rewriting and All That.* Cambridge, 1998.

[Berregeb *et al.*, 1998] Narjes Berregeb, Adel Bouhoula, and Michaël Rusinowitch. Observational proofs with critical contexts. In *Fundamental Approaches to Software Engineering*, volume 1382 of *Lecture Notes in Computer Science*, pp. 38–53. Springer, 1998.

[Bidoit and Hennicker, 2003] Michel Bidoit and Rolf Hennicker. Constructor-based observational logic. Technical Report LSV–03–9, Laboratoire Spcification et Verification, CNRS de Cachan, 2003.

[Birkhoff, 1935] Garrett Birkhoff. On the structure of abstract algebras. *Proceedings of the Cambridge Philosophical Society*, 31:433–454, 1935.

[Burstall and Goguen, 1977] Rod Burstall and Joseph Goguen. Putting theories together to make specifications. In R. Reddy, ed., *Proceedings, Fifth International Joint Conference on Artificial Intelligence*, pp. 1045–1058. Department of Computer Science, Carnegie-Mellon University, 1977.

[Burstall and Goguen, 1981] Rod Burstall and Joseph Goguen. An informal introduction to specifications using Clear. In R. Boyer and J. Moore, editors, *The Correctness Problem in Computer Science*, pp. 185–213. Academic, 1981. Reprinted in *Software Specification Techniques*, N. Gehani and A. McGettrick, eds., Addison-Wesley, 1985, pp. 363–390.

[Buss and Roşu, 2000] Samuel Buss and Grigore Roşu. Incompleteness of behavioral logics. In H. Reichel, ed., *Proceedings, Coalgebraic Methods in Computer Science (CMCS'00)*, volume 33 of *Electronic Notes in Theoretical Computer Science*, pp. 61–79. Elsevier Science, 2000.

[Căzănescu and Roşu, 1997] Virgil Emil Căzănescu and Grigore Roşu. Weak inclusion systems. *Mathematical Structures in Computer Science*, 7(2), 1997.

[Căzănescu and Roşu, 2000] Virgil Emil Căzănescu and Grigore Roşu. Weak inclusion systems, part 2. *Journal of Universal Computer Science*, 6(1):5–21, 2000.

[Cerioli *et al.*, 1998] Maura Cerioli, Till Mossakowski, and Horst Reichel. From total equational to partial first order. In *State of the Art in Algebraic Specification*. IFIP, to appear 1998.

[Clavel *et al.*, 2001] Manuel Clavel, Francisco Durán, Steven Eker, Patrick Lincoln, Narciso Martí-Oliet, José Meseguer, and José F. Quesada. Maude: Specification and programming in rewriting logic. *Theoretical Computer Science*, 2001.

[Clavel *et al.*, 1996] Manuel Clavel, Steven Eker, Patrick Lincoln, and José Meseguer. Principles of Maude. In J. Meseguer, ed., *Proceedings, First International Workshop on Rewriting Logic and its Applications*. Elsevier Science, 1996. Volume 4, *Electronic Notes in Theoretical Computer Science*.

[Diaconescu, 2002] Răzvan Diaconescu. Grothendieck institutions. *Applied Categorical Structures*, 10:383–402, 2002.

[Diaconescu and Futatsugi, 1998a] Răzvan Diaconescu and Kokichi Futatsugi. *CafeOBJ Report: The Language, Proof Techniques, and Methodologies for Object-Oriented Algebraic Specification*. World Scientific, 1998. AMAST Series in Computing, Volume 6.

[Diaconescu and Futatsugi, 1998b] Răzvan Diaconescu and Kokichi Futatsugi. Behavioural coherence in object-oriented algebraic specification. *Journal of Universal Computer Science*, 6(1):74–96, 2000.

[Duran and Meseguer, 2003] Francisco Duran and José Meseguer. Structured theories and institutions. *Theoretical Computer Science*, 309:357–380, 2003.

[Ehrig and Mahr, 1990] Hartmut Ehrig and Bernd Mahr. *Fundamentals of Algebraic Specification 2: Module Specifications and Constraints*. Springer, 1990. EATCS Monographs on Theoretical Computer Science, Vol. 21.

[Fiadeiro *et al.*, 2005] José Fiadeiro, Neil Harman, Markus Roggenbach, and Jan Rutten (Eds.). *Algebra and Coalgebra in Computer Science*. Springer, 2005. Lecture Notes in Computer Science, vol. 3629.

[Fiadeiro, 2004] José Luiz Fiadeiro. *Categories for Software Engineering*. Springer, 2004.

[Futatsugi *et al.*, 1985] Kokichi Futatsugi, Joseph Goguen, Jean-Pierre Jouannaud, and José Meseguer. Principles of OBJ2. In B. Reid, ed., *Proceedings, Twelfth ACM Symposium on Principles of Programming Languages*, pp. 52–66. Association for Computing Machinery, 1985.

[Futatsugi *et al.*, 2005] Kokichi Futatsugi, Joseph Goguen, and Kazuhiro Ogata. Verifying design with proof scores. In B. Meyer, ed., *Proceedings, Verified Software: Theories, Tools, Experiments*. Springer, 2005.

[Garland, 1999] Steven Garland. LP – the Larch Prover: version 3.1b, 1999. MIT, Laboratory for Computer Science, `http://www.sds.lcs.mit.edu/larch/LP`.

[Gaudel and Privara, 1991] Marie-Claude Gaudel and Igor Privara. Context induction: an exercise. Technical Report 687, LRI, Université de Paris-Sud, 1991.

[Goguen, 1971] Joseph Goguen. Mathematical representation of hierarchically organized systems. In E. Attinger, ed., *Global Systems Dynamics*, pp. 112–128. S. Karger, 1971.

[Goguen, 1977] Joseph Goguen. Abstract errors for abstract data types. In E. Neuhold, ed., *Proceedings, First IFIP Working Conference on Formal Description of Programming Concepts*, pp. 21.1–21.32. MIT, 1977. Also in *Formal Description of Programming Concepts*, P. Neuhold, Ed., North-Holland, pp. 491–522, 1979.

[Goguen, 1986] Joseph Goguen. Suggestions for using and organizing libraries in software development. In S. Kartashev and S. Kartashev, eds., *Proceedings, First International Conference on Supercomputing Systems*, pages 349–360. IEEE Computer Society, 1985. Also in *Supercomputing Systems*, S. and S. Kartashev, Eds., Elsevier, 1986.

[Goguen, 1989] Joseph Goguen. Principles of parameterized programming. In T. Biggerstaff and A. Perlis, eds., *Software Reusability, Volume I: Concepts and Models*, pp. 159–225. Addison Wesley, 1989.

[Gpgiem, 1988] Joseph Goguen. Higher-order functions considered unnecessary for higher-order programming. In D. Turner, editor, *Research Topics in Functional Programming*, pp. 309–352. Addison Wesley, 1990. University of Texas at Austin Year of Programming Series; preliminary version in SRI Technical Report SRI-CSL-88-1, 1988.

[Goguen, 1989] Joseph Goguen. Types as theories. In G. M. Reed, A. W. Roscoe, and R. Wachter, editors, *Topology and Category Theory in Computer Science*, pp. 357–390. Oxford, 1991.

[Goguen and Burstall, 1992] Joseph Goguen and Rod Burstall. Institutions: Abstract model theory for specification and programming. *Journal of the Association for Computing Machinery*, 39(1):95–146, 1992.

[Goguen and Ginali, 1978] Joseph Goguen and Susanna Ginali. A categorical approach to general systems theory. In G. Klir, ed., *Applied General Systems Research*, pp. 257–270. Plenum, 1978.

[Goguen and Lin, 2001] Joseph Goguen and Kai Lin. Web-based support for cooperative software engineering. *Annals of Software Engineering*, 12:25–32, 2001.

[Goguen and Lin, 2003] Joseph Goguen and Kai Lin. Behavioral verification of distributed concurrent systems with BOBJ. In H.-D. Ehrich and T. H. Tse, eds., *Proceedings, Conference on Quality Software*, pp. 216–235. IEEE Press, 2003.

[Goguen *et al.*, 2000a] Joseph Goguen, Kai Lin, and Grigore Roşu. Circular coinductive rewriting. In *Automated Software Engineering '00*, pp. 123–131. IEEE, 2000. Proceedings of a workshop held in Grenoble, France.

[Goguen *et al.*, 2000b] Joseph Goguen, Kai Lin, and Grigore Roşu. Behavioral and coinductive rewriting. In *Proceedings, Rewriting Logic Workshop, 2000.* Elsevier, 2001. Electronic Notes on Theoretical Computer Science, Volume 36, at `www.elsevier.nl/locate/entcs/volume36.html`.

[Goguen and Malcolm, 1999] Joseph Goguen and Grant Malcolm. Hidden coinduction: Behavioral correctness proofs for objects. *Mathematical Structures in Computer Science*, 9(3):287–319, June 1999.

[Goguen and Malcolm, 2000] Joseph Goguen and Grant Malcolm. More higher order programming in OBJ3. In J. Goguen and G. Malcolm, eds., *Software Engineering with OBJ: Algebraic Specification in Action*, pp. 397–408. Kluwer, 2000.

[Goguen and Malcolm, 1997] Joseph Goguen and Grant Malcolm. A hidden agenda. *Theoretical Computer Science*, 245(1):55–101, August 2000. Also UCSD Dept. Computer Science & Eng. Technical Report CS97–538, May 1997.

[Goguen and Meseguer, 1982] Joseph Goguen and José Meseguer. Rapid prototyping in the OBJ executable specification language. *Software Engineering Notes*, 7(5):75–84, December 1982. Proceedings of Rapid Prototyping Workshop.

[Goguen and Meseguer, 1985a] Joseph Goguen and José Meseguer. Completeness of many-sorted equational logic. *Houston Journal of Mathematics*, 11(3):307–334, 1985.

[Goguen and Meseguer, 1985b] Joseph Goguen and José Meseguer. Order-sorted algebra I: Equational deduction for multiple inheritance, overloading, exceptions and partial operations. *Theoretical Computer Science*, 105(2):217–273, 1992. Drafts exist from as early as 1985.

[Goguen and Meseguer, 1986] Joseph Goguen and José Meseguer. Eqlog: Equality, types, and generic modules for logic programming. In D. DeGroot and G. Lindstrom, eds., *Logic Programming: Functions, Relations and Equations*, pp. 295–363. Prentice-Hall, 1986.

[Goguen and Roşu, 1999] Joseph Goguen and Grigore Roşu. Hiding more of hidden algebra. In J. Wing, J. Woodcock, and J. Davies, eds., *FM'99 – Formal Methods*, pp. 1704–1719. Springer, 1999. Lecture Notes in Computer Sciences, Volume 1709, Proceedings of World Congress on Formal Methods, Toulouse, France.

[Goguen *et al.*, 2003] Joseph Goguen, Grigore Roşu, and Kai Lin. Conditional circular coinductive rewriting with case analysis. In M. Wirsing, D. Pattinson, and R. Hennicker, eds., *Recent Trends in Algebraic Development Techniques, 16th International Workshop, WADT'02*, pp. 216–232. Springer, Lecture Notes in Computer Science, volume 2755, 2003.

[Goguen and Tardo, 1979] Joseph Goguen and Joseph Tardo. An introduction to OBJ: A language for writing and testing software specifications. In M. Zelkowitz, ed., *Specification of Reliable Software*, pp. 170–189. IEEE, 1979. Reprinted in *Software Specification Techniques*, N. Gehani and A. McGettrick, eds., Addison Wesley, 1985, pp. 391–420.

[Goguen and Tracz, 1999] Joseph Goguen and William Tracz. An implementation-oriented semantics for module composition. In G. Leavens and M. Sitaraman, eds., *Foundations of Component-based Systems*, pp. 231–263. Cambridge, 2000.

[Goguen *et al.*, 2000] Joseph Goguen, Timothy Winkler, José Meseguer, Kokichi Futatsugi, and Jean-Pierre Jouannaud. Introducing OBJ. In J. Goguen and G. Malcolm, eds., *Software Engineering with OBJ: Algebraic Specification in Action*, pp. 3–167. Kluwer, 2000.

[Gray, 1989] John Gray. The category of sketches as a model for algebraic semantics. In J. Gray and A. Scedrov, eds., *Categories in computer science and logic*, volume 92 of *Contemporary Mathematics*, pp. 109–135. American Mathematical Society, 1989.

[Hennicker, 1990] Rolf Hennicker. Context induction: a proof principle for behavioural abstractions. In A. Miola, ed., *Proceedings, International Symposium on the Design and Implementation of Symbolic Computation Systems*, volume 429 of *Lecture Notes in Computer Science*, pp. 101–110. Springer, 1990.

[Hennicker and Bidoit, 1999] Rolf Hennicker and Michel Bidoit. Observational logic. In *Algebraic Methodology and Software Technology (AMAST'98)*, volume 1548 of *Lecture Notes in Computer Science*, pp. 263–277. Springer, 1999.

[Hudak *et al.*, 1992] Paul Hudak, Simon Peyton Jones, Philip Wadler, Arvind, et al. Report on the functional programming language Haskell. *ACM SIGPLAN Notices*, 27, May 1992. Version 1.2.

[Jacobs and Rutten, 1997] Bart Jacobs and Jan Rutten. A tutorial on (co)algebras and (co)induction. *Bulletin of the European Association for Theoretical Computer Science*, 62:222–259, June 1997.

[Lloyd, 1987] John Wilcox Lloyd. *Foundations of Logic Programming*. Springer, 1987. Second edition.

[Meseguer, 1989] José Meseguer. General logics. In H.-DĖbbinghaus et al., eds., *Proceedings, Logic Colloquium 1987*, pp. 275–329. North-Holland, 1989.

[Meseguer, 1997] José Meseguer. Membership algebra as a logical framework for equational specification. In F. Parisi-Presicce, ed., *Proceedings, WADT'97 – Workshop on Abstract Data Types*, pp. 18–61. Springer, 1998. Lecture Notes in Computer Science, Volume 1376.

[Meseguer and Goguen, 1993] José Meseguer and Joseph Goguen. Order-sorted algebra solves the constructor selector, multiple representation and coercion problems. *Information and Computation*, 103(1):114–158, March 1993. Revision of a paper presented at LICS 1987.

[Milner, 1980] Robin Milner. *A Calculus of Communicating Systems*. Springer, 1980. Lecture Notes in Computer Science, Volume 92.

[Milner *et al.*, 1997] Robin Milner, Mads Tofte, Robert Harper, and David MacQueen. *The Definition of Standard ML (Revised)*. MIT, 1997.

[Mossakowski *et al.*, 2005] Till Mossakowski, Joseph Goguen, Razvan Diaconescu, and Andrzej Tarlecki. What is a logic? In J.-Y. Beziau, ed., *Logica Universalis*, pp. 113–133. Birkhauser, 2005. Selected papers from the First World Conference on Universal Logic.

[Mosses, 2004] Peter Mosses, editor. *CASL Reference Manual*. Springer, 2004. Lecture Notes in Computer Science, Volume 2960.

[Park, 1980] David M.R. Park. *Concurrency and Automata on Infinite Sequences*. Springer, 1980. Lecture Notes in Computer Science, Volume 104.

[Pierce, 1991] Benjamin C. Pierce. *Basic Category Theory for Computer Scientists*. MIT, 1991.

[Reichel, 1981] Horst Reichel. Behavioural equivalence – a unifying concept for initial and final specifications. In *Proceedings, Third Hungarian Computer Science Conference*. Akademiai Kiado, 1981. Budapest.

[Reichel, 1995] Horst Reichel. Behavioural validity of conditional equations in abstract data types. In *Contributions to General Algebra 3*. Teubner, 1985. Proceedings of the Vienna Conference, June 21-24, 1984.

[Roşu, 2000] Grigore Roşu. *Hidden Logic*. PhD thesis, University of California at San Diego, 2000.

[Roşu and Goguen, 1998] Grigore Roşu and Joseph Goguen. Hidden congruent deduction. In R. Caferra and G. Salzer, eds., *Proceedings, 1998 Workshop on First Order Theorem Proving*, pp. 213–223. Technische Universität Wien, 1998. (Schloss Wilhelminenberg, Vienna, November 23-25, 1998).

[Roşu and Goguen, 2001] Grigore Roşu and Joseph Goguen. Circular coinduction. In *Proceedings, Int. Joint Conf. Automated Deduction*. Springer, 2000. Sienna, June 2001.

[Sannella *et al.*, 1992] Donald Sannella, Stefan Sokolowski, and Andrzej Tarlecki. Toward formal development of programs from algebraic specifications: parameterisation revisited. *Acta Informatica*, 29:689–736, 1992.

[Sannella and Tarlecki, 1986] Donald Sannella and Andrzej Tarlecki. Extended ML: an institution independent framework for formal program development. In D. Pitt, S. Abramsky, A. Poigné, and D. Rydeheard, eds., *Proceedings, Summer Workshop on Category Theory and Computer Programming*, pp. 364–389. Springer, 1986. Lecture Notes in Computer Science, Volume 240.

[Tracz, 1993] William Tracz. LILEANNA: a parameterized programming language. In *Proceedings, Second International Workshop on Software Reuse*, pp. 66–78, March 1993. Lucca, Italy.

[Ullman, 1998] Jeffrey Ullman. *Elements of ML Programming*. Prentice Hall, 1998.

[Whitehead, 1998] Alfred North Whitehead. *A Treatise on Universal Algebra, with Applications, I*. Cambridge, 1898. Reprinted 1960.

[Yatsu and Futatsugi, 1995] Hirokazu Yatsu and Kokichi Futatsugi. Modular specification in CafeOBJ. In K. Futatsugi, ed., *Proceedings, Workshop on Tenth Anniversary of OBJ2, Itoh, Japan, 21–22 October 1995*. Unisys, 1995.

Formal Approaches to Teamwork

John Grant, Sarit Kraus and Don Perlis

1 Introduction

Cooperation matters. Many everyday tasks cannot be done at all by a single agent, and many others are done more effectively by multiple agents. Moving a very heavy object is an example of the first sort, and moving a very long (but not heavy) object can be of the second. Many researchers have investigated the automation of cooperative behavior, producing a large body of published work with philosophical, formal, and implementational aspects. In this paper we focus on the second of these; our aim is to provide a comparative overview of a (hopefully somewhat representative) variety of approaches. Toward this end we have chosen a simple example of cooperative behavior (two agents working together to move a heavy object) and attempted to treat this example in some detail in each of the six distinct formal approaches, in order to reveal their differences and similarities, and noting where we needed to make some changes either to the example or to the formalism.

It is useful initially to ask just what it takes for agents to cooperate. Several authors provide answers stressing different aspects of cooperation. Bratman [1992] gives four criteria that a multi-agent activity must meet to be a "shared cooperative activity:" mutual responsiveness, commitment to the joint activity, commitment to mutual support, and formation of subplans that mesh with one another. Two of the papers we will consider later in detail deal with some key concepts about teams and cooperation. [Sonenberg *et al.*, 1992] states that a team of agents must have mutual beliefs, joint goals, joint plans, and joint intentions. [Wooldridge and Jennings, 1999] give four stages in cooperative behavior: recognition of the possibility for cooperation, team formation, plan formation, and execution.

There is now a substantial literature on teamwork. Some papers present non-formal philosophical discussion on what constitute teams (e.g., [Tuomela and Miller, 1988; Tuomela, 1991; Searle, 1990; Bratman, 1992]). Some of these non-formal models served as the basis for the formal ones. Others take a practical approach and provide techniques to implement teams on top of formal approaches (e.g., [Jennings, 1995; Tambe, 1997]). There are many other proposed interesting formalizations including [Singh, 1991; Castelfranchi, 1995; Singh, 1998]. For space reasons we have chosen to deal in detail with just six approaches (in some cases we combined several papers of the authors). We leave for future work an extended survey of formal teamwork models.

The formal models of teamwork build on formal models of mental states of individuals. There is substantial work on models of knowledge and belief such as [Megiddo, 1989; Konolige, 1986; Levesque, 1984; Shoham, 1989; Moore, 1985; Kraus and Lehmann, 1988; Gmytrasiewicz *et al.*, 1992; Haass, 1983; Dubois and Prade, 1994; Fagin *et al.*, 1995; Grant *et al.*, 2000]. Others discussed models of goals and intentions including [Singh and Asher, 1993; Rao and Georgeff, 1991b; Cohen and Levesque, 1990; Perrault, 1990; Cohen *et al.*, 1990; Moore, 1985; Konolige and Pollack, 1993; Rao and Georgeff, 1991a; Schut *et al.*, 2001; Kraus *et al.*, 1998; Kumar *et al.*, 2000; Alonso, 1997; Wooldridge, 2000]. To survey and compare all these works is not in the scope of our paper.

In Section 2 we describe an example task for two agents. We tried to find an example that is neither trivial nor difficult and that requires the cooperation of two agents as a team. The agents must go to a location where they find a large block they have to push to a new location. They must push together and do this twice in order to avoid an obstacle.

Section 3 contains the main part of the paper: we show how to express the example in the six formalisms. In each case we give names to the approaches based on the terminology used by the authors. Except for *SharedPlans* these are not official names for the formalisms discussed. We order the formalizations based on the order in which the key papers appeared. We call the approach of Cohen and Levesque (sometimes with co-authors) *Joint Intentions* [Levesque *et al.*, 1990]. The approach of the Sonenberg group we call *Team Plans* [Sonenberg *et al.*, 1992]. We use the term *SharedPlans* for the work of Grosz and Kraus [Grosz and Kraus, 1996]. We refer to Wooldridge and Jennings's model as *Cooperative Problem Solving* [Wooldridge and Jennings, 1999] and to Dunin-Keplicz and Verbrugge's by *Collective Intentions* [Dunin-Keplicz and Verbrugge, 2002]. Finally, we call the Grant, Kraus and Perlis approach *Cooperative Subcontracting* [Grant *et al.*, 2005].

In Section 4 we compare the six approaches based on several criteria, including their focus, formalism, representation of time, and facility for expressing complex plans. The paper ends with a brief conclusion and suggestion for future work in Section 5.

2 Example Description

We illustrate the six approaches we have chosen by expressing the same example in all of them. We chose an example that is neither particularly difficult, nor trivial, and that shows some important aspects of the formalizations of teamwork.

Agents A_1 and A_2 have the joint goal of pushing block Bl from location $locO$ (origin) to $locD$ (destination), starting at time t. There is an obstacle in the way so they have to push the block twice, once in the north direction, and then to the west. First the agents need to (plan to) be at $locO$, to be on the south side of Bl, and to do a push together; then they need to be on the west side of Bl and do another push. Bl will be at location $locT$ (temporary location) after the first push. The two agents cannot be in the same place simultaneously.

We assume that all the actions needed for performing the task are specified in a recipe. Hence, there is no need to plan in order to satisfy preconditions. In some systems the subactions are not specified in recipes and we will need to adjust the notation. Not surprisingly, as different authors stress different aspects of teamwork, the particular details are not equally well suited to all the approaches.

We assume that before the agents start the task, A_1 is at $loc1$ and A_2 is at $loc2$. Each agent is associated with a specified speed at which it moves. We do not deal with the speed separately. Thus the time it takes to move between two places is the distance between these places divided by the speed. Both the distance between $loc1$ and $locO$ and the distance between $loc2$ and $locO$ are known.

Then A_1 goes to the west corner of the south side of $locO$, $WC\_SS(locO)$ and A_2 goes to the east corner of the south side of $locO$, $EC\_SS(locO)$. For simplicity we assume that the "Go" action is a basic-level action. We assume basic-level actions are executable at will if appropriate situational conditions hold, and do not define this further (see Pollack's argument that this is a reasonable assumption in a computational setting [Pollack, 1986]).

The "Push" action is a multi-agent action (and thus a complex action). It requires two agents, one doing LPush, push from the left, and one doing RPush, push from the right. The push action takes one time unit.

Then agent A_1 goes to the north corner of the west side of $locT$ and A_2 goes to the south corner of the west side of $locT$ and they push again.

Notations:

- The action to be performed is $\alpha = Move(\{A_1, A_2\}, Bl, locO, locD)$, i.e., A_1 and A_2 will move block Bl from $locO$ to $locD$.
- We denote the group of the agents $GR = \{A_1, A_2\}$

- There are six subactions in the recipe for α.
  $\beta_1 = Go(A_1, loc1, WC\_SS(locO))$
  $\beta_2 = Go(A_2, loc2, EC\_SS(locO))$
  $\beta_3 = Push(\{A_1, A_2\}, Bl, locO, locT)$
  $\beta_4 = Go(A_1, WC\_SS(locT), NC\_WS(locT))$
  $\beta_5 = Go(A_2, EC\_SS(locT), SC\_WS(locT))$
  $\beta_6 = Push(\{A_1, A_2\}, Bl, locT, locD)$

- The recipe for α consists of the six subactions and a set of order constraints. We use the predicate $Before(\beta, \beta')$ to denote that β should be performed before β'. In the example, β_1 and β_2 should be performed before β_3; β_3 should be performed before β_4 and β_5 and the last two should be performed before β_6. The instantiated recipe for $\alpha = Move(\{A_1, A_2\}, Bl, locO, locD)$ is: $R_\alpha = \{\{\beta_1, \beta_2, \beta_3, \beta_4, \beta_5, \beta_6\}$, $\{Before(\beta_1, \beta_3), Before(\beta_2, \beta_3), Before(\beta_3, \beta_4), Before(\beta_3, \beta_5)$, $Before(\beta_4, \beta_6), Before(\beta_5, \beta_6)\}\}$

The recipe for the action $\beta_i = Push(GR, Bl, locX, locY)$ (with arbitrary locations $locX$ and $locY$) consists of two actions: $\gamma_i^1 = LPush(A_1, Bl, locX, locY)$ and $\gamma_i^2 = RPush(A_2, Bl, locX, locY)$. Denote by $Sim(\beta, \beta')$ the constraint that β and β' should be performed simultaneously. The recipe for such an action is: $\{\{\gamma_i^1, \gamma_i^2\}, \{Sim(\gamma_i^1, \gamma_i^2)\}\}$.

We denote by t_p the time when the plans, intentions, and beliefs are formed and by t the time when the agents start performing α.

For each action β, T_β denotes the time interval during which β is performed. We write k_1 and k_2 for the amount of time it takes for agent A_1 and A_2 respectively to go from their initial location to $locO$, and $k = max(k_1, k_2)$.

1. $T_{\beta_1} = [t, t + k_1]$.

2. $T_{\beta_2} = [t, t + k_2]$.

3. β_3 will start after β_1 and β_2. Thus
   $T_{\beta_3} = [t + k + 1, t + k + 2]$.

4. $T_{\beta_4} = [t + k + 3, t + k + 4]$.

5. $T_{\beta_5} = [t + k + 3, t + k + 4]$.

6. β_6 will start after β_4 and β_5. Thus $T_{\beta_6} = [t + k + 5, t + k + 6]$.

7. Thus $T_\alpha = [t, t + k + 6]$.

3 Formal Approaches

This is the main part of the paper. Here we express the example presented in the previous section in six formalizations, also making some comments about these approaches along the way. We present mainly those aspects that are directly relevant to the example and omit many important details discussed by the authors. In the presentations we use the same symbol α as a generic symbol for actions in definitions as well as for the specific action to move the block *Bl* from *locO* to *locD*. The meaning should always be clear from the context. We also make some changes to notation for the sake of uniformity: in particular we change Lisp notation, (Pxy) to P(x,y). In the following section we will also compare the six approaches in various ways.

3.1 Joint Intentions

Cohen and Levesque with several of their colleagues presented a series of papers on models for teamwork ([Levesque *et al.*, 1990; Cohen and Levesque, 1991b; Cohen and Levesque, 1991a; Cohen *et al.*, 1997]. Here we consider mainly the first paper of the series by Levesque, Cohen and Nunes [Levesque *et al.*, 1990] which is the most formal one. The other papers build on it. That paper focuses on the persistence of joint intentions and when agents will drop joint intentions.

They use constructs of dynamic logic to describe sequences of actions and modal operators to express time associated propositions. In their logic, $\alpha_1;\alpha_2$ is action composition and $\alpha_1||\alpha_2$ is the concurrent occurrence of α_1 and α_2. $p?$ is a test action: if p is true, the action succeeds, but if p is false it fails.

They do not deal with groups of agents performing complex subactions; in fact, they do not deal with expressing actions at different levels. Therefore we consider only basic-level actions when expressing our example in their system.

We denote $\alpha = [(\beta_1||\beta_2);(\gamma_3^1||\gamma_3^2);(\beta_4||\beta_5);(\gamma_6^1||\gamma_6^2)]$ that is:

$$
\begin{aligned}
&\alpha = \\
&\quad [(Go(A_1, loc1, WC\_SS(locO))||Go(A_2, loc2, EC\_SS(locO)));\\
&\quad (LPush(A_1, Bl, locO, locT)||RPush(A_2, Bl, locO, locT));\\
&\quad (Go(A_1, WC\_SS(locT), NC\_WS(locT))||\\
&\quad\ Go(A_2, EC\_SS(locT), SC\_WS(locT)));\\
&\quad (LPush(A_1, Bl, locT, locD)||RPush(A_2, Bl, locT, locD))]
\end{aligned}
$$

Levesque, Cohen and Nunes [Levesque *et al.*, 1990] do not show how to express exact times. In particular, they do not describe how to specify joint intentions for specific times. However, in another paper [Cohen and Levesque,

1990] that considers single agent intentions, they do present a way to specifying exact time. They introduce the concept of a time proposition such as $2:30PM/3/6/85$. Using a time proposition we can express tests of the form $2:30PM/3/6/85?$. Writing $2:30PM/7/20/05?;\beta_1;2:35PM/7/20/05?$ will express the sequence of events where β_1 will be performed July 20 between $2:30PM$ and $2:35PM$. In particular, suppose that in our example the agents should move the block on July 20, 2005. So if we say that t is 2PM and have specific values for k_1 and k_2, say 5 minutes and 10 minutes respectively, and we assume that the time unit is 1 minute, we can write the following:

$\alpha =$ [2:00PM/7/20/05?;
(β_1 2:05PM/7/20/05?||β_2 2:10PM/7/20/05?);
2:11PM/7/20/05?($\gamma_3^1||\gamma_3^2$);
2:13PM/7/20/05?; ($\beta_4||\beta_5$);
2:15PM/7/20/05?; ($\gamma_6^1||\gamma_6^2$);
2:16PM/7/20/05?]

To avoid the complexity of dealing with time, we consider the α without the time expression in the rest of this section.

In addition to dynamic logic constructs, [Levesque *et al.*, 1990] use temporal modal logic operators:

- UNTIL(p, q) specifies that until p is true q will remain true.
- $\Box p$ indicates that p is true from now on.
- $\diamond p$ indicates that p is true at some point in the future.
- DONE(x, y, α) indicates that α has just happened and x and y are its agents.
- DOING(x, y, α) indicates that x and y are doing α.

They also apply modal operators for mental states:

- GOAL(x, p) says that x has p as a goal.
- BEL(x, p) says that x has p as a belief.
- MB(x, y, p) says that p is mutually believed by x and y.

Using these operators they define the concept of a persistent goal:

- $PGOAL(x, p, q)$ means that x has a persistent goal that p will be true because of q.

$$\begin{aligned}&\mathrm{PGOAL}(x,p,q) =_{def}\\&\quad \mathrm{BEL}(x,\neg p)\wedge \mathrm{GOAL}(x,\diamond p)\wedge\\&\quad UNTIL([\mathrm{BEL}(x,p)\vee \mathrm{BEL}(x,\Box\neg p)\vee \mathrm{BEL}(x,\neg q)]\\&\qquad \mathrm{GOAL}(x,\diamond p))\end{aligned}$$

Intuitively this means that x has a persistent goal that p become true if it believes that p is false but wants it to become true until x comes to believe that p is true or p will never be true or that q is false.

They also define two more complex notions with respect to two agents:

- $\mathrm{MG}(x, y, p) =_{def} \mathrm{MB}(x, y, \mathrm{GOAL}(x, \diamond p) \wedge \mathrm{GOAL}(y, \diamond p))$. Intuitively, MG(mutual goal) means that both agents would like p to become true eventually and this is mutually believed.

- $$\begin{aligned}&\mathrm{WG}(x,y,p) =_{def}\\&[\neg\mathrm{BEL}(x,p)\wedge \mathrm{GOAL}(x,\diamond p)]\vee\\&[\mathrm{BEL}(x,p)\wedge \mathrm{GOAL}(x,\diamond \mathrm{MB}(x,y,p))]\vee\\&[\mathrm{BEL}(x,\Box\neg p)\wedge \mathrm{GOAL}(x,\diamond \mathrm{MB}(x,y,\Box\neg p))]\end{aligned}$$

  This form of "weak goal" (WG) involves three mutually exclusive cases: either x has the goal that p will be true sometime in the future, or believes that p is true and wants to make it mutually believed or believes that p will never be true from now on and wants to make that mutually believed.

- They also define WMG expressing that both agents have a "weak goal" that is mutually believed:

  $\mathrm{WMG}(x, y, p) =_{def} \mathrm{MB}(x, y, \mathrm{WG}(x, y, p) \wedge \mathrm{WG}(y, x, p))$.

Suppose q is a reason why the agents want to perform α. JPG, joint persistent goal, in our example with respect to q would be expressed as follows:

$$\begin{aligned}&\mathrm{JPG}(A_1, A_2, \mathrm{DONE}(A_1, A_2, \alpha), q) =^{def}\\&\ \mathrm{MB}(A_1, A_2, \neg\mathrm{DONE}(A_1, A_2, \alpha)) \wedge \mathrm{MG}(A_1, A_2, \mathrm{DONE}(A_1, A_2, \alpha))\wedge\\&\ \mathrm{UNTIL}([\mathrm{MB}(A_1, A_2, \mathrm{DONE}(A_1, A_2, \alpha)) \vee \mathrm{MB}(A_1, A_2, \Box\neg\mathrm{DONE}(A_1, A_2, \alpha))\\&\qquad \vee \mathrm{MB}(A_1, A_2, \neg q)],\\&\quad \mathrm{WMG}(A_1, A_2, \mathrm{DONE}(A_1, A_2, \alpha)))\end{aligned}$$

The two clauses indicate that A_1 and A_2 mutually believe that α has not been done yet and that they both have the mutual goal for α to be done eventually. The clause that starts with UNTIL specifies when the agents will stop having the persistent goal and what goals and beliefs A_1 and A_2 will have until then. In particular, the agents will stop having the JPG when either they mutually believe that α has been done, or they will mutually believe that α will never be done or they will mutually believe that the reason for doing α, namely q, is not true any more. Until then, they will have a weak mutual goal, which means that they mutually believe that both of them have a weak goal with respect to the other agent. A_1 having a WG with respect to A_2 that α will be done means that one of the following is true: (i) A_1 does not believe that the α has been done and has a goal that it will be eventually done. (ii) A_1 believes that α has been done and has a goal that this belief will become a mutual belief with A_2. (iii) A_1 believes that α will never be done and has a goal that this belief will become jointly believed with A_2.

Finally, we are ready to present their notion of joint intentions:

$$
\begin{array}{l}
\mathrm{JI}(A_1, A_2, \alpha, q) =^{def} \\
\quad \mathrm{JPG}(A_1, A_2, \\
\quad\quad \mathrm{DONE}(A_1, A_2, \\
\quad\quad\quad \mathrm{UNTIL}(DONE(A_1, A_2, \alpha), \\
\quad\quad\quad \mathrm{MB}(A_1, A_2, \mathrm{DOING}(A_1, A_2, \alpha)))?; \alpha), \\
\quad\quad q)
\end{array}
$$

A_1 and A_2 having joint intentions to do α means that they have a joint persistent goal to do α while mutually believing (throughout the execution of α) that they are doing it.

The main question is what are the individual intentions and beliefs that A_1 and A_2 have when they have joint intentions. Levesque, Cohen and Nunes prove that each agent has a persistent goal that α will be done, i.e., $PGOAL(A_1, \alpha, q)$ and $PGOAL(A_2, \alpha, q)$, but it is not clear what will be the intentions and beliefs toward the subactions, β_i. They say that if an agent is committed to doing $\beta_1; \beta_2$, it does not follow that the agent is committed to each action separately; for example, it does not make sense to be committed to push the block if going to *locO* failed. More non formal discussion on these issues are presented in [Cohen and Levesque, 1991b; Cohen *et al.*, 1997]. In their paper [Cohen and Levesque, 1991a] they prove that if the agents have a joint persistent goal that $\beta_1; \beta_2$ will be done, if they mutually believe that β_2 has not been done, then they will keep having a joint persistent goal that β_2 will be done at least as long they have the persistent goal that $\beta_1; \beta_2$ will be done. In our example, if A_1 and A_2

mutually believe that the push has not been done they will keep on having a persistent joint goal to do it at least while they have a persistent goal to do α.

3.2 Team Plans

Sonenberg et al in a series of papers [Sonenberg *et al.*, 1992; Kinny *et al.*, 1994; Rao *et al.*, 1992; Tidhar *et al.*, 1996] focus on groups performing complex plans to achieve a goal. They discuss how to find a suitable team for a task; how to synchronize the establishment of joint goals and the adoption of intentions; how to assign roles; and how to maintain proper temporal relations while executing different parts of the plan by different members of the team.

Here we consider the most extended version [Sonenberg *et al.*, 1992]. They apply three modal operators:

- BEL(A, ϕ) means that an agent A believes that ϕ is true.
- GOAL(A, ϕ) means that A has the goal ϕ.
- INTEND(A, π) means that A intends to do π.

Teams play an important role in their model. An individual agent is a team and a set of agents (either ordered or not) is a team. They also refer to team roles that we denote using $a, b, \ldots$. We use τ to denote actual teams. Using the individual agent operators they define team operators:

- MBEL(τ, ϕ) means that the team τ jointly believe ϕ.
- JGOAL(τ, ϕ) means that the team τ has the joint goal ϕ.
- JINTEND(τ, π) means that the team τ jointly intend to do π.

They reduce all the multi-agent attitudes to single agent attitudes and provide mechanisms for expressing which part of a joint plan is the responsibility of each member of the team.

As in the Joint Intentions approach, they also apply dynamic logic for simple plan expressions. However, their notation is slightly different. If α is a primitive action, then $*(\tau, \alpha)$ denotes the performance of α by the team τ; $!(\tau, \phi)$ the achievement of ϕ by τ; $(?\tau, \phi)$ the testing of ϕ by τ; and $\hat{}(\tau, \phi)$ that τ is waiting for ϕ to become true. They use ; for sequencing, & for and-parallelism and | for non deterministic choice; $ in front of a string indicates that the string is a variable. (This notation is slightly different from the one in [Levesque *et al.*, 1990] where parallelism is denoted by ||.) They do not discuss the representation of explicit times.

A plan is a tuple $(p, \phi_{purpose}, \phi_{precondition}, \rho_{body})$ where p is a unique plan name, $\phi_{purpose}$ is a goal formula that motivates the plan, $\phi_{precondition}$ is the preconditions that must be true in order to perform the plan and ρ_{body} is a plan graph or a plan expression. The plan includes team variables that represent the roles of the plan. There may be a library of plans that are available with non-assigned roles. Once a team is formed, the roles are assigned to specific individual agents or specific teams. For readability we simplified the notation of teams and roles. The syntax of [Sonenberg *et al.*, 1992] is much richer and allows dealing with very complex role structures and teams.

In our Move example, $\phi_{purpose}$ is
$\text{JGOAL}((a, b, (a, b)), \text{done}(!((a, b, (a, b))Move(Bl, locO, locD))))$.

In the team there are the individual agent roles a and b as well as a team (a, b) associated with the multi-agent actions.

The preconditions are $at(loc1, a, t)$ and $at(loc2, b, t)$.

The body is as follows:

$$\begin{array}{l}(*(a, Go(a, loc1, WC\_SS(locO)))\& *(b, Go(b, loc2, EC\_SS(locO))));\\ \quad *((a, b), Push((a, b), Bl, locO, locT));\\ (*(a, Go(a, WC\_SS(locT), NC\_WS(locT)))\&\\ \quad *(b, Go(b, EC\_SS(locT), SC\_WS(locT))));\\ \quad *((a, b), Push((a, b), Bl, locT, locD))\end{array}$$

When the actual team is formed, the role a is assigned to A_1 and b to A_2. Any agent may form a team. Assume that A_1 forms the team of our example and serves as the team leader. It needs to send A_2 the joint goal to move the block, as well as the joint plan and role assignments. A_1 is assigned the role a and A_2 the role b. The authors describe two protocols A_1 and A_2 can use to form the relevant beliefs and intentions (commit-and-cancel and agree-and-execute). If the agents agree on the plan p and the role assignments, they form a joint intention of the form

$$\begin{array}{l}\text{JINTEND}((A_1, A_2, (A_1, A_2)), p) \equiv\\ \quad \text{INTEND}(A_1, p) \wedge \text{BEL}(A_1, \text{JINTEND}((A_1, A_2, (A_1, A_2)), p)) \wedge\\ \quad \text{INTEND}(A_2, p) \wedge \text{BEL}(A_2, \text{JINTEND}((A_1, A_2, (A_1, A_2)), p)) \wedge\\ \quad \text{JINTEND}((A_1, A_2), p) \wedge \text{MBEL}((A_1, A_2), \text{JINTEND}((A_1, A_2, (A_1, A_2)), p))\end{array}$$

The first two lines specify that each agent intends to perform the complex plan p and believes that A_1, A_2 and the team (A_1, A_2) have a joint intention for doing p. The third line indicates that the team (A_1, A_2) jointly intend p and mutually believe that they all have a joint intention to do the plan p.

For the execution of the plan, each agent transforms the plan to the subplan that it needs to perform based on the agent's role. The function *try* attempts to execute a plan expression and returns the value true if successful, and false otherwise. For example, for the action *Go* that A_1 needs to perform, it will have the following in its executable plan:

$$\begin{array}{l} *(A_1, \$r = try(Go(A_1, loc1, WC\_SS(locO))));\\ (?(A_1, \$r); !(A_1, broadcast(succeeded(Go(A_1, loc1, WC\_SS(locO)))))|\\ (?(A_1, \neg\$r); !(A_1, broadcast(failed(Go(A_1, loc1, WC\_SS(locO)))));\\ \quad *(A_1, fail))). \end{array}$$

That is, A_1 will try to perform the *Go* action. The result of this attempt will be stored in $\$r$. If A_1 succeeds, it will broadcast this success; if A_1 fails it will broadcast its failure.

For the part of the plan where A_1 is not involved, such as the *Go* action for A_2, it will include the following:

$$\begin{array}{l} \hat{}(A_1, failed(Go(A_2, loc2, EC\_SS(locO))) \vee\\ \quad succeeded(Go(A_2, loc2, EC\_SS(locO)))));\\ ?(A_1, succeeded(Go(A_2, loc2, EC\_SS(locO)))) \end{array}$$

That is, A_1 will wait to know whether A_2 succeeded or failed and only then will it continues.

The transformed plan mimics the structure of the original plan, but it includes communications guaranteeing that each role-player knows when a given subplan has succeeded or failed. This common knowledge ensures that the temporal order of the actions and the behavior in response to failure in the original plan are preserved in the execution of the transformed plans. For this a new conjunction has been added to the JINTEND definition. In particular in our example the following will be added to the specification of JINTEND$((A_1, A_2, (A_1, A_2)), p)$:

$$\begin{array}{l} \text{BEL}(A_1,\\ \quad done(!(A_1, transform(a, p))) \wedge done(!(A_2, transform(b, p))) \wedge\\ \quad done(!((A_1, A_2), transform((a, b), p))) \rightarrow done(!((A_1, A_2, (A_1, A_2)), p))) \wedge\\ \text{BEL}(A_2,\\ \quad done(!(A_1, transform(a, p))) \wedge done(!(A_2, transform(b, p))) \wedge\\ \quad done(!((A_1, A_2), transform((a, b), p))) \rightarrow done(!((A_1, A_2, (A_1, A_2)), p))) \wedge\\ \text{MBEL}((A_1, A_2),\\ \quad done(!(A_1, transform(a, p))) \wedge done(!(A_2, transform(b, p))) \wedge\\ \quad done(!((A_1, A_2), transform((a, b), p))) \rightarrow done(!((A_1, A_2, (A_1, A_2)), p))) \end{array}$$

Intuitively this means that each agent believes that if each agent and subteam do their part of the plan as specified by the "transform" function (including the broadcast and the waiting parts) then the team will perform the plan. Similarly, the subteam (A_1, A_2) has such a mutual belief.

3.3 SharedPlans

The SharedPlan formalization of collaboration [Grosz and Kraus, 1996; Grosz and Kraus, 1999; Grosz and Kraus, 1993] is based on a mental-state view of plans [Bratman, 1987; Pollack, 1990]: agents are said to have plans when they have a particular set of intentions and beliefs. Both *shared plans* and *individual plans* are considered. Shared plans are constructed by groups of collaborating agents and include subsidiary shared plans [Lochbaum, 1994] formed by subgroups as well as subsidiary individual plans formed by individual participants in the group activity. The formalization distinguishes between complete plans—those in which all the requisite beliefs and intentions have been established—and partial plans.

The plan definitions apply intention operators.

- $\text{Int.To}(G, \alpha, T_i, T_\alpha, \text{con}(\alpha), IC_\alpha)$ represents agent's (G) intention at time T_i to do action α at time T_α under the constraints con(α) in the context IC_α.
- $\text{Int.Th}(G, prop, T_i, T_{prop}, \text{IC}_{prop})$ represents an agent's (G) intention at time T_i that a certain proposition *prop* hold in the intentional context IC_{prop} at time T_{prop}.

The significant distinction between intentions-to and intentions-that is not in the types of objects each relates, but in their connection to means-ends reasoning and in their different presumptions about an agent's ability to act in service of the intention. An Int.To commits an agent to means-ends reasoning [Bratman, 1987] and, at some point, to acting. In contrast, an Int.Th does not directly engender such behavior. Int.Ths form the basis for meshing subplans, helping one's collaborators, and coordinating status updates, all of which play an important role in collaborative plans; any of these activities may lead to the adoption of an Int.To and thus indirectly to means-ends reasoning. In addition, an agent can only adopt an Int.To toward an action for which it is the agent and that it believes it can do.

SharedPlan applies also the belief operators:

- $\text{Bel}(G, prop, T_{bel})$ means that G believes at time T_{bel} that *prop* is true.
- $\text{MB}(\text{GR}, prop, T_{bel})$ means that the group GR mutually believe at time T_{bel} that *prop* is true.

In addition, two meta-predicates are defined to represent the beliefs agents have about their own and their collaborators' abilities:

- $\text{CBA}(G, \alpha, R_\alpha, T_\alpha, \Theta)$ ("can bring about") represents an agent's (G) ability to do the action α using the recipe R_α at time T_α and under constraints Θ.

- CBAG(GR, $\alpha, R_\alpha, T_\alpha, \Theta$) (Can bring about group) represents the ability of the group GR to do the action α using the recipe R_α at time T_α and under constraints Θ.

The two main meta-predicates that are defined in the *SharedPlans* formalization are:

- FSP(P, GR, $\alpha, T_p, T_\alpha, R_\alpha, \Theta_\alpha, IC_\alpha$) means that the group GR has at time T_p a full plan P to do the multi-agent action α at time T_α using the recipe R_α under the constraint Θ_α in the context of IC_α.

- PSP(P, GR, $\alpha, T_p, T_\alpha, \Theta_\alpha, IC_\alpha$) means that the group GR has at time T_p a partial plan P to do the multi-agent action α at time T_α under the constraint Θ_α in the context of IC_α.

The meta-predicate FSP is used to represent the situation in which a group of agents has completely determined the recipe by which they are going to do some group activity, and members of the group have adopted Int.Tos toward all the basic-level actions in the recipe as well as Int.Ths toward the actions of the group and its other members. Most typically a group of agents will not have such a complete plan until after they have done some of the actions in the recipe. Groups, like the individual agents they comprise, typically have only partial plans for which the meta-predicatePSP is used. meta-predicate FIP, similarly defined, for individual agents.

Now we describe the full SharedPlan of our example. Here α is $Move(\{A_1, A_2\}, Bl, locO, locD)$, and for the rest of this subsection we use α for this specific action. $Recipes(\alpha)$ are all the possible recipes for α and R_{Move} is the recipe presented in Section 2 that consists of $\beta_1 - \beta_6$ with the order constraints described there.

Next we describe the constituents of $\mathrm{FSP}(P, \{A_1, A_2\}, \alpha, t_p, [T, T^e_\alpha], R_{Move}, \theta_\alpha, IC_\alpha)$, i.e., the full shared plan P of A_1,A_2 at time t_p to move the block Bl.

0. The group $GR = \{A_1, A_2\}$ has mutual belief that both of them are committed to the success of the group's doing α; in particular each agent has Int.Th at time t_p that they do α in the context of the FSP.

$$\begin{aligned}&\mathrm{MB}(\{A_1, A_2\}, (\forall G \in \{A_1, A_2\})\\ &\qquad \mathrm{Int.Th}(G, \mathrm{Do}(\{A_1, A_2\}, \alpha, [t, t+k+6]),\\ &\qquad\qquad\qquad t_p, [t, t+k+6], FSP(\alpha)), t_p)\end{aligned}$$

1. The group $\{A_1, A_2\}$ has mutual belief that the recipe R_α that they agreed to use is actually a recipe for α.

$\text{MB}(\{A_1, A_2\}, R_\alpha \in \textit{Recipes}(\alpha), t_p)$

For each of the actions of the recipe R_α, the agents have agreed who will do it and they adopt intentions and form beliefs. There are two cases: single-agent actions and multi-agent actions. In the recipe there are four single-agent actions, $\beta_1 = Go(A_1, loc1, WC\_SS(locO))$, $\beta_2 = Go(A_2, loc2, EC\_SS(locO))$, $\beta_4 = Go(A, WC\_SS(locT), NC\_WS(locT))$, $\beta_5 = Go(A_2, EC\_SS(locT), SC\_WS(locT))$ and two multi-agent actions $\beta_3 = Push(GR, Bl, locO, locT)$ and $\beta_6 = Push(GR, Bl, locT, locD)$.

We first consider one of the single-agent actions, β_1 only; the others are similar.

- A_1 has the intention to do β_1 at time $T_{\beta_1} = [t, t+k_1]$ under the relevant constraints from the recipe, i.e., $\{Before(\beta_1, \beta_3)\}$. Its context for doing it is the FSP of α. Formally,
  $\text{Int.To}(A_1, \beta_1, t_p, T_{\beta_1}, \{Before(\beta_1, \beta_3)\}, FSP(\alpha))$.
- Both agents have the mutual belief that A_1 intends to do β_1
  $\text{MB}(\{A_1, A_2\}, \text{Int.To}(A_1, \beta_1, t_p, T_{\beta_1}, \{Before(\beta_1, \beta_3)\}, FSP(\alpha)), t_p)$.
- Both agents mutually believe that A_1 can do β_1. As β_1 is a basic-level action, R_{β_1} is empty.
  $\text{MB}(\{A_1, A_2\}, \text{CBA}(A_1, \beta_1, R_{Empty}, T_{\beta_1}, \{Before(\beta_1, \beta_3)\}), t_p)$.
- A_2 is committed to the success of A_1 to perform β_1 by having an Int.Th that it will be able to perform it. This intention is mutually believed by the two agents.
  $\text{MB}(\{A_1, A_2\},$
  $\quad \text{Int.Th}(A_2, (\exists R_{\beta_1})\text{CBA}(A_1, \beta_1, R_{\beta_1}, T_{\beta_1},$
  $\qquad \{Before(\beta_1, \beta_3)\}), t_p, T_{\beta_1}, \beta_1), t_p)$
  If β_1 is a basic level action, R_{β_1} is empty.
- If β_1 were not a basic level action, A_1 should have a full individual plan for doing it and both agents should have mutual belief about it.
  - $\neg\text{basic.level}(\beta_1)$
  - There should be a recipe for the *Go* action, R_{β_1}, and full individual plan P_{β_1} of A_1 to do β_1 at T_{β_1}. The context is the full shared plan of α and the action should be done under the constraint $Before(\beta_1, \beta_3)$. Formally,
    $(\exists P_{\beta_1}, R_{\beta_1})$
    $\text{FIP}(P_{\beta_1}, A_1, \beta_1, t_p, T_{\beta_1}, R_{\beta_1}, \{Before(\beta_1, \beta_3)\}, FSP(\alpha))$.

– A_1 and A_2 mutually believe that such a plan and a recipe exists and that A_1 can bring about the performance of α according to the recipe. Note that A_2 does not need to know the recipe nor the details of the plan. It just needs to believe that A_1 has them.
$$\text{MB}(\{A_1, A_2\}, (\exists P_{\beta_1}, R_{\beta_1}) \\ [\text{CBA}(A_1, \beta_1, R_{\beta_1}, T_{\beta_1}, \{Before(\beta_1, \beta_3)\}) \wedge \\ \text{FIP}(P_{\beta_1}, A, \beta_1, t_p, T_{\beta_1}, R_{\beta_1}, \{Before(\beta_1, \beta_3)\}, FSP(\alpha))], t_p).$$

The clauses associated with the other single-agent actions of the recipe R_α, namely β_2 β_4 β_5, are similar to those of β_1. The intentions and beliefs associated with the multi-agent action $\beta_3 = Push(\{A_1, A_2\}, Bl, locO, locT)$ are different from that of the single-agent actions. They include the following:

- There should be a full SharedPlan, P_{β_3}, for the agents to do the Push using a recipe R_{β_3}. This recipe consists of $\gamma_3^1 = LPush(A_1, Bl, locX, locY)$ and $\gamma_3^2 = RPush(A_2, Bl, locX, locY)$ and the constraint $\{Sim(\gamma_i^1, \gamma_i^2)\}$. The relevant constraints for the Push are $\{Before(\beta_3, \beta_4), Before(\beta_3, \beta_5), Before(\beta_1, \beta_3), Before(\beta_2, \beta_3)\}$ and the context is the full SharedPlans for α:
$$(\exists P_{\beta_3}, R_{\beta_3})\ \text{FSP}(P_{\beta_3}, \{A_1, A_2\}, \beta_3, t_p, T_{\beta_3}, R_{\beta_3}, \{Before(\beta_3, \beta_4), \\ Before(\beta_3, \beta_5), Before(\beta_1, \beta_3), Before(\beta_2, \beta_3)\}, \text{FSP}(\alpha)).$$

- A_1 and A_2 mutually believe that they can perform β_3 and that they have a full SharedPlan to do it. Note that if the group for the Move included more agents, the entire group would have had this mutual belief. In our case since the subgroup and the entire group are the same, this clause can be inferred from the previous one.
$$\text{MB}(\{A_1, A_2\}, (\exists P_{\beta_3}, R_{\beta_3}) \\ [\text{CBAG}(\{A_1, A_2\}, \beta_3, R_{\beta_3}, T_{\beta_3}, \{Before(\beta_3, \beta_4), Before(\beta_3, \beta_5), \\ Before(\beta_1, \beta_3), Before(\beta_2, \beta_3)\}) \wedge \\ \text{FSP}(P_{\beta_3}, \{A_1, A_2\}, \beta_3, t_p, T_{\beta_3}, R_{\beta_3}, \{Before(\beta_3, \beta_4), Before(\beta_3, \beta_5), \\ Before(\beta_1, \beta_3), Before(\beta_2, \beta_3)\}, FSP(\alpha))], t_p)$$

- If there had been additional members in the group, they should have had Int.Th that the subgroup will succeed in performing β_3.

An important aspect of the *SharedPlans* model is that it motivates helpful behavior of the group members. If something goes wrong and one of the agents is not able to perform its action, other members in the group will consider helping. For example, if A_1 is running out of fuel and is not able to perform a *Go* action, A_2 will try to help it. This is motivated by the Int.Th of A_2 that A_1 will succeed in performing the *Go*. In addition, this Int.Th (via axioms included in the *SharedPlans* model) also motivates the following: if A_2 is able to reduce the costs of A_1 doing the Go action, without bearing a big loss, it will help A_1 too. For example, if A_2 finds out that there is an obstacle in A_1's way, it will let A_1 know.

3.4 Cooperative Problem Solving

The paper [Wooldridge and Jennings, 1999] provides a model of cooperative problem solving starting with the recognition of the potential for cooperation to the team action itself. The notation is based on a branching-time tree formed of states connected by arcs representing primitive actions. Paths through the tree can be characterized by sequences of actions. Terms may denote agents, groups of agents, sequences of actions, and other objects in the environment. Action expressions are formed by using ";" for sequencing, "—" for non-deterministic choice, "*" for iteration, and "?" for test actions.The important modal operators are as follows:

- $Bel(a, \phi)$ means that agent a has belief ϕ.
- $Goal(a, \phi)$ means that agent a has goal ϕ.
- $Agts(\alpha, g)$ means that g is an agent group required to do α.
- $Agt(\alpha, a)$ means that a is the only agent for α.
- $Happens(\alpha)$ means that action α is the first thing that happens on a given path.
- $A\ \phi$ means that ϕ holds on all paths (inevitably ϕ).

The following are important examples of some of the operators that can be defined using these concepts.

- Mutual belief, *M-Bel* is defined using the everyone believes, *E-Bel*, operator as follows:
  $E\text{-}Bel(g, \phi, 0) =_{def} \phi$
  $E\text{-}Bel(g, \phi, \ell + 1) =_{def} \forall a(a \in g \rightarrow Bel(a, E\text{-}Bel(g, \phi, \ell)))$, where ℓ represents the level of nesting for *E-Bel*
  $M\text{-}Bel(a, \phi) =_{def} E\text{-}Bel(g, \phi, \ell)$ for all levels ℓ.
- $Does(\alpha) =_{def} A\ Happens(\alpha)$ meaning that α happens on all paths.
- $Achieves(\alpha, \phi) =_{def}$
  $$E(Happens(\alpha)) \wedge A((Happens(\alpha) \rightarrow Happens(\alpha; \phi?))$$
  meaning that α achieves the goal ϕ if α happens on some path and on all paths if α happens then ϕ is true afterwards. (Note: E is the dual of A.)

- $Able_1(\alpha, \phi) =_{def} \exists\alpha(Bel(a, (Agt(\alpha, a) \wedge Achieves(\alpha, \phi))\wedge$
  $Agt(\alpha, a) \wedge Achieves(\alpha, \phi))$
  An agent has type 1 ability to achieve ϕ if it believes that some action can achieve ϕ and it can do the action.
  $Able(a, \phi) =_{def} Able_1(a, \phi) \vee Able_1(a, Able_1(a, \phi))$
  An agent has the ability to achieve ϕ if either it has type 1 ability for it or it has type 1 ability to bring about a situation where it has type 1 ability to achieve it.

- $J\text{-}Able_1(g, \phi) =_{def} \exists\alpha$
  $(M\text{-}Bel(g, Agts(\alpha, g) \wedge Achieves(\alpha, \phi)) \wedge Agts(\alpha, g) \wedge Achieves(\alpha, \phi))$
  Type 1 joint ability of a group of agents g means that there is an action that achieves ϕ that the group is required to do and this is mutually believed.
  $J\text{-}Able(g, \phi) =_{def} J\text{-}Able_1(g, \phi) \vee J\text{-}Able_1(g, J\text{-}Able_1(g, \phi))$
  Joint ability is obtained from type 1 joint ability analogously as ability is obtained from type 1 ability.

- $J\text{-}Commit(g, \phi, \psi, \chi, c)$ has a fairly complex definition. It defines joint commitment with the parameters g for the group, ϕ for the goal, ψ for the motivation, χ for the pre-condition, and c for the convention. We do not give the details here.

- $J\text{-}Intend(g, \alpha, \psi) =_{def} M\text{-}Bel(g, Agts(\alpha, g))\wedge$
  $J\text{-}Commit(g, A \diamond Happens(M{-}Bel(g, A\ Happens(\alpha)); \alpha), \psi, \chi_{soc}, c_{soc})$
  J-Intend is the joint intention of the group g with respect to action α and motivation ψ. Here $\diamond\ \phi$ means that ϕ is eventually satisfied; χ_{soc} and c_{soc} involve social pre-conditions and conventions. We can read joint intention to mean that the group g has a joint commitment that eventually g will believe that α will happen next, and then α happens next.

- $Team(g, \phi, , a) =_{def}$
  $\exists\alpha(M\text{-}Bel(g, Achieves(\alpha, \phi)) \wedge J\text{-}Intend(g, \alpha, Goal(a, \phi))$
  The agents g form a team to accomplish agent a's goal ϕ if the agents in g mutually believe that some action sequence α achieves ϕ and they jointly intend to do it.

We did not include here the details of various other steps including individual commitments, intentions, and the potential for cooperation.

Consider now how our example can be formalized using these concepts. Initially an agent, say A_1, is presented with the goal of having the block Bl at location $locD$. We can write this as $Goal(A_1, At(Bl, locD))$. As

[Wooldridge and Jennings, 1999] does not explicitly deal with parallel actions, we add parallelism by using the extra symbol &. There is also no discussion of complex subactions, so we consider only basic-level actions and write $\alpha = (\beta_1 \& \beta_2); (\gamma_3^1 \& \gamma_3^2); (\beta_4 \& \beta_5); (\gamma_6^1 \& \gamma_6^2)$. This formalism also does not have explicit time representation. In this case the group $g = (A_1, A_2)$. So we would get to $M\text{-}Bel((A_1, A_2), Achieves(\alpha, At(Bl, locD)))$. The preconditions here include $At(A_1, loc1)$, $At(A_2, loc2)$, $At(Bl, locO)$. If everything goes well, we get to $Team((A_1, A_2), At(Bl, locD), \alpha)$ at which point A_1 and A_2 form a team to apply the sequence of actions α to move the block from *locO* to *locD*. There will also be individual intentions to do the various subactions: for example A_1 to do β_1, A_2 to do β_2, and so on.

3.5 Collective Intentions

In a series of three papers, [Dunin-Keplicz and Verbrugge, 2002], [Dunin-Keplicz and Verbrugge, 2003], and [Dunin-Keplicz and Verbrugge, 2004] Dunin-Keplicz and Verbrugge formally characterize the concept of a team having a collective intention and collective commitment towards a common goal. Many aspects of teamwork get a thorough formal treatment. They give a modal logic axiomatization for the predicates and prove the completeness of an important portion of their formalization, collective intentions, by the use of Kripke models.

The language contains propositional symbols, symbols for agents and atomic actions, and a large number of predicate symbols. Individual actions are built up from atomic actions by using sequential composition: $(\alpha_1; \alpha_2)$, non-deterministic choice: $(\alpha_1 \cup \alpha_2)$, iteration: (α^*), as well as several other predicate symbols. Using individual actions, social plan expressions are formed by writing $< \alpha, a >$ (agent a associated with action α), $stit(G, \phi)$ (group G sees to it that ϕ), $confirm(\phi)$, as well as by combining social plan expressions Γ and Δ as $< \Gamma; \Delta >$ (sequential composition) and $< \Gamma || \Delta >$ (parallelism).

Various modalities are also defined. A partial list includes:

epistemic - versions of belief

motivational - versions of goals, intentions and commitments

temporal - versions of *done*, *succeeded*, *failed*

abilities and opportunities - $able(a, \alpha)$, $opp(a, \alpha)$

Now we list some of the important epistemic and motivational predicate symbols with their meanings; however we change the subscripts to parameters:

- $BEL(a, \phi)$ means that agent a believes ϕ.
- $E\text{-}BEL(G, \phi)$ means that every agent in group G believes ϕ.
- $C\text{-}BEL(G, \phi)$ means that group G collectively believes ϕ.
- $GOAL(a, \phi)$ means that agent a has goal ϕ.
- $INT(a, \phi)$ means that agent a has the intention to make ϕ true.
- $E\text{-}INT(G, \phi)$ means that every agent in G has an individual intention to make ϕ true.
- $M\text{-}INT(G, \phi)$ means that G has a mutual intention to make ϕ true.
- $C\text{-}INT(G, \phi)$ means that G has a collective intention to make ϕ true.
- $COMM(a, b, \phi)$ means that agent a commits to agent b to make ϕ true.
- $R\text{-}COMM(G, P, \phi)$ means that group G has a robust collective commitment to achieve ϕ by plan P.
- $S\text{-}COMM(G, P, \phi)$ means that group G has a strong collective commitment to achieve ϕ by plan P.
- $W\text{-}COMM(G, P, \phi)$ means that group G has a weak collective commitment to achieve ϕ by plan P.
- $T\text{-}COMM(G, P, \phi)$ means that group G has a team commitment to achieve ϕ by plan P.
- $D\text{-}COMM(G, P, \phi)$ means that group G has a distributed commitment to achieve ϕ by plan P.

E-BEL refers to the individual belief of each agent in the group. Collective belief, *C-BEL*, is a recursive extension of *E-BEL*: every agent believes that every agent believes ... ϕ. For intentions, again *E-INT* refers to the individual intention of each agent in a group. Mutual intention, *M-INT*, is a recursive extension of *E-INT*: every agent intends that every agent intends ... ϕ. Collective intention, *C-INT*, extends mutual intention by requiring also the collective belief of the group in the mutual intention. Commitment, $COMM$, involves an intention, a goal, and a collective belief in the intention and the goal. The various types of group commitments differ in some subtle details involving differences in the awareness of the responsibilities of the various agents in the group. The group commitments listed above go from the strongest to the weakest versions. For example, in the strongest version, robust commitment, essentially every agent is aware of all the intentions and commitments of all the other agents; while in the weakest version, distributed commitment, there is not even a collective intention and agents are aware only of their subactions.

To indicate the approach of the papers, which rely heavily on axiomatization of the concepts, we include several representative axioms. For introspection axioms we give only the positive versions; there are also negative introspection axioms.

A4 $BEL(a, \phi) \rightarrow BEL(a, BEL(a, \phi))$ (positive introspection)

R2 From ϕ infer $BEL(a, \phi)$ (all agents believe the true propositions)

C1 $E\text{-}BEL(G, \phi) \leftrightarrow \bigwedge_{a \in G} BEL(a, \phi)$ (the meaning of *E-BEL*)

C2 $C\text{-}BEL(G, \phi) \leftrightarrow E\text{-}BEL(G, \phi \wedge C\text{-}BEL(G, \phi))$ (collective belief in terms of *E-BEL*)

A2$_G$ $GOAL(a, \phi) \wedge GOAL(a, \phi \rightarrow \psi) \rightarrow GOAL(a, \psi)$ (goal distribution)

A7$_{GB}$ $GOAL(a, \phi) \rightarrow BEL(a, GOAL(a, \phi))$ (positive introspection for goals)

A2$_I$ $INT(a, \phi) \wedge INT(a, \phi \rightarrow \psi) \rightarrow INT(a, \psi)$ (intention distribution)

A7$_{IB}$ $INT(a, \phi) \rightarrow BEL(a, INT(a, \phi))$ (positive introspection for intentions)

A9$_{IG}$ $INT(a, \phi) \rightarrow GOAL(a, \phi)$ (intention implies goal)

M1 $E\text{-}INT(G, \phi) \leftrightarrow \bigwedge_{a \in G} INT(a, \phi)$ (the meaning of *E-INT*)

M2 $M\text{-}INT(G, \phi) \leftrightarrow E\text{-}INT(G, \phi \wedge M\text{-}INT(G, \phi))$ (mutual intention in terms of *E-INT*)

M3 $C\text{-}INT(G, \phi) \leftrightarrow M\text{-}INT(G, \phi) \wedge C\text{-}BEL(G, M\text{-}INT(G, \phi))$ (collective intention in terms of collective belief and mutual intention)

SC1 $COMM(a, b, \phi) \leftrightarrow INT(a, \phi) \wedge GOAL(b, stit(a, \phi))$
$\wedge C\text{-}BEL((a, b), INT(a, \phi) \wedge GOAL(b, stit(a, \phi)))$ (commitment for a proposition in terms of intentions, goals, and collective belief)

SC2 $COMM(a, b, \alpha) \leftrightarrow INT(a, \alpha) \wedge GOAL(b, done(a, \alpha)) \wedge$
$C\text{-}BEL((a, b), INT(a, \phi) \wedge GOAL(b, done(a, \alpha)))$ (commitment for an action in terms of intentions, goals, and collective belief)

Consider now how our example can be formalized in this framework. A social plan for it can be expressed as

$P =<<<<< \beta_1, A_1 > || < \beta_2, A_2 >>; << \gamma_3^1, A_1 > || < \gamma_3^2, A_2 >>>;$
$<< \beta_4, A_1 > || < \beta_5, A_2 >>>; << \gamma_6^1, A_1 > || < \gamma_6^2, A_2 >>>$

This will lead to the various individual commitments, such as $COMM(A_1, A_2, \beta_1)$, $COMM(A_2, A_1, \beta_2)$, and so on and at some point to the group commitment, probably $S\text{-}COMM((A_1, A_2), P, \alpha)$ in this case. The execution of α will follow the commitments; there is no explicit time representation. The last two papers go into substantial detail about what happens if some action fails during execution.

3.6 Cooperative Subcontracting

Grant, Kraus, and Perlis [Grant *et al.*, 2005; Grant *et al.*, 2002] provide a formal theory for the handling of agent intentions in a framework that allows multiple agents doing various tasks together. We briefly describe those aspects of the language that are needed for the formalization of our example; we omit many important details, such as how agents communicate. There are, in fact, several languages. The part of the language described here is really a metalanguage that has names for formulas in the language of the agents. This (meta)language is a sorted first-order language with different sorts for time, agents, actions, recipes, and formulas. To conform to the example given for this paper in the different formalisms we slightly change the symbols used in the original paper. The following symbols are used: for time: t, i, j, k; for agents: A_i; for actions: α, etc.; for recipes: r; for formulas: f.

Actions are either basic-level or complex; for a complex action there is always a recipe.

- $BL(\alpha, d)$ means that α is basic level and takes d units of time.
- $Rec(\alpha, r)$ means that α has recipe r. We only deal with the single recipe case here.
- $Mem(r, \beta, i, j, k)$ means that in recipe r the subaction β is one of the ith subactions (there may be more than one) starting at time j and ending at time k. These times are offsets relative to the beginning of the action.

Agents have beliefs, such as that they can do an action.

- $Bel(t, A_i, f)$ means that at time t agent A_i believes in the statement expressed by the formula f.
- $CanDo(t, A_i, \alpha)$ means that at time t agent A_i can do (at least start to do) action α.

We are primarily interested in the concept of intention. There are three main predicates for this.

- $ATD(t, A_i, A_j, \beta, \alpha, t')$ means that at time t agent A_i asks agent A_j to do action β as a subaction of action α at time t'.

- $PotInt(t, A_i, A_j, \beta, \alpha, t')$ means that at time t agent A_i directly assisting agent A_j has the potential intention to do action β as a subaction of action α at time t'.

- Int is defined exactly as $PotInt$, but it stands for the actual intention.

The axioms of this system show what happens in a time step given various initial conditions; for example, how an agent gains potential intentions and intentions.

The general process is as follows. The agent is asked to do an action at a particular time. This triggers a potential intention for the agent. This potential intention may lead to an intention assuming certain conditions are satisfied. In particular, the agent must believe that it can do the action at the right time, possibly with help from other agents. The parts of the action the agent cannot do, it can subcontract to other agents. A subcontracting means that an agent is asked to do an action leading again to a potential intention. The time for each action is explicitly represented by a time value. The recipes for the actions are known by the agents, as needed.

In our example, the action α is the action of moving the block from $locO$ to $locD$. There are two agents, A_1 and A_2. We must use subcontracting for the joint action; hence initially A_1 is asked to do this move and A_1 must subcontract some subactions to A_2. Some of the actions are done in parallel. Since our system requires that the first subaction of every action must be a single action, we add an essential dummy first subaction, as needed. In the case of the agents going to the block, this first subaction may be "prepare to go".

Later we will only sketch the attainment of intentions and the actual actions, but for clarity we write the recipe out in detail.

- $Rec(\alpha, r)$: r is the recipe for α: "move the block from $locO$ to $locD$".

- $Mem(r, \beta_1, 1, 1, 1)$: The "prepare to go" subaction starts and ends at time 1.

- $Mem(r, \beta_{21}, 2, 2, k_1 + 2)$: The "go to $locO$" subaction for agent A_1 takes k_1 time steps.

- $Mem(r, \beta_{22}, 2, 2, k_2 + 2)$: The "go to *locO*" subaction for agent A_2 takes k_2 time steps. Note that both β_{21} and β_{22} are considered the second subaction. Next recall that $k = max(k_1, k_2)$.
- $Mem(r, \beta_3, 3, k+3, k+6)$: The "push" subaction is a complex action.
- $Mem(r, \beta_{41}, 4, k + 7, k + 8)$: The first agent goes to the new position.
- $Mem(r, \beta_{42}, 4, k+7, k+8)$: The second agent goes to the new position.
- $Mem(r, \beta_3, 5, k + 9, k + 12)$: Another "push" is done.
- $Rec(\beta_3, r')$: r' is the recipe for "push".
- $Mem(r', \gamma_1, 1, 1, 1)$: The "prepare to push" subaction is analogousto the "prepare to go" subaction.
- $Mem(r', \gamma_{21}, 2, 2, 3)$: The "push from left" subaction takes one time step.
- $Mem(r', \gamma_{22}, 2, 2, 3)$: The "push from right" subaction takes one time step.

Next we briefly sketch some of the steps involved in the getting and keeping of intentions by the two agents. In our system we allow the retraction of intentions for various reasons but assume it doesn't happen here. In order to indicate clearly the time value, we write it first separately and leave it out of the formula itself. We intersperse the timed actions with various explanations.

time 0 $ATD(\_, A_1, \alpha, \_, t)$
A_1 is asked to do action α at time t.

time 1 $PotInt(A_1, \_, \alpha, \_, t)$
$Bel(A_1, \text{“}CanDo(t, A_1, \alpha)\text{”})$
The ATD triggers a $PotInt$ for A_1 to do action α at time t. We also assume that A_1 believes it can do α, possibly with help from other agents. This belief triggers potential intentions for the subactions of α while the potential intention for α is inherited.

time 2 $PotInt(A_1, \_, \alpha, \_, t)$
$PotInt(A_1, A_1, \beta_1, \alpha, t + 1)$
$Bel(A_1, \text{“}CanDo(t + 1, A_1, \beta_1)\text{”})$
The first formula shows the inheritance of $PotInt$. The second formula is the $PotInt$ for β_1. There is a similar $PotInt$ for all the βs. The third formula shows A_1's belief about β_1.

time 3 $Int(A_1, A_1, \beta_1, \alpha, t+1)$
$ATD(A_1, A_2, \beta_{22}, \alpha, t+2)$
The first formula gives the first intention for A_1. A similar intention is obtained for all subactions that A_1 believes it can do. The second formula shows A_1 asking A_2 to do β_{22}. This is a subaction A_1 cannot do. Similarly A_1 asks A_2 to do β_{42}.

time 4 $PotInt(A_2, A_1, \beta_{22}, \alpha, t+2)$
$Int(A_1, A_1, \gamma_1, \beta_3, t+k+4)$
$ATD(A_1, A_2, \gamma_{22}, \beta_3, t+k+5)$
The first formula shows that A_2 obtains a potential intention for β_{22}. The second formula shows the intention of A_1 to do γ_1. A_1 must ask A_2 to do γ_{22}. Continuing this process we just show the main formula at time 8.

time 8 $Int(A_1, \_, \alpha, \_, t)$
This is the point where A_1 gets an intention for α as all subactions have been assigned and for each subaction of α either A_1 or A_2 has an intention to do it.

Finally, we sketch the initiation of the first subactions:

time t $Ini(A_1, \_, \alpha)$
Given the continuing intention of A_1 to do α, then at time t A_1 actually initiates the action.

time t+1 $Ini(A_1, A_1, \beta_1)$
$Bel(A_1, \text{“}Done(t+1, \beta_1)\text{”})$
A_1 initiates β_1 which is just the preparation subaction and gets done immediately.

time t+2 $Ini(A_1, A_1, \beta_{21})$
$Ini(A_2, A_1, \beta_{22})$
A_1 initiates its subaction β_{21} and A_2 initiates its subaction β_{22}.

Assuming the agents can do all the subactions at the right time, they all get done. We do not show the details.

4 Comparison

There are many ways in which the formal approaches presented in the previous section can be compared. We have chosen several criteria for comparison and present it here. We do not claim that this is an exhaustive list.

We start by considering the focus of these approaches. We have found that this difference in focus leads in a natural way to differences in the way that different authors handle many issues. The focus of Joint Intentions is the persistence of joint intentions. The focus of Team Plans is on the complexity of actions and the roles of agents. SharedPlans emphasizes individual intentions and actions needed for teamwork. Cooperative Problem Solving emphasizes joint commitment to a goal and cooperation. Collective Intentions emphasizes the formalization of collective intentions in as multi-modal logical framework that allows for a proof of the completeness of the logic. Cooperative Subcontracting focuses on subcontracting between cooperative agents.

Consider now the formalisms themselves. Along this dimension we can distinguish approaches that use a modal logic and ones that have a syntactic approach. The modal logic approach uses various modalities in the syntax and Kripke models for the semantics. Typically, in dealing with plans and actions, some aspects of dynamic logic are included. Among the six systems, Joint Intentions, Team Plans, Cooperative Problem Solving, and Collective Intentions use the modal logic approach. SharedPlans and Cooperative Subcontracting both use the syntactic approach; the latter also deals with semantics via minimal models.

Another issue separating the various formalizations is whether in the system teamwork is based on the intentions and actions of single agents (working together) or primarily on mutual beliefs and joint intentions. This is basically a matter of degree because in all cases at some point agents must have their own beliefs and intentions. Here we find that Joint Intentions, Cooperative Problem Solving, and Collective Intentions emphasize collective/joint intentions, while Team Plans, SharedPlans, and Cooperative Subcontracting have more emphasis on the intentions of single agents.

We can also consider the representation of time. Joint Intentions, SharedPlans, and Cooperative Subcontracting allow for the explicit representation of time in the language. Team Plans has time in the semantics in the form of a time tree. Cooperative Problem Solving also does not have explicit time values in the syntax and deal with it through path formulas, expressing properties of a single path through a branching time structure. Collective Intentions does not deal with time.

Systems differ also in their facility for expressing complex plans. Team Plans places great emphasis on this issue. SharedPlans and Cooperative Subcontracting also possess specific constructs to express complex plans. Joint Intentions, Cooperative Problem Solving, and Collective Intentions do not deal with this issue.

Now we recall the four stages of cooperative behavior as given in the Introduction from the Cooperative Problem Solving Approach: recognition for the possibility of cooperation, team formation, plan formation, and execution. We modify some of the explanation there to provide a more comprehensive formulation of these four steps. Recognition is about agent abilities and beliefs, as well as the potential for cooperation. Team formation involves mutual beliefs, intentions, and commitments. Plan formation consists of possibly negotiation between agents as well as formulating the actions and their sequence that need to be done. Finally, team action should be the actual activity of the agents in accomplishing their goal. Using these definitions, we find that except for Joint Intentions, which emphasizes mainly team formation, the other approaches appear to contain at least some aspects of each of the four stages.

We also recall that Bratman [1992] gives four criteria that a multi-agent activity must meet to be a "shared cooperative activity": mutual responsiveness, commitment to the joint activity, commitment to mutual support, and formation of subplans that mesh with one another. All works consider the issue of the commitment to the joint activity. The requirements from an agent to be considered being committed to the team activity varied from one model to the other. With respect to the other requirements, SharedPlans considers all the other three of Bratman's requirement explicitly (Section 7 of [Grosz and Kraus, 1996]). Team Plans studies carefully the issue of coordination and synchronization of team activities and thus handles the issue of meshing subplans. It also handles mutual responsiveness in action, namely the responsiveness required when plans must be modified to cope with problems in execution. This may also lead to mutual support, but the details are left to future work (Section 5.3 of [Sonenberg *et al.*, 1992]). Collective Intentions provides a formal model for the dynamic change in collective commitments and intentions and thus handles mutual responsiveness and, in some ways, the meshing of subplans and mutual support. The original Joint Intentions paper [Levesque *et al.*, 1990] considers only the issue of the commitments to the joint activity and defines how agents disband a team. In a later paper, [Cohen *et al.*, 1997], issues such as commitment to mutual support and mutual responsiveness are discussed. The Cooperative Problem Solving considers implicitly the issue of meshing subplans and mutual support in their plan formation process. Cooperative Subcontracting deals with mutual responsiveness and with the meshing of subplans both in the planning phase and in the execution phase.

5 Conclusions and Future Work

We compared six approaches to teamwork from a substantial literature on the subject. Our primary technique was to see how these approaches handle a specific task of two agents pushing a block. Teamwork is a highly complex issue and we used the example to illustrate some important aspects of it.

We did not deal with some other issues associated with teamwork such as the details of preconditions, communication between agents, and intention reconsideration. Our examination of the six formalisms with respect to our example of two cooperating agents indicates that there are (at least) two sub-areas of emphasis in the literature: formation of the preconditions for cooperative action, and enactment based upon the cooperative preconditions. Some approaches tend to cover both of these, but none is entirely adequate for an end-to-end treatment that can describe the entire agent processing. Clearly, there is room for more work, especially integrative work; in particular, two interrelated components in need of more attention are explicit representations of time and primitive actions.

Acknowledgement

We wish to thank Artur Garcez for many helpful comments.This research was supported by NSF under grant IIS-0222914 and by AFOSR and ONR.

BIBLIOGRAPHY

[Alonso, 1997] E. Alonso. A formal framework for the representation of negotiation protocols, 1997.

[Bratman, 1987] M. E. Bratman. *Intention, Plans, and Practical Reason.* Harvard University Press, Cambridge, MA, 1987.

[Bratman, 1992] M. E. Bratman. Shared cooperative activity. *The Philosophical Review*, 101:327–341, 1992.

[Castelfranchi, 1995] C. Castelfranchi. Commitments: From individual intentions to groups and organizations. In *ICMAS 95*, 1995.

[Cohen and Levesque, 1990] P. R. Cohen and H. J. Levesque. Intention is choice with commitment. *Artificial Intelligence*, 42:263–310, 1990.

[Cohen and Levesque, 1991a] P. R. Cohen and H. J. Levesque. Confirmation and joint action. In *Proce. of IJCAI 1991*, 1991.

[Cohen and Levesque, 1991b] P. R. Cohen and H. J. Levesque. Teamwork. *Noûs*, pages 487–512, 1991.

[Cohen *et al.*, 1990] P. R. Cohen, J. Morgan, and M. E. Pollack (editors). *Intentions in Communication.* MIT Press, 1990.

[Cohen *et al.*, 1997] P. R. Cohen, H. J. Levesque, and I. Smith. On team formation. In J. Hintikka and R. Tuomela, editors, *Contemporary Action Theory Synthese.* 1997.

[Dubois and Prade, 1994] D. Dubois and H. Prade. A survey of belief revision and updating rules in various uncertainty models. *International Journal of Intelligent Systems*, 9:61–100, 1994.

[Dunin-Keplicz and Verbrugge, 2002] B. Dunin-Keplicz and R. Verbrugge. Collective intentions. *Fundamenta Informatica*, 49:271–295, 2002.

[Dunin-Keplicz and Verbrugge, 2003] B. Dunin-Keplicz and R. Verbrugge. Evolution of collective commitments during teamwork. *Fundamenta Informaticae*, 56(4):329–371, 2003.

[Dunin-Keplicz and Verbrugge, 2004] B. Dunin-Keplicz and R. Verbrugge. A tuning machine for cooperative problem solving. *Fundam. Inform.*, 63(2-3):283–307, 2004.

[Fagin *et al.*, 1995] R. Fagin, J. Halpern, Y. Moses, and M. Y. Vardi. *Reasoning About Knowledge.* MIT Press, 1995.

[Gmytrasiewicz *et al.*, 1992] P. Gmytrasiewicz, E. H. Durfee, and D. Wehe. A logic of knowledge and belief for recursive modeling: Preliminary report. In *Proc. of AAAI-92*, pages 628–634, California, 1992.

[Grant *et al.*, 2000] J. Grant, S. Kraus, and D. Perlis. A logic for characterizing multiple bounded agents. *Autonomous Agents and Multi-Agent Systems Journal*, 3(4):351–387, 2000.

[Grant *et al.*, 2002] J. Grant, S. Kraus, and D. Perlis. A logic-based model of intentions for multi-agent subcontracting. In *Proc. of AAAI-2002*, pages 320–325, 2002.

[Grant *et al.*, 2005] J. Grant, S. Kraus, and D. Perlis. A logic-based model of intention formation and action for multi-agent subcontracting. *Artificial Intelligence Journal*, 163(2):163–201, 2005.

[Grosz and Kraus, 1993] B. J. Grosz and Sarit Kraus. Collaborative plans for group activities. In *Proceedings of IJCAI-93*, pages 367–373, Chambery, France, August 1993.

[Grosz and Kraus, 1996] B. J. Grosz and S. Kraus. Collaborative plans for complex group activities. *Artificial Intelligence Journal*, 86(2):269–357, 1996.

[Grosz and Kraus, 1999] B. J. Grosz and S. Kraus. The evolution of sharedplans. In A. Rao and M. Wooldridge, editors, *Foundations and Theories of Rational Agency*, pages 227–262. Kluwer Academic Publishers, 1999.

[Haass, 1983] A. Haass. The syntactic theory of belief and knowledge. *Artificial Intelligence*, 28(3):245–293, 1983.

[Jennings, 1995] N. R. Jennings. Controlling cooperative problem solving in industrial multi-agent systems using joint intentions. *Artificial Intelligence Journal*, 75(2):1–46, 1995.

[Kinny *et al.*, 1994] D. Kinny, M. Ljungberg, A. S. Rao, E. Sonenberg, G. Tidhar, and E. Werner. Planned team activity. In C. Castelfranchi and E. Werner, editors, *Artificial Social Systems, Lecture Notes in Artificial Intelligence (LNAI-830)*, Amsterdam, The Netherlands, 1994. Springer Verlag.

[Konolige and Pollack, 1993] K. Konolige and M. E. Pollack. A representationalist theory of intention. In *Proc. of IJCAI-93*, pages 390–395, Chambery, France, August 1993.

[Konolige, 1986] K. Konolige. *A Deduction Model of Belief.* Pitman, London, 1986.

[Kraus and Lehmann, 1988] S. Kraus and D. Lehmann. Knowledge, belief and time. *Theoretical Computer Science*, 58:155–174, 1988.

[Kraus *et al.*, 1998] S. Kraus, K. Sycara, and A. Evenchik. Reaching agreements through argumentation: a logical model and implementation. *Artificial Intelligence*, 104(1-2):1–69, 1998.

[Kumar *et al.*, 2000] S. Kumar, M. J. Huber, D.R. McGee, P.R. Cohen, and H.J. Levesque. Semantics of agent communication languages for group interaction. In *AAAI 2000*, pages 42–47, 2000.

[Levesque *et al.*, 1990] H. J. Levesque, P. R. Cohen, and J. Nunes. On acting together. In *Proceedings of AAAI-90*, pages 94–99, Boston, MA, July 1990.

[Levesque, 1984] H. J. Levesque. A logic of implicit and explicit belief. In *Proc. of AAAI-84*, pages 198–202, Austin, TX, 1984.

[Lochbaum, 1994] K. Lochbaum. *Using Collaborative Plans to Model the Intentional Structure of Discourse.* PhD thesis, Harvard University, 1994. Available as Tech Report TR-25-94.

[Megiddo, 1989] N. Megiddo. On computable beliefs of rational machines. *Games and Economic Behavior*, 1:144–169, 1989.

[Moore, 1985] R. Moore. A formal theory of knowledge and action. In J. Hobbs and R. Moore, editors, *Formal Theories of the Commonsesnse World.* ABLEX publishing, Norwood, N.J., 1985.

[Perrault, 1990] R. Perrault. An application of default logic to speech act theory. In P.R. Cohen, J.L. Morgan, and M.E. Pollack, editors, *Intentions in Communication*, pages 161–185. Bradford Books at MIT Press, 1990.

[Pollack, 1986] M. E. Pollack. A model of plan inference that distinguishes between the beliefs of actors and observers. In *Proceedings of the 24th Annual Meeting of the Association for Computational Linguistics*, pages 207–214, New York, 1986.

[Pollack, 1990] M. E. Pollack. Plans as complex mental attitudes. In P. R. Cohen, J. Morgan, and M. E. Pollack, editors, *Intentions in Communication*, pages 77–103. MIT Press, 1990.

[Rao and Georgeff, 1991a] A. Rao and M. Georgeff. Deliberation and its role in the formation of intention. In *Proceedings of the Seventh Conference on Uncertainty in Artificial Intelligence*, San Mateo, California, 1991. Morgan Kaufmann Publishers, Inc.

[Rao and Georgeff, 1991b] A. Rao and M. P Georgeff. Asymmetry thesis and side-effect problems in linear-time and branching-time intention logics. In *Proc. of IJCAI-91*, pages 498–504, Sydney, Australia, August 1991.

[Rao *et al.*, 1992] A. Rao, M. P. Georgeff, and E. A Sonenberg. Social plans: A preliminary report. In *Decentralized Artificial Intelligence, Volume 3*, pages 57–76. Elsevier Science Publishers, 1992.

[Schut *et al.*, 2001] M. Schut, M. Wooldridge, and S. Parsons. Reasoning about intentions in uncertain domains. In *Proceedings of the Sixth European Conference on Symbolic and Quantitative Approaches to Reasoning with Uncertainty (ECSQARU-2001)*, Toulouse, France, September 2001.

[Searle, 1990] J. R. Searle. Collective intentions and actions. In *Intentions in Communication*, chapter 19. The MIT Press, 1990.

[Shoham, 1989] Y. Shoham. Belief as defeasible knowledge. In *Proc. Eleventh Int'l Joint Conf. on Artificial Intelligence*, Detroit, MI, 1989. Int'l Joint Conferences on Artificial Intelligence, Inc.

[Singh and Asher, 1993] M. P. Singh and N. M. Asher. A logic of intentions and beliefs. *Journal of Philosophical Logic*, 22:513–544, 1993.

[Singh, 1991] M. P. Singh. Group ability and structure. In Y. Demazeau and J.-P. Müller, editors, *Decentralized AI 2 — Proceedings of the Second European Workshop on Modelling Autonomous Agents in a Multi-Agent World (MAAMAW-90)*, pages 127–146, Saint-Quentin en Yvelines, France, 1991. Elsevier Science B.V.: Amsterdam, Netherland.

[Singh, 1998] M. P. Singh. The intentions of teams: Team structure, endodeixis, and exodeixis. In *Proceedings of the 13th European Conference on Artificial Intelligence (ECAI)*, pages 303–307. Wiley, 1998.

[Sonenberg *et al.*, 1992] E. Sonenberg, G. Tidhar, E. Werner, D. Kinny, M. Ljungberg, and A. Rao. Planned team activity. Technical Report 26, Australian Artificial Intelligence Institute, Australia, 1992.

[Tambe, 1997] M. Tambe. Towards flexible teamwork. *Journal of Artificial Intelligence Research*, 7:83–124, 1997.

[Tidhar *et al.*, 1996] G. Tidhar, A. Rao, and E. Sonenberg. Guided team selection. In *Proceedings of Second International Conference on Multi-Agent Systems*, pages 369–376, 1996.

[Tuomela and Miller, 1988] R. Tuomela and K. Miller. We-intentions. *Philosophical Studies*, 53:367–389, 1988.

[Tuomela, 1991] R. Tuomela. We will do it: An analysis of group-intentions. *Philosophy and Phenomenological Research*, 51:249–277, 1991.

[Wooldridge and Jennings, 1999] M. Wooldridge and N. R. Jennings. The cooperative problem-solving process. *Journal of Logic and Computation*, 9(4):563–592, 1999.

[Wooldridge, 2000] M. Wooldridge. *Reasoning about Rational Agents*. The MIT Press, 2000.

Detecting the Logical Content: Burley's 'Purity of Logic'

WILFRID HODGES

In one of his inaugural lectures Dov Gabbay demonstrated God's providence as follows—I quote from memory. God had seen fit to send his revelation to mankind in a language with no verb tenses. (In classical Hebrew the verb forms are distinguished as perfective, imperfective, passive, reduplicative, etc., but not as past, present or future.) So there was a problem to explain how facts about a temporal world can be expressed in a language with no tenses. Logicians could not only solve this problem; they could also apply for research grants for solving it. Thus God ensured that logicians would not go short of spare cash.

As a hardened atheist, what can I rescue from this story? Actually, quite a lot. In the history of logic we study texts that are written in a language generally quite different from our own, on subjects that supposedly have something to do with what we understand as logic. The links are often hard to make.

One thing that historians of logic do is to describe the deductive practices of earlier thinkers. In this sense Netz's [1999] close study of Euclid's procedures is an important contribution to the history of Greek logic. Here I try to make a small step in the same direction with a medieval logician. But *reconstructing practices is not enough.* Netz remarks [Netz, 1999, p. 216]:

> One of the most impressive features of Greek mathematics is its being practically mistake-free.

The point is obvious but still worth making: in mathematics some things are right and some are wrong, and we can often tell which. For example any procedure, using any system of concepts and representations, is just wrong if it yields the conclusion that π is exactly 22/7. This makes a sharp division between the history of mathematics and the history of philosophy.

Logic comes somewhere between the two, and another purpose of this essay is to test the question: Are logician X's procedures correct for what he was aiming to do? To answer the question we need to be able to discern the aims behind the procedures, and then we need to be able to tie these aims to something objective. Of course the next generation's notion of what is objective in logic may differ from ours, just as it sometimes does in mathematics. So this kind of analysis of the past needs to be redone in each generation.

To make things as concrete as possible, I chose a particular text, 'On the Purity of the Art of Logic' (*De Puritate Artis Logicae Tractatus Longior*), PL for short, written by Walter Burley in the late 1320s. Theophilus Boehner [Burleigh, 1955] has edited the Latin original, and Paul Vincent Spade gives a translation [Burley, 2000]. Spade helpfully numbers the paragraphs (1) to (1053), and my references follow this numbering.

I chose Burley because he is the beneficiary of the preceding century and a half's work by terminist logicians, but he is mercifully free from the lapses of common sense and the metaphysical irrelevances that disfigure much of later Western medieval logic.

This essay was written in a hurry against a publication deadline, so that some details will certainly be wrong. I intend to put on my website at `www.maths.qmul.ac.uk/~wilfrid` the full reference list mentioned in Section 1, together with corrections as I find them.

1 The book

The backbone of Burley's PL consists of a large number of inferences expressed by Latin sentences; each inference is labelled either good or not good. Burley calls the inferences 'consequences' (*consequentiae*). Listing them I found 128 consequences labelled 'good' and 178 labelled 'not good', making 306 in all; below I refer to this as the *reference list*. The number is not exact, for several reasons. Some consequences are parts of Burley's own argumentation, not inferences that he is discussing. Some consequences are expressed in irregular ways. Some consequences are broken up into more than one consequence during their discussion. In some consequences Burley uses a letter as an abbreviation for a certain piece of Latin text; I included these. But I left out all consequences that contain schematic letters, since these are strictly not consequences but rules defining classes of consequences.

Burley never mentions that the sentences in his consequences are in Latin. (Neither of the words *Latinus* or *lingua* appears in PL.) Nevertheless we need to mention it. Some of Burley's rules rely on facts or alleged facts about Latin that don't hold for other languages. These facts are of three kinds: (a) the use of word order for determining scope, (b) the use of noun cases for determining subjects, (c) the possible antecedents of anaphoric pronouns.

(a) Burley's rule of thumb for scopes is that an expression comes at the start of its scope. Thus (107)

A negation doesn't include in its scope what precedes it.
... negatio non habet dominium supra praecedens.

This is false in Burley's own first language of English, where we negate a finite verb by putting 'not' *after* it. Thus for example ([Baugh and Cable, 1978] p. 143) from about 1300:

Bot al men can noht, I-wis,
Understand Latin and Frankis.

Here the 'noht' follows not only the verb 'can' but also the quantifier 'al' which clearly lies within its semantic scope. (John Marenbon suggested to me that Burley might have dismissed English as a low status language.)

Translating *dominium* as 'scope' may convey the impression that Burley has a technical term here. But in fact this is the one occurrence of the word *dominium* in PL, and mostly we have to infer Burley's view of scope from his practice. For example at (98) he distinguishes between

All day someone is indoors here.
tota die aliquis homo est hic intus.

Someone is all day indoors here.
aliquis homo tota die est hic intus.

He reads the sentences in the way that we would describe by saying that the earlier quantifier has wider scope.

(b) Among the many logical 'rules' that Burley lists, one (662) refers explicitly to nominative case. The rule says that 'Only an A is a B' is equivalent to 'Every B is an A' when the subject A is in the nominative.

His explanation (643) is more puzzling than the rule it explains. The explanation is that without the restriction 'nominative' (*rectus*) the rule would give us that 'Any man's is a donkey' (*cuiuslibet hominis est asinus*) is equivalent to 'Only a donkey is man's', which he says is wrong because 'Any man's is a donkey' would be true if each man has both a donkey and a bull. Clearly he is reading 'Any man's is a donkey' as meaning 'Each man has a donkey'; so the logical predicate is not 'donkey' but 'has a donkey', and the correct converse would be 'Only a thing having a donkey is a man'.

One might guess that he is using the genitive case of 'man' (*hominis*) as a heuristic to warn us that the logical subject is not necessarily the grammatical subject; at (169ff) he uses the case endings in this way. But of course this has only an indirect connection with the rules for 'only'. Moreover it doesn't show that his rule needs 'nominative', since the word in the genitive is not in the proposition with 'only'. His reasoning here is too loose to allow any definitive correction.

At (701) he notes correctly that Latin nominalises sentences by putting the subject into the accusative (and the verb into the infinitive). Thus a sentence with another sentence as the subject can have what appears to be a subject in the accusative. English works differently; Burley's *Deum esse Deum* comes into English as 'that God is God' or 'for God to be God'.

(c) At (123) Burley states that a reflexive pronoun can have its antecedent either in the same clause or in an another clause. This is so far from the behaviour of English reflexives that I was surprised to find that Burley is right. There are classical examples, for instance Plautus [Plautus, 1910?] *Poenulus* I.ii:

> The lady can make a lump of flint fall in love with herself [sic].
> *Illa mulier lapidem silicem subigere ut se amet potest.*

Similar things are reported in Korean and Chechen. (Spade's explanatory example 'Socrates looked in the mirror and he saw himself' [Burley, 2000] p. 113 is off the point; the antecedent of 'himself' is 'he' in the same clause.)

One sometimes hears it said that the later medieval Western logicians adopted a rigid form of Latin that was intended to serve as a formal language. I found nothing whatever in PL to support this view. Burley never once suggests that his readings are anything other than the correct readings of normal Latin. He does acknowledge that some of his readings disagree with linguistic usage (*usus loquendi*), but in such cases he insists that his reading is correct literally (*de virtute sermonis*) (for example (191), (192), (731), (741)).

In any case it's hard to see what place a formal language would have in Burley's scheme of things. It's even harder to see how he could have saddled himself with a formal language as confusing as his rather stilted Latin. For example at (673) he reads the sentence

> *Tantum sciens grammaticam est homo.*

as false in the case where everybody knows both grammar and logic. This is mysterious. Spade ([Burley, 2000] p. 226 note 250) may have it right when he translates as

> Someone who knows nothing but grammar is a man.

But this includes *tantum* within the scope of *sciens*, contrary to the order of the words. If this is how to invent a formal language, it puts Burley in the same class as the man who allegedly first invented the vacuum cleaner except that he made it blow instead of suck.

2 Consequences

A consequence consists of one or more Latin sentences (called the 'antecedents', *antecedentes* or 'premises', *praemissae*), followed by 'Therefore' (*ergo*), followed by one Latin sentence (called the 'consequent', *consequens* or 'conclusion' *conclusio*). The parts of the consequence can be run together as a single sentence when convenient.

Among the 128 good consequences in the reference list, Burley phrases 65 in the style:

> It follows: A therefore B.
> *Sequitur* A *ergo* B.

Sometimes he adds 'well', apparently not meaning anything different:

> It follows well: A therefore B.
> *Sequitur bene:* A *ergo* B. (12 times)

Sometimes he says that the conclusion follows:

> If A, it follows that B.
> *Si* A, *sequitur quod* B. (1 time)
>
> With reference to A, B follows.
> *Ad* A *sequitur* B. (1 time)
>
> When A, it follows that B.
> *Cum* A, *sequitur quod* B. (1 time)
>
> From this: A, B follows.
> *Ex ista* A *sequitur* B. (1 time)
>
> From these premises: A, this conclusion follows well: B.
> *Ex istis praemissis* A *ista conclusio bene sequitur* B. (1 time)

(In the last two, 'this' plays the role of quotation marks, which Burley doesn't have.) When the antecedents have been stated, he sometimes says:

> The conclusion follows well: B.
> *Bene sequitur conclusio* B. (8 times)

Sometimes he says about a consequence:

> It is a good consequence.
> *Est consequentia bona.* (5 times)
>
> The consequence holds.
> *Tenet consequentia.* (3 times)

A syllogism is a particular kind of consequence; sometimes Burley uses the same language as above but with 'syllogism' in place of 'consequence'. For example we have *Syllogismus est bonus* 4 times.

Apart from the restriction to syllogisms, all the locutions above seem to be stylistic variants. When any of them are true of a consequence, I will say that the consequence is *good*. Burley describes the remaining consequences as 'not good', but I will shorten that to *bad*.

Compare the three items:

(i) If A then B. (*Si* A, B.)

(ii) With respect to A, B follows. (*Ad* A *sequitur* B.)

(iii) A, therefore B. (A *ergo* B.)

Here item (i) is not a consequence but a conditional. Burley is clear that conditionals and consequences are not the same thing. However, he notes that for each conditional (i) there is a corresponding sentence (ii), and he says (353f) that the sentence follows from the conditional and vice versa. Also since (ii) and (iii) seem to be stylistic variants, it appears that Burley takes all three forms as equivalent for purposes of reasoning. In (353) he describes the relationship between (i) and (ii) as being that (i) performs an act which is meant by (ii). This is less revealing than it might be, because in (706) he describes the relationship between (ii) and (iii) in exactly the same terms. (In Latin the distinction is between *actus significatus* and *actus exercitus*; see Nuchelmans [Nuchelmans, 1996] for the history of this distinction. I think the word *actus* here refers to the action performed by 'If' in relating two clauses, not to a speech act.)

In fact Burley's text shows some slippage between conditionals and consequences. For example at (308) he applies the word *sequitur* to a conditional, though the conditional is restated as a consequence a few lines later. At (311) he makes a statement about conditionals, but his example to illustrate it is a consequence. At (319) he speaks of an 'inference' being made in a conditional. At (69) and (153), *si* is answered by *ergo*.

3 'Follows'

Burley's text doesn't distinguish between primitive and derived notions. But he uses the word 'follows' (*sequitur*) hundreds of times, with no attempt to reduce it to any more basic notion. In particular he doesn't paraphrase it in such terms as 'If the premises are true then the conclusion must be true'. (He does state this as a necessary condition for a consequence to be good (258), but he never suggests it is sufficient.)

In fact a good deal of what he says about truth seems to presuppose some notion of following. For example he uses the goodness or badness of certain consequences in order to explain the circumstances in which a sentence is true. Section 14 below will report some bad consequences used this way. Sometimes Burley uses good consequences of the form

> 'If we suppose that' (*posito quod, supposito quod*) A; then (*tunc*) the sentence B is true.

to clarify the conditions under which a sentence is true. (Thus at (557), (593), (595), (643), (886); likewise at (98), (163), (507), (535), (594), (673), (735), (767), (776), (947) without *tunc*.) In one place (735) he says 'If we suppose that' A, 'it follows that' (*sequitur*) B is true. So this talk of 'supposing' is a way of expressing conditionals or consequences; but Burley seems to use it only in cases where the main interest lies in B, while A appears only as an example of circumstances in which B is true.

Another explanation of truth apparently in terms of consequence is Burley's theory of descent. At (82) he explains that the term 'man' supposits determinately (*supponit determinate*) in 'Any man runs', because from 'Any man runs' there follow 'Socrates runs' and 'Plato runs' and so on. We must come back to this later; but for the moment we note that the notion of suppositing determinately is defined in terms of the notion of certain things following. So if (as is sometimes claimed) the notion of suppositing determinately is intended to explain the conditions under which certain sentences are true, then for Burley the notion of 'following' is prior to the notion of being 'true under certain conditions'.

For 'prior' perhaps we should read 'not posterior'. There is little indication that Burley has his definitions in a hierarchy. Holism rules.

In any case we have to suppose that Burley expects his readers to come ready equipped with the notion of 'follows'. A corollary is that at least in simple cases, he must expect readers to be able to see for themselves that a consequence is good.

Many times through the book, Burley gives a rule for good consequences and then a consequence that is an instance of the rule. He doesn't tell us how to read the text, but two possible ways suggest themselves. One way is to read it forwards: we learn the rule, then we see how to use it to construct good consequences. In a classroom the teacher might invite the class to construct other instances of the rule, using the given instance as a template. Elementary mathematical texts often use a similar format today.

The second way of reading is backwards, and this is more interesting. We can suppose that the abler readers, already having the notion of 'follows', will be able to see for themselves that the example consequences are good. They can then generalise from these examples to see how the rules arise.

This way lies danger. Can we really infer a general rule from study of a single instance? That question looms large over the rest of this paper, because Burley often proceeds as if the answer is yes. To understand what he is doing, we will need to look at his procedures for proving universally quantified statements (and we will do this in Section 9).

Burley is not above justifying rules, or indeed anything else, by barefaced appeals to authority. For example at (345) he answers the objection that Boethius didn't mention a certain rule by retorting that it's in Aristotle; but this could be *ad hominem* against some old fogey. At (284) he says that a certain method of argument is correct because Aristotle used it. At (510) and (753) he says that something is his own opinion (*mihi videtur*, *ego dico tibi*), but calls in Aristotle to back him up. Generally his uses of Aristotle are benign—they give credit or aim to pull things together. For example at (32) and (957) he uses notions that he credits to Aristotle; at (763) he mentions that a method he uses is also in Aristotle.

In (228) he claims to report Aristotle's views on the meaning of the Latin word *est*. Historically this is bizarre. He may only mean that Aristotle's discussion of the Greek *estí* transfers to the Latin word; but one would have been happier if he had said so. (Al-Fārābī would have done.)

4 Propositions

For Burley every true proposition is true because of (*secundum quod*) something, known as its 'cause of truth' (*causa veritatis*). Whether or not this notion is prior to that of a good consequence, we can usefully treat it next, if only to introduce Burley's notions of sentence structure.

Just as an event can have several possible causes, a true proposition can have several possible causes of truth; any one of them is enough to make the proposition true. Burley confuses actual causes of truth with possible causes of truth. For example at (868) he says that the proposition

> It's not the case that every man except Socrates is running.

has two causes of truth, namely that somebody other than Socrates is not running, and that Socrates himself is running. Since the truth of either of these conditions would make the displayed sentence true, and these are the only things that could make it true, the sentence is equivalent to (*valet*) the disjunction

> Either some man other than Socrates is [not] running, or Socrates is running.

In this example it's clear that the second of the 'causes of truth' is not in fact true. In what follows I will speak of 'possible causes of truth' when the sentences in question are not assumed to be true. (Wittgenstein at 5.101 of the *Tractatus* [Wittgenstein, 1961] avoids Burley's potential muddle by defining explicitly: 'Diejenigen Wahrheitsmöglichkeiten seiner Wahrheitsargumente, welche den Satz bewahrheiten, will ich seine Wahrheitsgründe nennen.')

Some propositions are true because of two things. For example according to Burley (651) the proposition 'Only Socrates is running' is made true by the truth of 'Socrates is running' and 'Nothing other than Socrates is running'. Likewise 'Socrates and Plato are running' is made true by the truth of 'Socrates is running' and 'Plato is running'. In such cases Burley normally describes the analysing sentences as 'exponents' (*exponentes*). Just once (868) he gets in a muddle and calls them causes of truth. The reason is that he is discussing negations of conjunctions; such a negation is equivalent to a disjunction, so it has two possible causes of truth, namely the negations of the exponents of the conjunction, a point he makes again at (964).

For future reference, I note that the discussion at (651) and at (757f) seems to imply that for Burley the sentence 'Every A is a B' has exponent 'There is an A', but 'Nothing other than a B is an A' doesn't. If this is right, then presumably for Burley 'Every A is a B' and 'No A is a non-B' are not equivalent. (But he should have told us explicitly.)

We can catalogue propositions according to their possible causes of truth. Burley talks of propositional combinations as having two propositions and a 'principal' (*principale*), namely the connective that joins them (328), (506); the two joined propositions are the 'principal parts' (*partes principales* (522)). Presumably we find the possible causes of truth by looking first at the principal, though Burley doesn't spell this out. For example a sentence 'S or T' has two possible causes of truth, namely S and T. Probably the possible causes of truth of 'S and T' are the propositions formed by conjoining a possible cause of truth of S with a possible cause of truth of T. Probably similar ideas work for all propositional combinations. Modal propositions 'It is necessary that ... ' etc. have the modal operator as principal (327).

There remain two kinds of proposition, namely atomic propositions like 'Socrates runs', and propositions that are compound but neither modal nor propositional compounds. Burley treats these two types together. In each case he looks for two 'terms', normally noun phrases, which are respectively subject and predicate. The sentence expresses something about how many or which of the objects described by the subject term have the property expressed by the predicate term. The things described by the subject term are called the *supposita*. The required connection between the supposita and the property expressed by the predicate is determined by the other words of the proposition (the *syncategoremata*), such as 'is' (the principal in such sentences) or 'not', or 'every' or 'some' (when they are attached to the subject). Propositions of this general form are called 'categoricals' (*categoricae*). Propositions superficially of this form but with other pieces attached are treated separately. Examples are exceptives like 'Some man besides Socrates can laugh', and exclusives like 'Only a man is a donkey'.

For example a necessary condition for a universally quantified affirmative categorical to be true is (165) that the predicate 'is in' all the things contained under the subject (i.e. the supposita); I guess it becomes necessary and sufficient if we add 'there is at least one suppositum'. This possible cause of truth is at meta-level since it talks about the subject and predicate; the possible causes of truth of a disjunction by contrast were at the same conceptual level as the disjunction itself. This may be one of the reasons for the theory of 'descent to supposita', which essentially resolves a universal affirmative into a family of exponents at the same conceptual level (though Burley doesn't put it this way). In descent we remove the quantifier and replace the subject term by a name of one of its supposita; the resulting proposition is called a 'singular' (*singularis*) of the categorical. For example one of the singulars of 'Every man is running' is 'Socrates is running'. If the categorical is true then so are all its singulars. The converse holds too; Burley doesn't state this explicitly in PL, though he refers to it at (143), (163f), (175), (179) and (183).

Existentially quantified categoricals also allow descent, but this time each of the singulars is a possible cause of truth, so that the categorical is equivalent to their disjunction. Thus for example from 'Some man runs' we reach 'Socrates runs or Plato runs or etc.' (82). I will call this 'disjunctive descent' to distinguish it from the simple descent associated with universal quantifiers.

Burley generalises the idea of descent to other terms besides the subject, and he often talks of whether it's possible to deduce the singulars (*contingit descendere*), either separately or their disjunction. The possible answers are described as forms of 'supposition' (*suppositio*, strictly *suppositio personalis*, though this refinement is irrelevant to the present essay). Burley uses these forms of supposition to classify occurrences of terms in propositions. The forms of supposition are defined in terms of whether certain inferences hold. For example (100) the statement that 'man' in 'Every man runs' has 'confused and distributive supposition' means that from this sentence we can deduce that Socrates runs, that Plato runs etc. (Again I ignore some subtleties about mobile and absolute supposition.)

When we apply descent to terms that aren't the subject, we generally have to forget about dropping the attached quantifier, because there isn't one. Burley passes over this in silence. There is a more serious problem when we apply descent to the subject term but the determiner is something more complicated than a simple quantifier. When we drop the quantifier, the rest of the determiner may make no sense attached to a proper name. For example at (101) and (873) Burley happily descends from 'Every man except Socrates is running' to 'Plato except Socrates is running'. Here it looks as if the notion of descent has overrun its usefulness. (Burley can't handle complicated determiners anyway. See at (732) his inability to make sense of the sentence 'Only three men are running', *tantum tres homines currunt.*)

Altogether this system of possible causes of truth, exponents and descent might look like the beginnings of a Tarski-Montague-style truth definition for medieval Latin. But there are some major differences. One is Burley's lack of any systematic ideas on how sentences are built up. Another is that in spite of appearances, possible causes of truth are not compositional—it's not true that each piece makes its own separate contribution to the possible cause of truth. For example (118):

> For the sentences 'A man is running' and 'He is debating' to be true [in 'A man is running and he is debating'], it's required that 'A man is running' should be verified for one of the supposita of 'man', and that the second part should be verified for the same suppositum.
> *Ad hoc enim quod istae sint verae: 'Homo currit', et: 'Ille disputat', oportet quod ista: 'Homo currit', verificetur pro aliquo supposito hominis, et quod secunda pars verificetur pro eodem supposito.*

Or for another example (207f), in a present tense sentence the supposita of the subject term are limited to things that the term describes now; but in a past tense sentence the supposita can include things that the term used to describe. Burley warns of this at (4) when he says that the supposition of a term is a property that it has in relation to another term in the proposition.

5 When are consequences good?

Just as true sentences are true because of some *causa veritatis*, so good consequences are good because of some rule. Burley says (341):

> Every good consequence holds through some place, i.e. maximal proposition. A maximal proposition is simply a rule through

which a consequence holds.
Omnis consequentia bona tenet per aliquem locum qui est propositio maxima. Nam propositio maxima non est nisi regula, per quam consequentia tenet.

As Burley says, rules (in the relevant sense) are propositions, so they are statements that are either true ((422) *regulae verae*) or false ((301) *regulam falsam*). (But see (309) *regulam bonam.*) Burley gives many examples of such rules; they are general statements saying that all consequences with certain features are good.

Burley says in many places—I counted 54—that some consequence holds 'through' (*per*) a certain rule. I found no evidence that he means any more than that the rule is true and the consequence is an instance of it.

For example at (300) he says that 'it's clear' (*patet*) that a particular consequence is through a certain rule; since we are given no information of any kind about the context in which the consequence is being used, the rule must be something recognisable from the form of the consequence itself.

Another piece of evidence in the same direction is this. In a few places Burley gives a consequence schema, with letters in place of expressions, and says that it holds 'through a rule'. Thus at (433):

> The first mood is when both premises have affirmative antecedents. For example 'If A then B. If C then not B. Therefore if C then not A.' This is argued by the following rule: Whatever entails the opposite of the consequent entails the opposite of the antecedent.
> *Primus modus est, quando antecedens utriusque praemissae est propositio affirmativa. Verbi gratia: 'Si* A *est,* B *est; si* C *est,* B *non est; ergo si* C *est,* A *non est'. Et arguitur per hanc regulam: Quidquid antecedit ad oppositum consequentis, antecedit ad oppositum antecedentis ...*

(At (441) he explains that 'A *est*' stands for a proposition and 'A *non est*' stands for its negation.) The 'mood' here should be read as a rule, saying that every instance of the schema is a good consequence. This is exactly the content of the 'rule' through which Burley says the schema holds. The 'mood' and the 'rule' say the same thing, and the difference between them is that the 'mood' is written as a schema whereas the 'rule' is written as a statement about consequences. (There are similar examples at (434), (443f), (464), (465), (467), (468), (469). In some of these cases Burley refers to the schemas not as moods but as syllogisms.)

In four places ((263), (293), (296), (321)) Burley talks of consequences being 'based on' (*fundatae supra*) some rule. Any impressions that he might mean that the rule is the reason why the consequence is good are dispelled by (263), where he says that the rule in question is false and all consequences based on it are fallacious.

So the passage in (341) probably says rather less than one might have guessed. It doesn't say that for every good consequence there is a rule which is the *reason why* the consequence holds. It doesn't say that our *knowledge* of good consequences is always derived from general principles of reason. There are good Aristotelian authorities for both these views, but they are not in PL. Perhaps this is part of what Burley meant by entitling his book 'The Purity of Logic'.

One possible way to read (341) is as a methodological statement. It's a task of logic to classify good consequences; given a good consequence, a logician should always seek to catalogue it with other good consequences. But a phrase in (289) suggests a more explicit reading of (341):

> But these rules are enough for making syllogisms in conditional hypotheticals by skill.
> *Sed istae regulae sufficiunt ad artificialiter syllogizandum in hypotheticis conditionalibus.*

To do a thing *artificialiter* is to do it by using learned skills like those of a craftsman (*artifex*). Logicians do seek to catalogue good consequences by the rules that they obey; but they have a specific practical reason for doing this. If you learn the rules, you can use them to construct good arguments to a professional standard. Without these rules you have to rely on such common sense as you have.

Burley spells this out more fully in his late Commentary on the Ars Vetus ([Burlaeus, 1967] p. 2):

> And one should know that logic is useful as a faculty or power of distinguishing by skill between true and false in the separate branches of knowledge. For in these branches of knowledge one distinguishes true from false by reasoning from premises known to be true, to conclusions that follow from them, and logic teaches this kind of reasoning, so that it is through logic that in every branch of knowledge, true is distinguished from false by skill. Nor can one have any knowledge by skill except through logic. ... One should know that there are two ways to have logic, namely by instinct and by skill. A person who uses a syllogism without knowing that he is doing so has instinctive logic, but he doesn't have logic as a skill since he doesn't know that he is using syllogisms and has logic. ... But a person who

uses a syllogism and knows that he is doing so has logic as a skill, since he knows the nature of syllogisms.

Et est sciendum quod utilitas logicae est facultas seu potestas discernendi artificialiter verum a falso in singulis scientiis. Nam in singulis scientiis distinguit verum a falso per discursum factum a praemissis notis ad conclusiones sequentes ex illis, et talem discursum docet logica, et ideo per logicam distinguitur artificaliter verum a falso in omni scientia. Nec potest aliqua scientia artificialiter haberi sine logica. ... Sciendum quod logica potest haberi dupliciter, scilicet usualiter vel artificialiter. Utens enim syllogismo nesciens se sillogizare habet logicam usualem sed non habet artificialem quia nescit se syllogizare nec habere logicam. ... Sed utens syllogismo sciens se syllogizare habet logicam artificialem quia novit naturam syllogismorum.

6 Types of rule of good consequences

Burley refers to some consequences as 'syllogisms' (*syllogismi*); this word occurs 232 times in PL. He certainly doesn't mean just the familiar 'syllogisms' of Aristotle; he calls these 'categorical syllogisms' (*syllogismi categorici*). Other kinds of 'syllogism' are catalogued according to the kinds of proposition that occur in them. Thus we meet hypothetical syllogisms (249), syllogisms in causals (614), syllogisms of exclusives (931), syllogising in hypothetical conditionals (289), syllogising in disjunctives (545), syllogising in exceptives (931), syllogising in reduplicatives (935).

Burley describes the types of syllogism sometimes with consequence schemas ('moods') and sometimes with rules. Sometimes (as we saw above) he uses both. I think he is consistent in restricting the word 'syllogism' to actual consequences; a mood is not a syllogism but a way of forming syllogisms. Also syllogisms are good or not good, just as consequences are good or not good. This is slightly confused by his usage 'a syllogism is made' (*fit syllogismus*, e.g. at (496)), by which he always means that a *good* syllogism is made.

From the many examples that he gives, it seems that for Burley a syllogism is always a consequence with two premises and one conclusion. (At (283) he says explicitly that syllogisms, unlike some 'non-syllogistic consequences', have two premises.) In most of his examples the premises and conclusion of a syllogism all have a similar form, for example they might all be exclusives, though some of these exclusives might be negated. (A small exception is at (483ff), where he considers syllogisms that have one conditional premise and one categorical.) In most cases three terms are involved, and each proposition involves two of them. This allows Burley to catalogue most syllogisms into three 'figures' copying Aristotle's classification: in first figure the term appearing in both premises is in different positions in the two premises, in second figure it's in second position in both premises, and in third figure it's in first position in both. In hypothetical syllogisms there are three sub-propositions instead of three terms, and in reduplicatives the propositions each have three terms—though sometimes two are the same. (When one premise is conditional and the other is categorical, the classification into figures breaks down, so that at (486) he says 'This syllogism is not made in any figure'.)

What does Burley mean to convey when he describes a particular consequence or rule as a 'syllogism'? I don't think PL supports anything stronger than this: *A syllogism is a consequence that belongs in a class of consequences that can be handled systematically in a way analogous to Aristotle's treatment of categorical syllogisms.* But this leaves it open what he counts as analogous to Aristotle's treatment of categorical syllogisms.

At (767), commenting on propositions with certain features F, he says:

> But if [the premises have features F], no conclusion follows syllogistically and formally. This is clear from counterexamples in the terms.
> *Sed si in utraque praemissa ..., nulla conclusio sequitur syllogistice formaliter. Quod patet per instantias in terminis.*

This passage is problematic and we will come back to it in Sections 16 and 17. But it seems clear that he is discussing syllogisms with premises of a certain form. The fact that he limits himself here to *syllogisms* allows him to assume that in each case there is only a small number of possible conclusions, so that he can run through the possibilities. For each possible conclusion, he shows that it doesn't follow 'formally', by giving a 'counterexample in terms'; i.e. he shows that for each possible form of syllogistic conclusion, there is a bad consequence where the premises and the conclusion have the specified forms. The role of syllogisms here is to allow Burley to restrict himself to arguments of a certain type; a modern analogue would be to ask what follows 'by first-order arguments' from premises of a certain form.

Burley says 'syllogistically and formally'. I've said what I think he means by 'syllogistically'; what about 'formally'? Burley discusses this notion in several places, not always connected with syllogisms. The idea is that a consequence C is good 'formally' if all consequences derived from C by replacing the terms of C in certain ways are good too; to say that a consequence is formally good is to say that it is one of a class of good consequences that are related to each other in some way by substitutions. The problem is to pin down what kind of substitutions.

The reader can consult (363f), (380), (387)–(389), (767) and (995) and make some guesses. (Unfortunately (380) contains some textual corruption. Brunellus is a donkey, and the manuscripts disagree about where the arguments are about braying and where they are about laughing. Boehner's ([Burleigh, 1955] p. 84) preferred reading makes Burley say in consecutive sentences that the consequence

> Brunellus can laugh. Therefore some man can laugh.
> *Brunellus est risibilis, ergo homo est risibilis.*

is not good, and that it is good. Spade ([Burley, 2000] p. 171) removes this anomaly but still saddles Burley with the view that this consequence is good not because of any relationship between Brunellus and man, but because of the relationship between 'Brunellus', 'man' and 'can laugh'.)

For what it's worth, my impression is that Burley describes a good consequence as 'formally good' when it remains good under substitution for terms (the same substitution for each occurrence of the same term, of course), provided that all *conceptual* truths of the forms 'All A's are B's', 'No A's are B's', and their negations, are preserved by the substitution. For example 'All men are animals' is a conceptual truth, but 'All men can laugh' is just an 'accidental' truth. (See particularly (387).)

Burley allows a good consequence to depend on conceptual relationships of these kinds even when the relationships aren't spelled out as premises. I am guessing that when he says 'syllogism', part of what he means is that any such relationships, if they are needed, *are* stated explicitly as premises. So to say that a consequence holds 'syllogistically and formally' means that it holds under all systematic substitutions for terms, provided only that the substitutions take conceptually true premises to conceptually true premises. This is not the same notion as 'logically true' in the sense of Tarski [Tarski, 1936], a notion that never appears in PL.

Because of their restricted forms, syllogisms have a better chance of being 'obviously good' than some other arguments. Burley speaks of some syllogisms as 'obvious (in itself)' (*(de se) evidens*) or 'perfect' (*perfectus*), which seems to mean something similar. See (170), (172), (213), (236), (415), (499).

However, at (212), (213) and (236) he speaks of a syllogism being 'perfect and regulated' (*perfectus et regulatus*), and 'regulated' here means that the syllogism is derivable from a more basic rule of argument. We know this because at (212) and (236) he tells us what that more basic rule is, namely 'dici de omni'. This is important, because it means that even the most self-evident syllogism rules need not be the most basic argument steps. One of my main aims in this paper is to describe what (for Burley) those most basic steps are. Burley makes no attempt to list them.

7 Shallow and deep rules

It will be helpful to make a distinction between shallow and deep rules. A *shallow* rule is one that depends on only a bounded amount of unpacking of the premises and conclusion (the bound depending on the rule, not the consequence). All other rules are *deep*.

For example Burley (409) gives the rule:

> If A is the case, B is the case. If B is the case, C is not the case. Therefore if A is the case, C is not the case.
> *Si* A *est*, B *est; si* B *est* C *non est; ergo si* A *est*, C *non est.*

To see that this rule applies to a particular consequence, we need to analyse the premises and conclusion down to the forms

> If (A is the case), (B is the case).
> If (B is the case), (C is not the case).
> If (A is the case), (C is not the case).

and then match up the bracketed parts. So the rule is shallow. (Some applications of it might hide their structure, so that they need to be paraphrased before the analysis. This is a separate issue.)

Most rules of propositional logic are shallow. One deep rule is the Replacement rule saying that if a proposition p is logically equivalent to q, then we can replace any occurrence of p by an occurrence of q inside any proposition, and the resulting proposition will be logically equivalent to the original proposition. Burley uses a (non-propositional) rule of this type at (667), though he doesn't spell out the rule he is using.

Here is a more exotic propositional example (I believe from Leśniewski):

> If the conclusion of a consequence uses no connectives except 'if and only if', and each atomic sentence in the conclusion occurs an even number of times, then the consequence is good.

Since the connective 'if and only if' can occur any number of times in the conclusion, the atomic sentences can lie at any depth in the analysis. So the rule is deep.

More familiar examples of deep rules are the quantifier rules of first order logic. For example:

> Let $\phi(x)$ be a formula with just x free, and c a constant. Then the sequent $\phi(c) \vdash \exists x\phi(x)$ is valid.

This rule is deep because the occurrences of c may lie arbitrarily deep in the syntactic structure of ϕ. But note that a shallow version of this rule copes with most ordinary language applications. Take dirty face sentences:

> Socrates believes that everybody else believes that he believes that everybody else believes that ... that his face is dirty. Therefore someone believes that everybody else believes that he believes that everybody else believes that ... that his face is dirty.

Here the anaphoric pronouns 'he' and 'his' take over the role of the free variables. We can quantify the sentence by putting 'someone' in place of 'Socrates', and for this we need only analyse the sentence as far down as its subject.

In PL Burley explicitly states about eighty rules of good consequences, usually labelled 'rule' or 'way of forming syllogisms'. All of these rules are shallow. In view of the previous paragraph, this is not very surprising. However, there are various pieces of indirect evidence for the use of at least potentially deep rules in Latin logic even before Burley.

My first witness is John of Salisbury, whose Metalogicon [of Salisbury, 1929] dates from the mid twelfth century. Describing a certain logician whom he names by a pseudonym, John says ([of Salisbury, 1929] p. 10f, 829ab):

> So he needed a calculus whenever he had to dispute, so as to be able to recognise affirmative force and negative force. For in many cases two negations have the force of an affirmation, and likewise an odd number of negations creates a negative force. ... So in order to tell whether he was dealing with an odd or even number, he found it a prudent policy to take to debates a handful of beans and peas that he would call on.
> *... ita ut calculo opus esset, quotiens fuerat disputandum; alioquin vis affirmationis et negationis erat incognita. Nam plerumque vim affirmationis habet geminata negatio; itemque vis negatoria ab impari numero convalescit; ... Ut ergo pari loco an impari versetur deprehendi queat, ad disceptationes collectam fabam et pisam deferre, que conveniebatur, consilio prudenti consueverat.*

(So perhaps bean = odd, pea = even. There is a mild pun on *calculus*, which means both algorithm and pebble. The beginning of Chaucer's Miller's Tale relates that the clerk took with him on his travels a set of algorithm stones, *augrim-stones*.)

The implication of this passage is that logicians of this period were familiar with deeply nested negations and had some calculus of positive and negative occurrences. But we have to be cautious, because the nesting need not have been within single propositions. In a debate one might say 'I claim that the following argument, assuming the truth of your last response, refutes the contradictory of my previous statement'. Here there are two nested negations in one statement, but still each statement in the debate might have no nesting deeper than two.

My second witness is the De Probationibus Terminorum [Billingham, 1982] of Richard Billingham, composed possibly in the 1340s. Billingham writes ([Billingham, 1982] p. 51):

> From inferior to its superior with subject held fixed, and with any expression with the force of negation put later than the inferior and superior, the consequence is good.
> *ab inferiori ad suum superius cum constantia subiecti et cum dictione habente vim negationis postposita inferior et superiori, tenet consequentia.*

In Section 11 we will see how one can read this as a sound rule, at least for suitably regimented languages. The rule is deep because nothing is said about how deep the inferior and superior may lie in the structure of the proposition. The caution this time is that although he doesn't say so, Billingham may have intended the rule to be used only to replace the entire predicate in subject-predicate sentences, and this is shallow. His mention of the subject suggests this.

Billingham's rule has antecedents as far back as the Abbreviatio Montana ([Rijk, 1967] p. 86) from the late twelfth century. The Abbreviatio certainly has in mind replacement of subject or predicate in subject-predicate sentences, since it spells out the relevant sentence structures in detail.

8 Burley's quantifier rules

No doubt there is scope for someone to write a dissertation reducing Burley's stated rules to valid sequents in some appropriate formal calculus; I haven't pursued the idea. His propositional rules are standard, and they include a form of reductio ad absurdum (499). Most of these rules are explicitly stated.

Burley shows no awareness of any logical rules for sentences not built up by quantifiers or propositional connectives. At (371) he says that 'singulars' are 'equivalent to particulars' (*aequivalens particulari*) and claims that 'Nothing follows from particulars'. Counterexamples are easy to find. For example, taking as first premise one of the 'singulars' that Burley is discussing here,

> Socrates is running. Socrates is the husband of Xanthippe. Therefore the husband of Xanthippe is running.

Why does Burley think such consequences are bad? We aren't told. Perhaps he means only that no such consequence is a syllogism in any sense of 'syllogism' that he allows. (See Example 3 in Section 17 below.) But the net effect is that if we have to deal with a consequence where the premises explicitly or implicitly involve neither quantifiers nor sentence connectives, nothing in PL will help us.

The lack of such rules wouldn't necessarily hinder a medieval mathematician. For example the first of Euclid's Common Notions is the universal statement that things equal to the same thing are also equal to each other. Appeals to transitivity of identity could be seen as applications of this universally quantified law.

After rules for propositions, and the nonexistent rules for singulars, there come rules for quantifiers. Burley has names for some of the procedures involved; but the names describe a kind of move, not the conditions under which the move is valid. In this section I take some of the simpler cases. Universal quantification and monotonicity need closer treatment and will come in the sections that follow.

8.1 Dici de omni

This rule appears at (213), (236) and (983). It also appears at (210), (212) and again at (236) under the name *dici de omni vel de nullo*. The name *dici de omni* means 'saying about every'.

At (213) and (236) Burley applies this name to argument steps of the form

> From 'All A's are B's' and 'c is an A' infer 'c is a B'.

Burley correctly notes at (213) that if A is a class whose membership varies from one situation to another, then the inference only holds where the premise 'c is an A' holds in the same situation that is intended in the premise 'All A's are B's'. To illustrate he offers

> Everything white was black. Socrates was white. Therefore Socrates was black.
> *Omne album fuit nigrum, Sortes fuit albus, ergo Sortes fuit niger.*

The inference holds only if 'Everything white' means everything that was white, not everything that is white. (Of course Burley's details are dead wrong here. The relevant class of white things depends on the context in which the inference is used, and the possibilities are a great deal more complicated than just 'what was white' and 'what is white'. But the medieval Latins were serially blind to questions of context; we just have to live with that.)

At (981) Burley gives a bad consequence that appears to be of the *dici de omni* form. The consequence fails when we realise that the term in the '*c*' position needs analysis, and after analysis the inference no longer has the *dici de omni* form. The term in question is 'Socrates by virtue of the fact that he is an animate substance' (*Sortes inquantum est substantia animata*).

At (983) Burley seems to apply the name *dici de omni* to a different form of inference:

> Every man has perceptions by virtue of the fact that he is an animal. Everything that can laugh is a man. Therefore everything that can laugh has perceptions by virtue of the fact that it is an animal.
> *Omnis homo est sensibilis inquantum animal, omne risibile est homo, ergo omne risibile est sensibile inquantum animal.*

There is not enough here to tell whether he intends a different form of *dici de omni*, or whether he means that the argument holds by *dici de omni* together with other appropriate quantifier rules.

8.2 Descent

Descent allows us to infer, from a proposition S containing a term T, any sentence S′ got from S by replacing T by the name of an object described by T in the context of the sentence S. If T is the subject term, then we have to drop any quantifier expression attached to the term when we replace it by the name.

We discussed descent briefly in Section 4. A term T that allows descent as above is said to have confused and distributive supposition in S. So any statement of the form "Under such-and-such conditions a term has confused and distributive supposition" is in fact a rule of good consequences. The main rule of this kind is that in universally quantified categorical propositions the subject term has confused and distributive supposition. This statement is equivalent to the rule of *dici de omni* that we have just discussed.

> From 'Every A is a B' and '*c* is an A' there follows '*c* is a B'.

Another case (87) is a categorical proposition where the subject term carries a negative universal quantifier such as 'none of' or 'neither of'; in such a proposition the predicate has confused and distributive supposition. Thus:

From 'No A is a B' and '*c* is a B' there follows 'No A is *c*'.

In Section 10 we will see how to derive this by a monotonicity argument.

8.3 Existential generalisation

Burley certainly knows the move made in inferences such as

Socrates is running. Therefore a man is running.

But he seems to have no name for it. At (85) he describes it rather clumsily as 'The proposition is inferred from any one of the supposita of the term', i.e. the term 'man' in the case above. Note that since this is the subject term, English demands an explicit existential quantifier. (In Latin one can add *quidam*, but it's not required.) A term that allows this inference rule is said to have 'determinate' *determinata*) supposition if it also allows disjunctive descent, and 'simply confused' (*confusa tantum*) supposition if it doesn't.

Just as with descent, any rule saying that certain terms have one of these kinds of supposition yields rules of good consequence. For example when Burley tells us at (164) that in 'Each of them said something true', 'something true' has simply confused supposition, he is (among other things) licensing the inference

Each of them said the two times table. Therefore each of them said something true.

This example is quite interesting because the term 'something true' is a proper part of the predicate. So any justification of Burley's claim would need something tending towards a deep rule. But he gives none. Again we will see (in Sections 10f) how he could have done.

9 Universal generalisation

From the reference list, 25 of Burley's good consequences have universally quantified conclusions. In all of these the conclusion begins with a sign of quantification ('every' *omnis*, 'no' *nullus*, 'nothing' *nihil*, 'only' *tantum*). Sometimes Burley takes 'Every A is a B' as implying that 'There is an A', which he proves separately.

In some of these twenty-five consequences ((127), (543), (645), (667), (764), (921), (985)) Burley states the conclusion without any indication of how it's derived, except perhaps the general rule which it illustrates. In (750) his argument is unclear to me.

The rest of the twenty-five fall into two classes. First there are those inferences where the conclusion is derived by propositional rules from some other proposition or propositions that already contain either it or its exponents or a paraphrase of it ((642), (816), (818), (833), (895)). This includes those cases where he derives the negation of the premise from the negation of the consequent, as at (280), (486), (638) with a variation at (996). It also includes the cases where he derives the conclusion by transitivity of entailment, as at (305) and (990); (988) is a slight variation of this.

Second there are those cases where a premise already contains a corresponding quantifier, and Burley carries out some manipulation of the proposition while holding the quantifier fixed. In (667) he replaces the phrase 'man' inside a universally quantified sentence by the phrase 'distinct from any non-man'. At (85) he uses existential generalisation inside the scope of a universal quantifier. The remaining cases (648), (648) and (996) are basically monotonicity arguments. We will come to monotonicity arguments in the next section.

In some of these cases, Burley proves the consequence by invoking a categorical syllogism and then deriving the conclusion by propositional rules or paraphrase. One could ask how he proves the categorical syllogisms, but PL is largely silent on this.

Now all the proofs in these two classes have an interesting feature in common. The universal quantifier never appears or disappears; we can trace it from the conclusion, through intermediate steps, back into the premises. Frege in his Begriffsschrift of 1893 officially kept to this style too: he had no rules for adding universal quantifiers. But in practice Frege found the restriction intolerable, and he introduced a convention that allows one to drop universal quantifiers in favour of latin letters ([Frege, 1893] §17). Then we can compute using the latin letters, and restore the quantifiers later. Today we find it more natural to regard his rules for adding and removing latin letters as quantifier rules of his system.

One can distinguish at least three ways of reaching a universally quantified conclusion without maintaining the quantifier from premises to conclusion:

Full Enumeration: To prove that all A's are B's, list all the A's and check that each of them is a B.

Sample: To prove that all A's are B's, look at a suitable sample of A's and check that all A's in the sample are B's. (This method can lead to false results, for example if through lack of imagination you miss an important sort of A.)

Universal Generalisation: The method is to introduce a letter, say a, and proceed to make deductions using the formal proposition that a is an A, but no other assumptions using the letter a. If you succeed in deducing that a is a B, then it follows that all a's are B's. (This is the standard mathematicians' method, used by geometers since at least the fourth century BC.)

There is no evidence in PL that Burley understands any of Full Enumeration, Sample or Universal Generalisation well enough to use them reliably. There is some evidence that he doesn't.

For example at (672) he gives what I think must be intended as an example of Full Enumeration:

> And when it is proved: Everything distinct from any non-man is an animal, therefore everything distinct from any donkey is an animal; for the singulars of the antecedent imply the singulars of the consequent. For it follows: This thing differing from any non-man is an animal, therefore this thing distinct from any donkey is an animal, and likewise with the rest.
> *Et cum probatur: Omne differens a non-homine est animal, ergo omne differens ab asino est animal; nam singulares antecedentis inferunt singulares consequentis; sequitur enim: Hoc differens a non-homine est animal, ergo hoc differens ab asino est animal, et sic de aliis.*

The 'and likewise with the rest' makes it clear that Burley is talking about an enumeration, not a formal argument as in Universal Generalisation. But for this he should be enumerating the items in the range of the universal quantifier of the consequent, namely the things distinct from any donkey; he has gone to the wrong quantifier. A second problem is that it's not plausible to list all things distinct from any non-man (or from any donkey). The consequence is in fact bad, but Burley never gets round to explaining why. Is that perhaps because he didn't have the matter clear in his own mind? If he really had a clear understanding of the moves involved, one suspects he would have chosen a less confusing example to illustrate them.

There is a rule for making deductions from existentially quantified propositions 'Some A is a B':

Existential Instantiation: The method is to introduce a letter, say a, and proceed to make deductions using the formal proposition that a is an A and a B, but no other assumptions using the letter a. If you succeed in deducing a conclusion not mentioning a, then the conclusion follows already from 'Some A is a B'.

This rule is a kind of dual to Universal Generalisation.

Burley shows at (150) that he has come across some version of Existential Instantiation. But he thinks that the method works by taking one of the

A's and naming it a, an operation that he calls 'signing' (*signat*). Thus he considers an argument which starts 'For every magnitude there is a smaller magnitude; let A be this smaller magnitude'. He rightly objects to this argument: he says that it is not legitimate to replace the term 'a smaller magnitude' by a sign naming a particular magnitude (*non licet ponere aliquod suppositum eius*). But a is not a sign naming a particular magnitude. If the problem were what Burley says it is, then we would be inhibited from making deductions from 'There is a grain of sand on the Siberian coast' until we'd identified and named a grain of sand on the Siberian coast.

Burley's misunderstandings of Universal Generalisation and Existential Instantiation come together at (761f). Here he gives a proof of a universal 'Every B is an A' from an exclusive. He turns it around so as to derive the negation of the exclusive from the sentence 'Some B is not an A'. He correctly reasons: Let c be a B that is not an A, etc. We would suppose that he had correctly understood Existential Instantiation if he hadn't gone on to add:

> And the same goes for any other singular.
> *Et eodem modo est de qualibet alia singulari.*

There are two mistakes here. First, he thinks that a singular has been mentioned; it hasn't. Second, he seems to think that part of the argument is to generalise from one singular to all singulars. To me this suggests he has confused Existential Instantiation with Universal Generalisation, and then Universal Generalisation in turn with Full Enumeration.

John Peckham's *Perspectiva Communis*, written a few decades before Burley's book, proves a number of universally quantified geometrical statements. His arguments in general are certainly thin, fitting his intention to write a popular book. Also some of his physical or physiological views are indefensible, and there are a few slips in the geometry. But his geometrical demonstrations of universally quantified statements are normally squeaky clean examples of Universal Generalisation. I found nothing corresponding to the logical misunderstandings in Burley. True, Peckham introduces many of these proofs with 'For example' (*verbi gratia*); but this seems to be a turn of phrase, meaning 'Here's a way of seeing why this is true'. Did Peckham understand universal quantification better than Burley, or was it a feature of the age to fail to transfer to logic what they understood in mathematics?

I haven't mentioned Sample yet. We will see later that Burley's lack of fluency with basic techniques comes home to roost when he has to prove statements about all consequences with a certain form. When he can't deduce these statements from already known general laws, he is reduced to giving examples. Thus when he says that consequences with a certain form are bad, and gives a single example, we can't tell from the form of his argument whether he is meaning to show (a) that not all consequences with that form are good, by giving a counterexample (*instantia*) or (b) that all consequences with that form are bad, by giving an example that he wants us to recognise as typical. We will see that there are some fairly severe problems of interpretation, but (b) seems to be part of the story. And (b) is a case of Sample.

10 Ab inferiori ad superius

Burley refers many times to a move called 'from lower to higher' (*ab inferiori ad superius*). Though he talks of it as a way of arguing, perhaps the best way to think of it is as a class of consequences. The modern name is 'upwards monotonicity', and a modern description might go as follows (making allowance for some looseness about the grammar).

Write $S(T)$ for a sentence in which we mark an occurrence of a noun phrase T. Let T' be another noun phrase, and write $S(T'/T)$ for the sentence got by replacing T by T' at the marked occurrence.

We say that the marked occurrence of T in $S(T)$ is *upwards monotone* if for all noun phrases T' the consequence

Every T is a T'. $S(T)$. Therefore $S(T'/T)$.

is good. Burley describes a consequence of this form, whether or not it's good, as 'from lower to higher'.

We say that the marked occurrence of T in $S(T)$ is *downwards monotone* if for all noun phrases T' the consequence

Every T is a T'. $S(T'/T)$. Therefore $S(T)$.

is good. Burley describes a consequence of this form, whether or not it's good, as 'from higher to lower'. (Recall that for Burley 'Every T is a T'' implies there is a T, so downwards monotonicity doesn't go to empty terms. This will cause some technical nuisances in Section 17.)

A number of Burley's remarks on 'from lower to higher' could be paraphrased by saying that occurrences of noun phrases in certain situations are (or are not) upwards monotone. For example at (102) he shows that in 'No animal except one of these is a man', 'man' is not upwards monotone. At (304) he shows that in 'If a man is running, a thing that can laugh is running' the word 'man' is not upwards monotone. At (305) he remarks that in 'If a man is running, an animal is running', 'animal' is upwards monotone. At (648f) we read that 'man' is upwards monotone in 'Only a man is running', but not in 'Only something that can laugh is a man'. An example at (927) has two occurrences of the relevant term: in 'Whatever is true is true at this instant', 'true' is upwards monotone at its first occurrence. At (981) there is an example involving 'by virtue of'.

Burley knows that when we build up compound sentences, terms that are upwards monotone in the component sentences may stay upwards monotone in the compound, and that we can sometimes show this by climbing up through the construction. For example at (568) he correctly notes that in a disjunction 'P or Q', an upwards monotone occurrence in one of the disjuncts remains upwards monotone in the disjunction (though he doesn't prove this in detail). In the same paragraph he shows he knows that negating a sentence reverses the monotonicities of occurrences of terms in it.

An argument at (302) is revealing:

> From the same rule, viz., that whatever follows from the consequent [follows from the antecedent], it's clear that in a conditional whose antecedent is a particular or indefinite proposition, the subject of the antecedent has confused and distributive supposition with respect to the consequent, so that ... there follows a conditional in which the subject of the antecedent is inferior to the subject of the first conditional. For example it follows: If an animal is running then a substance is running, therefore if a man is running, a substance is running.
> *Ex eadem regula, scilicet quidquid sequitur ad consequens in etc., patet quod in conditionali, cuius antecedens est propositio particularis vel indefinita, subiectum antecedentis supponit confuse et distributive respectu consequentis, ita quod ... ad talem conditionalem, cuius antecedens est propositio particularis vel indefinita, sequitur conditionalis, in cuius antecedente subiicitur aliquod inferius ad subiectum primae conditionalis. Verbi gratia, sequitur enim: Si animal currit, substantia currit, ergo si homo currit, substantia currit.*

Here Burley is showing that 'animal' is downwards monotone in 'If an animal is running then a substance is running'. The argument he uses is the transitivity of 'following'. We can reconstruct as follows. First, 'man' in 'A man is running' is obviously upwards monotone. It follows that the consequence

> If a man is running then an animal is running.

is good. But this together with 'If an animal is running then a substance is running' yields the stated conclusion. Here Burley moves the monotonicity one step deeper in a compound sentence by using an argument rule directly related to the principal of the compound sentence. It's clear that this could be iterated any number of times, and so we would discover monotone occurrences at arbitrary depth inside sentences.

Burley also pushes the matter forward in another direction: we can build up compound *terms*. The appropriate definition now is where $P(T)$ is a term and $P(T'/T)$ the result of substituting. We say that the marked occurrence of T in $P(T)$ is *upwards monotone* if for all noun phrases T' the consequence

Every T is a T'. Therefore every $P(T)$ is a $P(T'/T)$.

is good. We say that the marked occurrence of T in $P(T)$ is *downwards monotone* if for all noun phrases T' the consequence

Every T' is a T. Therefore every $P(T)$ is a $P(T'/T)$.

is good. Burley has no technical term to cover these, but he discusses the phenomenon. For example at (671) he notes that in 'distinct from any donkey' the word 'donkey' is downwards monotone. At (194) he notes that a certain argument would work if the word 'going' in 'going to Rome' was upwards monotone; to show that it isn't, he cites 'existing to Rome'. (This example raises important issues not connected with monotonicity, of course.)

Burley is also aware of constructions that block monotonicity. He knows that sentences about knowledge provide examples: 'man' is not upwards or downwards monotone in 'You know whether a man is running' (383), (385). At (1029) he knows that something subtler than 'Every T is a T'' may be needed in temporal contexts. But rather than analyse what really is needed, he opts for an idle solution and requires that 'Every T is a T'' is a *necessary* truth.

Burley tends to connect monotonicity and distribution. For example at (302) he explains the downwards monotonicity of 'animal' in 'If an animal is running, a substance is running' as a case of confused and distributive supposition. At (304) he describes a case of upwards monotonicity as 'ascent'. At (168) he refers to a case of failure of upwards monotonicity as a matter of distribution. At (648f) he uses confused and distributive supposition as the reason why a certain term isn't upwards monotone.

In fact downwards monotonicity and descent to singulars are quite different phenomena, and Burley loses information by confusing them. Descent to singulars replaces a noun phrase by a proper noun, and deletes any quantifier attached to the noun phrase. Downwards monotonicity replaces a noun phrase by another noun phrase, and leaves the quantifier in place. (At (382) and (384) Burley discusses a move that replaces a noun phrase by a noun phrase *and* removes the quantifier. This is neither descent nor monotonicity; Burley makes it a matter of distribution.)

Downwards monotonicity implies descent to singulars, but not the other way round. To derive descent from downwards monotonicity, the simplest route is to use a noun phrase that is true of just the one individual. For example 'man' is downwards monotone in 'Every man is running'. We deduce that Dov Gabbay is running by carrying out the consequence

> Every person who is Dov is a man. Every man is running. Therefore every person who is Dov is running.

Since Dov is the only person who is Dov, the conclusion says simply that Dov is running. The same device works with 'There is' (but clearly not with quantifiers implying the existence of more than one thing, like 'For at least two').

In a kind of dual operation, upwards monotonicity implies ascent from singulars. For example the position T in 'Some T writes books' is upwards monotone, and this licenses the consequence

> Every person who is Dov is a logician. Some person who is Dov writes books. Therefore some logician writes books.

In other words, Dov is a logician, Dov writes books, so some logician writes books.

I think the main cause of the conflation of descent with distribution is that confused and distributive supposition was taken—long before Burley—to be the characteristic effect of universal quantifiers. (In PL see (87), (97), (111), (377), (383), (384), (752).) But universal quantifiers have both these properties: they generate downwards monotonicity and they allow descent to singulars. Since the former property implies the latter and not vice versa, it would have been more sensible if the medievals had taken the former as the characteristic property throughout, instead of shifting tacitly between them as Burley does.

11 A calculus of monotonicity

Burley's treatment of monotonicity almost amounts to a calculus. In this section let me set it out more systematically. Since we are talking about natural language sentences, the word DEFAULT should be up in neon lights throughout—or maybe John of Salisbury's judicious *plerumque*.

First consider simple affirmative categoricals. In 'Every A is a B', A is downwards monotone and B is upwards monotone. In 'Some A is a B', both terms are upwards monotone.

More complicated subject-predicate sentences are probably best treated as generalised quantifiers on two or more terms. For example in 'Ignoring A's, all B's are C's', A and C are upwards monotone while B is downwards.

In 'Only A's are B's' (read as Burley reads *tantum*), A is upwards monotone and B downwards. In 'At least five A's are B's', A and B are both upwards monotone.

Negating a sentence reverses all monotonicities in it.

Conjunction with 'and' and disjunction with 'or' preserve all monotonicities. This holds even where there are anaphoric pronouns, but of course we should avoid stupidities like trying to assign upwards monotonicity to a pronoun.

In 'If P then Q', monotonicities are reversed in P and preserved in Q. We should add that if conditionals are read intentionally, as Burley usually reads them, then the licensing 'Every T is a T'' should be true necessarily.

One can add further clauses in a similar spirit, for example to cope with monotonicity of noun phrases inside other noun phrases. The language covered by these constructions, with 'If ... then' interpreted so as to allow unrestricted monotonicities, can reasonably be called Monotone Latin.

Monotone Latin excludes constructions that block monotonicity. For example the constructions 'X knows whether P' or 'X is pleased that P' block monotonicities in P. The generalised quantifier 'A's are B's by virtue of the fact that they are C's' is arguably upwards monotone at B and downwards at A, but C is neither upwards nor downwards monotone.

This calculus yields a high proportion of the good consequences discussed by Burley, and infinitely many more not discussed by him. The rule of deducing $S(T'/T)$ from $S(T)$ and 'Every T is a T'', where T is upwards monotone in $S(T)$, is of course a deep rule. (Jan van Eijck [van Eijck, 2005] implements in Haskell a calculus which proves all good categorical syllogisms by two rules: monotonicity and symmetry.)

I think Burley and his colleagues could reasonably claim credit for the calculus of monotonicity. Granted, today's style demands greater rigour. Victor Sánchez Valencia in his PhD thesis [Valencia, 1991] builds a calculus of monotonicity based on the λ-calculus, with a fragment of natural language to illustrate. In one direction Burley goes further: he includes temporal and modal phenomena.

In fact I think it's fair to say that Burley understands the theory behind the calculus of monotonicity better than he understands how to apply it. He is not always reliable in recognising upwards and downwards monotonicity. One can extend the notion to occurrences of sentences (so that for example in 'If P then Q', the occurrence of P is downwards monotone and the occurrence of Q is upwards). Burley certainly has this idea; it was in the background in his discussion at (302). But for example at (452) he considers second figure conditional syllogisms with one premise an affirmed conditional and the other a negated one, and he says that in this case the proposition common to both premises is affirmed in both or denied in both. This is exactly what he should *not* be saying; the crucial fact that he misses here is that in these syllogisms one of the occurrences is upwards monotone and the other is downwards. In Section 17 below, we prove on general principle that if both had the same monotonicity, there would be no valid syllogisms of this type.

In [Spade, 1988] Spade describes what is essentially a fragment of Monotone Latin, and observes that in all categoricals in his language, every term allows either descent to singulars or ascent from singulars (his Theorem 2 on page 199). He sketches an argument in terms of quantifiers. In fact a stronger statement follows immediately from the construction of the calculus: Throughout Monotone Latin, every occurrence of a term is either upwards or downwards monotone.

12 Invalid consequences

Of the 306 consequences in the reference list, 178 are not good. This is 58% of the total. A rough count on a modern textbook (Kalish and Montague [Kalish and Montague, 1964], in some ways a modern counterpart of Burley) found 49 valid English arguments and 29 invalid; here 37% of the total are invalid.

The relative importance of invalid arguments in medieval Latin logic is an acknowledged fact. De Rijk ([Rijk, 1962] p. 22) says:

> ... the doctrine of fallacy forms the basis of terminist logic. For this logic developed as a result of the fact that ... the proposition was beginning to be subjected to a strictly linguistic analysis. The first impulse to this was given by the discovery of Aristotle's *Sophistici Elenchi* and especially by the circumstance that scholars made themselves familiar with this work.

Looking at De Rijk's evidence, I query only his phrase 'strictly linguistic'. One of the chief morals of the *Sophistici Elenchi* was that you can't distinguish valid from invalid arguments by strictly syntactic criteria; you have to look at the meanings. So far as there are valid argument forms, these forms are at least partly semantic. The terminists set out to describe the features of meaning that count towards validity. Burley inherited their work and many of their attitudes. For example he inherited their shallow grasp of syntax, and to this extent both he and they were anti-linguists. We have seen examples of this.

Sixteenth and seventeenth century thinkers were apt to complain that scholastic logicians had refined the art of proving invalidity to a point where they refused to accept refutation even by valid arguments. Thus Locke ([Locke, 1884] §189) in 1690:

> Is there any thing more inconsistent with Civil Conversation, and the End of all Debate, than not to take an Answer, though never so full and satisfactory, but still to go on with the Dispute as long as equivocal Sounds can furnish (a *medius terminus*) a Term to wrangle with on the one Side, or a Distinction on the other ...?

There is an uncomfortable amount of truth in the charge.

For example Paul of Venice ([of Venice, 1978] p. 20f) in the early fifteenth century argues that the following consequence is not good:

> There is no chimera. Therefore it is the case that there is no chimera.
> *Nulla chimaera est. Igitur ita est quod nulla chimaera est.*

He has two arguments. The first is that 'an affirmative proposition without any modal term does not follow from a non-pregnant negative proposition'. The second is that it's imaginable (*imaginabile*) that nothing exists or is the case (*nihil nec aliqualiter esset*), in which case the premise is true and the conclusion is false.

The second argument is an absurd appeal to introspection; happily Burley is free of this kind of nonsense. The first argument is an attempt to batter the reader with technology; but it makes no sense, because the manifest goodness of the consequence invalidates any theory that claims it's bad. *Eppur si muove.* This is highly relevant to Burley, because he makes several attempts at general rules guaranteeing the invalidity of consequences. As we go, we must ask how Burley hopes to justify these rules.

13 Showing that a consequence is bad

Burley's 'first general and principal rule of consequences' (258) is that

> if in some possible circumstances it is possible that there is a time when the antecedents are true and the consequent is false, then the consequence is not good.
> *... si aliquo casu possibili posito possit antecedens aliquando esse verum sine consequente, tunc non fuit consequentia bona.*

There are two nested possibles and one temporal operator here; I don't know whether Burley intended this or he just wrote carelessly. In fact for all his applications of the rule a simpler form suffices:

> If in some possible circumstances the antecedents are true and the consequent is false, then the consequence is not good.

He never claims the converse.

At least, I think he never means to claim the converse. At (499) he wants to show that a consequence is good. His method is to deduce a contradiction from the premises and the negation of the conclusion, and then appeal to *per impossibile*. But at the start of this argument he says 'If we suppose that the conclusion doesn't follow, let the negation of the conclusion be assumed'. This suggests that he thinks that badness of a consequence allows us to assume the premises and the negation of the conclusion. But a closer look shows that the words 'If we suppose that the conclusion doesn't follow' (*si datur, quod conclusio non sequitur*) are not a part of the argument at all, and it would have been clearer if he had left them out.

For 47 of the bad consequences in the reference list, Burley claims that the premises are true and the conclusion false. For a further 4 in the list he claims that this is the case under a posit. There are a large number of other cases where he leaves it to us to see that a given consequence is bad, and for most of these cases the true-premise false-conclusion test works. The test is logically sound.

More puzzling are the other arguments that Burley seems to use in order to show that certain consequences are bad. In later sections we will look at the systematic methods that he uses. Here I note one case where he uses a special argument. Burley is mistaken, but in an interesting way.

Burley argues at (239):

> Some are predicates that determinately include nonexistence, for example to be dead, to be decomposed and so on. And when it's argued from a proposition in which such a predicate is predicated, to simple existence, this is a fallacy of relatively/absolutely. Thus it doesn't follow: Caesar is dead, therefore Caesar exists.
> *... quaedam sunt praedicata, quae determinate includunt nonesse, sicut esse mortuum, esse corruptum et sic de aliis. Et quando arguitur a propositione, in qua praedicatur tale praedicatum, ad esse simpliciter, est fallacia secundum quid et simpliciter. Et ideo non sequitur: Caesar est mortuus, ergo Caesar est.*

Here's a counterexample to Burley's claim:

> The current Pope is dead. Therefore the current Pope exists.

If there is such a person as the current Pope, then he must exist. So the consequence is perfectly sound, and not an example of any fallacy.

Burley has made the following mistake. It's correct that if P is a predicate of the kind that he describes, then from 'A is P' (under reasonable assumptions on the form of A) we can correctly deduce 'A doesn't exist'. But it in no way follows that we can't also deduce 'A exists'. Burley has confused 'We can infer not-q' with 'We can't infer q'.

This confusion is still very common in the subcultures of logic, for example among those many people who kindly send me their refutations of Cantor's diagonal argument [Hodges, 1998a]. In the most sophisticated example to reach me, my correspondent quoted an inference rule from Mostowski and restated it as a non-inference rule in the Burley fashion.

Burley's mistake would rule out arguments *per impossibile.* Not all logicians accept arguments *per impossibile*, but we have seen that Burley did.

14 Bad consequences and false rules

Why does Burley give examples of bad consequences?

For many of his examples the reason is clear. First, there are consequences that are given to illustrate some point about the meanings of words in the consequence.

For example at (56) he clarifies the use of the aristotelian notion of *perfectio* with the help of some inferences involving it; (69) does the same for the notion of 'in the first instance' (*primo*). A consequence in (194) is to explain the meaning of 'being' (*ens*), and (242)–(248) perform the same service for 'is' (*est*), and likewise (690) and (694) for the notion 'one' (*unum*). At (118) and (184) he illustrates the behaviour of anaphoric pronouns in inferences, and at (126) he illustrates reflexive pronouns in the same way. An inference in (630) makes a point about truth conditions of sentences about the past, and one at (377) illuminates truth conditions for statements involving 'knows that'. Inferences at (921) and (924) illustrate truth conditions for sentences containing 'unless' (*nisi*), and (884) illustrates the use of numerical expressions. At (360) he explains the difference between 'following from A or (from) B' and 'following from A-or-B'.

Second, there are examples that illustrate or prove that some rule is false. Recall that a rule of good consequences is 'true' if and only if every consequence obeying the rule is good. So a false rule is one that is obeyed by at least one bad consequence. Burley's word for a bad consequence proving that a rule is false is *instantia.*

For example, if our account of formal consequence in Section 6 is correct, then to say that a consequence is not formally good is to say that a certain class of consequences contains at least one bad consequence. There are examples at (102), (160), (364) and (380). I suspect that (162) is another example, though Burley doesn't mention formal consequence here. He claims that the consequence 'It's impossible that a person who is standing is sitting; Socrates is standing; therefore it's impossible that Socrates is sitting' is bad because 'the premises don't have any terms in common, as is clear if the propositions are unpacked' (*praemissae non communicant in aliquo termino, ut patet, si istae propositiones resolvantur*). He presumably means that the first premise should be read as '(The proposition that a person who is standing is sitting) is impossible', so that its terms are 'the proposition that a person who is standing is sitting' and 'impossible'. This suggests that he has in mind a class of inferences which is closed under some kind of replacement of entire terms.

Sometimes Burley uses examples to show that there is no syllogism of a certain form. We saw that the notion of a syllogism doesn't have a sharp definition, so there is an element of vagueness about this kind of argument. Examples are at (420) and (497). Also (421) and (445) are in the middle of discussions of syllogisms and should probably be understood this way.

Sometimes Burley simply wants to point out that some rules that some people might think are good are in fact bad. Probably his examples call on his teaching experience.

For example at (277) he gives an example to show that while the goodness of 'Not Q, therefore not P' ensures the goodness of 'P, therefore Q', the implication fails if we replace 'Not Q' by a sentence that is merely inconsistent with Q. At (658) he points out that the equivalence of '*a* is not a B' and 'It's not true that *a* is a B' (where *a* is a proper name) breaks down if we replace '*a*' by 'Only *a*'. Other examples are at (300), (334), (659) and (878).

Some of the rules that Burley refutes are very silly. For example at (417) he seems to be attacking the rule 'If P then Q; if R then S; therefore if P then S' (though his example puts in place of the second S a sentence obviously implied by S). At (382) and (383) I was unable to see any remotely plausible rule that fits his description.

15 Fallacies

There are a number of false rules that Burley describes as 'fallacies' (*fallaciae*) and gives names to:

Figure of Speech (*figurae dictionis*) (93), (94), (95), (145), (147), (149), (153), (155), (157), (164), (1030).

Consequent (*consequentis*) (168), (241), (250), (263), (376), (420), (421),

(514), (652), (656), (671), (927).

Relative and Absolute (*secundum quid et simpliciter*) (239), (241), (242), (245), (684), (902).

Varying the Common Part (*accidentis ex variatione medii*) (310), (608), (610), (613)

Ambiguity (*aequivocationis*) (229)

For each fallacy Burley indicates a feature of a consequence that would make it an instance of the fallacy. These features need not be definitions of the fallacies; the feature that he indicates for Figure of Speech is one that particularly interests him, but the name itself suggests a wider class of features.

Now there are three things that we might say about a consequence C in connection with a particular fallacy Fa and corresponding feature Fe:

(I) C has feature Fe.

(II) C commits fallacy Fa.

(III) C is a bad consequence.

Burley is consistent about the relationships between (I), (II) and (III). They entail each other as follows:

$$(\mathrm{I}) \Rightarrow (\mathrm{II}) \Rightarrow (\mathrm{III}).$$

We can deduce this from the way he describes fallacious consequences. For example he says

> The consequence C is bad and a fallacy, because (usually *quia*, sometimes *nam*) it has the feature.
> (93), (155), (164), (168), (229), (376), (671), (927).
>
> The consequence C is bad because it commits the fallacy.
> (239), (657).
>
> The consequence C commits the fallacy because (usually *quoniam*, *quia*) it has the feature.
> (153), (1957), (613). (94) is similar.
>
> Every consequence with the feature commits the fallacy.
> (95), (245), (263). (420), (421), (514) and (611) are similar.

> Every consequence with the feature commits the fallacy and is bad.
> (241), (310), (652).

A comparison with the list at the head of this section confirms that these comments are spread fairly evenly across the types of fallacy.

One implication of all this is that there are various features, each of which guarantees the badness of every consequence that has it. Such features are rare on the ground in modern logic, but Burley seems to find them all over the place. The obvious candidates are that the premises are true and the conclusion false, or (if we want a formally recognisable property) that the premises are propositional tautologies and the conclusion is an explicit contradiction. But Burley's features go way beyond these. What has he found that we are missing?

In the case of Figure of Speech I think I know the answer. The feature that constitutes this fallacy is that we infer

> For every A there is a B such that .... Therefore there is a B such that for every A, ...

For example at (93)

> Twice (i.e. on each of two occasions) you ate some bread. Therefore there is some bread that you ate twice.

Now certainly not every consequence with this feature is bad. For example

> Everybody here has heard of someone who is Alfred Hitchcock. Therefore there is someone who is Alfred Hitchcock, and whom everybody here has heard of.

From Burley's detailed discussions of this fallacy, it seems clear that he would have ruled this out as a counterexample, because in this case no word 'imports multiplicity'. He appears to say at (89) that 'every', 'each', 'three times' etc. import multiplicity, as if we could check it from the word alone. But we already know from Section 4 that his semantic analyses are not compositional, and it would have been open to him to say that in this particular example 'Everybody' fails to import multiplicity because of the fact that the description 'is Alfred Hitchcock' can only hold of one thing.

If this is right, then it's part of the definition of the feature constituting Figure of Speech that the relevant consequence is bad. Since the implication from (I) to (III) goes by way of (II), it would follow that it's part of the definition of Figure of Speech that the consequence is bad; so the implication from (II) to (III) is part of the definition of (II).

Unfortunately this explanation is implausible for the fallacy of Consequent. First, Burley claims at (263) that a certain two rules (*regulae*) 'always produce a fallacy of the Consequent'. This claim is obtuse if the only reason that these rules produce a fallacy is that good consequences don't count as instances of them. Moreover at (300f), discussing a bad consequent, he says that it is clear (*patet*) that the consequence is argued by one of the false rules of (263), and the implication is that we can see that the conference is fallacious by seeing that it has a certain syntactic form. (I think this knocks out one possible soft-option escape route: namely that Burley means only that if the only reason we have for accepting a consequence is that it's an instance of a particular rule, then we don't have any good reason for accepting it.)

The 'false' rule that Burley says produces the fallacy at (300) is

> If q follows from p and r follows from p, then r follows from q.

Burley himself at (254), just a few pages earlier, has said

> From an impossibility anything follows.
> *Ex impossibili sequitur quodlibet.*

Thus if p and q are impossible, then q follows from p and r follows from p, so by the false rule, r follows from q; but in this case r really does follow from q, contrary to Burley's statement that the false rule always produces bad consequences. Another way to get good consequences out of the false rule is to start from any good consequence 'q therefore r' and take p arbitrarily.

There is no evidence in PL that Burley is aware of the contradiction between (254) and (263); so it's an idle question how he would have resolved it. However, this is one place where we can call on some logical facts. In some sense of 'follow', (254) is certainly correct. So the 'false' rule doesn't in fact always create bad consequences, and Burley is mistaken to say that it does.

But note that the two counterexamples above have an interesting feature in common. Going by way of (254), we finish with a consequence 'r follows from q' which is good regardless of what r is. Starting with a good consequence 'q therefore r', we satisfy the false rule regardless of what p is. So in both cases the counterexample depends on a redundancy. Similar redundancies will be important for us in Section 17.

With Varying the Common Part the situation is different yet again. Burley gives as an example the following ((310) slightly simplified):

> If it's no time then it's not day. If it's not day and it's some time, then it's night. Therefore if it's no time then it's night.

The fault in the argument is that 'it's not day' in the first premise is matched against 'it's not day and it's some time' in the second, and they are different. His point is that there is no good rule of the form 'If P then Q. If R then S. Therefore if P then S.' Now manifestly not every example of this rule is bad. For example the consequence 'If it's no time then it's not day but it is some time. If it's not day, then it's night. Therefore if it's no time then it's night.' is good and has this form. Examples are so easy to find that Burley can't conceivably mean that this form has any tendency to create bad consequences. His examples, here and elsewhere ((172), (608)), fit the following pattern:

> If we take a true syllogistic rule, and in one of the premises we replace the middle term by a new term, then the resulting rule is no longer true.

If this is what Burley means, then he is quite correct, as one easily checks.

We have here a *class of rules* in which every rule is false. This is a different matter altogether from a *class of consequences* in which every consequence is bad. Burley has done a disservice by lumping together the two kinds of class under the common name of 'fallacy'. But classes of false rules are an important notion in PL, and we devote our final two sections to them.

16 Classes of false rules

At (789) Burley makes the following very revealing remarks:

> But if each premise is exclusive affirmative and the principal word is negated in each, no conclusion follows by rules of syllogisms. For if any conclusion followed by rules of syllogisms, a negative conclusion would follow since each premise is negative. But no negative does follow. For it doesn't follow: Only an intelligent being is not a non-animal, only a non-man is not a non-animal, therefore only a non-man is not intelligent. Neither does it follow that a non-man is not intelligent, since the premises are true and the conclusion false. But if any negative conclusion followed, one of these would follow.
> *Si vero utraque praemissa sit exclusiva affirmativa et verbum principale negetur in utraque, nulla conclusio sequitur per regulas syllogismorum. Quia si aliqua conclusio sequeretur per regulas syllogismorum, conclusio negativa sequeretur, cum utraque praemissa sit negativa; sed nulla negativa sequitur. Non enim sequitur: Tantum intelligibile non est non-animal, tantum non-homo non est non-animal, ergo tantum non-homo non est in-*

> *telligibilis. Nec etiam sequitur, quod non-homo non est intelligibilis, quia praemissae sunt verae et conclusio falsa. Si tamen aliqua conclusio negativa sequeretur, altera istarum sequeretur.* (I follow Spade's text against Boehner's here; otherwise the example wouldn't fit Burley's description.)

Here Burley is talking about a certain class of second-figure rules of syllogistic type. He argues that every such rule is false, for the following reasons. (i) Since the premises are negative, the conclusion must be negative. (ii) No negative conclusion follows by rules of syllogisms.

How does Burley justify this argument? He presents no case at all for (i). It happens to be an established fact for categorical syllogisms, and for this case I don't have a neat proof either. But the syllogisms under discussion here are not categorical. As a general rule of argument (i) is a non-starter:

> Catullus never fails to delight me. Therefore Catullus delights me.
>
> 2 is nothing other than $1 + 1$. Therefore 2 is $1 + 1$.

And so on. It's curious that at (419), (769) and (820) Burley states a more general rule that 'Nothing follows from negatives'. If he had had the confidence to use this rule here it would have made (ii) unnecessary. I don't think we can absolve Burley of assuming that a rule which works for categorical syllogisms works for all other syllogisms. (The rule does have a folksy kind of plausibility: you can't get something for nothing.)

His proof of (ii) is that (iii) any negative conclusion would have to have one or other of two given forms; but (iv) for each of these forms he has counterexamples. Again he offers no argument for (iii). It certainly isn't true that the two forms exhaust the possible forms that he has been considering in this section of PL. He should have tried the two terms in either order, allowing at least the subject term to be negated, and this yields four forms.

Whether Burley is right about (iv) is a matter of interpretation. If not being a non-animal is the same as being an animal, then the existential exponents of the two premises say that there is an animal, and it follows that some intelligent being is a non-man, not a form that Burley bothers to consider. In this case Burley's statement is wrong. On the other hand if—as is probable—the 'not a non-animal' formulation is meant to cancel the existential exponents, then Burley's statement is correct and we will prove it in Example One of Section 17.

As a sample of Burley's reasoning style, (789) is comparatively mild. Elsewhere he explicitly claims to prove a general rule from a single instance. Thus for example at (769) he says

> But if the exclusion is negated in each premise, no conclusion follows, since nothing follows from negatives. And this is clear from a counterexample in the terms.
> *Si vero in utraque praemissa negetur exclusio, nulla conclusio sequitur, quia ex negativis nihil sequitur. Et patet per instantiam in terminis.*

He makes the same claim for other general rules at (814), (836) and (838). (At (838) he leaves it to the reader to find the counterexample.) At (767) and (799) he claims that 'counterexamples' make the truth of certain general statements clear.

We are in a situation we discussed in Section 9: Burley is aiming to prove a universally quantified statement. Unless he can find a way of converting some established universally quantified fact into the form required, he is stuck with considering examples, the method we called Sample. For exclusives he does have the method of translating into categoricals and then checking the rules for categorical syllogisms. But for instance 'Only A's are not B's' translates into 'All non-B's are A's', which is not a classical form of categorical. Yet he does allow exclusive sentences with negated predicate terms to occur in syllogisms, as at (787) 'Only an intelligent being is not a non-animal'.

A number of Burley's results, though correct, depend crucially on the existential exponents. For example at (797) he claims that there is no exclusive syllogism in second figure where one premise is negated exclusive and the other is categorical. Thus by implication he rejects the syllogism

> Not only A's are not B's. Every C is a B. Therefore not only C's are not A's.

This syllogism has counterexamples, but in all of them there are no non-B's. The fact that Burley's discussion never mentions this is witness to the fact that his methods are insensitive to existential assumptions.

Understandably he does make mistakes. At (767) he claims that from exclusive premises in first figure, where one premise is affirmative and the other is negative, no conclusion follows syllogistically. He misses the following good syllogism mood:

> Not only not-B's are A's. Only C's are B's. Therefore not only not-C's are A's.

Probably he was checking against categorical syllogisms, and the 'not-C's' in the subject of the conclusion threw him. At (827) he claims that if the premise containing the predicate term of the conclusion is a negated exclusive, then there is no good syllogism in third figure. A counterexample is the mood

> Not only a non-A is a B. Every A is an C. Therefore not only a non-C is a B.

I can't account for his oversight here, except that it's quite late in his discussion of exclusives and his sampling method does require an undue amount of concentration.

In fact the interesting thing is that Burley makes as few mistakes as he does. Clearly he has intuitions that are sounder than his methods. We can't profitably guess how those intuitions went; but we can at least report logical facts that yield most of the conclusions that he wanted, using tools closely related to his.

17 The Medieval Interpolation Theorem

In [Hodges, 1998b] I showed how the Lyndon interpolation theorem would have come to the rescue of medieval logicians if they had known it. That theorem is quite sophisticated, and it's wholly unrealistic to imagine any medieval thinking in those terms. So here let me rephrase the basic result in a way that bypasses most of the machinery of Lyndon's theorem. I came on it by first using Hintikka sets to prove Lyndon's theorem, and then stripping down to bare essentials. Obviously nobody in the middle ages knew the result, but I believe the account below is entirely in terms that they could have understood.

For simplicity I will ignore the difference between the medieval notion of formal inference and the modern notion of an inference rule valid under all substitutions for its function or relation symbols (two occurrences of the same symbol being replaced by the same symbol in both places). Today we know that for a wide range of argument forms, if there is a counterexample at all then there is one in the natural numbers, where the relationships all hold for purely conceptual reasons. (This is due to Kurt Gödel in 1930 for first-order logic.) So I will speak of a set of sentences as *formally consistent* when there is some replacement of terms which turns it into a set of sentences that can be simultaneously true.

We consider two sets of sentences, Φ and Ψ. We say that the pair Φ, Ψ is *formally consistent* if the set consisting of all the sentences in either Φ or Ψ is formally consistent. If T is a term, we say that the pair Φ, Ψ is *formally consistent under variation of terms* if some pair $\Phi(T'/T)$, Ψ is formally consistent, where $\Phi(T'/T)$ is the result of replacing all occurrences of T in sentences of Φ by occurrences of a new term T' not present in sentences

of either Φ or Ψ. (Since we are talking of formal consistency, it would be equivalent to say 'every pair' instead of 'some pair'.)

THEOREM 1 (Medieval Interpolation Theorem). *Suppose Φ and Ψ are sets of sentences of Monotone Latin, and T is a noun phrase which occurs in sentences of Φ and Ψ only with upward monotonicity, or only with downward monotonicity. Then if the pair Φ, Ψ is not formally consistent, it is not formally consistent under variation of T either.*

Sketch proof The proof goes by recursion on the construction of sentences of Monotone Latin. Since the syntax of Monotone Latin is not completely determined, the proof has to be a bit vague. With a formal language there would be no difficulty in tightening it up to a rigorous argument.

First suppose that Φ and Ψ consist of simple categoricals with singular subjects, for example 'Socrates is running' or 'Brussels is not a village'. Here the terms are the predicates; they are upwards monotone in affirmative categoricals and downwards in negated ones. So if T is everywhere upwards monotone, all categoricals using it are affirmative, and clearly these can't lead to a formal contradiction. Hence in this case, if Φ, Ψ is formally inconsistent, this must be entirely because of categoricals in which T doesn't occur. So the formal inconsistency will still be there if we vary the term T in Φ. Essentially the same argument applies when T is everywhere downwards monotone.

The remaining cases consider more complex sentences in terms of their exponents or possible causes of truth.

For example suppose Φ contains a sentence 'P or Q', which has two possible causes of truth, namely P and Q. If Φ, Ψ is formally consistent under variation of T, then there is some possible situation in which all the sentences of $\Phi(T'/T)$ and Ψ are true. Such a situation must make at least one of the sentences P, Q true, say P; and we note that every term that occurs in P also appears in 'P or Q' *with the same monotonicity*. So if Φ' is Φ with 'P or Q' replaced by P, then Φ', Ψ is still formally consistent under variation of T. But the sentences in Φ', Ψ are simpler than those in Φ, Ψ, so we can assume that the theorem has been proved for Φ', Ψ, and hence Φ', Ψ is formally consistent. Now P was a possible cause of truth of 'P or Q', and *this remains the case when T is replaced by T' throughout* P *and* Q. Hence Φ, Ψ is formally consistent too.

Other cases are the same in principle. The hardest are those involving quantifiers, and here I grant that some familiarity with modern methods would help. Suppose Φ contains a sentence S beginning with a universal quantifier: 'No A's are B's'. If Φ, Ψ is formally consistent under variation of T, then there is some possible situation in which all the sentences of

$\Phi(T'/T)$ and Ψ are true. This situation makes true all the sentences got from S by descent to singulars of A in the form 'Either c is not an A or c is not a B', and then replacing T by T' in them. The term T' (or T) has the same monotonicities in these sentences as it did in the sentence 'No A's are B's'. Replacing that sentence in Φ by the new sentences (before T has been replaced by T' in them) gives a new pair Φ', Ψ which is still formally consistent under variation of T. But the sentences in this pair are less complex than those in Φ, Ψ, so we can assume that the theorem holds for Φ', Ψ, and it tells us that Φ', Ψ is formally consistent. It follows that Φ, Ψ was formally consistent too. □

To apply the result in Burley's context, we need to draw out the existential assumptions explicitly. I don't know exactly what they are, but for example it seems we have the following monotonicities:

- Some A$\uparrow$ is a B$\uparrow$.
- No A$\downarrow$ is a B$\downarrow$. (Negation swaps the monotonicities.)
- Every A is a B $\equiv$ No A$\downarrow$ is a non-B$\uparrow$, and there is an A$\uparrow$.
- Only A's are B's $\equiv$ No non-A$\uparrow$ is a B$\downarrow$, and there is a B$\uparrow$.
- Not only A's are B's $\equiv$ Some non-A$\downarrow$ is a B$\uparrow$, or there is no B$\downarrow$.

Here are a few applications of the theorem. In each case we show that if a certain sort of consequence was formally valid, then it would remain formally valid if one of the terms was replaced, at one occurrence and not the other, by a new term. It's normally clear at once that none of the resulting consequences could be formally valid.

Example One. We saw in Section 16 that at (789) Burley claims that no syllogistic conclusion follows from a pair of premises of the form

Only an A is not a C. Only a B is not a C.

We also saw that Burley's claim is false if we read the premises as implying that something is not a C. So suppose we drop that implication. There remains:

No non-A$\uparrow$ is not a C$\uparrow$. No non-B$\uparrow$ is not a C$\uparrow$.

Suppose a conclusion P follows syllogistically. Let Φ consist of the first premise, and let Ψ consist of the second premise together with the contradictory negation of P. Since we are talking of syllogisms, the term C occurs nowhere in P, and hence it occurs with only upward monotonicity in both Φ and Ψ. So by the Medieval Monotonicity Theorem, if P follows formally from the two premises, then it already follows formally from the two premises

No non-A is not a D. No non-B is not a C.

But clearly the two premises don't interact, and nothing follows except what follows separately from each premise.

Example Two. This easy example uses monotonicity of sentences. At (423) Burley maintains that there is no syllogism with premises

If P$\downarrow$ then R$\uparrow$. If Q$\downarrow$ then R$\uparrow$.

or the same with 'not-R' in place of R. He says this is obvious, but gives a reason; the reason is obscure to me, but it seems to be an appeal to an analogous fact about categorical syllogisms. However, R is upwards monotone in both occurrences, so that by the Medieval Interpolation Theorem any syllogistic consequence would follow also from 'If P then S', 'If Q then R', which is absurd.

Example Three. At (372) and elsewhere, Burley quotes an old rule that nothing follows from two particular (i.e. existentially quantified) premises. In categorical syllogisms there are two possible forms of particular proposition, 'Some A$\uparrow$ is a B$\uparrow$' and 'Some A$\uparrow$ is not a B$\downarrow$'. In both of these, A is upwards monotone. The Medieval Interpolation Theorem allows us to deduce quickly that there is no valid syllogism of this type in third figure, where both premises have the same subject. If we are not in third figure, then at least one term appearing in the conclusion must be the subject of a premise, and so must be upwards monotone in the premise. In second figure both terms of the conclusion must therefore be upwards monotone in premises, and so the Medieval Interpolation Theorem yields that they must both be upwards monotone in the conclusion. There is only one such consequence, namely

Some A is a C. Some B is a C. Therefore some A is a B.

We find counterexamples at once. There remains the first figure; here there are two possibilities not ruled out by the Medieval Interpolation Theorem, namely

Some A is not a B. Some B is a C. Therefore some A is a C.
Some A is not a B. Some B is not a C. Therefore some A is not a C.

Both have obvious counterexamples. So the rule is correct for categorical syllogisms.

But note that in both the two first-figure examples there is a nontrivial conclusion about A and C that we could have derived from the same premises. For the first it's 'Some A is distinct from some C'. One could say that the reason why the Medieval Interpolation Theorem failed to rule out these two cases is that they are not intrinsically invalid, it's just that the syllogistic calculus is too limited in the types of proposition that it allows. There is an obvious moral, namely that one should *never* take for granted that a rule which works for categorical syllogisms works for any other class of consequences. (Burley is particularly open to the charge of playing fast and loose here, since at (667) he is perfectly willing to allow consequences containing 'Some A is distinct from all C'.)

Example Four. We saw that in (767) Burley claimed wrongly that nothing follows syllogistically from

Not only not-B's$\uparrow$ are A's. Only C's are B's.

Even ignoring the existential implications, B occurs with different monotonicities in the two premises, and this makes it likely that something will follow validly. To see whether any proposition P relating A and C does follow, it's reasonable to start with consequences that don't depend on the existential implications. Ignoring them, A has upwards monotonicity in the first premise and C has upwards in the second. So the Medieval Interpolation Theorem advises us to look for a conclusion where A and C both have upwards monotonicity. This greatly simplifies the set of examples that we need to search through. To find the counterexample that Burley missed took about a minute by this route.

BIBLIOGRAPHY

[Baugh and Cable, 1978] Albert C. Baugh and Thomas Cable. *A History of the English Language.* Routledge and Kegan Paul, London, 1978.

[Billingham, 1982] Richard Billingham. De probationibus terminorum (first recension). In L. M. De Rijk, editor, *Some 14th Century Tracts on the Probationes Terminorum,* pages 45–76. Ingenium, Nijmegen, 1982.

[Burlaeus, 1967] G. Burlaeus. *Super Artem Veterem.* Minerva, Frankfurt, 1967. Reprint of edition printed in Venice 1497.

[Burleigh, 1955] Walter Burleigh. *De Puritate Artis Logicae Tractatus Longior, with a revised edition of the Tractatus Brevior.* The Franciscan Institute, St. Bonaventure, New York, 1955. Ed. Philotheus Boehner.

[Burley, 2000] Walter Burley. *On the Purity of the Art of Logic: The Shorter and the Longer Treatises.* Sheridan Books, Chelsea Michigan, 2000. Translated by Paul Vincent Spade.

[Frege, 1893] Gottlob Frege. *Grundgesetze der Arithmetik Begriffsschriftliche Abgeleitet, I.* Pohle, Jena, 1893.

[Hodges, 1998a] Wilfrid Hodges. An editor recalls some hopeless papers. *Bulletin of Symbolic Logic*, 4:1–16, 1998.

[Hodges, 1998b] Wilfrid Hodges. The laws of distribution for syllogisms. *Notre Dame Journal of Formal Logic*, 39:221–230, 1998.

[Kalish and Montague, 1964] Donald Kalish and Richard Montague. *Logic: Techniques of Formal Reasoning.* Harcourt, Brace and World Inc., New York, 1964.

[Locke, 1884] John Locke. *Some Thoughts concerning Education.* Clay, London, 1884.

[Netz, 1999] Reviel Netz. *The Shaping of Deduction in Greek Mathematics.* Cambridge University Press, Cambridge, 1999.

[Nuchelmans, 1996] Gabriel Nuchelmans. *The distinction of* actus exercitus/actus significatus *in medieval semantics*, chapter vi. Variorum Reprints, Aldershot, Hampshire, 1996.

[of Salisbury, 1929] John of Salisbury. *Metalogicon.* Clarendon Press, Oxford, 1929. Ed. Clemens C. I. Webb.

[of Venice, 1978] Paul of Venice. *Logica Magna ii fasc. 6, Tractatus de Veritate et Falsitate Propositionis et Tractatus de Significato Propositionis.* British Academy, Oxford, 1978. Ed. Francesco del Punta.

[Plautus, 1910?] Titus Maccius Plautus. *Comoediae, vol. ii.* Clarendon Press, Oxford, 1910? Ed. W. M. Lindsay.

[Rijk, 1962] L. M. De Rijk. *Logica Modernorum i: On the Twelfth Century Theories of Fallacy.* Van Gorcum, Assen, 1962.

[Rijk, 1967] L. M. De Rijk. *Logica Modernorum ii Part Two: The Origin and Early Development of the Theory of Supposition.* Van Gorcum, Assen, 1967.

[Spade, 1988] Paul Vincent Spade. The logic of the categorical: The medieval theory of descent and ascent. In Normann Kretzmann, editor, *Meaning and Inference in Medieval Philosophy*, pages 187–224. Kluwer, Dordrecht, 1988.

[Tarski, 1936] Alfred Tarski. O pojęciu wynikania logicznego. *Przegląd Filozoficzny*, 39:58–68, 1936. translated as [Tarski, 1983].

[Tarski, 1983] Alfred Tarski. *Logic, Semantics, Metamathematics: papers from 1923 to 1938*, John Corcoran, ed., pp. 409–420. Hackett Publishing Company, Indianapolis, Indiana, 1983.

[Valencia, 1991] Víctor Manuel Sánchez Valencia. *Studies on Natural Logic and Categorial Grammar.* PhD thesis, University of Amsterdam, 1991.

[van Eijck, 2005] Jan van Eijck. Syllogistics = monotonicity + symmetry + existential import. Technical Report SEN-R0512, CWI, Amsterdam, July 2005. Available from `http://db.cwi.nl/rapporten/`.

[Wittgenstein, 1961] Ludwig Wittgenstein. *Tractatus Logico-Philosophicus.* Routledge and Kegan Paul, London, 1961.

Separation — Past, Present, and Future

IAN HODKINSON AND MARK REYNOLDS

1 Introduction

Separation is a remarkable concept, invented by Dov Gabbay in [Gabbay, 1981a], and elaborated in [Gabbay, 1989; Gabbay *et al.*, 1994]. Roughly, a temporal logic has the separation property if its formulas can be equivalently rewritten as a boolean combination of formulas, each of which depends only on the past, present, or future. This seemingly innocent and technical definition has had some far-reaching consequences, and has taken on a life of its own. Surprisingly, separation is closely connected to the important topic of *expressive completeness,* and is one of the main methods for proving expressive completeness. Separation has applications in *executable temporal logic,* and parts of this have been implemented. Separation has found recent uses in simplifying normal form theorems and axiomatising Ockhamist branching time logic, and in analysing the W3C language XPath.

Separation has attracted attention from the time it was introduced. Its simplicity and naturalness have made it appealing to many, and its technical intricacies are still providing employment today.

Reynolds first came across separation through Gabbay in around 1984, and gave a talk on it to Wilfrid Hodges' group at Queen Mary College London, thereby introducing it to Hodkinson who was a Ph.D. student there. Hodkinson and especially Reynolds worked on it with Gabbay during the preparation of the monograph [Gabbay *et al.*, 1994], and it has caught our attention several times since. In this short article, we are happy to return to the topic. We will discuss the original application of separation, some more recent activity on it, and some open problems about it. We thank Carsten Lutz for his ideas and interest, which have improved the 'Future' section.

We would like to take this opportunity to wish Dov a very happy birthday and convey our hopes that his own future will be as prosperous as his past and present.

2 Separation past

Here, we recall the definition of separation itself, and its connection to expressive completeness — the context in which it was first introduced.

2.1 Basic temporal logic

We have to begin with some definitions and notation. They are well known and mostly taken from [Gabbay *et al.*, 1994]. Probably, they can be skipped by those who are familiar with the field and willing to work out the notation at sight. (Temporal logic is the province of philosophers, linguists, and computer scientists, and there seems to be no *standard* notation in the field yet.)

A *flow of time* is a pair $(T, <)$, where T is a non-empty set (the 'time points') and $<$ is an irreflexive transitive relation on T. Examples: $(\mathbb{N}, <)$, where $\mathbb{N} = \{0, 1, 2, \ldots\}$; $(\mathbb{Z}, <)$, where $\mathbb{Z} = \{\ldots -1, 0, 1, 2, \ldots\}$; $(\mathbb{Q}, <)$, where $\mathbb{Q}$ is the set of rational numbers; $(\mathbb{R}, <)$, where $\mathbb{R}$ is the set of real numbers; and various non-linear flows such as trees. Flows of time are (special) Kripke frames. The idea of $<$ is that $t < u$ means that u is a *later* time than t. We will be using *classes* of flows of time, such as the linear flows, dense linear flows, and Dedekind complete linear flows.

We fix a set L of propositional atoms; we write $p, q, r, \ldots$ for atoms. An *assignment* is a map $h : L \to \wp(T)$. For a flow of time $(T, <)$ and an assignment h, the triple $(T, <, h)$ is called a *temporal structure*. The idea of the assignment is that the atom $p \in L$ is *true in* $(T, <, h)$ at the time points in $h(p)$, and *false* at all other times.

We can create *temporal formulas* from the atoms using the boolean operations $\neg$, $\wedge$, $\vee$, $\rightarrow$, $\leftrightarrow$, $\top$, $\bot$, and *temporal connectives.* The main connectives are F, P, G, H, U, S, T, Y, standing for *at some future (past) time, always in the future (past), Until, Since, Tomorrow,* and *Yesterday,* respectively. The use of the symbols F, P, G, H in this context is due to Prior, and U, S to Kamp; we are not sure about the origin of T, Y. All but U, S are unary (they take a single formula as argument); U, S are binary (they connect two formulas). So, if α, β are temporal formulas, then $F\alpha$, $G\alpha$, $T\alpha$, $U(\alpha, \beta)$, $\neg U(\neg S(\alpha \wedge \beta, \beta \rightarrow H\neg\alpha), T(\alpha \vee \beta))$, etc., are also formulas.

The semantics of formulas is defined by induction as follows. Let $\mathcal{M} = (T, <, h)$ be a temporal structure, and $t \in T$ a time point.

1. $\mathcal{M}, t \models p$ iff $t \in h(p)$, for $p \in L$,

2. the booleans are handled as usual,

3. $\mathcal{M}, t \models F\alpha$ iff $\mathcal{M}, u \models \alpha$ for some $u \in T$ with $u > t$,

4. $P\alpha$ is treated similarly using $u < t$ — i.e., the *mirror image* of the preceding clause,

5. $\mathcal{M}, t \models G\alpha$ iff $\mathcal{M}, u \models \alpha$ for all $u > t$,

6. $H\alpha$ — the mirror image clause,

7. $\mathcal{M}, t \models U(\alpha, \beta)$ iff there is $u > t$ with $\mathcal{M}, u \models \alpha$ and $\mathcal{M}, v \models \beta$ for all $v \in T$ with $t < v < u$,

8. $S(\alpha, \beta)$ — mirror image,

9. $\mathcal{M}, t \models T\alpha$ iff there is $u > t$ with $\mathcal{M}, u \models \alpha$ and there is no $v \in T$ with $t < v < u$,[1]

10. $Y\alpha$ — mirror image.

If desired, we can add more connectives with 'first-order-style' definitions as above. On the other hand, sometimes we want to restrict to certain sets $\mathcal{T}$ of connectives. For simplicity, we will generally call such a $\mathcal{T}$ a *temporal logic*. For example, we refer to 'the temporal logic US', by which we mean the formulas written with the booleans and the connectives U and S.

In this article we are only concerned with syntax and semantics, not with any kind of reasoning. But it will already be obvious that there are some connections between formulas. $H\alpha$ and $\neg P \neg \alpha$ 'mean the same'; so do $F\alpha$ and $U(\alpha, \top)$, and $T\alpha$ and $U(\alpha, \bot)$. On the other hand, $YU(\alpha, \beta)$ will mean the same as $\alpha \vee (\beta \wedge U(\alpha, \beta))$ in flows of time like $(\mathbb{Z}, <)$, but not in dense flows like $(\mathbb{Q}, <)$. We formalise this 'relativity' as follows.

DEFINITION 1. Given a class $\mathcal{C}$ of flows of time, we say that temporal formulas α, β are *equivalent over* $\mathcal{C}$ if $(T, <, h), t \models \alpha \leftrightarrow \beta$ for all $(T, <) \in \mathcal{C}$, all assignments $h : L \to \wp(T)$ and all $t \in T$.

We write 'equivalent over linear time' to abbreviate 'equivalent over the class of all linear flows of time', etc.

2.2 Standard translation and expressive completeness

As in modal logic, temporal formulas can be translated to first-order ones.

DEFINITION 2. Let L' be the first-order signature consisting of $<$, and unary relation symbols $P, Q, \ldots$ twinned with the atoms $p, q, \ldots \in L$, respectively.

[1]Note that in dense flows of time like $(\mathbb{Q}, <)$, $T\alpha$ is always false; but it is not 'meaningless'.

Each temporal formula α has a 'standard translation', which is a first-order L'-formula $\alpha^x(x)$ with free variable x (for any first-order variable x), as follows.

1. Each atom p translates to $p^x(x) = P(x)$.
2. The translation $-^x$ commutes with the booleans.
3. $(F\alpha)^x = \exists y(y > x \wedge \alpha^y)$.
4. $(G\alpha)^x = \forall y(y > x \rightarrow \alpha^y)$.
5. $U(\alpha, \beta)^x = \exists y(y > x \wedge \alpha^y \wedge \forall z(x < z < y \rightarrow \beta^z))$.
6. The others are similar.

It is clear that we can view a temporal structure $(T, <, h)$ as an L'-structure by interpreting $P, Q, \ldots$ as $h(p), h(q), \ldots$, respectively, and that by so doing, we have

$$(T, <, h), t \models \alpha \text{ iff } (T, <, h) \models \alpha^x[t]$$

for all $t \in T$ and all formulas α. *The translation $\alpha \mapsto \alpha^x$ is meaning-preserving.* So it makes good sense to extend definition 1:

DEFINITION 3. A temporal formula α is said to be *equivalent* to an L'-formula $\varphi(x)$ over a class $\mathcal{C}$ of flows of time if for all $(T, <) \in \mathcal{C}$ and all $h : L \rightarrow \wp(T)$, we have $(T, <, h) \models \forall x(\alpha^x \leftrightarrow \varphi(x))$.

Not all first-order formulas $\varphi(x)$ will have the form α^x for some temporal formula α. Surprisingly however, sometimes every $\varphi(x)$ is *equivalent* to some α.

DEFINITION 4. We say that a temporal logic $\mathcal{T}$ is *expressively complete over a class $\mathcal{C}$ of flows of time* if for every L'-formula $\varphi(x)$, there is a $\mathcal{T}$-formula α that is equivalent to φ over $\mathcal{C}$.

Kamp proved in the very important [Kamp, 1968] that U and S are expressively complete over Dedekind-complete linear time. This first expressive completeness theorem led to a canon of results, continuing today, and it brings our story finally to separation and Gabbay's contribution.

2.3 Separation

We now explain formally what separation is. Some jargon will be handy for this. Given a flow of time $(T, <)$, a time $t \in T$, and assignments $g, h : L \rightarrow \wp(T)$, we say that

- *g, h agree on t* if for all $p \in L$, we have $t \in g(p)$ iff $t \in h(p)$.

- *g and h agree on the past of t* if for all $u \in T$ with $u < t$, and all $p \in L$, we have $u \in g(p)$ iff $u \in h(p)$.
- *'g and h agree on the future of t'* is defined by the mirror image of the preceding clause.

DEFINITION 5. [Gabbay, 1981a] Let $\mathcal{C}$ be a class of flows of time. A temporal formula α is said to be

- *pure past over $\mathcal{C}$*, if for any $(T,<) \in \mathcal{C}$ and any $t \in T$, whenever $g, h : L \to \wp(T)$ are assignments that agree on the past of t, we have $(T,<,g), t \models \alpha$ iff $(T,<,h), t \models \alpha$;
- *pure present over $\mathcal{C}$*, if for any $(T,<) \in \mathcal{C}$, $t \in T$, and assignments $g, h : L \to \wp(T)$ that agree on t, we have $(T,<,g), t \models \alpha$ iff $(T,<,h), t \models \alpha$;
- *pure future over $\mathcal{C}$*, if for any $(T,<) \in \mathcal{C}$, any $t \in T$, and any assignments $g, h : L \to \wp(T)$ agreeing on the future of t, we have $(T,<,g), t \models \alpha$ iff $(T,<,h), t \models \alpha$;
- *pure over $\mathcal{C}$*, if it is pure past, pure present, or pure future over $\mathcal{C}$,
- *separated over $\mathcal{C}$*, if it is a boolean combination of formulas that are pure over $\mathcal{C}$.

$\mathcal{T}$ is said to have the *separation property over $\mathcal{C}$* if every $\mathcal{T}$-formula is equivalent over $\mathcal{C}$ to a formula that is separated over $\mathcal{C}$.

So a formula is pure past if its truth value at any time depends only on the values of its atoms in the past; and similarly for pure present and pure future. Hq and $S(p, q \to S(q,r))$ are pure past formulas. The formula $F(q \wedge Hr)$ is not pure, but it is equivalent over linear time to the separated formula $Hr \wedge r \wedge U(q,r)$.

2.4 Gabbay's theorem

We are now ready to state the important result of Gabbay that relates separation to expressive completeness. It was stated in [Gabbay, 1981a, theorem 11] and proved in [Gabbay, 1989] and [Gabbay *et al.*, 1994, §9.3]. The theorem is surprising because it connects two seemingly different conditions.[2]

THEOREM 6. *Let $\mathcal{C}$ be a class of linear flows of time, and $\mathcal{T}$ a temporal logic able to express F and P over $\mathcal{C}$. Then $\mathcal{T}$ is expressively complete over $\mathcal{C}$ iff it has the separation property over $\mathcal{C}$.*

[2]The legend goes that when Kamp heard this theorem in a seminar, he went out and bought Dov a cake.

Proof '$\Rightarrow$' is proved by showing that any first-order L'-formula can be separated over linear time. Here, we sketch a proof running along standard model-theoretic lines.[3] It is entirely 'classical', involving no temporal logic.

First observe that we can define equivalence, purity, and separatedness for *L'-formulas* $\varphi(x)$ semantically (over linear time). The definitions are analogous to those for temporal formulas (definitions 1 and 5).

Fix an L'-formula $\varphi(x)$ with free variable x, and let $L_\varphi \subseteq L'$ be the finite relational signature consisting of $<$ and the unary relation symbols that occur in φ. An L_φ-structure has the form $(T, <, h)$, where h is a map that assigns each unary relation symbol occurring in φ to a subset of T.

We want to show that there is a separated L_φ-formula that is equivalent to $\varphi(x)$ over linear time (that is, over all L_φ-structures $(T, <, h)$ where $(T, <)$ is a linear flow of time). We do this using compactness and games.

Let λ be a sentence saying that $<$ is an irreflexive linear order. Let $\Sigma(x)$ be the set of all separated L_φ-formulas $\sigma(x)$ implied by φ over linear time: formally, those such that $\lambda \vdash \forall x(\varphi(x) \rightarrow \sigma(x))$.

We first show that $\Sigma \models \varphi$ over linear time. So suppose that $(T, <, h) \models \Sigma[t]$, where $(T, <)$ is a linear flow of time, and $t \in T$. We show that $(T, <, h) \models \varphi[t]$.

Let $\Theta(x)$ be the set consisting of

- all pure L_φ-formulas $\pi(x)$ such that $(T, <, h) \models \pi[t]$,
- $\lambda \wedge \varphi(x)$.

We claim that Θ is consistent. If it were not, then by first-order compactness, there would be pure $\pi_1(x), \ldots, \pi_n(x) \in \Theta$ with $(T, <, h) \models \pi_i[t]$ for each i, such that $\lambda \vdash \forall x(\varphi(x) \rightarrow \neg \bigwedge_{i \leq n} \pi_i(x))$. But then, $\neg \bigwedge_{i \leq n} \pi_i(x) \in \Sigma$ and so $(T, <, h) \models \neg \bigwedge_{i \leq n} \pi_i[t]$. This is a contradiction, and establishes the consistency of Θ.

So there is a model $(T', <', h') \models \Theta[t']$, where $(T', <')$ is a linear flow of time and $t' \in T'$. In short, $(T, <, h, t)$ and $(T', <', h', t')$ are linear, agree on all pure L_φ-formulas, and φ is true in the latter.

But now, if $\pi(x)$ is a pure past L_φ-formula then $(T, <, h) \models \pi[t]$ iff $(T', <', h') \models \pi[t']$. Since we can relativise (the quantifiers in) any L_φ-sentence to the points in the past of t, so obtaining a pure past formula, this means that the substructures of $(T, <, h)$ and $(T', <', h')$ consisting of all time points in the past of t, t', respectively, are elementarily equivalent. The same holds for pure present and pure future formulas, so the substructures based on t, t' are elementarily equivalent, and the substructures based on points in the future of t, t' are elementarily equivalent too. Now an Ehrenfeucht–Fraïssé game

[3] For a more effective proof, see [Gabbay *et al.*, 1994, 9.3.2].

argument (which is legal because L_φ is finite), or the Feferman–Vaught theorem (cf. [Hodges, 1993, A.6.2]), will easily show that $(T, <, t, h)$ and $(T', <', t', h')$ are themselves elementarily equivalent. Since φ is true in the second structure, we have $(T, <, h) \models \varphi[t]$ as required.

We know now that $\Sigma \models \varphi$ over linear time. By compactness, and as linearity is first-order definable, there is a conjunction $\sigma(x)$ of separated formulas in Σ that implies φ over linear time. Hence, $\varphi(x)$ and $\sigma(x)$ are equivalent over linear time. Of course, σ is separated.

Having got through this first stage, it is now easy to show that $\mathcal{T}$ has the separation property over $\mathcal{C}$. Let α be a $\mathcal{T}$-formula. By the above, its standard translation α^x is equivalent over linear time, and so over $\mathcal{C}$, to a separated formula — a boolean combination of pure formulas. But $\mathcal{T}$ is expressively complete over $\mathcal{C}$, so each of these pure formulas is equivalent over $\mathcal{C}$ to a $\mathcal{T}$-formula, which is obviously pure as well. The corresponding boolean combination of these $\mathcal{T}$-formulas is separated and equivalent to α over $\mathcal{C}$. This completes the proof.

For '$\Leftarrow$', assume that $\mathcal{T}$ has the separation property over $\mathcal{C}$. We need to show that any first-order L'-formula $\varphi(x, P_1, \ldots, P_n)$, with one free variable x and unary relation symbols $P_1, \ldots, P_n$, is equivalent over $\mathcal{C}$ to some $\mathcal{T}$-formula. We go by induction on the quantifier depth k of φ.

If k is 0, then φ is a boolean combination of formulas of the form $P_i(x)$, $x = x$, and $x < x$, so the result is clear. Assume the result for k. It suffices to express $\exists y \varphi(x, y, P_1, \ldots, P_n)$ as a $\mathcal{T}$-formula, where φ has quantifier depth k. We can suppose that x and y do not occur bound in φ.

First, we want to remove x from φ somehow, to obtain a formula $\varphi'(y)$ to which we can apply the inductive hypothesis. How can we do it? Let us start by thinking about the atomic subformulas of φ involving x. They are of the form $P_i(x)$, $x = x$, $x < x$, $z = x$, $x = z$, $z < x$, and $x < z$, where z is some other variable than x. We can replace $x = z$ by $z = x$ and leave it for later. We replace $x = x$ by $\top$, and $x < x$ by $\bot$. This leaves $z < x$, $z = x$, $x < z$, and the $P_i(x)$.

Next, we root out the $P_i(x)$. The idea here is based on a simple observation about propositional logic, exemplified in the statement that an arbitrary propositional formula like $\neg p \wedge q$ is equivalent to

$$\begin{array}{rcl} (p & \to & \neg\top \wedge q) \\ \wedge\ (\neg p & \to & \neg\bot \wedge q). \end{array}$$

The right-hand sides here do not involve p.

Proceeding in this way, for each $S \subseteq \{1, 2, \ldots, n\}$, let φ^S be the result of replacing each atomic subformula $P_i(x)$ of φ by $\top$ if $i \in S$, and by $\bot$ if $i \notin S$. Then φ^S still has the same quantifier depth, k, but has no occurrences of any $P_i(x)$. (It may involve $P_i(z)$ for other variables z.) We see that $\exists y\varphi$ is equivalent to

$$\bigwedge_{S\subseteq\{1,\dots,n\}} \Big(\bigwedge_{i\in S} P_i(x) \wedge \bigwedge_{j\notin S} \neg P_j(x) \to \exists y\varphi^S(x,y,P_1,\dots,P_n)\Big).$$

The formulas $P_i(x)$ are equivalent over $\mathcal{C}$ to p_i, of course. So it is enough if we can express the $\exists y\varphi^S(x,y,P_1,\dots,P_n)$ as $\mathcal{T}$-formulas.

Hence, we can assume that the original formula φ is such a formula. That is, we suppose that x does not occur in atomic subformulas of φ of the form $P_i(x)$, but only in ones of the form $z = x$, $z < x$, and $x < z$.

It remains to get rid of these too. To do it, we temporarily introduce new atoms $n_=$, $n_<$, $n_>$, with corresponding unary relation symbols $N_=$, etc. We replace each atomic subformula $z = x$ in φ (where z is any variable) by $N_=(z)$. Similarly, replace $x < z$ by $N_>(z)$, and $z < x$ by $N_<(z)$. This yields a formula $\varphi'(y, P_1,\dots,P_n, N_<, N_=, N_>)$ of quantifier depth k and with no occurrences of x at all. For a while, we will restrict ourselves to structures interpreting $N_=$ as $\{x\}$, $N_<$ as $\{t : t < x\}$, and $N_>$ as $\{t : t > x\}$. Then each $N_*(z)$ is equivalent to the formula it replaced, so $\exists y\varphi(x,y,P_1,\dots,P_n)$ will be equivalent to $\exists y\varphi'(y,P_1,\dots,P_n,N_<,N_=,N_>)$.

Inductively, $\varphi'(y,P_1,\dots,P_n,N_<,N_=,N_>)$ is equivalent over $\mathcal{C}$ to some $\mathcal{T}$-formula $\alpha(p_1,\dots,p_n,n_<,n_=,n_>)$. The flows in $\mathcal{C}$ are linear, so $\exists y\varphi'$ is equivalent over $\mathcal{C}$ to $\beta = \alpha \vee F\alpha \vee P\alpha$ (here we use the hypothesis that F and P are $\mathcal{T}$-expressible). Thus, *under our restriction*, the original formula $\exists y\varphi(x,y,P_1,\dots,P_n)$ is equivalent over $\mathcal{C}$ to $\beta(p_1,\dots,p_n,n_<,n_=,n_>)$.

Finally, we come to the key step. We remove the temporary n-atoms from β. We do this using the separation property! We separate β, obtaining a boolean combination $\gamma(p_1,\dots,p_n,n_<,n_=,n_>)$ of *pure* $\mathcal{T}$-formulas. Consider, say, a *pure past* formula $\delta(p_1,\dots,p_n,n_<,n_=,n_>)$ from this boolean combination. As δ is pure past, its truth value at x only depends on the values of its atoms at points $t < x$. It is independent of their values at points $t \geq x$. But under our restriction, the values of the n-atoms at points $< x$ is entirely predictable: $n_<$ is true, and the others are false. Accordingly, let us replace $n_<$ in δ by $\top$, and replace $n_=$ and $n_>$ by $\bot$. We obtain $\delta^*(p_1,\dots,p_n) = \delta(p_1,\dots,p_n,\top,\bot,\bot)$. Then, always under our restriction, the truth values of δ^* and δ at x are the same.

We adjust each pure formula δ in γ in a similar way. If δ is pure present, $n_=$ is replaced by $\top$, and the others by $\bot$. For pure future δ, $n_>$ is replaced by $\top$ instead. The result is a boolean combination $\gamma^*(p_1,\dots,p_n)$, which, subject to our restriction, is equivalent to $\exists y\varphi(x,y,P_1,\dots,P_n)$.

But the restriction concerned atoms that do not appear in γ^*. So it has no force. Without any restriction on assignments to atoms, γ^* is equivalent over $\mathcal{C}$ to $\exists y\varphi(x, y, P_1, \ldots, P_n)$. This completes the induction and the proof. ❑

Given an algorithm to separate a formula over $\mathcal{C}$, the proof above can be elaborated into an algorithm that will translate any given first-order formula $\varphi(x, P_1, \ldots, P_n)$ to a $\mathcal{C}$-equivalent $\mathcal{T}$-formula. However, it appears that the complexity of this algorithm is high. Over $(\mathbb{N}, <)$, the problem is known to have non-elementary complexity: that is, there is no translation algorithm that runs in exponential time, double exponential time, triple exponential time, etc. A stronger result was proved in [Etessami and Wilke, 2000]: that there is no elementary bound on the *length* of a US-formula equivalent to a first-order one. See section 4 for some related problems on the separation process.

2.5 Separation and expressive completeness

Not all temporal logics have the separation property. For example, FP over linear time does not: the formula $F(q \wedge Hr)$ cannot be separated using only F and P. We now discuss some temporal logics that do have the separation property.

THEOREM 7 (Gabbay; see [Gabbay, 1989], [Gabbay *et al.*, 1994, 10.2.9]). *Until and Since have the separation property over* $(\mathbb{N}, <)$ *(i.e., over the class* $\{(\mathbb{N}, <)\}$*).*

Hence, Until and Since are expressively complete over $(\mathbb{N}, <)$. The proof involves an induction showing that each formula can be separated.

THEOREM 8 (Gabbay–Reynolds; see [Gabbay *et al.*, 1994, §§10.3–11]).

1. *Until and Since have the separation property over* $(\mathbb{R}, <)$.
2. *Until, Since, and the Stavi connectives have the separation property over the class of all linear flows of time.*

Hence, U and S are expressively complete over $(\mathbb{R}, <)$, and U, S, and the Stavi connectives are expressively complete over all linear flows. The *Stavi connectives* were mentioned (without definition) in [Gabbay *et al.*, 1980] and are versions of U, S oriented towards Dedekind cuts in the flow of time; we refer to [Gabbay, 1981a; Gabbay *et al.*, 1994] for details.

The proof of theorem 8 is broadly similar to that of theorem 7, but more complicated. Both proofs are effective. The spirit of the proofs suggests a 'heuristic' to tell whether a given set of connectives is expressively complete: as Gabbay said, *'try to separate and see where you get stuck!'* This may inspire the addition of extra connectives that are expressively complete. For example, trying to separate $F(q \wedge Hr)$ over linear time shows the need for

Until. We already said that this formula cannot be separated using only F and P. But a look at what the formula says suggests the addition of $U(q, r)$ to express what it says about the future. Then $F(q \wedge Hr)$ is equivalent to $Hr \wedge r \wedge U(q, r)$, which is separated over linear time.

The concept of separation is not confined to linear time; generalisations to non-linear flows have been known since the 1980s. The idea is to break up the flow of time into suitable 'regions' on which the values of the atoms $n_<, n_=, n_>$ are constant, and that allow the Ehrenfeucht–Fraïssé game argument of theorem 6 to go through. Various separation and expressive completeness results for trees were proved in [Amir, 1982a; Amir, 1982b; Amir, 1985; Amir and Gabbay, 1987; Immerman and Kozen, 1987; Schlingloff, 1992].

2.6 Syntactic separation

The proof of theorem 7 actually shows that over $(\mathbb{N}, <)$, every US-formula α is equivalent to a boolean combination β of U-formulas and S-formulas. We call such a β *syntactically separated.* This is formally stronger than a simple 'semantic' separation result, and it can be very useful to have. But on the theoretical level, it is only a marginal improvement. This is for two reasons.

First, while recognising a *syntactically* separated formula is trivial, it is not so hard to tell whether a formula is *semantically* pure or separated. The complexity is often the same as for the validity problem, and temporal logic is frequently advertised in terms of the relatively low complexity of its validity problem.

LEMMA 9. *The problems of deciding whether, over* $(\mathbb{N}, <)$, *a given* US-*formula is (a) pure past, (b) pure present, (c) pure future, (d) separated, are* PSPACE-*complete.*

Proof The proof relies on the PSPACE-completeness of the problem of validity of US-formulas over $(\mathbb{N}, <)$ [Sistla and Clarke, 1985].

A US-formula $\alpha(p_1, \ldots, p_n)$ is pure future over $(\mathbb{N}, <)$ iff the formula

$$\Big(G \bigwedge_{i=1}^{n} (p_i \leftrightarrow q_i)\Big) \to \Big(\alpha(p_1, \ldots, p_n) \leftrightarrow \alpha(q_1, \ldots, q_n)\Big)$$

is valid in $(\mathbb{N}, <)$. This formula is constructible from α in polynomial time. By checking its validity, we decide pure future-ness of α in PSPACE. The other two cases are dealt with similarly. This algorithm is due to Lutz (personal communication, 2005).

Whether α is separated over $(\mathbb{N}, <)$ can then be checked by searching for a boolean decomposition of α into pure formulas; the search can be conducted in PSPACE.

The proof of PSPACE-hardness uses the rather counter-intuitive fact that for any US-formula α and any atom q not occurring in α, the following five conditions are equivalent (all taken over $(\mathbb{N}, <)$): (i) α is valid, (ii) $\alpha \vee GHq$ is pure past, (iiv) $\alpha \vee GHq$ is pure present, (iv) $\alpha \vee GHq$ is pure future, (v) $\alpha \vee GHq$ is separated. This is easily checked. If α is valid, so is $\alpha \vee GHq$, so it is trivially pure and separated. If α is not valid, we can choose $h : L \to \wp(\mathbb{N})$ and $t \in \mathbb{N}$ with $(\mathbb{N}, <, h), t \models \neg\alpha$. Then the value of $\alpha \vee GHq$ at t depends on whether $h(q) = \mathbb{N}$ or not. As this can be varied without changing the value of q in the past of t, we see that $\alpha \vee GHq$ is not pure past. Similarly, it is not pure present or pure future. Nor is $\alpha \vee GHq$ separated, since GHq is not pure. Thus, we have reduced validity in polynomial time to being pure or separated, showing PSPACE-hardness of the latter. ❏

The second reason is that whilst a syntactic separation result is available over $(\mathbb{N}, <)$, it is not possible to obtain one over general flows of time. Over $(\mathbb{R}, <)$ for example, $P\neg U(\top, q)$ is pure past, but is not equivalent to any S-formula. However, lemma 9 generalises easily to $(\mathbb{R}, <)$ using the PSPACE-completeness of validity of US-formulas over $(\mathbb{R}, <)$ proved in [Reynolds, 1999].

In short, we have to live with semantic separation in many cases, and the algorithmic costs of doing so are not too bad.

3 Separation: present

In this section we consider recent and current work involving separation. We also mention briefly some earlier uses of separation which have stood the test of time particularly well: these are cases where a proof or technique which relies on separation still attracts some current research activity.

3.1 Expressive Completeness

As outlined above, separation can demonstrate expressive completeness of temporal languages. It gives a theoretical justification for using a particular language. Separation, along with the construction in the proof of theorem 6 stands as one of the main ways of translating from first-order logic into temporal logic in such important cases as natural numbers and real numbers time.

Other ways of approaching such a task do exist but seem to be even less well developed towards actual implementation. These include the direct syntactical arguments in Kamp [Kamp, 1968], the outline by Stavi in [Gabbay *et al.*, 1980] and the game-theoretic version of this in chapter 12 of [Gabbay *et al.*, 1994].

Of course, this translation is not done as part of any practical reasoning application as it seems to be too expensive.

More recently, Marx noticed a connection with temporal logic and used an original but also quite traditional separation approach to analyse the expressiveness of XPath, the W3C-standard node addressing language for XML documents [Marx, 2004]. He was able to add some natural extra relations to XPath to achieve an expressive completeness result.

3.2 Removing past-time operators

Separation is also called upon to justify *not* using past-time operators in temporal languages which specify systems. It has been well argued [Lichtenstein *et al.*, 1985] that specifications using past-time operators are more natural. It is also clear that it is often the case that including the past-time operators adds no complexity to reasoning tasks [Sistla and Clarke, 1985]. However, across the wide ranging uses of temporal languages we find many examples where past-time operators are neglected. These include reasoning about branching time where the future-only CTL and CTL* are long established. Reasoning about evolving knowledge using temporal-epistemic combinations is another example: the basic languages are set out in [Fagin *et al.*, 1995] and past-time operators are only mentioned briefly to note that they might complicate the exposition.

The justification for not having past-time operators is often not explicit. We would agree with [Lichtenstein *et al.*, 1985] in supposing that researchers are relying on the claim in [Gabbay *et al.*, 1980] that the past-time operators add nothing to the expressivity of the language with Until over the natural numbers time.

The claim (Theorem 2.1 in [Gabbay *et al.*, 1980]), which follows easily from theorem 7, is just that for every formula $\phi(x)$ of the first-order language of order (L' from definition 2) there is a temporal formula from the language with just U which is equivalent to ϕ over $(\mathbb{N}, <)$ at time 0. It follows that

LEMMA 10. *For any US-formula α there is a U-formula β that is equivalent to α at time* 0. *That is, for any $h : L \to \wp(\mathbb{N})$, we have* $(\mathbb{N}, <, h), 0 \models \alpha \leftrightarrow \beta$.

To see this, just syntactically separate α, and replace all maximal subformulas $S(X, Y)$ of the result by **false**. We will call such an equivalence at time zero, *initial equivalence* between formulas.

We should note that, in many computer science oriented publications since [Gabbay *et al.*, 1980], the temporal languages considered use what are called the non-strict versions of U and S. To distinguish these from the strict versions which we introduced in section 2 we write them in an infix manner and as $U_=$ and $S_=$. For example, $M, t \models pU_=q$ iff there is $u \geq t$

with $M, u \models q$ and for all $v \in T$, if $t \leq v < u$ then $M, v \models p$. The language with $U_=$ and T is known as LTL, and if $S_=$ and Y are included then we might call this LTL+PAST (but the notation is not everywhere consistent).

It is clear that on natural number time structures, $\mathcal{T} = (\mathbb{N}, <, h)$, strict until can be defined by $U(\beta, \alpha) \equiv T(\alpha U_= \beta)$ and, equally, the LTL operators can be defined from U: $\alpha U_= \beta \equiv \beta \vee (\alpha \wedge U(\beta, \alpha))$ and $T\alpha \equiv U(\alpha, \mathbf{false})$. Similarly for strict since.

The claim from [Gabbay *et al.*, 1980] thus tells us that the past-time operators $S_=$ and Y do not add anything to the expressivity of the language with $U_=$ and T, provided we are just interested in the satisfiability of formulas at time zero. For specifying the behaviour of a system, of course, it is natural to just state its desired behaviour as a formula to hold at time zero.

It is not clear that there are any other ways, apart from separation, to remove past-time operators from US-formulas over the natural numbers to make an initially equivalent U-formula.

3.3 Executable temporal logic

Separation stands as one of the foundations on which the intuitively appealing "declarative past implies imperative future" idea of [Gabbay, 1989] is used to allow arbitrary temporal (declarative) specifications to be executed as (imperative) programs. This idea, and executable temporal logics more generally, are used for rapid proto-typing in fields from system modelling [Finger *et al.*, 1993] to multimedia [Bowman *et al.*, 2003] and AI [Jonker and Wijngaards, 2003].

Considering the particular Metatem approach [Barringer *et al.*, 1996] to executable temporal logic, we see that separation gives a theoretical justification for only dealing with specifications given in a "past implies future" normal form. Alternatively, it gives a process for converting an arbitrary specification into this normal form. A Metatem program is of the form

$$G(\bigwedge_{i=1}^{n}(\xi_i \to \psi_i))$$

where the ξ_i are pure past and the ψ_i are boolean combinations of pure present and pure future formulas. Note that there are further syntactic restrictions. The idea is that on the basis of the declarative truth of the past time ξ_i the program should go on to "do" ψ_i.

The process given in chapter 4 in [Barringer *et al.*, 1996] shows how an arbitrary formula of the propositional language with $U_=S_=YT$ can be converted into an equally satisfiable formula in this normal form. This process starts by assuming that the formula will be separated: it is not easy to see any alternative than to use the separation process described in [Gabbay *et al.*, 1994].

3.4 The safety-liveness normal form

A more recent application for the separation process is in achieving another normal form for temporal formulas. This is the safety-liveness form of [Lichtenstein *et al.*, 1985].

One of the many interesting observations made in [Lichtenstein *et al.*, 1985] is that by using some transformations based on separation, automata and regular expressions one can find, for any temporal formula (possibly involving past operators) an equivalent formula in which the only future operators are unnested and of the form "infinitely often" or "eventually always". This form is used to show that any temporal formula is equivalent to a positive boolean combination of what [Lichtenstein *et al.*, 1985] define as safety and simple liveness properties— safety properties being those that must continue to hold forever, and liveness properties being those which must eventually happen. We thus call this the safety-liveness form.

From [Lichtenstein *et al.*, 1985] we have:

LEMMA 11. *Every formula of LTL+Past is initially equivalent to a formula of the form:*

$$\bigvee_{i=1}^{n}(GF\alpha_i) \wedge (FG\beta_i)$$

where each α_i and β_i are formulas without $U_=$ or X.

The proof in [Lichtenstein *et al.*, 1985] first uses separation to find a LTL equivalent. It then proceeds via standard translations of a LTL formula into an equivalent Büchi automaton (i.e. an automaton that accepts exactly the structures which are models of the formula). The standard translations use a result from [McNaughton, 1966] which allows us to find a deterministic equivalent to any Büchi automaton and so, via a recursive construction, allows us to find a negation of a given automaton.

The automaton will be what is known as counter-free and this allows us (via results from [McNaughton and Papert, 1971]) to find star-free regular expressions which describe various possible limiting behaviours in terms of finite prefixes of the structure. Finally, these are converted into equivalent past-time formulas via a reverse translation.

A newer proof of lemma 11 is given in [Reynolds, 2000]. It also uses separation but relies on a less varied range of other powerful techniques.

We use the stronger separation result appropriate for any Dedekind complete linear order, i.e. $(T, <)$ such that every non-empty subset of T with an upper bound has a least upper bound. This is used to establish theorem 8 above. The result (Theorem 10.3.20, [Gabbay *et al.*, 1994]) involves semantic separation and concepts of purity. Here we just need to extract some ideas from the proof.

We introduce new operators
$K^+(\alpha) = \neg U(\mathbf{true}, \neg\alpha)$,
$K^-(\alpha) = \neg S(\mathbf{true}, \neg\alpha)$,
$\Gamma^+(\alpha) = \neg K^+(\neg\alpha) \wedge K^-(\neg\alpha)$, and
$\Gamma^-(\alpha) = \neg K^-(\neg\alpha) \wedge K^+(\neg\alpha)$
and then add these to the US language to define an extended language, which we call ESTL. $K^+(\alpha)$ holds at a point when α holds arbitrarily soon afterwards and the other connectives have similarly intuitive readings.

In the proof of the theorem, some dozens of ESTL equivalences, or *acceptable rewrites*, are presented. The general idea is that formulas with past operators within the scope of future operators (or vice versa) are rewritten with one less level of such nesting. A recursive procedure is given for eventually eliminating all such nesting via a series of these rewrites. See [Gabbay *et al.*, 1994] for details.

In ESTL, we say that a formula is *syntactically separated* iff it is a boolean combination of formulas that are either atoms or of the form $U(\alpha, \beta)$ and not containing S, $S(\alpha, \beta)$ and not containing U, $K^+(\alpha)$ and not containing S, or $K^-(\alpha)$ and not containing U. Such subformulas are semantically pure (as in definition 5) and we will see that over a certain flow of time, which is nearly everywhere discrete, they have a particularly simple form.

From the proof of Theorem 10.3.20 in [Gabbay *et al.*, 1994] we have the following:

LEMMA 12. *Over Dedekind complete time, each formula in ESTL can be acceptably rewritten as an equivalent formula which is syntactically separated.*

It must be pointed out that despite only involving straightforward syntactic rewrites, the general separation procedure may be rather computationally complex. In fact, the time complexity of the procedure (and hence the size blow-up in formulas) is probably nonelementary.

The new proof from [Reynolds, 2000] of lemma 11 based solely on separation is as follows.

Proof The idea is to add a point ∞ after all the natural numbers to get a new Dedekind complete linear order $\mathbb{N}^\infty$. We will work in the temporal logic with US over propositional structures with $\mathbb{N}^\infty$ as the flow of time.

Our first step is to relativize the US version of our formula ϕ to places where $\kappa = U(\mathbf{true}, \mathbf{true})$ holds (via recursive use of such transformations of $U(\alpha, \beta)$ to $U(\kappa \wedge \alpha, \kappa \rightarrow \beta)$). Say that the result is ϕ^+. It is clear that

ϕ is true at 0 in a natural number structure iff ϕ^+ is true at 0 in any $\mathbb{N}^\infty$ extension.

To say that ϕ^+ holds at time 0 is clearly equivalent to saying that $\beta^+ = S(\neg S(\mathbf{true}, \mathbf{true}) \wedge \phi^+, \mathbf{true})$ holds at time ∞. By using the separation technique Lemma 12, we can find, (effectively via the straightforward syntactic transformation via the set of rewrite rules), an equivalent formula γ^+, which is a boolean combination of syntactically pure ESTL formulas. Obviously we can dispense with the present and future parts of this combination (eg, assume that all atoms are false at time ∞ and any formula $U(\alpha, \beta)$ or $K^+(\alpha)$ can be replaced by falsity).

The next step is to get rid of $K^\pm$ and $\Gamma^\pm$. Note that we can rewrite $K^-(\alpha)$ as $\neg S(\mathbf{true}, \neg\alpha)$ so we need only consider maximal subformulas of γ^+ of the form $S(\alpha, \beta)$ in which α and β contain no U. The fact that $\mathbb{N}^\infty$ is discrete at all points $< \infty$ at which we evaluate such α and β (and their subformulas) allows us to equivalently rewrite each $K^+(\psi)$ and $\Gamma^-(\psi)$ subformula as **false** and each $\Gamma^+(\psi)$ formula as $\neg S(\mathbf{true}, \neg\psi)$. An induction on the construction of each α or β is actually needed here to guarantee that we do only need to evaluate formulas at points $< \infty$ but this is straightforward. Thus we may assume that γ^+ is a boolean combination of US formulas of the form $S(\alpha, \beta)$ containing no U.

So consider a pure past boolean component of γ^+. It will be in the form $\delta = S(\eta, \theta)$ with η and θ being formulas without U. To say that δ holds in the extended structure at time ∞ is just to say that at time 0 in the original structure, $\theta \wedge S(\eta, \theta)$ eventually always holds. To say that δ does not hold in the extended structure at time ∞ is just to say that at time 0 in the original structure, $\neg(\theta \wedge S(\eta, \theta))$ holds infinitely often. By rewriting γ^+ in disjunctive normal form we can extract the required α_i and β_i from conjunctions of such $\theta \wedge S(\eta, \theta)$ and their negations: use the equivalences $FG\alpha \wedge FG\beta \leftrightarrow FG(\alpha \wedge \beta)$ and $GF\alpha \vee GF\beta \leftrightarrow GF(\alpha \vee \beta)$ to collect the conjunctions together. ❑

Obviously the new proof calls on a more limited variety of machinery for finding the formulas: the procedure is just a series of syntactic rewrites. However, we should again warn the reader that the complexity of the separation process (used in both the new proof and the original) and the blow-up in the size of formulas may be non-elementary. Note that there may even be an exponential blow-up in translating from TL to the strict language.

An example transformation is set out in [Reynolds, 2000]. The formula $\phi = qU_{=}p$ rewritten as $p \vee (q \wedge U(p, q))$ in strict form is translated (eventually) to the safety-liveness form, $(FGYP(H\mathbf{false} \wedge p) \wedge GF\mathbf{true}) \vee (FGYP(p \wedge Y(qS(H\mathbf{false} \wedge q))) \wedge GF\mathbf{true})$.

Safety-liveness form and automata

As shown in [Reynolds, 2000], the safety-liveness form allows us to easily and naturally find an equivalent deterministic automaton, i.e. one that accepts exactly the models of the formula. There are immediate advantages in reasoning tasks if one happens to already have a formula in safety-liveness form. For example, the number of states needed in an equivalent automaton is exponential in the length of the formula: normally, without safety-liveness, a double exponential blow-up is required.

Alternatively, many of the applications of automata to theorem-proving, decision procedures, synthesis of models, and executable temporal logic, can be recast in terms of the properties of formulas in safety-liveness form. There is no need to translate to automata to carry out these reasoning tasks. See [Reynolds, 2000] for details.

*Using safety-liveness to axiomatize PCTL**

Perhaps the most useful applications of the safety-liveness form are for those who are developing axiomatizations and theorem-proving methods for extensions of LTL. The important observation here is that automata are increasingly being used in such proofs (see, for example, [Walukiewicz, 1995], [Kesten and Pnueli, 1995], [Kaivola, 1996], and [Reynolds, 2001]) but, as in the last of these references, having an automaton in a completeness proof may involve problems with describing it within the object language. However, past operators and the safety-liveness form allow many of the advantages of automata but stay within the confines of the original language.

Here we mention one such application concerning the widely used branching time logic CTL* from [Emerson and Halpern, 1986]. This logic has been particularly hard to develop reasoning tools for. The relatively recent axiomatization in [Reynolds, 2001] makes great use of automata and needs to rely on a special rule of inference to allow the introduction of new atoms into a proof, which can be used to represent states of an automaton. We end this section by seeing how the safety-liveness form can be used so that if past operators are available then a lot of this extra machinery is not needed. Below we will briefly look at the work in [Reynolds, 2005] where there is a complete axiomatization for the logic PCTL*, CTL* with the past operators: an axiomatization involving only the standard rules of inference and a few dozen simple and intuitive axioms.

It is possible that similar applications may be found in axiomatizations and theorem-proving methods for many of the large number of currently interesting extensions of LTL including branching-time logics, logics with quantified propositions and combinations of temporal and epistemic logics.

PCTL* was defined in [Laroussinie and Schnoebelen, 1994], as the full branching-time computational tree logic CTL* with the addition of (linear) past operators. This was also studied in [Zanardo and Carmo, 1993] and in [Kupferman and Pnueli, 1995]. The latter paper also introduces a different way of adding past operators to CTL* (and CTL): the semantics of branching past. See [Kupferman and Pnueli, 1995] or [Laroussinie and Schnoebelen, 2000] for details.

CTL* has been hard to axiomatize despite its recognized wide applicability. The problem is seen as its limit closure property: roughly the idea that any increasing sequence of prefixes of paths through a structure is part of one path. The solution in the axiomatization [Reynolds, 2001] is to bring in automata and a related special rule of inference. PCTL*, incorporating CTL*, and all the advantages of past operators, is even more useful but exhibits the same limit closure problem.

The formulas of PCTL* are built from atomic propositions recursively using classical connectives, the LTL+Past temporal connectives Y, $S_=$, T, and $U_=$ and the path-switching modality E: if α and β are formulas then so are $Y\alpha$, $\alpha S_= \beta$, $T\alpha$, $\alpha U_= \beta$, and $E\alpha$. We use the usual LTL+Past abbreviations plus $A\alpha \equiv \neg E \neg \alpha$.

Formulas are evaluated in *(total) Kripke structures*, $M = (S, R, g)$ where:

S is the non-empty set of *states*

R is a total binary relation $\subseteq S \times S$
(i.e. for every $s \in S$, there is some $t \in S$ such that $(s,t) \in R$) and

g $: S \to \wp(\mathcal{L})$ is a labelling of the states with sets of atoms.

A *fullpath* in M is an infinite sequence $b = \langle b_0, b_1, b_2, ...\rangle$ of states of M such that for each i, $(b_i, b_{i+1}) \in R$.

Truth of formulas is evaluated at indices in fullpaths in structures. We write $M, b, i \models \alpha$ iff the formula α is true of the fullpath b at the index (time) i in the structure $M = (S, R, g)$. This is defined recursively by:

$M, b, i \models \mathbf{true}$		
$M, b, i \models p$	iff	$p \in g(b_i)$, any $p \in \mathcal{L}$
$M, b, i \models T\alpha$	iff	$M, b, i+1 \models \alpha$
$M, b, i \models \alpha U_= \beta$	iff	there is some $j \geq i$ such that $M, b, j \models \beta$ and for each k, if $i \leq k < j$ then $M, b, k \models \alpha$
$M, b, i \models Y\alpha$	iff	$i > 0$ and $M, b, i-1 \models \alpha$
$M, b, i \models \alpha S_= \beta$	iff	there is some $j \leq i$ such that $M, b, j \models \beta$ and for each k, if $j < k \leq i$ then $M, b, k \models \alpha$
$M, b, i \models E\alpha$	iff	there is a fullpath b' such that $\langle b_0, \ldots, b_i\rangle = \langle b'_0, \ldots, b'_i\rangle$ and $M, b', i \models \alpha$

We say that α is valid in PCTL* iff for all Kripke structures M, for all

fullpaths b in M, for all indices i, we have $M, b, i \models \alpha$.

To find a Hilbert system capable of deriving exactly these validities, [Reynolds, 2005] extends the axiom system for LTL (given in [Lichtenstein *et al.*, 1985]) by the usual S5 rules and axioms for path-switching, plus that propositional atoms only depend on states:

APS $p \rightarrow Ap$, for each atomic proposition p

plus some interaction between modalities:

AT $AT\alpha \rightarrow TA\alpha$

AY $AY\alpha \leftrightarrow YA\alpha$

plus the limit closure axiom from [Reynolds, 2001]:

LC $AG(E\alpha \rightarrow ET((E\beta)U(E\alpha))) \rightarrow (E\alpha \rightarrow EG((E\beta)U(E\alpha)))$

THEOREM 13 ([Reynolds, 2005]). *The Hilbert system is sound and complete for PCTL*.*

The completeness proof is very similar to that in [Reynolds, 2001] which shows completeness for a Hilbert system for CTL*. The interesting part is not the addition of the past operators: these can be handled by standard linear temporal logic techniques for past operators given that the axioms virtually define them in terms of the future time operators.

The interesting part of this completeness proof is to show that we do not need to call on the extra, unusual rule (called AA) which is used in [Reynolds, 2001]. Space limitations prevent us from introducing AA properly here. The rule allows new atoms to be brought into a proof provided that their truth values in a tree structure are determined functionally by the truth-values of atoms (both new and original) in the past. There are similarities with the IRR rule of [Gabbay, 1981b]. In using the CTL* proof system to make a derivation and in giving its completeness proof, this special rule allows us to bring a deterministic Rabin automaton into the proof. The new atoms can be used to represent the states of the automaton and the automaton can tell us where we are up to in trying to satisfy LTL formulas along branches of the tree. By making sure certain states do not come up very often if other states do not (using the acceptance pairs of the automaton) we can guarantee that all branches satisfy a particular LTL formula: even the uncountable number of branches which appear in the limit of a construction of such a tree do.

The observation that allows us to modify the completeness proof to cope with PCTL* without the AA rule is that when past operators are available in the language and we have the safety-liveness form of the original formula (and it will be in the original signature) then no new atoms are needed to record our progress along the branches of the tree. We can build a deterministic automaton accepting exactly the models of a given LTL+Past formula by only using (sets or conjunctions of) LTL+Past formulas in the original signature to make up the states.

In fact, with the syntactic method of finding the safety-liveness form there need be no mention of automata at all in the proof. All we have to do, after translating the formula into the right form, is to make sure that eventually some β_i stays true in our construction (as we build a particular branch from the root towards the leaf) and make sure α_i is true infinitely often.

See [Reynolds, 2005] for details.

4 Separation: future

As we have seen, separation has proven theoretically interesting and both theoretically and practically useful. However, it is far from fully understood, even in the most common situations. We end with some (as far as we know) open problems about it, which may provide work for future researchers. We focus on *complexity* and *succinctness,* which are of some current interest and for which there is relevant recent work. We thank Carsten Lutz for interesting discussions.

4.1 Initial equivalence

We begin with initial equivalence: in $(\mathbb{N}, <)$, at time 0. Recall from section 3.2 that temporal formulas α, β are said to be *initially equivalent* if $(\mathbb{N}, <, h), 0 \models \alpha \leftrightarrow \beta$ for all assignments $h : L \to \wp(\mathbb{N})$.

REMARK 14.

(i) By lemma 10, we know that for any US-formula α, there is a U-formula β that is initially equivalent to α.

(ii) If we combine lemma 10 with the separation algorithm obtained from the proof of theorem 7, we obtain an algorithm to construct β from α, but its complexity appears to be non-elementary. However, [Schnoebelen, 2003, footnote 22] notes that an *elementary upper bound* on the minimal size of an U-formula initially equivalent to a given US-formula can be obtained by combining the standard translation from US-formulas to counter-free Büchi automata and the elementary translation from these automata to U-formulas using results from [Wilke, 1999]. This also gives an *elementary algorithm* to perform the translation.

(iii) [Laroussinie *et al.*, 2002, theorem 3.1] showed that US can be *exponentially more succinct* than U over $(\mathbb{N}, <)$ at time 0. That is, obtaining β from α above can introduce an exponential blow-up in size. Hence, any algorithm to construct β from α requires at least exponential time in the worst case.

This prompts the following questions:

Q1. What is the exact *complexity* of the problem of constructing β from α in remark 14?

Q2. *(Succinctness)* What is the minimum length of β in terms of the length of α?

4.2 Separation

Similar questions can be asked about separation itself. As we said, the proof of theorem 7 provides an algorithm that constructs, for any US-formula α, a US-formula β that is (syntactically) separated and equivalent to α over $(\mathbb{N}, <)$. However, very little is known about the complexity of separation algorithms and the succinctness of their output.

We pose our questions in the general context of a temporal logic $\mathcal{T}$ with the separation property over a class $\mathcal{C}$ of flows of time. Assuming that validity of $\mathcal{T}$-formulas over $\mathcal{C}$ is decidable, it follows from the proof of lemma 9 that there exists an algorithm to separate any $\mathcal{T}$-formula α over $\mathcal{C}$. We just enumerate all $\mathcal{T}$-formulas, stopping when a separated one equivalent to α is found. Separatedness and equivalence are decidable (cf. lemma 9 for the former), so this is an effective process; and since $\mathcal{T}$ has the separation property, it will terminate.

Q3. *Complexity of separation.* What is the optimal complexity of algorithms that, given a $\mathcal{T}$-formula α, output a separated $\mathcal{T}$-formula equivalent to α over $\mathcal{C}$?

This problem has many versions, depending on circumstances. The chief concrete instances of it are

(a) for U, S over $(\mathbb{N}, <)$,

(b) for U, S over $(\mathbb{N}, <)$, but requiring the output formula to be *syntactically* separated,

(c) for U, S over Dedekind-complete time,

(d) for U, S, and the Stavi connectives over linear time.

It follows from remark 14(iii) that over $(\mathbb{N}, <)$, all separation algorithms require at least exponential time.

Q4. *Succinctness of separation.* What can one say about the length of a shortest separated $\mathcal{T}$-formula equivalent over $\mathcal{C}$ to a given $\mathcal{T}$-formula?

This is asking whether there is an inevitable loss of succinctness when separating a formula, and if so, how much? Again, there are various concrete instances of the problem. For U and S over $(\mathbb{N}, <)$, it follows from remark 14(iii) that separation sometimes incurs an exponential increase in length. It would be interesting if non-separated formulas were non-elementarily more succinct than separated ones in this case, since it follows from [Sistla and Clarke, 1985] that satisfiability over $(\mathbb{N}, <)$ for both arbitrary US-formulas and for U-ones (which are certainly separated) is PSPACE-complete, and both have the exponential-size model property. It would also mark a difference from the special case discussed in remark 14(ii).

REMARK 15. Questions Q3 and Q4 are connected. The following are equivalent:

(a) There is an elementary algorithm to separate US-formulas over $(\mathbb{N}, <)$,

(b) There is an elementary upper bound on the length of a shortest separated US-formula equivalent over $(\mathbb{N}, <)$ to a given formula.

$(a) \Rightarrow (b)$ is trivial, since it takes at least as much time to compute the separated formula as to output it. For $(b) \Rightarrow (a)$, recall from [Sistla and Clarke, 1985] that there is an elementary (PSPACE) algorithm to decide equivalence of US-formulas over $(\mathbb{N}, <)$. So assuming (b) and given a US-formula α, we can enumerate all separated US-formulas of length at most the elementary bound obtained from α, using lemma 9 to verify separatedness, and checking each for equivalence to α. This is an elementary algorithm to separate α.

Obviously, there are more refined results, bounding the complexity of each side in terms of the other.

4.3 Other sets of connectives

Now we consider what might happen when we change the set of connectives.

Consider any temporal logic $\mathcal{T}$ with finitely many connectives which are all first-order-definable. As US is expressively complete over $(\mathbb{N}, <)$, each $\mathcal{T}$-formula α can be translated into a US-formula $\alpha^\dagger$ that is equivalent to α over $(\mathbb{N} <)$. The translation is straightforward. For each n-ary $\mathcal{T}$-connective $\sharp$, we fix a US-formula $\beta_\sharp(p_1, \ldots, p_n)$ that is equivalent over $(\mathbb{N}, <)$ to $\sharp(p_1, \ldots, p_n)$. Such a formula $\beta_\sharp$ exists by expressive completeness of US over $(\mathbb{N}, <)$. Now we put

- $p^\dagger = p$ for any atom p,
- $(\alpha \wedge \beta)^\dagger = \alpha^\dagger \wedge \beta^\dagger$ and $(\neg\alpha)^\dagger = \neg\alpha^\dagger$,
- $\sharp(\alpha_1, \ldots, \alpha_n)^\dagger = \beta_\sharp(\alpha_1^\dagger, \ldots, \alpha_n^\dagger)$.

The translation $\alpha \mapsto \alpha^\dagger$ *entails at most an exponential increase in size* (and such an increase can be attained). To see this, let $d(\alpha)$ be the depth of nesting of boolean and temporal connectives in the formula α. Formally, $d(p) = 0$ for an atom p, $d(\neg\alpha) = 1 + d(\alpha)$, $d(\alpha \wedge \beta) = 1 + \max(d(\alpha), d(\beta))$, and $d(\sharp(\alpha_1, \ldots, \alpha_n)) = 1 + \max\{d(\alpha_1), \ldots, d(\alpha_n)\}$, for each temporal connective $\sharp$. Here, α can be either a $\mathcal{T}$-formula or a US-formula. Let k be the maximal value of $d(\beta_\sharp)$ (running over all $\mathcal{T}$-connectives $\sharp$). We will assume that $k \geq 1$. Then every nested occurrence of a boolean or temporal connective in α raises the depth of $\alpha^\dagger$ by at most k. So we see that

$$d(\alpha^\dagger) \leq k \cdot d(\alpha) \quad \text{for all } \mathcal{T}\text{-formulas } \alpha,$$

and indeed, this is easily proved formally by induction on α.

Write $|\alpha|$ for the length of α, ignoring commas and brackets for simplicity. By considering formation trees of formulas, we see that for each US-formula β we have $|\beta| \leq 2^{d(\beta)}$. For $\mathcal{T}$-formulas α, we trivially have $d(\alpha) \leq |\alpha|$. So the length $|\alpha^\dagger|$ of the translation is bounded in terms of the length $|\alpha|$ of the original formula by

$$|\alpha^\dagger| \leq 2^{d(\alpha^\dagger)} \leq 2^{k \cdot d(\alpha)} \leq 2^{k|\alpha|}.$$

Clearly, then, the translation $\alpha \mapsto \alpha^\dagger$ is effective and takes at most exponential time. The main requirements were that the connectives in $\mathcal{T}$ are finite in number and have first-order definitions, and that US is expressively complete over $(\mathbb{N}, <)$. So if $\mathcal{T}$ is expressively complete as well, the translation process can be reversed. The translation obviously preserves purity and being separated, since these are defined semantically. It follows that

PROPOSITION 16. *Let $\mathcal{T}$ be a finite set of first-order-definable connectives that is expressively complete over* $(\mathbb{N}, <)$. *Then the following are equivalent:*

- *There is an elementary algorithm to separate $\mathcal{T}$-formulas over* $(\mathbb{N}, <)$.
- *There is an elementary algorithm that separates US-formulas over* $(\mathbb{N}, <)$.

A similar result holds for succinctness. As in remark 15, the complexity estimates can be refined. Still, they do not give exact bounds, so the following questions are perhaps worth answering:

Q5. Find a finite set $\mathcal{T}$ of connectives that is expressively complete over $(\mathbb{N}, <)$, and such that the complexity of the problem of separating $\mathcal{T}$-formulas is least possible.

Q6. Find a finite set $\mathcal{T}$ of connectives that is expressively complete over $(\mathbb{N}, <)$, and such that the length of the smallest separated $\mathcal{T}$-formula equivalent to a given $\mathcal{T}$-formula is as small as possible.

There are special problems for other flows of time.

Q7. Over linear time, or over $(\mathbb{Q}, <)$:

(a) Is it decidable whether a given US-formula can be separated?

(b) [Rabinovich] Is it decidable whether a given first-order formula $\varphi(x, P_1, \ldots, P_n)$ is equivalent to a US-formula?

In each case, if the answer is positive, one can ask about complexity.

We hope that these questions will find interesting solutions before too long. At any rate, they show that the topic of separation, after more than 20 years, is still very much alive.

BIBLIOGRAPHY

[Amir and Gabbay, 1987] A. Amir and D. M. Gabbay. Preservation of expressive completeness in temporal models. *Information and Computation*, 72:66–83, 1987.

[Amir, 1982a] A. Amir. Expressive completeness failure in branching time structures. *Journal of Computer and System Sciences*, 34(1), 1982.

[Amir, 1982b] A. Amir. *Functional Completeness in Tense Logic*. PhD thesis, Bar-Ilan University, Ramat-Gan, Israel, 1982.

[Amir, 1985] A. Amir. Separation in nonlinear time models. *Information and Control*, 66:177–203, 1985.

[Barringer *et al.*, 1996] H. Barringer, M. Fisher, D. Gabbay, R. Owens, and M. Reynolds, editors. *The Imperative Future*. Research Studies Press, Somerset, 1996.

[Bowman *et al.*, 2003] H. Bowman, H. Cameron, P. King, and S. J. Thompson. Mexitl: Multimedia in executable interval temporal logic. *Formal Methods in System Design*, 22:5–38, January 2003.

[Emerson and Halpern, 1986] E. Emerson and J. Halpern. 'Sometimes' and 'not never' revisited: on branching versus linear time. *J. ACM*, 33, 1986.

[Etessami and Wilke, 2000] K. Etessami and T. Wilke. An until hierarchy and other applications of an Ehrenfeucht–Fraïssé game for temporal logic. *Information and Computation*, 160:88–108, 2000.

[Fagin *et al.*, 1995] R. Fagin, J. Halpern, Y. Moses, and M. Vardi. *Reasoning about Knowledge*. The MIT Press, 1995.

[Finger *et al.*, 1993] M. Finger, M. Fisher, and R. Owens. Metatem at work: Modelling reactive systems using executable temporal logic. In *6th Intl. Conf. on Industrial and Engineering applications of AI and Expert Systems*, 1993.

[Gabbay *et al.*, 1980] D. M. Gabbay, A. Pnueli, S. Shelah, and J. Stavi. On the temporal analysis of fairness. In *7th ACM Symposium on Principles of Programming Languages, Las Vegas*, pages 163–173, 1980.

[Gabbay *et al.*, 1994] D. Gabbay, I. Hodkinson, and M. Reynolds. *Temporal logic: mathematical foundations and computational aspects, Vol. 1.* Clarendon Press, Oxford, 1994.

[Gabbay, 1981a] D. Gabbay. Expressive functional completeness in tense logic (preliminary report). In U. Monnich, editor, *Aspects of Philosophical Logic*, pages 91–117. Reidel, Dordrecht, 1981.

[Gabbay, 1981b] D. M. Gabbay. An irreflexivity lemma with applications to axiomatizations of conditions on tense frames. In U. Monnich, editor, *Aspects of Philosophical Logic*, pages 67–89. Reidel, Dordrecht, 1981.

[Gabbay, 1989] D. Gabbay. Declarative past and imperative future: Executable temporal logic for interactive systems. In B. Banieqbal, H. Barringer, and A. Pnueli, editors, *Proceedings of Colloquium on Temporal Logic in Specification, Altrincham, 1987*, volume 398 of *Lecture Notes in Computer Science*, pages 67–89. Springer-Verlag, 1989.

[Hodges, 1993] W. Hodges. *Model theory*, volume 42 of *Encyclopedia of mathematics and its applications.* Cambridge University Press, 1993.

[Immerman and Kozen, 1987] N. Immerman and D. Kozen. Definability with bounded number of bound variables. In *LICS87, Proceedings of the Symposium on Logic in Computer Science, Ithaca, New York*, pages 236–244, Washington, 1987. Computer Society Press.

[Jonker and Wijngaards, 2003] C. M. Jonker and W. C. A. Wijngaards. A temporal modelling environment for internally grounded beliefs, desires and intentions. Technical Report 040, Utrecht: UU, owi CKI (Artificial intelligence preprint series), 2003.

[Kaivola, 1996] R. Kaivola. Axiomatising extended computation tree logic, in trees in algebra and programming. In *CAAP'96, 21st International Colloquium, Proceedings*, volume 1059, pages 87–101. Springer, 1996.

[Kamp, 1968] H. Kamp. *Tense logic and the theory of linear order.* PhD thesis, University of California, Los Angeles, 1968.

[Kesten and Pnueli, 1995] Y. Kesten and A. Pnueli. A complete proof system for QPTL. In *Proceedings, Tenth Annual IEEE Symposium on Logic in Computer Science*, pages 2–12, San Diego, California, 26–29 June 1995. IEEE Computer Society Press.

[Kupferman and Pnueli, 1995] O. Kupferman and A. Pnueli. Once and for all. In *Proc. 10th IEEE Symposium on Logic in Computer Science*, pages 25–35, San Diego, June 1995.

[Laroussinie and Schnoebelen, 1994] F. Laroussinie and Ph. Schnoebelen. A hierarchy of temporal logics with past. In *Proc. STACS'94, Caen, France*, volume 775 of *LNCS*, pages 47–58. Springer–Verlag, 1994.

[Laroussinie and Schnoebelen, 2000] F. Laroussinie and Ph. Schnoebelen. Specification in CTL+Past for verification in CTL. *Information and Computation*, 156:236–263, 2000.

[Laroussinie *et al.*, 2002] F. Laroussinie, N. Markey, and Ph. Schnoebelen. Temporal logic with forgettable past. In *Proc. 17th Annual IEEE Symposium on Logic in Computer Science (LICS'02)*, pages 383–392. IEEE Computer Society Press, 2002. Available at `www.lsv.ens-cachan.fr/Publis/PAPERS/PDF/LMS-lics2002.pdf`.

[Lichtenstein *et al.*, 1985] O. Lichtenstein, A. Pnueli, and L. Zuck. The glory of the past. In R. Parikh, editor, *Logics of Programs* (*Proc. Conf. Brooklyn USA 1985*), volume 193 of *Lecture Notes in Computer Science*, pages 196–218. Springer-Verlag, Berlin, 1985.

[Marx, 2004] M. Marx. Conditional XPath, the first order complete XPath dialect. In *Proc. ACM SIGMOD/PODS (Principles of Database Systems)*, 2004. Available at `www.sigmod.org/pods/proc04/`.

[McNaughton and Papert, 1971] R. McNaughton and S. Papert. *Counter Free Automata.* MIT Press, 1971.

[McNaughton, 1966] R. McNaughton. Testing and generating infinite sequences by finite automata. *Information and Control*, 9:521–530, 1966.

[Reynolds, 1999] M. Reynolds. The complexity of temporal logic over the reals. 1999. Submitted.

[Reynolds, 2000] M. Reynolds. More past glories. In *Fifteenth Annual IEEE Symposium on Logic in Computer Science (LICS'2000), Santa Barbara, California, USA, June 26-28, 2000*, pages 229–240. IEEE, 2000.

[Reynolds, 2001] M. Reynolds. An axiomatization of full computation tree logic. *J. Symbolic Logic*, 66(3):1011–1057, 2001.

[Reynolds, 2005] M. Reynolds. An axiomatization of PCTL*. *Information and Computation*, 201:72–119, 2005.

[Schlingloff, 1992] B-H. Schlingloff. Expressive completeness of temporal logic over trees. *Journal of Applied Non-classical Logics*, 2:157–180, 1992.

[Schnoebelen, 2003] Ph. Schnoebelen. The complexity of temporal logic model checking. In *Proc. 4th Workshop on Advances in Modal Logic (AiML'02)*, pages 393–436. King's College Publications, 2003.

[Sistla and Clarke, 1985] A. P. Sistla and E. M. Clarke. The complexity of propositional linear temporal logics. *J. ACM*, 32:733–749, 1985.

[Walukiewicz, 1995] I. Walukiewicz. A complete deductive system for the μ-calculus. Research Series RS-95-6, BRICS, Department of Computer Science, University of Aarhus, January 1995. 39 pp.

[Wilke, 1999] T. Wilke. Classifying discrete temporal properties. In *Proc. 16th Ann. Symp. Theoretical Aspects of Computer Science (STACS'99)*, volume 1563 of *Lecture Notes in Computer Science*, pages 32–46. Springer-Verlag, 1999.

[Zanardo and Carmo, 1993] A. Zanardo and J. Carmo. Ockhamist computational logic: Past-sensitive necessitation in CTL. *J. Logic Computat.*, 3(3):249–268, June 1993.

Kripke's Modal Objection to the Description Theory of Reference

Dale Jacquette

1 Names and Descriptions

The description theory was once the most widely accepted account of the reference of proper names. As developed by Russell, and in later versions by P. F. Strawson and John R. Searle, the description theory implements Frege's distinction between sense (*Sinn*) and reference (*Bedeutung*).[1] Russell interprets nonlogically proper names as disguised definite descriptions that unpack a name's Fregean sense by specifying the uniquely identifying properties of the particular object that a proper name denotes. The referent of a proper name is thereby determined as the object uniquely picked out by its associated definite description, as the only object to which the description truly applies. In a simplified example, if the proper name 'Aristotle' abbreviates the definite description, 'the student of Plato and teacher of Alexander the Great', then the referent of the name 'Aristotle' is the one and only object in the world that fits the description as being both the student of Plato and teacher of Alexander the Great. If the description theory works, then it further instantiates Frege's central semantic thesis that sense determines reference, or, equivalently, that intension determines extension.[2]

Saul A. Kripke, in his lectures on *Naming and Necessity*, challenges the description theory by means of a modal objection. The criticism is that we can intelligibly consider the object designated by a proper name in counterfactual circumstances where not all or even none of the properties attributed to it by a contingently true definite description apply in nonactual alternative logically possible situations or worlds. We presumably refer to the

[1]Bertrand Russell, "On Denoting", *Mind*, 14, 1905, pp. 479–493. P. F. Strawson, *Individuals: An Essay in Descriptive Metaphysics* (London: Methuen & Co. Ltd., 1959), pp. 184–187. John R. Searle, "Proper Names", *Mind*, 67, 1958, pp. 166–173.

[2]Gottlob Frege, "*Über Sinn und Bedeutung*", *Zeitschrift für Philosophie und philosophische Kritik*, 100, 1892, pp. 25–50; translated as, "On Sense and Reference", in P. T. Geach and Max Black, *Translations from the Philosophical Writings of Gottlob Frege*, (Oxford: Basil Blackwell, 1970), pp. 56–78.

same thing by the use of the same proper name in these worlds in which by hypothesis their associated definite descriptions do not hold true of them. Kripke concludes that as a result the description theory of the referential meaning of proper names cannot be explicated as contingently corresponding definite descriptions. The present essay critically evaluates Kripke's argument in light of difficulties raised by Brian Loar alleging a fallacy of equivocation between wide and narrow modal scope readings of the proposition that a named object might not have had the properties predicated of it by means of an associated definite description in at least some non-actual logically possible worlds. I further examine Kripke's reflections on the irrelevance of modal scope distinctions for his objection, arguing that Kripke does not adequately establish that ordinary language is incapable of disambiguating modal scope, and that potential equivocations of modal scope persist even in contexts from which Kripke claims to have eliminated them. I formalize the distinction between wide and narrow modal scope and pursue a reply to Loar and defense of Kripke's objection to the description theory, developing a suggestion by Kent Bach that enforces a narrow modal scope interpretation univocally on the argument's assumptions and conclusion.

After arguing that Kripke's modal objection to the description theory of proper names avoids Loar's criticism, I turn next to a different problem in Kripke's picture of the reference of proper names as rigid designators. Kripke holds that proper names refer to the same entity in every logically possible world in which they refer at all, and that their referential meaning is transmitted historically from the equivalent of an initial baptism or naming ceremony to each link in a causal chain of language users who intend to refer to the same thing by their use of the name as the intended referent. The referential meaning of names is thereby socially disseminated among the members of a spatiotemporally extended linguistic community in what Kripke sees as a preferable way of accounting for how it is that the use of a proper name like 'Aristotle' refers to a particular Greek man in the ancient world than any speaker's ability to substitute a definite description of properties that uniquely apply to the name's referent. As a more damaging critique, I identify a circularity in Kripke's reasoning. Although Kripke appears to reject the description theory of the reference of proper names in order to clear the way for his alternative causal-historical picture of the reference of names as rigid designators, I maintain that Kripke's modal objection instead presupposes from the outset that proper names are rigid designators, without which assumption the modal objection to the description theory cannot be upheld. Kripke's concept of proper names as rigid designators is thereby rescued from Loar's frying-pan but promptly deposited in the fire. It is preserved from the criticism that it commits the fallacy of equivocation involving modal scope ambiguities, but charged with vicious circularity in its strategy of insisting that proper names should be regarded as rigid designators. Finally, I argue that Kripke's modal objection is avoided altogether by world-indexing the properties attributed to the objects designated by proper names in an expanded more complete definite description that is not limited only to the properties the objects possess in the actual world but in principle in every logically possible world.

2 Kripke's Modal Objection and Loar's Countercriticism

Kripke's modal objection challenges the theory that the reference of proper names can be understood in terms of a uniquely referring definite description. If a proper name can be used to refer to an object under logically possible circumstances in which the properties attributed to it by the associated description do not hold true of the object to which the name refers, then, he argues, the name's referential meaning is not correctly analyzed as the corresponding description:

> Let me return to the question about names which I raised. As I said, there is a popular modern substitute for the theory of Frege and Russell; it is adopted even by such a strong critic of many views of Frege and Russell, especially the latter, as Strawson. The substitute is that, although a name is not a disguised description it either abbreviates, or anyway its reference is determined by, some cluster of descriptions. The question is whether this is true...Such a suggestion [referring to John R. Searle's version of the theory in his essay "Proper Names"], if 'necessary' is used in the way I have been using it in this lecture, must clearly be false...Most of the things commonly attributed to Aristotle are things that Aristotle might not have done at all. In a situation in which he didn't do them, we would describe that as a situation in which *Aristotle* didn't do them...Not only is it true *of* the man Aristotle that he might not have gone into pedagogy; it is also true that we use the term 'Aristotle' in such a way that, in thinking of a counterfactual situation in which Aristotle didn't go into any of the fields and do any of the achievements we commonly attribute to him, still we would say that was a situation in which *Aristotle* did not do these things.[3]

[3] Saul A. Kripke, *Naming and Necessity* (Cambridge: Harvard University Press, 1980), pp. 60–62.

The force of Kripke's complaint against the description theory is that we can in fact and often do use proper names to refer to objects in counterfactual circumstances in which the properties attributed to them by definite descriptions do not hold true of them. If this is possible, then it is hard to see how we could accept a description theory of the referential meaning of proper names by which names refer to whatever objects a description uniquely applies. If what I referentially *mean* by the name 'Aristotle' is the student of Plato and teacher of Alexander the Great, then, contrary to established practice, I will not be able to refer to Aristotle as possibly not having studied with Plato or possibly not having taught Alexander the Great. In those logically possible worlds where the description does not apply the name 'Aristotle' cannot refer to that same object that contingently possesses those properties in the actual world.

Kripke's modal objection is criticized by Loar in his essay, "The Semantics of Singular Terms". Loar describes the explication of a proper name N by a definite description *the* F by the form, 'N means *the* F'. He schematizes Kripke's argument in this way:

Loar's Analysis of Kripke's Objection

(a) If N were used to mean *the* F, then 'N might not have been the F' would be false.

(b) But 'N might not have been the F' is true.

(c) Therefore, N is not used to mean *the* F.[4]

Loar concludes that Kripke's argument is 'defective', when he maintains:

> Names are normally read as having wider scope than modal operators — and that is why premise (b) is true for the relevant Fs. For example, 'Aristotle might not have been the pre-eminent ancient philosopher called "Aristotle"' is true because it asserts of Aristotle that he might not have had that contingent property. But if the N-position in the sentence mentioned in premise (a) is read similarly as having widest scope, then premise (a) is false, as may be seen by substituting 'the F' for N. That 'the pre-eminent ancient philosopher called "Aristotle"' means *the pre-eminent ancient philosopher called 'Aristotle'* does not entail that it is false that the pre-eminent ancient philosopher called Aristotle is such that he might not have had that property.[5]

[4] Brian Loar, "The Semantics of Singular Terms", *Philosophical Studies*, 30, 1976, p. 373.

[5] Ibid.

The modality of meaning is center stage in both Kripke's objection and Loar's counter-objection. Kripke raises the problem of the modal status of names and definite descriptions, in the usual semantic idiom, across alternative logically possible worlds. He argues that the meaning of a name 'N' is not the definite description '*the* F', if it is logically possible that N is not the F. To explain the *meaning* of the name 'Aristotle' as 'the student of Plato and tutor of Alexander the Great' is unsatisfactory, according to Kripke, because Aristotle need not have been the student of Plato or the tutor of Alexander the Great.

3 Kripke on the Irrelevance of Modal Scope

In the 'Preface' to the (1980) book version of *Naming and Necessity*, Kripke rejects problems of modal scope ambiguity as irrelevant to his theory of meaning. The issues he raises are immediately pertinent to the question of whether Loar's criticism or anything resembling it can possibly succeed. This makes it worthwhile to consider Kripke's reaction to Loar-style efforts to identify equivocations of modal scope in his modal objection to descriptivism. Kripke maintains in a passage worth quoting at length that modal scope is a red herring:

> Another misconception concerns the relation of rigidity to scope, which apparently I treated too briefly. It seems often to be supposed that all the linguistic intuitions I adduce on behalf of rigidity could just as well be handled by reading names in various sentences as nonrigid designators with wide scopes, analogously to wide scope descriptions. It would, indeed, be possible to interpret *some* of these intuitions as results of scope ambiguities instead of rigidity — this I recognize in the monograph. To this extent the objection is justified, but it seems to me to be wrong to suppose that *all* our intuitions can be handled in this way...It has even been asserted that my own view itself reduces to a view about scope, that the doctrine of rigidity simply *is* the doctrine that natural language has a convention that a name, in the context of any sentence, should be read with a large scope including all modal operators. This latter idea is particularly wide of the mark; in terms of modal logic, it represents a technical error. Let me deal with it first. [Sentences] (1) and (2) [(1) = 'Aristotle was fond of dogs'; (2) = 'The last great philosopher of antiquity was fond of dogs'] are 'simple' sentences. Neither contains modal or other operators, so there is no room for any scope distinctions. No scope convention about more complex sentences affects the interpretation of *these* sentences. Yet the

> issue of rigidity makes sense as applied to both. My view is that 'Aristotle' in (1) is rigid, but 'the last great philosopher of antiquity' in (2) is not. No hypothesis about scope conventions for modal contexts expresses this view; it is a doctrine about the truth conditions, with respect to counterfactual situations, of (the propositions expressed by) *all* sentences, including *simple* sentences. / This shows that the view that *reduces* rigidity to scope in the manner stated is simply in error.[6]

Kripke may well be right to maintain that rigid designation contexts cannot simply be reduced to nonrigid designation contexts with wide modal scope. The question in any case does not directly address the problem of whether differences in modal scope are relevant to his modal objection to the description theory of reference. Kripke does not deny that there are differences between narrow and wide modal scope, but insists only that rigid designation cannot be reduced to nonrigid designation with wide modal scope. Such a claim might even be understood as presupposing rather than discrediting the distinction. A more serious consideration is the problem of whether Kripke's modal objection can be recast in such a way as to eliminate modal operators altogether. Kripke gestures toward a similar method of discounting modal scope distinctions in his critique of the description theory, when he continues:

> Another remark, not so directly relying on counterfactual situations, may illuminate matters. In the monograph I argued that the truth conditions of 'It might have been the case that Aristotle was fond of dogs' conform to the rigidity theory: no proof that some person *other* than Aristotle might have been both fond of dogs and the greatest philosopher of antiquity is relevant to the truth of the quoted statement. The situation is unchanged if we replace 'the greatest philosopher of antiquity' by any other (nonrigid) definite description thought of as identifying Aristotle. Similarly, I held, 'It might have been the case that Aristotle was not a philosopher' expresses a truth, though 'It might have been the case that the greatest philosopher of antiquity was not a philosopher' does not, contrary to Russell's theory...Now the last quoted sentence would express a truth if the description used were read, contrary to my intent, with wide scope. So perhaps it might be supposed that the problem simply arises from an (unaccountable!) tendency to give 'Aristotle' a wide scope reading while the descriptions are given a small scope

[6]Kripke, *Naming and Necessity*, pp. 10–12.

> reading; sentences with both names and descriptions, however, would be subject in principle to both readings.[7]

Here Kripke acknowledges the kind of counter-criticism Loar had raised to his original modal objection to the description theory of reference. Let us leave aside the question of whether there could be a principled basis for attributing narrow modal scope to definite descriptions and wide modal scope to names. The question is whether the ambiguity, principled or unprincipled, in its own right, so to speak, as a fact of language, invalidates the key inference in Kripke's modal objection. Kripke's next move is to anticipate precisely the joint narrow modal scope reading of both assumptions. Now he adds:

> My point, however, was that the contrast would hold if all the sentences involved were explicitly construed with small scopes (perhaps by inserting a colon after 'that'). Further, I gave examples (referred to above) to indicate that the situation with names was not in fact parallel to that with large scope descriptions. Proponents of the contrary view often seem to have overlooked these examples, but this is not my point here. The contrary view must hold that our language and thought are, somehow, impotent to keep the distinction straight, that it is this which is responsible for the difficulty. It is hard to see how this can be: how did we make the distinction if we cannot make it? If the formulation with a that clause really is so tangled that we are unable to distinguish one reading of it from another, what about:
>
> (4) What (1) expresses might have been the case.
>
> Doesn't this express the desired assertion, with no scope ambiguities? If not, what would do so? (The formulation might be a bit more natural in a dialogue: 'Aristotle was fond of dogs.' 'That's not the case, though it might have been.') Now my claim is that our understanding of (4) conforms to the theory of rigidity. No possible situation in which anyone but Aristotle himself was fond of dogs can be relevant.[8]

It is worth remarking again that Kripke recognizes that his modal objection to the description theory goes through, contrary to Loar-style counter-criticisms, provided that both of the relevant assumptions are given narrow

[7] Ibid, pp. 12–13.
[8] Ibid., pp. 13–14.

modal scope readings. Such a stipulation avoids Loar's critique, as we have seen in a more sympathetic reconstruction of Kripke's original argument.

What remains controversial in Kripke's defense is his further assertion. He claims that in order to suggest that modal scope distinctions are relevant to his modal objection to the description theory depends on the implausible view that ordinary language has no way of observing modal scope distinctions. A critic of Kripke's argument, however, does not need to suppose that language generally has no devices for marking such distinctions. As Kripke observes, on the contrary, Loar-style criticisms appear to presuppose that the distinctions can be drawn, informally as well as formally. The failure of Loar's objection, as Kripke also notes, seems instead to result from the fact that he simply overlooks Kripke's intended interpretation of all sentences in the original objection as requiring narrow modal scope.

The problem is not with language generally, if Loar's criticism is correct, but with Kripke's use of it to express his modal objection to the description theory. A critic like Loar need only argue that Kripke adopts assumptions in which the disambiguating devices available in the language are not clearly or correctly applied, thereby contributing to a fatal equivocation in the modal scopes of the objection's assumptions. This is what Loar tries to argue — unsuccessfully, I would say, in agreement with Kripke. The counter-criticism, correct or incorrect, by no means constitutes an insupportable indictment of language generally or its ability to distinguish between narrow and wide modal scopes, and so by itself does not render all efforts to invoke modal scope distinctions as irrelevant to Kripke's modal objection to the description theory of reference.

Similar qualifications can be made in response to Kripke's reply in the Preface. Issues of modal scope might be said not to arise if Kripke can reformulate his objection to the description theory of reference without explicitly invoking scope-ambiguous modal operators. Yet it is not at all clear that the strategy is appropriate here. Kripke argues that (4) and its dialogue counterpart involves no modal scope ambiguities, and that therefore 'our understanding' of (4) 'conforms to the theory of rigidity'.

Several replies are in order. Kripke attempts a kind of Quinean semantic ascent in both attempts to eliminate modal scope ambiguity. Such efforts bring in further ambiguities and possibilities for equivocation of their own. Sentence (4), for example, speaks of 'What (1) expresses'. But what, exactly, does (1) express? It states that Aristotle is fond of dogs, which (4) says 'might have been the case'. Thus, we have, in effect, as an implication of Kripke's paraphrase, (4*) 'It might have been the case that Aristotle is fond of dogs'. Why would (4*) not be subject to the same ambiguity of modal scope as the original sentences in Kripke's modal objection? The dialogue form introduces the further problem of indexicals, involving the term 'that' in 'That's not the case' and 'it' in 'although it might have been'.

Again, it is difficult to see how, when these constructions are given and interpreted to produce the sentences Kripke needs for his conclusion, they are supposed to avoid all modal scope ambiguities. For then we have: 'It is not the case that Aristotle was fond of dogs although it might have been the case that Aristotle was fond of dogs'. An ambiguity and threat of equivocation in that case holds for any Kripke-style modal objection to the description theory of reference that attempts to define the name 'Aristotle' in part as involving the true predication by which Aristotle is fond (or not fond) of dogs. Shall we then define 'Aristotle' as denoting the individual who, among other things, is fond of dogs, for whom it is impossible as a result, relative to the respective language, not to be identical with the individual who, etc., is fond of dogs? Or shall we say that it is impossible for 'Aristotle' to denote the individual who, among other things, is fond of dogs only if Aristotle is not identical to the individual who, etc., is fond of dogs?

The modal ambiguity does not go away, but attaches at a higher theoretical level in these contexts to multiple exact interpretations of the description theory of reference whenever we consider the relevant identities and predications of properties to an individual whose name is defined in terms of a definite description as possible or impossible relative to a fixed language. The only way to confront the problem is to clarify the respective modal scope of each predication, and to refuse to settle for any interpretation in which modal scope ambiguities and equivocations are preserved. Kripke does this well enough in remarking that his original modal objection to the description theory of reference was meant to involve only narrow or 'small' modal scope, thereby scotching Loar's counter-criticism. The possibility of ramified modal scope ambiguities, unfortunately, is not eliminated by Kripke's use of semantic ascent.

Kripke entertains these reductions exclusively in connection with the issue of rigid designation, and not in application to the modal scope ambiguities alleged to plague his modal objection to the description theory of reference. This arguably puts the cart before the horse. We first need a convincing refutation of the description theory of reference as a reason for adopting the theory of rigid designation. That, at least, is how Kripke presents and encourages us to expect the proper development of the distinction, and that strategically makes the most sense. We shall later see that Kripke's conclusions about rigid designation are presupposed by rather than derived from his modal objection to the description theory of reference. The main point of the present discussion is that Kripke's efforts to dismiss the problem of equivocal modal scope in the argument are not persuasive, so that if the modal objection is to carry weight against descriptivism, Loar's counter-criticism will need to be met in another way.

4 Formalization of Modal Scope and Defense of Kripke Against Loar

Loar argues that Kripke's argument is equivocal. The proper name 'N' can be understood as having typically wide scope relative to the modal operator in assumption (b), by which the assumption is true. Under the same wide scope interpretation of 'N', however, assumption (a) is false. The resulting dilemma is that Kripke's modal objection to the description theory of reference is either unsound or invalid by virtue of committing the fallacy of equivocation.

The difference between wide and narrow scope readings of a proper name in Loar's criticism can be formalized. It is useful to distinguish the meaning relation '$=_{\mathrm{df}}$' from identity '$=$' in Kripke's argument. The meaning relation '$=_{\mathrm{df}}$' is intensional, and identity '$=$' is extensional; a distinction that turns out to be crucial to Kripke's objection to the description theory of reference. Without trying formally to define these concepts, it is standard to regard '$=_{\mathrm{df}}$' and similar devices as indicating the codesignation of a *definiendum* or *analysandum* in terms of its necessary and sufficient properties specified in the *definiens* or *analysans*. Identity '$=$' is interpreted on the model of familiar interpretations of Leibniz's Law, such that $a = b$ iff every nonintensional property of a is a property of b, and conversely.

The analysis of scope ambiguity in the proposition that N means *the* F but N might not have been or it is logically possible that N is not *the* F can then be formally represented. The definite description '*the* F' is standardly rendered as '$\imath x Fx$'. Throughout, we suppose that the alethic modalities in question regardless of scope are relative to a fixed language L; thus, in all that follows, we silently interpret $\Diamond$ as $\Diamond^L$ and $\Box$ as $\Box^L$. The ambiguity of modal scope for these two interpretations of names in terms of definite descriptions can now be made explicit:

Modal Scope Ambiguity

$N =_{\mathrm{df}} \imath x Fx \wedge \Diamond[N \neq \imath x Fx]$	wide modal scope for N
$\Diamond[N =_{\mathrm{df}} \imath x Fx \wedge N \neq \imath x Fx]$	narrow modal scope for N

The difference between identity of meaning in '$=_{\mathrm{df}}$' and ordinary identity as '$=$' is crucial in preserving the narrow scope interpretation from trivial falsehood as a substitution instance of $\Diamond[p \wedge \neg p]$. The distinction captures the intuitive sense of wide and narrow scope readings for 'N' in Kripke's argument. The wide scope interpretation of proper name 'N' relative to the modal operator allows the meaning of 'N' to be fixed by identifying its reference with that of the definite description in the actual world, while in another logically possible world, N need not be the same definitely described object. This reading is modally wide in the sense that it allows the name to have different application from one logically possible world to another. The narrow scope interpretation of 'N' relative to the possibility operator requires that the meaning of the name be identified with the definite description despite being nonidentical with the definitely described object in at least one logically possible world. This reading is modally narrow in the sense that it restricts the meaning identification and identity denial to an individual limitedly circumscribed logically possible world.[9]

These conventions facilitate the formalization of Loar's critique of Kripke's modal argument. The dilemma in Loar's objection is represented by comparing Kripke's assumption (a) in narrow and wide scope interpretations. The second horn of the dilemma is reflected in the attempt to interpret both assumptions (a) and (b) under a single wide scope interpretation, by which assumption (a) is false and the argument as a whole unsound. An obvious difference in the two forms can then be appreciated.

Formalization of Loar's Criticism

(a^w)	$N =_{\text{df}} \imath x Fx \rightarrow \neg\Diamond[N \neq \imath x Fx]$	(wide scope reading)
(a^n)	$\neg\Diamond[N =_{\text{df}} \imath x Fx \rightarrow N \neq \imath x Fx]$	(narrow scope reading)
(b)	$\Diamond[N \neq \imath x Fx]$	
(c)	$N \neq_{\text{df}} \imath x Fx$	

The argument appears to be a valid inference in *modus tollendo tollens*, but only if the major premise is the wide scope reading in (a^w), which Loar maintains is false. If the narrow scope reading in (a^n) is offered instead, the derivation is invalid.

Loar remarks that Kripke's assumption (b) is "normally read as having the wider scope". The wide scope reading of (a) is logically unsound. If

[9]Quantifier scope distinctions in Russellian definite descripitons are discussed by Stephen Neale, *Facing Facts* (Oxford: The Clarendon Press, 2001), pp. 100–108. See Neale, "Term Limits", *Philosophical Perspectives*, 7, *Language and Logic*, edited by James E. Tomberlin (Atascadero: Ridgeview Publishing Company, 1993), pp. 89–123; especially p. 109–114. Neale is not specifically concerned with ambiguities of modal scope, but his treatment of substitutivity within descripiton contexts is instructive with respect to related questions of Quine's objections to intensional operator scopes. As Neale writes, p. 117: "Certainly a clear account of substitution and the functioning of terms is going to be a prerequisite to clear-headed philosophical inquires that make use of locutions that appear to say things about possibility, time, events and causation."

$N =_{\mathrm{df}} \imath xFx$ *within* the modal context of a single logically possible world, then within that same modal context we can substitute '$\imath xFx$' for 'N'. Doing so on the present assumptions results in the evident logical falsehood: $N =_{\mathrm{df}} \imath xFx \rightarrow \imath xFx \neq \imath xFx$. As Kent Bach in *Thought and Reference* observes, although Loar argues that the wide scope reading of (a) is false, and the mixed narrow scope reading of (a) and wide scope reading of (b) is equivocal, Loar does not consider what happens when both (a) and (b) are given a narrow scope reading. Bach does not address the question of whether or not Kripke's assumption (b) in Loar's reconstruction would be true on a narrow scope interpretation. This is the alternative to be explored in what follows. I will argue that the distinction between the wide or narrow scope reading of Kripke's assumption (b) is irrelevant to the validity of Kripke's modal argument, because the inference goes through with a narrow scope reading of Kripke's assumption (a) on both the wide and narrow scope readings of assumption (b), indifferently.

To substantiate this claim, I begin by revising Loar's reconstruction of Kripke's argument as a *reductio ad absurdum*, rather than as an inference in *modus tollendo tollens*. The central thesis of the description theory of reference is introduced as an hypothesis for *reductio*, and rejected when it is shown to entail a contradiction.

Revised Reconstruction of Kripke's Objection

(1)	$N =_{\mathrm{df}} \imath xFx$	hypothesis
(a^n)	$\neg\Diamond[N =_{\mathrm{df}} \imath xFx \rightarrow N \neq \imath xFx]$	narrow scope reading of (a)
(b^n)	$\Diamond[N \neq \imath xFx]$	narrow scope reading of (b)
(2)	$\Box\neg[N =_{\mathrm{df}} \imath xFx \rightarrow N \neq \imath xFx]$	(a^n) modal duality
(3)	$\Box[N =_{\mathrm{df}} \imath xFx \wedge N = \imath xFx]$	(2) propositional logic
(4)	$\Box[N = \imath xFx]$	(3) weak modal entailment
(c)	$N \neq_{\mathrm{df}} \imath xFx$	(b^n),(4) *reductio ad absurdum*

The narrow scope reading of Kripke's assumption (a), moreover, need not be independently assumed, but can be derived by a similar style of inference from relatively unproblematic propositions, on the assumption of the description theory of reference, offered again for the sake of argument, that $N =_{\mathrm{df}} \imath xFx$. The proof has this logical structure:

Derivation of Narrow Scope Interpretation

(1)	$N =_{\text{df}} \imath xFx$	definition
(2)	$\forall x \Box [x = x]$	necessary self-identity
(3)	$\forall x \neg \Diamond [x \neq x]$	modal duality
(4)	$\neg \Diamond [\imath xFx \neq \imath xFx]$	(3) instantiation
(5)	$\neg \Diamond [N =_{\text{df}} \imath xFx \rightarrow \imath xFx \neq \imath xFx]$	(4) propositional logic
(6)	$\neg \Diamond [N =_{\text{df}} \imath xFx \rightarrow N \neq \imath xFx]$	(1),(5) uniform substitution

In the previous derivation, I referred to Kripke's assumption (b) in the proposed formalization as having a narrow scope reading, which I denoted (b^n). It is difficult to see any basis for modal scope ambiguity in such a truth-functionally simple proposition. The negation in $\Diamond[N \neq \imath xFx]$, construed as $\Diamond\neg[N = \imath xFx]$, in the denial of an identity predication, does not lend itself to distinct modal scopes for 'N'. The only possibility for imposing a wide scope reading on (b), despite Loar's assertion that (b) is 'normally' read as having wide scope, is to interpret (b) intuitively as saying that $\Diamond N \neq \imath xFx]$ *where* $N =_{\text{df}} \imath xFx$, and formalizing the reading with wide modal scope as $N =_{\text{df}} \imath xFx \wedge \Diamond[N \neq \imath xFx]$, by contrast with an imaginable narrow scope reading, of the form, $\Diamond[N =_{\text{df}} \imath xFx \wedge N \neq \imath xFx]$. It is appropriate in this sense to examine the consequences of reinterpreting Kripke's original assumption (b) as having wide as opposed to narrow scope.

The wide scope reading of assumption (b) can first be seen as logically entailing the narrow scope reading in the deductively valid version of Kripke's argument already presented. Here is the reduction:

Reduction of Narrow from Wide Scope Interpretation

(b^w)	$\Diamond[N =_{\text{df}} \imath xFx \wedge N \neq \imath xFx]$	wide scope reading of (b)
(1)	$\Diamond[N =_{\text{df}} \imath xFx] \wedge \Diamond[N \neq \imath xFx]$	(b^w) weak modal $\Diamond$-distribution
(b^n)	$\Diamond[N =_{\text{df}} \imath xFx]$	(1) propositional logic

The modal principle required for the inference of proposition (1) from the wide scope reading of Kripke's assumption (b) in (b^w) is the modal commutative principle for conjunction, $\vdash [\Diamond[p \wedge q] \rightarrow [\Diamond p \wedge \Diamond q]]$, which is a theorem of even the weakest modal systems. Intuitively, if there is a logically possible world in which both propositions p and q are true, then there is sure to be at least one logically possible world in which p is true and at least one logically possible world (at least the same logically possible world in which p is true) in which q is true. (Not conversely, however: $\nvdash [[\Diamond p \wedge \Diamond q] \rightarrow \Diamond[p \wedge q]]$. From the fact that there is a logically possible world in which a proposition p is true and a (perhaps different) logically possible world in which proposition q is true, it does not follow that there is at least one single logically possible world in which both propositions p and q are true.)

5 Naming and Necessary Referential Meaning

If Loar's criticism of Kripke's modal objection is answered in this way, then Kripke's reinterpreted argument appears not only deductively valid but logically sound. By resolving the modal scope ambiguity in Kripke's inference, showing that scope is irrelevant to the interpretation of Kripke's assumption (b), and establishing the logical truth of Kripke's assumption (a), the argument as a whole can no longer be dismissed by force of Loar's dilemma, but must be more seriously reconsidered as a tenable refutation of the description theory of reference. In what follows, I evaluate Kripke's argument reconstructed as above with univocal narrow modal scope in both assumptions, as an inference that, whatever its other advantages or disadvantages, at least avoids Loar's objection.

To begin, it is worthwhile to call attention to another difficulty in Kripke's argument. Why does Kripke not simply distinguish between contingent and necessary reference? The difference is that necessary reference is the univocal designation of the numerically identical object by a proper name in every logically possible world in which the name refers at all. It is thus, in effect, Kripke's concept of proper names as rigid designators. Contingent reference makes no such stipulation about proper names as having univocal transwordly identical reference, but allows that a name might not refer to the very same object in different logically possible worlds. A proper name might then contingently refer equivocally to different individuals in different logically possible worlds, or refer only to the objects in a particular logically possible world, each world requiring its own set of strictly logically proper names. Contingent reference is evidently the modal semantics of choice for the referential meaning of proper names in a counterpart modal logic. Counterpart theory stands opposed to Kripke's commitment to transworld identity. That, however, while it is Kripke's firm intuition, does not represent a monopoly of opinion, and does not provide an argument for preferring necessary referential meaning or the account of proper names as rigid designators over the alternative model of contingent referential meaning.

Where $N = \imath x Fx \wedge \Diamond[N \neq \imath x Fx]$, it might be reasonable for the description theory of proper names to assert that $N =_{\mathrm{df}} \imath x Fx \wedge \neg\Box[N \neq_{\mathrm{df}} \imath x Fx]$. Such a distinction makes it possible to say that while the proper name 'Aristotle' (*in fact*) *means* 'the student of Plato and tutor of Alexander the Great', 'Aristotle' *does not necessarily mean* 'the student of Plato and tutor of Alexander the Great'. A question that naturally arises if the distinction between meaning and necessary meaning is invoked is whether the *necessary* meaning of names can be articulated by *any* definite description. The answer seems to be that necessary meaning can easily be accommodated by requiring that a definite description that gives the necessary meaning of a name do so in terms of the named entity's essential properties only, to the exclusion of any merely accidental logically contingent predications. What obtains in that case is an analysis of necessary meaning as opposed to ordinary or modally unqualified (contingent) meaning. On this interpretation, the necessary meaning equivalence of 'N' and '$\imath xFx$' implies that N necessarily or essentially has property F, $\Box[N =_{\text{df}} \imath xFx] \rightarrow \Box F(N)$. As candidates for necessary definite descriptions, we might consider *haecceities* or individual essences, or intensionally essentially individuating abstractions of the complete set of an object's properties. Kripke allows for the necessity of such definite descriptions when he speculates that genetic parenthood might be essential to the identity of persons, and so presumably also for other nonhuman animals.[10] Additional categories of logically necessary definite descriptions should similarly be available for names of nonliving things, involving essential properties derived from Kripke's concept of natural kinds.[11] The question is whether even such necessary descriptions articulate the *meanings* of names.

The fact that Kripke does not avail himself of a distinction between meaning and necessary meaning is significant. Kripke appears to regard meaning as inherently necessary, so to speak, by definition. Such a theory interprets the 'meaning' of proper names as elliptical for 'necessary meaning', and 'necessary meaning' as redundant. The assumption is explicit in Kripke's statements about proper names as rigid designators, and rigid designators as referring to the same object in every logically possible world in which they refer.[12] The commitment to meaning as logically necessary is also reflected in assumption (a) of Loar's reconstruction of Kripke's argument. Definite descriptions, when they are not limited to an object's essential properties, cannot provide logically necessary meaning because they are satisfied by different individuals in different logically possible worlds. It is the *reference* rather than the descriptive *sense* of a name, the extension rather than the intension, that Kripke seems to regard as logically necessary in the concept of rigid designation. Thus, in another nonactual logically possible world, Speucippus rather than Aristotle is the student of Plato and tutor of Alexander the Great, while in another world it is Xenophon, and so on.

[10] Kripke, *Naming and Necessity*, pp. 48, 110–115, 140–142.
[11] Ibid., especially pp. 116–129, 134–144.
[12] Ibid., pp. 48–54.

Bach, again in *Thought and Reference*, raises a related problem for Kripke's argument, that is similar to and inspired by Loar's. Bach writes:

> In order for [Kripke's assumption] (a) [in Loar's reconstruction] to be true, 'N' must take narrow scope relative to the modal operator. So the question (which Loar did not ask) is whether (b) is true on *its* narrow scope reading. Now how we assess it depends on what, according to the description theory in question, particular instances of 'N' are supposed to mean. We can accept (b) only if we have independent reason to reject that theory; otherwise, we are just begging the question.[13]

Bach then considers a revision of Kripke's argument, recast in terms of Bach's Nominal Description Theory (NDT) of reference. Bach's NDT says in effect that "a name 'N' makes the same contribution ['to the meanings of the sentences in which it occurs'] as the metalinguistic description 'the bearer of "N"'".[14] The NDT reformulation of Kripke's modal objection to the description theory of reference is given as:

Bach's Analysis of Kripke's Objection

(a_n) If 'N' meant 'the bearer of "N"', then 'N might not have been the bearer of "N"' would be true.

But

(b_n) 'N might not have been the bearer of "N"' is true.

Therefore,

(c_n) 'N' does not mean 'the bearer of "N"'.[15]

Bach claims that this is the strongest version of Kripke's argument, but that it does not refute NDT. The reason is that NDT makes a specifically semantic assertion about the meaning of name 'N', whereas (b_n) in the revised inference, according to Bach, even on the narrow reading of its modal scope, is not about the semantics of naming, but at most about the pragmatics of how names are or might actually be used. Thus, for Bach, in defense of the purely semantic thesis of his NDT, the meaning of name 'N' remains constant in every logically possible world as 'the bearer of name "N"', although, as (b_n) states, language users might have decided not to use

[13]Kent Bach, *Thought and Reference* (Oxford: The Clarendon Press, 1987), p. 150.

[14]Ibid., p. 34. See also Bach, "Georgione was So-Called Because of his Name", *Philosophical Perspectioves*, 16, *Language and Mind*, edited by James E. Tomberlin (Malden: Blackwell Publishing Co., 2002), pp. 73–103.

[15]Bach, *Thought and Reference*, pp. 150–151.

'N' but some other name instead to refer to whatever object they happen to have conventionally named by 'N'.[16]

Without entering into the merits of Bach's NDT as opposed to Russell's or Searle's description theory of reference, I find Kripke's unamended version of the modal objection perfectly capable of withstanding Loar's criticism just as it stands, provided that a narrow scope reading of both assumptions (a) and (b) is given to avoid modal scope equivocation in the interpretation of proper names whose meaning is hypothetically identified with an ordinary (nonessentialist) definite description. The question is whether Kripke's modal argument successfully defeats the description theory of reference, or whether there is a further objection other than Loar's scope ambiguity problem that constitutes a potentially more damaging criticism of Kripke's modal argument against the description theory of reference. In this light, Bach capitulates too easily to Loar's assault on Kripke, and should not even need to consider whether a strengthened version of Kripke's modal argument could threaten NDT. Bach again does not consider the advantages of an essentialist description theory, which is different still from both the ordinary description theory on the one hand, and his NDT.

6 Rigid Designation versus Descriptivism

It is important to see that Kripke's modal objection proceeds from the assumption that the *meaning* construed specifically as the *sense* of proper names like 'Aristotle' or 'N' according to the description theory that he wants to refute is always logically necessary. If it is reasonable to distinguish between contingent meaning and necessary meaning, and to implement the

[16]See ibid., p. 151: "The truth of (b_n) cannot be ruled out simply on the grounds that with narrow scope we first consider a counterfactual situation and then determine reference with respect to that situation. For it might be that where names are involved, the effect of narrow scope is the same as that of wide scope, that is, that scope makes no difference. Along these lines it could be argued against NDT and in support of (b_n) that NDT mistakenly identifies the referent of 'N' *with respect to* a counterfactual situation with the bearer of 'N' *in* that situation. Even though 'N' takes narrow scope rather than wide, its referent relative to that situation is the actual N (this is, of course, just what RDT [Rigid Designator Theory] says). But the referent of 'the bearer of "N"' relative to that siutation is not the actual N. So (b_n) does not mean 'the bearer of "N"'. / However plausible this defence of (b_n) may be, it does not have adverse consequences for NDT as a semantic thesis about sentences containing names. To the extent that the referent of a name, even when it has narrow scope, is constant across counterfactual situations, this is not the semantic fact that RDT says it is. I contend that it really describes merely a pragmatic phenomenon involving *uses* of names, and that it seems significant semantically, as in the context of Kripke's appeal to intuition, only because of a certain method he adotps for individuating uses of names. I will argue that this seemingly innocuous method has no semantic basis." Bach introduces his distinction between semantics and pragmatics on pp. 4–6, 61–69, 82–88.

distinction in the description theory of reference by definite descriptions respectively involving nonessential versus essential predications, then Kripke's argument is unsound unless the logical necessity of meaning as either sense or reference can be satisfactorily upheld.[17]

For Kripke's modal objection to work, it must be shown that it is part of the meaning of 'meaning' that the properly understood meaning of names is logically necessary. If this assumption is false, then there is no justification for not invoking the distinction between contingent and necessary meaning to disprove Kripke's crucial assumption (a). To maintain on the narrow modal scope reading of assumption (a) that it is logically impossible for 'N' to *mean* '*the F*', but for N not to *be* the F, $\neg\Diamond[N =_{\mathrm{df}} \imath xFx \rightarrow N \neq \imath xFx]$, requires a concept of necessary meaning of the *reference* if not the *sense* of proper names. If not, then there is no logical obstacle to a name's having a definite description as its meaning even when the object named is not identical to a definitely described object.

It is hard to imagine the sort of argument that could be given to uphold the meaning of proper names as logically necessary. Without such a rationale, there may be no alternative to the distinction between logically contingent and logically necessary meaning. If such an argument were forthcoming, it would play directly into the second horn of a dilemma. Kripke can only prove that the meaning of names is logically necessary by arguing that definite descriptions contingently true of named objects are not or do not constitute the meaning of an object's name. Kripke, in that case, seems to be involved in a vicious circularity, in which he needs to prove that the (sense) meanings of names are not definite descriptions, in order to prove that the (reference) meanings of names are logically necessary, in order to prove that the (sense) meanings of names are not definite descriptions.

[17]Scott Soames in *Beyond Rigidity: The Unfinished Semantic Agenda of Naming and Necessity* (Oxford: Oxford University Press, 2002), pp. 24–50, argues on intuitive grounds for the complementary conclusion that descriptivist efforts to overcome Kripke's modal argument by univocal appeal to wide-scope readings of the relevant modal operators are unsuccessful. See also, Soames, "The Modal Argument: Wide Scope and Rigidified Descriptions", *Noûs*, 32, 1998, pp. 1–22. For an effort to defend the description theory by appeal to a wide-scope reading of Kripke's modal operators, David Sosa, "Rigidity in the Scope of Russell's Theory", *Noûs*, 35, 2001, pp. 1–38; with a detailed criticism of Soames in the 'Appendix', pp. 24–34. See Jaakko Hintikka and Gabriel Sandu, "The Fallacies of the New Theory of Reference", in *Paradigms for Language Theory and Other Essays* (Dordrecht–London–Boston: Kluwer Academic Publishers, 1997) *Jaakko Hintikka Selected Papers*, vol. 4, pp. 175–218. The authors defend a version of descriptivism within the framework of Hintikka's Independence Friendly (IF) logic, arguing that 'direct reference' by-passing all descriptive content in the reference of proper names is not a matter of the *de dicto–de re* distinction, but of 'rule-ordering', which is to say of combined quantifier and modal scopes. See also Delia Graff, "Descriptions as Predicates", *Philosophical Studies*, 102, 2001, pp. 1–42.

If Kripke resorts to the unargued assumption that the referential meanings of names are logically necessary as somehow intuitively obvious or the like, then the conclusion he later wants to derive, that proper names are rigid designators that denote the identical entity in every logically possible world in which they refer at all, is caught up in a similar circularity. The problem is that Kripke needs to show that names are rigid designators because the meanings of names are not explained by the description theory of reference, and that the description theory of reference does not explain the meaning of names because of the necessity of meaning. A commitment to this thesis is clearly tantamount to the interpretation of proper names as logically rigid designators. In asserting assumption (a) (in Loar's reconstruction), in other words, Kripke simply announces his prior acceptance of the analysis of names as rigid designators.[18]

The overall structure of Kripke's reasoning in the lectures is that names must be understood as rigid designators because their meanings cannot be adequately interpreted in the Frege–Russell mode as disguised definite descriptions. Indeed, when the distinction between contingent and necessary meaning is considered, the drift of Kripke's argument is revealed as a question-begging inference, in which it is tacitly assumed from the outset that names are rigid designators. We must assume that names are rigid designators in order to accept Kripke's premise that if a name 'N' were used to mean '*the* F', then 'N is *the* F', according to the description theory of reference, is logically necessary, or that 'N might not have been the F' would be false, contrary to the logically contingent alethic modality of referentially nonrigid definite descriptions.

[18]Kripke claims that all names are rigid, but does not maintain that all descriptions are nonrigid. An exception he mentions is the description of the number π as 'the ratio of the circumference to the diameter of a circle' (p. 60). He nevertheless resists the suggestion that the meaning of 'π' is the above description. He holds that a difference in the modality of a name and description is sufficient to prove that the description does not contstitute the meaning of the name, but he further holds that their sharing in the same modality is not sufficient for the description to constitute the name's meaning. See Richard L. Mendelsohn, "Rigid Designation and Informative Identity Sentences", *Midwest Studies in Philosophy*, 4 (Minneapolis: University of Minnesota Press, 1979), pp. 307–320. Melvin Fitting and Mendelsohn, in *First-Order Modal Logic*, (Dordrecht–Boston: Kluwer Academic Publishers, 1998), p. 217, define rigid designation as requiring (Proposition 10.2.4) that: "a singular term c is rigid if all instances of the following are valid $\langle\lambda x.\Box\Phi(x)\rangle(c) \equiv \Box\langle\lambda x.\Phi(x)\rangle(c)$. That is, a rigid designator is a designator for which scope does not matter." But compare Baruch A. Brody, "Kripke on Proper Names", *Midwest Studies in Philosophy*, 2 (Mineapolis: University of Minnesota Press, 1977), p. 68: "...in light of his own account of rigid designation, it is unclear what rigid designation can mean over and above the description's lying outside the scope of the model [*sic*, 'modal'] operator (when that happens, after all, we will be referring, in talking about a possible world or a counterfactual situation, to the same object that we would be referring to when we are talking about the actual world)."

The scope ambiguity for names relative to modal operators to which Loar in his counter-objection refers provides another ground for disputing Kripke's modal objection to the description theory of reference. The same point can be made by means of the *de dicto–de re* distinction, where the meaning of name 'N' is understood *de re* rather than *de dicto* as '*the* F' (as the *reference* of 'N'). It is only on the basis of such a wide modal scope or *de re* interpretation that it is intelligible for Loar to entertain the substitution of 'the F' for 'N' in assumption (a) of his reconstruction of Kripke's argument, in trying to prove that on the wide scope reading the proposition is incorrect. This method of discrediting Kripke's assumption (a) nevertheless seems irrelevant. The point of requiring that names and descriptions have narrow rather than wide scope for names with *de dicto* rather than *de re* illocutionary force is rather to advance a conditional claim about the modality of statements of meaning expressed as a relation between names and definite descriptions specifically *as distinct terms.*[19]

The problem torpedoes Loar's criticism of Kripke's argument by limiting its disproof of assumption (a) to an irrelevant wide scope or *de re* interepretation of the meaning of names and descriptions as *different* terms of putatively identical meaning. To say as Loar does above, that "Names are *normally* read as having wider scope than modal operators...", may be true, but does not imply anything definitive about the particular reading required for the interpretation of names in Loar's reconstruction of Kripke's argument, where their modal scope might be assumed to be relatively abnormal. Loar does not leave things hanging here, of course. He argues that the wide scope reading seems to be presupposed by the truth of Kripke's premise (b), when he adds: "...that is why premise (b) is true for the relevant Fs." Yet I doubt that the wide-scope reading of 'N' is presupposed by the truth of Kripke's assumption (b). Suppose, on the contrary, that 'N' is interpreted as having narrow scope. It remains true even in that case that N *might not* have been the F, provided that the narrow-scope reading of the name does not automatically entail that the meaning of 'N' is logically necessary. Loar's observations seem finally to overlook the subjunctive mood of Kripke's assumption (a), which as a conditional need not be false on the same wide scope reading of 'N' as Loar attributes to its occurrence in Kripke's assumption (b). Assumption (a) states only *subjunctively* that *if* (*per impossibile*) 'N' *were* used to mean '*the* F', then (*per impossibile*) 'N might not have been the F' *would* be false. Embedded in the subjunctive context, the sentence 'N might not have been the F' need not be inter-

[19] Bach, *Thought and Reference*, pp. 60, 62–63 and *passim.*

preted with narrow scope in order to sustain Kripke's modal objection to the description theory.[20]

There is, moreover, another solution to the problem Kripke raises. Why not world-index the properties attributed to an object in a much-expanded definite description? Thus, if Aristotle was the student of Plato and teacher of Alexander the Great in the actual world $w_@$, but the student of Gorgias and teacher of Glaucon in an alternative logically possible world w_1, or not a philosopher or sophist at all, but a shepherd or soldier in distinct worlds w_2 and w_3, then, instead of saying simply that the referential meaning of 'Aristotle' is the student of Plato and teacher of Alexander the Great, we can avoid Kripke's modal objection altogether by specifying more precisely instead that the referent of the proper name 'Aristotle' is the student of Plato and teacher of Alexander the Great in world $w_@$ *and* the student of Gorgias and teacher of Glaucon in world w_1 *and* the sheperd of Mount Lycabettus in world w_2 *and* the soldier who distinguished himself at the battle of X in world w_3 *and* ...etc., etc. If we pack into the description by which a proper name is analyzed all of the information Kripke requires in order to do justice to the counterfactual referential uses we may wish to make of the name, then the modal objection is met on its own terms within the framework of the description theory and there is no need to go beyond the theory in order to account for the possibility of naming the same things in other logically possible worlds where they do not happen to possess the same properties as they do in the actual world. We must adopt much the same method in order to avoid trivial objections to the indiscernibility of identicals when we consider the same objects whose properties change over time. We do not say that Milo is simply bald and not bald, but that Milo

[20]Loar, "The Semantics of Singular Terms", p. 373: "Kripke anticipates that such a point [about the scope of the name] might be made, and he replies [quoting the final sentence of the citation from Kripke's *Naming and Necessity* with which this essay begins] ...Now we would indeed say that, and we would then be giving 'Aristotle' wider scope than the quantifier over counterfactual situations. And if we substitute 'the pre-eminent ancient philosopher called "Aristotle"' and gave *it* widest scope we would be entitled to assert the result as well. Ordinary particulars are the referents of our utterances always by virtue of their having contingent individuating properties which are intrinsic to what we mean." Loar nevertheless does not consider the possibilities of avoiding equivocation in Kripke's argument by giving both occurrences of 'N' narrow modal scope. Kripke explicitly tries to forestall counter-objections involving appeal to distinctions of scope. In *Naming and Necessity*, immediately after presenting the modal objection to interpreting the meaning of the name 'Aristotle' on a description theory of reference, pp. 60–61, he writes: "This is not a distinction of scope, as happens sometimes in the case of descriptions, where someone might say that the man who taught Alexander might not have taught Alexander; though it could not have been true that: the man who taught Alexander didn't teach Alexander. This is Russell's distinction of scope ...It seems to me clear that this is not the case here."

is bald at time t_1 and not bald at time t_2. Why not, then, avail ourselves of the comparable device for specifying the worlds rather than the times within a world at which an object has a given property in a more complete description of its properties in order to avoid Kripke's modal objection to descriptivism?[21]

7 Conclusion: What's Wrong With This Picture?

Kripke attacks the description theory as the leading analysis of the reference of proper names in its day. His objection resiliently resists Loar's counter-criticism, although Kripke seems to misstate key aspects of the problem of modal scope in his efforts to neutralize scope ambiguity issues in the modal objection. Loar's objection is more competently met by Bach's recommendation of recognizing differences in modal scope both in symbolic and colloquial languages while sustaining a univocal narrow scope reading in the assumptions and conclusion of Kripke's argument.

Kripke's alternative picture of the referential meaning of proper names is nevertheless flawed. What is wrong with the picture Kripke offers stems from two problems in the modal objection. First, there is a circularity in the reasoning whereby Kripke proposes to introduce the concept of rigid designators as the best remedy for the limitations of the description theory of proper names in the wake of the modal objection, when in fact the modal objection presupposes that proper names are rigid designators and cannot be advanced unless that assumption is made. Second, and perhaps more importantly, Kripke's modal objection is entirely avoidable by a more pedestrian solution than the concept of rigid designation by expanding the descriptions that unpack the sense of a proper name and determine its referent not only in the actual world but by indexing its properties as they obtain in every logically possible world.

What now if someone responds to the circularity objection to Kripke's modal argument by saying that Kripke is merely tapped into a common set of pretheoretical intuitions that support both the modal objection and the theory of proper names as rigid designators? There does not immediately appear to be anything necessarily amiss in relying on the same gut feeling

[21] A similar but still significantly different objection is considered by Michael Nelson in "Descriptivism Defended", *Noûs*, 36, 2002, pp. 409: "One classic response to this argument is the Rigidified Descriptivist's responses. *Rigidified Descriptivism* (RD) is the thesis that for every proper name n there is some definite description $\ulcorner$the actual $F\urcorner$ that gives the content of n. The sentence $\ulcorner n$ is $G\urcorner$ does *not* differ in modal profile from $\ulcorner$The actual F is a $G\urcorner$. For any context c and world of evaluation w, $\ulcorner$ the actual $F\urcorner$ designates relative to c and wrt w the entity that is designated by the $\ulcorner F\urcorner$ relative to c and wrt the world of c, if it designates anything. So [the assumption that the modal profile of $\ulcorner n$ is a $G\urcorner$ differs from the modal profile of $\ulcorner$The F is a $G\urcorner$] is false when restricted to rigidified definite descriptions. Thus the modal argument goes flaccid against RD."

that proper names must be able to function unequivocally, in effect as rigid designators, in the actual world and when considering the same objects in counterfactual circumstances. This, unfortunately, is precisely where the interesting questions get begged in trying to criticize descriptivism. We find at the beginning such intuitions colliding with those of counterpart modal logicians as to whether it is the numerically identical individuals we are projecting into counterfactual worlds or instead their close counterparts. Kripke is unmistakably on the transworld strict identity side of the dispute in the semantics of counterfactuals, so it is worthwhile to see to what extent we can preserve Kripke's insights in this regard while still questioning the impact of his modal objection to the description theory. The problem then is whether there need be any vicious circularity in Kripke's presupposing that proper names function as rigid designators in order to advance the modal objection. Perhaps he depends on his intuitions about proper names designating the same individuals in every logically possible world despite failing to satisfy the same descriptions, and then later declares that the concept of proper names to which he had appealed all along can afterward be characterized terminologically as rigid designators.

That does not sound so bad. It is only a matter, then, in the end, of what is essential to Kripke's argument as opposed to what may be only accidental to its presentation. Let us accordingly try to assess what is really at stake and how much damage, if any, is done by the fact that Kripke in his modal objection to the rival description theory seems to presuppose that proper names are rigid designators. We are confronted in that case with an argument in which the description cannot be supposed to fail on grounds that are independent of the account by which it is replaced. Kripke's alternative picture appears gratifying tailor-made to supply the defect that led to the description theory's downfall, which is not such an unusual situation when theories are refuted and supplanted by competitors. Arguably, on the contrary, what we find in Kripke's modal objection is a very different dynamic in which the description theory is made to appear defective when challenged by a putatively opposing account of the reference of proper names that requires names to name the same thing in every logically possible world in which they name anything existent at all. Kripke creates the impression that the description theorists had just shortsightedly lost track of the fact that proper names need to function referentially not only with respect to the actual world, but with respect to nonactual worlds.

There is consequently much more presupposed by Kripke's modal objection than merely the assumption that proper names designate the same objects in nonactual logically possible worlds as they do in the actual world. We can readily imagine a descriptivist admitting that Aristotle might not have been Plato's student or Alexander the Great's tutor, or as failing to have any of the properties included in his associated definite description. Such allowances need nevertheless not make any dent in descriptivism unless we also assume, as Kripke apparently does, that it is *true* that Aristotle in the relevant counterfactual worlds has properties other than those he has in the actual world. Why, however, should a decriptivist who has not already accepted Kripke's picture of proper names as rigid designators admit anything of the kind? Why should the descriptivist not say that only existent entities (truly) have properties, and offer a different account of the acknowledged fact that Aristotle, student of Plato and tutor of Alexander the Great, might not have been or done these things? Kripke in mounting the modal objection to descriptivism assumes that there is no other way to explain the truth of counterfactuals than to posit the same objects as those that inhabit the actual world residing also in nonactual merely logically possible worlds where they truly have the properties that we say relative to their dispositions in the actual world they might have had instead.

The descriptivist is certainly not committed to all of this. Without such agreement, however, the description is untouched or anyway not decisively refuted by Kripke's modal objection. It seems to follow, as a result, that Kripke in presupposing the package of propositions that goes along with assuming that proper names are rigid designators does in fact beg the question against the description theory of the referential meaning of proper names in the usual sense of what is intended as viciously circular reasoning. The argument has no leverage against anyone who has not already bought into the concept of names as rigid designators, and yet it is the defeat of descriptivism that is supposed to incline open-minded semantic theorists away from descriptivism and toward Kripke's alternative picture of proper names as rigid designators whose referential meaning is socially transmitted along the links of a causal-historical chain from speaker to speaker. Otherwise, there is nothing more dramatic in what purports to be an argument against the description theory than a statement of an opposing set of intuitions about the function of proper names, with no compelling reason to turn away from the received description theory in favor of the nondescriptivist picture of the reference of proper names represented by Kripke's concept of names as rigid designators. We may or may not prefer Kripke's alternative picture of names, but we look in vain to Kripke's modal objection as providing a solid reason for rejecting descriptivism or justification for casting about for another way of thinking about the referential meaning of proper names that will compensate for demonstrated limitations or explanatory lacunae in the description theory.[22]

---

[22]An abbreviated version of this paper was presented at the Sixteenth International

Dedication

I am enormously pleased to dedicate this paper to Dov Gabbay on the occasion of his 60th birthday. The quality of his extraordinary productivity in many fields of logic, computing theory and its applications in linguistics, and semantic philosophy, including modal logic and its model set theoretical semantics, have inspired so many of his admirers, including myself, to venture into areas of philosophical analysis with the tools of formal symbolic logic. Congratulations, Dov, and many happy returns and best wishes for the continuation of all your research projects.

Symposium, LOGICA 2002, Zahradky Castle, Czech Republic, June 18–21, 2002, and published under the title, "Kripke on Identity and the Description Theory of Reference", *The Logica Yearbook 2002*, edited by Timothy Childers and Ondrej Majer (Prague: Filosofia, 2003) (Institute of Philosophy, Academy of Sciences of the Czech Republic), pp. 109–116.

Infinitary Normalization

JAN WILLEM KLOP AND ROEL DE VRIJER

Dedicated to Dov Gabbay, in celebration of his 60th anniversary, with fond memories of the first author of numerous meetings around 1990 at Coseners House in Abingdon, England, where Dov (together with Tom and Samson) was getting the Handbook of Logic in Computer Science in shape.

In infinitary orthogonal first-order term rewriting the properties confluence (CR), Uniqueness of Normal forms (UN), Parallel Moves Lemma (PML) have been generalized to their infinitary versions CR^∞, UN^∞, PML^∞, and so on. Several relations between these properties have been established in the literature.

Generalization of the termination properties, Strong Normalization (SN) and Weak Normalization (WN) to SN^∞ and WN^∞ is less straightforward. We present and explain the definitions of these infinitary normalization notions, and establish that as a global property of orthogonal TRSs they coincide, so at that level there is just one notion of infinitary normalization. Locally, at the level of individual terms, the notions are still different. In the setting of orthogonal term rewriting we also provide an elementary proof of UN^∞, the infinitary Unique Normal form property.

1 Outline

We work in the framework of infinitary first-order term rewriting, dealing with transfinite rewrite sequences that may converge to a limit and the fundamental notion of an infinite normal form. Infinite normal forms can e.g. be seen to arise quite naturally as the limits of infinite processes generating "streams" of natural numbers, for example the primes or the fibonacci numbers.

After recapitulating some of the basic definitions and facts, we consider the question of how to generalize the notions of weak normalization (WN) and strong normalization (SN) to the infinite setting, obtaining WN^∞ and SN^∞, respectively. It turns out, somewhat surprisingly, that when applied to orthogonal term rewriting systems (OTRSs), these notions coincide. Moreover, although CR^∞ is no longer a general property of infinitary orthogonal rewriting, we still have that—finite or infinite—normal forms are unique (UN^∞). The proofs of these facts use classical techniques of orthogonal term rewriting, such as the parallel moves lemma, generalized to the infinitary setting and an analysis of infinitary head normalization.

The notion of SN^∞ was first introduced in [Kennaway, 1992]. A slightly different definition than ours is given there, in topological terms, in the context of abstract rewriting. Given our concrete approach to the notion of infinitary reduction the two definitions appear to amount to the same. Also UN^∞ for orthogonal TRSs is not new here; in [Terese, 2003] it is shown as a consequence of an analysis of meaningless terms and Böhm trees in a broader setting, including infinitary lambda calculus. For first-order term rewriting systems a simpler proof of UN^∞ can be given, making use of the infinitary parallel moves lemma. Note that PML^∞ fails for the lambda calculus. The main technical contribution of the present paper is that WN^∞ and SN^∞ coincide as properties of infinitary orthogonal TRSs.

2 Introduction to infinitary term rewriting

We are concerned with the framework of first-order term rewriting and we assume familiarity with that area. For general background reading on term rewriting the reader may consult any standard text, for example [Baader and Nipkow, 1998], [Terese, 2003], [Klop, 1992], [Dershowitz and Jouannaud, 1990] and for more specific information on infinitary rewriting [Kennaway *et al.*, 1995], [Klop and de Vrijer, 1991] or the chapter Infinitary Rewriting by Kennaway and de Vries in [Terese, 2003]. Some of the basic notions will be recapitulated when and where needed, and the same for notation. We will also assume some familiarity with ordinal numbers.

2.1 Finitary and infinitary perspectives on term rewriting

One aspect of term rewriting is that it can be used to model computations with normal forms as the intended outcomes.

As a simple example consider the TRS $\mathcal{N}$ specifying the natural numbers with zero, successor and addition, with the reduction rules of Figure 1, which go back to [Dedekind, 1888].

$$\begin{array}{lcl} A(x,0) & \to & x \\ A(x,S(y)) & \to & S(A(x,y)) \end{array}$$

Figure 1. Dedekind's rules

Closed terms in this TRS represent arithmetical expressions involving the addition operator and the outcomes are the closed normal forms 0, $S(0)$, $S(S(0))$, etcetera, the *numerals*. Thus we have e.g. the reduction

$$\begin{aligned} A(A(S(S(0)),0),S(0)) &\rightarrow S(A(A(S(S(0)),0),0)) \\ &\rightarrow S(A(S(S(0)),0)) \\ &\rightarrow S(S(S(0))) \end{aligned}$$

modeling a computation of $(2 + 0) + 1$ with outcome 3. This example illustrates the interest in terminating reductions with finite terms (normal forms) as outcomes.

There is also an infinitary aspect to term rewriting. Again consider the TRS $\mathcal{N}$, its signature now expanded with a binary symbol P and a unary E and with the additional reduction rule $E(x) \rightarrow P(x, E(S(S(x))))$. Think of P as pairing, or a list-forming `cons` operator, which we will also denote by the infix symbol : for better readability. Beginning with the term $E(0)$ we now have the following infinite reduction

$$E(0) \rightarrow 0 : E(S(S(0))) \rightarrow 0 : S(S(0)) : E(S(S(S(S(0))))) \rightarrow \cdots$$

The consecutive terms in this reduction appear to converge in the limit to an infinite "term" representing the *stream* of even natural numbers, namely

$$0 : S(S(0)) : S(S(S(S(0)))) : S(S(S(S(S(S(0)))))) : \ldots$$

Likewise the stream of all naturals is generated from the term $N(0)$ using a unary symbol N with the reduction rule $N(x) \rightarrow x : N(S(x))$ and for example the constant stream of zeros $0 : 0 : 0 : 0 : \ldots$ can be obtained as the limit of an infinite reduction

$$Z \rightarrow 0 : Z \rightarrow 0 : 0 : Z \rightarrow 0 : 0 : 0 : Z \rightarrow \cdots$$

starting from a constant Z and using the single reduction rule $Z \rightarrow 0 : Z$

The finitary and infinitary perspectives on term rewriting give rise to a difference in emphasis on properties of term rewriting systems. If the finite normal forms are considered as the outcomes of computations, strong normalization is an attractive property: regardless of your reduction strategy, an outcome will always be found. The objective of generating streams as outcomes, on the other hand, is incompatible with strong normalization.

2.2 Convergence and limits defined

The structure of the *infinite terms* that arise as limits is most clearly displayed by representing them as infinite term trees. In Figure 2 infinite term trees are drawn for the streams of zeros and naturals, respectively generated by the terms Z and $N(0)$. Note that the function symbols Z and N themselves do not occur anymore in the limit terms, although they did occur in all finite approximations.

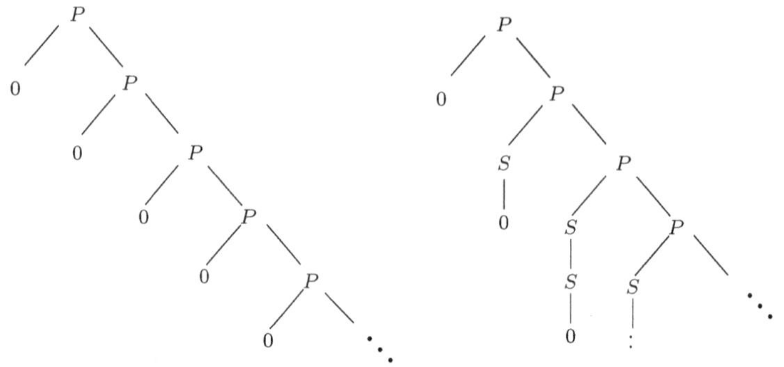

Figure 2. Infinite term trees

The question arises what the formal status of infinite terms is. We just offer a quick and informal answer. For a more extensive treatment we refer to [Terese, 2003]. First note that each finite term corresponds one-one to its *term tree*: a finite set of labeled positions satisfying the following three requirements:

1. the set of positions is closed under prefixes

2. each position is labeled by a function symbol or a variable

3. the arity of the function symbol at a position equals the number of outgoing edges at that position

So the constants (function symbols of arity 0) and the variables are at the endpoints of the term tree. Now we take an *infinite term tree* to be just a possibly infinite set of labeled positions, satisfying the same requirements 1-3. So in particular in an infinite term each position has a finite distance to the root.

It is sometimes convenient to use recursion equations to characterize infinite term trees. The example of the stream of zeros is then given by the equation $t = P(0, t)$, or shorter $t = 0 : t$, with the obvious semantics.[1]

Having extended our domain of rewriting to the infinite terms, we keep the notion of rewriting itself as it was: rewrite rules are pairs of finite

[1]Of course one thinks here of a least fixed-point semantics. We note that it could in fact be implemented by the very techniques of infinitary term rewriting that this paper is about.

terms.[2] However, now also infinite terms can be rewritten. A redex $C[l^\sigma]$ is still identified as a pattern occurring at a finite position with prefix C, but now the substitution σ may involve also infinite terms. So for example with the reduction rule $I(x) \to x$, the infinite term $I(I(I(I\ldots)))$ characterized by the equation $t = I(t)$ has a redex at each of its positions. Note that in this particular case all these redexes and their reducts happen to be the same term, all identical to the original term itself, which we will henceforth denote by I^ω.

In order to explain the notion of convergence we use, we consider again the example of the stream of the naturals. We have the infinite reduction

$$N(0) \to 0 : N(S(0)) \to 0 : S(0) : N(S(S(0))) \to \cdots$$

In Figure 3 we get a clear picture by drawing the term trees again.

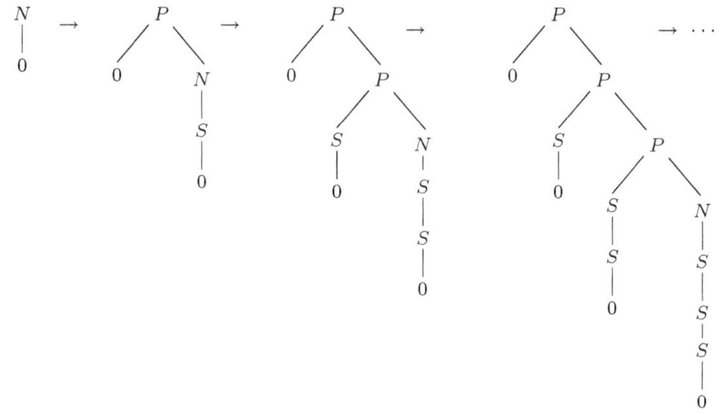

Figure 3. Converging infinite reduction

Looking at the terms as they evolve during this reduction, two things stand out. First, after each step a larger and larger prefix remains fixed throughout the rest of the reduction and, secondly, the depth (i.e. the distance to the root) of the contracted redexes increases. In accordance with these observations we formulate the following two conditions:

1. For any depth n the prefix of positions up to depth n eventually gets fixed throughout the rest of the reduction.

[2]One may also consider rewrite rules involving infinite terms. Most of the existing theory of infinitary rewriting extends to rewrite rules with infinite righthand side. For rules with infinite lefthand sides the situation is less clear.

2. The depth (i.e. the distance to the root) of the contracted redexes eventually grows beyond any finite value.

Note that condition 1 alone already would be sufficient to guarantee the existence of a well-defined limit, any position has a finite depth and hence will eventually be fixated. In the literature this is called *weak convergence* and it was used in the ground-breaking paper on infinitary term rewriting [Dershowitz *et al.*, 1991]. However, later developments have made it clear that it is better to require the stronger property 2, which is easily seen to imply 1. The reason is that with the stronger notion the resulting theory of infinitary term rewriting is much better behaved.[3] For details of this consult the background literature, e.g. [Terese, 2003]. Convergence according to the stronger notion 2 is usually called *strong convergence*, but since we will not be concerned anymore with alternative notions we will also call it *convergence* without more.

An example af a weakly but not strongly convergent reduction sequence can be given for example with the reduction rule $A(x) \to A(B(x)$:

$$A(x) \to A(B(x)) \to A(B(B(x))) \to A(B(B(B(x)))) \to \cdots$$

The infinite term A(B(B(B(. . .)))) could in principle be considered as the limit, but we will not do so because, as all redex contractions occur at the root, the reduction sequence is not strongly convergent.

It is illustrative to contrast this example of non-convergence with a "mirror" example, the TRS with the rule $A \to B(A)$. The reduction

$$A \to B(A) \to B(B(A)) \to B(B(B(A))) \to \cdots$$

is convergent, with as limit the infinite normal form B^ω. To indicate a convergent reduction of length ω we write $A \to^\omega B^\omega$.

2.3 Transfinite reductions

Reduction can proceed beyond the ordinal ω. An easy way to see this is by adding a passive pairing operator to the signature of the last example and considering the convergent reduction

$$P(A, A) \to P(B(A), A) \to P(B(B(A)), A) \to P(B(B(B(A))), A) \to \cdots$$

We have that $P(A, A) \to^\omega P(B^\omega, A)$, but now the limit $P(B^\omega, A)$ is not a normal form, as it contains the redex A. So after the first ω steps another one is possible and so on, and by taking another limit we reach the normal form $P(B^\omega, B^\omega)$, in $\omega + \omega$ steps. Notation: $A \to^{\omega+\omega} P(B^\omega, B^\omega)$.

[3]More specifically, adopting 2 enables us to extend the notion of descendant of a subterm after a reduction to infinite reductions, where the critical point in the definition is of course the limit case.

Now it is not difficult to construct longer reductions, for example the infinite term characterized by $t = P(A, t)$ has a reduction of length ω^2 to $t = P(B^\omega, t)$. As we will see in Example 3 below, convergent transfinite reductions can be constructed of any countable ordinal length.

DEFINITION 1. We sum up the notion of a transfinite reduction ρ of length β. It consists of rewrite steps $t_\alpha \to_{s_\alpha} t_{\alpha+1}$:

$$\rho: \; t_0 \to_{s_0} t_1 \to_{s_1} \; \cdots \; t_\omega \to_{s_\omega} t_{\omega+1} \to_{s_{\omega+1}} \; \cdots$$

or in a more compact notation $\rho = (s_\alpha)_{\alpha<\beta}$. This only makes sense if for each limit ordinal $\lambda < \beta$ the *prefix* $\rho_\lambda = (s_\alpha)_{\alpha<\lambda}$ of ρ is convergent, with limit t_λ. Let d_α be the depth—in the term t_α—of the redex that is contracted in step s_α. Then for each limit ordinal $\lambda < \beta$ we must have that $(d_\alpha)_{\alpha<\lambda}$ tends to infinity.

Figure 4 depicts the course of the redex depths in a convergent reduction of length ω^2.

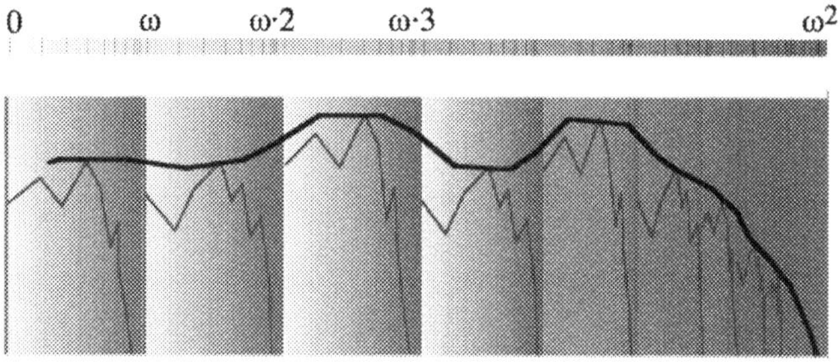

Figure 4. Convergence at each limit ordinal $\leq \omega^2$

Note that in Definition 1 we do not require convergence at the ordinal β, that is, convergence of the whole reduction. We get back to the topic of divergent transfinite reductions in Section 2.5.

2.4 Compression

Continuing the example of the last section, there is also a convergent reduction $P(A, A) \to^\omega P(B^\omega, B^\omega)$, by alternating between performing steps in the left and right argument positions of P:

$$P(A, A) \to P(B(A), A) \to P(B(A), B(A)) \to P(B(B(A)), B(A)) \to \; \cdots$$

The transition from the reduction of length $\omega \cdot 2$ to that of length ω is called *compression*. A key result in the theory of infinitary rewriting is that for

OTRSs compression of transfinite reductions to length ω is always possible. As a matter of fact, left-linearity of the TRS already suffices for having compression.

2.5 Convergence and divergence

For a transfinite reduction ρ of length λ with λ a limit ordinal there are two possibilities: either ρ is convergent or it is not. In the first case there is a limit term t_λ and we have that $\rho : t_0 \twoheadrightarrow^\lambda t_\lambda$. In the second case we say that ρ is *divergent*. Note again that for any reduction of length λ, convergent or divergent, all prefixes must be well-defined, meaning convergence at each limit ordinal $< \lambda$.

Now we look into the question of what constitutes a divergent reduction ρ of length λ. Let ρ again consist of steps s_α. Divergence of ρ means that there exists a finite number n such that for every $\alpha < \lambda$ there exists a $\beta > \alpha$ such that the step s_β has depth $\leq n$. If we then take N to be the smallest such n, we have that infinitely many steps s_β have depth N.

Conversely, if for some N there are infinitely many steps of depth N, then we have divergence. For let $X = \{\alpha \mid d_\alpha = N\}$ be an infinite set. Then we construct an infinite sequence of ordinals $\alpha_1, \alpha_2, \ldots$ such that for all i, $d_{\alpha_i} = N$ by taking α_1 the smallest element of X, α_2 the next smallest and so on. At the limit β of this increasing sequence we do not have convergence of ρ_β. We must then have $\beta = \lambda$, as otherwise a prefix of ρ would be ill-defined, and hence the reduction ρ is divergent.

We proved:

THEOREM 2. *A transfinite reduction is divergent if and only if for some N there are infinitely many steps at depth N.*

An immediate consequence of this theorem is that all convergent transfinite reductions have countable length. Namely for each N there can only be finitely many steps of depth N and a countable union of finite sets is countable.

But we can also directly prove a stronger result, namely that all reductions, no matter whether they are convergent or divergent, must be countable. For assume not. Then there is certainly a reduction $\rho = (s_\alpha)_{\alpha<\omega_1}$, where ω_1 is the first uncountable ordinal. Infinitary pigeon holing yields an $N \in \omega$ such that ρ has infinitely (even uncountably) many steps at depth N. As before we consider the infinite set $X = \{\alpha \mid d_\alpha = N\}$ and construct an infinite increasing sequence $\alpha_1, \alpha_2, \ldots$ of ordinals inside X. At the limit β of this increasing sequence we do not have convergence of the prefix ρ_β, but since β is a countable limit ordinal, $\beta < \omega_1$ and hence ρ is not well-defined.

EXAMPLE 3. It is instructive and also entertaining to play some more with the infinite term I^ω, which, as we already remarked, reduces only to itself and has at each position an identical redex. So, as a rewrite step is determined by its depth, a transfinite reduction $\rho = (s_\alpha)_{\alpha<\lambda}$ of I^ω can be identified with the sequence $(d_\alpha)_{\alpha<\lambda}$, where d_α is the depth of s_α.

The constant sequence $(0, 0, 0, \ldots)$, for example, codes the divergent reduction of length ω consisting of only root steps, whereas in the reduction $(0, 1, 2, \ldots)$ the depth of the steps tends to infinity, hence it converges. An example of a divergent reduction of length $\omega \cdot 2$ is $(0, 1, 2, \ldots\ 0, 0, 0, \ldots)$. We find converging and diverging reductions $(d_\alpha)_{\alpha<\omega^2}$ and $(d'_\alpha)_{\alpha<\omega^2}$ of length ω^2 by taking $d_{\omega \cdot n+m} = n + m$ and $d'_{\omega \cdot n+m} = m$, respectively.

We will now show that (1) the term I^ω admits a convergent reduction of length any countable ordinal λ, and (2) if λ is a limit ordinal also a divergent reduction.

1. Take a countable λ. So there is a bijection $d : \lambda \to \omega$. Then the reduction $(d(\alpha))_{\alpha<\lambda}$ is a convergent reduction of length λ by Theorem 2.

2. We can use 1 to construct also a divergent reduction of length λ, if it is a limit ordinal. For consider a convergent reduction $\rho = (d_\alpha)_{\alpha<\lambda}$ and let $\alpha_1, \alpha_2, \ldots$ be an increasing sequence of ordinals with limit λ. Define $\rho' = (d'_\alpha)_{\alpha<\lambda}$, where $d'_{\alpha_i} = 0$ for all $i \in \omega$ and $d'_\alpha = d_\alpha$ for all other $\alpha < \lambda$. Then one easily sees that ρ'_β still converges at any limit ordinal $\beta < \lambda$, as it differs from ρ_β in at most finitely many places. Hence ρ' is well-defined and it obviously diverges at λ.

3 Normal forms and normalization properties

3.1 WN and SN in the finitary setting

Consider the TRS $\mathcal{T} = \{f(x) \to b,\ a \to f(a)\}$. In $\mathcal{T}$ we have an infinite reduction originating from the term a:

$$a \to f(a) \to f(f(a)) \to f(f(f(a))) \to \cdots$$

but on the other hand any term can be reduced to a normal form, in particular we have $a \twoheadrightarrow b$ via the reduction

$$a \to f(a) \to b$$

This is an example of a TRS that is *weakly normalizing* (WN), every term reduces to a normal form, but not *strongly normalizing* (SN), there are reductions that go on indefinitely, without ever reaching a normal form.[4]

[4]Term rewriting systems that are SN are also called *terminating*.

Of course there is the trivial implication SN $\Longrightarrow$ WN and the above example shows that the converse does not hold. In this respect it is worth mentioning a classic result of [O'Donnell, 1977], specifying a situation where the converse *does* go through. It goes back to [Church, 1941], where it is proved that in the λI-calculus WN and SN are equivalent. A TRS is called *non-erasing* if in each rewrite rule all variables in the left-hand side occur also in the right-hand side.

THEOREM 4 (O'Donnell). *For non-erasing OTRSs we have WN $\Longleftrightarrow$ SN.*

For the sake of completeness we add a recently discovered fact from [Ketema *et al.*, 2005]. Here AC is the property of *acyclicity*, there are no reduction cycles.

THEOREM 5. *For OTRSs we have WN $\Longrightarrow$ AC.*

3.2 Infinite normal forms

A normal form, finite or infinite, is a term from which no rewrite step is possible, that is, a term without redex occurrences. Typical examples of infinite normal forms are the terms representing the streams of zeros and naturals, depicted in Section 2.2. Less standard examples in the TRS $\mathcal{N}$ of addition are the infinite term S^ω and the infinite binary tree labeled with A's, defined by the recursion equation $t = A(t, t)$.

A typical example of an infinite term that is not a normal form is the term I^ω from Section 2.2. As we already indicated this term does not reduce to a normal form either, as it can only reduce to itself. So I^ω is an example of a term that is not WN$^\infty$, where WN$^\infty$ is the property of reducing by a possibly transfinite reduction to an (infinitary) normal form.

3.3 The notion of infinitary strong normalization

Now we want to consider the question what SN$^\infty$ could mean. To keep the analogy with finitary SN it should be something like: no matter how you reduce, if you just keep going, in the end a normal form will always be reached. Naturally it might take any (countable) transfinite number of steps. This analysis leads us to the following provisional definition.[5]

> A term t has the property SN$^\infty$ if any maximal transfinite reduction from t reduces t to a normal form.

[5] In [Kennaway, 1992] a similar phrasing is rejected, apparently due to a subtle difference in perspective. Our transfinite reductions of length α presuppose (strong) convergence at all limit ordinals $\lambda < \alpha$ from the outset, whereas Kennaway also considers weak convergence, only to be eliminated later on. The resulting notions of SN$^\infty$ are the same.

The question that then has to be answered is of course: what are maximal reductions? But that is not difficult. There are just two types of transfinite reductions from t that cannot be prolonged:

1. The reductions that reduce t to a normal form.

2. The reductions that diverge.

Now the first possibility satisfies the provisional criterion, and only the second violates it. Hence it is clear what the definition of SN^∞ should be.

DEFINITION 6. A (finite or infinite) term is SN^∞ if it has no divergent reductions.

The notions of finitary SN and infinitary SN^∞ are independent, as we will point out in 1 and 2 below. Here especially the failure of SN $\Longrightarrow$ SN^∞, although easy to understand, may come as a surprise.

1. SN $\not\Rightarrow$ SN^∞. Consider the two-rule TRS $\mathcal{N}$ for addition from Section 2.1. It is clearly SN. However SN^∞ fails as witnessed by the infinite term recursively defined by $t = A(t, 0)$. We have $t \to t$, yielding a divergent reduction, i.e. $\neg\mathrm{SN}^\infty$.

   For another, even simpler counterexample consider again the one-rule TRS $I(x) \to x$, trivially SN, and the infinite term I^ω that has no normal form.

2. $\mathrm{SN}^\infty \not\Rightarrow$ SN. There is a far from obvious couterexample here: take the fragment of Combinatory Logic (CL) consisting of the terms solely built by application from the combinator S, with the reduction rule $Sxyz \to xz(yz)$. This is an orthogonal TRS which is not SN, as e.g. the term $SSS(SSS)(SSS)$ has an infinite reduction. But it has the property SN^∞, according to [Waldmann, 2000].

3.4 Newman's Lemma

A typical application of SN in the finitary case is in Newman's Lemma: WCR & SN $\Longrightarrow$ CR. However, infinitary Newman Lemma fails, that is, we do not have the implication WCR & SN^∞ $\Longrightarrow$ CR^∞. The following is an easy counterexample. An alternative counterexample can be found in [Kennaway, 1992].

EXAMPLE 7. Consider the TRS R with the three rules:

$$\begin{array}{lcl} C & \to & A(C) \\ C & \to & B \\ A(B) & \to & B \end{array}$$

So R is not orthogonal. All reductions from C are depicted in Figure 5. There are two normal forms, A^ω and B. Hence UN$^\infty$ does not hold and

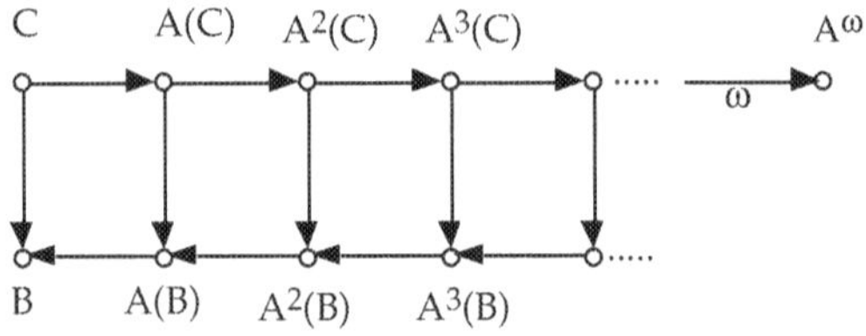

Figure 5. Reduction graph of C

neither does CR$^\infty$. As all relevant terms of R are shown, it is clear that R is CR and also WCR. We also have SN$^\infty$, as one easily sees.

4 Infinitary orthogonal rewriting

It is by now well-known that even for orthogonal TRSs infinitary confluence may fail. This is shown by the following example.

EXAMPLE 8. Consider the TRS with rules

$$\begin{array}{lcl} C & \to & A(B(C)) \\ A(x) & \to & x \\ B(x) & \to & x \end{array}$$

Then CR$^\infty$ fails since C reduces in ω steps to A$^\omega$ and B$^\omega$, infinite terms that both only reduce to themselves, so having no common reduct.

In this example it is essential that there are two collapsing rewrite rules and conditions can be given under which CR$^\infty$ does go through, but we will not pursue this matter here, see [Kennaway *et al.*, 1995]. An important reason for the interest in confluence is that it implies uniqueness of normal forms. Below we will prove that despite the failure of CR$^\infty$, in infinitary orthogonal rewriting we do have UN$^\infty$, uniqueness of normal forms.

4.1 The Parallel Moves Lemma

A *parallel step* consists of the contractions of a possibly infinite set of disjoint redexes. This can be done in any order and will always result in a convergent reduction.

Fundamental is the infinitary parallel moves lemma (PML$^\infty$) for OTRSs in this form:

THEOREM 9. *It is always possible to construct the finite or transfinite reduction diagram of a convergent reduction ρ against a parallel step p. The projection p/ρ is again a parallel step and ρ/p a convergent reduction.*

By way of example we give a more microscopic view of this construction for the case that ρ has length $\omega \cdot 2$. So we have

$$\rho: \ t_0 \rightarrow_{s_0} t_1 \rightarrow_{s_1} \cdots \quad t_\omega \rightarrow_{s_\omega} t_{\omega+1} \rightarrow_{s_{\omega+1}} \cdots \quad t_{\omega \cdot 2}$$

Define $p_0 = p$ and $p_\alpha = p/\rho_\alpha$ (where ρ_α is the prefix of ρ of length α). Note that then also $p_{\alpha+1} = p_\alpha/s_\alpha$.) Finally let $S_\alpha = s_\alpha/p_\alpha$.

We then have locally at each t_α the diagram construction with initial term t_α of the single step $t_\alpha \rightarrow_{s_\alpha} t_{\alpha+1}$ against the parallel step p_α. The bottom and right residual steps are the S_α and $p_{\alpha+1}$. The complete diagram construction consists of collating the local diagrams. It is essential here that at the limits we have convergence. See Figure 6.

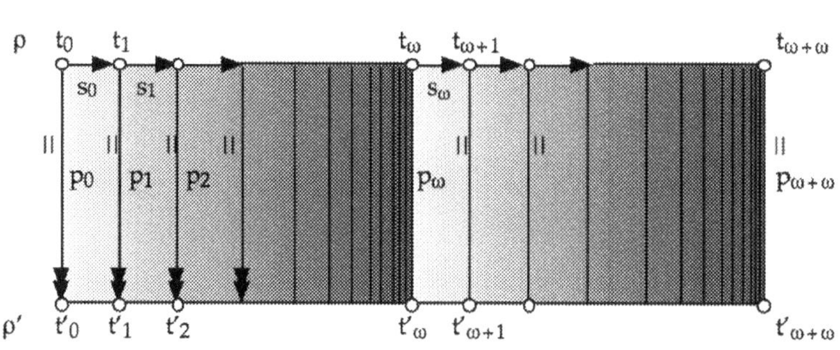

Figure 6. PML$^\infty$ for a transfinite reduction of length $\omega \cdot 2$

To appreciate that PML$^\infty$ is non-trivial we mention that for infinitary lambda calculus (λ^∞), this fundamental lemma fails.

One can make sense also of the projection of a divergent reduction ρ over a convergent reduction σ, given that the projections of ρ's prefixes exist. Let the ρ_α's ($\alpha < \lambda$) be the prefixes of ρ. Then, as the prefixes of a divergent reduction are convergent reductions, we can take ρ/σ as the union of the ρ_α/σ, $\alpha < \lambda$. A priori ρ/σ can be either convergent or divergent.

In fact, already the projections of convergent reductions σ and ρ over each other may be divergent. An example is the reduction diagram in Figure 7. It is obtained from the reductions $C \twoheadrightarrow^\omega A^\omega$ and $C \twoheadrightarrow^\omega B^\omega$ in the counterexample to CR$^\infty$ above. In Figure 7 the steps crossing a light layer are empty steps. The reduction $A^\omega \rightarrow A^\omega \rightarrow \cdots$ in the righthand side is a root reduction, hence divergent; likewise the reduction from B^ω at the bottom.

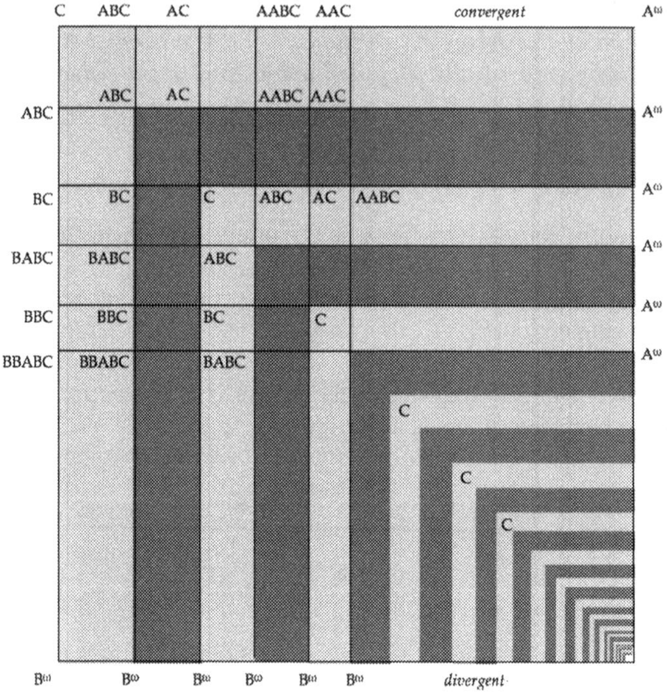

Figure 7. Non-confluent infinite reduction diagram

5 Unique normal forms: UN^∞ for OTRSs

In this section we assume orthogonality for all TRSs and under this restriction we prove uniqueness of infinitary normal forms (UN^∞).

There is the well-known distinction between *root* (or *head*) reduction and *non-root* reduction, also called *internal* reduction. In a root (head) step the root of the term is part of the contracted redex, in an internal step the root is left untouched. We will generalize this notion of internal reduction relative to the root position to internal reduction relative to an arbitrary prefix C. This is called *C-stable reduction* or ι_C-reduction and it leaves all of C untouched.

DEFINITION 10.

1. Let C be a prefix of t. A rewrite step from t is *C-stable* (a ι_C-step) if the contracted redex lies below C. Idem for parallel steps.

2. *C-stable* reduction or *ι_C-reduction* is reduction consisting of only C-stable steps.

3. A term is called *C-stable* if it allows only C-stable reduction.

4. If C is the "full" prefix up to depth n then C-stable reduction, is also called *n-stable* reduction. Accordingly a term allowing only n-stable reduction is called *n-stable.*

So internal or root-stable reduction is the same as 1-stable reduction, or C-stable reduction where C is just the root. A root-stable term is also called a *head normal form.*

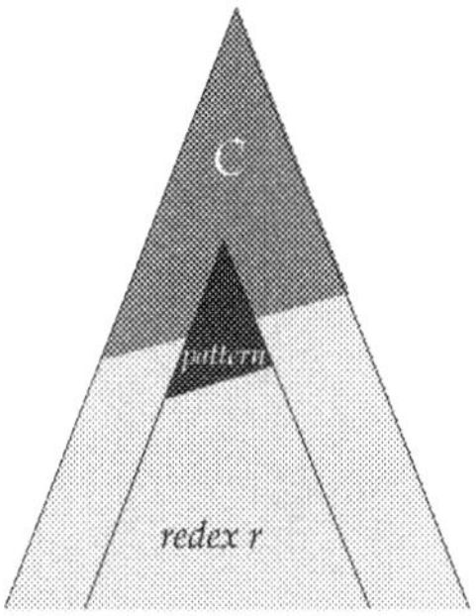

Figure 8. Redex r overlapping with prefix C

LEMMA 11.

1. *If redex r overlaps with C and $t \to s$ by a ι_C-step, then r has a unique residual in s, at the same position and still a redex.*

2. *The same for $\rho : t \twoheadrightarrow s$.*

3. *The same for $\rho : t \to^{\alpha} s$.*

Proof.

1. This is the crux of orthogonal rewriting. Not only C is stable, but also $C \cup \pi(r)$, the union of C with the pattern $\pi(r)$ of r.

2. Repetition of 1.

3. If the prefix $C \cup \pi(r)$ is stable in ρ, it will also be present in the limit (by the very definition of what a limit is).

■

PROPOSITION 12. *If t reduces to an infinite normal form by ι_C-reduction, then no redex in t overlaps with C.*

Proof. By Lemma 11(3). ■

It is easy to see that projection of (parallel) ι_C-steps over each other yields ι_C-steps as residuals.

LEMMA 13. *Projection of a ι_C-reduction over a (parallel) ι_C-step yields a ι_C-reduction again.*

PROPOSITION 14.

1. *If t reduces to an infinite normal form by ι_C-reduction ρ and $t \to s$, then C is a prefix of s and no redex in s overlaps with C.*
2. *If t reduces to an infinite normal form by ι_C-reduction, then t is C-stable.*
3. *If t reduces to an infinite normal form by n-stable reduction, then t is n-stable.*

Proof.

1. No redex in t overlaps with C because of Proposition 12. Hence $t \to_p s$ is a ι_C-step. Then by Lemma 4 the projection ρ/p is a ι_C-reduction again, by PML$^\infty$ to a normal form. Apply Proposition 12 again.
2. By Proposition 12 no redex in t itself overlaps with C and by repeating (1) we see that this property is preserved under reduction.
3. This is a special case of (2).

■

THEOREM 15. *For OTRSs we have UN$^\infty$.*

Proof. Consider two reductions ρ and σ from the same term t, both converging to infinitary normal forms, say nf_1 and nf_2, respectively.

$$\begin{array}{ll}\rho: & t_0 \to_{s_0} t_1 \to_{s_1} \cdots nf_1 \\ \sigma: & t'_0 \to_{s'_0} t'_1 \to_{s'_1} \cdots nf_2\end{array}$$

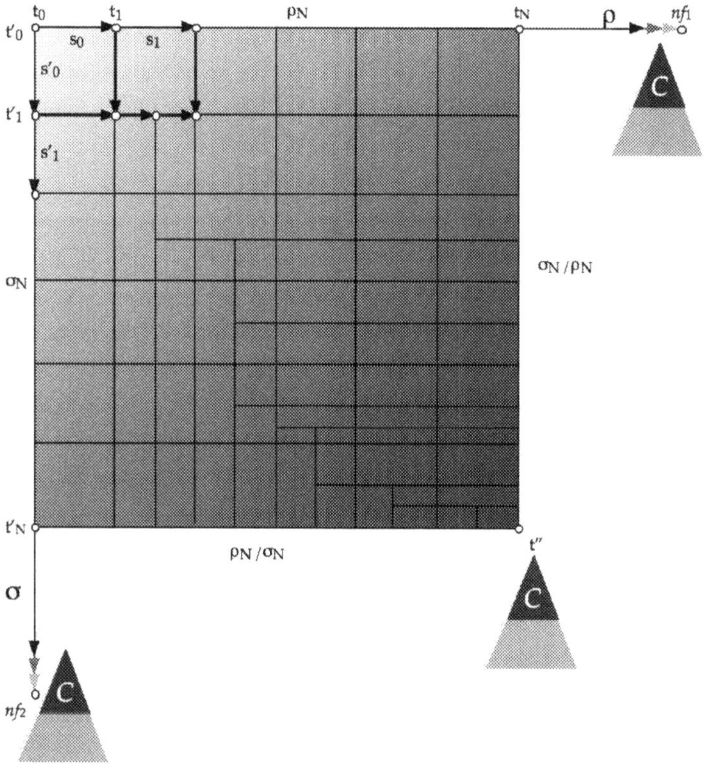

Figure 9. Proof of UN^{∞}

By compression we may assume that ρ and σ have length at most ω. We show that all finite prefixes of the normal forms nf_1 and nf_2 are identical, hence $nf_1 = nf_2$. The proof is depicted in Figure 9.

Let C be a finite prefix of nf_1, say of depth n. Both in ρ and in σ all redex activity will eventually be below depth n, say after N_1 and N_2 steps respectively, with $N = \max(N_1, N_2)$. Now note that by Proposition 14(3), the terms t_N and t'_N are n-stable, as both reduce to infinite normal form by n-stable reduction. And in particular t_N will be C-stable.

Projecting the finite reductions ρ_N and σ_N yields a common reduct t'' of t_N and t'_N. Because of the n-stability of t_N and $t_{N'}$, t'' will share its prefix up to depth n with both these terms. So in particular C will also be a prefix of t'_N, and hence of nf_2. ■

As a corollary we now have the following theorem from [Dershowitz *et al.*, 1991].[6] The proof is immediate, analogous to that of UN & SN $\Longrightarrow$ CR for the finitary case.

COROLLARY 16. *For OTRSs the implication* $\mathrm{SN}^\infty \Longrightarrow \mathrm{CR}^\infty$ *holds.*

6 The equivalence of SN^∞ and WN^∞

Also in this section we assume orthogonality. We show that as properties of an OTRS the notions SN^∞ and WN^∞ are equivalent.

PROPOSITION 17. *Consider a transfinite divergent reduction ρ of length λ, containing infinitely many head steps and a coinitial parallel step p. Then the projections $\sigma_\alpha = \rho_\alpha/p$ $(\alpha < \lambda)$ are the prefixes of a divergent reduction σ of length λ, also containing infinitely many head steps.*

Proof. Let s_α be the step in ρ performed at ordinal α. So we have

$$\rho: \; t_0 \rightarrow_{s_0} t_1 \rightarrow_{s_1} \cdots \; t_\omega \rightarrow_{s_\omega} t_{\omega+1} \rightarrow_{s_{\omega+1}} \cdots$$

Define $p_0 = p$ and $p_\alpha = p/\rho_\alpha$ (Note that then also $p_{\alpha+1} = p_\alpha/s_\alpha$.) Finally let $S_\alpha = s_\alpha/p_\alpha$.

There are two possibilities:

1. For all ordinals α such that s_α is a head step, p_α is internal. Then each of the corresponding S_α's is a head step as well: infinitely many.

2. For some α such that s_α is a head step, p_α is a head step too. Then $p_{\alpha+1} = \varnothing$ and hence $p_\beta = \varnothing$ for all $\beta > \alpha$. Hence $S_\beta = s_\beta$ for all $\beta > \alpha$. Since the set $\{s_\beta \mid \beta > \alpha\}$ contains infinitely many head steps, so does $\{S_\beta \mid \beta > \alpha\}$.

In both cases σ contains infinitely many head steps, hence diverges. ■

COROLLARY 18. *Projection of a ρ containing infinitely many head steps over any finite reduction yields a σ containing infinitely many head steps.*

PROPOSITION 19. *Assume t has a divergent reduction ρ containing infinitely many head steps. Then t does not reduce to a head normal form.*

[6] As a matter of fact, in [Dershowitz *et al.*, 1991] instead of SN^∞ a notion is used that is called there *top termination* and which corresponds to our notion SHN^∞ (see Section 7). However, a result of the present paper is that as properties of OTRSs the notions SN^∞ and SHN^∞ are equivalent.

Proof. For a proof by contradiction suppose t has a transfinite reduction to head normal form h. Claim: t can be reduced to h in a finite reduction φ. For, suppose not. Then by compression we may assume that $t \twoheadrightarrow^{\omega} h$. For this reduction to be convergent the head must be in rest after finitely many steps. By Corollary 18 projection of ρ over φ yields a divergent reduction from h containing infinitely many head steps. This contradicts the assumption that h is a head normal form. ∎

COROLLARY 20. *A term t cannot both have a normal form and a reduction containing infinitely many head steps.*

PROPOSITION 21. *Consider an OTRS $\mathcal{T}$ and suppose there exists a divergent reduction ρ in $\mathcal{T}$. Then there is in $\mathcal{T}$ also a divergent reduction σ containing infinitely many head steps.*

Proof. Let $\rho = (s_\alpha)_{\alpha<\lambda}$, as in the display in the proof of Proposition 17 above. Divergence of ρ implies that λ is a limit ordinal and that for some n we have that for every $\alpha < \lambda$ there exists a $\beta > \alpha$ such that the step s_β has depth $\leq n$. Let N be the smallest such n.

If $N = 0$, then we are done, take $\sigma = \rho$.

Otherwise $N > 0$ and then, as ρ has only finitely many steps with depth $< N$, there exists an ordinal $\Gamma < \lambda$ such that for all $\beta > \Gamma$ the depth of step s_β is $\geq N$. Beyond Γ the prefix of the term t_Γ consisting of all positions up to depth $N - 1$ will be fixed throughout the rest of the reduction ρ. So there is a fixed finite set of positions of depth N, uniform for all terms t_β, $\beta > \Gamma$. At these positions all infinitely many steps s_β, $\beta > \Gamma$ of depth N must take place and by pigeon holing at least one of these positions, say P, then will have infinitely many of these steps. Now consider the reduction from the term $t_\Gamma|_P$ consisting of all steps s_β, $\beta > \Gamma$ that take place at or below P. That will be a transfinite reduction σ containing infinitely many root steps, which is therefore also divergent. ∎

THEOREM 22. *For OTRSs we have* $\mathrm{SN}^\infty(\mathcal{T}) \iff \mathrm{WN}^\infty(\mathcal{T})$.

Proof. ($\Rightarrow$) is trivial.

($\Leftarrow$) Assume that $\mathcal{T}$ is not SN^∞. That is, in $\mathcal{T}$ there exists a divergent reduction. Then by Proposition 21 there exists in $\mathcal{T}$ also a reduction with infinitely many head steps. By Corollary 20 it then follows that $\mathcal{T}$ is not WN^∞. ∎

On reflection, recalling O'Donnell's Theorem 4 (OD) which states that in the finitary case the equivalence of WN and SN follows from non-erasingness, it is remarkable that according to Theorem 22 in the infinitary setting the equivalence holds without more. Here it is crucial to keep in mind that Theorem 22 is about WN^∞ and SN^∞ as properties of an OTRS $\mathcal{T}$. At the level of terms the infinitary case deviates from finitary as well, but quite in the opposite direction: for terms O'Donnell's Theorem does not hold at all. Abbreviating the property of non-erasingness by NE, the infinitary version of O'Donnell's Theorem for terms would read:

$$\mathrm{OD}^\infty: \quad \mathrm{NE} \Longrightarrow (\,\mathrm{WN}^\infty(t) \Longleftrightarrow \mathrm{SN}^\infty(t)\,)$$

Failure of OD^∞ is demonstrated by the following example.

EXAMPLE 23. Consider the term $a(c)$ in the non-erasing orthogonal TRS $\mathcal{T} = \{c \to c,\ a(x) \to b(a(x))\}$; it violates OD^∞ by being WN^∞ but not SN^∞.

WN^∞: The infinite reduction $a(c) \to b(a(c)) \to b(b(a(c))) \to \cdots$ reduces $a(c)$ to its infinitary normal form b^ω.

$\neg\mathrm{SN}^\infty$: There is also the divergent reduction $a(c) \to a(c) \to a(c) \to \cdots$

For an intuitive explanation first note that $a(c)$ is a counterexample to OD^∞ in particular in the sense that it does not satisfy the implication

$$\mathrm{WN}^\infty(t)\ \&\ \neg\mathrm{SN}^\infty(t) \quad \Longrightarrow \quad \neg\mathrm{NE}$$

In the finitary case, erasure ($\neg$NE) is "needed" for getting rid of the part of a $\neg$SN-term that generates an infinite reduction, in order to pass to a normal form. One wonders why the same would not be needed as well in the infinitary case. Well, look at the example again. The c is not erased literally, but in the infinite reduction it is pushed over the edge of infinity, so to say, with the same effect as erasure: the potentially divergent part c has disappeared in the infinite normal form b^ω.

7 Weak and Strong Head Normalization

We also consider the notions of *weak* and *strong head normalization* (WHN and SHN). A head normal form is a term which is root-stable, as defined in Section 5. Then we have for a term t:

WHN: There is a reduction of t to a head normal form.

SHN: In each infinite reduction of t after a finite number of steps a head normal form is reached.

The infinitary versions then follow naturally.

WHN$^\infty$: There is a possibly transfinite reduction of t to a head normal form.

SHN$^\infty$: In each maximal transfinite reduction of t, no matter whether converging or diverging, at some point a head normal form is reached.

We restrict attention to orthogonal systems again. Then, if a head normal form can be reached by an infinite reduction, by compression it can be reached already in a reduction of length $\leq \omega$. In this reduction the root becomes stable after finitely many steps. So for finite terms there is no difference between finitary and infinitary WHN. As to SHN$^\infty$, again one easily sees that for finite terms it is equivalent to finitary SHN. By contrast note that I^ω is an infinite term that is neither SHN$^\infty$ nor WHN$^\infty$ in the TRS with rule $I(x) \to x$, notwithstanding the TRS being both WHN and SHN.

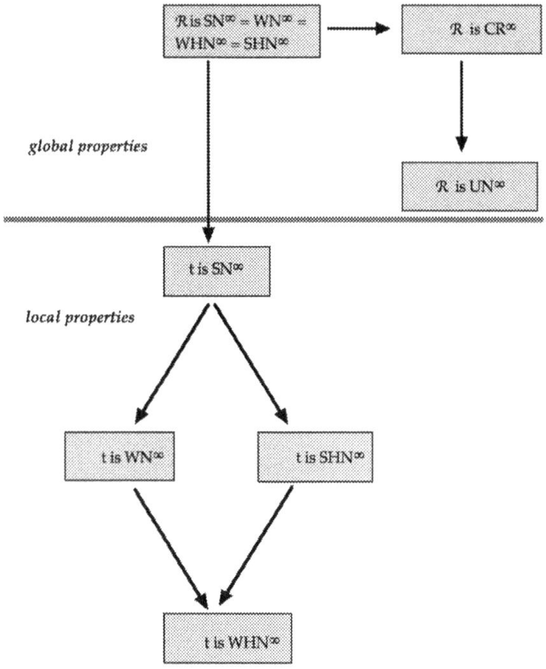

Figure 10. Survey of infinitary properties

The following two examples show that the four implications between the local properties SN$^\infty$, WN$^\infty$, SHN$^\infty$ and WHN$^\infty$ as in Figure 10 are strict.

EXAMPLES 24.

1. In the TRS with rules $\{c \to c, a(x) \to b\}$ the term $a(c)$ is WN^∞, $\neg\mathrm{SN}^\infty$, WHN^∞ and $\neg\mathrm{SHN}^\infty$.

2. In the TRS with the single rule $c \to c$ and additional unary function symbol e the term $e(c)$ is SHN^∞, $\neg\mathrm{SN}^\infty$, WHN^∞ and $\neg\mathrm{WN}^\infty$

This was all at the level of terms. At the global level of orthogonal TRSs the notions of WHN^∞ and SHN^∞ both coincide with WN^∞ and SN^∞, which we showed to be the same.

We conclude this section on head normalization by noting that the implication WN $\Longrightarrow$ AC (Theorem 5) can be strengthened to WHN $\Longrightarrow$ AC. This was proved in [Ketema *et al.*, 2004].

8 Concluding remarks

There are two directions in which we would like to see the subject of this paper extended. The first is to relax the requirement of orthogonality to weak orthogonality. A TRS is weakly orthogonal if the reduction rules are left-linear and critical pairs $\langle t, s\rangle$ generated by overlapping rules are trivial, i.e. of the form $\langle t, t\rangle$. An example of a weakly orthogonal TRS is obtained by adding to the TRS for addition as in this paper a function symbol P for predecessor, with extra reduction rules $S(P(x)) \to x$ and $P(S(x)) \to x$.

The second direction is the extension to the higher-order case, where bound variables are present, as in λ-calculus. For $\lambda\beta$-calculus the property UN^∞ does hold, see [Terese, 2003], but for general orthogonal higher-order systems UN^∞ is a conjecture only. The infinitary λ-calculus itself is fully covered in [Terese, 2003]. Work on general infinitary higher-order rewriting has already begun in [Ketema and Simonsen, 2005].

A combination of these two directions is found in the $\lambda\beta\eta$-calculus, which is a higher-order weakly orthogonal TRS. Unfortunately, because the η-rule tests for the absence of a variable[7], something which may happen only in the limit, fundamental theorems such as compression do not hold for the infinitary $\lambda\beta\eta$-calculus. Recently, Severi and de Vries have considered both η-reduction and η-expansion with the aim of generating and studying models by means of transfinite rewriting techniques [Severi and de Vries, 2002; Severi and de Vries, 2005].

Figure 11 summarizes the situation (OCRS stands for Orthogonal Combinatory Reduction System).

[7]In higher-order terminology: the rule is not fully-extended.

	PML	CR	UN	PML^∞	CR^∞	UN^∞
OTRS	*yes*	*yes*	*yes*	*yes*	*no*	*yes*
w.o. TRS	*yes*	*yes*	*yes*	?	*no*	?
$\lambda\beta$	*yes*	*yes*	*yes*	*no*	*no*	*yes*
OCRS	*yes*	*yes*	*yes*	*no*	*no*	?

Figure 11. Some open questions in infinitary rewriting

Acknowledgements

We would like to thank Clemens Grabmayer, Jeroen Ketema and Vincent van Oostrom for stimulating conversations and helpful comments.

BIBLIOGRAPHY

[Baader and Nipkow, 1998] F. Baader and T. Nipkow. *Term Rewriting and All That.* Cambridge University Press, 1998.

[Church, 1941] A. Church. *The Calculi of Lambda-Conversion*, volume 6 of *Annals of Mathematics Studies.* Princeton University Press, 1941.

[Dedekind, 1888] R. Dedekind. *Was sind und was sollen die Zahlen?* Brunswick, 1888.

[Dershowitz and Jouannaud, 1990] N. Dershowitz and J.-P. Jouannaud. Rewrite systems. In J. van Leeuwen, editor, *Handbook of Theoretical Computer Science, Vol. B, Formal Models and Semantics*, pages 243–320. Elsevier, 1990.

[Dershowitz *et al.*, 1991] Nachum Dershowitz, Stéphane Kaplan, and David A. Plaisted. Rewrite, rewrite, rewrite, rewrite, rewrite, ... *Theoretical Computer Science*, 83(1):71–96, 1991.

[Kennaway *et al.*, 1995] J.R. Kennaway, J.W. Klop, M.R. Sleep, and F.J. de Vries. Transfinite reductions in orthogonal term rewriting systems. *Information and Computation*, 119(1):18–38, 1995.

[Kennaway, 1992] J.R. Kennaway. On transfinite abstract reduction systems. Technical Report CS-R9205, CWI, January 1992.

[Ketema and Simonsen, 2005] Jeroen Ketema and Jakob Grue Simonsen. Infinitary combinatory reduction systems. In Jürgen Giesl, editor, *Proceedings of the 16th International Conference on Rewriting Techniques and Applications (RTA 2005), Nara, Japan, April 19-21, 2005*, volume 3467 of *Lecture Notes in Computer Science*, pages 438–452. Springer-Verlag, 2005.

[Ketema *et al.*, 2004] Jeroen Ketema, Jan Willem Klop, and Vincent van Oostrom. Vicious circles in rewriting systems. Technical Report SEN-E0427, Centre for Mathematics and Computer Science (CWI), Amsterdam, 2004.

[Ketema *et al.*, 2005] Jeroen Ketema, Jan Willem Klop, and Vincent van Oostrom. Vicious circles in orthogonal term rewriting systems. In Sergio Antoy and Yoshihito Toyama, editors, *Proceedings of the 4th International Workshop on Reduction Strategies in Rewriting and Programming (WRS'04)*, volume 124(2) of *Electronic Notes in Theoretical Computer Science*, pages 65–77. Elsevier Science, 2005.

[Klop and de Vrijer, 1991] J.W. Klop and R.C. de Vrijer. Extended term rewriting systems. In S. Kaplan and M. Okada, editors, *Conditional and typed rewriting systems*, volume 516 of *Lecture Notes in Computer Science*, pages 26–50. Springer-Verlag, Berlin, 1991.

[Klop, 1992] J.W. Klop. Term rewriting systems. In S. Abramsky, D.M. Gabbay, and T.S.E. Maibaum, editors, *Handbook of Logic in Computer Science*, volume 2, pages 1–116. Oxford University Press, New York, 1992.

[O'Donnell, 1977] M.J. O'Donnell. *Computing in Systems Described by Equations*, volume 58 of *Lecture Notes in Computer Science*. Springer-Verlag, 1977.

[Severi and de Vries, 2002] P. Severi and F.-J. de Vries. An extensional Böhm model. In Sophie Tison, editor, *Proc. of the 13th Int. Conf. on Rewriting Techniques and Applications, RTA'02*, volume 2378 of *Lecture Notes in Computer Science*, pages 159–173. Springer-Verlag, 2002.

[Severi and de Vries, 2005] P. Severi and F.-J. de Vries. Order structures on Böhm-like models. In *Proceedings of CSL'05*, 2005. To appear.

[Terese, 2003] Terese. *Term Rewriting Systems*. Cambridge University Press, 2003.

[Waldmann, 2000] J. Waldmann. The combinator **S**. *Information and Computation*, 159(1–2):2–21, 2000.

Modal Logics for Metric Spaces: Open Problems

AGI KURUCZ, FRANK WOLTER AND
MICHAEL ZAKHARYASCHEV

Modal logics and their models have been used to speak about and represent *topological spaces* since the 1940s [Tarski, 1938; Tsao Chen, 1938; McKinsey, 1941; McKinsey and Tarski, 1944]. Examples include Tarski's programme of algebraisation of topology ("*of creating an algebraic apparatus for the treatment of portions of point-set topology,*" to be more precise) which involved modal logic **S4** [McKinsey and Tarski, 1944], and the use of the extension of **S4** with the universal modality (and its fragments) for spatial representation and reasoning; see, e.g., [Bennett, 1994; Nutt, 1999; Egenhofer and Franzosa, 1991; Egenhofer and Herring, 1991; Aiello and van Benthem, 2002; Gabelaia *et al.*, 2005] and references therein.

Metric spaces are even more important mathematical structures that are fundamental for many areas of mathematics and computer science (recent examples include classification in bioinformatics, linguistics, botany, etc. using various similarity measures). A natural research programme is then to find out to which extent modal-like formalisms can be useful for speaking about metric spaces. Such a programme was launched in 2000 [Sturm *et al.*, 2000; Kutz *et al.*, 2002; Kutz *et al.*, 2003].

The aim of this note is to attract attention to the most important open problems and new directions of research in this exciting and promising area.

1 Distance spaces

Recall that a *metric space* is a pair (Δ, d), where Δ is a nonempty set (of points) and d is a function from $\Delta \times \Delta$ into the set $\mathbb{R}^{\geq 0}$ (of non-negative real numbers) satisfying the following axioms

identity of indiscernibles:	$d(x, y) = 0 \text{ iff } x = y,$	(1)
triangle inequality:	$d(x, z) \leq d(x, y) + d(y, z),$	(2)
symmetry:	$d(x, y) = d(y, x)$	(3)

for all $x, y, z \in \Delta$. The value $d(x, y)$ is called the *distance* from the point x to the point y. Given a metric space (Δ, d), a point $x \in \Delta$ and a nonempty $Y \subseteq \Delta$, define the *distance* $d(x, Y)$ *from* x *to* Y by taking

$$d(x, Y) \quad = \quad \inf\{d(x, y) \mid y \in Y\},$$

and put $d(y, \emptyset) = \infty$. The distance $d(X, Y)$ between two nonempty sets X and Y is

$$d(X, Y) \quad = \quad \inf\{d(x, y) \mid x \in X, \ y \in Y\}.$$

Although acceptable in many cases, the defined concept of metric space is not universally applicable to all interesting measures of distance between points, especially those used in everyday life. Consider, for instance, the following two examples:

(i) If $d(x, y)$ is the flight-time from x to y then, as we know it too well, d is not necessarily symmetric, even approximately (just take a plane from London to Tokyo and back).

(ii) Often we do not measure distances by means of real numbers but rather using more fuzzy notions such as 'short,' 'medium' and 'long.' To represent these measures we can, of course, take functions d from $\Delta \times \Delta$ into the subset $\{1, 2, 3\}$ of $\mathbb{R}^{\geq 0}$ and define *short* := 1, *medium* := 2, and *long* := 3. So we can still regard these distances as real numbers. However, for measures of this type the triangle inequality (2) does not make sense (short plus short can still be short, but it can also be medium or long).

Spaces (Δ, d) satisfying only axiom (1) will be called *distance spaces.*

Recall also that a *topological space* is a pair $(U, \mathbb{I})$ in which U is a nonempty set, the *universe* of the space, and $\mathbb{I}$ is the *interior operator* on U satisfying the *Kuratowski axioms*: for all $X, Y \subseteq U$,

$$\mathbb{I}(X \cap Y) \ = \ \mathbb{I}X \cap \mathbb{I}Y, \quad \mathbb{I}X \ \subseteq \ \mathbb{I}\mathbb{I}X, \quad \mathbb{I}X \ \subseteq \ X \quad \text{and} \quad \mathbb{I}U \ = \ U.$$

The operator dual to $\mathbb{I}$ is called the *closure operator* and denoted by $\mathbb{C}$: for every $X \subseteq U$, we have $\mathbb{C}X = U - \mathbb{I}(U - X)$. Thus, $\mathbb{I}X$ is the *interior* of a set X, while $\mathbb{C}X$ is its *closure.* X is called *open* if $X = \mathbb{I}X$ and *closed* if $X = \mathbb{C}X$.

Each metric space (Δ, d) gives rise to the *interior operator* $\mathbb{I}_d$ on Δ: for all $X \subseteq \Delta$,

$$\mathbb{I}_d X \quad = \quad \{x \in X \mid \exists \varepsilon > 0 \ \forall y \ (d(x, y) < \varepsilon \to y \in X)\}.$$

The pair $(\Delta, \mathbb{I}_d)$ is called the *topological space induced by* the metric space (Δ, d). The dual *closure operator* $\mathbb{C}_d$ in this space can be defined by the equality

$$\mathbb{C}_d X \quad = \quad \{x \in W \mid \forall \varepsilon > 0 \ \exists y \in X \ d(x, y) < \varepsilon\}.$$

Examples. We briefly remind the reader of a few standard examples of metric and topological spaces that will be used in what follows.

1 The *one-dimensional Euclidean space* is the set of real numbers $\mathbb{R}$ equipped with the following metric on it

$$d_1(x, y) \;=\; |x - y|.$$

Let $X \subseteq \mathbb{R}$. A point $x \in \mathbb{R}$ is said to be *interior* in X if there is some $\varepsilon > 0$ such that the whole open interval $(x - \varepsilon, x + \varepsilon)$ belongs to X. The interior $\mathbb{I}X$ of X is defined then as the set of all interior points in X. It is not hard to check that $(\mathbb{R}, \mathbb{I})$ is the topological space induced by the Euclidean metric d_1. Open sets in $(\mathbb{R}, \mathbb{I})$ are (possibly infinite) unions of open intervals (a, b), where $a \leq b$. The closure of (a, b), for $a < b$, is the closed interval $[a, b]$, with the end points a and b being its boundary.

2 In the same manner one can define *n-dimensional Euclidean spaces* based on the universes $\mathbb{R}^n$ with the metric

$$d_n(x, y) \;=\; \sqrt{\sum_{i=1}^{n}(x_i - y_i)^2}$$

(in the definition of interior points x one should take n-dimensional ε-neighbourhoods of x).

3 Further well-known examples are *metric spaces on graphs*: the distance between two nodes of a graph is defined as the length of the shortest path between them. Special cases are the *tree metric spaces.*

4 A topological space is called an *Aleksandrov space* [Alexandroff, 1937] if arbitrary (not only finite) intersections of open sets are open. Aleksandrov spaces are closely related to *quasi-ordered sets*, that is, pairs $\mathfrak{G} = (V, R)$, where V is a nonempty set and R a transitive and reflexive relation on V. Every such quasi-order $\mathfrak{G}$ induces the interior operator $\mathbb{I}_{\mathfrak{G}}$ on V: for $X \subseteq V$,

$$\mathbb{I}_{\mathfrak{G}}X \;=\; \{x \in X \mid \forall y \in V\ (xRy \rightarrow y \in X)\}.$$

In other words, the open sets of the topological space $\mathfrak{T}_{\mathfrak{G}} = (V, \mathbb{I}_{\mathfrak{G}})$ are the *upward closed* (or *R-closed*) subsets of V. It is well-known (see, e.g., [Bourbaki, 1966]) that $\mathfrak{T}_{\mathfrak{G}}$ is an Aleksandrov space and, conversely, every Aleksandrov space is induced by a quasi-order.

2 Modal logics of distance spaces

The intended *distance models* we would like to talk about with the help of modal-like formalisms are structures of the form

(4) $\mathfrak{I} = (\mathfrak{D}, \ell_1^{\mathfrak{I}}, \ell_2^{\mathfrak{I}}, \dots, p_1^{\mathfrak{I}}, p_2^{\mathfrak{I}}, \dots)$

where $\mathfrak{D} = (\Delta, d)$ is a distance space, the $\ell_i^{\mathfrak{I}}$ are some elements (or *locations*) of Δ and the $p_i^{\mathfrak{I}}$ are subsets of Δ. Distance models with the underlying distance space being a metric space will be called *metric models.*

We divide our languages designed for talking about distance models into two groups: those without *quantification over distances* and those that do use (explicitly or implicitly) such quantification.

2.1 Logics without quantification over distances

We introduce the following 'parameterised modalities' or 'bounded quantifiers:'

- $\exists^{=a}$ meaning 'somewhere at distance a,'
- $\exists^{<a}$ meaning 'somewhere at distance $< a$,'
- $\exists^{>a}$ meaning 'somewhere at distance $> a$,' and
- $\exists^{<b}_{>a}$ meaning 'somewhere at distance d with $a < d < b$,'

where a and b are some numbers from $\mathbb{R}^{\geq 0}$ (or rather $\mathbb{Q}^{\geq 0}$ to avoid the problem of representing real numbers and keep the language countable). Then one can also define duals like $\forall^{<b}_{>a}$ meaning 'everywhere within distance d for $a < d < b$,' etc. (As the expressive completeness result below shows, once we restrict ourselves to the 'modal' paradigm, our choice of operators is rather natural.)

More precisely, given a distance model $\mathfrak{I}$ of the form (4), we interpret our operators as

$$\begin{aligned}
(\exists^{=a}\tau)^{\mathfrak{I}} &= \{x \in \Delta \mid \exists y\, (d(x,y) = a \;\wedge\; y \in \tau^{\mathfrak{I}})\},\\
(\exists^{<a}\tau)^{\mathfrak{I}} &= \{x \in \Delta \mid \exists y\, (d(x,y) < a \;\wedge\; y \in \tau^{\mathfrak{I}})\},\\
(\exists^{>a}\tau)^{\mathfrak{I}} &= \{x \in \Delta \mid \exists y\, (d(x,y) > a \;\wedge\; y \in \tau^{\mathfrak{I}})\},\\
(\exists^{<b}_{>a}\tau)^{\mathfrak{I}} &= \{x \in \Delta \mid \exists y\, (a < d(x,y) < b \;\wedge\; y \in \tau^{\mathfrak{I}})\},
\end{aligned}$$

where $\tau^{\mathfrak{I}} \subseteq \Delta$.

The full 'modal' language of distance spaces. The full language $\mathcal{MS}$ of distance spaces with the operators $\exists^{=a}$, $\exists^{<a}$, $\exists^{>a}$, $\exists^{<b}_{>a}$ (and their duals $\forall^{=a}$, $\forall^{<a}$, etc.) interpreted as defined above was introduced and analysed in [Kutz *et al.*, 2003]. Formally, the expressions of this language are defined as follows:

(5) $\tau \;::=\; p_i \mid \{\ell_i\} \mid \neg\tau \mid \tau_1 \sqcap \tau_2 \mid \exists^{=a}\tau \mid \exists^{<a}\tau \mid \exists^{>a}\tau \mid \exists^{<b}_{>a}\tau,$

where $a, b \in \mathbb{Q}^{\geq 0}$ with $a < b$, and the ℓ_i are *location constants* (or *nominals*) interpreted by singleton sets. As expressions of the form τ are interpreted as subsets of distance spaces, we will call them (*spatial*) *terms*. Given some terms, we allow the language to say some simple things about them by means of *formulas* that are defined as follows:

$$\varphi \ ::= \ \tau_1 \sqsubseteq \tau_2 \ \mid \ d(\ell_1, \ell_2) = a \ \mid \ d(\ell_1, \ell_2) < a \ \mid \ \neg\varphi \ \mid \ \varphi_1 \wedge \varphi_2$$

(in particular, we can express $\ell_i \in \tau$ and $\tau_1 = \tau_2$). Formulas are interpreted in distance models as *true* or *false* in the natural way. Various *logics* in (fragments of) this language can be obtained by restricting the class of distance spaces underlying our models, say, to the class of metric spaces.

Below we summarise what is known about the language $\mathcal{MS}$ interpreted over various classes of distance spaces. Perhaps the most important result is the following *expressive completeness theorem* [Kutz *et al.*, 2003] which describes precisely how the modal language $\mathcal{MS}$ is related to first-order logic over *metric models*:

> *Over the class metric models, the language $\mathcal{MS}$ is expressively complete for (or has the same expressive power as) the two-variable fragment of first-order logic with countably many unary predicates and binary predicates of the form $d(x, y) < a$ and $d(x, y) = a$ for $a \in \mathbb{Q}^{\geq 0}$.*

A rather transparent axiomatisation of formulas of $\mathcal{MS}$ that are valid in metric models has been given in [Kutz, 2005] using some 'Gabbay-style' rules known from hybrid logic. To give the reader some idea of the axiomatisation, we observe first that $\exists^{\leq a}$ and $\exists^{>a}$ behave like normal modal 'diamonds,' $\forall^{\leq a}\tau \sqcap \forall^{>a}\tau$ is the *universal modality* $\boxtimes$, while $\exists^{>0}\tau$ is the *difference operator* (so in fact, nominals are expressible in $\mathcal{MS}$). Typical 'non-modal' axioms look as follows:

$$\tau \ \sqsubseteq \ \forall^{\leq a}\exists^{\leq a}\tau,$$
$$\exists^{\leq a}\forall^{>b}\tau \ \sqsubseteq \ \forall^{>a+b}\tau.$$

It is also shown in [Kutz, 2005] that $\mathcal{MS}$ does not have the Craig interpolation property over metric models.

The satisfiability[1] problem for $\mathcal{MS}$ over metric models was proved to be *undecidable* in [Kutz *et al.*, 2003]. It turns out that the 'doughnut' operators $\exists^{\leq a}_{>0}$ and the propositional constants $\top$ and $\bot$ are already enough for obtaining undecidability: this relatively small fragment can 'enforce' the

[1] Throughout, it does not matter whether we consider term or formula satisfiability.

$\mathbb{N} \times \mathbb{N}$ grid using the 'punctured' centers of circles, and so we can encode in it the undecidable $\mathbb{N} \times \mathbb{N}$ tiling problem.

An important class of distance models consists of those that are based on the *one-dimensional Euclidean space* $\mathbb{R}$. The situation here can be understood by embedding into quantitative temporal logics: the full language $\mathcal{MS}$ without the operator $\exists^{=a}$ (but with operators $\exists^{\leq a}_{\geq b}$ for $a \neq b$) turns out to be *decidable* over $\mathbb{R}$. It is EXPSPACE-complete under the binary coding and PSPACE-complete under the unary coding of parameters [Hirshfeld and Rabinovich, 2004; Alur *et al.*, 1996]. Note, however, that the fragment of $\mathcal{MS}$ with the operators $\exists^{=a}$ only is *undecidable* over $\mathbb{R}$; see [Alur *et al.*, 1996].

These undecidability results have motivated the study of 'well-behaved' fragments of $\mathcal{MS}$ over various classes of models. In particular, two such 'reasonable' fragments have been discovered.

The $(\exists^{\leq a}, \exists^{>a})$-fragment. The terms of this fragment are formed as follows:

$$\tau \quad ::= \quad p_i \mid \{\ell_i\} \mid \neg\tau \mid \tau_1 \sqcap \tau_2 \mid \exists^{\leq a}\tau \mid \exists^{>a}\tau.$$

This fragment (together with several others without the doughnut operators) turns out to be *decidable* over various classes of distance models (over *metric models*, in particular), and even has the finite model property with respect to intended models (e.g., a term is satisfiable in a metric model iff it is satisfiable in a finite metric model) [Kutz *et al.*, 2003]. The computational complexity of these satisfiability problems was proved to be in non-deterministic exponential time in [Kutz *et al.*, 2003]. Using a different, carefully crafted 'Fisher–Ladner closure,' one can actually prove EXPTIME-completeness of these problems, provided that the numerical parameters are coded in *unary* [Wolter and Zakharyaschev, 2005b].

PROBLEM 1. What is the complexity of the satisfiability problem for the $(\exists^{\leq a}, \exists^{>a})$-fragment over metric models under the binary coding of parameters?

Note that the $(\exists^{\leq a}, \exists^{>a})$-fragment is *undecidable* over models based on Euclidean spaces $\mathbb{R}^n$, for $n \geq 2$ [Kutz *et al.*, 2003; Wolter and Zakharyaschev, 2005a].

Hilbert-style axiomatisations for the $(\exists^{\leq a}, \exists^{>a})$-fragment over arbitrary distance models (and several subclasses) are provided in [Kutz *et al.*, 2002]. Observe that the universal modality and the difference operator (and so nominals) are still expressible in this fragment.

The ($\exists^{\leq a}$, $\exists^{<a}$)-fragment. The terms of this fragment are formed as follows:

$$\tau \ ::= \ p_i \mid \{\ell_i\} \mid \neg\tau \mid \tau_1 \sqcap \tau_2 \mid \exists^{\leq a}\tau \mid \exists^{<a}\tau$$

(observe that the universal modality and nominals are no longer expressible, and so the $\{\ell_i\}$ are not just syntactic sugar). In many contexts—e.g., if we represent a similarity measure between objects of a certain domain by means of a metric—we may not need operators of the form $\exists^{>a}$. The ($\exists^{\leq a}$, $\exists^{<a}$)-fragment extended with the *universal* and *existential modalities* ⊠ and ⟐ was considered in [Wolter and Zakharyaschev, 2003]. The satisfiability problem for this language over *metric models* is EXPTIME-complete even if the numerical parameters are coded in *binary*, and enjoys the finite model property in the same sense as above. The crucial observation in the proof of this result is that the logic turns out to be complete with respect to *tree metric spaces*, a feature not shared by the richer languages considered above. An intriguing fact is that the fragments with only strict operators $\exists^{<a}$ and only non-strict ones $\exists^{\leq a}$ behave similarly, which perhaps reflects our everyday life disregard of the borders. Note that using both these operators we can say that the distance between two sets p and q is precisely a:

$$(p \sqcap \exists^{\leq a} q \neq \bot) \quad \wedge \quad (p \sqcap \exists^{<a} q = \bot)$$

The ($\exists^{\leq a}$, $\exists^{<a}$)-fragment is PSPACE-complete over models based on $\mathbb{R}$ under binary coding of parameters; see [Lutz *et al.*, 2005]. However, for $n \geq 2$, the ($\exists^{\leq a}$, $\exists^{<a}$)-fragment becomes *undecidable* over $\mathbb{R}^n$, even without nominals [Kutz *et al.*, 2003; Wolter and Zakharyaschev, 2005a]. Thus, no interesting decidable fragment of $\mathcal{MS}$ over $\mathbb{R}^2$ is known so far. Interesting candidates which might be decidable are given in the next open problem:

PROBLEM 2. Is the language with operators $\exists^{<a}$ only decidable over $\mathbb{R}^2$? What about the fragment with operators $\exists^{\leq a}$?

Besides the full $\mathbb{R}^2$, natural and useful spaces to consider are bounded subspaces like $[0,1] \times [0,1]$. It is to be noted that the undecidability proofs mentioned above do not go through in this case.

PROBLEM 3. Investigate the satisfiability problem for fragments of $\mathcal{MS}$ over bounded subspaces of $\mathbb{R}^n$ (such as $[0,1] \times [0,1]$).

Of course, the choice of the two fragments of $\mathcal{MS}$ discussed above is rather *ad hoc*. There are many open questions along these lines:

PROBLEM 4. Give a complete classification of the fragments of $\mathcal{MS}$ over various classes of distance models with respect to their satisfiability and axiomatisation problems. Given a class $\mathcal{C}$ of distance models, are there natural 'maximal' decidable fragments of $\mathcal{MS}$ over $\mathcal{C}$? If so, what is their computational complexity (under unary and binary coding of parameters)?

We conjecture that a natural candidate for a 'maximal' decidable fragment of $\mathcal{MS}$ over various classes is the language with the operators $\exists^{\leq a}$, $\exists^{<a}$, $\exists^{>a}$, and $\exists^{\geq a}$.

2.2 Logics with quantification over distances

The language $\mathcal{MS}$ does not allow any *quantification over distances*. In particular, we can neither reason about the topology induced by a metric space nor compare distances without fixing their absolute values. A natural extension $\mathcal{QMS}$ of $\mathcal{MS}$ with quantification over distances can be obtained by allowing *individual variables* $x, y, z, \dots$ over $\mathbb{R}^{>0}$ or $\mathbb{Q}^{>0}$ in distance operators as well as quantification over these variables. Formally, the $\mathcal{QMS}$-terms are defined by adding to (5) terms of the form $\exists x\, \tau$ and by allowing *variables* $x, y, z, \dots$ along with concrete parameters in the distance operators, for example,

$$\exists^{=x}\tau, \quad \exists^{<x}\tau, \quad \exists^{>x}\tau, \quad \exists^{<x}_{>y}\tau, \quad \exists^{<x}_{>0}\tau.$$

To interpret $\mathcal{QMS}$-terms in distance models of the form (4), we also need *assignments* $\mathfrak{a}$ of *positive* real numbers $\mathfrak{a}(x) \in \mathbb{R}^{>0}$ to the individual variables x.[2] Then we have

$$\begin{aligned}
(\exists^{=x}\tau)^{\mathfrak{I},\mathfrak{a}} &= (\exists^{=\mathfrak{a}(x)}\tau)^{\mathfrak{I}} \\
(\exists^{<x}\tau)^{\mathfrak{I},\mathfrak{a}} &= (\exists^{<\mathfrak{a}(x)}\tau)^{\mathfrak{I}} \\
(\exists^{>x}\tau)^{\mathfrak{I},\mathfrak{a}} &= (\exists^{>\mathfrak{a}(x)}\tau)^{\mathfrak{I}} \\
(\exists^{<x}_{>y}\tau)^{\mathfrak{I},\mathfrak{a}} &= (\exists^{<\mathfrak{a}(x)}_{>\mathfrak{a}(y)}\tau)^{\mathfrak{I}} \\
(\exists x\, \tau)^{\mathfrak{I},\mathfrak{a}} &= \bigcup\{\tau^{\mathfrak{I},\mathfrak{b}} \mid \mathfrak{b}(y) = \mathfrak{a}(y), \text{ for } y \neq x\}\ .
\end{aligned}$$

Not much is known about the expressive power of this language. We conjecture, in particular, that the following problem can be solved in a positive way:

PROBLEM 5. Is the language $\mathcal{QMS}$ expressively complete for the two-sorted first-order logic where one sort is over $\mathbb{R}^{>0}$ and the other over the metric space underlying a given metric model, with only two variables of the second sort being allowed?

[2]We quantify over *positive* real numbers rather than *non-negative* ones in order to obtain short and transparent definitions of standard topological operators; see (6). The expressivity of the language does not depend on this assumption.

However, we do have a number of interesting results for some fragments of $\mathcal{QMS}$.

'Modal' languages of metric and topological spaces. The terms of the fragment $\mathcal{MT}$ of $\mathcal{QMS}$ can be formed as follows:

$$\tau \;::=\; p_i \mid \neg\tau \mid \tau_1 \sqcap \tau_2 \mid \exists^{<a}\tau \mid \exists^{\leq a}\tau \mid \exists x\forall^{<x}\tau \mid \forall x\exists^{<x}\tau \mid \forall x\forall^{<x}\tau \mid \exists x\exists^{<x}\tau,$$

where $a \in \mathbb{Q}^{\geq 0}$. (Observe the similarities between these 'directly closed' terms and expressions of Computational Tree Logic $\mathcal{CTL}$.) It is not hard to see that by adding similar 'non-strict' operators like $\forall x\exists^{\leq x}$ and $\exists x\forall^{\leq x}$ we do not increase the expressive power of the language. So in fact, what we obtain this way is an extension of the *nominal-free* ($\exists^{\leq a}$, $\exists^{<a}$)-fragment of $\mathcal{MS}$ above with the *topological interior* and *closure operators*

$$(6) \quad \mathbf{I}\tau = \exists x\forall^{<x}\tau, \qquad \mathbf{C}\tau = \forall x\exists^{<x}\tau,$$

and the *universal* and *existential modalities*

$$\boxed{\forall}\tau = \forall x\forall^{<x}\tau, \qquad \diamondsuit\tau = \exists x\exists^{<x}\tau,$$

The intended meanings of these terms of course only 'work' in *metric models*:

$$(\exists x\,\forall^{<x}p)^{\mathfrak{I}} = \bigcup_{a\in\mathbb{R}^{>0}} (\forall^{<a}p)^{\mathfrak{I}}, \qquad (\forall x\,\exists^{<x}p)^{\mathfrak{I}} = \bigcap_{a\in\mathbb{R}^{>0}} (\exists^{<a}p)^{\mathfrak{I}},$$

$$(\boxed{\forall}\tau)^{\mathfrak{I}} = \begin{cases} \Delta, & \text{if } \tau^{\mathfrak{I}} = \Delta, \\ \emptyset, & \text{otherwise,} \end{cases} \qquad (\diamondsuit\tau)^{\mathfrak{I}} = \begin{cases} \Delta, & \text{if } \tau^{\mathfrak{I}} \neq \emptyset, \\ \emptyset, & \text{otherwise.} \end{cases}$$

The language $\mathcal{MT}$ over metric models can be also regarded as an extension of the modal logic $\mathbf{S4}_u$ of topological spaces with the metric operators $\exists^{<a}$ and $\exists^{\leq a}$. Such a view was taken in [Wolter and Zakharyaschev, 2005a] where this logic was first introduced and investigated.

Note first that this logic does not have the finite model property with respect to metric models because the topology induced by a finite metric space is trivial. For example, the term $p \sqcap \mathbf{C}\neg p$ is not satisfiable in any finite metric model, yet is satisfiable in every Euclidean space. Moreover, the logic in question is not compact in the sense that there is an infinite set Γ of terms such that, for every finite $\Gamma' \subseteq \Gamma$, there exists a model $\mathfrak{I}$ with $\bigcap_{\tau\in\Gamma'} \tau^{\mathfrak{I}} \neq \emptyset$, but there exists no model $\mathfrak{I}$ for which $\bigcap_{\tau\in\Gamma} \tau^{\mathfrak{I}} \neq \emptyset$. An example is given by the set of terms $\{\neg\mathbf{C}p\} \cup \{\exists^{<\frac{1}{n}}p \mid n \in \mathbb{N}^+\}$.

It turns out, however, that the intended metric models for this logic can be represented in the form of relational structures *à la* Kripke frames, which can be regarded as partial descriptions of scenarios that can be realised in metric models. This representation theorem—in fact, a generalisation of the McKinsey–Tarski [McKinsey and Tarski, 1944] representation theorem for topological spaces—reduces reasoning with almost always infinite metric models to reasoning with finite relational models, which can be shown to be EXPTIME-complete even for the binary coding of the numerical parameters.

The formulas of $\mathcal{MT}$ that are valid in metric models can be axiomatised in a natural way (bearing in mind that both distance and topological operators are in fact normal modalities): we have the **S4**-axioms for **I** and **C**, standard axioms for $\forall^{<a}$ and $\exists^{<a}$ reflecting, in particular, the triangle inequality

$$\begin{aligned} \tau &\sqsubseteq \forall^{<a}\exists^{<a}\tau, \\ \exists^{\leq a}\exists^{\leq b}\tau &\sqsubseteq \exists^{\leq a+b}\tau, \\ &\text{etc.} \end{aligned}$$

and only two axioms connecting metric and topology

$$\begin{aligned} \mathbf{C}\tau &\sqsubseteq \exists^{<a}\tau, \\ \exists^{<a}\mathbf{C}\tau &\sqsubseteq \exists^{<a}\tau, \end{aligned}$$

see [Wolter and Zakharyaschev, 2005a] for more details.

PROBLEM 6. Investigate axiomatisation and satisfiability problems for $\mathcal{MT}$ over metric spaces whose induced topological spaces are connected.

We conjecture that this logic can be axiomatised by adding the connectivity axiom

$$\diamondsuit\mathbf{I}p \sqcap \diamondsuit\mathbf{I}q \sqcap \boxdot(\mathbf{I}p \sqcup \mathbf{I}q) \quad \sqsubseteq \quad \diamondsuit(\mathbf{I}p \sqcap \mathbf{I}q)$$

to the axioms over arbitrary metric models, and that results on the satisfiability problem are similar to those for the arbitrary metric case.

PROBLEM 7. Investigate axiomatisation and satisfiability problems for $\mathcal{MT}$ over other interesting classes of metric and topological metric spaces.

PROBLEM 8. What happens if we extend $\mathbf{S4}_u$ with the operators $\exists^{\leq a}$ and $\exists^{>a}$?

PROBLEM 9. What happens if we extend $\mathcal{MT}$ with nominals?

The satisfiability problem for $\mathcal{MT}$ over the Euclidean space $\mathbb{R}$ is *decidable*; see [Hirshfeld and Rabinovich, 2004]. It becomes undecidable over models based on $\mathbb{R}^2$ (or over its various subspaces), as it contains the undecidable $(\exists^{\leq a}, \exists^{<a})$-fragment of $\mathcal{MS}$, see above. However, the following questions are open:

PROBLEM 10. Is there a transparent axiomatisation of $\mathcal{MT}$ over $\mathbb{R}$? What is the computational complexity of the satisfiability problem?

PROBLEM 11. Is the satisfiability problem for $\mathcal{MT}$ over $\mathbb{R}^2$ (or its subspaces) recursively enumerable? What happens if we omit the operators $\exists^{\leq a}$?

The language of comparative similarity. We can be a bit more 'liberal' regarding the quantifier patterns of $\mathcal{MT}$ and (similarly to Computational Tree Logic $\mathcal{CTL}^+$) only require that the operators $\exists^{<x}$ and $\exists^{\leq x}$ cannot occur nested without an $\exists x$ in between. We then end up with the *similarity language* $\mathcal{SL}$ containing the '*closer operator*'

$$\tau_1 \leftleftarrows \tau_2 \;=\; \exists x\,(\exists^{\leq x}\tau_1 \sqcap \neg\exists^{\leq x}\tau_2).$$

Its semantical meaning in distance models of the form (4) is defined as follows:

$$(7)\quad (\tau_1 \leftleftarrows \tau_2)^{\mathfrak{I}} \;=\; \{x \in \Delta \mid d(x, \tau_1^{\mathfrak{I}}) < d(x, \tau_2^{\mathfrak{I}})\}\,.$$

In other words, $\tau_1 \leftleftarrows \tau_2$ is (interpreted by) the set containing those objects of Δ that are 'closer' (or 'more similar') to τ_1 than to τ_2.

This allows us to represent and reason about predicates like 'X is closer to Y than it is to Z' which are quite common in our everyday life ('the body was in the middle of the room, rather closer to the door than to the window').

The closer operator itself turns out to be quite powerful. Using it we can express (in metric spaces) the interior (and so the closure) operator by taking

$$\mathbf{I}\tau \;=\; \top \leftleftarrows \neg\tau.$$

Indeed, by the definition above, we have

$$(\mathbf{I}\tau)^{\mathfrak{I}} \;=\; \{x \in \Delta \mid d(x, \Delta - \tau^{\mathfrak{I}}) > 0\}.$$

We can also express the existential (and so the universal) modality:

$$\diamondsuit\tau \;=\; \tau \leftleftarrows \bot$$

because $d(x, \emptyset) = \infty$. Thus, the similarity language having the sole closer operator interpreted in metric models results in a logic that contains full $\mathbf{S4}_u$, and can again be regarded as a qualitative spatial formalism for reasoning about metric spaces with their induced topologies. We call it the *language of comparative similarity* and denote by $\mathcal{CSL}$.

One more interesting operator is

$$\tau_1 \leftrightarrows \tau_2 \;=\; \neg(\tau_1 \leftleftarrows \tau_2) \sqcap \neg(\tau_2 \leftleftarrows \tau_1)$$

which defines the set of points located at the same distance from τ_1 and τ_2.

As a small illustrating example consider the formula

$$(8) \quad p \sqsubseteq (q \leftleftarrows r) \;\wedge\; q \sqsubseteq (r \leftleftarrows p) \;\wedge\; r \sqsubseteq (p \leftleftarrows q) \;\wedge\; p \neq \bot.$$

One can readily check that it is satisfiable in a three-point non-symmetrical 'graph model,' say, in the one depicted below where the distance from x to y is the length of the shortest directed path from x to y.

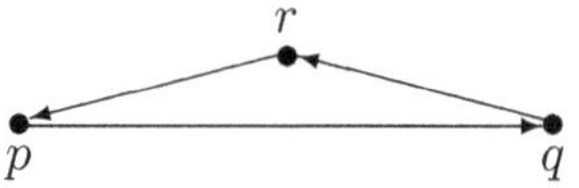

On the other hand, it can be satisfied in the following subspace of $\mathbb{R}$

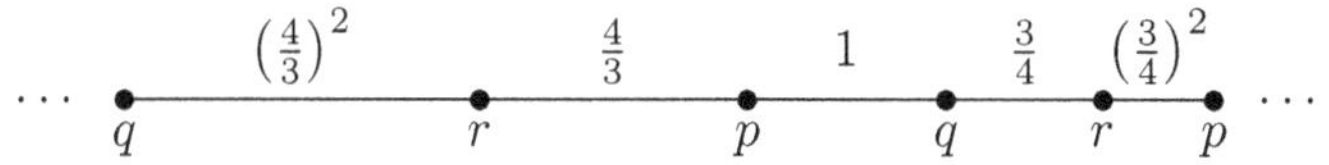

The price we have to pay for the expressivity of the closer operator is that the satisfiability problem for $\mathcal{CSL}$ over natural classes of metric spaces becomes EXPTIME-hard (remember that $\mathbf{S4}_u$ is PSPACE-complete).

Special classes of distance spaces are the so-called *min-spaces* that satisfy the *min-condition*

$$d(X,Y) \;=\; \inf\{d(x,y) \mid x \in X, y \in Y\} \;=\; \min\{d(x,y) \mid x \in X, y \in Y\},$$

for all sets X and Y. Over min-spaces, $\mathcal{CSL}$ is actually EXPTIME-complete (even if we extend it with distance operators $\exists^{<a}$ and $\exists^{\leq a}$, code the numerical parameters in binary, and allow nominals as well) [Sheremet *et al.*, 2005]. Actually, the complexity remains the same no matter whether we assume symmetry (3) and the triangle inequality (2). Note that the term (8) is not satisfiable in any symmetric model satisfying the min-condition.

Rather unexpectedly, valid formulas of $\mathcal{CSL}$ are not even recursively enumerable when interpreted over finite subspaces (or arbitrary min-subspaces) of $\mathbb{R}$ or in $\mathbb{R}$ itself [Sheremet *et al.*, 1999; Sheremet *et al.*, 2005]. This can be proved by a reduction of Hilbert's 10th problem on the unsolvability of Diophantine equations; see, e.g., [Barwise, 1977] and references therein. The same holds for min-subspaces of $\mathbb{R}^n$, where $n \geq 2$. To give the reader some impression of what structures can be enforced on such subspaces of $\mathbb{R}^2$ by terms with the closer operator, consider the following formula

$$(p_0 \neq \bot) \;\wedge\; (p_1 \neq \bot) \;\wedge \bigwedge_{\substack{i,j<7\\ j\neq i,i\oplus 1}} (p_i \sqsubseteq (p_j \leftrightarrows p_{i\oplus 1})),$$

where $\oplus$ is addition modulo 7. One can show that to satisfy this formula, a subspace of $\mathbb{R}^2$ must contain an infinite grid of the form

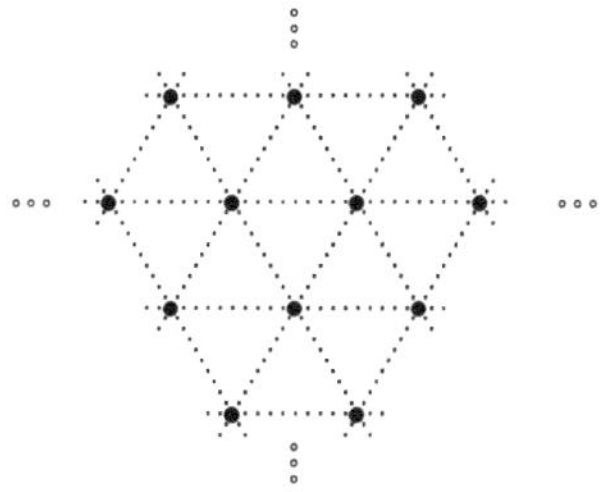

However, finding axiomatisations for $\mathcal{CSL}$ over other classes of models is open:

PROBLEM 12. Axiomatise the formulas of $\mathcal{CSL}$ that are valid in various classes of models.

As concerns evaluating formulas of $\mathcal{CSL}$ in arbitrary (not necessarily min) metric spaces, we only know that satisfiability is decidable (and EXPTIME-hard).

PROBLEM 13. What is the computational complexity of the satisfiability problem for $\mathcal{CSL}$ over arbitrary metric models? What happens if we extend the language with nominals?

PROBLEM 14. Gärdenfors [Gärdenfors, 2000] suggests that atomic terms of similarity languages should be interpreted by convex subsets of $\mathbb{R}^n$. Investigate the computational behaviour of $\mathcal{CSL}$ and its extensions under such or similar 'valuation restrictions.'

PROBLEM 15. Characterise (un)decidable fragments of full $\mathcal{QMS}$ over various classes of models.

Our similarity languages are 'crisp' in the sense that they operate with precise distances like 'the distance between two proteins is 3.1415926...' In practice such distances are only computed *approximately*. An interesting and important problem is the following:

PROBLEM 16. Develop logical formalisms capable of dealing with non-crisp distances, e.g., using vagueness/fuzziness/probability.

Acknowledgments

The work on this paper was partially supported by U.K. EPSRC grants no. GR/S61973/01, GR/S61966/01, GR/S63175/01 and GR/S63182/01.

BIBLIOGRAPHY

[Aiello and van Benthem, 2002] M. Aiello and J. van Benthem. A modal walk through space. *Journal of Applied Non-Classical Logics*, 12(3–4):319–364, 2002.

[Alexandroff, 1937] P. S. Alexandroff. Diskrete Räume. *Matematicheskii Sbornik*, 2 (44):501–518, 1937.

[Alur *et al.*, 1996] R. Alur, T. Feder, and T. Henzinger. The benefits of relaxing punctuality. *Journal of the ACM*, 43:116–146, 1996.

[Barwise, 1977] J. Barwise, editor. *Handbook of Mathematical Logic*. North-Holland, Amsterdam, 1977.

[Bennett, 1994] B. Bennett. Spatial reasoning with propositional logic. In *Proceedings of the 4th International Conference on Knowledge Representation and Reasoning*, pages 51–62. Morgan Kaufmann, 1994.

[Bourbaki, 1966] N. Bourbaki. *General topology, Part 1*. Hermann, Paris and Addison-Wesley, 1966.

[Egenhofer and Franzosa, 1991] M. Egenhofer and R. Franzosa. Point-set topological spatial relations. *International Journal of Geographical Information Systems*, 5:161–174, 1991.

[Egenhofer and Herring, 1991] M. Egenhofer and J. Herring. Categorizing topological relationships between regions, lines and point in geographic databases. Technical report, University of Maine, 1991.

[Gabelaia *et al.*, 2005] D. Gabelaia, R. Kontchakov, A. Kurucz, F. Wolter, and M. Zakharyaschev. Combining spatial and temporal logics: expressiveness vs. complexity. *Journal of Artificial Intelligence Research (JAIR)*, 23:167–243, 2005.

[Gärdenfors, 2000] P. Gärdenfors. *Conceptual Spaces. The Geometry of Thought*. The MIT Press, 2000.

[Hirshfeld and Rabinovich, 2004] Y. Hirshfeld and A. Rabinovich. Logics for real time: decidability and complexity. *Fundamenta Informaticae*, 62:1–28, 2004.

[Kutz *et al.*, 2002] O. Kutz, H. Sturm, N.-Y. Suzuki, F. Wolter, and M. Zakharyaschev. Axiomatizing distance logics. *Journal of Applied Non-Classical Logic*, 12:425–440, 2002.

[Kutz *et al.*, 2003] O. Kutz, H. Sturm, N.-Y. Suzuki, F. Wolter, and M. Zakharyaschev. Logics of metric spaces. *ACM Transactions on Computational Logic*, 4:260–294, 2003.

[Kutz, 2005] O. Kutz. Notes on logics of metric spaces. *Studia Logica*, 2005. (In print).

[Lutz *et al.*, 2005] C. Lutz, D. Walther, and F. Wolter. Quantitative temporal logics: PSPACE and below. In *Proceedings of the Twelfth International Symposium on Temporal Representation and Reasoning*, Burlington, VT, USA, 2005. IEEE Computer Society Press.

[McKinsey and Tarski, 1944] J.C.C. McKinsey and A. Tarski. The algebra of topology. *Annals of Mathematics*, 45:141–191, 1944.

[McKinsey, 1941] J.C.C. McKinsey. A solution of the decision problem for the Lewis systems **S2** and **S4**, with an application to topology. *Journal of Symbolic Logic*, 6:117–134, 1941.

[Nutt, 1999] W. Nutt. On the translation of qualitative spatial reasoning problems into modal logics. In W. Burgard, T. Christaller, and A. Cremers, editors, *Advances in Artificial Intelligence. Proceedings of the 23rd Annual German Conference on Artificial Intelligence (KI'99)*, volume 1701 of *Lecture Notes in Computer Science*, pages 113–124. Springer, 1999.

[Sheremet *et al.*, 1999] M. Sheremet, D. Tishkovsky, F. Wolter, and M. Zakharyaschev. 'Closer' representation and reasoning. In I. Horrocks, U. Sattler, and F. Wolter, editors, *International Workshop on Description Logics, (DL 2005)*, pages 25–36, 1999.

[Sheremet *et al.*, 2005] M. Sheremet, D. Tishkovsky, F. Wolter, and M. Zakharyaschev. Comparative similarity, tree automata, and Diophantine equations. Submitted; available at `http://www.dcs.kcl.ac.uk/staff/mz/`, 2005.

[Sturm *et al.*, 2000] H. Sturm, N.-Y. Suzuki, F. Wolter, and M. Zakharyaschev. Semi-qualitative reasoning about distances: A preliminary report. In M. Ojeda-Aciego, I.P. de Guzmán, G. Brewka, and L. Moniz Pereira, editors, *JELIA*, volume 1919 of *Lecture Notes in Computer Science*, pages 37–56. Springer, 2000.

[Tarski, 1938] A. Tarski. Der Aussagenkalkül und die Topologie. *Fundamenta Mathematicae*, 31:103–134, 1938.

[Tsao Chen, 1938] T. Tsao Chen. Algebraic postulates and a geometric interpretation of the Lewis calculus of strict implication. *Bulletin of the AMS*, 44:737–744, 1938.

[Wolter and Zakharyaschev, 2003] F. Wolter and M. Zakharyaschev. Reasoning about distances. In *Proceedings of the 18th International Joint Conference on Artificial Intelligence (IJCAI 2003)*, pages 1275–1280. Morgan Kaufmann, 2003.

[Wolter and Zakharyaschev, 2005a] F. Wolter and M. Zakharyaschev. A logic for metric and topology. *Journal of Symbolic Logic*, 70:795–828, 2005.

[Wolter and Zakharyaschev, 2005b] F. Wolter and M. Zakharyaschev. On the computational complexity of metric logics. Manuscript, 2005.

Partial Correctness Assertions Provable in Dynamic Logic

Daniel Leivant

1 Introduction

Hoare-style logics prove partial-correctness assertions (PCAs) about imperative programs. Their prominent role in program verification is due in part to their being syntax-directed: the inference rules follow the inductive buildup of programs. As a result, proofs can be converted into program annotations, and inference rules can guide program derivation and transformation. In contrast, some commonly-used inference rules of Dynamic Logic are not syntax-directed, allowing reasoning that intertwines formulas and programs in complex ways. This added complexity is obviously necessary when proving program-properties that are themselves more complex than PCAs, such as $\langle\alpha^*\rangle\top \rightarrow [\alpha^*]\,\varphi$.[1]

But does Dynamic Logic yield PCAs that Hoare's Logic fails to prove? That is, are there PCAs that are proved in first-order logic augmented with Pratt-Segerberg's **DL** rules for program modalities, but are not provable using only PCAs along the way?.

The answer might depend, of course, on the background first-order theory **T**, which in the case of Hoare's Logic manifests itself via the implicational first-order formulas used in the Rule of Consequence,

$$\frac{\varphi \rightarrow \varphi' \quad \varphi'\,[\alpha]\,\psi' \quad \psi' \rightarrow \psi}{\varphi\,[\alpha]\,\psi}$$

If a Hoare-style logic **H** is relatively complete for a structure $\mathcal{S}$, and **T** consists of the entire first-order theory $Th(\mathcal{S})$ of $\mathcal{S}$, then Cook's Relative Completeness Theorem [Cook, 1978] applies, and already **H**(**T**) (i.e. **H** based on **T**) proves all PCAs that are true in $\mathcal{S}$. Unfortunately, this observation is not particularly helpful, because $Th(\mathcal{S})$ is not effectively axiomatizable for most structures $\mathcal{S}$ of interest.

[1] I.e., the formula stating that if some iteration of α terminates then φ is true after all iterations.

Consider then the opposite extreme case, where **T** is empty, i.e. the Consequence Rule invokes only implications that are provable in first-order logic, no axioms added. Even though Dynamic Logic has a far more complex proof theory than Hoare Logic, it is not immediately obvious that this difference must manifest itself in more PCAs being proved. In general, extending a formalism with extra expressive or deductive power need not imply that new theorems of a simple form are proved.[2]

We shall show (Theorem 7) that, in fact, **DL** based on the empty theory proves far more PCAs than Hoare's Logic based on the empty theory. To focus on the essentials, we formulate this and subsequent theorems for a simple programming language, namely regular programs with first-order tests and assignments as atomic actions. (Guarded iterative programs, i.e. **while** programs, are definable in terms of these regular programs.)

Since **DL** is conservative over **H** when the background theory **T** is as strong and possible, and not conservative when **T** is empty, it is natural to ask whether there is a theory $\mathbf{T}_0$ that demarcates a transition between these two states of affair. In Theorems 8 and 9 we consider this question for programs over the natural numbers, and prove that Peano Arithmetic **PA** is the transition point: **DL**(**T**) is conservative over **H**(**T**) for all extensions **T** of **PA**, but is not when **T** is a sound theory whose axioms are of bounded complexity. A preliminary version of these results appeared as [Leivant, 2004a].

2 Dynamic Logic and Arithmetic

2.1 Dynamic Logic over Regular Programs

To focus on the essentials, we refer to the simplest non-trivial imperative programming language, namely regular programs over assignments, with first-order tests [Segerberg, 1977; Harel *et al.*, 2000]. We posit a vocabulary V (a finite set of identifiers for functions and relations, each assigned a non-negative integer as an arity). Let A be the set of V*-assignments,* that is expressions of the form $x := \mathbf{t}$, where $\mathbf{t}$ is a V-term; here x is said to be an *activated* variable. The set P of regular programs over V (with quantifier-free tests) is generated inductively by the following abstract-syntax closure conditions.

$$
\begin{array}{lllll}
A & \ni & a & & (V\text{-assignments}) \\
\Xi & \ni & \xi & & (\text{quantifier-free } V\text{-formulas}) \\
P & \ni & \alpha & ::= & a \mid ?\xi \mid \alpha;\alpha \mid \alpha \cup \alpha \mid \alpha^*
\end{array}
$$

[2]For example, extending Peano Arithmetic with all true Π^0_1 sentences yields a dramatic increase in proof theoretic power, but without any new provably recursive functions [Kreisel, 1965].

Given a V-structure $\mathcal{S}$, the semantics of programs α is a binary relation $\xrightarrow{\alpha}$ between valuations (where a valuation is a function from the set of variables to elements of $\mathcal{S}$). That relation is defined by a straightforward recurrence on the complexity of α (see e.g. [Harel *et al.*, 2000]).

As usual, guarded iterative programs ("while programs") are definable by programs in P: $\texttt{skip} \equiv ?\top$, $\texttt{abort} \equiv ?\bot$, $(\texttt{if}\ \xi\ \texttt{then}\ \alpha)\ \texttt{else}\ \beta \equiv (?\xi;\alpha) \cup (?\neg\xi;\beta)$, and $(\texttt{while}\ \xi\ \texttt{do}\ \alpha) \equiv (?\xi;\alpha)^*;\ (?\neg\xi)$. When α is a guarded iterative program, the relation $\xrightarrow{\alpha}$ is univalent, i.e. a partial function. This is not the case for arbitrary regular α, since the union and star constructs are nondeterministic.

The *DL-formulas* are generated from atomic V-formulas by propositional connectives, quantifiers binding variables that are not activated in their scope, and the modal formation rule: if φ is a DL-formula, and α is a program, then $[\alpha]\,\varphi$ is a formula (pronounced "box alpha phi"). The latter is intended to convey that φ is true following any execution of α; that is, for a V-structure $\mathcal{S}$ and valuation η therein, $\mathcal{S},\,\eta \models [\alpha]\,\varphi$ iff $\mathcal{S},\,\eta' \models \varphi$ whenever $\eta\xrightarrow{\alpha}\eta'$. A construct dual to the box is the diamond, $\langle\alpha\rangle\varphi$, definable as $\neg[\alpha]\neg\varphi$. Thus $\mathcal{S},\,\eta \models \langle\alpha\rangle\,\varphi$ just in case $\mathcal{S},\,\eta' \models \varphi$ for some valuation η' where $\eta\xrightarrow{\alpha}\eta'$. The semantics of formulas φ is defined by recurrence on the syntax, just as for first-order formulas, but using the above for formulas of the form $[\alpha]\,\varphi$.

A DL formula of the form $\varphi \rightarrow [\alpha]\psi$, with φ and ψ first-order, is said to be a *partial-correctness assertion (PCA)*. It is often useful to abbreviate the formula above by $\varphi[\alpha]\psi$. The first-order formula φ is dubbed the PCA's *pre-condition*, and ψ its *post-condition*.

More general notions of programs and formulas are possible in Dynamic Logic, notably allowing first-order tests, or even DL-formulas as tests.[3] However, these are unrelated to our present purpose of comparing DL to Hoare's logic.

2.2 A Deductive Calculus for Dynamic Logic

A natural deductive calculus **DL** for Dynamic Logic is obtained by augmenting first-order logic with the rules of Table 1. This formalization is due primarily to Pratt [Pratt, 1976; Harel *et al.*, 1977]. The assignment rule is Hoare's, and the others are related to Segerberg's Propositional Dynamic Logic for regular programs [Segerberg, 1977]. (This is similar to the formalism 14.12 of [Harel *et al.*, 2000], but with the Convergence Rule omitted.[4] For first-order quantifier rules, the definition of "free occurrence of variable

[3] In the latter case programs and formulas are generated in tandem by a simultaneous inductive definition, see [Harel *et al.*, 2000].

[4] See [Leivant, 2004b] for a discussion of that rule.

First-order logic		
Modality	Generalization:	$\dfrac{\vdash \varphi}{[\alpha]\varphi}$ (no open assumptions)
	Box:	$[\alpha](\varphi \to \psi) \to ([\alpha]\varphi \to [\alpha]\psi)$
Atomic Programs	Assignment:	$[x := \mathbf{t}]\,\varphi \leftrightarrow \{\mathbf{t}/x\}\varphi$ ($\mathbf{t}$ free for x in φ)
	Test:	$[?\chi]\varphi \leftrightarrow (\chi \to \varphi)$
Program constructs	Composition:	$[\alpha;\beta]\varphi \leftrightarrow [\alpha][\beta]\varphi$
	Union:	$[\alpha \cup \beta]\varphi \leftrightarrow [\alpha]\varphi \wedge [\beta]\varphi$
	Iteration:	$[\alpha^*]\varphi \leftrightarrow \varphi \wedge [\alpha][\alpha^*]\varphi$
Limit	Invariance:	$\dfrac{\varphi \to [\alpha]\,\varphi}{\varphi \to [\alpha^*]\,\varphi}$ (no open assumptions)

Table 1. Pratt-Segerberg's Deductive System

x in formula φ" is amended to exclude the scope of $[\alpha]$ and $\langle\alpha\rangle$ in case x is activated in α.

From the Invariance Rule we obtain the Schema of Induction:

$$\psi \wedge [\alpha^*](\psi \to [\alpha]\,\psi) \to [\alpha^*]\,\psi)$$

Indeed, taking $\varphi \equiv \psi \wedge [\alpha^*](\psi \to [\alpha]\psi)$, we have $\vdash \varphi \to [\alpha]\varphi$ using the remaining axioms and rules, and so $\varphi \to [\alpha^*]\varphi$ by Iteration. The Induction template above readily follows. Note, however, that even if ψ is a first-order formula, the formula φ above is not.

A *V-theory* is a set of closed first-order V-formulas. Given a V-theory $\mathbf{T}$ we write $\mathsf{DL}(\mathbf{T})$ for the deductive formalism DL augmented with the formulas in $\mathbf{T}$ as axioms. We refer to $\mathbf{T}$ as the *background theory*.

By a straightforward induction on derivations, we conclude that DL is sound:

THEOREM 1. (Soundness of DL) *Let* $\mathbf{T}$ *be a V-theory,* φ *a DL V-formula. Suppose that* $\mathbf{DL}(\mathbf{T}) \vdash \varphi$. *Then* $\mathbf{T} \models \varphi$; *that is* φ *is true in every model of* $\mathbf{T}$.

2.3 An Interpretation of Dynamic Logic in Peano Arithmetic

The interpretation of logics of programs (executed over $\mathbb{N}$) in PA has been considered repeatedly, e.g. in [Bergstra and Tucker, 1983; Hajek, 1983; Hajek, 1986]. We rephrase it for reference below. Recall that *Peano Arithmetic* (**PA**) is the first-order theory over the vocabulary consisting of identifiers for $\mathbf{0}$, s (the successor function), $+$ and $\times$. There are three groups of axioms:

1. The two separation axioms for $\mathbb{N}$, i.e. Peano's Third and Fourth Axioms
$$\forall x.\, \mathsf{s}(x) \neq \mathbf{0} \qquad \text{and} \qquad \forall x, y.\, \mathsf{s}(x) = \mathsf{s}(y) \rightarrow x = y$$
2. The defining recurrence equations for addition and multiplication.
3. All instances of the schema of Induction,
$$\forall x.\, (\varphi[x] \rightarrow \varphi[\mathsf{s}x]) \rightarrow (\varphi[\mathbf{0}] \rightarrow \forall x.\, \varphi[x])$$

We will be interested in Π^0_1 formulas, that is formulas in the vocabulary above of the form $\forall \vec{x}.\, \psi$, where all quantifiers in ψ are bounded. It is well-known that these formulas are expressed in the form $\forall \vec{x}.\, \psi$, ψ an equation $\mathbf{t} = 0$, provided we refer also to a finite number of certain primitive-recursive (indeed, Kalmar-elementary) functions. We thus augment the vocabulary above with identifiers for these functions, and extend the theory with the recurrence equations defining them. In particular, we posit that the following functions are included:

1. A pairing function p (with $0 \notin \mathit{Range}(p)$); a sequence $x_0 \ldots x_n$ can then be coded unambiguously as a list: $\langle x_0, \ldots, x_n \rangle \equiv p(x_0, d(x_1, \cdots p(x_n, 0) \cdots))$.
2. Pair-projection functions j_0 and j_1, such that $j_i(d(x_0, x_1)) = x_i$, and $j_i(0) = 0$. A sequence-projection function can then be defined such that $J(i, x) = j_0(j_1^{[i]}(x))$; thus $J(i, \langle x_0, \ldots, x_n \rangle) = x_i$ if $i \leq n+1$, and 0 otherwise.

We write V_{PA} for the extended vocabulary, and **PA** for the extended theory. (Note that this theory is interpretable in standard Peano Arithmetic.) Also, ν will stand for the conjunction of the axioms of **PA** other than instances of Induction.

For each program α, and list of variables $\vec{x} = (x_1 \ldots x_k)$ that includes all variables mentioned in α, we define a V_{PA}-formulas $\mu^{\vec{x}}_{\alpha}$, with free variables $\vec{u} = (u_1 \ldots u_k)$ and $\vec{v} = (v_1 \ldots v_k)$ (not used in α), intended to define the

program α	formula $\mu_\alpha^{x_0 \ldots x_k}[u_0 \ldots u_k, v_0 \ldots v_k]$
$x_i := t[\vec{x}]$	$v_i = t[\vec{u}] \wedge \bigwedge_{j \neq i} v_j = u_j$
$?\xi[\vec{x}]$	$\xi[\vec{u}] \wedge \vec{v} = \vec{u}$
$\beta;\gamma$	$\exists \vec{w}.\ \mu_\beta^{\vec{x}}[\vec{u}, \vec{w}] \wedge \mu_\gamma^{\vec{x}}[\vec{w}, \vec{v}]$
$\beta \cup \gamma$	$\mu_\beta^{\vec{x}}[\vec{u}, \vec{v}] \vee \mu_\gamma^{\vec{x}}[\vec{u}, \vec{v}]$
β^*	$\exists \ell, c_0, \ldots, c_k \quad \wedge_i\, u_i = p(0, c_i)$ $\wedge \quad \forall j < k.\ \mu_\beta^{\vec{x}}[X_j, X_{j+1}]$ $\wedge \quad \wedge_i\, v_i = p(\ell, c_i)$ where X_j stands for $(p(j, c_0), \ldots, p(j, c_k))$

Table 2. Formulas defining program semantics over $\mathbb{N}$

semantics of α over $\mathbb{N}$: the formula holds iff $\eta \xrightarrow{\alpha} \eta'$ whenever $\eta(\vec{x}) = \vec{u}$ and $\eta'(\vec{x}) = \vec{v}$. The definition is by recurrence on α, as displayed in Table 2.

Of course, if $\vec{x}, \vec{y}$ are two different variable lists that include all program variables in α, then the formulas $\mu_\alpha^{\vec{x}}$ and $\mu_\alpha^{\vec{y}}$ are equivalent. We therefore write simply μ_α for $\mu_\alpha^{\vec{x}}$, where $\vec{x}$ is a listing of the variables occurring in α (say ordered as in some given ordering of all variables).

Using the formulas μ_α we interpret DL formulas in **PA**. If φ is a formula where all variables in programs are among $\vec{x}$, then V_{PA}-formula $\varphi^{\mu,\vec{x}}$ (with the same free variables as φ) is defined recursively as displayed in Table 3. (The cases for disjunction and $\exists$ are analogous.) Again, we write simply φ^μ when $\vec{x}$ above is a listing of the variables actually occurring in φ.

PROPOSITION 2. *If* $\mathbf{DL}(\mathbf{PA}) \vdash \varphi$, *then* $\mathbf{PA} \vdash \varphi^\mu$. *More generally, if* $\mathbf{T}$ *is an extension of* $\mathbf{PA}$, *and* $\mathbf{DL}(\mathbf{T}) \vdash \varphi$, *then* $\mathbf{T} \vdash \varphi^\mu$.

Proof. We show that if P is a proof in $\mathsf{DL}(\mathbf{T})$ of φ from assumptions $\psi_1, \ldots, \psi_k$, then φ^μ is provable in $\mathbf{T}$ from $\psi_1^\mu, \ldots, \psi_k^\mu$. The proof is a straightforward by induction on the length of P. The Induction Rule of **PA** is used to deal with the Invariance Rule of **DL**. ∎

2.4 An Interpretation of Peano Arithmetic in Dynamic Logic

While Dynamic Logic is intended as a logical formalism, that is with arbitrary relational structures as potential models, the semantic of iteration is based on the standard natural numbers. This makes it possible

DL formula φ	interpreting formula φ^μ
$\mathbf{t} = \mathbf{q}$	$\mathbf{t} = \mathbf{q}$
$\neg\psi$	$\neg(\psi^\mu)$
$\psi \wedge \chi$	$\psi^\mu \wedge \chi^\mu$
$\forall u.\varphi$	$\forall u.\ \varphi^\mu$ (Note: u not in $\vec{x}$)
$[\alpha]\varphi$	$\forall \vec{v}.\ (\ \mu_\alpha[\vec{x}, \vec{v}] \rightarrow \{\vec{v}/\vec{x}\}\varphi^\mu\ )$

Table 3. Interpretation of **DL**(**PA**) in **PA**

to generate "numeric variables" by programs. Let $\mathbf{N}(x)$ be the program $x := \mathbf{0}; (x := \mathsf{s}(x))^*$ (with s denoting the successor function). Then $\mathcal{S}, \eta \models [\mathbf{N}(x)]\varphi$ iff the formula $\forall x.\ N(x) \rightarrow \varphi$ is true in $\mathcal{S}, \eta$, with the unary identifier N interpreted as the set of denotations in $\mathcal{S}$ of the numerals $\mathbf{0}, \mathsf{s}(\mathbf{0}), \mathsf{s}(\mathsf{s}(\mathbf{0})), \ldots, \mathsf{s}^{[n]}(\mathbf{0}), \ldots$.

Thus, we interpret **PA** in **DL** as follows, writing φ^D for the **DL** formula that interprets a V_{PA}-formula φ. An equation $\mathbf{t} = \mathbf{t}'$ is interpreted as itself; the interpretation commutes with the propositional connectives: $(\neg\varphi)^D \equiv_{\mathrm{df}} \neg(\varphi^D)$, etc.; finally $(\forall x.\varphi)^D$ is $[\mathbf{N}(x)](\varphi^D)$, and $(\exists x.\varphi)^D$ is $\langle\mathbf{N}(x)\rangle(\varphi^D)$.

THEOREM 3. *Let φ be a closed V_{PA}-formula. The following are equivalent, where $\mathcal{N}$ is the standard model of* **PA**.

1. $\mathcal{N} \models \varphi$
2. $\mathcal{N} \models \varphi^D$
3. $\models \nu \rightarrow \varphi^D$

Proof. We prove a more general statement, referring to formulas φ that are not necessarily closed. For a valuation η in $\mathbb{N}$ and a V_{PA}-structure $\mathcal{S}$ write $\eta_{\mathcal{S}}$ for the valuation in $\mathcal{S}$ that assigns to a variable x the value $[\![\overline{\eta(x)}]\!]_{\mathcal{S}}$, i.e. the value in $\mathcal{S}$ of the n'th numeral, where $n = \eta(x)$. We now prove that for every valuation η in $\mathcal{N}$, the following are equivalent.

1. $\mathcal{N}, \eta \models \varphi$
2. $\mathcal{N}, \eta \models \varphi^D$

3. In every V_{PA}-structure $\mathcal{S}$ $\quad \mathcal{S}, \eta_{\mathcal{S}} \models \nu \rightarrow \varphi^D$.

(1) and (2) are equivalent by a straightfoward induction on φ, and (3) implies (2), since $\eta_{\mathcal{N}}$ is η, and $\mathcal{N} \models \nu$.

We prove, by induction on the number of logical operators in φ, that (2) implies (3). If φ is an equation $\mathbf{t} = \mathbf{t}'$, then φ^D is φ. Then (2) implies that $\mathbf{t}$ and $\mathbf{t}'$ evaluate under η to the same natural number n, and so there are calculations, based on the recurrence equations in ν, that derive the equations $\mathbf{t} = \bar{n}$ and $\mathbf{t}' = \bar{n}$ from the equations $x = \overline{\eta(x)}$ for variables x occurring in $\mathbf{t}$ or $\mathbf{t}'$. Thus $\mathcal{S}, \eta_{\mathcal{S}} \models \nu \rightarrow (\mathbf{t} = \mathbf{t}')$ for any model $\mathcal{S}$ of ν.

If φ is $\neg\psi$, consider first the case where ψ is an equation $\mathbf{t} = \mathbf{t}'$. Then (2) implies that $\mathbf{t}$ and $\mathbf{t}'$ evaluate under η to distinct natural numbers n, n'. Therefore, if $\mathcal{S}$ is a model of ν, then $\mathcal{S}, \eta_{\mathcal{S}} \models \mathbf{t} = \bar{n} \;\wedge\; \mathbf{t}' = \overline{n'}$, whereas $\nu \models \bar{n} \neq \overline{n'}$ since the Separation Axioms are among the conjuncts in ν. Thus $\quad \mathcal{S}, \eta_{\mathcal{S}} \models \nu \rightarrow \neg(\mathbf{t} = \mathbf{t}')$.

If φ is $\neg(\psi_0 \wedge \psi_1)$, then φ^D is $\neg(\psi_0^D \wedge \psi_1^D)$. So (2) implies that $\quad \mathcal{N}, \eta \models \neg\psi_i^D$ for $i = 0$ or $i = 1$. By IH this implies, for any model $\mathcal{S}$ of ν, that $\mathcal{S}, \eta_{\mathcal{S}} \models \neg\psi_i^D$, so $\quad \mathcal{S}, \eta_{\mathcal{S}} \models \varphi^D$.

The remaining cases for a negated φ, as well as the cases for other propositional connectives, are similar.

Finally, suppose that φ is $\forall x\, \psi$, so φ^D is $[\mathbf{N}(x)]\, \psi^D$. Then (2) implies $\mathcal{N}, \eta \models \{\bar{n}/x\}\psi$ for every $n \in \mathbb{N}$, which by IH implies that, for all models $\mathcal{S}$ of ν, $\mathcal{S}, \eta_{\mathcal{S}} \models \{\bar{n}/x\}\psi$. But in every $\mathcal{S}$, every terminating excution of the program $\mathbf{N}(x)$ yields a valuation that differs from the initial valuation $\eta\mathcal{S}$ only in assigning to x the value in $\mathcal{S}$ of some numeral $\bar{n}$. So $\mathcal{S}, \eta_{\mathcal{S}} \models [\mathbf{N}(x)]\, \psi$. ■

Note that our interpretation of **PA** in **DL** does not use quantification in **DL**. This is because all basic data is generated inductively, and so can be referred to as the output of a program. This does not imply, however, that quantification is generally redundant in Dynamic Logic. For example, **DL** specifications for programs over graphs would naturally use quantifiers over vertices and edges.

We could also interpret **PA** in **DL** without recourse to the nondeterministic * operator, by using insteach of the program $\mathbf{N}(x)$ the deterministic program

$$\mathbf{N}'(x) \quad \equiv \quad y := x;\ \mathbf{while}\ y \neq 0\ \mathbf{do}\ y := y - 1\ \mathbf{end}$$

(where y is a variable not used elsewhere). We would then interpret $\forall x.\varphi$ by $\forall x.\ [\mathbf{N}'(x)]\, \varphi^D$. Here the use of a quantifier in φ^D is indispensible of course, since nondeterminism is not available as a way to refer simultaneously to an unbounded number of values.

2.5 Proof Theoretic Equivalences

PROPOSITION 4. *If φ is a V_{PA}-formula, then* $\mathbf{PA} \vdash \varphi \leftrightarrow (\varphi^D)^\mu$.

Proof. By induction on φ. The only non-trivial case is for quantification. If φ is $\forall x.\,\psi$, then φ^D is $[N(x)]\psi^D$, and $(\varphi^D)^\mu$, i.e. $([N(x)]\psi^D)^\mu$, is trivially equivalent to $\forall u\,((\exists \ell. u = 0 + \ell) \to \{u/x\}\,(\psi^D)^\mu)$, which is equivalent to $\forall x.\varphi$. ■

LEMMA 5. *If a V_{PA}-formula φ is provable in* $\mathbf{PA}+\mathbf{T}$, *then the DL formula $\nu \to \varphi^D$ is provable in* $\mathbf{DL}(\mathbf{T})$.

Proof. By induction on the length of the proof of φ in $\mathbf{PA} + \mathbf{T}$. The only interesting case is instances of Induction, i.e. where φ is of the form

$$\forall x\,(\psi[x] \to \psi[\mathbf{s}x]) \to \psi[\mathbf{0}] \to \forall x\,\psi[x]$$

Then φ^D is

$$[x := \mathbf{0};\ (x := \mathsf{s}(x))^*](\psi^D[x] \to \psi^D[\mathsf{s}(x)]) \quad \to \\ \psi^D[\mathbf{0}] \to [x := \mathbf{0};\ (x := \mathsf{s}(x))^*]\psi[x]$$

By the Box-Distribution and Assignment rules of **DL**, it suffices to prove

$$[(x := \mathsf{s}(x))^*](\varphi^D[x] \to \varphi^D[\mathsf{s}(x)]) \quad \to \quad \varphi^D[x] \to [(x := \mathsf{s}(x))^*]\varphi[x]$$

or equivalently

$$[(x := \mathsf{s}(x))^*](\varphi^D[x] \to [x := \mathsf{s}(x)]\varphi^D[x]) \quad \to \quad \varphi^D[x] \to [(x := \mathsf{s}(x))^*]\varphi[x]$$

But this is an instance of the Induction Schema for DL, which we derived above in **DL**. ■

THEOREM 6. *Let* $\mathbf{T}$ *be a V_{PA}-theory, and φ a closed V_{PA}-formula. The following are equivalent.*

1. $\mathbf{PA} + \mathbf{T} \vdash \varphi$.
2. $\mathbf{DL}(\mathbf{PA} + \mathbf{T}) \vdash \varphi^D$.
3. $\mathbf{DL}(\mathbf{T}) \vdash \nu \to \varphi^D$.

Proof. Since ν is provable in $\mathbf{PA}$, (3) implies (2).

By Proposition 2 (2) implies $\mathbf{PA} + \mathbf{T} \vdash (\varphi^D)^\mu$, which by Proposition 4 implies (1).

Finally, (1) implies (3) by Lemma 5. ■

2.6 Interpreting Inductive Algebras

We can use DL modalities to force variables to range over any given inductively generated algebra, rather than the natural number as above. For example, the set Σ^* of words over a finite alphabet Σ can be identified with the free algebra generated from the constant ε, denoting the empty word, and, for each $a \in \Sigma$, a unary function identifier a. For example, if $\Sigma = \{0,1\}$ the constructors are the 0-ary ε and the unary $\mathbf{0}$ and $\mathbf{1}$. A word such as 011 is represented as $\mathbf{0}(\mathbf{1}(\mathbf{1}(\varepsilon)))$.

Define now a program analogous to $\mathbf{N}(x)$:

$$\mathbf{W}_{\{0,1\}}(x) \quad \equiv \quad x := \varepsilon;\ ((x := \mathbf{0}(x)) \cup (x := \mathbf{1}(x)))^*$$

Then $\mathcal{S}, \eta \models [\mathbf{W}_{\{0,1\}}(x)]\ \varphi$ exactly when φ is true for all denotations of the terms above.

The definition is similar for arbitrary word algebras Σ, and, indeed, for any free algebra. Multi-sorted free algebras can also be represented by such iterative programs. For example, to have x range over the algebra of lists over $\mathbb{N}$, with Λ denoting NIL and c denoting **cons**, we use the program

$$\mathbf{L}_N(x) \quad \equiv \quad x := \Lambda\ ;\ (\ y := \mathbf{0}\ ;\ (y := \mathsf{s}(y))^*;\ x := \mathsf{c}(y, x))^*$$

It is not hard to prove for every inductive algebra (possibly multi-sorted) statements analogous to Theorems 3 and 6. We shall not have use for these generalizations here.

3 When is Dynamic Logic conservative over Hoare's Logic?

3.1 Hoare's Logic for Regular Programs

Let V be a vocabulary, and $\mathbf{T}$ a V-theory. We define Hoare's Logic $\mathbf{H}^*(\mathbf{T})$ for reasoning about PCAs for regular V-programs with assignments. A crucial property of this logic is that it refers to PCAs and first-order V-formulas only, and not to more complex DL formulas. The inference rules are displayed in Table (4).

A formalism $\mathbf{H}(\mathbf{T})$ for reasoning about PCAs for guarded iterative programs is obtained by replacing the rules for Branching, Query, and Iteration by rules for the remaining program constructs of guarded iterative programs.

3.2 The Cases of Maximal and Minimal Background Theories

The largest possible sound first-order theory $\mathbf{T}$ for the structure $\mathcal{N}$ is the set *Th*($\mathcal{N}$) of <u>all</u> V_{PA}-formulas that are true in $\mathcal{N}$. For this choice of $\mathbf{T}$, $\mathbf{H}(\mathbf{T})$ proves exactly the PCAs that are true in $\mathcal{N}$, by Cook's Relative

ASSIGNMENT	$\{\mathbf{t}/x\}\varphi\ [x := \mathbf{t}]\ \varphi$	
COMPOSITION	$\dfrac{\psi\ [\alpha]\ \chi \qquad \chi\ [\beta]\ \varphi}{\psi\ [\alpha;\beta]\ \varphi}$	
BRANCHING	$\dfrac{\psi\ [\alpha]\ \varphi \qquad \psi\ [\beta]\ \varphi}{\psi\ [\alpha \cup \beta]\ \varphi}$	
ITERATION	$\dfrac{\varphi\ [\alpha]\ \varphi}{\varphi\ [\alpha^*]\ \varphi}$	
QUERY	$\dfrac{\psi \wedge \xi \rightarrow \varphi}{\psi\ [?\xi]\ \varphi}$	ξ quantifier-free
PRE-CONSEQUENCE	$\dfrac{\mathbf{T} \vdash \psi' \rightarrow \psi \qquad \psi\ [\alpha]\ \varphi}{\psi'\ [\alpha]\ \varphi}$	
POST-CONSEQUENCE	$\dfrac{\psi\ [\alpha]\ \varphi \qquad \mathbf{T} \vdash \varphi \rightarrow \varphi'}{\psi\ [\alpha]\ \varphi'}$	

Table 4. Hoare's Logic for regular programs

Completeness Theorem [Cook, 1978]. Since $\mathbf{DL}(\mathbf{T})$ is sound for $\mathcal{N}$, it cannot possibly prove additional PCAs.

At the other extreme we have as $\mathbf{T}$ the empty theory. The proof theory of **DL** is far richer and more complex than that of Hoare's Logic, but (as discussed in the Introduction) this by itself does not necessarily imply that more PCAs are proved in **DL**. The needed link between proof theoretic power and PCAs is provided by the following.

THEOREM 7. $\mathbf{DL}(\emptyset)$ *is not conservative over* $\mathbf{H}(\emptyset)$*: there are PCAs that are provable in Dynamic Logic, but not in Hoare's Logic.*

Proof. Let **PRA** be Primitive Recursive Arithmetic.[5] Let χ be a universal sentence $\forall x.\, \mathbf{t} = \mathbf{0}$, where $\mathbf{t} \equiv \mathbf{t}[x]$ is a term with x as the only variable, which is provable in **PA** but not in **PRA**; for example a sentence expressing the consistency of **PRA**. (The latter is expressible as a purely-universal formula since we use an extended vocabulary for **PA**.) Referring to the

[5] See e.g. [Hájek and Pudlák, 1993]. We can take here even Elementary Arithmetic.

interpretation above of **PA** in **DL**, χ^D is the PCA $\top\ [\mathbf{N}(x)]\ (\mathbf{t} = \mathbf{0})$, and so, by Theorem 6, the PCA

$$\pi \quad \equiv_{\mathrm{df}} \quad \nu\ [x := 0;\ (x := \mathsf{s}(x))^*]\ (\mathbf{t} = \mathbf{0})$$

is provable in $\mathbf{DL}(\emptyset)$.

Towards contradiction, suppose that π is provable in $\mathbf{H}(\emptyset)$. Given the syntax-directed form of the rules of **H**, we must have first-order formulas $\varphi(x)$ and $\psi(x)$ such that ν implies the following formulas in first-order logic:

$$\begin{array}{rl} & \varphi(0) \\ & \varphi(x) \to \psi(x) \\ & \psi(x) \to \psi(\mathsf{s}x) \\ \text{and} & \psi(x) \to \mathbf{t} = \mathbf{0} \end{array} \tag{1}$$

Reasoning within **PRA**, we thus have, for each $n \in \mathbb{N}$, a first-order proof of $\mathbf{t}[\bar{n}] = \mathbf{0}$ from ν. But **PRA** proves cut-elimination for first-order logic, so **PRA** proves that for all n there is a cut-free proof of $\mathbf{t}(\bar{n}) = \mathbf{0}$ from ν, whence a quantifier free proof of $\mathbf{t}(\bar{n}) = \mathbf{0}$ from instances of ν. It follows that **PRA** proves $\forall x.\, \mathbf{t} = \mathbf{0}$, contradicting the choice of **t**. ■

Note. The use of the nondeterministic program $\mathbf{N}(x)$ permit a streamlined proof, but is inessential. Had we used $\mathbf{N}'$ instead, the truth of $\forall x\, \mathbf{t} = \mathbf{0}$ would be equivalent to the validity of the DL formula

$$\nu \to \forall x\ [\mathbf{N}'(x)]\ \mathbf{t} = \mathbf{0}$$

i.e. to the validity of the PCA

$$\nu\ [\mathbf{N}'(x)]\ \mathbf{t} = \mathbf{0}$$

The rest of the argument is similar to the proof above.

3.3 The Boundary of Conservativeness is Peano's Arithmetic

We have seen that **DL** is conservative over **H** in the presence of the complete first-order theory *Th*($\mathcal{N}$) of $\mathcal{N}$, but is not conservative over **H** when the background theory is empty. It is natural to ask for a transition point.

Since the proof theoretic power of **DL** itself is akin to that of Peano Arithmetic, as illustrated by the results of [Bergstra and Tucker, 1983; Hajek, 1983; Hajek, 1986], it is not surprising that that point is Peano Arithmetic.

THEOREM 8. *Let* **T** *be a subtheory of Th*($\mathcal{N}$). *If* **T** *contains* **PA**, *then* **DL(T)** *is conservative over* **H(T)**.

Proof. Let the formulas $\mu_\alpha[\vec{u},\vec{v}]$ be as above. We prove first, by induction on α, that for all first-order formulas φ,

$$\mathbf{H(PA)} \vdash (\forall \vec{v}.\, \mu_\alpha[\vec{x},\vec{v}] \rightarrow \varphi[\vec{v}]) \ \ [\alpha] \ \ \varphi[\vec{x}] \tag{2}$$

If α is an assignment $x_i := \mathbf{t}$ then the pre-condition above is equivalent in first-order logic to $\{\mathbf{t}/x_i\}\varphi[\vec{x}]$. But $\{\mathbf{t}/x_i\}\varphi[\vec{x}]\ [\alpha]\ \varphi$ is an instance of the Assignment Rule, and so (2) follows by Pre-Consequence.

If α is a query $?\xi$, then $(\xi \rightarrow \varphi)\ [?\xi]\ \varphi$ follows from $(\xi \rightarrow \varphi) \wedge \xi \rightarrow \varphi$. Since the pre-condition of (2) is equivalent to $\xi \rightarrow \varphi$, we get (2) by Pre-Consequence.

If α is $\beta;\gamma$, then the precondition above is equivalent to

$$(\forall \vec{w}.\, (\, \mu_\beta[\vec{x},\vec{w}] \ \rightarrow \ \forall \vec{v}.\, (\, \mu_\gamma[\vec{w},\vec{v}] \rightarrow \varphi[\vec{v}]\,)\,)$$

By Pre-Consequence and IH for β we therefore have

$$\mathbf{H(PA)} \vdash (\forall \vec{v}.\, \mu_\alpha[\vec{x},\vec{v}] \rightarrow \varphi) \ \ [\beta] \ \ \forall \vec{v}.\, (\, \mu_\gamma[\vec{x},\vec{v}] \rightarrow \varphi[\vec{v}]\,)\,)$$

By IH for γ we have

$$\forall \vec{v}.\, (\, \mu_\gamma[\vec{x},\vec{v}] \rightarrow \varphi[\vec{v}]\,)\,) \ \ [\gamma] \ \ \varphi[\vec{x}]$$

We thus get (2) by the Composition Rule.

If α is $\beta \cup \gamma$, then the precondition above implies both $\forall \vec{v}.\ \mu_\beta[\vec{x},\vec{v}] \rightarrow \varphi$ and $\forall \vec{v}.\ \mu_\gamma[\vec{x},\vec{v}] \rightarrow \varphi$. So by IH and Pre-Consequence we have $\forall \vec{v}.\, \mu_\alpha[\vec{x},\vec{v}]\ \ [\beta]\ \ \varphi$ as well as $\forall \vec{v}.\, \mu_\alpha[\vec{x},\vec{v}]\ \ [\gamma]\ \ \varphi$, which together imply (2) by Branching.

Finally, if α is β^*, then the precondition of (2) implies

$$\forall \vec{v}.\, (\, \mu_\beta[\vec{x},\vec{v}] \rightarrow \forall w.(\mu_\alpha[\vec{v},\vec{w}] \rightarrow \varphi[\vec{w}]\,)\,) \tag{3}$$

as well as $\varphi[\vec{v}]$. From (3) we get, by Pre-Consequence and IH,

$$\forall \vec{v}.\, (\, \mu_\alpha[\vec{x},\vec{v}] \rightarrow \varphi[\vec{v}]) \ \ [\beta] \ \ \forall \vec{v}.\, (\, \mu_\alpha[\vec{x},\vec{v}] \rightarrow \varphi[\vec{v}])$$

By the Invariance Rule this implies

$$\forall \vec{v}.\, (\, \mu_\alpha[\vec{x},\vec{v}] \rightarrow \varphi[\vec{v}]) \ \ [\alpha] \ \ \forall \vec{v}.\, (\, \mu_\alpha[\vec{x},\vec{v}] \rightarrow \varphi[\vec{v}])$$

and so

$$\forall \vec{v}.\, (\, \mu_\alpha[\vec{x},\vec{v}] \rightarrow \varphi[\vec{v}]) \ \ [\alpha] \ \ \varphi$$

by Post-Consequence. This concludes the induction.

Now suppose that $\mathbf{DL}(\mathbf{T}) \vdash \varphi \to [\alpha]\,\psi$, where φ and ψ are first-order V_{PA}-formulas. Then, by Theorem 6,

$$\mathbf{PA} \vdash \varphi \wedge \mu_\alpha[\vec{x}, \vec{v}] \to \{\vec{v}/\vec{x}\}\psi \tag{4}$$

But $\mathbf{T}$ contains $\mathbf{PA}$, so the formulas above is provable in $\mathbf{T}$. Combining this with (2), we obtain by the rule of Pre-Consequence of $\mathbf{H}$ that $\varphi[\alpha]\psi$ is provable in $\mathbf{H}(\mathbf{PA})$. ■

We prove a result dual to Theorem 8, showing that if $\mathbf{T}$ is virtually any subtheory of $\mathbf{PA}$ of interest, then $\mathsf{DL}(\mathbf{T})$ is not conservative over $\mathsf{H}(\mathbf{T})$. We do not know whether the result remains true when its axioms have unbounded complexity. Let $\mathbf{PR}_k$ be $\mathbf{PA}$ with induction restricted to Π_k formulas.

THEOREM 9. *Let* $\mathbf{T}$ *be a* V_{PA}*-theory whose axioms are all of complexity* $\leq k$*, and which is consistent with* $\mathbf{PA}_k$*.*[6] *Then* $\mathbf{DL}(\mathbf{T})$ *is not conservative over* $\mathbf{H}(\mathbf{T})$*.*

Proof. The proof is similar to that of Theorem 7. Let χ be a universal sentence $\forall x.\, \mathbf{t}[x] = \mathbf{0}$ which is provable in $\mathbf{PA}{+}\mathbf{T}$, but not in $\mathbf{PA}_k + \mathbf{T}$, for instance a sentence expressing the consistency of $\mathbf{PA}_k + \mathbf{T}$. Then χ^D is the PCA $\top\ [\mathbf{N}(x)]\ (\mathbf{t} = \mathbf{0})$, and so, by Theorem 6, the PCA

$$\boldsymbol{\pi} \quad \equiv_{\mathrm{df}} \quad \nu\ [x := 0;\ (x := \mathsf{s}(x))^*]\ (\mathbf{t} = \mathbf{0})$$

is provable in $\mathbf{DL}(\mathbf{T})$.

However, if $\boldsymbol{\pi}$ were provable in $\mathbf{H}(\mathbf{T})$, then the formulas of (1) above would be provable from $\mathbf{T}$. Reasoning within $\mathbf{PA}_k + \mathbf{T}$, we obtain for each $n \in \mathbb{N}$ a first-order proof of $\mathbf{t}(\bar{n}) = \mathbf{0}$ from Π_k instances of induction. Since $\mathbf{PA}_1$ proves cut-elimination for first-order logic, all formulas in a cut-free proof of $\mathbf{t}(\bar{n}) = \mathbf{0}$ from $\mathbf{T}$ are (at most) Π_k. Using a partial truth definition and induction on proofs [Kreisel, 1965; Troelstra, 1973], it follows that $\mathbf{PA}_k + \mathbf{T}$ proves $\forall n.\, \mathbf{t}(\bar{n}) = \mathbf{0}$, contradicting the choice of $\mathbf{t}$. ■

Acknowledgements

Research partially supported by NSF grant CCR-CCR-0105651.

BIBLIOGRAPHY

[Bergstra and Tucker, 1983] J.A. Bergstra and J.V. Tucker. Hoare's Logic and Peano's Arithmetic. *Theoretical Computer Science*, 22:265–284, 1983.

[Cook, 1978] Stephen A. Cook. Soundness and completeness of an axiom system for program verification. *SIAM J. Computing*, 7(1):70–90, 1978.

[6] For the latter it suffices, of course, that $\mathbf{T}$ be sound for $\mathcal{N}$.

[Hájek and Pudlák, 1993] Petr Hájek and Pavel Pudlák. *Metamathematics of First-Order Arithmetic*. Perspectives in Mathematical Logic. Springer Verlag, Berlin, 1993.

[Hajek, 1983] Petr Hajek. Arithmetical interpretations of Dynamic Logic. *Jourmal of Symbolic Logic*, 48:704–713, 1983.

[Hajek, 1986] Petr Hajek. A simple dynamic logic. *Theoretical Computer Science*, 46:239–259, 1986.

[Harel *et al.*, 1977] D. Harel, A. Meyer, and V. Pratt. Computability and completeness in logics of programs. In *Proceedings of the ninth symposium on the Theorey of Computing*, pages 261–268, Providence, 1977. ACM.

[Harel *et al.*, 2000] David Harel, Dexter Kozen, and Jerzy Tiuryn. *Dynamic Logic*. MIT Press, Cabridge, MA, 2000.

[Kreisel, 1965] G. Kreisel. Mathematical logic. In T. Saaty, editor, *Lectures on Modern Mathematics*, volume III, pages 95–195. John Wiley, New York, 1965.

[Leivant, 2004a] Daniel Leivant. Partial correctness assertions provable in dynamic logics. In *Seventh Conference on Foundations of Software Science and Computation Structures (FOSSACS'04)*, pages 304–317. Springer-Verlag, LNCS 2987, 2004.

[Leivant, 2004b] Daniel Leivant. Proving termination assertions in dynamic logics. In *Proceedings of the Nineteenth IEEE Conference on Logic in Computer Science*, pages 89–99, Washington, 2004. IEEE Computer Society Press.

[Pratt, 1976] V. Pratt. Semantical considerations on Floyd-Hoare logic. In *Proceedings of the seventeenth symposium on Foundations of Computer Science*, pages 109–121, Washington, 1976. IEEE Computer Society.

[Segerberg, 1977] Krister Segerberg. A completeness theorem in the modal logic of programs (preliminary report). *Notices of the American Mathematical Society*, 24(6):A–552, 1977.

[Troelstra, 1973] A. S. Troelstra. *Metamathematical Investigation of Intuitionistic Arithmetic and Analysis*. Volume 344 of LNM. Springer-Verlag, Berlin, 1973.

Abduction and Cognition in Human and Logical Agents

Lorenzo Magnani

1 Abduction and Cognition

The development of human society has now reached a technological level in which issues concerning the creation and dynamics of information - especially in science - are absolutely crucial. Gradually, philosophical methods and problems are studied and understood in terms of the new information-theoretic notions. A new paradigm, aimed at unifying the different perspectives and providing some new design insights, arises by emphasizing the significance of the concept of *abduction*, in order to illustrate the problem-solving process and to propose a unified and rational epistemological model of scientific discovery, diagnostic reasoning, and other kinds of creative reasoning [Magnani, 2001c]. The concept of abduction nicely ties together both issues related to the dynamics of information and its systematic embodiment in segments of various types of knowledge.

Abduction is the process of *inferring* certain facts and/or laws and hypotheses that render some sentences plausible, that *explain* or *discover* some (eventually new) phenomenon or observation; it is the process of reasoning in which explanatory hypotheses are formed and evaluated. There are two main epistemological meanings of the word abduction [Magnani, 2001c]: 1) abduction that only generates "plausible" hypotheses ("selective" or "creative") and 2) abduction considered as inference "to the best explanation", which also evaluates hypotheses (cf. Figure 1). An illustration from the field of medical knowledge is represented by the discovery of a new disease and the manifestations it causes which can be considered as the result of a creative abductive inference. Therefore, "creative" abduction deals with the whole field of the growth of scientific knowledge. This is irrelevant in medical diagnosis where instead the task is to "select" from an encyclopedia of pre-stored diagnostic entities. We can call both inferences ampliative, selective and creative, because in both cases the reasoning involved amplifies, or goes beyond, the information incorporated in the premises.

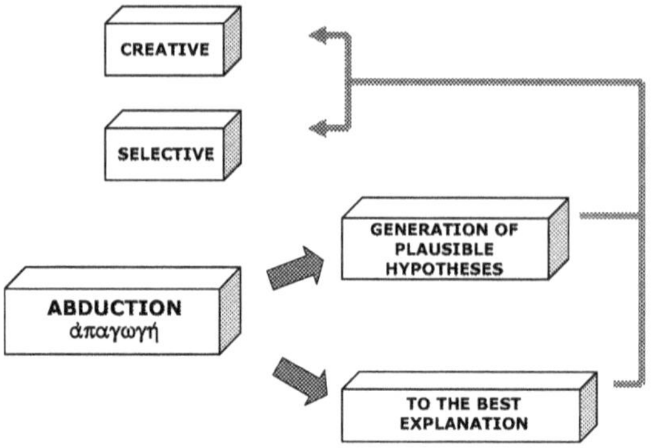

Figure 1. Creative and selective abduction.

Theoretical abduction[1] certainly illustrates much of what is important in creative abductive reasoning, in humans and in computational programs, but fails to account for many cases of explanations occurring in science when the exploitation of environment is crucial. It fails to account for those cases in which there is a kind of "discovering through doing", cases in which new and still unexpressed information is codified by means of manipulations of some external objects (*epistemic mediators*, cf.. below in this paper). The concept of *manipulative abduction*[2] captures a large part of scientific thinking where the role of action is central, and where the features of this action are implicit and hard to be elicited: action can provide otherwise unavailable information that enables the agent to solve problems by starting and by performing a suitable abductive process of generation or selection of hypotheses.

1.1 The "internal" side of creative reasoning

Throughout his career Peirce defended the thesis that, besides deduction and induction[3], there is a third mode of inference that constitutes the only method for really improving scientific knowledge, which he called *abduction*.

[1] [Magnani, 2001c] introduces the concept of theoretical abduction as a form of internal processing. He maintains that there are two kinds of theoretical abduction, "sentential", related to logic and to verbal/symbolic inferences, and "model-based", related to the exploitation of models such as diagrams, pictures, etc, cf. below in this paper.

[2] Manipulative abduction and epistemic mediators are introduced and illustrated in [Magnani, 2001b] and [Magnani, 2001c].

[3] Peirce clearly contrasted abduction with induction and deduction, by using the famous syllogistic model. More details on the differences between abductive and inductive/deductive inferences can be found in [Flach and Kakas, 2000] and [Magnani, 2001c].

Science improves and grows continuously, but this continuous enrichment cannot be due to deduction, nor to induction: deduction does not produce any new idea, whereas induction produces very simple ideas. New ideas in science are due to *abduction*, a particular kind of non-deductive[4] inference that involves the generation and evaluation of explanatory hypotheses.

I and others [Ramoni *et al.*, 1992] have developed an epistemological model of medical reasoning, called the Select and Test Model (ST-MODEL) which can be described in terms of the classical notions of abduction, deduction and induction. It describes the different roles played by such basic inference types in developing various kinds of medical reasoning (diagnosis, therapy planning, monitoring) but can be extended and regarded also as an illustration of scientific theory change. The model is consistent with the Peircian view regarding the various stages of scientific inquiry in terms of "hypothesis" generation, deduction (prediction), and induction.

As previously illustrated, I have introduced a distinction between "creative" and "selective" abduction. All we can expect of our "selective" abduction, is that it tends to produce hypotheses for further examination that have some chance of turning out to be the best explanation. Selective abduction will always produce hypotheses that give at least a partial explanation and therefore have a small amount of initial plausibility. In the syllogistic view advocated by Peirce (see below) concerning abduction as inference to the best explanation one might require that the final chosen explanation be the most "plausible".

Since the time of John Stuart Mill, the name given to all kinds of non deductive reasoning has been induction, considered as an aggregate of many methods for discovering causal relationships. Consequently induction in its widest sense is an ampliative process of the generalization of knowledge. Peirce distinguished various types of induction: a common feature of all kinds of induction is the ability to compare individual statements: by using induction it is possible to synthesize individual statements into general laws – inductive generalizations – in a defeasible way, but it is also possible to confirm or discount hypotheses.

Following Peirce, I am clearly referring here to the latter type of induction: abduction creates or selects hypotheses; from these hypotheses consequences are derived by deduction that are compared with the available data by induction. This perspective on hypothesis testing in terms of induction is also known in philosophy of science as the "hypothetico-deductive method" [Hempel, 1966] and is related to the idea of confirmation of scientific hypotheses, predominant in neopositivistic philosophy but also present in the anti-inductivist tradition of falsificationism [Popper, 1959].

[4]Non-deductive if we use the attribute "deductive" as designated by classical logic.

Deduction is an inference that refers to a logical implication. Deduction may be distinguished from abduction and induction on the grounds that the truth of the conclusion of the inference is guaranteed by the truth of the premises on which it is based only in deduction. Deduction refers to the so-called non-defeasible arguments. It should be clear that, on the contrary, when we say that the premises of an argument provide partial support for the conclusion, we mean that if the premises were true, they would give us good reasons – but not conclusive reasons – to accept the conclusion. That is to say, although the premises, if true, provide some evidence to support the conclusion, the conclusion may still be false (arguments of this type are called inductive, or abductive, arguments).

From the deductive perspective of classical logic the abductive inference rule corresponds to the well-known fallacy called affirming the consequent (simplified to the propositional case $\varphi \rightarrow \psi$, ψ , then φ).

1.2 Sentential abduction

Many attempts have been made to model abduction by developing some formal tools in order to illustrate its computational properties and the relationships with the different forms of deductive reasoning [see, for example, [Bylander *et al.*, 1991]]. Some of the formal models of abductive reasoning are based on the theory of the *epistemic state* of an agent [Boutilier and Becher, 1995], where the epistemic state of an individual is modeled as a consistent set of beliefs that can change by expansion and contraction (*belief revision framework*).

Deductive models of abduction may be characterized as follows. An explanation for β relative to background theory T will be any α that, together with T, entails β (normally with the additional condition that $\alpha \cup T$ be consistent). Such theories are usually generalized in many directions: first of all by showing that explanations entail their conclusions only in a *defeasible* way (there are many potential explanations), thus joining the whole area of so-called nonmonotonic logic or of probabilistic treatments; second, trying to show how some of the explanations are relatively implausible, elaborating suitable technical tools (for example in terms of modal logic) able to capture the notion of preference among explanations.

The idea of consistency that underlies some of the more recent deductive consistency-based models of selective abduction (diagnostic reasoning) is the following: any inconsistency (anomalous observation) refers to an aberrant behavior that can usually be accounted for by finding some set of components of a system that, if behaving abnormally, will entail or justify the actual observation. The observation is anomalous because it contradicts the expectation that the system involved is working according to specification.

This types of deductive model go beyond the mere treatment of selective abduction in terms of preferred explanations and include the role of those components whose abnormality makes the observation (no longer anomalous) consistent with the description of the system [Boutilier and Becher, 1995; Magnani, 2001a].

This kind of *sentential frameworks* exclusively deals with selective abduction (diagnostic reasoning)[5] and relates to the idea of preserving *consistency*. Exclusively considering the sentential view of abduction does not enable us to say much about creative processes in science, and, therefore, about the nomological and most interesting creative aspects of abduction. It mainly refers to the *selective* (diagnostic) aspects of reasoning and to the idea that abduction is mainly an inference *to the best explanation* [Magnani, 2001c]: when used to express the creative events it is either empty or replicates the well-known *Gestalt* model of radical innovation. It is empty because the sentential view stops any attempt to analyze the creative processes: the event of creating something new is considered so radical and instantaneous that its irrationality is immediately involved.

For Peirce abduction is an *inferential process* that includes all the operations whereby hypotheses and theories are constructed. Hence abduction has to be considered as a kind of *ampliative* inference that, as already stressed, is not logical and truth preserving: indeed valid deduction does not yield any new information, for example new hypotheses previously unknown.

From the point of view of computational philosophy the sentential models of theoretical abduction are limited, because they do not capture various reasoning tasks [Magnani, 1999]:

1. the role of statistical explanations, where what is explained follows only probabilistically and not deductively from the laws and other tools that do the explaining;

2. the sufficient conditions for explanation;

3. the fact that sometimes the explanations consist of the application of schemas that fit a phenomenon into a pattern without realizing a deductive inference;

4. the idea of the existence of high-level kinds of *creative* abductions;

5. the existence of model-based abductions (cf. the following section);

[5]As previously indicated, it is important to distinguish between *selective* (abduction that merely selects from an encyclopedia of pre-stored hypotheses), and *creative* abduction (abduction that generates new hypotheses).

6. the fact that explanations usually are not complete but only furnish *partial* accounts of the pertinent evidence [Thagard and Shelley, 1997];

7. the fact that one of the most important virtues of a new scientific hypothesis (or of a scientific theory) is its power of explaining *new*, previously *unknown* facts: "[...] these facts will be [...] unknown at the time of the abduction, and even more so must the auxiliary data which help to explain them be unknown. Hence these future, so far unknown explananda, cannot be among the premises of an abductive inference" [Hintikka, 1998], observations become real and explainable only by means of new hypotheses and theories, once discovered by abduction.

Important developments in the fields of logical models of abduction – also touching some related problems in artificial intelligence (AI) and devoted to overcome the limitations above – are illustrated in [Flach and Kakas, 2000] and in [Gabbay and Kruse, 2000; Gabbay and Woods, 2005; Gabbay and Woods, 2006]. Cf. also the recent papers contained in the collections [Magnani *et al.*, 2002; Magnani, 2006].

1.3 Model-based abduction and its external dimension

Computational philosophy taught us how to provide a suitable framework for constructing actual models of the most interesting cases of conceptual changes in science: we do not have to limit ourselves to the *sentential* view of theoretical abduction but we have to consider a broader *inferential* one: the *model-based* sides of creative abduction.

From Peirce's philosophical point of view, all thinking is in signs, and signs can be icons, indices or symbols. Moreover, all inference is a form of sign activity, where the word sign includes "feeling, image, conception, and other representation" [Peirce, CP, 5.283], and, in Kantian words, all synthetic forms of cognition. That is, a considerable part of the thinking activity is model-based. Of course model-based reasoning acquires its peculiar creative relevance when embedded in abductive processes, so that we can individuate a *model-based abduction*. Hence, we must think in terms of model-based abduction (and not in terms of sentential abduction) to explain complex processes like scientific conceptual change. Different varieties of *model-based abductions* [Magnani, 1999] are related to the high-level types of scientific conceptual change [see, for instance, [Thagard, 1992]].

Following Nersessian [Nersessian, 1995; Nersessian, 1999], the term "model-based reasoning" is used to indicate the construction and manipulation of various kinds of representations, not mainly sentential and/or formal, but mental and/or related to external mediators.

Obvious examples of model-based reasoning are constructing and manipulating visual representations, thought experiment, analogical reasoning, but also for example the so-called "tunnel effect" [Cornuéjols *et al.*, 2000], occurring when models are built at the intersection of some operational interpretation domain – with its interpretation capabilities – and a new ill-known domain.

Manipulative abduction [Magnani, 2001c] - contrasted with theoretical abduction - happens when we are thinking through doing and not only, in a pragmatic sense, about doing. So the idea of manipulative abduction goes beyond the well-known role of experiments as capable of forming new scientific laws by means of the results (nature's answers to the investigator's question) they present, or of merely playing a predictive role (in confirmation and in falsification). Manipulative abduction refers to an extra-theoretical behavior that aims at creating communicable accounts of new experiences to integrate them into previously existing systems of experimental and linguistic (theoretical) practices. As I said above, the existence of this kind of extra-theoretical cognitive behavior is also testified by the many everyday situations in which humans are perfectly able to perform very efficacious (and habitual) tasks without the immediate possibility of realizing their conceptual explanation. In the following two sections manipulative abduction will be considered from the perspective of the relationship between unexpressed knowledge and external representations.

2 Unexpressed Knowledge and External Mediators

The power of model-based abduction mainly depends on its ability to render explicit a certain amount of important information, unexpressed at the level of available data. It also has a fundamental role in the process of transformation of knowledge from its *tacit* to its *explicit* forms, and in the subsequent elicitation and use of knowledge. Let us describe how this happens.

As pointed out by Polanyi in his epistemological investigation, a large part of knowledge is not explicit, but tacit: we know more than we can tell and we can know nothing without relying upon those things which we may not be able to tell [Polanyi, 1966]. Polanyi's concept of knowledge is based on three main theses: first, discovery cannot be accounted for by a set of articulated rules or algorithms; second, knowledge is public and also to a very great extent personal (i.e. it is constructed by humans and therefore contains emotions, "passions"); third, an important part of knowledge is tacit.

As Polanyi contends, human beings acquire and use knowledge by actively creating and organizing their own experience: for example tacit knowledge is the practical knowledge used to perform a task. The existence of this kind of not merely theoretical knowing behavior is also testified by the many everyday situations in which humans are perfectly able to perform very efficacious (and habitual) tasks without the immediate possibility of realizing their conceptual explanation. In some cases the conceptual account for doing these things was at one point present in memory, but now has deteriorated, and it is necessary to reproduce it, in other cases the account has to be constructed for the first time, like in creative experimental settings in science.

[Hutchins, 1995] illustrates the case of a navigation instructor that performed an automatized task for 3 years involving a complicated set of plotting manipulations and procedures. The insight concerning the conceptual relationships between relative and geographic motion came to him suddenly "as lay in his bunk one night". This example explains that many forms of learning can be represented as the result of the capability of giving conceptual and theoretical details to already automatized manipulative executions. The instructor does not discover anything new from the point of view of the objective knowledge about the involved skill, however, we can say that his conceptual awareness is new from the local perspective of his individuality.

We can find a similar situation also in the process of scientific creativity. In the cognitive view of science, it has been too often underlined that conceptual change just involves a *theoretical* and "internal" replacement of the main concepts. But usually researchers forget that a large part of these processes are instead due to *practical* and "external" *manipulations* of some kind, prerequisite to the subsequent work of theoretical arrangement and knowledge creation. When these processes are creative we can speak of manipulative abduction (cf. above). Scientists sometimes need a first "rough" and concrete experience of the world to develop their systems, as a *cognitive-historical* analysis of scientific change [Nersessian, 1992; Gooding, 1990] has carefully shown.

Traditional examinations of how problem-solving heuristics create new representations in science have analyzed the frequent use of analogical reasoning, imagistic reasoning, and thought experiment, from an internal point of view. However, attention has not been focalized on those particular kinds of heuristics that resort to the existence of *extra-theoretical* ways of thinking – *thinking through doing* [Magnani, 2002b]. Indeed many cognitive processes are centered on *external representations*, as a means to create communicable accounts of new experiences ready to be integrated into previously existing systems of experimental and linguistic (theoretical) practices.

Interesting insights can arise regarding these problems studying them from a different contrasting approach, which moves away from Simon's paradigm, but which can offer a rational solution to the problem of creativity and conceptual change in terms of mathematical models: the *dynamic* approach [Port and van Gelder, 1995]. The traditional computational view treats cognition as a process that computes internal symbolic representations of the external world. But this approach is considered too reductive, since it is based on the functionalist hypothesis (which cannot render the *external dimension* of cognition), and on a computation of static entities. It is useful to integrate it with a dynamical modeling of cognition, which is able to describe abductive processes as *dynamical entities* "unfolding" in real time (we can also gain a better cognitive-historical perspective) [Magnani and Piazza, forthcoming]. From this point of view it is possible to model the terms (objects or propositions) that constitute abduction by considering the *attractors* in a dynamical system. This can be achieved by topologically specifying the semantic content of the inferential process through the spatial relations between its defining attractors. We can therefore consider the process of progressive development of "new" concepts and replacement of old ones in terms of temporal evolving patterns defined by interactions between topological configurations of attractors.

Moreover, a central point in the dynamical approach is the importance assigned to the "whole" cognitive system: cognitive activity is in fact the result of a complex interplay and simultaneous coevolution, in time, of the states of mind, body, and external environment. Even if, of course, a large portion of the complex environment of a thinking agent is internal, and consists in the proper software composed of the knowledge base and of the inferential expertise of the individual, nevertheless a "real" cognitive system is composed by "distributed cognition" among people and some "external" objects and technical artifacts [Hutchins, 1995; Norman, 1993].

For example, in the case of the construction and examination of diagrams in geometrical reasoning, specific experiments serve as states and the implied operators are the manipulations and observations that transform one state into another. The geometrical outcome depends upon practices and specific sensory-motor activities performed on a non-symbolic object, which acts as a dedicated external representational medium supporting the various operators at work. There is a kind of an epistemic negotiation between the sensory framework of the geometer and the external reality of the diagram [Magnani, 2002a]. This process involves an external representation consisting of written symbols and figures that for example are manipulated "by hand". The cognitive system is not merely the mind-brain of the person performing the geometrical task, but the system consisting of the whole body (cognition is *embodied*) of the person plus the external physical representation. In geometrical discovery the whole activity of cognition is located in the system consisting of a human together with diagrams.

An external representation can modify the kind of computation that a human agent uses to reason about a problem: the Roman numeration system eliminates, by means of the external signs, some of the hardest parts of the addition, whereas the Arabic system does the same in the case of the difficult computations in multiplication. The capacity for inner reasoning and thought results from the internalization of the originally external forms of representation. In the case of the external representations we can have various objectified knowledge and structures (like physical symbols – e.g. written symbols, and objects – e.g. three-dimensional models, shapes and dimensions), but also external rules, relations, and constraints incorporated in physical situations (spatial relations of written digits, physical constraints in geometrical diagrams and abacuses) [Zhang, 1997]. The external representations are contrasted with the internal representations that consist in the knowledge and the structure in memory, as propositions, productions, schemas, neural networks, models, prototypes, images.

The external representations are not merely memory aids: they can give people access to knowledge and skills that are unavailable to internal representations, help researchers to easily identify aspects and to make further inferences, they constrain the range of possible cognitive outcomes in a way that some actions are allowed and others forbidden. The mind is limited because of the restricted range of information processing, the limited power of working memory and attention, the limited speed of some learning and reasoning operations; on the other hand the environment is intricate, because of the huge amount of data, real time requirement, uncertainty factors. Consequently, we have to consider the whole system, consisting of both internal and external representations, and their role in optimizing the whole cognitive performance of the distribution of the various subtasks. It is well-known that in the history of geometry many researchers used internal mental imagery and mental representations of diagrams, but also self-generated diagrams (external) to help their thinking.

3 The Manipulative Framework

I have illustrated above the notion of tacit knowledge and I have proposed an extension of that concept. From the perspective of a more adequate information-theoretic account of cognition surely there is something more important beyond the tacit knowledge "internal" to the subject – considered by Polanyi as personal, embodied and context specific. We can also speak of a sort of tacit information "embodied" into the whole relationship between our mind-body system and suitable external representations. An information we can extract, explicitly develop, and transform in knowledge contents, to solve problems, as it was already manifest, for instance, in the geometrical problem contained in the *Meno* [Plato, 1977], even if philosophers know perfectly that Plato considered this activity to be just the result of reminiscence and not of discovery [Magnani, 2001c, chapter 1].

As I have already stressed, Peirce considers inferential any cognitive activity whatever, not only conscious abstract thought; he also includes perceptual knowledge and subconscious cognitive activity. For instance in subconscious mental activities visual representations play an immediate role. Peirce gives an interesting example of model-based abduction related to sense activity: "A man can distinguish different textures of cloth by feeling: but not immediately, for he requires to move fingers over the cloth, which shows that he is obliged to compare sensations of one instant with those of another" [Peirce, CP, 5.221]. This surely suggests that abductive movements have also interesting extra-theoretical characters and that there is a role in abductive reasoning for various kinds of manipulations of external objects. *All* knowing is *inferring* and inferring is not instantaneous, it happens in a process that needs an activity of comparisons involving many kinds of models in a more or less considerable lapse of time.

All these considerations suggest, then, that there exist a creative form of thinking through doing,[6] fundamental as much as the theoretical one. I call this kind of reasoning *manipulative abduction* (cf. above). As already said *manipulative* abduction happens when we are thinking *through* doing and not only, in a pragmatic sense, about doing. Of course the study of this kind of reasoning is important not only in delineating the actual practice of abduction, but also in the development of programs computationally adequate to rediscover, or discover for the first time, for example, scientific hypotheses and mathematical theorems or laws.

Various *templates* of manipulative behavior exhibit some regularities. The activity of manipulating external things and representations is highly conjectural and not immediately explanatory: these templates are "hypotheses of behavior" (creative or already cognitively present in the scientist's mind-body system, and sometimes already applied) that abductively enable a kind of epistemic "doing". Hence, some templates of action and manipulation can be selected in the set of the ones available and pre-stored, others have to be created for the first time to perform the most interesting creative cognitive accomplishments of manipulative abduction.

Some common features of the tacit templates of manipulative abduction (cf. Figure 3), that enable us to manipulate things and experiments in science are related to: 1. sensibility towards the aspects of the phe-

[6]In this way the cognitive task is achieved on *external* representations used in lieu of internal ones. Here action performs an *epistemic* and not a merely performatory role, relevant to abductive reasoning.

nomenon which can be regarded as *curious* or *anomalous*; manipulations have to be able to introduce potential inconsistencies in the received knowledge (Oersted's report of his experiment about electromagnetism is devoted to describe some anomalous aspects that did not depend on any particular theory of the nature of electricity and magnetism); 2. preliminary sensibility towards the *dynamical* character of the phenomenon, and not to entities and their properties, common aim of manipulations is to practically reorder the dynamic sequence of events into a static spatial one that should promote a subsequent bird's-eye view (narrative or visual-diagrammatic); 3. referral to experimental manipulations that exploit *artificial apparatus* to free new possible stable and repeatable sources of information about hidden knowledge and constraints (Davy set-up in terms of an artifactual tower of needles showed that magnetization was related to orientation and does not require physical contact); 4. various contingent ways of epistemic acting: *looking* from different perspectives, *checking* the different information available, *comparing* subsequent events, *choosing*, *discarding*, *imaging* further manipulations, *re-ordering* and *changing relationships* in the world by implicitly *evaluating* the usefulness of a new order (for instance, to help memory).

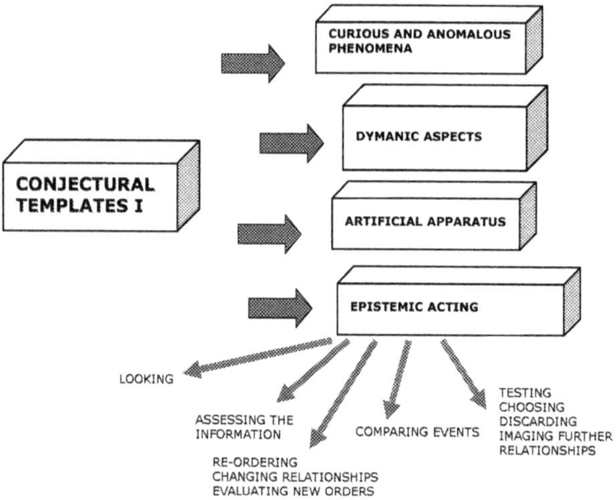

Figure 2. Conjectural templates I.

[Gooding, 1990] refers to this kind of concrete manipulative reasoning when he illustrates the role in science of the so-called "construals" that embody tacit inferences in procedures that are often apparatus and machine based. The embodiment is of course an expert manipulation of objects in a highly constrained experimental environment, and is directed by abductive movements that imply the strategic application of old and new *templates* of behavior mainly connected with extra-theoretical components, for instance emotional, esthetical, ethical, and economic.

The whole activity of manipulation is in fact devoted to building various external *epistemic mediators*[7] that function as an enormous new source of information and knowledge. Therefore, manipulative abduction represents a kind of redistribution of the epistemic and cognitive effort to manage objects and information that cannot be immediately represented or found internally (for example exploiting the resources of visual imagery).[8]

From the point of view of everyday situations manipulative abductive reasoning and epistemic mediators exhibit very other interesting templates (we can find the first three in geometrical constructions)(cf. Figure 3): 1. action elaborates a *simplification* of the reasoning task and a redistribution of effort across time [Hutchins, 1995], when we need to manipulate concrete things in order to understand structures which are otherwise too abstract [Piaget, 1974]], or when we are in presence of *redundant* and unmanageable information; 2. action can be useful in presence of *incomplete* or *inconsistent* information – not only from the "perceptual" point of view – or of a diminished capacity to act upon the world: it is used to get more data to restore coherence and to improve deficient knowledge; 3. action enables us to build *external artifactual models* of task mechanisms instead of the corresponding internal ones, that are adequate to adapt the environment to the agent's needs. 4. action as a *control of sense data* illustrates how we can change the position of our body (and/or of the external objects) and how to exploit various kinds of prostheses (Galileo's telescope, technological instruments and interfaces) to get various new kinds of stimulation: action provides some tactile and visual information (e.g., in surgery), otherwise unavailable.

[7]This expression, introduced by Magnani [Magnani, 2001c], is derived from the cognitive anthropologist Hutchins [Hutchins, 1995], who coined the expression "mediating structure" to refer to various external tools that can be built to cognitively help the activity of navigating in modern but also in "primitive" settings. Any written procedure is a simple example of a cognitive "mediating structure" with possible cognitive aims, so mathematical symbols and diagrams: "Language, cultural knowledge, mental models, arithmetic procedures, and rules of logic are all mediating structures too. So are traffic lights, supermarkets layouts, and the contexts we arrange for one another's behavior. Mediating structures can be embodied in artifacts, in ideas, in systems of social interactions [...]" [Hutchins, 1995, pp. 290–291].

[8]It is difficult to preserve precise spatial and geometrical relationships using mental imagery, in many situations, especially when one set of them has to be moved relative to another.

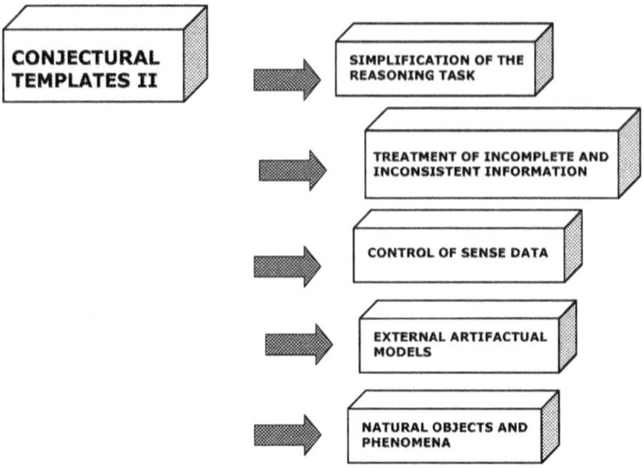

Figure 3. Conjectural templates II.

Also natural phenomena can play the role of external artifactual models: under Micronesians' manipulations of their images, the stars acquire a structure that "becomes one of the most important structured representational media of the Micronesian system" [Hutchins, 1995, p. 172]. The external artifactual models are endowed with functional properties as components of a memory system crossing the boundary between person and environment (for example they are able to transform the tasks involved in allowing simple manipulations that promote further visual inferences at the level of model-based abduction). The cognitive process is *distributed* between a person (or a group of people) and external representation(s), and so obviously *embedded* and *situated* in a society and in a historical culture.

4 Mimetic and Creative Representations

4.1 External and internal representations

After having illustrated the main features of my cognitive-epistemological analysis of abduction, it is necessary to further stress the dynamics involved in the interplay between internal and external representations. This will furnish a tool for exploring the relationship between human and logical agents in section 5.

We can say that

- *external representations* are formed by external materials that re-express (through reification) concepts and problems that are already present in the mind or concepts and problems that do not have a *natural home* in the brain;

- *internalized representations* are internal re-projections, a kind of recapitulations, (learning) of external representations in terms of neural patterns of activation in the brain. They can sometimes be "internally" manipulated like external objects and can originate new internal reconstructed representations through the neural activity of *transformation* and *integration*.

This process explains why human beings seem to perform both computations of a *connectionist* type such as the ones involving representations as

- (I Level) *patterns of neural activation* that arise as the result of the interaction between body and environment (and suitably shaped by the evolution and the individual history): pattern completion or image recognition,

  and computations that use representations as

- (II Level) *derived combinatorial syntax and semantics* dynamically shaped by the various external representations and reasoning devices found or constructed in the environment (for example geometrical diagrams); they are neurologically represented contingently as patterns of neural activations that "sometimes" tend to become stabilized structures and to fix and so *to permanently belong to the I Level* above.

The I Level originates those *sensations* (they constitute a kind of "face" we think the world has), that provide room for the II Level to reflect the structure of the environment, and, most important, that can follow the computations suggested by these external structures. It is clear we can now conclude that the growth of the brain and especially the synaptic and dendritic growth are profoundly determined by the environment.

When the fixation is reached the patterns of neural activation no longer need a direct stimulus from the environment for their construction. In a certain sense they can be viewed as *fixed internal records* of *external structures* that *can exist* also in the absence of such external structures. These patterns of neural activation that constitute the I Level Representations always keep record of the experience that generated them and, thus, always carry the II Level Representation associated to them, even if in a different form, the form of *memory* and not the form of a vivid sensorial experience. Now, the human agent, via neural mechanisms, can retrieve these II Level Representations and use them as *internal* representations or use parts of them to construct new internal representations very different from the ones stored in memory (cf. also [Gatti and Magnani, 2005]).

Human beings delegate cognitive features to external representations because in many problem solving situations the internal computation would be impossible or it would involve a very great effort because of human mind's limited capacity. First a kind of alienation is performed, second a recapitulation is accomplished at the neuronal level by re-representing internally that which was "discovered" outside. Consequently only later on we perform cognitive operations on the structure of data that synaptic patterns have "picked up" in an analogical way from the environment. We can maintain that internal representations used in cognitive processes like many events of *meaning creation* meaning have a deep origin in the experience lived in the environment.

I think there are two kinds of artefacts that play the role of *external objects* (representations) active in this process of disembodiment of the mind: *creative* and *mimetic*. Mimetic external representations mirror concepts and problems that are already represented in the brain and need to be enhanced, solved, further complicated, etc. so they sometimes can creatively give rise to new concepts and meanings.[9]

From this perspective the expansion of the minds is in the meantime a continuous process of *disembodiment* of the minds themselves into the *material world* around them. In this regard the evolution of the mind is inextricably linked with the evolution of many kinds of large, integrated, material cognitive systems. In the following sections I will illustrate some features of this extraordinary interplay between human brains and the *logical cognitive systems* they make. We acknowledge that material artefacts like logical agents are tools for thoughts as is language: tools for exploring, expanding, and manipulating our own minds.

4.2 Manipulating meanings

If the structures of the environment play such an important role in shaping our representations and, hence, our cognitive processes, we can expect that physical manipulations of the environment deserve a cognitive relevance.

Several authors have pointed out the role that physical actions can have at a cognitive level. In this sense [Kirsh and Maglio, 1994] distinguish actions into two categories, namely *pragmatic actions* and *epistemic actions*. Pragmatic actions are the actions that an agent performs in the environment in order to bring itself physically closer to a goal. In this case the action modifies the environment so that the latter acquires a configuration that

[9]Following this perspective it is at this point evident that the "mind" transcends the boundary of the individual and includes parts of that individual's environment.

helps the agent to reach a goal which is understood as physical, that is, as a desired state of affairs. Epistemic actions are the actions that an agent performs in the environment in order to discharge the mind of a cognitive load or to extract information that is hidden or that would be very hard to obtain only by internal computation.

In this section I want to focus specifically on the relationship that can exist between manipulations of the environment and representations. In particular, I want to examine whether external manipulations can be considered as means to construct external representations.

If a manipulative action performed upon the environment is devoted to create a configuration of elements that carries relevant information, that action will well be able to be considered as a cognitive process and the configuration of elements it creates will well be able to be considered an external representation and mediators. In this case, we can really speak of an *embodied* cognitive process in which an action constructs an external representation by means of manipulation. We define *cognitive manipulating* as any manipulation of the environment devoted to construct external configurations that can count as representations.

An example of cognitive manipulating is the diagrammatic demonstration illustrated in figure 5, taken from the field of geometry. In this case a simple manipulation of the triangle ABC gives rise to an external configuration ABCDE that carries relevant information about the internal angles of a triangle. It is the Euclid's Proposition XXXII, *Elements*, Book I. Given the properties of the parallel lines (Proposition XIX), we have $\alpha = \alpha'$ and $\beta = \beta'$. Hence the exterior angle $\alpha' + \beta'$ is equal to $\alpha + \beta$, and because of $\gamma + \alpha + \beta = 2R$, we have $\alpha + \beta + \psi = 2R$ (from [Euclid, 1846, p. 30]).

The entire process through which an agent arrives at a physical action that can count as cognitive manipulating can be understood by means of the concept of manipulative abduction (cf. above section 3 and [Magnani, 2001b]). Manipulative abduction is a specific case of cognitive manipulating in which an agent, when faced with an external situation from which it is hard or impossible to extract new meaningful features of an object, selects or creates an action that structures the environment in such a way that it gives information which would be otherwise unavailable and which is used specifically to infer explanatory hypotheses.

5 Beyond Peirce: Human Agents, Logical Agents

When searching for a satisfying logic of abduction (see above section 1.2) Gabbay and Woods [2005] disagree with the Peircian rejection of the possibility of a practical logic and consequently of a logic of abductive reasoning in practical contexts. We have to overcome Peirce's limitations that are

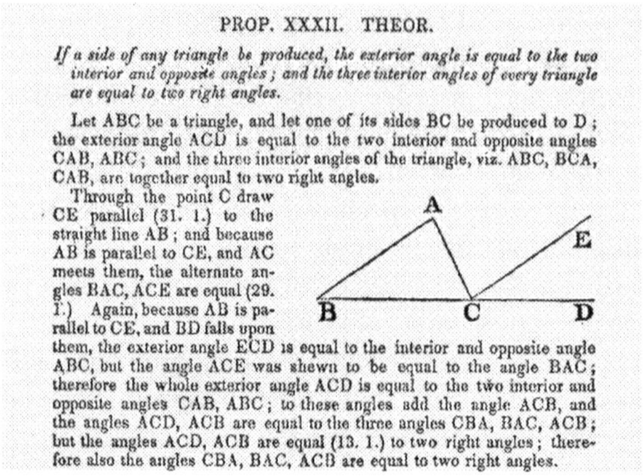
PROP. XXXII. THEOR.

If a side of any triangle be produced, the exterior angle is equal to the two interior and opposite angles; and the three interior angles of every triangle are equal to two right angles.

Let ABC be a triangle, and let one of its sides BC be produced to D; the exterior angle ACD is equal to the two interior and opposite angles CAB, ABC; and the three interior angles of the triangle, viz. ABC, BCA, CAB, are together equal to two right angles.

Through the point C draw CE parallel (31. 1.) to the straight line AB; and because AB is parallel to CE, and AC meets them, the alternate angles BAC, ACE are equal (29. 1.) Again, because AB is parallel to CE, and BD falls upon them, the exterior angle ECD is equal to the interior and opposite angle ABC, but the angle ACE was shewn to be equal to the angle BAC; therefore the whole exterior angle ACD is equal to the two interior and opposite angles CAB, ABC; to these angles add the angle ACB, and the angles ACD, ACB are equal to the three angles CBA, BAC, ACB; but the angles ACD, ACB are equal (13. 1.) to two right angles; therefore also the angles CBA, BAC, ACB are equal to two right angles.

Figure 4. Diagrammatic demonstration that the sum of the internal angles of any triangle is 180°. (a) Triangle: ABC. (b) Diagrammatic manipulations: ABCDE.

stated in these terms:

> My proposition is that logic, in the strict sense of the term, has nothing to do with how you think [...]. Logic in the narrower sense is that science which concerns itself primarily with distinguishing reasonings into good and bad reasonings, and with distinguishing probable reasonings into strong and weak reasonings. Secondarily, logic concerns itself with all that it must study in order to draw those distinctions about reasoning, and with nothing else [Peirce, 2005, p. 143]

Gabbay and Woods clearly maintain that Peirce's abduction, depicted as both (a) a surrender to an idea, and (b) a method for testing its consequences, perfectly resembles practical reasoning. Anyway, Peirce says "In everyday business reasoning is tolerably successful but I am inclined to think that it is done as well without the aid of theory as with it" (p. 109).

Facing the problem of "reaching" a logic of abduction Gabbay and Woods contend it is just a formal and somewhat idealized description of an abductive agent. Following my cognitive description of abductive reasoning illustrated above, an abductive human agent can naturalistically be seen in the perspective of the role of manipulations and of the interplay between internal – neural – representations and external ones. We have also seen that in this framework both conscious and unconscious inferences are important. Gabbay and Woods [2005, p. 27–29] seem to agree with me: they indeed stress the function of consciousness and indicate both its narrow bandwidth and its slow processing of information: an extraordinary quantity of information processed by the human system cannot be accessed to by consciousness. Analogously, the bandwidth of natural language is narrower than the bandwidth of sensation (model-based level, in our terms), a great quantity of what we know we are not able to communicate one to another – linguistic intercourse is a series of exchanges whose bandwidth is 16 bits per second. Moreover, a side effect of consciousness is that is suppresses information.

The importance of abduction from the evolutionary point of view is clear: as Dennett [2003, p. 248] points out, it is better to kill a hypothesis in the mind than to be killed ourselves because we precipitously experienced the real environment. It is in this way, he says, that we are "Popperian creatures". I would think more appropriate to say we are "Peircian creatures", but what it is important here is that we do not have to know we are Popperian creatures to be one, as is the case with my consciousness-free – but hypotheses-making cat – as well as with computers programmed to simulate planning and mean-ends reasoning.

Clearly logic is historically related to conscious and propositional thinking and it seems to disregard the subconscious and prelinguistic levels of thinking. This fact leads to the following dilemma: rules of logic are thought of as having something to do with how human beings actually think as practical agents, then by and large they are too complex for conscious deployment; on the other hand, unconscious performance or tacit knowledge is a matter of certain things happening under the appropriate conditions and the right order, but it is unlikely to suppose that this is a matter of following rules (an inclination which seems embedded in contemporary computer science): "Given the cognitive goals typically set by practical agents, validity and inductive strength are typically not appropriate (or possible) standards for their attainment. [...] This, rather than computational costs, is the deep reason that practical agents do not in the main execute systems of deductive or inductive logic as classically conceived" [Gabbay and Woods, 2005, p. 25]

In the following paragraphs I plan to offer a new clarification of this dilemma. Frequent complains claim that logic provides rules which humans cannot conform their conscious thinking to except for very few conscious exceptions, so that logicality cannot be consider just a matter of following rules: indeed the intervention in reasoning of implicit knowledge sometimes renders performances relatively effortless, as I have illustrated in section 2.

If, as Gabbay and Woods contend, a logic is a formalized *idealization* of a type of agent, a *logical agent*, which features of a human agent will a logic of abduction describe? They maintain that a real human agent is a kind of biological realization of a nonmonotonic paraconsistent base logic and surely the strategies provided by classical logic and some strictly related non standard logics form a very small part of the individual cognitive skills given the fact that human agents are not in general dedicated to error avoidance like "classical" logical agents. They say: "A formal model is an idealized description of what an abductive agent does. As such, it reflects some degree of departure from empirical accuracy. Thus an ideal model I is distinct from a descriptive model D" (p. 22). Immediately they add that in this logic of a practical agent basic questions of *relevance* and *plausibility* are central.

Gabbay and Woods describe the features of "real" human thinking agents in the following way, consistent with the cognitive architecture I have illustrated in the previous sections: a human abductive agent certainly operates at two levels, conscious and unconscious, and at *both* levels it engages or it is influenced by truth conditions on propositional structures, state conditions on belief structures and their fixation, and sets of rules defined for various argumentative structures, for instance for evaluating arguments. These three capacities cut across explicit and implicit thinking.

I think that many of the most important inference skills of a human agent are endowed with a *story* which varies with the multiple propositional relations she finds in her environment and which she takes into account, and with various cognitive reasons to change her mind or to think in a different way, and with multiple motivations to deploy various tactics of argument.

Facing the problem of logical modeling this kind of practical (abductive) human agent, logical systems can be considered *mimetic*, in the sense of the "mimetic representations" I introduced in the previous section, (to make an example, nonmonotonic systems seem to "mime" better – they are more psychologically adequate – human beings' reasoning performances).

First of all a good mimetic abductive ideal logical agent is embedded in a situation of *nescience* and it is characterized by an *ignorance preserving* (but also *ignorance mitigating*) character [Gabbay and Woods, 2005] and by the following general distinct levels

- a *base logic* L_1 with proof procedures Π;
- an *abductive algorithm* which deploys Π to look for missing premises and other formulas to be abduced;
- a *logic* L_2 for deciding which abduced formulas to choose, which criteria of selection apply, etc. This logic is related to the specifica-

tion of suitable problems of plausibility, relevance (topical, full-use, irredundancy-oriented, probabilistic) and economy, making the ideal agent able to discount and select information which does not resolve the task at hand.

The second and the third components together – endowed with what it is called a *filtrating* power – form *the logic of discovery*. It is in this new formal framework (GW model) that Gabbay and Woods criticize the so called AKM model of abduction[10] and also present a *non-explanationist* abduction, considered as not intrinsically consequentialist.

6 Logical Agents as Mimetic and Creative Representations and Mediators

Following our previous considerations it would seem that logical systems can be fruitfully seen as *external representations* expressed through artificial languages (in part mathematical) and through ordinary language, to the aim of *mimicking* various human reasoning performances in an idealized and rigorous way.

I think this view is partial and does not unveil other important features of logical agents. Indeed it is important to stress that logical agents can also play the role of *creative* representations human beings externalize and manipulate not just to mirror the internal ways of thinking of human agents but to find room for concepts and new ways of inferring which cannot – at a certain time – be found internally "in the mind".

In summary, we can say that

- logical systems as *external representations* are formed by external materials that either *mimic* (through reification) concepts and problems already internally present in the brain or *creatively* express concepts and problems that do not have a "natural home" in the brain;[11]

- *internalized logical representations* are internal re-projections, a kind of recapitulations, (learning) of external logical representations in terms of neural patterns of activation in the brain. In some simple cases – cf. for example some precise rules of inferences of classical logic in the case of ordinary people or more complex logical rules and structures in the case of expert logicians – can be "internally" manipulated like external objects and can originate new internal reconstructed representations through the neural activity of *transformation* and *integration*.

[10]Among its supporters are [Aliseda-Llera, 1997; Kuipers, 1999; Magnani, 2001c; Meheus *et al.*, 2002; Meheus and Batens, 2006] .

[11]Also Gabbay and Woods consider mimetic the logical agent: "An abductive logic describes or mimics what an abductive agent does" [2005, p. 346].

We have already stressed that this process explains – from a cognitive point of view – why human agents seem to perform both computations of a *connectionist* type such as the ones involving representations as

- (I Level) *patterns of neural activation* that arise as the result of the interaction between body and environment (and suitably shaped by the evolution and the individual history): pattern completion or image recognition,

  and computations that use representations as

- (II Level) *derived combinatorial syntax and semantics* dynamically shaped by the various artificial external representations and reasoning devices found or constructed in the environment (for example logical systems); they are – more or less completely – neurologically represented contingently as patterns of neural activations that "sometimes" tend to become stabilized structures and to fix and so *to permanently belong to the I Level* above.

As already stressed the I Level originates those *sensations* (they constitute a kind of "face" we think the world has), that provide room for the II Level to reflect the structure of the environment, and, most important, that can follow the computations suggested by the *logical external structures* available. It is clear that in this case we can conclude that the growth of the brain and especially the synaptic and dendritic growth are profoundly determined by the environment. Consequently we can hypothesize a form of co-evolution between the logical brain and the development of the external logical systems. Brains build logic manipulating external symbols and structures and learn from them after having manipulated them to discover new concepts and features.

When the fixation is reached – imagine for instance the precise rule *modus tollens* as an internally explicit and stabilized form of thinking – the pattern of neural activation no longer needs a direct stimulus from the external logical representation in the environment for its construction. It can be viewed as a *fixed internal record* of an *external structure* that *can exist* also in the absence of such external structure. The pattern of neural activation that constitutes the I Level Representation has kept record of the experience that generated it and, thus, carries the II Level Representation associated to it, even if in a different form, the form of *memory* and not the form of the vivid sensorial experience of *modus tollens* written externally, over there, for instance in a blackboard. Now, the human agent, via neural mechanisms, can retrieve that II Level Representation and use it as an *internal* representation (and can use it to construct new internal representations less complicated than the ones previously available and stored in memory).

At this point we can easily understand the particular mimetic and creative role played by external logical representations:

1. some "new ways of inferring" performed by the biological human agents appear hidden and tacit (see the considerations on unconscious thinking above) and can be rendered explicit by building external logical *mimetic* models and structures; later on the agent will be able to pick up and use what suggested by the constraints and features intrinsic to their external materiality and the relative established conventionality: artificial languages, proofs, examples, etc.;[12]

2. some inferences can be *discovered* only through a problem solving process occurring in a distributed interplay between brains and external representations. I have called this process *disembodiment of the mind* [Magnani, 2005]: the representations are *mediators* of results obtained and allow human beings

   (a) to re-represent in their brains new reasoning devices picked up outside, externally, previously absent at the internal level and thus impossible: first, a kind of alienation is performed, second, a recapitulation is accomplished at the neuronal level by re-representing internally that which has been "discovered" outside. We perform cognitive logical operations on the structure of data that synaptic patterns have "picked up" in an analogical way from the explicit logical representations in the environment;

   (b) to re-represent in their brains portions of reasoning devices which, insofar as explicit, can facilitate inferences that previously involved a very great effort because of human brain's limited capacity. In this case the thinking performance is not completely processed internally but in a *hybrid* interplay between internal (both tacit and explicit) and external logical representations. In some cases this interaction is between the internal level and a computational tool which in turn can exploit logical representations to perform inferences (cf. above section 3 "The manipulative framework").

[12]A simple mathematical example: it is relatively neurologically easy to perform an addition of numbers by depicting in our *mind* – thanks to that brain device that is called visual buffer – the images of that addition *thought* as it occurs concretely, with paper and pencil, taking advantage of external diagrammatic representations.

An evolved mind is unlikely to have a *natural home* for complicated concepts like the ones a logical agent introduces, as such concepts do not exist in the natural world: so whereas evolved minds could perform some trivial deductions or abductions in a more or less tacit way by exploiting modules shaped by natural selection,[13] how could one think exploiting explicit concepts without having picked them up outside, after having produced them? A mind consisting of different separated implicit templates of thinking and modes of inferences exemplified in various exemplars expressed through natural language cannot come up with certain logical entities without the help of the external representations. The only way is *to extend* the mind into the *material world*, exploiting paper, blackboards, symbols, artificial languages, and other various semiotic tools, to provide anchors for ways of inferring that have no natural home within the mind, that is for ways of inferring that take us beyond those that natural selection could enable us to possess at a certain moment.

Hence, we can hypothesize – for example – that many truth preserving reasoning habits which in human agents are performed internally have a deep origin in the past experience lived in the interplay with logical systems at first represented in the environment. As I have just illustrated other recorded thinking habits only partially occur internally because they are hybridized with the exploitation of already available external logical artefacts.

7 Externalization in Demonstrative Environments

It is well-known that there are external representations that are representations of other external representations. In some cases they carry new scientific knowledge. To make an example, Hilbert's *Grundlagen der Geometrie* is a "formal" representation of the geometrical problem solving through diagrams: in Hilbertian systems solutions of problems become proofs of theorems in terms of an axiomatic model. In turn a calculator is able to re-represent (through an artifact) (and to perform) those geometrical proofs with diagrams already performed by human beings with pencil and paper. In this case we have representations that *mimic* – as I have illustrated above – particular cognitive performances that we usually attribute to our minds.

I have contended that our brains delegate cognitive (and epistemic) roles to externalities and then tend to "adopt" and recapitulate what they have

[13]Logicians express this idea in terms of the first of two basic low resources principles: "[...] 1. low resources individual agents perceive *natural kinds* around them, 2. low resources individual agents are *hasty generalizers* of *generic rules*" [2005, p. 352]. Indeed Gabbay and Woods characterize human agents as those that command few cognitive resources, cf. also my considerations on the main features of practical and theoretical agents below in section 7.

checked occurring outside, over there, after having manipulated – often with creative results – the external invented structured model. We have said that mind representations are also over there, in the environment, where mind has objectified itself in various structures that *mimic* and *enhance* its internal representations.

Let us consider the example of the externalizations of cognitive abilities generated by the computational turn. Turing adds a new structure to the list of external objectified reasoning devices: an abstract tool (the Logical Computing Machine, LCM) endowed with powerful mimetic properties. We have concluded the subsection 4.1 remarking that the creative "mind" is in itself extended and, so to say, both internal and external: the mind transcends the boundary of the individual and includes parts of that individual's environment. Turing's LCM, which is an externalized device, is able to *mimic* human cognitive operations that occur in that interplay between the internal mind and the external one. Indeed Turing already in 1950 maintains that, taking advantage of the existence of the LCM, "Digital computers [...] can be constructed, and indeed have been constructed, and [...] they can in fact mimic the actions of a human computer very closely" [Turing, 1950].

In the light of my perspective both (Universal) Logical Computing Machine (LCM) (the theoretical artifact) and (Universal) Practical Computing Machine (PCM) (the practical artifact) are whole *mimetic minds* because they are able to mimic the mind in a kind of universal way. LCM and PCM are able to re-represent and perform in a very powerful way plenty of cognitive skills of human beings. Universal Turing machines are discrete-state machines, DMS, "with a Laplacian behavior" [Longo, 2002; Lassègue, 1998; Lassègue, 1999]: "it is always possible to predict all future states") and they are equivalent to all formalisms for computability (what is thinkable is calculable and mechanizable), and because universal they are able to simulate – that is to *mimic* – any human cognitive function, that is what is usually called mind.

Universal Turing machines are just a further extremely fruitful step of the disembodiment of the mind I have described above. A natural consequence of this perspective is that they do not represent (against classical AI and modern cognitivist computationalism) a "knowledge" of the mind and of human intelligence. Turing is perfectly aware of the fact that brain is not a DSM, but as he says, a "continuous" system, where instead a mathematical modeling can guarantee a satisfactory scientific intelligibility (cf. his studies on morphogenesis).

We have seen that our brains delegate meaningful cognitive (and epistemic) roles to externalities and then tend to "adopt" what they have checked occurring outside, over there, in the external invented structures and models. And a large part of meaning formation takes advantage of the exploitation of external representations and mediators. We have said that PCMs can be considered mimetic minds (they are ideal "practical" – in Turing's sense – agents): what is in turn the cognitive status of "logical agents" from the point of view of their *demonstrative* aspect?

It is at the level of the external representations which originate the logical ideal agent that we firstly realize the *demonstrative* character of a reasoning process. For example this is the case of Hilbertian formal systems, Gentzen's proofs, semantic tableaux, etc., but also of heterogeneous systems. Let us consider the amazing case of the heterogeneous systems: they produce representations in a demonstrative framework which originate from a number of different representation systems, sentential, but also model-based, diagrammatic, usually considered non-demonstrative: the advantage is that they allow a reasoner to bridge the gaps among various formalisms and to construct threads of proof which cross the boundaries of the systems of representations [Swoboda and Allwein, 2002].

In doing this, heterogeneous systems allow the reasoner to take advantage of each component system's ability to express information in that component's area of expertise. For example "recast" rules are clearly elicited as rules of inference that allow the exchange of information between the various representations (cf. Figure 5). We have two ways to use them: one for the extraction of information from a diagram to be expressed in a sentential form and another that allows the extraction of information from a formula to be incorporated into a diagram [Swoboda and Allwein, 2002].[14]

Gabbay and Woods propose a hierarchy of cognitive agents: "An agent is a practical agent to the extent that it commands comparatively few cognitive resources in relation to comparatively modest cognitive goals. [...] An agent is theoretical to the extent that it commands comparatively much in the way of cognitive resources, directed at comparatively strict goals" [2005, pp. 12-13]. They also add that it is typical of human individuals to function as practical agents and that it is typical of what they call "institutions" to function as theoretical agents (p. 14); moreover, agents tend toward enhancement of cognitive assets when this makes possible the achievement of cognitive goals previously unaffordable or anattainable. Of course the ideal logical agents I am considering here are theoretical agents.

The externalization of inferential skills in demonstrative systems presents interesting cognitive features (cf. also [Longo, 2005]) which I believe deserve to be further analyzed and which can further develop the distinction above

[14]Clarifications of exact processes and semantic requirements of manipulative inferences and distributed cognition are given in [Shimojima, 2002].

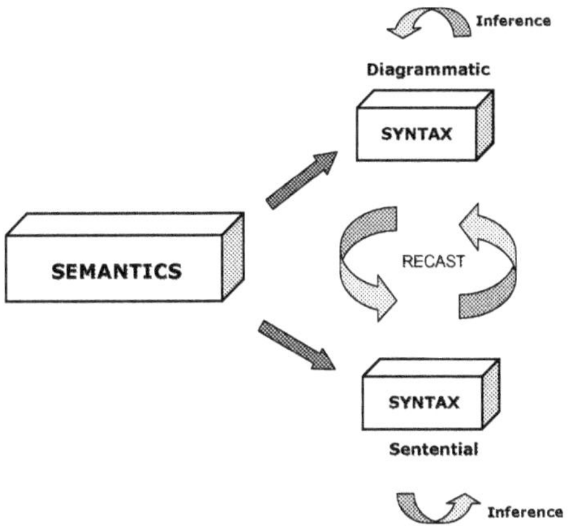

Figure 5. A heterogeneous logical system [Swoboda and Allwein, 2002].

between theoretical and practical agents:

1. *symbolic*: they activate and "anchor" meanings in material communicative and intersubjective *mediators* in the framework of the phylogenetic, ontogenetic, and cultural reality of the human being and its language. It seems these logical agents originated in embodied cognition, gestures, and manipulations of the environment we share with some mammals but also non mammal animals (cf. the case of monkeys' knots and pigeons' categorization, in [Grialou *et al.*, 2005]).[15]

2. *abstract*: they are based on a *maximal independence* regarding sensory modality; they strongly stabilize experience and common categorization. The maximality is especially important: it refers to their practical and historical invariance and stability;

3. *rigorous*: the rigor of proof is reached through a difficult practical experience. For instance, in the case of mathematics, as the maximal place for convincing and sharable reasoning. Rigor lies in the stability of proofs and in the fact they can be iterated. Following this perspective mathematics is the best example of maximal stability and conceptual invariance. Logic is in turn a set of proof invariants, a set of structures that are preserved from one proof to another or which

[15]Cf. also the cognitive analysis of the origin of the mathematical continuous line as a pre-conceptual invariant of three cognitive practices [Theissier, 2005], and of the numeric line [Châtelet, 1993; Dehaene, 1997; Butterworth, 1999].

are preserved by proof transformations. As the externalization and result of a distilled praxis, the praxis of proof, it is made of maximally stable regularities;

4. I also say that a *maximization of memorylessness*[16] "variably" characterizes demonstrative reasoning. This is particularly tangible in the case of the vast idealization of classical logic and related approaches. The inferences described by classical logic do not yield sensitive information – so to say – about their real past life in human agents' use, contrarily to the "conceptual" – narrative – descriptions of human non-demonstrative processes, which variously involve "historical", "contextual", and "heuristic" memories. Indeed many thinking behaviors in human agents – for examples abductive inferences, especially in their generative part – are context-dependent. As already noted their *stories* vary with the multiple propositional relations the human agent finds in her environment and which she is able to take into account, and with various cognitive reasons to change her mind or to think in a different way, and with multiple motivations to deploy various tactics of argument:

> Good reasoning is always good in relation to a goal or an agenda which may be tacit. [...] Reasoning validly is never *itself* a goal of good reasoning; otherwise one could always achieve it simply by repeating a premiss as conclusion, or by entering a new premiss that contradicts one already present. [...] It is that the reasoning actually performed by individual agents is sufficiently reliable not to kill them. It is reasoning that precludes neither security not prosperity. This is a fact of fundamental importance. It helps establish the fallibilist position that it is not unreasonable to pursue modes of reasoning that are known to be imperfect [Gabbay and Woods, 2005, pp. 19-20].

As we have already illustrated in section 5 human agents, as practical agents, are hasty inducers and bad predictors, unlike ideal agents. In conclusion, we can say abductive inferences in human agents have a memory, a story: consequently, an abductive ideal logical agent has to variably weaken many of the aspects of classical logic and to overcome the relative demonstrative limitations.

[16] I derive this expression from [Leyton, 2001] that introduces a very interesting new geometry where forms are no longer memoryless like in classical approaches such as the Euclidean and the Kleinian in terms of groups of transformations.

I think that a great contribution given to logic by Gabbay is the creation of the *labelled deductive systems* (and their application to the logic of abduction), where data is structured and labelled and different insertion policies can be formulated [Gabbay, 2002; Gabbay and Woods, 2006]. The labelled deductive systems fulfil the request of weakening the rigidity of classical logic but also of many non standard logics strictly related to it, opening a new era in logic: the attention to the role of *meta-levels* – for instance in the logic of abduction – formalizes the flexibility and "historicity" of many kinds of human thinking which are meaningful in certain application areas they address. Gabbay and Woods' conclusion about psychologism is clear and leads to a new conception of logic:

> If [...] it is legitimate to regard logic as furnishing formal models of certain aspects of the cognitive behavior of logical agents, then not only do psychological considerations have a defensible place, they cannot reasonably be excluded [Gabbay and Woods, 2005, p. 2].[17]

We can conclude by stressing the fact that human non-demonstrative inferential processes are more and more externalized and objectified at least in three ways:

1. through Turing's Universal Practical Computing Machines we can have running programs that are able to mimic "the actions of a human computer very closely" [Turing, 1950], and so - amazingly - also those human agents' "actions" that correspond to the complicated inferential performances like abduction (cf. the whole area of artificial intelligence);

2. human non-demonstrative processes are more and more externalized and made available in forms of explicit narratives and learnable templates of behavior (cf. also the study of fallacies as important tools of the human "kit" that provides evolutionary advantages, in this sense a fallacy of the affirming the consequent – which depicts abduction in classical logic – is better than nothing [Woods, 2004]).[18]

[17]An analogous example of the new modeling flexibility of recent logic is represented by the work in the dynamic logics of reasoning of van Benthem [1996]. This logic offers a distinction between inferences that are dependent on short term representations and those that depend on long-term memory, which involves the processing of representations of greater abstraction. In this way it is possible to formally and flexibly reproduce the interplay that occurs in human agents' thinking both at the level of short-term memory – more inclined to be damaged by inconsistencies – and at the level of the long-term memory, where inconsistencies can be inert.

[18]Cf. also Gabbay and Woods [2005, pp. 33-36].

3. new demonstrative systems – ideal logical agents – are created able to model in a deductive way many non-demonstrative thinking processes, like abduction, analogy, creativity, spatial and visual reasoning, etc.[19]

8 Conclusions

The main thesis of this paper is that the externalization of the mind is a significant cognitive perspective able to unveil some basic features of abduction and its logical systems. We have illustrated that an evolved mind is unlikely to have a natural home for explicit logical concepts and inferences, as such concepts and inferences did not exist in an already known – non human-modified – natural world: the cognitive referral to the central role of the relation between meaningful behavior and dynamical interactions with the environment becomes critical to the problem of modeling up-to-date logical systems. I think that the issue could further suggest research on the role played by symbolisms, abstractness, and rigor, in the perspective of their capacity to characterize externalized demonstrative systems.

Moreover, I think the cognitive role of what I call "mimetic" and "creative" representations can be further studied also taking advantage of the research on practical logic. The continuous construction of new logical systems of practical reasoning (also related to the field of artificial intelligence) will constitute important and interesting new "mimetic inferences" externalized and available over there, in the environment. They can also provide new tools for creating new reasoning devices in areas like abduction, analogical, visual, and spatial inferences, both in science and everyday situations, so that this can extend the epistemological and the psychological theory.

The perspectives above, resorting to the exploitation of a very interdisciplinary interplay, will further shed light on how concrete manipulations of external objects influence the generation of hypotheses and thus on the characters of what I call manipulative abduction, showing how we can find methods of constructivity – and their logical and computational counterparts – in scientific and everyday reasoning based on external models and "epistemic mediators" [Magnani, 2002a].

---

[19] A skeptical conclusion about the superiority of demonstrative over non-demonstrative reasoning is provided by the following philosophical argumentation of Cellucci [2005] I agree with, which seems to emphasize the role of *ignorance preservation* in logic: "To know whether an argument is demonstrative one must know whether its premises are true. But knowing whether they are true is generally impossible", as Gödel teaches. So they have the same status of the premises of non-demonstrative reasoning. Moreover: demonstrative reasoning cannot be more cogent than the premises from which its starts; the justification of deductive inferences in any absolute sense is impossible, they can be justified as much, or as little, as non-deductive - ampliative - inferences. Also checking soundness is a problem.

Some results in the specific domain of calculus are presented in [Magnani and Dossena, 2004] were diagrams which play an *optical* role – microscopes and "microscopes within microscopes", telescopes, windows, a *mirror* role (to externalize rough mental models), and an *unveiling* role (to help create new and interesting mathematical concepts, theories, and structures) are studied. They play the role of epistemic mediators able to perform the explanatory abductive task of endowing with new "intuitive" meanings difficult mathematical concepts and of providing a better understanding of the calculus, through a non-standard model of analysis. I also maintain they can be used in many other different epistemological and cognitive situations. Another interesting application is given in the area of chance discovery [Magnani and Piazza, forthcoming]: concrete manipulations of the external world constitute a fundamental passage in chance discovery: by a process of manipulative abduction it is possible to build prostheses that furnish a kind of embodied and unexpressed knowledge that holds a key role in the subsequent processes of scientific comprehension and discovery but also in ethical thinking and in moral deliberation.

BIBLIOGRAPHY

[Aliseda-Llera, 1997] A. Aliseda-Llera. *Seeking Explanations: Abduction in Logic, Philosophy of Science and Artificial Intelligence.* PhD thesis, Amsterdam: Institute for Logic, Language and Computation, 1997.

[Boutilier and Becher, 1995] C. Boutilier and V. Becher. Abduction as belief revision. *Artificial Intelligence*, 77:43–94, 1995.

[Butterworth, 1999] B. Butterworth. *The Mathematical Brain.* MacMillan, New York, 1999.

[Bylander *et al.*, 1991] T. Bylander, D. Allemang, M.C. Tanner, and J.R. Josephson. The computational complexity of abduction. *Artificial Intelligence*, 49:25–60, 1991.

[Cellucci, 2005] C. Cellucci. Mathematical discourse vs. mathematical intuition. In C. Cellucci and D. Gillies, editors, *Mathematical Reasoning and Heuristics*, pages 138–166, London, 2005. King's College Publications. Special Issue of the *Logic Journal of IJPL.*

[Châtelet, 1993] G. Châtelet. *Les enjeus du mobile.* Seuil, Paris, 1993. English transl. by R. Shore and M. Zagha, *Figuring Space: Philosophy, Mathematics, and Physics*, Kluwer, Dordrecht, 2000.

[Cornuéjols *et al.*, 2000] A. Cornuéjols, A. Tiberghien, and G. Collet. A new mechanism for transfer between conceptual domains in scientific discovery and education. *Foundations of Science*, 5(2):129–155, 2000. Special Issue on 'Model-based Reasoning in Science: Learning and Discovery', ed. by L. Magnani, N.J. Nersessian, and P. Thagard.

[Dehaene, 1997] S. Dehaene. *The Number Sense.* Oxford University Press, Oxford, 1997.

[Dennett, 2003] D. Dennett. *Freedom Evolve.* Viking, New York, 2003.

[Euclid, 1846] Euclid. *Elements of Geometry: Containing the First Six Books of Euclid, with a Supplement on the Quadrature of the Circle and the Geometry of the Solids: To Which are Added Elements of Plane and Spherical Trigonometry.* W.E. Dean, New York, 1846. Edited by John Playfair.

[Flach and Kakas, 2000] P. Flach and A. Kakas, editors. *Abductive and Inductive Reasoning: Essays on Their Relation and Integration*, Dordrecht, 2000. Kluwer Academic Publishers.

[Gabbay and Kruse, 2000] D.M. Gabbay and R. Kruse. *Abductive Reasoning and Learning*. Kluwer, Dordrecht, 2000. Edited by K.L. Ketner. In *Handook of Defeasible Reasoning and Uncertainty Management Systems*, vol. 4.

[Gabbay and Woods, 2005] D.M. Gabbay and J. Woods. *The Reach of Abduction*. North-Holland, Amsterdam, 2005. volume 2 of *A Practical Logic of Cognitive Systems*.

[Gabbay and Woods, 2006] D.M. Gabbay and J. Woods. A formal model of abduction. In L. Magnani, editor, *Abduction and Creative Inferences in Science*, 2006. Special Issue of the *Logic Journal of IJPL*.

[Gabbay, 2002] D.M. Gabbay. Abduction in labelled deductive systems. In D.M. Gabbay and R. Kruse, editors, *Handook of Defeasible Reasoning and Uncertainty Management Systems*, pages 99–153, Dordrecht, 2002. Kluwer. In *Handook of Defeasible Reasoning and Uncertainty Management Systems*, vol. 4.

[Gatti and Magnani, 2005] A. Gatti and L. Magnani. On the representational role of the environment and on the cognitive nature of manipulations. In L. Magnani, editor, *Computing, Philosophy, and Cognition*, 2005. Proceedings of the European Conference of Computing and Philosophy, Pavia, Italy, 3-4 June 2004.

[Gooding, 1990] D. Gooding. *Experiment and the Making of Meaning*. Kluwer, Dordrecht, 1990.

[Grialou *et al.*, 2005] P. Grialou, G. Longo, and M. Okada, editors. *Images and Reasoning*, Tokyo, 2005. Keio University.

[Hempel, 1966] C.G. Hempel. *Philosophy of Natural Science*. Prentice-Hall, Englewood Cliffs, NJ, 1966.

[Hintikka, 1998] J. Hintikka. What is abduction? The fundamental problem of contemporary epistemology. *Transactions of the Charles S. Peirce Society*, 34:503–533, 1998.

[Hutchins, 1995] E. Hutchins. *Cognition in the Wild*. MIT Press, Cambridge, MA, 1995.

[Kirsh and Maglio, 1994] D. Kirsh and P. Maglio. On distinguishing epistemic from pragmatic action. *Cognitive Science*, 18:513–49, 1994.

[Kuipers, 1999] T.A.F. Kuipers. Abduction aiming at empirical progress of even truth approximation leading to a challenge for computational modelling. *Foundations of Science*, 4:307–323, 1999.

[Lassègue, 1998] J. Lassègue. *Turing*. Les Belles Lettres, Paris, 1998.

[Lassègue, 1999] J. Lassègue. Turing entre formel et forme; remarque sur la convergence des perspectives morphologiques. *Intellectica*, 35(2):185–198, 1999.

[Leyton, 2001] M. Leyton. *A Generative Theory of Shape*. Springer, Berlin, 2001.

[Longo, 2002] G. Longo. Laplace, turing, et la géométrie impossible du "jeu de l'imitation": aléas, determinisme e programmes dans le test de turing. *Intellectica*, 35(2):131–61, 2002.

[Longo, 2005] G. Longo. The cognitive foundations of mathematics: human gestures in proofs and mathematical incompleteness of formalisms. In P. Grialou, G. Longo, and M. Okada, editors, *Images and Reasoning*, pages 105–134, Tokyo, 2005. Keio University.

[Magnani and Dossena, 2004] L. Magnani and R. Dossena. Perceiving the infinite and the infinitesimal world: unveiling and optical diagrams and the construction of mathematical concepts, 2004. Forthcoming in *Foundations of Science*.

[Magnani and Piazza, forthcoming] L. Magnani and M. Piazza. Morphodynamical abduction: causation by attractors dynamics of explanatory hypotheses in science, forthcoming. *Foundations of Science*.

[Magnani *et al.*, 2002] L. Magnani, N.J. Nersessian, and C. Pizzi, editors. *Logical and Computational Apsects of Model-Based Reasoning*, Dordrecht, 2002. Kluwer. Special Issue of the *Logic Journal of IJPL*.

[Magnani, 1999] L. Magnani. Inconsistencies and creative abduction in science. In *AI and Scientific Creativity. Proceedings of the AISB99 Symposium on Scientific Creativity*, pages 1–8, Edinburgh, 1999. Society for the Study of Artificial Intelligence and Simulation of Behaviour, University of Edinburgh.

[Magnani, 2001a] L. Magnani. Limitations of recent formal models of abductive reasoning. In A. Kakas and F. Toni, editors, *Abductive Reasoning. KKR-2, 17th International Joint Conference on Artificial Intelligence (IJCAI 2001)*, pages 34–40, Berlin, 2001. Springer.

[Magnani, 2001b] L. Magnani. *Philosophy and Geometry. Theoretical and Historical Issues*. Kluwer Academic Publisher, Dordrecht, 2001.

[Magnani, 2001c] L. Magnani. *Abduction, Reason, and Science. Processes of Discovery and Explanation*. Kluwer Academic/Plenum Publishers, New York, 2001.

[Magnani, 2002a] L. Magnani. Epistemic mediators and model-based discovery in science. In L. Magnani and N.J. Nersessian, editors, *Model-Based Reasoning: Science, Technology, Values*, pages 305–329, New York, 2002. Kluwer Academic/Plenum Publishers.

[Magnani, 2002b] L. Magnani. Thinking through doing, external representations in abductive reasoning. In *AISB 2002 Symposium on AI and Creativity in Arts and Science*, London, 2002. Imperial College.

[Magnani, 2005] L. Magnani. Creativity and the disembodiment of mind. In P. Gervas, A. Pease, and T. Veale, editors, *Proceedings of CC05, Computational Creativity Workshop, IJCAIO2005, Edinburgh*, 2005.

[Magnani, 2006] L. Magnani, editor. *Abduction and Creative Inferences in Science*, 2006. Special Issue of the *Logic Journal of IJPL*.

[Meheus and Batens, 2006] J. Meheus and D. Batens. A formal logic for abductive reasoning. In L. Magnani, editor, *Abduction and Creative Inferences in Science*, 2006. Special Issue of the *Logic Journal of IJPL*.

[Meheus *et al.*, 2002] J. Meheus, L. Verhoeven, M. Van Dyck, and D. Provijn. Ampliative adaptive logics and the foundation of logic-based approaches to abduction. In L. Magnani, N.J. Nersessian, and C. Pizzi, editors, *Logical and Computational Aspects of Model-Based Reasoning*, pages 39–71, Dordrecht, 2002. Kluwer.

[Nersessian, 1992] N.J. Nersessian. How do scientists think? Capturing the dynamics of conceptual change in science. In R. Giere, editor, *Cognitive Models of Science*, Minnesota Studies in the Philosophy of Science, pages 3–44, Minneapolis, 1992. University of Minnesota Press.

[Nersessian, 1995] N.J. Nersessian. Should physicists preach what they practice? Constructive modeling in doing and learning physics. *Science and Education*, 4:203–226, 1995.

[Nersessian, 1999] N.J. Nersessian. Model-based reasoning in conceptual change. In L. Magnani, N.J. Nersessian, and P. Thagard, editors, *Model-based Reasoning in Scientific Discovery*, pages 5–22, New York, 1999. Kluwer Academic/Plenum Publishers.

[Norman, 1993] D.A. Norman. *Things that Make Us Smart. Defending Human Attributes in the Age of the Machine*. Addison-Wesley, Reading, MA, 1993.

[Peirce, 2005] C. S. Peirce. *Reasoning and the Logic of Things: The 1898 Cambridge Conferences Lectures by Charles Sanders Peirce*. Harvard University Press, Amsterdam, 2005. Edited by K.L. Ketner.

[Peirce, CP] C.S. Peirce. *Collected Papers*. Harvard University Press, Cambridge, MA, CP. 1–6, ed. by C. Hartshorne and P. Weiss, 7–8, ed. by A. W. Burks, 1931-1958.

[Piaget, 1974] J. Piaget. *Adaption and Intelligence*. University of Chicago Press, Chicago, 1974.

[Plato, 1977] Plato. *Plato in Twelve Volumes*. Harvard University Press, Cambridge, MA, 1977. vol. II, Laches, Protagoras, Meno, Euthydemus, with an English translation by W.R.M. Lamb.

[Polanyi, 1966] M. Polanyi. *The Tacit Dimension*. Routledge & Kegan Paul, London, 1966.

[Popper, 1959] K.R. Popper. *The Logic of Scientific Discovery*. Hutchinson, London, New York, 1959.

[Port and van Gelder, 1995] R.F. Port and T. van Gelder, editors. *Mind as Motion. Explorations in the Dynamics of Cognition*, Cambridge, MA, 1995. MIT Press.

[Ramoni *et al.*, 1992] M. Ramoni, M. Stefanelli, L. Magnani, and G. Barosi. An epistemological framework for medical knowledge-based systems. *IEEE Transactions on Systems, Man, and Cybernetics*, 22(6):1361–1375, 1992.

[Shimojima, 2002] A. Shimojima. A logical analysis of graphical consistency proof. In L. Magnani, N.J. Nersessian, and C. Pizzi, editors, *Logical and Computational Aspects of Model-Based Reasoning*, pages 93–116, Dordrecht, 2002. Kluwer.

[Swoboda and Allwein, 2002] N. Swoboda and G. Allwein. A case study of the design and implentation of heterogeneous reasoning systems. In L. Magnani, N.J. Nersessian, and C. Pizzi, editors, *Logical and Computational Aspects of Model-Based Reasoning*, pages 3–20, Dordrecht, 2002. Kluwer.

[Thagard and Shelley, 1997] P. Thagard and C.P. Shelley. Abductive reasoning: logic, visual thinking, and coherence. In M.L. Dalla Chiara, K. Doets, D. Mundici, and J. van Benthem, editors, *Logic and Scientific Methods*, pages 413–427, Dordrecht, 1997. Kluwer.

[Thagard, 1992] P. Thagard. *Conceptual Revolutions.* Princeton University Press, Princeton, 1992.

[Theissier, 2005] B. Theissier. Protomathematics, perception and the meaning of mathematical objects. In P. Grialou, G. Longo, and M. Okada, editors, *Images and Reasoning*, pages 135–45, Tokyo, 2005. Keio University.

[Turing, 1950] A.M. Turing. Computing machinery and intelligence. *Mind*, 49:433–460, 1950.

[van Benthem, 1996] J. van Benthem. *Exploring Logical Dynamics.* CSLI Publications, Stanford, CA, 1996.

[Woods, 2004] J. Woods. *The Death of Argument.* Kluwer, Dordrecht, 2004.

[Zhang, 1997] J. Zhang. The nature of external representations in problem-solving. *Cognitive Science*, 21(2):179–217, 1997.

Friendliness for Logicians

DAVID MAKINSON

1 Friendliness

For all of us who know him, Dov Gabbay is the friendliest of logicians. For this Festschrift, therefore, it is a great pleasure to dedicate this essay to him.

1.1 Rationale, Definition, Notation

Recall the definition of classical consequence in propositional logic. Let A be any set of formulae, and x any individual formula. Then x is said to be a classical consequence of A, written $A \vdash x$, iff for every valuation v on all letters of the language, if $v(A) = 1$ then $v(x) = 1$.

Trivially, the only letters that count here are those occurring in A or in x. So the definition may be rephrased as: $A \vdash x$ iff for every partial valuation v on $E(A,x)$, if $v(A) = 1$ then $v(x) = 1$. Equivalently again, $A \vdash x$ iff for every partial valuation v on $E(A)$, if $v(A) = 1$ then $v^+(x) = 1$ for every extension v^+ to $E(A, x)$.

Expressed in this last way, classical consequence is a $\forall\forall$ concept. It is natural to ask: what does the corresponding $\forall\exists$ concept look like, and how does it behave? This simple question, born of no more than curiosity, is the starting point of our investigation.

The definition is straightforward:

- We say that A is *friendly* to x and write $A \approx x$, iff every partial valuation v on $E(A)$ with $v(A) = 1$ may be extended to a partial valuation v^+ on $E(A, x)$ with $v^+(x) = 1$.

- Equivalently: iff for every partial valuation v on $E(A)$ with $v(A) = 1$ there is a partial valuation w on $E(x)$ agreeing with v on letters in $E(A) \cap E(x)$, with $w(x) = 1$.

- Equivalently: iff for every valuation v on the set E of all elementary letters of the language with $v(A) = 1$ there is a valuation w (on all letters) agreeing with v on letters in $E(A)$, with $w(x) = 1$.

The notation used in these definitions is fairly straightforward, but we state it explicitly for reference. We use lower case $a, b, \ldots, x, y, \ldots$ to range over *formulae* of classical propositional logic. It will be convenient to include the zero-ary falsum $\bot$ among the primitive connectives. *Sets of formulae* are denoted by upper case letters $A, B, \ldots, X, Y, \ldots$, reserving L for the set of all formulae, E for the set of all elementary letters, and $F, G, \ldots$ for subsets of the elementary letters. For any formula a, we write $E(a)$ to mean the set of all *elementary letters* occurring in a. Similarly for sets A of formulae. For any set A of formulae, L_A stands for the sub-language generated by $E(A)$, i.e. the set of all formulae y with $E(y) \subseteq E(A)$. Thus $L_A = L_{E(A)}$.

Classical consequence is written as $\vdash$ when treated as a relation, Cn when viewed as an operation. The relation of *classical equivalence* is written $\dashv\vdash$. When we speak of a *valuation*, we always mean a Boolean valuation, i.e. a function into {0,1} defined on the entire set E of elementary letters of the language and extended to cover all formulae in the usual way. A *partial valuation* is a restriction of a valuation to a subset of E.

To lighten notation, we follow the common convention of usually writing A, x for $A \cup \{x\}$. $A \vdash B$ is short for '$A \vdash b$ for all $b \in B$'. Also, $v(A) = 1$ is short for '$v(a) = 1$ for all $a \in A$', while $v(A) = 0$ is short for '$v(a) = 0$ for some $a \in A$'.

1.2 Remarks on the Definition

Of the three equivalent ways of defining friendliness, we will usually be working with the first. Thus throughout the paper (except for the appendix) we will be talking about *partial* valuations rather than full ones. In this context, it is essential to keep in mind some fine distinctions, which are easy to overlook because they are without much significance for classical consequence.

- $E(a)$ is the set of all elementary letters *actually occurring* in a, rather than the least set of letters needed to get a formula classically equivalent to a. For example, if $a = p \wedge (q \vee \neg q)$ then $E(a)$ is $\{p, q\}$, not $\{p\}$. We will look at least letter-sets and a corresponding notion of *sympathy* later, in section 3.

- When we speak of a partial valuation v on a set F of elementary letters, we mean one with *exactly* F as domain. Any valuation on a proper superset F^+ of F, agreeing with v over F, will be called an *extension* of v.

It will sometimes shorten formulations to apply the notion of friendliness to partial valuations themselves. Let F be any set of elementary letters, and let v be any partial valuation on F. Let x be any formula. We say that v is *friendly* to x iff it may be extended to a partial valuation v^+ on $F \cup E(x)$ with $v^+(x) = 1$. Clearly, whenever a partial valuation is friendly to a formula then so too are all its restrictions. In other words, whenever a partial valuation is not friendly to a formula, none of its extensions are friendly to it. The first definition of $A \models\!\approx x$ may thus be expressed concisely as follows:

- $A \models\!\approx x$ iff every partial valuation v on $E(A)$ with $v(A) = 1$ is friendly to x.

Similar definitions of friendliness may be made for first-order logic, speaking of (partial) models rather than partial valuations. It should be noted, however, that in the first-order case there are several ways of understanding the notion of an extension of a model, which give rise to variant concepts of friendliness. On the one hand, we could require that when we extend a partial model the domain of discourse must remain fixed, as well as the interpretations into it of the already given predicate letters; in the literature this is usually called an 'expansion'. On the other hand, we may allow the domain to increase. In this case we have sub-options to choose from, according to whether we keep the interpretations of the already given predicate letters fixed, or allow them to flow out into the enlarged domain in some way.

But for simplicity, in this paper *we will remain within the propositional context.* We will not discuss the question of what would be the most interesting way of generalizing the definition of friendliness to the first-order context. Nor, apart from some passing negative observations, will we tabulate which among our results for the propositional context carry over to which among the first-order notions.

In section 2 we discuss links between the notion of friendliness and several other operations and concepts in the literature. Readers of a historical bent may prefer to start there and return, but we begin by clarifying the behaviour of the friendliness relation itself.

1.3 Properties that Fail

At first sight, the relation of friendliness seems to be hopelessly ill behaved. It fails many familiar features of classical consequence. In particular:

- It is not closed under substitution for elementary letters. Example: $p \models\!\approx p \wedge q$ where p, q are (here and always) distinct elementary letters, but $p \not\models\!\approx p \wedge \neg p$.
- It fails monotony and left strengthening. Example: $p \models\!\approx p \wedge q$, but $\{p, \neg q\} \not\models\!\approx p \wedge q$ and similarly $p \wedge \neg q \not\models\!\approx p \wedge q$.

- It fails cautious monotony and cautious left strengthening. Example: $p \mathrel{|\!\approx} q$ and $p \mathrel{|\!\approx} \neg q$, but $\{p,q\} \mathrel{|\!\not\approx} \neg q$ and likewise $p \wedge q \mathrel{|\!\not\approx} \neg q$.
- It fails left classical equivalence. Example: $p \mathrel{|\!\approx} p \wedge q$ but $p \wedge (q \vee \neg q) \mathrel{|\!\not\approx} p \wedge q$.
- It fails conjunction in the conclusion. Example: $p \mathrel{|\!\approx} q$, $p \mathrel{|\!\approx} \neg q$, but $p \mathrel{|\!\not\approx} q \wedge \neg q$.
- For essentially the same reason, it fails a general form of cumulative transitivity. Example: $p \mathrel{|\!\approx} q, p \mathrel{|\!\approx} \neg q$, and $p \wedge q \wedge \neg q \mathrel{|\!\approx} \neg p$, but $p \mathrel{|\!\not\approx} \neg p$.
- It fails plain transitivity. Example: $p \mathrel{|\!\approx} q, q \mathrel{|\!\approx} \neg p$, but $p \mathrel{|\!\not\approx} \neg p$.
- It fails disjunction in the premisses. Example: $p \mathrel{|\!\approx} p \leftrightarrow q, q \mathrel{|\!\approx} p \leftrightarrow q$, but $p \vee q \mathrel{|\!\not\approx} p \leftrightarrow q$.

Nevertheless, friendliness does have positive properties including 'local' versions of some of the above, which we now describe.

1.4 Relationship of Friendliness to Classical Consequence

We begin by clarifying the relation of friendliness to classical consequence.

Supraclassicality. Whenever $A \vdash x$ then $A \mathrel{|\!\approx} x$. Briefly: $\vdash \subseteq \mathrel{|\!\approx}$

Verification. Immediate from the definition of $\mathrel{|\!\approx}$. □

The inclusion is proper; for example, when p, q are distinct elementary letters then $p \mathrel{|\!\approx} q$ but not $p \vdash q$. Friendliness is not the trivial relation over the language; for example, when a is a tautology and x a contradiction, $a \mathrel{|\!\not\approx} x$. For a less extreme example, $p \vee q \mathrel{|\!\not\approx} p \wedge q$ where p, q are distinct elementary letters.

However, there are special cases where friendliness collapses into classical consequence, and others where it collapses into non-consequence of the negation.

First Reduction case. Whenever $E(x) \subseteq E(A)$ then $A \mathrel{|\!\approx} x$ iff $A \vdash x$.

Verification. Right to left is given unconditionally by supraclassicality, so we need only show left to right. Suppose $E(x) \subseteq E(A)$ and $A \mathrel{|\!\approx} x$. Let v be any partial valuation on $E(A)$ with $v(A) = 1$. We need to show that $v^+(x) = 1$ for every extension v^+ of v to $E(A, x)$. Since $A \mathrel{|\!\approx} x, v^+(x) = 1$ for some extension v^+ of v to $E(A, x)$. But since $E(x) \subseteq E(A), E(A, x) = E(A)$, so the unique extension of v to $E(A, x)$ is v itself. Thus $v(x) = 1$ and indeed $v^+(x) = 1$ for every extension v^+ of v to $E(A, x)$. □

Second Reduction Case. Suppose A is consistent and for each elementary letter $p \in E(A)$, either $A \vdash p$ or $A \vdash \neg p$. Then $A \mathrel{|\!\approx} x$ iff $A \nvdash \neg x$.

Verification. Under the hypotheses, suppose first that $A \mathrel{|\!\approx} x$. Since A is consistent, there is some partial valuation v on $E(A)$ with $v(A) = 1$. Choose any one such v. Since $A \mathrel{|\!\approx} x$, we have $v^+(x) = 1$ for some extension v^+ of v to $E(A, x)$. Thus $v^+(\neg x) = 0$ while $v^+(A) = 1$, so $A \nvdash \neg x$.

For the converse, suppose $A \nvdash \neg x$. Then there a partial valuation v on $E(A)$ with $v(A) = 1$ that can be extended to a partial valuation v^+ on $E(A, x)$ with $v^+(x) = 1$. Since either $A \vdash p$ or $A \nvdash \neg p$, for each elementary letter $p \in E(A)$, v is the only partial valuation on $E(A)$ with $v(A) = 1$. Hence every partial valuation w on $E(A)$ with $w(A) = 1$ can be extended to a partial valuation w^+ on $E(A, x)$ with $w^+(x) = 1$. □

We also have the following important characterization of friendliness in terms of classical consistency.

Characterization in terms of consistency. $A \mathrel{|\!\approx} x$ iff every set B of formulae in L_A that is consistent with A, is consistent with x.

Verification. Suppose first that $A \mathrel{|\!\approx} x$. Let B be any set of formulae in L_A that is consistent with A. Then there is a partial valuation v on $E(A)$ with $v(A) = 1, v(B) = 1$. From the supposition, v may be extended to a partial valuation v^+ on $E(A, x)$ with $v^+(x) = 1$. Since v^+ extends v and $v(B) = 1$ we have $v^+(B) = 1$. Hence B is consistent with x, as desired.

For the converse, suppose that $A \mathrel{|\!\not\approx} x$. Then there is a partial valuation v on $E(A)$ with $v(A) = 1$, such that $v^+(x) = 0$ for every extension v^+ of v to $E(A, x)$. Put B to be the state-description (set of literals) in L_A that corresponds to v; in the limiting case that $E(A) = \emptyset$ put $B = \{\top\}$.

We complete the verification by showing that B is consistent with A but not consistent with x. The former is immediate from the fact that $v(A) = 1$ and by construction also $v(B) = 1$. For the latter, we observe that by construction, v is the only partial valuation on $E(B) = E(A)$ with $v(B) = 1$, and by hypothesis $v^+(x) = 0$ for every extension v^+ of v to $E(A, x)$. Thus there is no partial valuation w on $E(B, x) = E(A, x)$ with $w(B) = 1$ and $w(x) = 1$. In other words, B is inconsistent with x. □

This characterization can be refined. Our first refinement says, in effect, that in the characterization individual formulae c can do all the work of sets B of formulae.

First Refinement. $A \mathrel{|\!\approx} x$ iff $A \vdash c$ for every $c \in L_A$ with $x \vdash c$.

Verification. Suppose first $A \mid\!\approx x$. Applying the characterization from left to right, we have that every formula in L_A that is consistent with A, is consistent with x. Contrapositively, whenever $c \in L_A$ and $x \vdash c$ then $A \vdash c$.

In the other direction, suppose $A \not\mid\!\approx x$. Applying the characterization from right to left, there is a set B of formulae in L_A that is consistent with A, but is not consistent with x. Since B is not consistent with x, compactness tells us that is has a finite subset C that is not consistent with x. Then $x \vdash c$, where $c = \neg \bigwedge C$. But $A \nvdash c$, since A is consistent with B and so with its subset C. □

A second refinement will be useful for proving compactness for friendliness. In effect, in the characterization it suffices to consider only formulae $c \in L_A \cap L_x$, i.e. with $E(c) \subseteq E(A) \cap E(x)$.

Second Refinement. $A \mid\!\approx x$ iff $A \vdash c$ for every $c \in L_A \cap L_x$ with $x \vdash c$.

Verification. Left to right is immediate from the first corollary. For the converse, suppose $A \not\mid\!\approx x$. Then by the first corollary, there is a $d \in L_A$ with $x \vdash d$ but $A \nvdash d$. Since $x \vdash d$, classical interpolation tells us that there is a $c \in L_d \cap L_x \subseteq L_A \cap L_x$ with $x \vdash c \vdash d$. Since $c \vdash d$ and $A \nvdash d$ we have $A \nvdash c$ as desired. □

We note in passing that in the first-order context, if we define friendliness in terms of expansions (see section 1.2), then the second reduction case, the characterization in terms of consistency, and its two refinements, all fail in their right-to-left part. A single example serves for the three. Consider the language L with just one unary predicate letter P (no equality symbol, no individual constants), and put $\Gamma = Cn(\forall x(Px))$ to be the complete and consistent theory in that language. Let φ be the formula $\exists x \exists y(Rxy \wedge \neg Ryx)$, containing the additional letter R not available in L. On the one hand $\Gamma \nvdash \neg\varphi$; also every set Δ of formulae in L that is consistent with Γ, is consistent with φ. On the other hand, there is a model that satisfies Γ which has no expansion satisfying φ. Take any model with a singleton domain interpreting P as the whole domain. This satisfies Γ, but it cannot be expanded to a model satisfying Γ,φ, which would require two elements in the domain.

1.5 Closure Properties of Friendliness

We now see which among the familiar properties of classical consequence remain for friendliness. We begin with two that carry over without restriction.

Right weakening. Whenever $A \mid\!\approx x \vdash y$ then $A \mid\!\approx y$.

Verification. Immediate from the definition of $\mid\!\approx$. □

It follows from this, of course, that the relation is syntax-independent in its right argument, i.e. satisfies right classical equivalence: whenever $x \dashv\vdash y$ then $A \mid\!\approx x$ iff $A \mid\!\approx y$. This contrasts with the already noted syntax-dependence on the left.

Singleton cumulative transitivity. Whenever $A \mid\!\approx x$ and $A, x \mid\!\approx y$ then $A \mid\!\approx y$.

Verification. Suppose $A \mid\!\approx x$ and $A,x \mid\!\approx y$. Let v be any partial valuation on $E(A)$ with $v(A) = 1$. By the first hypothesis, v may be extended to a partial valuation v^+ on $E(A, x)$ with $v^+(x) = 1$, so also $v^+(A, x) = 1$. By the second hypothesis, v^+ may be extended to a partial valuation v^{++} on $E(A, x, y)$ with $v^{++}(y) = 1$. Restrict v^{++} to $E(A, y)$, call it v^{++-}. Then v^{++-} is still an extension of v with domain $E(A)$, and $v^{++-}(y) = 1$. □

We now formulate some properties that carry over in a restricted form only. The following are straightforward; compactness and interpolation are subtler and will be discussed in the following sections.

Local left strengthening. Suppose $E(B) \subseteq E(A)$. Then $B \vdash A \mid\!\approx x$ implies $B \mid\!\approx x$.

Verification. Suppose $B \vdash A \mid\!\approx x$. Consider any partial valuation v on $E(B)$ with $v(B) = 1$; we need to show that v is friendly to x. Extend v to any partial valuation v^+ on $E(A) \supseteq E(B)$. Then $v^+(B) = v(B) = 1$, and so since $B \vdash A$ we have $v^+(A) = 1$. Since $A \mid\!\approx x$, there is an extension v^{++} of v^+ to $E(A, x)$ with $v^{++}(x) = 1$. Restrict v^{++} to $E(B, x)$, call it v^{++-}. Then clearly $v^{++-}(x) = v^{++}(x) = 1$. But v^{++-} is still an extension of v with domain $E(B)$. Hence v is friendly to x, as desired. □

Local left equivalence. Suppose $E(B) \subseteq E(A)$. Then $A \mid\!\approx x$ and $A \dashv\vdash B$ together imply $B \mid\!\approx x$.

Verification. When $A \dashv\vdash B$ then $B \vdash A$ so we can apply local left strengthening. □

Local monotony. Suppose $E(B) \subseteq E(A)$. If $A \mid\!\approx x$ and $A \subseteq B$ then $B \mid\!\approx x$.

Verification. When $A \subseteq B$ then $B \vdash A$; apply local left strengthening. □

Local disjunction in the premisses. Suppose $E(b_2) \subseteq E(A,b_1)$ and $E(b_1) \subseteq E(A, b_2)$. Then $A, b_1 \mid\!\approx x$ and $A, b_2 \mid\!\approx x$ together imply $A, b_1 \vee b_2 \mid\!\approx x$.

Verification. Suppose $A, b_1 \vee b_2 \not\approx x$. Then there is a partial valuation v on $E(A, b_1 \vee b_2)$ with $v(A, b_1 \vee b_2) = 1$ that is not friendly to x. By the hypotheses, $E(A, b_1 \vee b_2) = E(A, b_1) = E(A, b_2)$. Since $v(A, b_1 \vee b_2) = 1$ either $v(A, b_1) = 1$ or $v(A, b_2) = 1$. Hence either v is a partial valuation on $E(A, b_1)$ with $v(A, b_1) = 1$ but not friendly to x, or similarly with b_2. That is, either $A, b_1 \not\approx x$ or $A, b_2 \not\approx x$. □

Proof by exhaustion. $A, b \mid\!\approx x$ and $A, \neg b \mid\!\approx x$ together imply $A \mid\!\approx x$.

Verification. Clearly $E(\neg b) = E(b) \subseteq E(A, b)$ and conversely $E(b) = E(\neg b) \subseteq E(A, \neg b)$ so we may apply local disjunction in the premisses to get $A, b \vee \neg b \mid\!\approx x$. Clearly also $E(A) \subseteq E(A, b \vee \neg b)$ and also $A \vdash (A, b \vee \neg b) \mid\!\approx x$, so we may apply local left strengthening to get $A \mid\!\approx x$ as desired. □

The properties obtained so far lead to another characterization. In a broad sense of the term, it can be seen as an axiomatization of the relation of friendliness, modulo classical consequence. 'A broad sense', since the right-hand side of the third condition is not closed under substitution.

Observation. Friendliness is the least relation R between sets of formulae and individual formulae that satisfies the following three conditions:

1. $\vdash \subseteq R$,
2. $\langle A, x \rangle \in R$ whenever $\langle A \cup \{b\}, x \rangle \in R$ and $\langle A \cup \{\neg b\}, x \rangle \in R$,
3. $\langle A, x \rangle \in R$ whenever $A \nvdash \neg x$ and for each elementary letter $p \in E(A)$, either $A \vdash p$ or $A \vdash \neg p$.

Verification. First observe that the total relation between sets of formulae and individual formulae satisfies these three conditions, and so there is at least one such relation. Further, the intersection of any non-empty set of such relations is itself such a relation (despite the negative term $A \nvdash \neg x$ in the third condition, which negates classical consequence rather than the relation R). Thus there is a unique least such relation R, call it R_0.

We already know that $\mid\!\approx$ satisfies all three conditions (supraclassicality, proof by exhaustion, second reduction case). Thus $R_0 \subseteq \mid\!\approx$.

For the converse, suppose $\langle A, x \rangle \notin R_0$; we need to show that $A \not\approx x$. Let $p_1, \ldots, p_n$ be all the elementary letters in $E(A)$. Define sets $A_0, \ldots, A_n$ by setting $A_0 = A$ and putting $A_{i+1} = A_i \cup \{p_{i+1}\}$ if $\langle A_i \cup \{p_{i+1}\}, x \rangle \notin R_0$ and otherwise $A_{i+1} = A_i \cup \{\neg p_{i+1}\}$. By hypothesis, $\langle A_0, x \rangle \notin R_0$ and an easy induction using condition (2) gives us $\langle A_n, x \rangle \notin R_0$. But for each elementary letter $p \in E(A)$, either $A_n \vdash p$ or $A_n \vdash \neg p$, so condition (3) tells us that $A_n \vdash \neg x$. Also, since $\langle A_n, x \rangle \notin R_0$, condition (1) tells us that A_n is consistent, so there is at least one partial valuation v on $E(A_n) = E(A)$ with $v(A_n) = 1$. Since $A_n \vdash \neg x$, we have $v^+(x) = 0$ for every extension v^+ of v to $E(A, x)$, so $A \not\models\!\approx x$ as desired. □

1.6 Compactness

In the context of friendliness, some care must be taken with the formulation of compactness. When the property is formulated in exactly the same way as in classical logic, it tells us very little. For suppose $A \models\!\approx x$. Then:

- On the one hand, in the limiting case that x is inconsistent the definition of $\models\!\approx$ implies that A must also be inconsistent, so by classical compactness there is a finite inconsistent subset $B \subseteq A$, so that by the definition of $\models\!\approx$ again, $B \models\!\approx x$.
- On the other hand, in the principal case that x is consistent, we have immediately that $\emptyset \models\!\approx x$. This leaves us hungry, for while the empty set is certainly finite we would like something more substantial.

This motivates the following strengthened formulation. Bearing in mind that friendliness does not satisfy monotony, it is quite strong.

Compactness. Let A be a non-empty set with $A \models\!\approx x$. Then there is a finite subset $B \subseteq A$ such that $C \models\!\approx x$ for every C with $B \subseteq C \subseteq A$.

Proof. Suppose $A \models\!\approx x$. By the second refinement of the characterization of friendliness in terms of consistency, whenever $c \in L_A \cap L_x$ and $x \vdash c$ then $A \vdash c$. Hence by compactness for classical consequence, for every $c \in L_A \cap L_x$ with $x \vdash c$ there is a finite subset $B_c \subseteq A$ with $B_c \vdash c$. Since x is an individual formula, there are only finitely many $c \in L_A \cap L_x \subseteq L_x$ up to classical equivalence. Taking the finite union of the corresponding sets B_c, we conclude that there is a finite subset $B \subseteq A$ such that $B \vdash c$ for every $c \in L_A \cap L_x$ with $x \vdash c$.

Now let C be any set with $B \subseteq C \subseteq A$. We need to show that $C \models\!\approx x$. Since $B \subseteq C$, monotony for classical consequence gives us $C \vdash c$ for every $c \in L_A \cap L_x$ with $x \vdash c$. Also, since $C \subseteq A$, we have $L_C \subseteq L_A$ and so $C \vdash c$ for every $c \in L_C \cap L_x$ with $x \vdash c$. Applying again the second refinement of the characterization of friendliness, we have $C \models\!\approx x$ as desired. ■

1.7 Interpolation

As in the case of compactness, interpolation for friendliness is trivial when formulated in the way customary in classical logic. For suppose $A \models\!\approx x$; we want to show that there is a formula b with $E(b) \subseteq E(A) \cap E(x)$ such that both $A \mathrel{|\!\approx} b$ and $b \mathrel{|\!\approx} x$. On the one hand, if A is inconsistent, we can put $b = \bot$ giving us $A \vdash b \vdash x$ so $A \mathrel{|\!\approx} b \mathrel{|\!\approx} x$. On the other hand, if A is consistent then since $A \mathrel{|\!\approx} x$, x must also be consistent, so we can put $b = \top$, so that $A \vdash b$ and thus $A \mathrel{|\!\approx} b$, and also $b \mathrel{|\!\approx} x$ using the consistency of x.

The following formulation strengthens the property by guaranteeing that in suitable conditions, b can be chosen more informatively.

Interpolation. Whenever $A \mathrel{|\!\approx} x$ there is a finite set $F \subseteq E(A) \cap E(x)$ of elementary letters such that for every finite set G of elementary letters with $F \subseteq G \subseteq E(A)$ there is a formula b with the following properties:

1. $E(b) = G$
2. $A \mathrel{|\!\approx} b$ (indeed $A \vdash b$)
3. $b \mathrel{|\!\approx} x$
4. b is consistent, provided A is consistent
5. b is not a tautology, provided there is a non-tautology $y \in L_A \cap L_x$ with $A \vdash y$.

Remark. Before giving the proof, we note that the rather odd proviso in property (5) cannot be weakened to, say: A and x are not tautologous. Example: $A = p \vee q, x = q \vee r$. Then $A \mathrel{|\!\approx} x$, but the only formulae b with $E(b) \subseteq E(A) \cap E(x) = \{q\}$ and both $A \mathrel{|\!\approx} b$ and $b \mathrel{|\!\approx} x$ are the tautologies containing at most the letter q.

Proof. Suppose $A \mathrel{|\!\approx} x$. Since x is a single formula, $E(x)$ is finite, and thus so too is $E(A) \cap E(x)$. Hence, up to classical equivalence, there is a strongest formula a with $E(a) \subseteq E(A) \cap E(x)$ and $A \vdash a$. Take any such a and put $F = E(a)$, which is clearly finite. Let G be any finite set of letters with $F \subseteq G \subseteq E(A)$. Form b by conjoining with a the disjunctions $q \vee \neg q$ for the finitely many letters q in $G \backslash F$. We claim that b fulfils all requirements.

Property (1) is immediate by construction. Also by construction $A \vdash a \dashv\vdash b$ and so by supraclassicality, $A \mathrel{|\!\approx} b$, giving (2). For property (4), if A is consistent then since $A \vdash b$, b is also consistent. For (5), suppose there is a non-tautology $y \in L_A \cap L_x$ with $A \vdash y$. Then by its construction, a is not a tautology, and so since $a \dashv\vdash b$, b is not a tautology.

It remains to show (3). Suppose $b \mathrel{|\!\not\approx} x$; we derive a contradiction. Since $b \mathrel{|\!\not\approx} x$ there is a partial valuation v on $E(b) = G \subseteq E(A)$ with $v(b) = 1$, which is not friendly to x, i.e. such that $v^+(x) = 0$ for every extension v^+ of v to $E(b, x)$. Fix such a v for the remainder of the proof.

Write k for the state-description formula in L_b that corresponds to v. Then clearly $v(k) = 1$ and also $k \vdash \neg x$. Put $b^* = b \wedge \neg k$. We show that b^* is a formula in L_A with $A \vdash b^*$ and $b \nvdash b^*$, thus contradicting the construction of b.

For $b^* \in L_A$: This is immediate since both $b, \neg k \in L_A$.

For $b \nvdash b^*$: It suffices to show $b \nvdash \neg k$, i.e. that $k \nvdash \neg b$. We have by its construction that $k \vdash b$; and since $v(k) = 1, k$ is satisfiable, so $b \nvdash \neg k$ as desired.

For $A \vdash b^*$: Since $A \vdash b$ it suffices to show $A \vdash \neg k$. As a preliminary observation, we show that there is no extension w of v to $E(A)$ with $w(A) = 1$. For let w be such an extension. Since by hypothesis $A \mathrel{|}\!\!\approx x$, there is an extension w^+ of w to $E(A, x)$ with $w^+(x) = 1$. Clearly, w^+ is also an extension of v to $E(A, x)$. Now restrict w^+ to $E(b, x)$, which is possible since $E(b) \subseteq E(A)$ so that $E(b, x) \subseteq E(A, x)$, and call it w^{+-}. Clearly $w^{+-}(x) = 1$ and also w^{+-} is still an extension of v, which has domain $E(b)$. But this contradicts the fact that v is not friendly to x. This completes the preliminary step of showing that there is no extension w of v to $E(A)$ with $w(A) = 1$.

Now let w be any partial valuation on $E(A) \supseteq E(b) = E(k)$ with $w(\neg k) = 0$, i.e. $w(k) = 1$. It remains to show that $w(A) = 0$. Restrict w to $E(k) = E(b) = \text{domain}(v)$, call it w^-. Clearly $w^-(k) = 1$. Hence by the construction of k as a state-description in L_b corresponding to v, $w^- = v$. Thus w is an extension of v to $E(A)$. So by the preliminary observation, $w(A) = 0$ as desired. ■

1.8 Friendliness as an Operation

Up to now, we have treated friendliness as a relation between formulae (or sets of formulae) on the left and formulae on the right. But just as in the case of classical consequence and well-known nonmonotonic consequences, we can consider it as an operation, taking sets of formulae to sets of formulae, by defining $Fr(A) = \{x : A \mathrel{|}\!\!\approx x\}$.

However, this may not be a very useful perspective for friendliness, in contrast to the situation for the usual nonmonotonic consequence relations. The reason is that friendliness is much further from being a closure relation. It fails monotony but also, as we have seen in section 1.3, it fails both conjunction in the conclusion and general cumulative transitivity. Expressed as an operation, it also fails idempotence (the same counterexample can be used as for cumulative transitivity). These properties are all satisfied by the usual nonmonotonic consequence relations (see e.g. Makinson 2005), and their absence makes the operational notation much less convenient to use.

So, in this section we examine just one question regarding the operational version: when do we have $Fr(A) = Fr(B)$ for sets A, B of formulae?

Observation. $Fr(A) = Fr(B)$ iff either $A \dashv\vdash B$ and $E(A) = E(B)$ or else A, B are both contradictions.

Verification. In one direction, suppose RHS. We want to show $Fr(A) = Fr(B)$. In the limiting case that A,B are both contradictions, we have $Fr(A) = L = Fr(B)$ vacuously from the definition of friendliness. So consider the principal case that $A \dashv\vdash B$ and $E(A) = E(B)$. Then $Fr(A) = Fr(B)$ by two applications of local left equivalence (section 1.5).

For the other direction, suppose $Fr(A) = Fr(B)$. Suppose that A, B are not both contradictions. We need to show that $E(A) = E(B)$ and $A \dashv\vdash B$.

First, we observe that neither of A,B is a contradiction. For suppose A, say, is a contradiction. Then $A \vdash \bot$ and so by supraclassicality of friendliness, $A \mid\!\approx \bot$ and so since $Fr(A) = Fr(B)$ we have $B \mid\!\approx \bot$, so B is a contradiction.

Next, we show $E(A) = E(B)$. It suffices to show $E(A) \subseteq E(B)$; the converse is similar. Suppose $p \in E(A)$ but $p \notin E(B)$; we derive a contradiction. Since $p \notin E(B)$ clearly $B \mid\!\approx p$ and also $B \mid\!\approx \neg p$. Since $Fr(A) = Fr(B)$, this gives us $A \mid\!\approx p$ and also $A \mid\!\approx \neg p$. Since $p \in E(A)$, the first reduction case for friendliness tells us that $A \vdash p$ and also $A \vdash \neg p$ so that A is inconsistent, contradicting what has been shown.

Finally, we show $A \dashv\vdash B$. It suffices to show $A \vdash B$; the converse is similar. Take any $b \in B$. we need to show $A \vdash b$. Now $B \vdash b$ so by supraclassicality $B \mid\!\approx b$ so since $Fr(A) = Fr(B)$ we have $A \mid\!\approx b$. Since $E(A) = E(B)$ and $b \in B$ we have $E(b) \subseteq E(A)$ so by the first reduction case for friendliness, $A \vdash b$ as desired, and the proof is complete. □

1.9 Joint Friendliness: Two Notions

For classical consequence, we have followed the common convention of writing $A \vdash B$ to mean that $A \vdash b$ for all $b \in B$. For friendliness, it is tempting to write $A \mid\!\approx B$ analogously. But care is needed, for there is an important distinction that does not arise in the classical case. We must distinguish between two relationships:

- $A \mid\!\approx_{\forall\forall\exists} B$: for every partial valuation v on $E(A)$ with $v(A) = 1$ and every $b \in B$, there is an extension v^+ of v to $E(A, b)$ with $v^+(b) = 1$.
- $A \mid\!\approx_{\forall\exists\forall} B$: For every partial valuation v on $E(A)$ with $v(A) = 1$ there is an extension v^+ of v to $E(A, B)$ with $v^+(B) = 1$, i.e. with $v^+(b) = 1$ for every $b \in B$.

The former says the same as $A \mid\!\approx b$ for all $b \in B$. But the latter says more. For classical consequence, where conjunction in the conclusion is satisfied, no such distinction arose. We call $\approx_{\forall\forall\exists}$ *weak* joint friendliness, $\approx_{\forall\exists\forall}$ *strong*. When we refer to joint friendliness (sections 2.2 and 3.4), we will specify clearly which is intended.

1.10 Internalizing the Relation

It is natural to ask whether we can internalize the relation of friendliness as a conditional connective of the object language.

It can be done quite trivially by adding an iterable two-place connective $\rightsquigarrow$ to the object language and adding to the familiar Boolean rules the following one. To bring the formulation as close as possible to standard ones for propositional connectives, we state it with v,w,u understood as full valuations, i.e. defined on the set E of all elementary letters.

$v(a \rightsquigarrow x) = 1$ iff for every full valuation w with $v(w) = 1$ there is a full valuation u that agrees with w on all elementary letters in $E(a)$ and such that $u(x) = 1$.

The same effect can be achieved by means of indexed unary modal operators. Consider a language with operators $\Box_a$ and $\Diamond_a$ for all formulae a. This is a little unusual, as the set of connectives is not fixed in advance, but is defined inductively along with the formulae in which they occur; but that is not a problem. We read these connectives by the following rules:

$v(\Box_a x) = 1$ iff for every valuation w that agrees with v on all elementary letters in $E(a)$, we have $w(x) = 1$.

$v(\Diamond_a x) = 1$ iff for some valuation w that agrees with v on all elementary letters in $E(a)$, we have $w(x) = 1$.

We may then identify plain $\Box$ and $\Diamond$ as $\Box_\top$ and $\Diamond_\top$ (or equivalently $\Box_\bot$ and $\Diamond_\bot$), giving us the familiar evaluation rules:

$v(\Box_x) = v(\Box_\top x) = v(\Box_\bot x) = 1$ iff $w(x) = 1$ for every valuation w.

$v(\Diamond_x) = v(\Diamond_\top x) = v(\Diamond_\bot x) = 1$ iff $w(x) = 1$ for some valuation w.

With this equipment, we may represent $a \approx x$ in the object language by the formula $\Box(a \to \Diamond_a x)$. Given the rules given above for evaluating indexed modal operators, this formula will satisfy the same evaluation condition that we gave for the trivial internalization. It will come out as true under one valuation iff it does so under all valuations, and that iff the relation $a \approx x$ holds.

However, it should be understood that when we internalize the relation of friendliness (whether directly or via indexed modal operators) the resulting system is rather unusual. The set of all valid formulae (defined as those formulae that are true under every valuation) is not closed under substitution, for the very same reason as the relation of friendliness was not so closed. The same example can be used to illustrate the failure. On the one hand, the formula $(p \to \Diamond_p(p \wedge q)$ is valid, while its substitution instance

$(p \to \Diamond_p(p \wedge \neg p)$ is not.

Thus while internalization is perfectly possible, the propositional system that it gives us unlike most modal and other non-classical propositional logics, for which the set of valid formulae is closed under substitution. In the author's view, this difference is not a disqualification — see e.g the discussion in Makinson (2005). But it not clear that internalization provides any insights that are not already available when friendliness is treated as a relation between formulae.

2 Links with Familiar Notions

Friendliness has many friends: several other notions familiar from the literature are connected with it. Roughly speaking, the links are of two main kinds.

- Certain well-known operations from the history of logic, distant and recent, can be seen as *instances* of friendliness.
- There are also more general *conceptual links*, notably with Ramsey eliminability and related notions that have been studied in the context of first-order logic.

We begin with some instances of friendliness.

2.1 Forgetting Letters from Formulae

Consider any formula a and any subset F of its elementary letters, i.e. $F \subseteq E(a)$. Let $\sigma_1, \ldots, \sigma_k$ be the $k = 2^n$ substitutions of $\bot, \top$ for the n letters in F. Following Weber (1987) and later papers such as Lin and Reiter (1994) and Lang, Liberatore, Marquis (2003), we may define $f_F(a)$, the result of *forgetting* the letters in F from a, as $\sigma_1(a) \vee \ldots \vee \sigma_k(a)$. Equivalently, in recursive form, $f_\emptyset(a) = a$, and $f_{F,q}(a) = \sigma_\bot(f_F(a)) \vee \sigma_\top(f_F(a))$, where the functions $\sigma_\bot$ and $\sigma_\top$ substitute $\bot, \top$ for the letter q.

As is well known, $a \vdash f_F(a)$. The converse fails, i.e. $f_F(a) \nvdash a$; for example $f_F(p) = \bot \vee \top \nvdash p$. However, $f_F(a)$ is easily shown to be the strongest formula b in the language generated by $E(a) \backslash F$ such that $a \vdash b$.

In fact, the notion goes back to Boole, whose focus was however rather different. From his point of view, the central logical relation was equality, coresponding to classical equivalence. Accordingly, the most important fact for him about what we now call forgetting was the equality that he introduced under the name of 'development' in Boole (1847): $a \dashv\vdash (\neg p \wedge \sigma_\bot(a)) \vee (p \wedge \sigma_\top(a))$. The consequence $a \vdash \sigma_\bot(a) \vee \sigma_\top(a) = f_a(a)$ is however implicit (in dual form) in the discussion of the 'elimination' of a term in an equation, in Boole (1854).

Observation. $f_F(a) \approx\!\!\models a$.

Verification. Let v be any partial valuation on $E(f_F(a)) = E(a)\backslash F$ and suppose $v(f_F(a)) = 1$. Then $v(\sigma_i(a)) = 1$ for some $i \leq k$. Extend v to v^+ on $E(a)$ by putting $v^+(q) = 0{,}1$ according as $\sigma_i(q)) = \bot, \top$ for each $q \in F$. Then clearly by induction on length of formulae, $v^+(a) = v(\sigma_i(a)) = 1$ and we are done. □

2.2 Ejective Substitution

It is natural to ask whether this observation can be extended to a more general result linking friendliness and substitution. It cannot cover all substitutions, for we do not always have $\sigma(a) \approx\!\!\models a$, even when σ is a one-one correspondence on letters. Consider for example the formula $a = p \wedge \neg q$ and the substitution σ that simply interchanges the two letters, putting $\sigma(p) = q$ and $\sigma(q) = p$ so that $\sigma(a) = q \wedge \neg p \not\approx\!\!\models a = p \wedge \neg q$ (witness the only partial valuation that makes the premiss true).

Nevertheless, we do have a positive result for a certain class of substitutions. Let σ be any substitution on the set E of all elementary letters, and let A be any set of formulae. We call σ *ejective for A* iff for every letter $p \in E(A)$, either $\sigma(p) = p$ or $p \notin E(\sigma(A))$.

Observation. Let a be any formula, and let σ be any substitution that is ejective for a. Then $\sigma(a) \approx\!\!\models a$. More generally, when A is a set of formulae and σ is ejective for A then $\sigma(A) \approx\!\!\models_{\forall\exists\forall} A$.

Verification. The notation $\approx\!\!\models_{\forall\exists\forall}$ for strong joint friendliness is explained in section 1.9. Consider any partial valuation v on $E(\sigma(A))$ with $v(\sigma(A)) = 1$. We extend v to v^+ on $E(\sigma(A), A)$ by putting $v^+(q) = v(\sigma(q))$ for each letter q in $E(A)\backslash E(\sigma(A))$. We want to show that $v^+(A) = v(\sigma(A)) = 1$. It suffices to show by induction that for every subformula b of any formula in $A, v^+(b) = v(\sigma(b))$.

For the basis, if b is a letter p then either $\sigma(p) = p$ or $p \notin E(\sigma(A))$. In the former case $p \in E(\sigma(A))$, so $v(p)$ is defined, so since v^+ extends v we have $v^+(p) = v(p) = v(\sigma(p))$ as desired. In the latter case, $p \in E(A)\backslash E(\sigma(A))$, so that $v^+(p) = v(\sigma(p))$ by definition.

The induction step is then routine using the definitions of a substitution and of a Boolean valuation. □

This observation covers the 'friendly forgetfulness' property $f_F(a) \approx\!\!\models a$ as a special case. For when a function σ substitutes $\bot, \top$ for some of the elementary letters in a (and is the identity on all other letters) then it is ejective for a. Indeed, it is ejective *tout court*, in the stronger sense that for every letter p, either $\sigma(p) = p$ or $p \notin E(\sigma(L)) = E(\sigma(E))$. Thus we have

$f_F(a) = \sigma_1(a) \vee \ldots \vee \sigma_k(a)$ where each substitution σ_i is ejective, so that each $\sigma_i(a) \approx a$. But $E(\sigma_i(a)) = E(a) \backslash F = E(\sigma_j(a))$ for all $i, j \leq k$ and so we may apply local disjunction in the premisses (section 1.5) putting $A = \emptyset$ to conclude that $\sigma_1(a) \vee \ldots \vee \sigma_k(a) \approx a$ as desired.

2.3 Identifying Letters

The above observation has a further corollary. By an *identification of letters* we mean a substitution σ on E into E such that for every letter p, either $\sigma(p) = p$ or $p \neq \sigma(q)$ for all letters q. Equivalently: such that whenever $p = \sigma(q)$ for some letter q then $\sigma(p) = p$. Equivalently: such that for some partition of E and some choice function γ on that partition, $\sigma(p) = \gamma(|p|)$.

Corollary. $\sigma(A) \approx_{\forall\exists\forall} A$ for any identification σ of letters. In particular, when a is an individual formula and σ is an identification of letters, then $\sigma(a) \approx a$.

Verification. By the observation in section 2.2, it suffices to observe that every identification of letters is ejective *tout court*, and so ejective for A. Let σ be any identification of letters. Suppose $p \in E(A)$ and $\sigma(p) \neq p$. Since σ is an identification of letters, this gives us $p \neq \sigma(q)$ for all letters q. Since σ takes E into E this implies that $p \notin E(\sigma(E)) = E(\sigma(L))$. □

2.4 Existential Quantification

The concept of forgetting can also be expressed in the language of quantified Boolean formulae. Put $g_F(a) = \exists p_1 \ldots \exists p_n(a)$ where $F = \{p_1, \ldots, p_n\}$. Then under the standard semantics for quantified Boolean formulae, $g_F(a)$ has exactly the same truth conditions as $f_F(a)$. So, with the notion of friendliness suitably enlarged to cover such formulae (rather than just unquantified Boolean formulae, as in this paper), we can say that $g_F(a)$ is friendly to a.

More generally, it is clear that in any language admitting existential quantifiers over a syntactic category of items, the existential quantification $\exists i_1 \ldots \exists i_n(a)$ over selected variables from that category will, under a natural enlargement of the notion, be friendly to a.

However, it should also be observed that the forgetting function $f_F(a)$, its quantified Boolean analogue $g_F(a)$, and existentialization $\exists i_1 \ldots \exists i_n(a)$ all have a more intimate relation to their argument a than mere friendliness. For we have not only $f_F(a) \approx a, g_F(a) \approx a, \exists i_1 \ldots \exists i_n(a) \approx a$ but also the classical consequences in the reverse direction: $a \vdash f_F(a), a \vdash g_F(a), a \vdash \exists i_1 \ldots \exists i_n(a)(a)$. This contrasts with the fact that for friendliness in general we may have $b \approx a$ without $a \vdash b$: witness the example $p \approx q$ but $q \nvdash p$ where p, q are distinct elementary letters.

2.5 Skolemization

The process of Skolemization of a formula of first-order logic manifests friendliness in a very special way. Taking for example the formula $\alpha = \forall x \exists y(Rxy)$, we can introduce a function letter f and consider both the formula $sk(\alpha) = \forall x(Rxf(x))$ and its existential quantification $\exists f(sk(\alpha)) = \exists f \forall x(Rxf(x))$. These formulae belong respectively to first-order logic with function letters, and second-order logic.

As Skolem observed, we have $sk(\alpha) \vdash \alpha$ in first-order logic, and also $\alpha \dashv\vdash \exists f(sk(\alpha))$ in second-order logic (assuming the axiom of choice in our metalanguage). The equivalence between α and $\exists f(sk(\alpha))$ means that the relation between these two is much tighter than for plain existentialization.

While $sk(\alpha) \vdash \alpha$, the converse fails: $\alpha \nvdash sk(\alpha)$. But we do have $\alpha \models\!\!\approx sk(\alpha)$ where $\models\!\!\approx$ is the friendliness in the first-order context, understood in terms of expansions (section 1.2). For every (partial) model interpreting the predicate letter R in a domain, if that model satisfies α then it has an expansion also interpreting the function letter f in the same domain that satisfies $sk(\alpha)$.

Here again there is an especially close relationship. As is well known, a and $sk(\alpha)$ are equivalent for logical truth, i.e. a is true in all first-order models iff $sk(\alpha)$ is. This does not hold for friendliness in general. In our base territory of classical propositional logic, $p \vee \neg p \models\!\!\approx q$ but the left is a tautology while the right is not.

As is well known, the passage from α to $sk(\alpha)$ also contrasts with existentialization in this regard. For example $\exists x(\exists x(Px) \rightarrow Px)$ is friendly to $\exists x(Px) \rightarrow Px$, but the left is a logical truth while the right is not.

2.6 Ramsey Eliminability

As well as the above particular instances of friendliness, there are also more general connections with concepts that have arisen elsewhere. Of these, the closest is with Ramsey eliminability of a predicate or other term in a theory.

This notion takes its origin in the philosophy of science, and more specifically in discussions concerning the relation between the observational and theoretical components of empirical scientific theories. It was first sketched in rough terms by F. P. Ramsey in notes of 1929, published in the posthumous collection Ramsey (1931, chapter 'Theories'). It was taken up and given its name by Sneed (1971, chapter 3); and subsequently discussed in a number of books and papers including van Benthem (1978) and Rantala (1991). All of these are expressed in the context of first-order languages.

Formulations differ in subtle but significant respects. What they all have in common is that every model of one set Γ of (first-order) formulae should be capable of expansion to a model of a larger set Δ that possibly contains further letters (individual constants, predicates, or function signs). We recall that by an *expansion* of a model is meant another model with the same domain, same interpretations of the letters that were interpreted in the first model, plus interpretations of whatever new letters are concerned.

Where the formulations differ is in what Γ and Δ are taken to be; which of them is taken to be an arbitrary set of formulae while the other is taken as a function of it. The story is as follows.

- For Rantala (1991, pages 150–151): Γ is taken to be an arbitrary set of first-order formulae, and Δ is put at $\Gamma \cup \{\varphi\}$ where φ is a (likewise first-order) formula. Rantala focusses on the case that this formula has just one new letter beyond those occurring in Γ, thought of as a candidate for reduction; however the definition is meaningful without that restriction. The concept is envisaged as expressing a property of the new letter(s) in φ modulo the set $\Gamma \cup \{\varphi\}$, rather than a relation between Γ and$\Delta = \Gamma \cup \{\varphi\}$.

- By contrast, for van Benthem (1978, page 325), it is Δ that is is taken to be an arbitrary set of first-order formulae, while Γ is taken to be $Cn(\Delta) \cap L_0$, where L_0 is an arbitrarily chosen sublanguage of the language L of Δ. Again, the concept is envisaged as expressing a property of the omitted letter set $L \backslash L_0$ modulo the formula set Δ.

Typically, L_0 will be made up of all the letters in L except for one, which is thought of as a candidate for reduction. In that case, we have exactly a notion introduced by de Bouvère (1959, chapter II.2). He used the failure of this property of the omitted letter (say, a predicate P) modulo a theory Δ, as a method for showing that P is not explicitly definable in Δ. This contrasts with the better-known technique going back to Padoa (1901), which proceeds by showing that some model of Γ can be expanded in two distinct ways to a model of Δ. Unlike de Bouvère's method, that of Padoa is complete for the task, as shown in a celebrated theorem of Beth (1956).

As is well known, the formulations of Rantala and van Benthem are not equivalent. On the one hand, when $\Delta = \Gamma \cup \{\varphi\}$ and L_0 is the language of Γ, then $\Gamma \subseteq Cn(\Delta) \cap L_0$. Hence, if every model of Γ can be expanded to a model of Δ, then every model of $Cn(\Delta) \cap L_0$ can too. In other words, Ramsey eliminability in the sense of Rantala implies the same in the sense of van Benthem. But in general, Γ may be a proper subset of $Cn(\Delta) \cap L_0$. So it may happen that whilst every model of $Cn(\Delta) \cap L_0$ can be expanded to one of Δ, there is some model of Γ (but not satisfying $Cn(\Delta) \cap L_0$) that cannot be so expanded. Thus Ramsey eliminability in the sense of van Benthem does not imply the same in the sense of Rantala. Specific examples have been given in the literature.

To compare these two concepts with friendliness as studied in this paper, we extract the purely propositional content, and write it in the notation that we have been using. We write $L_{E(B)\backslash F}$ for the language generated by the letters that are in $E(B)\backslash F$.

- From Rantala: The letters in $E(x)\backslash E(A)$ are Ramsey eliminable from a set A, x of formulae iff every partial valuation v on $E(A)$ with $v(A) = 1$ can be extended to a partial valuation v^+ on $E(A, x)$ with $v^+(A, x) = 1$.
- From van Benthem: Consider any set B of formulae and any set F of elementary letters with $F \subseteq E(B)$. The letters in F are Ramsey eliminable from B iff every partial valuation v on $E(B)\backslash F$ with $v(Cn(B) \cap L_{E(B)\backslash F}) = 1$ can be extended to a partial valuation v^+ on $E(B)$ with $v^+(B) = 1$.

Of these, the Rantala-style concept is equivalent to friendliness of A to x, as defined and studied in this paper.

Observation. Let A be any set of propositional formulae and x a propositional formula. Then A is friendly to x iff the letters in $E(x)\backslash E(A)$ are Ramsey eliminable from A, x in the sense of Rantala.

Verification. The only difference between the definition of friendliness and the propositional reduction of Rantala's version of Ramsey eliminability is that whereas the former requires the extension v^+ to satisfy x, the latter requires it to satisfy A, x. But these are equivalent when v^+ extends v and $v(A) = 1$. □

We have already remarked that even in the first-order context, the formulation of van Benthem is weaker than that of Rantala. Indeed, it is very much weaker since, as is well-known, every *finite* model of $Cn(\Delta) \cap L_0$ can be expanded to a model of Δ. In the purely propositional context, it becomes so much weaker that it always holds, as we now show.

Observation. Let B be any set of propositional formulae and $F \subseteq E(B)$ any subset of its elementary letters. Then the letters in F are Ramsey eliminable from B in the sense of van Benthem.

Proof. We need to show that every partial valuation v on $E(B)\backslash F$ with $v(Cn(B) \cap L_{E(B)\backslash F}) = 1$ can be extended to a partial valuation v^+ on $E(B)$ with $v^+(B) = 1$.

Let v be a partial valuation on $E(B)\backslash F$ with $v(Cn(B) \cap L_{E(B)\backslash F}) = 1$. Suppose for reductio ad absurdum that v cannot be extended to a partial valuation v^+ on $E(B)$ with $v^+(B) = 1$. Let S be the state-description corresponding to v, i.e. the set of all literals in $L_{E(B)\backslash F}$ that are true under v. In the limiting case that $F = E(B)$ so that $E(B)\backslash F = \emptyset$, put $S = \{\top\}$.

We note first that $S \cup B$ is inconsistent. Reason: For any partial valuation w on $E(S \cup B) = E(B)$ with $w(S \cup B) = 1$ we have $w(S) = 1$ so w must must agree with v over F, so w is an extension of v to $E(B)$. Also $w(B) = 1$, contrary to the supposition.

Since $S \cup B$ is inconsistent, compactness tells us that there is a formula s that is the conjunction of finitely many elements of S, such that $\neg s \in Cn(B)$. But also by construction, $\neg s \in L_{E(B)\backslash F}$. Hence $\neg s \in Cn(B) \cap L_{E(B)\backslash F}$ and so by hypothesis $v(\neg s) = 1$, contradicting the fact that by the construction of S we have $v(s) = 1$. ■

This argument is along much the same lines as that for the characterization of friendliness in terms of consistency in section 1.4. Like that characterization, it does not carry over to first-order contexts; indeed, the counterexample given in section 1.4 also serves here.

Corollary. De Bouvère's method can never be used in purely propositional logic as a way of showing that an elementary letter is not explicitly definable given a set A of propositional formulae.

Verification. Apply the observation with F chosen to be a singleton subset of $E(B)$. □

2.7 Leśniewski's Criterion of Conservativity

Friendliness is also closely related to the criterion of conservativity (alias non-creativity) in the theory of definition.

In lectures of the early 1920s, Leśniewski articulated two criteria that we usually want definitions to satisfy: eliminability and conservativity. A published account was given in Leśniewski (1931), with an easily accessible exposition in Suppes (1957, chapter 8). It is conservativity that connects with friendliness. The concept is usually formulated in the context of first-order logic. To clarify the link with friendliness, we again extract the purely propositional content.

Let A be any set of propositional formulae and let x be a formula. A,x is said to be a *conservative extension* of A iff $Cn(A, x) \cap L_A \subseteq Cn(A)$, i.e. iff $A \vdash c$ for every $c \in L_A$ such that $A, x \vdash c$.

Observation. In the propositional context: $A \approx\!\!\!\!\mid\; x$ iff A,x is a conservative extension of A.

Proof. We already know from the first refinement of the characterization of friendliness in terms of consistency, in section 1.4, that $A \approx x$ iff (1) $A \vdash c$ for every $c \in L_A$ with $x \vdash c$. So we need only show the equivalence of this with (2) $A \vdash c$ for every $c \in L_A$ with $A, x \vdash c$.

One direction is immediate: by the monotony of classical consequence, (2) clearly implies (1). For the converse, suppose (1). Suppose $c \in L_A$ and $A, x \vdash c$; we need to show $A \vdash c$. Since $A, x \vdash c$ compactness tells us that $a, x \vdash c$ where a is the conjunction of some finite subset of A, and so also $x \vdash a \to c$. Clearly since $c \in L_A$ we also have $a \to c \in L_A$ So we may apply (1) to get $A \vdash a \to c$, and so since $A \vdash a$ we have $A \vdash c$ as desired. ■

Corollary. On the level of propositional logic: A, x is a conservative extension of A iff the letters in $E(x) \backslash E(A)$ are Ramsey eliminable from A, x in the sense of Rantala.

Verification. By the observation just established, A, x is a conservative extension of A iff $A \approx x$. By the first observation of section 2.6, $A \approx x$ iff the letters in $E(x) \backslash E(A)$ are Ramsey eliminable from A, x in the sense of Rantala. □

Again this corollary is known to fail in the first-order context, where only the right-to-left half holds. An equivalence does hold, but it is between the left and a weaker version of the right: Γ, φ is a conservative extension of Γ iff every model of Γ is elementary equivalent to (i.e. satisfies the same first-order formulae as) some model of Γ that can be expanded to a model of Γ, φ.

2.8 Information-Preserving Paraconsistent Consequence

A less intimate connection with friendliness can be found in the construction of a certain paraconsistent consequence relation, effected in Pietruszczak (2004). This relation, which is a subrelation of classical consequence, is defined by Pietruszczak using a notion of preservation of information. But he also gives it an alternative characterization (his theorem 6.1) that makes contact with friendliness, or more precisely, with its syntax-independent counterpart sympathy, which we will define below in section 3.1.

Specifically, Pietruszczak's relation of information-preserving consequence holds between a formula a and a formula x iff four conditions hold: a classically entails x; a is classically consistent; x is not a tautology; and a further condition, formulated in terms of valuations, also holds. This further condition is not given a name, but is exactly the relation of sympathy, holding in the reverse direction from x to a.

Thus, roughly speaking, the syntax-independent version of friendliness has been used as one of the ingredients to construct a certain kind of paraconsistent subrelation of classical consequence. We have, in other words, an application of the relation.

The present author would comment, however, that the paraconsistent consequence so defined has a rather mixed bag of properties. As well as failing certain consequences that the paraconsistent logician desperately seeks to avoid (e.g. implication from $a \wedge \neg a$ to any proposition whatsoever, and from any proposition to $x \vee \neg x$), and failing others that some are willing to lose in order to achieve this (e.g. from a to $a \vee x$ for any x) the relation fails certain other properties that few paraconsistent logicians would be happy to see depart.

One of these is closure of the consequence relation under uniform substitution (of arbitrary formulae for elementary letters). Others are implication from $p \wedge q$ to any of $p \vee q, p \leftrightarrow q, p \rightarrow q, q \rightarrow p$, and likewise from $p \leftrightarrow q$ to either of $p \rightarrow q, q \rightarrow p$. Verification of all these failures is straightforward: none of the right formulae is friendly to the left one.

2.9 Coupled Semantic Decomposition Trees

Finally, we mention a connection with the theory of semantic decomposition trees (alias semantic tableaux) in classical logic. Developed by Beth, Hintikka and others, these trees entered the arena of textbooks with Jeffrey [1967]. Designed to test formulae for satisfiablility, the trees can of course be used to test an inference for invalidity by checking the satisfiability of the set (or conjunction) consisting of the premisses and negation of the conclusion. But Jeffrey also suggested another technique for the purpose, which he called 'coupled trees'.

Roughly speaking, he constructed a (signed) tree for the premisses, and another one for the conclusion. If every open branch of the former tree contains all the signed elementary letters that occur on some open branch of the latter one, then the inference is valid. However, as Jeffrey noted, the converse is not true without qualification. This is due to the possible absence of elementary letters in branches of the first tree, as for the inference from p to $q \vee \neg q$, likewise from p to $(p \wedge q) \vee (p \wedge \neg q)$. For this reason, he introduced an additional rule allowing the introduction of new elementary letters (by branching to an arbitrary formula and to its negation) when constructing a tree.

In the revised version of the textbook, published in 1981, Jeffrey omitted the technique of 'coupled trees' altogether, presumably because of the inelegance of the additional rule. In the meantime, Dunn [1976] showed that it could be adapted neatly to the so-called first-degree entailments of relevance logics. One simply requires that every branch (even closed) of the former tree contains all the signed elementary letters that occur on some branch (even closed) of the latter tree. This characterizes first-degree entailment without the need for any additional rules.

We remark that the technique of 'coupled trees' is even more naturally suited to determining whether a set A of formulae is friendly to another formula x. Construct the two (signed) trees as before. Call two branches *compatible* iff they do not contain any elementary letter with opposite signs. To test whether A is friendly to x, we simply check whether every open branch of the tree for A is compatible with some open branch of the tree for x. This characterizes friendliness without additional rules. We omit the straightforward verification.

3 From Friendliness to Sympathy

3.1 Definitions

We now consider a normalized version of friendliness that is syntax-independent on the left as well as on the right.

It is well known that for any finite set A of Boolean formulae, there is a unique least set F of elementary letters such that A is classically equivalent to some set of formulae in the language generated by F.

Although this is usually stated and proven for finite sets A only, it also holds for infinite ones. More specifically, let A be any set of formulae:

- Put $E!(A)$ to be the set of all letters p that are *essential for* A, in the sense that there are two valuations v, w, on the set E of all elementary letters of the language, that agree on all letters other than p but disagree in the value they give to A. Clearly $E!(A) \subseteq E(A)$.

- Put A^* to be the set of all formulae x with both $A \vdash x$ and $E(x) \subseteq E!(A)$. Clearly $E(A^*) = E!(A)$.

Clearly, whenever $A \dashv\vdash B$ then $E!(A) = E!(B)$ and also $A^* = B^*$. Moreover, as we show in the Appendix:

Least letter-set theorem. $A \dashv\vdash A^*$, and for every set B of formulae with $A \dashv\vdash B, E(A^*) \subseteq E(B)$.

We say that a set A of formulae is *sympathetic* to x and write $A \mid\!\sim x$, iff $A^* \mid\!\approx x$. This notion can be seen as a normalized version of friendliness, making it syntax-independent in the left argument.

Unrestricted left classical equivalence for $\mid\!\sim$. Whenever $A \dashv\vdash B$, then $A \mid\!\sim x$ iff $B \mid\!\sim x$.

Verification. Whenever $A \dashv\vdash B$ then as noted $A^* = B^*$, so $A^* \models\!\approx x$ iff $B^* \models\!\approx x$, i.e. $A \mid\!\sim x$ iff $B \mid\!\sim x$. □

From the least letter-set theorem we have immediately the following useful criterion for membership in $E!(A)$.

Criterion for membership *in* $E!(A)$. Let p be any elementary letter. Then $p \in E!(A)$ iff $p \in E(B)$ for every set B of formulae with $B \dashv\vdash A$.

We also have the following four criteria for sympathy.

Criteria for sympathy. Each of the following is equivalent to $A \mid\!\sim x$:

(a) $B \models\!\approx x$ for every B with $A \dashv\vdash B$ and $E(B) = E!(A)$

(b) $A^* \models\!\approx x$

(c) $B \models\!\approx x$ for some B with $A \dashv\vdash B$ and $E(B) = E!(A)$

(d) $B \models\!\approx x$ for some B with $A \dashv\vdash B$.

Verification. $A \mid\!\sim x$ is defined as (b), and immediately (a) $\Rightarrow$ (b) $\Rightarrow$ (c) $\Rightarrow$ (d). So we need only show (d) $\Rightarrow$ (a). Suppose $B \models\!\approx x$ for some B with $A \dashv\vdash B$. Let $A \dashv\vdash C$ and $E(C) = E!(A)$. We need to show $C \models\!\approx x$. Let v be any partial valuation on $E(C)$ with $v(C) = 1$. We need to find an extension v^+ of v to $E(C, x)$ with $v^+(x) = 1$. Since $E(C) = E!(A) = E(A^*) \subseteq E(B)$ by the least letter-set theorem, we may fix an arbitrary extension w of v to $E(B)$. Since $C \dashv\vdash A \dashv\vdash B$, we have $w(B) = 1$. Since $B \models\!\approx x$ there is an extension w^+ of w to $E(B, x)$ with $w^+(x) = 1$. Then w^+ is an extension of v to $E(B, x)$. Since $E(C) \subseteq E(B)$ we also have $E(C, x) \subseteq E(B, x)$, so we may restrict w^+ to $E(C, x)$, call it w^{+-}. Clearly w^{+-} is still an extension of v and also $w^{+-}(x) = w^+(x) = 1$, so we may put $v^+ = w^{+-}$ and it has the desired properties. □

Corollary: broadening. Whenever $A \models\!\approx x$ then $A \mid\!\sim x$.

Verification. By criterion (d). □

Evidently, the inclusion converse to broadening fails. Example: $p \wedge (q \vee \neg q) \not\models\!\approx p \wedge q$ but $(p \wedge (q \vee \neg q)) \mid\!\sim p \wedge q$ since $(p \wedge (q \vee \neg q)) \dashv\vdash p \models\!\approx p \wedge q$.

3.2 Property Failures for Sympathy: Inherited and New

All of the property failures that we bulleted for $\approx$ in section 1.3 are also failures for $\mid\!\sim$. We can take the same counterexamples and observe that for each premiss a, $E!(a) = E(a)$. On the other hand and perhaps surprisingly, there are two important properties that succeeded for $\approx$ but fail for $\mid\!\sim$: local disjunction in the premisses and compactness.

The following example, due to Pavlos Peppas (personal communication) illustrates the failure of local disjunction in the premisses.

Counterexample to local disjunction in the premisses. Put $a = p \vee r$, $b_1 = p \wedge q, b_2 = \neg q$, and $x = \neg q \vee \neg r$. Then $E(b_2) \subseteq E(a, b_1); E(b_1) \subseteq E(a, b_2); a, b_1 \mid\!\sim x; a, b_2 \mid\!\sim x$; but $a, b_1 \vee b_2 \not\mid\!\sim x$.

Verification. Clearly $E(b_2) \subseteq E(a, b_1)$ and indeed $E!(b_2) \subseteq E!(a, b_1)$. Also $E(b_1) \subseteq E(a, b_2)$ and indeed $E!(b_1) \subseteq E!(a, b_2)$. Also $a, b_1 \mid\!\sim x$ since $\{a, b_1\} \dashv\vdash b_1 \approx x$, applying criterion (d) for sympathy. Also $a \wedge b_2 \vdash x$ so that $a \wedge b_2 \approx x$ and thus $a, b_2 \mid\!\sim x$. But $a, (b_1 \vee b_2) \not\mid\!\sim x$.

To check the last, note that $a \wedge (b_1 \vee b_2) = (p \vee r) \wedge ((p \wedge q) \vee \neg q) \dashv\vdash p \vee (r \wedge \neg q)$ so that $E!(a, (b_1 \vee b_2)) = \{p, q, r\}$. So by criterion (a) for sympathy, it suffices to check that $p \vee (r \wedge \neg q) \not\approx \neg q \vee \neg r$. Since every letter on the right already occurs on the left, it suffices to show $p \vee (r \wedge \neg q) \nvdash \neg q \vee \neg r$ by the reduction case for friendliness (section 1.4). But this is clear putting $v(p) = v(q) = v(r) = 1$. □

By suitably tweaking this example, we can turn it into one that illustrates the failure, for sympathy, of the closely related rule of proof by exhaustion.

Counterexample to proof by exhaustion. Put $a = p \vee \neg q \vee r$, $b = p \wedge q; x = \neg q \vee \neg r$. Then $a, b \mid\!\sim x; a, \neg b \mid\!\sim x$; but $a \not\mid\!\sim x$.

Verification. Similar to that of the preceding example, but we give the details. Again we have $a, b \mid\!\sim x$ since $\{a, b\} \dashv\vdash b \approx x$, applying criterion (d) for sympathy. Also $a, \neg b \mid\!\sim x$ since $a, \neg b \vdash x$. But $a \not\mid\!\sim x$ since $E!(a) = \{p, q, r\}$, so by criterion (a) for sympathy, it suffices to check that $p \vee \neg q \vee r \not\approx \neg q \vee \neg r$. Since every letter on the right already occurs on the left, it suffices to show $p \vee \neg q \vee r \nvdash \neg q \vee \neg r$ by the reduction case for friendliness. But this is clear putting $v(p) = v(q) = v(r) = 1$. □

The next example illustrates the failure of compactness for sympathy. Consider a language with countably many elementary letters $q, p_1, p_2, \ldots$.

Counterexample to compactness. Put A to be the set of all formulae a_n that are of the form $(p_1 \wedge \ldots \wedge p_n) \vee q$ for odd $n \geq 1$, or of the form $(p_1 \wedge \ldots \wedge p_n) \vee \neg q$ for even $n \geq 1$. Then $A \mid\!\sim q$ but $B \not\mid\!\sim q$ for every finite non-empty subset $B \subseteq A$.

Verification. To show $A \mathrel{|\!\sim} q$ it suffices, by criterion (d) for sympathy, to find an $X \dashv\vdash A$ with $X \mathrel{|\!\approx} q$. Putting $X = \{p_i : i \geq 0\}$ we clearly have the former, and since q does not occur in any formula in X we also have the latter.

Now let B be any finite non-empty subset of A. To complete the verification of the example, we need to show that $B \mathrel{|\!\not\sim} q$, i.e. that $B^* \mathrel{|\!\not\approx} q$.

First, we show that q is essential to B. Consider the largest n such that $a_n \in B$; this exists because B is finite and non-empty. We examine the case that n is odd, so that $a_n = (p_1 \wedge \ldots \wedge p_n) \vee q$; the case for even n is similar. Put $v(p_i) = w(p_i) = 1$ for all $i < n, v(p_n) = w(p_n) = 0$, and $v(q) = 1$ while $w(q) = 0$. Then $w(a_n) = 0$ so that $w(B) = 0$. On the other hand, $v(a_n) = 1$ (since $v(q) = 1$) and also $v(a_i) = 1$ for all $i < n$ (since p_n does not occur in any such a_i) so that $v(B) = 1$. Since v, w agree on all p_i for all $i \leq n$ while disagreeing on B, this shows that q is essential to B, as desired.

We can now show that $B^* \mathrel{|\!\not\sim} q$. Put $u(p_i) = 1$ for all $i \leq n$ and $u(q) = 0$. Then $u(a_i) = 1$ for all $i \leq n$ so that $u(B) = 1$ and hence $B \nvdash q$; so since $B \dashv\vdash B^*$ we have $B^* \nvdash q$. But since q is essential to B, q occurs in B^*. So by the first reduction case of section 1.4, since $B^* \nvdash q$ we have finally $B^* \mathrel{|\!\not\approx} q$ completing the verification of the example. □

3.3 Property Successes for Sympathy

Apart from disjunction in the premisses and compactness, all of the other properties that we noted as satisfied by friendliness also hold for sympathy. We consider them one by one. Whenever possible, we derive the property for $\mathrel{|\!\sim}$ from the one for $\mathrel{|\!\approx}$, rather than argue from scratch. Most of the verifications are straightforward; only singleton cumulative transitivity is rather tricky, needing some lemmas on least letter-sets.

Supraclassicality for $\mathrel{|\!\sim}$. Whenever $A \vdash x$ then $A \mathrel{|\!\sim} x$.

Verification. Suppose $A \vdash x$. Then $A \mathrel{|\!\approx} x$ by supraclassicality for $\mathrel{|\!\approx}$, so $A \mathrel{|\!\sim} x$ by broadening. □

Reduction case for $\mathrel{|\!\sim}$. Whenever $E(x) \subseteq E!(A)$ then $A \mathrel{|\!\sim} x$ iff $A \vdash x$.

Verification. Right to left is given by supraclassicality. For the converse, suppose $E(x) \subseteq E!(A)$. Suppose $A \mathrel{|\!\sim} x$. By definition, $A^* \mathrel{|\!\approx} x$. Recalling that $E!(A) = E(A^*)$ so that $E(x) \subseteq E(A^*)$, the reduction case for friendliness tells us $A^* \vdash x$. Since $A \dashv\vdash A^*$ we have $A \vdash x$ as desired. □

Characterization of $\mathrel{|\!\sim}$ in terms of consistency. $A \mathrel{|\!\sim} x$ iff every set of formulae in $L_{E!(A)}$ that is consistent with A, is consistent with x.

Verification. By definition, $A \mid\!\sim x$ iff $A^* \mid\!\approx x$. Applying the corresponding consistency characterization of $\mid\!\approx$ and the fact that $A^* \dashv\vdash A$, the desired equivalence follows. □

Right weakening for $\mid\!\sim$. Whenever $A \mid\!\sim x \vdash y$ then $A \mid\!\sim y$

Verification. From the definition of $\mid\!\sim$ and right weakening for $\mid\!\approx$. □

This implies right classical equivalence for sympathy: whenever $x \dashv\vdash y$ then $A \mid\!\sim x$ iff $A \mid\!\sim y$. The relation $\mid\!\sim$ is thus syntax-independent on both left and right.

Local left strengthening for $\mid\!\sim$. Suppose $E!(B) \subseteq E!(A)$. If $B \vdash A \mid\!\sim x$ then $B \mid\!\sim x$.

Verification. Immediate from the corresponding property of $\mid\!\approx$, the definition of $\mid\!\sim$, and the fact that $A^* \dashv\vdash A$. □

Local monotony for $\mid\!\sim$. Suppose $E!(B) \subseteq E!(A)$. If $A \mid\!\sim x$ and $A \subseteq B$ then $B \mid\!\sim x$.

Verification. If $A \subseteq B$ then $B \vdash A$. □

Note that in these two 'local' properties, the locality condition concerns $E!(A), E!(B)$ rather than $E(A), E(B)$.

3.4 Singleton Cumulative Transitivity for Sympathy

We have postponed consideration of singleton cumulative transitivity because its proof requires two lemmas about least letter-sets.

Lemma. $E!(A, B) \subseteq E!(A) \cup E!(B) \subseteq E!(A) \cup E(B)$.

Verification. The right inclusion is immediate from $E!(B) \subseteq E(B)$. For the left inclusion, suppose $p \in E!(A, B)$. Then there are partial valuations v_0, v_1 on $E(A, B)$ that agree on all letters in this domain other than p, with $v_0(A, B) = 0$ and $v_1(A, B) = 1$. Since $v_0(A, B) = 0$, either $v_0(A) = 0$ or $v_0(B) = 0$.

Suppose the former; the argument for the latter is similar. Restrict v_0, v_1 to $E(A)$, call them v_0^-, v_1^-. Then $v_0^-(A) = 0$ whilst $v_1^-(A) = 1$, but v_0^-, v_1^- agree on all letters in their common domain other than p. Hence $p \in E!(A) \subseteq E!(A) \cup E!(B)$ as desired. □

Lemma. If $A \mid\!\approx x$ then $E!(A) \subseteq E!(A, x)$. Indeed, more generally: If $A \mid\!\approx_{\forall\exists\forall} B$ then $E!(A) \subseteq E!(A, B)$.

Verification. Suppose $A \mid\!\approx_{\forall\exists\forall} B$ (defined in section 1.9) and $p \in E!(A)$. From the latter, there are partial valuations v_0, v_1 on $E(A)$ that agree on all letters in this domain other than p, with $v_0(A) = 0$ and $v_1(A) = 1$. Since $A \mid\!\approx_{\forall\exists\forall} B, v_1$ can be extended to a valuation v_1^+ on $E(A, B)$ with $v_1^+(B) = 1$, so $v_1^+(A, B) = 1$. Now extend v_0 to $E(A, B)$ by putting $v_0^+(q) = v_1^+(q)$ for every letter $q \in E(A, B) \backslash E(A)$. Then clearly v_0^+, v_1^+ agree on all letters in their common domain except p, and disagree on A, B since $v_1^+(A, B) = 1$ while $v_0^+(A, B) = 0$ since $v_0(A) = 0$. Hence $p \in E!(A, B)$ as desired. □

Singleton cumulative transitivity for $\mid\!\sim$. Whenever $A \mid\!\sim x$ and $A, x \mid\!\sim y$ then $A \mid\!\sim y$.

Proof. Suppose $A \mid\!\sim x$ and $A, x \mid\!\sim y$. From the hypotheses we have $A^* \mid\!\approx x$ and $(A, x)^* \mid\!\approx y$. We need to show $A^* \mid\!\approx y$.

Let v be any partial valuation on $E(A^*) = E!(A)$ with $v(A^*) = 1$. We need to find an extension w of v to $E(A^*, y) = E!(A) \cup E(y)$ with $w(y) = 1$.

Since $A^* \mid\!\approx x$ and $v(A^*) = 1, v$ can be extended to a v^+ on $E(A^*, x) = E!(A) \cup E(x)$ with $v^+(x) = 1$. By the first lemma, we may restrict v^+ to the subset $E!(A, x)$ of its domain, call it v^{+-}. By the second lemma, since $A^* \mid\!\approx x$ we have $E(A^*) = E!(A) \subseteq E!(A, x)$, so v^{+-} is an extension of v. Also, $v^{+-}((A, x)^*) = v^+((A, x)^*) = v^+(A^*, x)$. Also $v^+(A^*) = v(A^*) = 1$ and $v^+(x) = 1$. Putting this together, $v^+(A^*, x) = 1$ so $v^{+-}((A, x)^*) = 1$.

Hence, since $(A, x)^* \mid\!\approx y, v^{+-}$ may be extended from $E!(A, x)$ to a valuation v^{+-+} on $E!(A, x) \cup E(y)$ with $v^{+-+}(y) = 1$. Since v^{+-} is an extension of v it follows that v^{+-+} is also an extension of v. Finally, restrict v^{+-+} to $E!(A) \cup E(y)$, which by the second lemma again is a subset of $E!(A, x) \cup E(y)$; call it v^{+-+-}. This is still an extension of v, defined on $E!(A)$, and also $v^{+-+-}(y)$ is well defined with $v^{+-+-}(y) = v^{+-+}(y) = 1$. Put $w = v^{+-+-}$ and the proof is complete. ■

3.5 Interpolation for Sympathy

An interpolation property for sympathy follows readily from its counterpart for friendliness. We need to be careful, however, about where we can write A, versus A^*, in the formulation.

Interpolation for $\mid\!\sim$. Whenever $A \mid\!\sim x$ there is a finite set $F \subseteq E(A^*) \cap E(x) \subseteq E(A) \cap E(x)$ of elementary letters, such that for every finite set G of elementary letters with $F \subseteq G \subseteq E(A^*)$ there is a formula b with the following properties:

1. $E(b) = G$
2. $A \mid\!\sim b$ (indeed $A \vdash b$)

3. $b \mid\!\sim x$

4. b is consistent, provided A is consistent

5. b is not a tautology, provided there is a non-tautology $y \in L_A \cap L_x$ with $A \vdash y$.

Proof. Suppose $A \mid\!\sim x$. By definition, $A^* \mid\!\approx x$. So by interpolation for friendliness, we have the above but with A^* in place of A in properties (2), (4), (5). Since $A \dashv\vdash A^*$ we also have (2), (4) for A. It remains to check condition (5).

Suppose there is a non-tautology $y \in L_A \cap L_x$ with $A \vdash y$. We need to find a non-tautology $z \in L_{A^*} \cap L_x$ with $A^* \vdash z$. Consider the 2^k formulae that can be obtained from y by substituting $\top, \bot$ for the k letters ($k \geq 0$) in $E(y)$ that are not in $E(A^*)$. Since y is not a tautology, at least one of these 2^k formulae is not a tautology; choose one as z. Clearly $z \in L_{A*} \cap L_x$. Also, since $A \vdash y$ and $A \dashv\vdash A^*$ we have $A^* \vdash y$ and so since the substitution producing z is the identity on A^* we have $A^* \vdash z$ and the verification is complete. ■

3.6 Further Remarks on the Concept of an Essential Letter

Karl Schlechta (personal communication) has observed that it is possible to generalize the notion of an essential letter, making it relative to an arbitrary *set of valuations* rather than to a *set of formulae.* In detail: let W be an arbitrary set of valuations. We say that a letter p is *essential to* W iff there are two valuations that agree on all letters other than p, but one in and the other outside W.

As is often the case when we pass to arbitrary sets of valuations in place of sets of formulae (which correspond to definable sets of valuations), we get an equivalent notion in the finite case, but a more general one in the infinite case with loss of some properties. Without following this through systematically, we give one example. When dealing with sets of formulae, we have the following:

Observation. Let A be any set of formulae. Then A is contingent (neither a tautology nor a contradiction) iff at least one of its elementary letters is essential to it.

Verification. Right to left is immediate from the definition of an essential letter. For the converse, suppose that A contingent. Then there are two partial valuations v, w on $E(A)$, with $v(A) = 1$ and $w(A) = 0$. From the latter, there is a formula $a \in A$ with $w(a) = 0$. Let v_w be the partial valuation on $E(A)$ defined by putting $v_w(p) = w(p)$ for all letters in $E(a)$, and $v_w(p) = v(p)$ for all other letters. Then v, v_w disagree on only finitely many letters, and we have $v(A) = 1$ while $v_w(a) = 0$ so that $v_w(A) = 0$.

Since v, w_v disagree on only finitely many letters, there is a finite chain $v_1, \ldots, v_n$ of partial valuations on $E(a)$ beginning with $v_1 = v$ and ending with $v_n = v_w$, each disagreeing with its predecessor on just one letter. Take the last v_k in the chain with $v_k(A) = 1$. Then $k < n$ and $v_{k+1}(A) = 0$. Thus v_k, v_{k+1} are partial valuations on $E(A)$ that agree on all letters except one, but give A different values, so that letter is essential to A. □

This argument goes through no matter what the cardinality of the set of the elementary letters, and independently of whether they can be well ordered. But the observation fails for its counterpart in terms of sets of valuations, even for a countable language.

The counterpart says: Let W be any subset of the set of all valuations; then W is proper and non-empty iff at least elementary letter is essential to it. Right to left does hold: if at least one letter is essential to W, then immediately from the definition W is neither empty nor the set of all valuations. But left to right fails. Example: put W to be the set of all valuations that make only finitely many elementary letters true. This is neither empty nor the set of all valuations. But when a valuation is in W, so is every valuation that differs from it at exactly one letter.

4 Open Questions

4.1 Specific Problems

- Can we give an axiomatic characterization of friendliness (or for sympathy) that is more traditional in style than the one at the end of section 1.5?

- What is the most interesting way of defining friendliness in a first-order context, and which of its properties carry over?

- Which properties of the notion of an essential letter carry over when that notion is understood modulo an arbitrary set of valuations, as in section 3.6, rather than modulo a set of formulae?

4.2 Open-Ended Questions

- How much of the theory of friendliness remains if we generalize from the classical two-valued context to a many-valued one?

- Is it helpful to characterize friendliness and sympathy using appropriate three-valued possible worlds structures, with a relation between possible worlds representing the extension of one partial valuation by another?

- Are there any interesting connections between the theory of friendliness and possible-worlds semantics for intuitionistic logic?

5 Appendix

5.1 Proof of Least Letter-Set Theorem

As remarked in the text, proofs of the least letter-set theorem usually cover only the finite case. Perhaps the most elegant such proof, given for example by Parikh (1999), uses interpolation for classical logic. We recall it briefly.

Let A be any finite set of Boolean formulae. Since A is finite, $E(A)$ is also finite, so there is at least one minimal subset $F \subseteq E(A)$ with the property that A is classically equivalent to some set of formulae in the language generated by F. So we need only show that F is unique. Let G be any other such minimal set of letters. Then there are sets B, C of formulae in L_F, L_G respectively with $B \dashv\vdash A \dashv\vdash C$ so $B \vdash C$ so by interpolation for classical logic there is a set X of formulae in $L_{F \cap G}$ with $B \vdash X \vdash C$ so $A \vdash B \vdash X \vdash C \vdash A$ so $A \dashv\vdash X$. But since F, G were both minimal, it follows that $F = F \cap G = G$ and we are done.

Unfortunately, this elegant argument is not available in the infinite case, as we cannot assume that there is a minimal F with the property. We give a different proof covering the infinite as well as the finite case. We have not been able to ascertain whether such a proof already occurs in the literature.

We recall from section 3.1 the definitions that will be needed.

- $E!(A)$ is the set of all letters p that are *essential for* A, in the sense that there are two valuations v, w, on the set E of all elementary letters of the language, that agree on all letters other than p but disagree in the value they give to A. Clearly $E!(A) \subseteq E(A)$, and whenever $A \dashv\vdash B$ then $E!(A) = E!(B)$.

- A^* is the set of all formulae x with both $A \vdash x$ and $E(x) \subseteq E!(A)$. Clearly $E(A^*) = E!(A)$. Clearly, whenever $A \dashv\vdash B$ then $A^* = B^*$.

Clearly, it would be equivalent to formulate the definition of $E!(A)$ in terms of partial valuations on $E(A)$ rather than full valuations on the entire set E of elementary letters, but working with full valuations here streamlines the argument.

We proceed via a lemma. Roughly speaking, it says that letters that are individually inessential to a set of formulae, are also jointly so.

Lemma. Let v, w be any two valuations on E that agree on $E!(A)$. Then $v(A) = 1$ iff $w(A) = 1$.

Proof. First we use induction to show that the lemma holds whenever v, w disagree on only finitely many letters. Then we use this to show that it holds when they disagree on infinitely many letters.

For the basis of the induction put $n = 0$, i.e. suppose that v, w disagree on no letters. Then $v = w$ and we are done. For the induction step, suppose that the lemma holds whenever two valuations disagree on just n letters. Suppose v,w disagree on just n+1 letters $p_1, \ldots, p_n, p_{n+1}$. Let w' be a valuation that is just like w except that $w'(p_{n+1}) = v(p_{n+1})$. Then w' disagrees with v on just n letters, and so by the induction hypothesis $v(A) = 1$ iff $w'(A) = 1$. But also w' disagrees with w on just the one letter p_{n+1}. Since v, w agree on $E!(A)$ while disagreeing on p_{n+1} we know that $p_{n+1} \notin E!(A)$, i.e. p_{n+1} is not essential for A. Hence since w, w' agree on every letter other than p_{n+1} we have by the definition of essential letters that $w(A) = 1$ iff $w'(A) = 1$. Putting these together, $v(A) = 1$ iff $w(A) = 1$ as desired. This completes the induction.

Now suppose that v, w are any two valuations on L that agree on $E!(A)$ but differ on infinitely many letters. We want to show that $v(A) = 1$ iff $w(A) = 1$. Suppose otherwise; we obtain a contradiction. Then either $v(A) = 1$ while $w(A) = 0$, or $w(A) = 1$ while $v(A) = 0$. Consider the former; the latter case is similar.

Since $w(A) = 0$, we have $w(a) = 0$ for some $a \in A$. Let v_w be the valuation like v except for the letters in a, where it is like w. Then v_w disagrees with v on just finitely many letters. Moreover, none of those letters are in $E!(A)$. For suppose $v_w(p) \neq v(p)$. Then the letter p occurs in a, so $v_w(p) = w(p)$ so $w(p) \neq v(p)$ and thus $p \notin E!(A)$ by the supposition that v, w agree on $E!(A)$. Hence the finite part of the lemma gives us $v(A) = 1$ iff $v_w(A) = 1$. By supposition, $v(A) = 1$ so we have $v_w(A) = 1$. Since $a \in A$ this gives $v_w(a) = 1$. But $w(a) = 0$ and by the construction of v_w we have $v_w(a) = w(a)$. Hence $v_w(a) = 0$ giving us the desired contradiction. ■

Least letter-set theorem. $A \dashv\vdash A^*$, and for every set B of formulae with $A \dashv\vdash B, E(A^*) \subseteq E(B)$.

Proof. We need to show (1) $E!(A) \subseteq E(B)$ for every B with $A \dashv\vdash B$, and (2) $A \dashv\vdash A^*$.

For (1), suppose $A \dashv\vdash B$, $p \in E!(A)$, but $p \notin E(B)$; we obtain a contradiction. The diagram illustrates the argument that follows.

Since $p \in E!(A)$ there are valuations v,w on L with $v(q) = w(q)$ for all letters q with $q \neq p$, but $v(A) \neq w(A)$ (top row). Since $p \notin E(B)$ this implies $v(B) = w(B)$ (bottom row). But since $A \dashv\vdash B$ we have both $v(A) = v(B)$ and $w(A) = w(B)$ (side columns), giving a contradiction.

$v(A)$	$\neq$	$w(A)$
$=$		$=$
$v(B)$	$=$	$w(B)$

For (2), by construction, we have $A \vdash A^*$. Suppose $A^* \nvdash A$; we derive a contradiction. Since $A^* \nvdash A$ there is a valuation v with $v(A^*) = 1$ and $v(A) = 0$, i.e. $v(a) = 0$ for some $a \in A$. Let S be the set of all literals $\pm q$ with $q \in E(A^*)$ such that $v(\pm q) = 1$. Then clearly $S \vdash A^*$. We break the argument into two cases, deriving a contradiction in each.

Case 1. Suppose S is inconsistent with A. Then by classical compactness, some finite subset $S_f \subseteq S$ is inconsistent with A. Hence $A \vdash \neg \wedge S_f$. Since all letters in $\neg \wedge S_f$ are in $E(A^*)$ it follows that $\neg \wedge S_f \in A^*$, so since $v(A^*) = 1$ we have $v(\neg \wedge S_f) = 1$. But by the construction of S we also have $v(\wedge S_f) = 1$, giving us the desired contradiction.

Case 2. Suppose S is consistent with A. Then there is a valuation w with $w(S) = w(A) = 1$. Since $w(S) = 1$ it follows that w agrees with v on all letters in $E(A^*)$. So the lemma tells us that $v(A) = 1$ iff $w(A) = 1$. So since $w(A) = 1$ we have $v(A) = 1$. Since $a \in A$, this gives $v(a) = 1$, contradicting $v(a) = 0$ and completing the proof of (2). ■

Acknowledgements

Many friendly logicians helped in various ways. In particular, thanks to Pavlos Peppas for the counterexample to disjunction in the premiss for the relation of sympathy in section 3.2, Lloyd Humberstone for discussions on links in part 2, and Karl Schlechta for the concept of a letter essential to a set of valuations in section 3.6. Anatoli Degtyarev, Kurt Engesser, Maribel Fernández, Dov Gabbay, Jamie Gabbay, George Kourousias and Odinaldo Rodrigues also commented on various versions.

A preliminary version of this paper was published in *Logica Universalis*, ed J.-Y. Beziau (Basel: Birkhauser Verlag, 2005) pages 191-205. The present version adds several new sections (1.8-1.10, 3.5-3.6, and all of part 2) as well as additional material in other sections (notably the axiomatization of friendliness in 1.5, a much stronger version of compactness in 1.6, more information about interpolant formulae in 1.7 and 3.5, and counterexamples to proof by exhaustion and to compactness for sympathy in 3.2).

BIBLIOGRAPHY

[Beth, 1953] Beth, Evert W. 1953. On Padoa's method in the theory of definition, *Nederl. Akad. Wetensch. Proc. Ser. A* 56: 330-339; also *Indagationes Mathematicae* 15: 330-339.

[Boole, 1847] Boole, George. 1847. *The Mathematical Analysis of Logic*. Cambridge: Macmillan.

[Boole, 1854] Boole, George. 1854. *An Investigation into the Laws of Thought*. London: Walton.

[de Bouvère, 1959] de Bouvère, K.L. 1959. *A Method in Proofs of Undefinability*. Amsterdam: North Holland.

[Dunn, 1976] Dunn, J.M. 1976. Intuitive semantics for first-degree entailments and coupled trees, *Philosophical Studies*, **29**, 149–168.

[Jeffrey, 1967] Jeffrey, R.C. 1967. *Formal Logic: Its Scope and Limits* (second edition 1981). New York: McGraw-Hill.

[Lang *et al.*, 2003] Lang, J., P. Liberatore, P. Marquis. 2003. Propositional independence: formula-variable independence and forgetting, *Journal of Artificial Intelligence Reseach*, 18: 391–443.

[Leśniewski, 1931] Leśniewski, S. 1931. Über definitionen in der sogennanten Theorie der Deduktion. *Comptes Rendus des Séances de la Société des Sciences et des Lettres de Varsovie*, Classe 3, XXIV: 300-302.

[Lin and Reiter, 1994] Lin, F. and R. Reiter (1994). Forget it!. In R. Greiner and D. Subramanian, eds. *Working Notes on AAAI Fall Symposium on Relevance*. Menlo Park: AAAI Press. Also at `http://www.cs.toronto.edu/cogrobo/forgetting.ps.Z`

[Makinson, 2005] Makinson, David. 2005. *Bridges from Classical to Nonmonotonic Logic*. London: King's College Publications. Series: Texts in Computing, vol 5.

[Padoa, 1901] Padoa, A. 1901. Essai d'une théorie algébrique des nombres entiers, précédé d'une introduction logique à une théorie deductive quelconque. *Bibliothèque du Congrès International de Philosophie, Paris 1900*, vol 3: 309-365. Paris: Armand Colin.

[Parikh, 1999] Parikh, R. 1999. Beliefs, belief revision, and splitting languages. Pages 266–278 of L. Moss et al eds, *Logic, Language and Computation*, vol 2. CSLI Lecture Notes n° 96: 266-278. California: CSLI Publications.

[Pietruszczak, 2004] Pietruszczak, A. 2004. The consequence relation preserving logical information, *Logic and Logical Philosophy* 13: 89-120.

[Ramsey, 1931] Ramsey, F.P. 1931. *The Foundations of Mathematics and Other Logical Essays* ed. R.B. Braithwaite. London: Kegan Paul, Trench, Trubner.

[Rantala, 1991] Rantala, V. 1991. Definitions and definability, pages 135-159 of James H. Fetzer et al *Definitions and Definability: Philosophical Perspectives*. Dordrecht : Kluwer.

[Sneed, 1971] Sneed, J.D. 1971. *The Logical Structure of Mathematical Physics*. Dordrecht: Reidel.

[Suppes, 1957] Suppes, P. 1957. *Introduction to Logic*. Princeton: Van Nostrand.

[van Benthem, 1978] Van Benthem, J.F.A.K. 1978. Ramsey eliminability, *Studia Logica* 37: 321-336.

[Weber, 1987] Weber, A. 1987. Updating propositional formulae, pages 487–500 in L. Kerschberg, ed. *Proceedings of the First Conference on Expert Data Systems*. Benjamin Cummings.

Interpolation and Joint Consistency

LARISA MAKSIMOVA

Interpolation theorem proved by W.Craig [Craig, 1957] in 1957 for classical first order logic was a source of a lot of investigation devoted to interpolation problem in various logical theories [Barwise, 1985; Gabbay, 1976; Gabbay, 1981; Gabbay and Maksimova, 2005]. Interpolation is considered as desirable and "nice" property; also it has important practical applications in computer science [Bicarregui *et al.*, 2001].

Interpolation in intuitionistic and modal predicate logics was investigated by Dov Gabbay in [Gabbay, 1971; Gabbay, 1972; Gabbay, 1976; Gabbay, 1981]. In [Gabbay, 1972] D. Gabbay proved that in classical modal logics, just as in the classical predicate logic, the Craig interpolation property CIP is equivalent to the Robinson consistency property RCP that is an analog of the joint consistency theorem [Robinson, 1956]:

Let P, Q, R be disjoint lists of predicate symbols and individual constants, and Γ be an L-consistent L-theory of the language $\mathcal{F}(P, Q)$, Δ an L-consistent L-theory of $\mathcal{F}(P, R)$ such that $\Gamma \cap \Delta$ is a complete L-theory of the language $\mathcal{F}(P)$. Then the set $\Gamma \cup \Delta$ is L-consistent.

It was proved by D. Gabbay [Gabbay, 1971] that in the intuitionistic predicate logics the full version of RCP does not hold. But a weaker version of RCP is valid, and this weaker version is equivalent to CIP in all superintuitionistic predicate logics. By the way, in propositional intermediate logics RCP is equivalent to CIP [Maksimova, 1979a].

The most known systems of modal logic, for instance, Lewis' systems S4 and S5, the logic GL of provability, the logic K4 and the least normal modal logic K possess the interpolation property. On the other hand, it appeared that this important property is rather rare. For instance, there is a continuum of normal extensions of the propositional modal logic S4 but only finitely many of these logics possess the interpolation [Maksimova, 1979b]. Moreover, there are two natural variants CIP (interpolation for implication) and IPD (interpolation for deducibility) of the interpolation property, which are not equivalent on the class of modal logics, more exactly, IPD is weaker than CIP.

Due to great significance of the interpolation, its weaker versions may also be useful. In [Maksiomov, 2003] a restricted form of interpolation IPR was introduced, which appeared to be weaker than IPD. In the present paper we define a new variant WIP of interpolation that is equivalent to some form WRP of Robinson's consistency property in modal and superintuitionistic predicate logics. We prove that WIP is equivalent to CIP over S5, and so only three consistent extensions of the propositional S5 have WIP. On the contrast, WIP holds for all extensions of the propositional logics S4.1 and GL. All propositional superintuitionistic logics also have WIP. It does not hold for superintuitionistic predicate logics. Nevertheless, all logics with CIP also possess WIP and, therefore, have WRP. In particular, WRP is valid in the intuitionistic predicate logic.

It was proved in [Maksiomov, 2003] that IPR is equivalent to the restricted amalgamation RAP. Here we find an algebraic equivalent for WIP. We define a generalisation Pre-RAP of RAP and prove that WIP is equivalent to Pre-RAP.

1 Various versions of interpolation

If $\mathbf{p}$ is a list of predicate symbols and individual constants, let $A(\mathbf{p})$ denote a formula whose all non-logical symbols are in $\mathbf{p}$, and $\mathcal{F}(\mathbf{p})$ the set of all such formulas.

Let L be a logic, $\vdash_L$ the deducibility relation in L. Suppose that $\mathbf{p}$, $\mathbf{q}$, $\mathbf{r}$ are disjoint lists of non-logical symbols. We consider the languages which contain neither equality nor functional symbols but do contain at least one constant $\top$ ("truth") or $\bot$ ("false"). One can define two *interpolation properties CIP* and *IPD* as follows:

CIP. If $\vdash_L A(\mathbf{p},\mathbf{q}) \to B(\mathbf{p},\mathbf{r})$, then there exists a formula $C(\mathbf{p})$ such that $\vdash_L A(\mathbf{p},\mathbf{q}) \to C(\mathbf{p})$ and $\vdash_L C\mathbf{p}) \to B(\mathbf{p},\mathbf{r})$.

IPD. If $A(\mathbf{p},\mathbf{q}) \vdash_L B(\mathbf{p},\mathbf{r})$, then there exists a formula $C(\mathbf{p})$ such that $A(\mathbf{p},\mathbf{q}) \vdash_L C(\mathbf{p})$ and $C\mathbf{p}) \vdash_L B(\mathbf{p},\mathbf{r})$.

The most known modal logics such as Lewis' systems S4 and S5, Grzegorczyk's logic Grz, the logic GL of provability, the logic K4 and the least normal modal logic K have CIP.

The properties CIP and IPD are equivalent in classical and intuitionistic theories due to the deduction theorem. It is not true in modal logics, where there are two rules of inference R1: $A, A \to B/B$ and R2: $A/\Box A$. The well-known deduction theorem in normal modal logics says that

$$\Gamma, A \vdash_L B \iff \Gamma \vdash_L [n]A \to B \text{ for some } n \geq 0,$$

where $[n]A \rightleftharpoons A\&\Box A\&\ldots\&\Box^n A$.

In [Maksiomov, 2003] a *restricted interpolation property* was introduced:

IPR. If $A(\mathbf{p},\mathbf{q}), B(\mathbf{p},\mathbf{r}) \vdash_L C(\mathbf{p})$, then there exists a formula $A'(\mathbf{p})$ such that $A(\mathbf{p},\mathbf{q}) \vdash_L A'(\mathbf{p})$ and $A'(\mathbf{p}), B(\mathbf{p},\mathbf{r}) \vdash_L C(\mathbf{p})$.

By the deduction theorem, one can easily show that IPD implies IPR in modal logics. On the other hand, IPD does not follow from IPR [Maksiomov, 2003]. Now we define the *weak interpolation property*

WIP. If $A(\mathbf{p},\mathbf{q}), B(\mathbf{p},\mathbf{r}) \vdash_L \bot$, then there exists a formula $A'(\mathbf{p})$ such that $A(\mathbf{p},\mathbf{q}) \vdash_L A'(\mathbf{p})$ and $A'(\mathbf{p}), B(\mathbf{p},\mathbf{r}) \vdash_L \bot$.

Obviously, WIP is a particular case of IPR. So we have

$$CIP \Rightarrow IPD \Rightarrow IPR \Rightarrow WIP.$$

The converse arrows are, in general, not valid in normal modal logics, although it is easy to derive CIP from WIP in classical logic.

In the classical predicate logic CIP is equivalent to Robinson's consistency property

RCP. Let T_1, T_2 be two consistent L-theories in the languages $\mathcal{L}_1, \mathcal{L}_2$ respectively. If $T_1 \cap T_2$ is a complete L-theory in the common language $\mathcal{L}_1 \cap \mathcal{L}_2$, then $T_1 \cup T_2$ is L-consistent.

The same equivalence holds in all classical modal logics [Gabbay, 1976]. We recall the definitions. By $\Gamma \rightarrow_L A$ we denote deducibility of A from Γ in L by the rule R1. Then $\Gamma \rightarrow_L B$ holds if and only if there are $n \geq 0$ and some formulas $A_1, \dots, A_n \in \Gamma$ such that

$$L \vdash (A_1 \& \dots \& A_n) \rightarrow B.$$

We say that a set Γ is *L-consistent* if $\Gamma \not\rightarrow_L \bot$. A set T of formulas of the language $\mathcal{L}$ is said to be an *L-theory* of this language if it is closed under $\rightarrow_L$, i.e. $T \rightarrow_L A$ for $A \in \mathcal{L}$ implies $A \in T$. An L-theory T of the language $\mathcal{L}$ is *complete* in $\mathcal{L}$ if $A \in T$ or $\neg A \in T$ for any formula $A \in \mathcal{L}$.

We note that RCP is equivalent to

RCP'. Let T_1, T_2 be two L-theories in the languages $\mathcal{L}_1, \mathcal{L}_2$ respectively, $\mathcal{L}_0 = \mathcal{L}_1 \cap \mathcal{L}_2$, $T_{i0} = T_i \cap \mathcal{L}_0$. If the set $T_{10} \cup T_{20}$ in the common language $\mathcal{L}_0$ is L-consistent, then $T_1 \cup T_2$ is L-consistent.

Lemma 1.1. For any modal logic L, RCP is equivalent to RCP'.

Proof. It is proved in [Gabbay, 1976] that RCP is equivalent to CIP. We show that CIP implies RCP' and RCP' implies RCP.

Assume that L have CIP and prove RCP'. Take two L-theories T_1, T_2 in the languages $\mathcal{L}_1, \mathcal{L}_2$ respectively. If $T_1 \cup T_2$ is L-inconsistent, i.e. $T_1 \cup T_2 \rightarrow_L$

$\perp$, we denote by A and by B the conjunctions of formulas in T_1 and T_2 used in the derivation of $\perp$. Then $L \vdash A \to (B \to \perp)$. By CIP there is $A' \in \mathcal{L}_0$ such that $L \vdash A \to A'$ and $L \vdash A' \to (B \to \perp)$. Then $L \vdash B \to (A' \to \perp)$. It follows that $A' \in T_{10}$ and $\neg A' \in T_{20}$, so $T_{10} \cup T_{20}$ is inconsistent. Thus L has RCP'.

Now let L have RCP'. Take two consistent L-theories T_1, T_2 in the languages $\mathcal{L}_1, \mathcal{L}_2$ respectively such that $T_1 \cap T_2$ is complete in $\mathcal{L}_0 = \mathcal{L}_1 \cap \mathcal{L}_2$. We show that $T_{10} \cup T_{20}$ is L-consistent, where $T_{i0} = T_i \cap \mathcal{L}_0$.

Indeed, if $T_{10} \cup T_{20} \to_L \perp$, then there are $A \in T_{10}$ and $B \in T_{20}$ such that $L \vdash A \to (B \to \perp)$. It follows that $\neg B \in T_1$ and $B \notin T_1$. On the other hand, $B \in T_2$ and so $\neg B \notin T_2$. Hence $B \notin T_1 \cap T_2$ and $\neg B \notin T_1 \cap T_2$ in contradiction with $\mathcal{L}_0$-completeness of $T_1 \cap T_2$.

So $T_{10} \cup T_{20}$ is L-consistent, and by RCP' the set $T_1 \cup T_2$ is L-consistent. Thus L has RCP. ■

By $\Gamma \vdash_L A$ we denote deducibility of A from Γ in L by the rules R1 and R2. We say that a set Γ is *strongly L-consistent* if $\Gamma \nvdash_L \perp$.

A set $\Gamma \subseteq \mathcal{L}$ is called an *open L-theory of this language* whenever it is closed with respect to $\vdash_L$, i.e. for every formula $A \in \mathcal{L}$, $\Gamma \vdash_L A$ implies $A \in \Gamma$.

Now we define an analog of RCP' taking open L-theories instead of L-theories. We say that a logic L has the *weak Robinson property* if it satisfies

WRP. Let T_1, T_2 be two open L-theories in the languages $\mathcal{L}_1, \mathcal{L}_2$ respectively, $\mathcal{L}_0 = \mathcal{L}_1 \cap \mathcal{L}_2$, $T_{i0} = T_i \cap \mathcal{L}_0$. If the set $T_{10} \cup T_{20}$ in the common language is strongly L-consistent, then $T_1 \cup T_2$ is strongly L-consistent.

We easily see

Proposition 1.2. A logic L has WRP if and only if L has WIP.

Proof. Let L have WRP, A of the language $\mathcal{L}_1$, B of the language $\mathcal{L}_2$ and $\mathcal{L}_0 = \mathcal{L}_1 \cap \mathcal{L}_2$. Assume $A, B \vdash_L \perp$. Let us define two open L-theories

$$T_1 = \{A' \in \mathcal{L}_1 |\ A \vdash_L A'\},\ T_2 = \{B' \in \mathcal{L}_2 |\ B \vdash_L B'\}.$$

Then by WRP we obtain $T_{10} \cup T_{20} \vdash_L \perp$, where $T_{i0} = T_i \cap \mathcal{L}_0$. Since derivations are finitary, it follows that there is $A' \in T_1 \cap \mathcal{L}_0$ such that $A', T_{20} \vdash \perp$. Thus $A \vdash_L A'$ and $A', B \vdash_L \perp$.

Now assume L have WIP. Take two open L-theories T_1, T_2 in the languages $\mathcal{L}_1, \mathcal{L}_2$ respectively. Let the theory $T_{10} \cup T_{20}$ in the common language be strongly L-consistent. If $T_1 \cup T_2 \vdash_L \perp$, we denote by A and by B the conjunctions of formulas in T_1 and T_2 used in the derivation of $\perp$. Then $A, B \vdash_L \perp$, and we apply WIP. There is $A' \in \mathcal{L}_0$ such that $A \vdash_L A'$ and $A', B \vdash_L \perp$. Again by WIP there is $B' \in \mathcal{L}_0$ such that $B \vdash_L B'$ and

$A', B' \vdash_L \bot$. It means that $T_{10} \cup T_{20} \vdash_L \bot$. A contradiction. Thus L has WRP. ■

We will show in Section 2 that all propositional logics over $S4.1$ or over GL possess WIP. On the other hand, WIP is non-trivial over K and even over $S5$. Denote $\boxdot A := A \& \Box A$.

Proposition 1.3. Let L be an extension of $K4$ containing the formula $(p \to \boxdot\neg\boxdot\neg p)$. If L has WIP then L has IPD.

Proof. Let $A(\mathbf{p}, \mathbf{q}) \vdash_L B(\mathbf{p}, \mathbf{r})$. Then

$$A(\mathbf{p}, \mathbf{q}) \vdash_L \boxdot B(\mathbf{p}, \mathbf{r})$$

and

$$A(\mathbf{p}, \mathbf{q}), \neg\boxdot B(\mathbf{p}, \mathbf{r}) \vdash_L \bot.$$

By WIP there is a formula $A'(\mathbf{p})$ such that

$$A(\mathbf{p}, \mathbf{q}) \vdash_L A'(\mathbf{p}) \text{ and } A'(\mathbf{p}), \neg\boxdot B(\mathbf{p}, \mathbf{r}) \vdash_L \bot.$$

By deduction theorem we have

$$\neg\boxdot B(\mathbf{p}, \mathbf{r}) \vdash_L \neg\boxdot A'(\mathbf{p}),$$

and again by deduction theorem

$$L \vdash \boxdot\neg\boxdot B(\mathbf{p}, \mathbf{r}) \to \neg\boxdot A'(\mathbf{p}).$$

If $L \vdash (p \to \boxdot\neg\boxdot\neg p)$, we obtain

$$L \vdash \neg B(\mathbf{p}, \mathbf{r}) \to \neg\boxdot A'(\mathbf{p})$$

and

$$L \vdash \boxdot A'(\mathbf{p}) \to B(\mathbf{p}, \mathbf{r}).$$

So

$$A'(\mathbf{p}) \vdash_L B(\mathbf{p}, \mathbf{r}).$$

■

Theorem 1.4. For any propositional logic L extending $S5$, WIP implies IPD, and so $CIP \iff IPD \iff IPR \iff WIP$ over $S5$.

Proof. Assume that L contains $S5$ and has WIP. By Proposition 1.3 L has IPD. It was proved in [Maksimova, 1979b; Maksimova, 1980] that IPD implies CIP over S5. ■

Recall from [Maksimova, 1979b; Maksimova, 1980] that the propositional $S5$ has only 3 consistent extensions with CIP, namely, $S5$ itself, the logic of the two-element frame with the total relation R, and the logic of the one-element reflexive frame.

2 Weak interpolation and pre-RAP

In this section we investigate propositional normal modal logics. In [Maksiomov, 2003] the equivalence of IPR and RAP was proved. Now we find an algebraic equivalent of WIP. For the properties CIP and IPD it was done in [Maksimova, 1992; Gabbay and Maksimova, 2005].

A *normal modal logic* is any set of modal formulas containing all the tautologies of the two-valued propositional logic and the axioms $\Box(A \to B) \to (\Box A \to \Box B)$, and closed under the inference rules R1: $A, A \to B/B$ and R2: $A/\Box A$, and the substitution rule. The set of normal modal logics containing a normal modal logic L is denoted by $NE(L)$. If L is a normal modal logic, by $\vdash_L$ we denote deducibility in L by the rules R1 and R2. The well-known deduction theorem in normal modal logics says that

$$\Gamma, A \vdash_L B \iff \Gamma \vdash_L [n]A \to B \text{ for some } n \geq 0,$$

where $[n]A \rightleftharpoons A \& \Box A \& \ldots \& \Box^n A$. Recall some standard denotations for logics in $NE(\mathrm{K})$:

$\mathrm{K4} = \mathrm{K} + (\Box p \to \Box\Box p)$,
$\mathrm{S4} = \mathrm{K4} + (\Box p \to p)$,
$\mathrm{S4.1} = \mathrm{S4} + (\Box\Diamond p \to \Diamond\Box p)$,
$\mathrm{S5} = \mathrm{S4} + (p \to \Box\Diamond p)$,
$\mathrm{GL} = \mathrm{K4} + (\Box(\Box p \to p) \to \Box p)$.

It is well known that there exists a duality between normal modal logics and varieties of modal algebras. *A modal algebra* is an algebra $\mathbf{A} = (A, \to, \neg, \Box)$ that is a boolean algebra with respect to $\to$ and $\neg$ and, moreover, satisfies the conditions $\Box\top = \top$ and $\Box(x \to y) \leq \Box x \to \Box y$. A modal algebra $\mathbf{A}$ is called *transitive* if it satisfies an inequality $\Box x \leq \Box\Box x$; *a topoboolean algebra*, or *interior algebra* if it is transitive and satisfies $\Box x \leq x$; an *epistemic algebra* if it is an interior algebra satisfying $x \leq \Box\Diamond x$; a *diagonalizable* algebra if it satisfies $\Box(\Box x \to x) = \Box x$. It is well known that the modal logic K4 can be characterized by the variety of all transitive algebras, S4 by the variety of topoboolean algebras, GL by diagonalizable algebras.

If A is a formula, $\mathbf{A}$ a modal algebra, then $\mathbf{A} \models A$ denotes that the identity $A = \top$ is satisfied in $\mathbf{A}$. We write $\mathbf{A} \models L$ instead of $(\forall A \in L)(\mathbf{A} \models A)$. We denote $V(L) = \{\mathbf{A} | \mathbf{A} \models L\}$. Each normal modal logic L is characterized by the variety $V(L)$.

With each open L-theory T of the language $\mathcal{F}(\mathbf{p})$, one can associate an equivalence relation on the set $\mathcal{F}(\mathbf{p})$:

$$A \sim_T B \rightleftharpoons (A \leftrightarrow B) \in T.$$

Due to the replacement lemma, $\sim_T$ is a congruence on $\mathcal{F}(\mathbf{p})$. It makes possible to define *the Lindenbaum-Tarski algebra*

$$\mathbf{A}(\mathbf{p},T) = \mathcal{F}(\mathbf{p})/{\sim_T}$$

as a quotient-algebra of $\mathcal{F}(\mathbf{p})$. Let us denote $\|A\| = \{B|A{\sim_T}B\}$ for each formula $A = A(\mathbf{p})$. The following statement holds (see, for instance,[Rasiowa and Sikorski, 1963]):

Lemma 2.1. Let $L \in NE(K)$. For each open L-theory T of the language $\mathcal{F}(\mathbf{p})$, the algebra $\mathbf{A}(\mathbf{p},T)$ is in $V(L)$; the canonical mapping $\varkappa : \mathcal{F} \to \mathbf{A}(\mathbf{p},T)$, where for every formula $A = A(\mathbf{p})$

$$\varkappa(A) = \|A\| = A/{\sim_T},$$

is a homomorphism and, moreover, $A \in T \Leftrightarrow \|A\| = \top$ in $\mathbf{A}(\mathbf{p},T)$.

In particular, $\mathbf{A}(\mathbf{p},L)$ is a free algebra of $V(L)$, with free generators $\|p\|$, where $p \in \mathbf{p}$.

All varieties of modal algebras possess such important properties as congruence-distributivity and congruence extension property CEP: If $\mathbf{A}$ is a subalgebra of $\mathbf{B}$ then every congruence Φ on $\mathbf{A}$ can be extended to a congruence Ψ on $\mathbf{B}$ such that $\Psi \cap \mathbf{A}^2 = \Phi$.

Proposition 2.2. For any normal modal logic L:

1. L has CIP if and only if $V(L)$ is super-amalgamable,
2. IPD is equivalent to the amalgamation property AP,
3. IPR is equivalent to RAP.

Proof. (1,2) See [Maksimova, 1992]. (3) Proved in [Maksiomov, 2003]. ■

Recall that a class V has *Amalgamation Property* if it satisfies the condition

AP. For each $\mathbf{A}, \mathbf{B}, \mathbf{C} \in V$ such that $\mathbf{A}$ is a common subalgebra of $\mathbf{B}$ and $\mathbf{C}$, there exist an algebra $\mathbf{D}$ in V and monomorphisms $\delta : \mathbf{B} \to \mathbf{D}$, $\varepsilon : \mathbf{C} \to \mathbf{D}$, such that $\delta(x) = \varepsilon(x)$ for all $x \in \mathbf{A}$.

A class V has *Super-Amalgamation Property (SAP)* if it has AP and, moreover, satisfies the condition:

$$\delta(x) \leq \varepsilon(y) \iff (\exists z \in \mathbf{A})(x \leq z \text{ and } z \leq y),$$

$$\delta(x) \geq \varepsilon(y) \iff (\exists z \in \mathbf{A})(x \geq z \text{ and } z \geq y);$$

V has *Strong Amalgamation Property (StrAP)* if it satisfies AP and, moreover, $\delta(\mathbf{B}) \cap \varepsilon(\mathbf{C}) = \delta(\mathbf{A})$.

We say that a class V has *Restricted Amalgamation Property* if it satisfies the condition:

RAP. For each $\mathbf{A}, \mathbf{B}, \mathbf{C} \in V$ such that $\mathbf{A}$ is a common subalgebra of $\mathbf{B}$ and $\mathbf{C}$ there exist an algebra $\mathbf{D}$ in V and homomorphisms $\delta : \mathbf{B} \to \mathbf{D}$, $\varepsilon : \mathbf{C} \to \mathbf{D}$ such that $\delta(x) = \varepsilon(x)$ for all $x \in \mathbf{A}$ and the restriction δ' of δ onto $\mathbf{A}$ is a monomorphism.

We say that a class V has *Pre-RAP* if it satisfies the condition:

PRAP. For each $\mathbf{A}, \mathbf{B}, \mathbf{C} \in V$ such that $\mathbf{A}$ is a common subalgebra of $\mathbf{B}$ and $\mathbf{C}$ there exist an algebra $\mathbf{D}$ in V and homomorphisms $\delta : \mathbf{B} \to \mathbf{D}$, $\varepsilon : \mathbf{C} \to \mathbf{D}$ such that $\delta(x) = \varepsilon(x)$ for all $x \in \mathbf{A}$, where $\mathbf{D}$ is non-degenerate whenever $\mathbf{A}$ is non-degenerate.

By analogy with Theorem 2.2 of [Maksiomov, 2003] we prove

Theorem 2.3. A normal modal logic L has WIP iff L has WRP iff $V(L)$ has PRAP.

Proof. Equivalence of WIP and WRP was proved in Proposition 1.2. Assume that L has WRP. We prove that $V(L)$ has PRAP. Let $\mathbf{A}, \mathbf{B}, \mathbf{C} \in V(L)$ and $\mathbf{A}$ be a subalgebra of both $\mathbf{B}$ and $\mathbf{C}$. We take the sets of propositional variables $\mathbf{p} = \{p_a | a \in \mathbf{A}\}$, $\mathbf{q} = \{q_b | b \in \mathbf{B} - \mathbf{A}\}$, $\mathbf{r} = \{r_c | c \in \mathbf{C} - \mathbf{A}\}$ and define the valuations $v' : \mathbf{p} \cup \mathbf{q} \to \mathbf{B}$ and $v'' : \mathbf{p} \cup \mathbf{r} \to \mathbf{C}$ by

$$v'(p_a) = v''(p_a) = a, \; v'(q_b) = b, \; v''(r_c) = c.$$

We denote

$$\begin{aligned} T_1 &= \{B | B \in \mathcal{F}(\mathbf{p}, \mathbf{q}) \text{ and } v'(B) = \top_{\mathbf{B}}\}, \\ T_2 &= \{C | C \in \mathcal{F}(\mathbf{p}, \mathbf{r}) \text{ and } v''(C) = \top_{\mathbf{C}}\}, \\ T &= \{D | D \in \mathcal{F}(\mathbf{p}, \mathbf{q}, \mathbf{r}) \text{ and } T_1 \cup T_2 \vdash_L D\}. \end{aligned}$$

Then T_1, T_2 and T are open L-theories of the languages $\mathcal{F}(\mathbf{p}, \mathbf{q}), \mathcal{F}(\mathbf{p}, \mathbf{r})$ and $\mathcal{F}(\mathbf{p}, \mathbf{q}, \mathbf{r})$ respectively.

Let $\mathcal{F} = \mathcal{F}(\mathbf{p}, \mathbf{q}, \mathbf{r})$ and $\mathbf{D} = \mathbf{A}(\mathbf{p} \cup \mathbf{q} \cup \mathbf{r}, T) = \mathcal{F} / \sim_T$ be a Lindenbaum-Tarski algebra, where $D \sim D' \rightleftharpoons T_1, T_2 \vdash D \leftrightarrow D'$.

Due to $T_1, T_2 \subseteq T$, the mappings $\delta : \mathbf{B} \to \mathbf{D}$ and $\varepsilon : \mathbf{C} \to \mathbf{D}$ defined by $\delta(a) = \varepsilon(a) = \|p_a\|$ for $a \in \mathbf{A}$, $\delta(b) = \|q_b\|$ for $b \in \mathbf{B} - \mathbf{A}$, and $\varepsilon(c) = \|r_c\|$ for $c \in \mathbf{C} - \mathbf{A}$, are homomorphisms that coincide on $\mathbf{A}$.

Assume that $\mathbf{D}$ is degenerate. It means that $\bot_{\mathbf{D}} = \top_{\mathbf{D}}$ and so $T_1, T_2 \vdash_L \bot \leftrightarrow \top$, i.e.

$$T_1, T_2 \vdash_L \bot.$$

By WRP we obtain

$$T_{10}, T_{20} \vdash_L \bot,$$

where

$$T_{10} = \{B | B \in \mathcal{F}(\mathbf{p}) \text{ and } v'(B) = \top_{\mathbf{B}}\},$$
$$T_{20} = \{C | C \in \mathcal{F}(\mathbf{p}) \text{ and } v''(C) = \top_{\mathbf{C}}\}.$$

It follows that $v'(B) = v''(C) = \top_{\mathbf{A}}$ for all $B \in T_{10}, C \in T_{20}$ and so $v'(\bot) = \bot_{\mathbf{A}} = \top_{\mathbf{A}}$. Hence $\mathbf{A}$ is degenerate. Thus $V(L)$ has PRAP.

For the converse, assume that $V(L)$ has PRAP and prove WIP. Suppose that $A(\mathbf{p}, \mathbf{q}), B(\mathbf{p}, \mathbf{r}) \vdash_L \bot$. We consider

$$T = \{A'(\mathbf{p}) \mid A(\mathbf{p}, \mathbf{q}) \vdash_L A'(\mathbf{p})\}$$

and prove that

(1) $T, B(\mathbf{p}, \mathbf{r}) \vdash_L \bot$.

Let us define

$$T_2 = \{F(\mathbf{p}, \mathbf{r}) \mid T, B(\mathbf{p}, \mathbf{r}) \vdash_L F(\mathbf{p}, \mathbf{r})\},\ T_0 = T_2 \cap \mathcal{F}(\mathbf{p}),$$

$$T_1 = \{F(\mathbf{p}, \mathbf{q}) \mid A(\mathbf{p}, \mathbf{q}), T_0 \vdash_L F(\mathbf{p}, \mathbf{q})\}.$$

We show that $T_1 \cap \mathcal{F}(\mathbf{p}) = T_0$.

Assume $F(\mathbf{p}) \in T_1$. Then $A(\mathbf{p}, \mathbf{q}), T_0 \vdash_L F(\mathbf{p})$, so by the deduction theorem we have $A(\mathbf{p}, \mathbf{q}) \vdash_L [n]A_1(\mathbf{p}) \to F(\mathbf{p})$ for some $n \geq 0$ and some $A_1(\mathbf{p}) \in T_0$. It follows that $([n]A_1(\mathbf{p}) \to F(\mathbf{p})) \in T \subseteq T_0$, $[n]A_1(\mathbf{p}) \in T_0$, and so $F(\mathbf{p}) \in T_0$. Thus we have

$$F(\mathbf{p}) \in T_2 \iff F(\mathbf{p}) \in T_0 \iff F(\mathbf{p}) \in T_1$$

for any formula $F(\mathbf{p})$.

We set $\mathbf{B} = \mathcal{F}(\mathbf{p}, \mathbf{q})/\sim_{T_1}$, $\mathbf{C} = \mathcal{F}(\mathbf{p}, \mathbf{r})/\sim_{T_2}$; let $\mathbf{A}$ be a subalgebra of $\mathbf{B}$ generated by $\mathbf{p}/T_1$. Then $\mathbf{A}$ is embeddable into $\mathbf{C}$ by a monomorphism $\varphi(p/\sim_{T_1}) = p/\sim_{T_2}$ for $p \in \mathbf{p}$. It is clear from Lemma 2.1 that $\mathbf{A}, \mathbf{B}, \mathbf{C} \in V(L)$.

Suppose for contradiction that $T, B(\mathbf{p}, \mathbf{r}) \nvdash_L \bot$, i.e. $T_2 \nvdash_L \bot$. Then $\bot_{\mathbf{C}} \neq \top_{\mathbf{C}}$ and $\bot_{\mathbf{A}} \neq \top_{\mathbf{A}}$. By PRAP there exist a non-degenerate $\mathbf{D} \in V(L)$ and homomorphisms $\delta : \mathbf{B} \to \mathbf{D}$, $\varepsilon : \mathbf{C} \to \mathbf{D}$ such that for every $p \in \mathbf{p}$ we have $\delta(p/\sim_{T_1}) = \varepsilon\varphi(p/\sim_{T_1}) = \varepsilon(p/\sim_{T_2})$.

Let us consider the valuation

$$v(p) = \delta(p/\sim_{T_1}) \text{ for } p \in \mathbf{p},$$
$$v(q) = \delta(q/\sim_{T_1}) \text{ for } q \in \mathbf{q},$$
$$v(r) = \varepsilon(r/\sim_{T_2}) \text{ for } r \in \mathbf{r}.$$

We note that $A(\mathbf{p}, \mathbf{q}) \in T_1$, so $v(A(\mathbf{p}, \mathbf{q})) = \delta(A(\mathbf{p}, \mathbf{q})/\sim_{T_1}) = \top_{\mathbf{D}}$. In addition, $B(\mathbf{p}, \mathbf{r}) \in T_2$, so $v(B(\mathbf{p}, \mathbf{r})) = \varepsilon(B(\mathbf{p}, \mathbf{r})/\sim_{T_2}) = \top_{\mathbf{D}}$. Due to

$A(\mathbf{p},\mathbf{q}), B(\mathbf{p},\mathbf{r}) \vdash_L \bot$, we obtain $v(\bot) = \delta(\bot/\sim_{T_1}) = \top_{\mathbf{D}}$. It follows that $\bot_{\mathbf{D}} = v(\bot) = \top_{\mathbf{D}}$, i.e. $\mathbf{D}$ is degenerate in contradiction with the assumption. Thus $T, B(\mathbf{p},\mathbf{q}) \vdash_L \bot$, and we proved (1).

Since all derivations are finite, there exist a finite subset Γ of T such that $\Gamma, B(\mathbf{p},\mathbf{q}) \vdash_L \bot$. Taking the conjunction of all formulas in Γ as $A'(\mathbf{p})$, we obtain $A(\mathbf{p},\mathbf{q}) \vdash_L A'(\mathbf{p})$ and $A'(\mathbf{p}), B(\mathbf{p},\mathbf{q}) \vdash_L \bot$. ∎

We apply this theorem to propositional modal logics. First recall that there is a one-to-one correspondence between congruences and open filters on any modal algebra. For a modal algebra $\mathbf{A}$, its subset Φ is an *open filter* if it satisfies the conditions: (1) $(x\&y) \in \Phi$ for all $x, y \in \Phi$, (2) $x \in \Phi$ and $x \leq y$ imply $y \in \Phi$, (3) $x \in \Phi$ implies $\Box x \in \Phi$. For $x \in \mathbf{A}$ and an open filter Φ we denote $x/\Phi = \{y | \ (x \leftrightarrow y) \in \Phi\}$.

Corollary 2.4. All modal logics in $NE(S4.1)$ or in $NE(GL)$ have WIP.

Proof. First consider extensions of $S4.1$. We prove that for any $L \in NE(S4.1)$, its corresponding variety $V(L)$ has PRAP. Take algebras $\mathbf{A}, \mathbf{B}, \mathbf{C}$ in $V(L)$ such that $\mathbf{A}$ is a common subalgebra of $\mathbf{B}$ and $\mathbf{C}$. Since the logic $S4.1$ has CIP, $V(S4.1)$ is amalgamable, so there exist an algebra $\mathbf{D}$ in $V(S4.1)$ and monomorphisms $\delta : \mathbf{B} \to \mathbf{D}$, $\varepsilon : \mathbf{C} \to \mathbf{D}$ such that $\delta(x) = \varepsilon(x)$ for all $x \in \mathbf{A}$. If $\mathbf{A}$ is non-degenerate, $\mathbf{D}$ is also non-degenerate. Take the set

$$S = \{\Box x \vee \Box \neg x | \ x \in \mathbf{D}\}.$$

We note that for any n

$$S4.1 \vdash \Diamond((\Box p_1 \vee \Box \neg p_1) \& \ldots \& (\Box p_n \vee \Box \neg p_n)).$$

It follows that the filter Φ generated by S is proper, i.e. it does not contain $\bot$. Moreover, Φ satisfies the condition: $x \in \Phi \Rightarrow \Box x \in \Phi$, i.e. it is an open filter. Define $\mathbf{D}' = \mathbf{D}/\Phi$, $h(x) = \delta(x)/\Phi$ for $x \in \mathbf{B}$ and $g(y) = \varepsilon(y)/\Phi$ for $y \in \mathbf{C}$. Then $\mathbf{D}'$ satisfies the condition: $\Box x = \Diamond x = x$ for all $x \in \mathbf{D}'$. It follows that $\mathbf{D}' \in V(L)$. Moreover, $\mathbf{D}'$ is non-degenerate, and h and g are homomorphisms which coincide on $\mathbf{A}$, as required.

Let $L \in NE(GL)$. We act by analogy. For any $\mathbf{A}, \mathbf{B}, \mathbf{C} \in V(L)$ such that $\mathbf{A}$ is a common subalgebra of $\mathbf{B}$ and $\mathbf{C}$, we find an algebra $\mathbf{D}$ in $V(GL)$ and monomorphisms $\delta : \mathbf{B} \to \mathbf{D}$, $\varepsilon : \mathbf{C} \to \mathbf{D}$ such that $\delta(x) = \varepsilon(x)$ for all $x \in \mathbf{A}$. If $\mathbf{A}$ is non-degenerate, $\mathbf{D}$ is also non-degenerate. We note that $\Box\bot \neq \bot$ in $\mathbf{D}$. Indeed from $\Box\bot = \bot$ we derive $\Box\bot \to \bot = \top$, $\top = \Box(\Box\bot \to \bot) \leq \Box\bot$ and $\top = \Box\bot = \bot$, i.e. $\mathbf{D}$ would be degenerate. We take a filter Φ on $\mathbf{D}$ generated by $\Box\bot$. Then it is an open filter, and $\mathbf{D}' = \mathbf{D}/\Phi$ is a non-degenerate algebra in $V(L)$; $h(x) = \delta(x)/\Phi$ for $x \in \mathbf{B}$ and $g(y) = \varepsilon(y)/\Phi$ for $y \in \mathbf{C}$ are homomorphisms which coincide on $\mathbf{A}$. ∎

3 Superintuitionistic logics

As we already mentioned, D.Gabbay [Gabbay, 1981] proved that CIP in extensions of intermediate predicate logics is equivalent to a weaker version RCP$''$ of Robinson's consistency property. It was proved in [Gabbay, 1981] that the general form RCP of Robinson's property fails in the intuitionistic predicate logic. The notion of an intuitionistic theory was defined as a pair (T, F), where T was a set of "true" formulas and F a set of "false" formulas. And in RCP we wished to keep all true and all false formulas of both theories (T_1, F_1) and (T_2, F_2), which was not always possible. The weaker property RCP$''$, equivalent to CIP, required an additional condition $F_1 \subseteq F_2$, in particular, F_1 should be in the common language.

Now we define an *open L-theory* as a set T closed with respect to $\vdash_L$; in the case of superintuitionistic logics it is equivalent to T to be closed under $\rightarrow_L$. Also L-consistency and strong L-consistency are equivalent in superintuitionistic logics due to the deduction theorem. Thus we can define Weak Robinson's Property WRP as follows:

WRP. Let T_1 and T_2 be two open L-theories in the languages $\mathcal{L}_1$ and $\mathcal{L}_2$ respectively, $\mathcal{L}_0 = \mathcal{L}_1 \cap \mathcal{L}_2$, $T_{i0} = T_i \cap \mathcal{L}_0$. If the set $T_{10} \cup T_{20}$ in the common language is L-consistent, then $T_1 \cup T_2$ is L-consistent.

It is clear that an open L-theory T is the same as the theory $(T, \emptyset)$. The following theorem is a re-formulation of Gabbay's [Gabbay, 1981, Theorem 8.32].

Theorem 3.1. For any superintutionistic (predicate or propositional) logic L, WIP is equivalent to WRP.

Proof. By analogy with Proposition 1.2. ∎

Corollary 3.2. If a superintuitionistic predicate logic has CIP, then it has WRP.

Proof. It is clear that CIP implies WIP. Then the statement immediately follows from Theorem 3.1. ∎

In particular, the intuitionistic predicate logic possesses WRP. Of course, this corollary is also true for propositional logics. But we do not need it due to

Theorem 3.3. All intermediate propositional logics possess WIP.

Proof. Let L be a consistent superintuitionistic logic and

$$A(\mathbf{p}, \mathbf{q}), B(\mathbf{p}, \mathbf{r}) \vdash_L \bot.$$

Then

$$\vdash_L \neg(A(\mathbf{p},\mathbf{q})\&B(\mathbf{p},\mathbf{r})),$$

so the same formula is provable in the classical calculus, and also in the intuitionistic calculus Int by Glivenko's theorem. It follows that

$$Int \vdash (A(\mathbf{p},\mathbf{q}) \to \neg B(\mathbf{p},\mathbf{r})).$$

Since Int has CIP, there exists a formula $A'(\mathbf{p})$ such that

$$Int \vdash A(\mathbf{p},\mathbf{q}) \to A'(\mathbf{p}) \text{ and } Int \vdash A'(\mathbf{p}) \to \neg B(\mathbf{p},\mathbf{r}).$$

Hence

$$A(\mathbf{p},\mathbf{q}) \vdash_L A'(\mathbf{p}) \text{ and } A'(\mathbf{p}), B(\mathbf{p},\mathbf{r}) \vdash_L \bot.$$

■

As a consequence, we also obtain the following version of interpolation that is already known to be true over Int.

Corollary 3.4. For any propositional intermediate logic L,

if $L \vdash A(\mathbf{p},\mathbf{q}) \to B(\mathbf{p},\mathbf{r})$, then there is $C(\mathbf{p})$ such that $L \vdash A(\mathbf{p},\mathbf{q}) \to C(\mathbf{p})$ and $L \vdash C(\mathbf{p}) \to \neg\neg B(\mathbf{p},\mathbf{r})$.

Proof. Let $L \vdash A(\mathbf{p},\mathbf{q}) \to B(\mathbf{p},\mathbf{r})$. Then $A(\mathbf{p},\mathbf{q}), \neg B(\mathbf{p},\mathbf{r}) \vdash_L \bot$, and by Theorem 3.3 there is $C(\mathbf{p})$ such that

$$A(\mathbf{p},\mathbf{q}) \vdash_L C(\mathbf{p}) \text{ and } C(\mathbf{p}), \neg B(\mathbf{p},\mathbf{r}) \vdash_L \bot.$$

By deduction theorem we have

$$L \vdash A(\mathbf{p},\mathbf{q}) \to C(\mathbf{p}) \text{ and } L \vdash C(\mathbf{p}) \to \neg\neg B(\mathbf{p},\mathbf{r}).$$

■

Theorem 3.3 remains true for those predicate logics intermediate between the intuitionistic and the classical predicate logics, which contain the formula $\forall x \neg\neg A \to \neg\neg \forall x\, A$, since these logics satisfy Glivenko's theorem. On the other hand, there are superintuitionistic predicate logics without WIP. In [Suzuki, 2003] an example of a superintuitionistic predicate logic is found, which is not H^*-complete. A logic L is said to be *H^*-complete* if for all formulas A and B having no predicate symbols or free individual variables in common, $L \vdash A \vee B$ implies $L \vdash \neg\neg A$ or $L \vdash \neg\neg B$. We show that WIP implies H^*-completeness. Assume that L has WIP and $L \vdash A \vee B$. Then $\neg A, \neg B \vdash_L \bot$ and by WIP there is a variable-free formula C such that $\neg A \vdash_L C$ and $C, \neg B \vdash_L \bot$. If C is equivalent to $\bot$, then $\vdash_L \neg\neg A$; if C is equivalent to $\top$, then $\vdash_L \neg\neg B$. So the same counter-example works for WIP.

Acknowledgement

Supported by INTAS, grant no. 04-77-7080.

BIBLIOGRAPHY

[Barwise, 1985] *J.Barwise, S.Feferman, eds.* Model-Theoretic Logics. New York: Springer-Verlag, 1985.

[Bicarregui *et al.*, 2001] *J.Bicarregui, T.Dimitrakos, D.Gabbay, T.Maibaum.* Interpolation in Practical Formal Development. Logic Journal of the IGPL, 9, no. 2 (2001), 247-259.

[Craig, 1957] *W.Craig.* Three uses of Herbrand-Gentzen theorem in relating model theory and proof theory. J. Symbolic Logic, 22 (1957), 269-285.

[Gabbay, 1971] *D.M.Gabbay.* Semantic proof of Craig's interpolation theorem for intuitionistic logic and its extensions, Part I, Part II. In: Logic Colloquium '69, North-Holland, Amsterdam, 1971, 391-410.

[Gabbay, 1972] *D.M.Gabbay.* Craig's interpolation theorem for modal logics. Lecture Notes in Mathematics, 255 (1972), 111-127.

[Gabbay, 1976] *D.M.Gabbay.* Investigations in Modal and Tense Logics with Applications to Problems in Philosophy and Linguistics, D.Reidel Publ. Co., Dordrecht, 1976.

[Gabbay, 1981] *D.M.Gabbay.* Semantical Investigations in Heyting's Intuitionistic Logic, D.Reidel Publ. Co., Dordrecht, 1981.

[Gabbay and Maksimova, 2005] *D.M.Gabbay, L.Maksimova.* Interpolation and Definability: Volume 1. Modal and Intuitionistic Logics. Clarendon Press, Oxford, 2005.

[Hoogland, 2001] *E.Hoogland.* Definability and Interpolation. Model-theoretic investigations. ILLC Dissertation Series DS-2001-05, Amsterdam, 2001.

[Maksimova, 1979a] *L.Maksimova.* Interpolation properties of superintuitionistic logics. Studia Logica, 38 (1979), 419–428.

[Maksimova, 1979b] *L.L.Maksimova.* Interpolation theorems in modal logics and amalgamable varieties of topoboolean algebras. Algebra i Logika, 18, no. 5 (1979), 556–586.

[Maksimova, 1980] *L.L.Maksimova.* Interpolation theorems in modal logics: Sufficient conditions. Algebra i Logika, 19, no. 2 (1980), 194–213.

[Maksimova, 1992] *L.L.Maksimova.* Modal Logics and Varieties of Modal Algebras: the Beth Property, Interpolation and Amalgamation. Algebra and Logic, 31, no.2 (1992), 145-166.

[Maksimova, 2000] *L.L.Maksimova.* Intuitionistic Logic and Implicit Definability. Annals of Pure and Applied Logic, 105 (2000), 83–102.

[Maksiomov, 2003] *L. Maksimova.* Restricted interpolation in modal logics. In: P.Balbiani, N.-Y.Suzuki, F.Wolter, M.Zakharyaschev, eds. Advances in Modal Logics, Volume 4, King's College London Publications, London, 2003, 297-312.

[Maltsev, 1970] *A.I.Maltsev.* Algebraic systems. Moscow: Nauka, 1970.

[Ono, 1986] *H. Ono.* Interpolation and the Robinson property for logics not closed under the Boolean operations. Algebra Universalis, 23 (1986), 111–122.

[Rasiowa and Sikorski, 1963] *H.Rasiowa, R.Sikorski.* The Mathematics of Metamathematics. Warszawa: PWN, 1963.

[Robinson, 1956] *A. Robinson.* A result on consistency and its application to the theory of definition. Indagationes Mathematicae, 18 (1956), 47–58.

[Suzuki, 2003] *N.-Y.Suzuki.* Hallden-completeness in super-intuitionistic predicate logics. Studia Logica 73 (2003), 113–130.

Goal-Directed Methods for Fuzzy Logics

GEORGE METCALFE AND NICOLA OLIVETTI

1 Historical Overview

1.1 N-Prolog

The main ideas underlying the goal-directed methodology were put forward in the early 1980s by Gabbay and developed further by a number of researchers. Recognizing that the deductive mechanism of Prolog could be generalized to support more sophisticated forms of reasoning, Gabbay (together with Reyle) [Gabbay and Reyle, 1984; Gabbay, 1985] proposed an extension of the Horn clause language, called N-Prolog, capable of performing *hypothetical reasoning*. In N-Prolog, implicational goals of the form $D \to G$ are permitted that succeed from a program P iff the goal G succeeds from the program $P \cup \{D\}$, e.g. the goal

$$pass(john, course123) \to can\_graduate(john)$$

with intuitive meaning "if John passes course123, can he graduate?" succeeds from P iff we can conclude that John can graduate from P expanded with the fact that he has passed course123. Note that here the deduction theorem is used to define the *meaning* of an implicational goal, and indeed that the evaluation of such goals may involve a change of the program; in this case, a simple addition of data to the program but in general, a possibly more sophisticated *update* of the program. Hypothetical goals may occur also in the body of a clause, as in for example, Gabbay's formalization of the law for British citizenship:

$$\begin{array}{rcl} born\_in\_UK(X) \wedge father(X,Y) & & \\ \wedge\ (alive(Y,T) \to british\_citizen(Y,T)) & \to & british\_citizen(X,T) \\ british\_citizen(X,T_1) \wedge (T_1 < T_2) \wedge alive(Z,T_1) & \to & british\_citizen(Z,T_2) \\ dead(X,T) \wedge alive(X,T) & \to & \bot \end{array}$$

The first clause establishes the counterfactual claim that X is a British citizen at time T if X was born in the UK and has a father Y who if alive at time T would be a British citizen. The second clause claims that being a British citizen persists over time, while the third expresses the incompatibility of *dead* and *alive*.

The addition of hypothetical goals as described above is not simply a procedural trick, but rather has a well-understood logical basis; the goals that succeed from a program P in N-Prolog being exactly the logical consequences of P in *Intuitionistic logic*. Related work includes Hudelmaier's optimized version of N-Prolog [Hudelmaier, 1990], a comparison of various Prolog extensions by Reed and Loveland [Reed and Loveland, 1992], and a similar hypothetical extension proposed by McCarty [McCarty, 1988] in the context of deductive databases (see also [Bonner, 1990]). Also, further hypothetical extensions of logic programming were proposed by Gabbay and others [Gabbay *et al.*, 2000] in order to define a conditional extension of logic programming incorporating a revision mechanism.[1]

1.2 Algorithmic proofs

Some years later Gabbay [Gabbay, 1992] proposed a procedural *interpretation* of logics. Whereas the traditional view of logics is extensional, i.e. a logic is a set of theorems (valid formulas) in some language, Gabbay suggested identifying a logical system with a pair (SetOfTheorems, Procedure), where Procedure is a correct deduction method for SetOfTheorems, that is to say, it generates all and only the members of SetOfTheorems. According to this perspective, (ClassicalLogic, TruthTables) is for instance a different logical system from (ClassicalLogic, Tableaux) or (ClassicalLogic, Resolution).

This conceptual shift is motivated by a number of considerations:

1. A new geography of logical systems is created, providing an alternative perspective on the differences and similarities between logics (in the traditional sense). For instance Classical logic with truth tables may be viewed as being very close to finite many-valued logic with truth tables, and Classical logic with sequent calculus as being very close to Intuitionistic logic with sequent calculus, whereas Intuitionistic logic and (finite) many-valued logic do not seem to be closely related.

2. The identification of a logical system with a pair (SetOfTheorems, Procedure) allows the possibility of discovering (or re-discovering) logics (in the traditional sense) by changing the deduction procedure, e.g. by making it stronger or weaker.

[1] A revision mechanism is needed to preserve the consistency of a database containing integrity constraints since they may be violated by the (hypothetical) insertion of new data.

3. Other reasoning mechanisms may be defined on top of the deduction procedure, prominent examples being negation by (finite) failure and abduction. Note, moreover, that while two deduction procedures for some logic may coincide on derivable formulas, they may differ in finite-failure and/or lead to the computation of different solutions for abduction.

1.3 The goal-directed methodology

Goal-directed methods fall under the procedural perspective mentioned above; being an extension of the standard computation mechanism of logic programming. Denoting by $\Gamma \vdash^? A$, the query "does A follow from Γ?" where Γ is a database (collection) of formulas and A is a goal formula, a deduction method for queries is *goal-directed* in the sense that each step in a proof is determined by the form of the current goal. More precisely, a complex goal is decomposed until its atomic constituents are reached, while an atomic goal q is matched (if possible) with the "head" of a formula $G \rightarrow q$ in the database, and its "body" G asked in turn. This latter step may be viewed as a sort of resolution or backchaining step. We illustrate this process here with a simple example of propositional N-Prolog for Intuitionistic logic with only $\rightarrow$ and $\wedge$; goal-formulas (G-formulas) and database formulas (D-formulas) being defined as follows:

$$D = q \mid G \rightarrow q \mid D \wedge D$$
$$G = q \mid D \rightarrow G \mid G \wedge G$$

A database Γ is a set of D-formulas $\{D_1, \ldots, D_n\}$, every formula in the fragment being equivalent both to a database (interpreted as a conjunction) and to a G-formula. The following goal-directed deduction procedure is sound and complete for this fragment:

$$(success) \quad \Gamma \vdash^? q \ \text{ if } q \in \Gamma$$

$$(implication) \quad \text{From } \Gamma \vdash^? D \rightarrow G \text{ step to } \Gamma, D \vdash^? G$$

$$(and) \quad \text{From } \Gamma \vdash^? G_1 \wedge G_2 \text{ step to } \Gamma \vdash^? G_1 \text{ and } \Gamma \vdash^? G_2$$

$$(reduction) \quad \text{From } \Gamma \vdash^? q \text{ step to } \Gamma \vdash^? G \text{ if } G \rightarrow q \in \Gamma$$

The goal-directed model of deduction may be refined in several ways in order to capture (efficiently) a large variety of logical systems:

1. The database can be structured, e.g. as a multiset, a list, or some more complicated structure, or constrained, e.g. by allowing only D-formulas in a certain position to be matched with the current atomic

goal in the (*reduction*) step. More generally, databases may be structured according to Gabbay's theory of *Labelled Deductive Systems* (LDS) [Gabbay, 1996; Gabbay and Olivetti, 2002a], where a labelled goal is asked from a set of labelled data and the goal-directed rules impose constraints on label propagation and combination.

2. Goal-directed algorithms can be modified to ensure termination, either by loop-checking or by "diminishing resources" i.e. removing formulas "used" to match an atomic goal. Note that in the latter case, the deletion of used data should usually be compensated for in some way to maintain completeness.

3. Goals (possibly also states of the database) previously occurring in a deduction may be stored in a history, and re-asked (or re-used in some way) using *restart rules.*

For example, a (terminating) proof system for Classical logic is obtained by modifying the calculus above to allow the replacement of the current goal by any other goal previously occurring in the same derivation branch. To this end, queries are reformulated as structures of the form $\Gamma \vdash^? G; H$ where H is a set of (atomic) goals called the *history.* We revise the reduction rule as follows:

$$(reduction) \quad \text{From } \Gamma, G \rightarrow q \vdash^? q; H \text{ step to } \Gamma \vdash^? G; H \cup \{q\}$$

and add a rule allowing restarts from any goal in the history:

$$(restart) \quad \text{From } \Gamma \vdash^? q; H \text{ step to } \Gamma \vdash^? p; H \cup \{q\} \text{ if } p \in H$$

For instance Peirce's axiom (which is not valid in Intuitionistic logic) is derivable as follows:

$$\begin{array}{rcll}
 & \vdash^? & ((p \rightarrow q) \rightarrow p) \rightarrow p; \emptyset & (implication) \\
(p \rightarrow q) \rightarrow p & \vdash^? & p; \emptyset & (reduction) \\
 & \vdash^? & p \rightarrow q; \{p\} & (implication) \\
p & \vdash^? & q; \{p\} & (restart) \\
p & \vdash^? & p; \{p, q\} & (success)
\end{array}$$

Goal-directed presentations of a wide variety of logics have been developed in Gabbay and Olivetti's book [Gabbay and Olivetti, 2000], and also by several other authors, e.g. [Giordano *et al.*, 1992; Bollen, 1991; Harland, 1997]. Moreover it has been shown that goal-directed proof procedures are useful to obtain theoretical results on logics such as interpolation [Gabbay and Olivetti, 2002b].

1.4 Uniform proofs

The goal-directed approach is closely related to the Uniform Proof paradigm proposed by Miller and others at the beginning of 1990s, see e.g. [Miller *et al.*, 1991; Harland and Pym, 1991; Hodas and Miller, 1994], and used as the basis for logic programming in various non-classical logics. The methodology and underlying perspective is different, however. The starting point for developing a uniform proof calculus is an analysis of a sequent calculus for the relevant logic; a uniform proof of a sequent $\Gamma \vdash A$ being a sequent calculus proof where (reading upwards) the goal A is decomposed first and the rules are applied to formulas in Γ only when A is atomic. In this respect the connectives occurring in the goal A may be viewed as "instructions" for directing proof search. A uniform proof calculus for a logic is then obtained from an analysis of the *permutability* of the rules of the calculus. This analysis allows the identification of fragments of the logic, such that if a sequent expressed in this fragment has a proof, then it has a uniform proof. The approach has been applied most successfully to logics (or fragments of logics) having a single-conclusion sequent calculus, although the treatment of multiple-conclusion calculi is also possible [Nadathur, 1998].

2 t-Norm Based Fuzzy Logics

Fuzzy logics are many-valued logics that form a suitable basis for reasoning under *vagueness*, providing the core of systems formalising approximate reasoning in (the field of) Fuzzy Logic. Such logics may be defined in a number of ways. Here, we focus on the influential "t-norm based" approach of Hájek [Hájek, 1998], which makes the following two basic assumptions or "design choices":

1. The set of truth values for the logic is the *real unit interval* $[0, 1]$.

2. The logic is *truth-functional*, i.e. the truth value of a compound formula is a function of the truth values of its subformulas.

Hájek also proposes restrictions on truth functions interpreting conjunction, thereby arriving at the well-known class of *continuous t-norms*: that is; continuous binary functions $* : [0, 1]^2 \to [0, 1]$ such that for all $x, y, z \in [0, 1]$:

1. $x * y = y * x$ (Commutativity)

2. $(x * y) * z = x * (y * z)$ (Associativity)

3. $x \leq y$ implies $x * z \leq y * z$ (Monotonicity)

4. $1 * x = x$ (Identity)

Suitable functions for interpreting *implication* (satisfying e.g. a generalized modus ponens property) are then the *residua* of continuous t-norms, defined as functions $\Rightarrow_*$: $[0,1]^2 \to [0,1]$ such that for all $x, y \in [0,1]$:

$$x \Rightarrow_* y =_{def} max\{z \ : \ x * z \leq y\}$$

The most important examples of continuous t-norms and their residua are:

	t-Norm	Residuum
Łukasiewicz	$x *_{\mathbf{Ł}} y = max(0, x+y-1)$	$x \Rightarrow_{\mathbf{Ł}} y = min(1, 1-x+y)$
Gödel	$x *_{\mathbf{G}} y = min(x,y)$	$x \Rightarrow_{\mathbf{G}} y = \begin{cases} 1 & \text{if } x \leq y \\ y & \text{otherwise} \end{cases}$
Product	$x *_{\mathbf{\Pi}} y = x \cdot y$	$x \Rightarrow_{\mathbf{\Pi}} y = \begin{cases} 1 & \text{if } x \leq y \\ \frac{y}{x} & \text{otherwise} \end{cases}$

Apart from the historical and practical importance of the Łukasiewicz and Gödel t-norms and their associated logics, there is a further reason to consider the above t-norms fundamental: namely, that any continuous t-norm is an ordinal sum construction of these three, see e.g. [Hájek, 1998] for details.

Following Hájek, it can now be seen that each continuous t-norm determines a *propositional fuzzy logic* $\mathbf{L}_*$ based on a language with binary connectives $\odot$ and $\to$, and a constant $\bot$; *valuations* for $\mathbf{L}_*$ being functions v assigning to each propositional variable a truth value from the real unit interval $[0,1]$, uniquely extended to formulas by:

$$\begin{array}{rcl} v(A \odot B) & = & v(A) * v(B) \\ v(A \to B) & = & v(A) \Rightarrow_* v(B) \\ v(\bot) & = & 0 \end{array}$$

A is *valid* for $\mathbf{L}_*$, written $\models_{\mathbf{L}_*} A$, if $v(A) = 1$ for all valuations v for $\mathbf{L}_*$.

Note that for any continuous t-norm $*$, the min and max functions expressing the lattice properties of $[0,1]$ can be defined using just $*$ and $\Rightarrow_*$, i.e. for all $x, y \in [0,1]$:

1. $min(x,y) = x * (x \Rightarrow_* y)$.
2. $max(x,y) = min((x \Rightarrow_* y) \Rightarrow_* y, (y \Rightarrow_* x) \Rightarrow_* x)$.

Moreover, the function $x \Rightarrow_* 0$ gives suitable properties for interpreting *negation*, being e.g. anti-monotonic with $0 \Rightarrow_* 0 = 1$ and $1 \Rightarrow_* 0 = 0$. Accordingly, the following connectives are also defined:

- $A \wedge B =_{def} A \odot (A \to B)$.
- $A \vee B =_{def} ((A \to B) \to B) \wedge ((B \to A) \to A)$.
- $\neg A =_{def} A \to \bot$.

In [Hájek, 1998] Hájek gives the following axiomatization for a *Basic fuzzy logic* **BL** that he conjectures (proved in [Cignoli *et al.*, 2000]) axiomatizes the formulas valid in *all* logics based on continuous t-norms:

$$\begin{array}{rl}
(A1) & (A \to B) \to ((B \to C) \to (A \to C)) \\
(A2) & (A \odot B) \to A \\
(A3) & (A \odot B) \to (B \odot A) \\
(A4) & (A \odot (A \to B)) \to (B \odot (B \to A)) \\
(A5a) & (A \to (B \to C)) \to ((A \odot B) \to C) \\
(A5b) & ((A \odot B) \to C) \to (A \to (B \to C)) \\
(A6) & ((A \to B) \to C) \to (((B \to A) \to C) \to C) \\
(A7) & \bot \to A
\end{array}$$

$$\frac{A \quad A \to B}{B}\ (mp)$$

Axiomatizations for the three fundamental logics are obtained as follows:

- *Łukasiewicz logic* **Ł** is **BL** plus (INV) $\neg\neg A \to A$.
- *Gödel logic* **G** is **BL** plus (ID) $A \to (A \odot A)$.
- *Product logic* **Π** is **BL** plus (Π) $\neg\neg A \to ((A \to (A \odot B)) \to B)$ and (S) $\neg(A \wedge \neg A)$.

For **Ł**, we remark that an alternative axiomatization can be given based on a language with connectives $\to$ and $\bot$, i.e. (mp) with:

$$\begin{array}{rl}
(\text{Ł}1) & A \to (B \to A) \\
(\text{Ł}2) & (A \to B) \to ((B \to C) \to (A \to C)) \\
(\text{Ł}3) & ((A \to B) \to B) \to ((B \to A) \to A) \\
(\text{Ł}4) & ((A \to \bot) \to (B \to \bot)) \to (B \to A)
\end{array}$$

The t-norm approach described above has been generalized in a number of directions. In particular, since a sufficient and necessary condition for a t-norm to have a residuum, is that it be *left-continuous*, a logic called *Monoidal t-norm logic* has been introduced in [Esteva and Godo, 2001] that captures exactly the tautologies of all left-continuous t-norm logics [Jenei and Montagna, 2002].

3 Uniform Goal-Directed Methods for Fuzzy Logics

A variety of proof methods have been defined for fuzzy logics. In particular, calculi for many of the most important t-norm based logics have been presented in the framework of *hypersequents*, a generalization of sequents to multisets of sequents, introduced independently by Avron [Avron, 1987] and Pottinger [Pottinger, 1983]. The first calculus of this type was defined for **G** by Avron in [Avron, 1991] and consists of the same "standard" rules for connectives as for e.g. Intuitionistic logic, together with further structural rules characterizing the linearity of truth values for the logic. More recently, hypersequent calculi with "non-standard" rules, have been defined by the current authors and Gabbay for **Ł** [Metcalfe *et al.*, 2005b] and **Π** [Metcalfe *et al.*, 2004a]. Moreover, taking such calculi as a starting point, goal-directed methods for **G** [Metcalfe *et al.*, 2003] and **Ł** [Metcalfe *et al.*, 2004b] have been defined that have a number of appealing properties, including the subformula property, termination, and optimal complexity, and provide a suitable basis for fuzzy logic programming [Metcalfe *et al.*, 2005a].

One unsatisfactory feature of the mentioned calculi, however, is the lack of uniformity in the rules. Essentially a new calculus has to be defined (and indeed implemented) for each logic. This contrasts for example with the situation for substructural logics where differences between logics consist solely in the presence or absence of structural rules in the calculus. This issue has been tackled recently for **Ł**, **G**, and **Π** by Ciabattoni et al. in [Ciabattoni *et al.*, 2005] where calculi with uniform rules are defined in a framework of *relational hypersequents*: roughly speaking, a further generalization of hypersequents that allows two types of sequent to occur. In this section, we make use both of previous work on goal-directed calculi for **G** and **Ł**, and the new insights provided by the relational hypersequent approach, to give uniform goal-directed rules for all three fundamental fuzzy logics. Focussing for simplicity of presentation on the implicational fragments, we define uniform rules that are sound and invertible for **Ł**, **G**, and **Π**, then show how these can be used either as the basis for Co-NP decision procedures, or extended to give fully goal-directed algorithms.

3.1 Uniform implication rules

We begin with the notion of a goal-directed query for fuzzy logics, generalizing both the structures for Intuitionistic and Classical logic given above, and the definitions of previous work. Such queries consists of a database together with a *multiset* of goals (rather than just one), and a history of previous *states* of the database with goals (rather than just goals). As above, we also allow the possibility of *restarts*, limited however to at most one for each multiset of goals. More precisely, and noting that henceforth we assume all set notation to refer to multisets:

DEFINITION 1 (Goal-Directed Query). A *goal-directed query* (query for short) is a structure of the form:

$$\Gamma_1 \vdash^? \Delta_1; R_1; \{(\Gamma_2 \vdash^? \Delta_2; R_2), \ldots, (\Gamma_n \vdash^? \Delta_n; R_n)\} \quad \text{where}$$

- $\Gamma_1, \ldots, \Gamma_n$ are multisets of formulas called *databases*.
- $\Delta_1, \ldots, \Delta_n$ are multisets of formulas called *goals*.
- $R_1, \ldots, R_n$ are multisets of at most one atomic formula called *restarts*.

Intuitively, the meaning of a query is that for each valuation for the logic, there should be a state of the database where the associated goals "follow from" that database (possibly using restart), where "follow from" is particular to the logic.

DEFINITION 2 (Validity of Queries). Let Q be a query:

$$\Gamma_1 \vdash^? \Delta_1; R_1; \{(\Gamma_2 \vdash^? \Delta_2; R_2), \ldots, (\Gamma_n \vdash^? \Delta_n; R_n)\}$$

Q is *satisfied* for $\mathbf{S}$ by a valuation v if for some i, $1 \leq i \leq n$:

$$\text{either} \quad \#^v_{\mathbf{S}}\Gamma_i \leq \#^v_{\mathbf{S}}\Delta_i \quad \text{or} \quad \#^v_{\mathbf{S}}(\Gamma_i \cup R_i) < \#^v_{\mathbf{S}}\Delta_i$$

where $\#^v_{\mathbf{S}}\emptyset = 1$ for $\mathbf{S} \in \{\mathbf{Ł}, \mathbf{G}, \mathbf{\Pi}\}$ and:

- $\#^v_{\mathbf{Ł}}(\Gamma) = 1 + \sum_{A \in \Gamma}\{v(A) - 1\}$
- $\#^v_{\mathbf{G}}(\Gamma) = min_{A \in \Gamma}\{v(A)\}$
- $\#^v_{\mathbf{\Pi}}(\Gamma) = \prod_{A \in \Gamma}\{v(A)\}$

Q is *valid* for $\mathbf{S}$, written $\models_{\mathbf{S}} Q$, iff Q is satisfied by all valuations v for $\mathbf{S}$.

Observe that for $\mathbf{G}$ and $\mathbf{\Pi}$, the valuation $\#^v_{\mathbf{S}}$ of a multiset of formulas is defined using the relevant t-norm, min and $\cdot$, respectively. For $\mathbf{Ł}$, on the other hand, the valuation $\#^v_{\mathbf{Ł}}$ is defined using the "unbounded" part of the t-norm, i.e. instead of $min(1, x + y - 1)$, we use simply $x + y - 1$. Note moreover, that the restart formula gives a choice for each state: either the restart is absent from the database and the relation is "less than or equal to", or the restart is present, and the relation is "strictly less than". However we emphasize that despite the complicated interpretation, the crucial point is that a single formula A is valid for any of the three logics $\mathbf{S} \in \{\mathbf{Ł}, \mathbf{G}, \mathbf{\Pi}\}$ iff the query $\vdash^? A; \emptyset; \emptyset$ is valid for $\mathbf{S}$. Hence checking validity for queries includes checking validity for formulas as a special case.

We now define uniform rules for handling implication in these logics, writing $\{A_1, \ldots, A_n\} \to B$ as shorthand for $A_1 \to (A_2 \to \ldots (A_n \to B) \ldots)$:

DEFINITION 3 (Implication Rules).

(*implication*) From $\Gamma \vdash^? \Pi \to q, \Delta; R; H$ step to
$\Gamma \vdash^? \Delta; R; H$ and $\Gamma, \Pi \vdash^? q, \Delta; R; H$

(*l-reduction*) From $\Gamma, \Pi \to q \vdash^? q, \Delta; R; H$ step to
$\Gamma, q \vdash^? \Pi, q, \Delta; R; H \cup \{(\Gamma \vdash^? q, \Delta; R)\}$ and
$\vdash^? \Pi; \{q\}; H \cup \{(\Gamma \vdash^? q, \Delta; R)\}$

(*r-reduction*) From $\Gamma \vdash^? q, \Delta; R_1; H \cup \{(\Gamma', \Pi \to q \vdash^? \Delta'; R_2)\}$ step to
$\Gamma', q \vdash^? \Pi, \Delta'; R_2; H \cup \{(\Gamma' \vdash^? \Delta'; R_2), (\Gamma \vdash^? q, \Delta; R_1)\}$ and
$\vdash^? \Pi; \{q\}; H \cup \{(\Gamma' \vdash^? \Delta'; R_2), (\Gamma \vdash^? q, \Delta; R_1)\}$

The implication rule (*implication*) treats a query with an implicational goal $\Pi \to q$, and steps to *two* further queries: one where this goal is removed, and one where Π is added to the database and q replaces $\Pi \to q$ as a goal. However, note that if $\Pi \to q$ is the *only* goal, then a rule with one premise is sufficient, i.e.:

(1-*implication*) From $\Gamma \vdash^? \Pi \to q; R; H$ step to $\Gamma, \Pi \vdash^? q; R; H$

The presence of an implicational formula in one of the states of the database is treated by two rules. *Local reduction* (*l-reduction*) and *remote reduction* (*r-reduction*) treat the cases where a goal matches the head of a formula in the current database, and in a database of a state in the history, respectively.

EXAMPLE 4. We illustrate these rules with a simple example:

$$\vdash^? \{p, p \to q\} \to q; \emptyset; \emptyset \quad (1\text{-}implication)$$
$$p, p \to q \vdash^? q; \emptyset; \emptyset \quad (l\text{-}reduction)$$
$$/ \qquad \backslash$$
$$p, q \vdash^? p, q; \emptyset; \{(p \vdash^? q; \emptyset)\} \qquad \vdash^? p; \{q\}; \{(p \vdash^? q; \emptyset)\}$$

We now check that the implication rules are sound (i.e. if the premises are valid, then the conclusion is valid) and invertible (i.e. if the conclusion is valid, then the premises are valid) for each logic.

LEMMA 5. *The implication rules are sound and invertible for* **Ł**, **G**, *and* **Π**.

Proof. Note first that we may assume the common part H of the histories of the premises and rules to be empty, since if H is satisfied for a valuation then clearly both the premises and conclusion are satisfied. Let v be a valuation for $\mathbf{S} \in \{\mathbf{Ł}, \mathbf{G}, \mathbf{\Pi}\}$. We treat each rule in turn:

- (*implication*). The cases of **Ł** and **Π** follow almost immediately by elementary arithmetic, so we just check the case of **G**. As a first subcase, if $\#^v_{\mathbf{G}}\Pi \leq v(q)$, then $v(\Pi \to q) = 1$, and clearly the first premise is satisfied iff the conclusion is satisfied. Moreover, if the conclusion is satisfied, then either $\#^v_{\mathbf{G}}\Gamma \leq \#^v_{\mathbf{G}}\Delta$ or $min(\#^v_{\mathbf{G}}\Gamma, v(p)) < \#^v_{\mathbf{G}}\Delta$ where $R = \{p\}$. If the former, then clearly $min(\#^v_{\mathbf{G}}\Gamma, \#^v_{\mathbf{G}}\Pi) \leq min(\#^v_{\mathbf{G}}\Delta, v(q))$. If the latter and $\#^v_{\mathbf{G}}\Pi < v(q)$, then $min(\#^v_{\mathbf{G}}\Gamma, v(p), \#^v_{\mathbf{G}}\Pi) < min(v(q), \#^v_{\mathbf{G}}\Delta)$; otherwise $\#^v_{\mathbf{G}}\Pi = v(q)$ and $min(\#^v_{\mathbf{G}}\Gamma, \#^v_{\mathbf{G}}Pi) \leq min(v(q), \#^v_{\mathbf{G}}\Delta)$. Now for the second subcase, suppose that $\#^v_{\mathbf{G}}\Pi > v(q)$ and hence $v(\Pi \to q) = v(q)$. Clearly if the conclusion is satisfied, then both premises are satisfied. Also, if the first premise is satisfied, then it follows from the fact that $min(\#^v_{\mathbf{G}}\Pi, v(q)) = v(q)$ that the conclusion is also satisfied.
- (*l-reduction*). First, assume that $\#^v_{\mathbf{S}}\Pi \leq v(q)$, and hence that $v(\Pi \to q) = 1$. Clearly, both premises are satisfied if the conclusion is satisfied. Moreover, the conclusion is satisfied, for $\#^v_{\mathbf{S}}\Pi = 1$ if the first premise is satisfied, and for $\#^v_{\mathbf{S}}\Pi < 1$, if the second premise is satisfied. Now assume that $\#^v_{\mathbf{S}}\Pi > v(q)$. Clearly, the second premise is satisfied. For **Ł** and **Π**, it follows by simple arithmetic that the conclusion is satisfied iff the first premise is satisfied. For **G**, $v(\Pi \to q) = v(q)$, and it follows that the first premise is satisfied iff the conclusion is satisfied.
- (*r-reduction*). Very similar to the previous case. ■

However, to use these rules as a calculus for establishing the validity of formulas, we require a further rule allowing "switching" between the current database and states of the history.

DEFINITION 6 (Switch Rule).

$$(switch) \quad \text{From } \Gamma_1 \vdash^? \Delta_1; R_1; H \cup \{(\Gamma_2 \vdash^? \Delta_2; R_2)\} \text{ step to}$$
$$\Gamma_2 \vdash^? \Delta_2; R_2; H \cup \{(\Gamma_1 \vdash^? \Delta_1; R_1)\}$$

Observe that each implication rule reduces a formula occurring in the query into subformulas. Hence, assuming that a loop checking mechanism is used for (*switch*) to stop it repeating ad infinitum, these rules terminate with queries where all goals are atomic and fail to match the head of any non-atomic database formula. We call such queries *irreducible.*

DEFINITION 7 (Irreducible Queries). Let Q be a query:

$$\Gamma_1 \vdash^? \Delta_1; R_1; \{(\Gamma_2 \vdash^? \Delta_2; R_2), \ldots, (\Gamma_n \vdash^? \Delta_n; R_n)\}$$

Q is *irreducible* iff:

1. Δ_i is atomic for $i = 1, \ldots, n$.

2. $\Gamma_i = \Pi_i \cup \Sigma_i$ for $i = 1, \ldots, n$ where Σ_i is atomic.

3. $Head(\Pi_1 \cup \ldots \cup \Pi_n) \cap (\Delta_1 \cup \ldots \cup \Delta_n) = \emptyset$.

where $Head(\Pi \to q) = q$ and $Head(\Gamma) = \{Head(A) : A \in \Gamma\}$.

PROPOSITION 8. *Applying the implication rules with (switch) to queries (using loop-checking for applications of (switch)) terminates with irreducible queries.*

Proof. We define the following measures and well-orderings:

- $c(q) = 1$ for q atomic.
- $c(A \to B) = c(A) + c(B) + 1$ for formulas A, B.
- $mc(\Gamma) = \{c(A) : A \in \Gamma\}$ for a multiset of formulas Γ.

For a query $Q = \Gamma_1 \vdash^? \Delta_1; R_1; \{(\Gamma_2 \vdash^? \Delta_2; R_2), \ldots, (\Gamma_n \vdash^? \Delta_n; R_n)\}$:

- $mmc(Q) = \{mc(\Gamma_i) \cup mc(\Delta_i) : 1 \leq i \leq n\}$.

For multisets α, β of integers:

1. $\alpha <_m \beta$ if $\alpha \subset \beta$.

2. $\alpha <_m \beta$ if $\alpha <_m \gamma$ where $\gamma = (\beta - \{j\}) \cup \{i, \ldots, i\}$ and $i < j$.

For multisets ϕ, ψ of multisets of integers:

1. $\phi <_{mm} \psi$ if $\phi \subset \psi$.

2. $\phi <_{mm} \psi$ if $\phi <_{mm} \chi$ where $\chi = (\psi - \{\alpha\}) \cup \{\beta, \ldots, \beta\}$ and $\beta <_m \alpha$.

For each implication rule with premises $Q_1, \ldots, Q_n$ and conclusion Q, we have $mmc(Q_i) <_{mm} mmc(Q)$ for $i = 1, \ldots, n$. Hence we arrive (using (*switch*) to move to states in the history) at queries which, since (*implication*) cannot be applied, must have atomic goals, and, since (*l-reduction*) and (*r-reduction*) cannot be applied, do not have a database formula whose head matches a goal. ■

Moreover, if an irreducible query is valid, then by removing non-atomic database formulas we obtain an atomic query that is also valid.

LEMMA 9. *For* $\mathbf{S} \in \{\mathbf{Ł}, \mathbf{G}, \mathbf{\Pi}\}$, *if the following conditions hold:*

1. $\models_{\mathbf{S}} \Gamma_1, \Pi_1 \vdash^? \Delta_1; R_1; \{(\Gamma_2, \Pi_2 \vdash^? \Delta_2; R_2), \ldots, (\Gamma_n, \Pi_n \vdash^? \Delta_n; R_n)\}$.
2. Γ_i *and* Δ_i *are atomic for* $i = 1, \ldots, n$.
3. $Head(\Pi_1 \cup \ldots \cup \Pi_n) \cap (\Delta_1 \cup \ldots \cup \Delta_n) = \emptyset$.

Then: $\models_{\mathbf{S}} \Gamma_1 \vdash^? \Delta_1; R_1; \{(\Gamma_2 \vdash^? \Delta_2; R_2), \ldots, (\Gamma_n \vdash^? \Delta_n; R_n)\}$

Proof. Assume $\not\models_{\mathbf{S}} \Gamma_1 \vdash^? \Delta_1; R_1; \{(\Gamma_2 \vdash^? \Delta_2; R_2), \ldots, (\Gamma_n \vdash^? \Delta_n; R_n)\}$. This means that there is a valuation v for $\mathbf{S}$ such that $\#^v_{\mathbf{S}}\Gamma_i > \#^v_{\mathbf{S}}\Delta_i$ and $\#^v_{\mathbf{S}}(\Gamma_i \cup R_i) \geq \#^v_{\mathbf{S}}\Delta_i$ for $i = 1, \ldots, n$. We define a new valuation w for $\mathbf{S}$ as follows:

$$w(q) = \begin{cases} 1 & \text{if } q \in Head(\Pi_1 \cup \ldots \cup \Pi_n) \\ v(q) & \text{otherwise} \end{cases}$$

Since $Head(\Pi_1 \cup \ldots \cup \Pi_n) \cap (\Delta_1 \cup \ldots \cup \Delta_n) = \emptyset$ we have that $\#^w_{\mathbf{S}}\Delta_i = \#^v_{\mathbf{S}}\Delta_i$ for $i = 1, \ldots, n$. We also get that $\#^w_{\mathbf{S}}\Pi_i = 1$, $\#^w_{\mathbf{S}}\Gamma_i \geq \#^v_{\mathbf{S}}\Gamma_i$, and $\#^w_{\mathbf{S}}R_i \geq \#^v_{\mathbf{S}}R_i$ for $i = 1, \ldots, n$. Hence $\#^w_{\mathbf{S}}(\Pi_i \cup \Gamma_i) > \#^w_{\mathbf{S}}\Delta_i$ and $\#^w_{\mathbf{S}}(\Pi_i \cup \Gamma_i \cup R_i) \geq \#^w_{\mathbf{S}}\Delta_i$ for $i = 1, \ldots, n$, i.e. $\not\models_{\mathbf{S}} \Gamma_1, \Pi_1 \vdash^? \Delta_1; R_1; \{(\Gamma_2, \Pi_2 \vdash^? \Delta_2; R_2), \ldots, (\Gamma_n, \Pi_n \vdash^? \Delta_n; R_n)\}$, a contradiction as required. ■

3.2 Co-NP Decision Procedures

Since the implication rules are both sound and invertible for **Ł**, **G**, and **Π**, and terminating (modulo loop-checking for (*switch*)), we are able to reduce checking the validity of a query in these logics to checking the validity of irreducible queries, which reduces in turn (by Lemma 9) to checking the validity of *atomic* queries. This is really only a useful step if we then have that checking the validity of atomic queries is less complex than the original validity problem for each logic. In fact, while the validity problem for all these logics is Co-NP complete (see e.g., [Hájek, 1998] for proofs and references), it can be shown (following [Ciabattoni *et al.*, 2005]) that checking validity for atomic queries, and hence for irreducible queries, is in each case *polynomial*.

LEMMA 10. *For an atomic query Q:*

$$\Gamma_1 \vdash^? \Delta_1; R_1; \{(\Gamma_2 \vdash^? \Delta_2; R_2), \ldots, (\Gamma_n \vdash^? \Delta_n; R_n)\}$$

$\models_{\mathbf{S}} Q$ *iff* $C^Q_{\mathbf{S}}$ *is inconsistent over* $[0, 1]$ *for:*

$$C^Q_{\mathbf{S}} = \{\circ_{\mathbf{S}}\Gamma_i > \circ_{\mathbf{S}}\Delta_i, \circ_{\mathbf{S}}(\Gamma_i \cup R_i) \geq \circ_{\mathbf{S}}\Delta_i \ : \ 1 \leq i \leq n\}$$

where $\circ_{\mathbf{S}}\emptyset = 1$ *for* $\mathbf{S} \in \{\mathbf{Ł}, \mathbf{G}, \mathbf{\Pi}\}$, *and:*

- $\circ_{\mathbf{Ł}}(\Gamma) = 1 + \sum_{q \in \Gamma}\{x_q - 1\}$
- $\circ_{\mathbf{G}}(\Gamma) = min_{q \in \Gamma}\{x_q\}$
- $\circ_{\mathbf{\Pi}}(\Gamma) = \prod_{q \in \Gamma}\{x_q\}$

where x_q is a real-valued variable for any propositional variable q.

Proof. Immediate from Definition 2. ■

THEOREM 11. *Checking the validity of atomic queries for* $\mathbf{Ł}$, $\mathbf{G}$, *and* $\mathbf{\Pi}$ *is polynomial.*

Proof. By Lemma 10, the result follows if we can show for each atomic query Q that checking the inconsistency of $C^Q_{\mathbf{S}}$ over $[0, 1]$ is polynomial for $\mathbf{S} \in \{\mathbf{Ł}, \mathbf{G}, \mathbf{\Pi}\}$. For $\mathbf{Ł}$ this follows from the fact that linear programming is polynomial; for $\mathbf{G}$ this follows from a theorem on relations over a finite domain; while for $\mathbf{\Pi}$ we require a rather subtle argument involving linear programming. Details of these proofs may be found in [Ciabattoni *et al.*, 2005]. ■

Co-NP decision procedures are obtained by modifying the reduction rules in order to prevent an exponential growth in the size of queries. To this end we introduce *new propositional variables* during the reduction process, that may be thought of as marking different options for the database.

DEFINITION 12 (New Reduction Rules).

(*l-reduction*) From $\Gamma, \Pi \to q \vdash^? q, \Delta; R; H$ step to
$\Gamma, q \vdash^? q, \Delta; R; H$ and $\vdash^? p, \Pi; \{q\}; H \cup \{(\Gamma, p \vdash^? q, \Delta; R)\}$
where p is a new propositional variable

(*r-reduction*) From $\Gamma \vdash^? q, \Delta; R_1; H \cup \{(\Gamma', \Pi \to q \vdash^? \Delta'; R_2)\}$ step to
$\Gamma \vdash^? q, \Delta; R_1; H \cup \{(\Gamma', q \vdash^? \Delta'; R_2)\}$ and
$\vdash^? \Pi, p; \{q\}; H \cup \{(\Gamma \vdash^? q, \Delta; R_1), (\Gamma', p \vdash^? \Delta'; R_2)\}$
where p is a new propositional variable

LEMMA 13. *The new reduction rules are sound and invertible for* **Ł**, **G**, *and* **Π**.

Proof. We just consider (*l-reduction*) (as before disregarding common parts of the history), the case of (*r-reduction*) being very similar. For soundness, we treat several cases. Assume $\#^v_{\mathbf{S}}\Pi \leq v(q)$. If $\#^v_{\mathbf{S}}\Pi = 1$, then we are done by the first premise. If $\#^v_{\mathbf{S}}\Pi < 1$, then extend v with $v(p) = 1$ and the result follows from the second premise. If $\#^v_{\mathbf{S}}\Pi > v(q)$, then we take $v(p) = v(\Pi \to q)$, and we are done by the second premise. For invertibility, note that if the conclusion is satisfied by a valuation v, then clearly the first premise is satisfied. For the second premise, if $v(q) \geq \#^v_{\mathbf{S}}(\Pi \cup \{p\})$, then it follows for each logic that $v(p) \leq v(\Pi \to q)$ and hence from the conclusion that $\#^v_{\mathbf{S}}(\Gamma \cup \{p\}) \leq \#^v_{\mathbf{S}}(\Delta \cup \{q\})$. ■

THEOREM 14. *The new reduction rules,* (*implication*), *and* (*switch*) *provide Co-NP decision procedures for the validity problems for* **Ł**, **G**, *and* **Π**.

Proof. As in Proposition 8, for the premises $Q_1, \ldots, Q_n$ and conclusion Q of the new reduction rules, $mmc(Q_i) < mmc(Q)$ for $i = 1, \ldots, n$. Moreover each application of these rules and (*implication*) gives only a constant increase in the size (number of symbols) in the query. Hence, to show that a formula is not valid in **Ł**, **G**, or **Π**, we can apply the new reduction rules (sound and invertible by Lemma 13) and (*implication*) exhaustively (using (*switch*) to move between different databases) until we reach irreducible queries, making a non-deterministic choice of two branches where necessary. The result then follows from Theorem 11. ■

3.3 Rules for deciding irreducible queries

As an alternative to polynomial decision procedures, we provide goal-directed rules below that check validity *directly* for each logic, thereby obtaining an algorithmic interpretation of **Ł**, **G**, and **Π**. We begin by defining a "standard" stock of rules for calculi, including a new "mingle" rule that allows the combination of databases and goals.

DEFINITION 15 (Standard Rules). The *standard rules* consist of the implication rules, (*switch*), and:

(*mingle*) From $\Gamma_1 \vdash^? q, \Delta_1; R_1; H \cup \{(\Gamma_2, q \vdash^? \Delta_2; R_2)\}$ step to
$\Gamma_1, \Gamma_2 \vdash^? \Delta_1, \Delta_2; \emptyset; H \cup \{(\Gamma_1 \vdash^? q, \Delta_1; R_1), (\Gamma_2, q \vdash^? \Delta_2; R_2)\}$

We now define calculi for **Ł**, **G**, and **Π** by extending the standard rules with different restart and success rules, noting that we write $\subseteq_m$ and $\subseteq_s$ for the *multiset* and *set* subset relations respectively.

DEFINITION 16 (**GDŁ**). **GDŁ** consists of the standard rules plus:

($success_{\mathbf{Ł}}$) $\Gamma \vdash^? \Delta; R; H$ if $\Delta \subseteq_m \Gamma$

($restart_{\mathbf{Ł}}$) From $\Gamma_1 \vdash^? q, \Delta_1; R_1; H \cup \{(\Gamma_2 \vdash^? \Delta_2; R_2)\}$
where $R_1 \cup R_2 = R \cup \{q\}$, step to
$\Gamma_1, \Gamma_2 \vdash^? \Delta_1, \Delta_2; R; H \cup \{(\Gamma_1 \vdash^? q, \Delta_1; R_1), (\Gamma_2 \vdash^? \Delta_2; R_2)\}$

THEOREM 17. *Q succeeds in* **GDŁ** *iff* $\models_{\mathbf{Ł}} Q$.

Proof. The left-to-right direction requires proving the soundness of (*mingle*), ($success_{\mathbf{Ł}}$), and ($restart_{\mathbf{Ł}}$) for **Ł**, which we leave as exercises in elementary arithmetic. For the right-to-left direction we proceed as follows. Let $Q = \Gamma_1 \vdash^? \Delta_1; R_1; \{(\Gamma_2 \vdash^? \Delta_2; R_2), \ldots, (\Gamma_n \vdash^? \Delta_n; R_n)\}$ be a query. If $\models_{\mathbf{Ł}} Q$, then the set $C^Q_{\mathbf{Ł}}$ is inconsistent over $[0, 1]$. Hence by linear programming methods there exist $\lambda_1, \mu_1, \ldots, \lambda_n, \mu_n \in \mathbb{N}$ such that $\lambda_i > 0$ for some $1 \leq i \leq n$ and:

$$\bigcup_{i=1}^{n} (\lambda_i + \mu_i)\Delta_i \subseteq_m \bigcup_{i=1}^{n} ((\lambda_i + \mu_i)\Gamma_i \cup \mu_i R_i)$$

We show that Q succeeds in **GDŁ** by induction on $\gamma = \sum_{i=1}^{n}(\lambda_i + \mu_i)$. If $\gamma = 1$ then we have $\Delta_i \subseteq_m \Gamma_i$ for some $1 \leq i \leq n$, and hence that Q succeeds by an application of (*switch*) if necessary and ($success_{\mathbf{Ł}}$). For $\gamma > 1$ we consider i such that $\lambda_i > 0$. If $\Delta_i \subseteq_m \Gamma_i$, then we are done by (*switch*) if necessary and ($success_{\mathbf{Ł}}$). Otherwise, we have $q \in \Delta_i$ where one of the following cases occurs:

1. $q \in \Gamma_j$ for some $j \neq i$ with $\lambda_j > 0$. Since we can always apply (*switch*), we can assume without loss of generality that $i = 1$ and $j = 2$. Now by applying (*mingle*) to Q we obtain a query Q':

   $$\Gamma_1, \Gamma_2 - \{q\} \vdash^? \Delta_1 - \{q\}, \Delta_2; \emptyset; \{(\Gamma_1 \vdash^? \Delta_1; R_1), \ldots, (\Gamma_n \vdash^? \Delta_n; R_n)\}$$

   If $\lambda_1 \geq \lambda_2$, then:

   $$(\lambda_1 - \lambda_2)\Delta_1 \cup \lambda_2(\Delta_1 - \{q\} \cup \Delta_2) \cup \Delta' \subseteq_m (\lambda_1 - \lambda_2)\Gamma_1 \cup \lambda_2(\Gamma_1 \cup \Gamma_2 - \{q\}) \cup \Gamma'$$

   for $\Delta' = \bigcup_{i=3}^n \lambda_i \Delta_i \cup \bigcup_{i=1}^n \mu_i \Delta_i$ and $\Gamma' = \bigcup_{i=3}^n \lambda_i \Gamma_i \cup \bigcup_{i=1}^n \mu_i(\Gamma_i \cup R_i)$. Since $(\lambda_1 - \lambda_2) + \lambda_2 + \sum_{i=3}^n \lambda_i + \sum_{i=1}^n \mu_i < \lambda$, by the induction hypothesis, Q' succeeds in **GDŁ** and we are done. The case where $\lambda_2 \geq \lambda_1$ is very similar.

2. $q \in R_j$ for some j with $\mu_j > 0$. If $i \neq j$, then without loss of generality we assume that $i = 1$ and $j = 2$. Otherwise $i = j$, and observe that either $\Delta_i \subseteq_m \Gamma_i$ and we can apply (*success*$_{\mathbf{L}}$), or there must exist $k \neq i$ such that either $\lambda_k > 0$ or $\mu_k > 0$. If the latter, then assume without loss of generality that $i = 1$ and $k = 2$. Now, in both cases, by applying (*restart*$_{\mathbf{L}}$) we obtain:

   $$Q' = \Gamma_1, \Gamma_2 \vdash^? \Delta_1 - \{q\}, \Delta_2; R_1; \{(\Gamma_1 \vdash^? \Delta_1; R_1), \ldots, (\Gamma_n \vdash^? \Delta_n; R_n)\}$$

   We then proceed similarly to above, separating the possibilities that $\lambda_1 \leq \mu_2$ and $\mu_2 \leq \lambda_1$. ■

DEFINITION 18 (**GDG**). **GDG** consists of the standard rules plus:

(*success*$_{\mathbf{G}}$) $\Gamma \vdash^? \Delta; R; H$ if $\Delta \subseteq_s \Gamma$

(*restart*$_{\mathbf{G}}$) From $\Gamma_1 \vdash^? q, \Delta_1; R; H \cup \{(\Gamma_2 \vdash^? \Delta_2; \{q\})\}$ step to
$\Gamma_1, \Gamma_2 \vdash^? \Delta_1, \Delta_2; \emptyset; H \cup \{(\Gamma_1 \vdash^? q, \Delta_1; R), (\Gamma_2 \vdash^? \Delta_2; \{q\})\}$

To show the completeness of **GDG** it will be helpful to have the following lemma, which establishes (roughly speaking) that we can restrict our attention to queries with only one goal per database.

LEMMA 19. *For queries* $Q = \Gamma \vdash^? \Delta_1, \Delta_2, R; H$, $Q_1 = \Gamma \vdash^? \Delta_1, R; H$ *and* $Q_2 = \Gamma \vdash^? \Delta_2, R; H$:

1. *If* $\models_{\mathbf{G}} Q$, *then* $\models_{\mathbf{G}} Q_1$ *and* $\models_{\mathbf{G}} Q_2$.

2. *If* Q_1 *and* Q_2 *succeed in* **GDG**, *then* Q *suceeds in* **GDG**.

Proof. The first part follows from the definition of validity for **G**, the second is proved by induction on the joint lengths of derivations of Q_1 and Q_2 in **GDG**. ■

THEOREM 20. *Q succeeds in* **GDG** *iff* $\models_{\mathbf{G}} Q$.

Proof. The left-to-right direction requires checking the soundness of $(mingle)$, $(success_{\mathbf{G}})$, and $(restart_{\mathbf{G}})$ for **GDG** (an easy exercise). For the right-to-left direction, let $Q = \Gamma_1 \vdash^? \Delta_1; R_1; \{(\Gamma_2 \vdash^? \Delta_2; R_2), \ldots, (\Gamma_n \vdash^? \Delta_n; R_n)\}$ be a query, assuming by Lemma 19 that $|\Delta_i| = 1$ for $i = 1, \ldots, n$. If $\models_{\mathbf{G}} Q$ then the set $C_{\mathbf{G}}^Q$ is inconsistent over $[0, 1]$. By a result on linear orders (see e.g. [Baaz *et al.*, 2001] for details), this holds iff there are $p_1, \ldots, p_m$, and distinct j_i for $i = 1, \ldots, m$ such that:

(a) $p_i \in \Delta_{j_i}$ and $p_{i+1} \in \Gamma_{j_i} \cup R_{j_i}$ for $i = 1, \ldots, m - 1$.

(b) $p_m \in \Delta_{j_m}$ and $p_1 \in \Gamma_{j_m}$.

We show that such queries are derivable in **GDG** by induction on m. If $m = 1$, then using $(switch)$ if necessary, we succeed since $\{p_1\} = \Delta_{j_m} \subseteq_s \Gamma_{j_m}$. For $m > 1$, assuming without loss of generality that $j_m = 1$ and $j_{m-1} = 2$, we have $p_m \in \Delta_1$ and $p_1 \in \Gamma_1$, and $p_{m-1} \in \Delta_2$ and $p_m \in \Gamma_2 \cup R_2$. If $p_m \in \Gamma_2$, then we apply $(mingle)$ to obtain the query Q':

$$\Gamma_1, \Gamma_2 - \{p_m\} \vdash^? \Delta_2; \emptyset; \{(\Gamma_1 \vdash^? \Delta_1; R_1), \ldots, (\Gamma_n \vdash^? \Delta_n; R_n)\}$$

But now, since $\Delta_2 = \{p_{m-1}\}$, we have $p_1, \ldots, p_{m-1}$ meeting requirements (a) and (b) above, and hence by the induction hypothesis Q' succeeds in **GDG**. For the case where $p_m \in R_2$, we apply $(restart_{\mathbf{G}})$ to obtain Q'':

$$\Gamma_1, \Gamma_2 \vdash^? \Delta_2; \emptyset; \{(\Gamma_1 \vdash^? \Delta_1; R_1), \ldots, (\Gamma_n \vdash^? \Delta_n; R_n)\}$$

Again, since $\Delta_2 = \{p_{m-1}\}$, we have $p_1, \ldots, p_{m-1}$ meeting requirements (a) and (b) above, and hence by the induction hypothesis Q'' succeeds in **GDG**. ■

To obtain a calculus for **Π**, we adapt the $(restart_{\mathbf{L}})$ rule of **GDŁ** to check that the restart formula used is "zero-ok", i.e. that the query is satisfied whenever this formula takes value 0.

DEFINITION 21 (**GDΠ**). **GDΠ** is exactly the same as for **GDŁ** except that for $(restart_{\mathbf{\Pi}})$ we add the condition:

- If $q \in R_1$, then q is zero-ok for the conclusion of the rule.

where q is *zero-ok* for a query $Q = \Gamma_1 \vdash^? \Delta_1; R_1; \{(\Gamma_2 \vdash^? \Delta_2; R_2), \ldots, (\Gamma_n \vdash^? \Delta_n; R_n)\}$ if one of the following two conditions holds:

1. $q \in \Gamma_i$ for some i, $1 \leq i \leq n$.

2. $R_i = \{q\}$ for some i, $1 \leq i \leq n$, and p is zero-ok for all $p \in \Delta_i$.

LEMMA 22. *Let* $Q = \Gamma \vdash^? \Delta; \{q\}; H$ *be a query, and let* $Q' = \Gamma \vdash^? \Delta; \emptyset; H$*:*

1. *If* $\models_{\mathbf{\Pi}} Q$ *and* q *is not zero-ok, then* $\models_{\mathbf{\Pi}} Q'$.

2. *If* Q' *succeeds in* **GDΠ**, *then* Q *succeeds in* **GDΠ**.

Proof. The first part follows from the definition of validity for **Π**, the second via an induction on the height of a proof of Q' in **GDΠ**. ■

THEOREM 23. Q *succeeds in* **GDΠ** *iff* $\models_{\mathbf{\Pi}} Q$.

Proof. The left-to-right direction follows by checking the soundness of $(mingle)$, $(success_{\mathbf{\Pi}})$, and $(restart_{\mathbf{\Pi}})$ for **GDΠ** (a straightforward exercise). For the right-to-left direction, let $Q = \Gamma_1 \vdash^? \Delta_1; R_1; \{(\Gamma_2 \vdash^? \Delta_2; R_2), \ldots, (\Gamma_n \vdash^? \Delta_n; R_n)\}$ be a query. We can assume by Lemma 22 that all restarts occurring in Q are zero-ok. If $\models_{\mathbf{\Pi}} Q$, then $C^Q_{\mathbf{\Pi}}$ is inconsistent over $[0,1]$ and therefore also $(0,1]$. Hence, by linear programming methods there exist $\lambda_1, \mu_1, \ldots, \lambda_n, \mu_n \in \mathbb{N}$ such that $\lambda_i > 0$ for some i, $1 \leq i \leq n$, and:

$$\bigcup_{i=1}^{n} (\lambda_i + \mu_i)\Delta_i \subseteq_m \bigcup_{i=1}^{n} ((\lambda_i + \mu_i)\Gamma_i \cup \mu_i R_i)$$

We now proceed as for **GDŁ** with a proof by induction on $\gamma = \sum_{i=1}^{n}(\lambda_i + \mu_i)$ that Q succeeds in **GDΠ**, the only difference being that every application of $(restart_{\mathbf{\Pi}})$ requires (which holds by assumption) that the restart formula is zero-ok. ■

EXAMPLE 24. Consider the following atomic query:

$$p \vdash^? q; \{q\}; \{(q \vdash^? p, p; \emptyset)\}$$

In the case of **G** we apply $(mingle)$ and obtain:

$$p, q \vdash^? q, p, p; \emptyset; \{(p \vdash^? q; \{q\}), (q \vdash^? p, p; \emptyset)\}$$

which succeeds by $(success_{\mathbf{G}})$ for **GDG** since $\{q, p, p\} \subseteq_s \{p, q\}$. For **Ł** on the other hand, we apply the $(restart_{\text{Ł}})$ rule of **GDŁ**, giving:

$$p, q \vdash^? p, p; \emptyset; \{(p \vdash^? q; \{q\}), (q \vdash^? p, p; \emptyset)\}$$

We now apply $(mingle)$ to get:

$$p, q, p \vdash^? p, p, q; \emptyset; \{(p \vdash^? q; \{q\}), (q \vdash^? p, p; \emptyset), (p, q \vdash^? p, p; \emptyset)\}$$

which succeeds by $(success_{\text{Ł}})$ in **GDŁ** since $\{p, p, q\} \subseteq_m \{p, q, p\}$. Moreover, since in the application of $(restart)$ we have that q occurs in one of the databases and is hence zero-ok, we also obtain a proof in **GDΠ**.

4 Concluding Remarks

Uniform goal-directed rules have been defined for the implicational fragments of the three fundamental fuzzy logics **Ł**, **G**, and **Π**, that can be extended to both co-NP decision procedures and fully goal-directed calculi. There remain, however, a number of issues requiring further attention. In particular, we began this chapter by recalling logic-programming as a source and underlying motivation for the goal-directed paradigm. In order to develop logic programming languages for fuzzy logics, however, various extensions of the proof methods presented here are required. First of all, a richer propositional language should be defined using suitable notions of database and goal formulas. Note, however, that the importance of extending the language may differ depending on the particular logic; for example, **Ł** may be based on a language with just $\to$ and $\bot$, while for **G** and **Π** more connectives are necessary. These issues have been considered for **Ł** and **G** in [Metcalfe *et al.*, 2003] and [Metcalfe *et al.*, 2004b] respectively; for example, the following rules are suitable for treating conjunctive and disjunctive goals in all three logics:

(and) From $\Gamma \vdash^? A \wedge B, \Delta; R; H$ step to
$\Gamma \vdash^? A, \Delta; R; H$ and $\Gamma \vdash^? B, \Delta; R; H$

(or) From $\Gamma \vdash^? A \vee B, \Delta; R; H$ step to
$\Gamma \vdash^? A, \Delta; R; H \cup \{(\Gamma \vdash^? B, \Delta; R)\}$

For logic programming applications, the language should also be extended to the first-order setting, a natural compromise being the fragment containing all universally-quantified formulas. In this case, the reduction rules should incorporate a suitable *unification* mechanism. Also, *logical consequence* is crucial, being the basis for fuzzy logic programming methods found in the literature [Klawonn and Kruse, 1994; Vojtás, 2001; Smutná-Hlinená and Vojtás, 2004]. That is, queries should be permitted that express that a goal G is a logical consequence of a database Γ, rather than the internal consequence that $\Gamma^* \to G$ is valid, where Γ^* is a t-norm combination of the formulas of Γ. In the case of **Ł** and **Π** these two notions of consequence are different, although they may be related via the deduction theorem for the logics, see [Metcalfe *et al.*, 2004b] for details.

From a more theoretical perspective, observe that although we have defined uniform goal-directed methods for **Ł**, **G**, and **Π**, these results also extend to a number of other logics. In particular, the uniform rules are (obviously) sound and invertible, and provide the basis for decision procedures for *intersections* of the three main fuzzy logics discussed and axiomatized in [Cignoli *et al.*, 2000]. The rules are also sound and invertible for *finite-valued* Łukasiewicz and Gödel logics, and a number of related fuzzy logics such as Cancellative hoop logic [Esteva *et al.*, 2003]. Hence the question remains as to exactly which logics can be captured using these rules. In particular, it would be interesting to show that the rules are uniform for all t-norm logics, including Hájek's logic **BL** which still lacks a reasonable calculus. Finally, observe that the goal-directed calculi for **Ł**, **G**, and **Π**, are really very closely related, the only differences lying in the underlying data structure, sets or multisets, and slight alterations in the restart rules. Moreover, note that a goal-directed calculus for *Classical logic* is obtained simply by changing the $(mingle)$ rule in **GDG** to allow the combination of databases without the corresponding combination of goals. These observations give further evidence of the uniformity of the goal-directed methodology, and more generally of the algorithmic proof perspective pioneered by Gabbay.

BIBLIOGRAPHY

[Avron, 1987] A. Avron. A constructive analysis of RM. *Journal of Symbolic Logic*, 52(4):939–951, 1987.

[Avron, 1991] A. Avron. Hypersequents, logical consequence and intermediate logics for concurrency. *Annals of Mathematics and Artificial Intelligence*, 4(3–4):225–248, 1991.

[Baaz *et al.*, 2001] M. Baaz, A. Ciabattoni, and C. Fermüller. Cut-elimination in a sequents-of-relations calculus for Gödel logic. In *International Symposium on Multiple Valued Logic (ISMVL'2001)*, pages 181–186. IEEE, 2001.

[Bollen, 1991] A. W. Bollen. Relevant logic programming. *Journal of Automated Reasoning*, pages 563–585, 1991.

[Bonner, 1990] A. J. Bonner. Hypothetical datalog: Complexity and expressibility. *Theoretical Computer Science*, pages 3–51, 1990.

[Ciabattoni *et al.*, 2005] A. Ciabattoni, C. G. Fermüller, and G. Metcalfe. Uniform Rules and Dialogue Games for Fuzzy Logics. In *Proceedings of LPAR 2004*, volume 3452 of *LNAI*, pages 496–510. Springer, 2005.

[Cignoli *et al.*, 2000] R. Cignoli, F. Esteva, L. Godo, and A. Torrens. Basic fuzzy logic is the logic of continuous t-norms and their residua. *Soft Computing*, 4(2):106–112, 2000.

[Esteva and Godo, 2001] F. Esteva and L. Godo. Monoidal t-norm based logic: towards a logic for left-continuous t-norms. *Fuzzy Sets and Systems*, 124:271–288, 2001.

[Esteva *et al.*, 2003] F. Esteva, L. Godo, P. Hájek, and F. Montagna. Hoops and fuzzy logic. *Journal of Logic and Computation*, 13(4):532–555, 2003.

[Gabbay and Olivetti, 2000] D. Gabbay and N. Olivetti. *Goal-directed Proof Theory*. Kluwer Academic Publishers, 2000.

[Gabbay and Olivetti, 2002a] D. Gabbay and N. Olivetti. Goal oriented deductions. In D. Gabbay and F. Guenthner, editors, *Handbook of Philosophical Logic*, volume 9, pages 199–285. Kluwer Academic Publishers, second edition, 2002.

[Gabbay and Olivetti, 2002b] D. Gabbay and Nicola Olivetti. Interpolation in goal-directed proof systems 1. In *Logic Colloquium 2001*. A K Peters Ltd, 2002.

[Gabbay and Reyle, 1984] D. Gabbay and Uwe Reyle. N-Prolog: An extension of Prolog with hypothetical implication. *Journal of Logic Programming*, 4:319–355, 1984.

[Gabbay *et al.*, 2000] D. Gabbay, L. Giordano, A. Martelli, N. Olivetti, and M. L. Sapino. Conditional reasoning in logic programming. *Journal of Logic Programming*, 44(1–3):37–74, 2000.

[Gabbay, 1985] D. Gabbay. N-Prolog part 2. *Journal of Logic Programming*, pages 251–283, 1985.

[Gabbay, 1992] D. Gabbay. Elements of algorithmic proof. In S. Abramsky et al., editors, *Handbook of Logic in Computer Science*, volume 2, pages 311–413. Oxford University Press, 1992.

[Gabbay, 1996] D. Gabbay. *Labelled Deductive Systems*, volume 1—Foundations. Oxford University Press, 1996.

[Giordano *et al.*, 1992] L. Giordano, A. Martelli, and G. F. Rossi. Extending Horn clause logic with implication goals. *Theoretical Computer Science*, 95:43–74, 1992.

[Hájek, 1998] P. Hájek. *Metamathematics of Fuzzy Logic*. Kluwer, Dordrecht, 1998.

[Harland and Pym, 1991] J. Harland and D. Pym. The uniform proof-theoretic foundation of linear logic programming. In *Proc. of the 1991 International Logic Programming Symposium*, pages 304–318, 1991.

[Harland, 1997] J. Harland. On goal-directed provability in classical logic. *Computer Languages*, 23(2–4):161–178, 1997.

[Hodas and Miller, 1994] J. Hodas and D. Miller. Logic programming in a fragment of intuitionistic linear logic. *Information and Computation*, 110:327–365, 1994.

[Hudelmaier, 1990] J. Hudelmaier. Decision procedure for propositional n-prolog. In P. Schroeder-Heister, editor, *Extensions of Logic Programming*, pages 245–251. Springer-Verlag, 1990.

[Jenei and Montagna, 2002] S. Jenei and F. Montagna. A proof of standard completeness for Esteva and Godo's MTL logic. *Studia Logica*, 70(2):183–192, 2002.

[Klawonn and Kruse, 1994] F. Klawonn and R. Kruse. A Łukasiewicz logic based Prolog. *Mathware & Soft Computing*, 1(1):5–29, 1994.

[McCarty, 1988] L. T. McCarty. Clausal intuitionistic logic. II. Tableau proof procedures. *Journal of Logic Programming*, 5(2):93–132, 1988.

[Metcalfe *et al.*, 2003] G. Metcalfe, N. Olivetti, and D. Gabbay. Goal-directed calculi for Gödel-Dummett logics. In M. Baaz and J. A. Makowsky, editors, *Proceedings of CSL 2003*, volume 2803 of *LNCS*, pages 413–426. Springer, 2003.

[Metcalfe *et al.*, 2004a] G. Metcalfe, N. Olivetti, and D. Gabbay. Analytic proof calculi for product logics. *Archive for Mathematical Logic*, 43(7):859–889, 2004.

[Metcalfe *et al.*, 2004b] G. Metcalfe, N. Olivetti, and D. Gabbay. Goal-directed methods for Łukasiewicz logics. In J. Marcinkowski and A. Tarlecki, editors, *Proceedings of CSL 2004*, volume 3210 of *LNCS*, pages 85–99. Springer, 2004.

[Metcalfe *et al.*, 2005a] G. Metcalfe, N. Olivetti, and D. Gabbay. Lukasiewicz logic: From proof systems to logic programming. To appear in Logic Journal of the IGPL, 2005.

[Metcalfe *et al.*, 2005b] G. Metcalfe, N. Olivetti, and D. Gabbay. Sequent and hypersequent calculi for abelian and Łukasiewicz logics. *ACM Transactions on Computational Logic*, 6(3):578–613, 2005.

[Miller *et al.*, 1991] D. Miller, G. Nadathur, F. Pfenning, and A. Scedrov. Uniform proofs as a foundation for logic programming. *Annals of Pure and Applied Logic*, 51:125–157, 1991.

[Nadathur, 1998] G. Nadathur. Uniform provability in classical logic. *Journal of Logic and Computation*, 8:209–229, 1998.

[Pottinger, 1983] G. Pottinger. Uniform, cut-free formulations of T, S4 and S5 (abstract). *Journal of Symbolic Logic*, 48(3):900, 1983.

[Reed and Loveland, 1992] D. W. Reed and D. W. Loveland. A comparison of three Prolog extensions. *Journal of Logic Programming*, 12(1):25–50, 1992.

[Smutná-Hlinená and Vojtás, 2004] D. Smutná-Hlinená and P. Vojtás. Graded many-valued resolution with aggregation. *Fuzzy Sets and Systems*, 143(1):157–168, 2004.

[Vojtás, 2001] P. Vojtás. Fuzzy logic programming. *Fuzzy Sets and Systems*, 124:361–370, 2001.

'Still'

ALICE G. B. TER MEULEN

1 Introduction

If human communication is modeled as a highly structured, dynamic process of information exchange between two natural agents each having unique access to his own information state, messages may be interpreted as instructions to update the belief state of the recipient, while adjusting the speaker's beliefs about the information state of the recipient accordingly. This paper analyzes some specific aspects of the dynamic behavior of the aspectual adverb *still* in ordinary English, offering a case study of how preservation of information may be indicated, factual and subjective evaluative information may be mixed and background information may be adjusted by reasoning in context.[1]

Extensionally the aspectual adverb *still* may informally be considered to indicate the simple continuation of given state or a description of lack of change. For instance, *D is still on the phone* expresses at least the information that D was on the phone at an earlier point in time (t_1) and is on the phone at the current reference time (t_0), and at no time in between t_1 and $t_0 D$ was not on the phone. In terms of entailments, *D is still on the phone* entails *D has been on the phone and is on the phone now.* If the aspectual adverb no longer in *D is no longer on the phone* shares with *still* the entailment that D has been on the phone, the perfect tense *have been P* may be considered a presupposition of both *still be P* and *no longer be P*, related by internal negation of P at the current reference time. But this purely extensional account in terms of truth conditional content at two reference times is too simplistic, as it does not provide any means to account for the following issues and observations on the semantic properties of *still.*

States come in quite a few semantically different flavors: some may be as timeless and eternal as we like to believe our classical logical tautologies to be (e.g. *P or not P is true*), some may be ephemeral and known to be so (e.g. *It is raining*), and some are either overtly or covertly conditional in nature

[1] The interested reader is referred to Smessaert and ter Meulen [2004] for a more detailed analysis of aspectual adverbs in a dynamic semantic framework.

(e.g. *It is raining, if I am not mistaken*, and *Lions have manes*). Clearly, not all states can straightforwardly be modified by still, which indicates that there must be more to its meaning than conveying the simple continuation of a given state. Consider the examples in (1).

1. (a) still crazy after all those years
   (b) ∗ still born after all those years
   (c) ∗ still mailed to London

Typically, states, such as being crazy (1a), normally considered to endure only for a limited time, may be modified by *still*, whereas perfect states, resulting from a once-in-a- lifetime action, such as being born (1b), cannot be felicitously so modified. But even temporary states, such as being mailed, resulting from an action or event, and known to end at some point, cannot always be modified by *still*, as (1c) shows. Furthermore, its counterpart *still not*, relating a negative state in the past to its continuation at present, is often acceptable, even when still is not, as seen in (2).

2. (a) Still not born after all these days
   (b) Still not mailed to London

Entailments may be modified by *still*, even if its corresponding premise may not be, as we see from (3).

3. (a) The wall is (∗still) reddened
   (b) The wall is (still) red

If the wall is made red, someone has caused the wall to be red, as expressed in (3a), and such a perfect tense causative predicate is not modifiable by *still*. But its entailment (3b), consisting of a simple adjectival state predicate, may be easily accepted with such an aspectual adverb, although it is hard to see prima facie to what semantic difference in the meaning of (3a) and (3b) this should be attributed. This offers us a first indication that anticipated or future polarity reversals must enter into the interpretation of *still*. As a thought exercise, it may well be conceivable to interpret (3a) in a reverse time, as if a movie were projected backwards. In that case, *still* could be acceptable, because the causal event of the wall getting painted is future from the current state of the wall being reddened.

Secondly, a perhaps surprising property of the English aspectual adverbs is their effective usage with marked high pitch prosody indicating additional subjective information about the speaker's attitude towards the described state. For instance, in (4b) a marked high pitch pronunciation of the utterance, indicated here with the capital letters, adds to the information already contained in (4a) that the speaker is in some way or other surprised or had a contrary expectation that *D* would no longer be teaching.

4. (a) *D*. is still teaching
   (b) *D*. is STILL teaching

Although there is no direct information exactly which epistemic or evaluative attitude the speaker intends to convey with his special intonational meaning of (4b), it is clear to any competent speaker of English that the speaker had counterfactually expected or perhaps believed *D*. not to be teaching anymore. Of course, there is lots of room for misunderstandings between communicating agents here, for assumptions, prejudices, norms of any kind, or other background may be adduced by a recipient of (4b) to come to believe that the speaker is disappointed by the fact that D is teaching at the current reference time, or perhaps considers it an unusual demonstration of stamina or whatnot. It has already become evident that *still* seems to describe an imminent polarity transition from a current positive phase to a future negative one, while presupposing a past positive phase continuous with the current one. This apparent reference to a future polarity transition ending the current positive phase is even more tangible in the interrogatives in (5).

5. (a) Are you STILL here?
   (b) Who is still here?

Clearly, from (5a) the addressee is to infer that the questioner expected him to be elsewhere by now, whereas the wh-question in (5b) presupposes that some people, who were here earlier, have left and seeks information on the people currently at the location of the speaker, who are supposed to leave as well. This shows an interesting interaction with *wh*-quantification, modifying both the usual existential presupposition of such interrogatives, as well as the set of possible referents that provide values for the variable in the information solicited. This set is obviously restricted by the supposition that the referents are also to leave, like the ones who have already left.

There is a related special usage of *still* in a single word question, responding to a simple assertion in a dialogue, as in (6).

6. (a) Speaker 1: *D. is asleep.*
   (b) Speaker 2: *STILL?*

Obviously the second speaker does not intend with his question to question the continuation of John's sleeping state. On the contrary, he adopts, at least conditionally, as true what the first speaker asserts as a fact that *D* is currently asleep, presupposing the information that *D* was asleep earlier as well. But the second speaker seeks from his interlocutor agreement with his own subjective assessment that D is late to wake up, or should have been awake by now. In the next section it is made explicit just how this subjective speaker assessment may be derived from the basic meaning of *still*.

Aspectual adverbs interact in an interesting way with ordinary quantificational noun-phrases (NPs). For instance, (7a) and (7b) are logically equivalent, if we assume that the predicates *being asleep* and *waking up* are polar opposites.

7. (a) Everyone is still asleep

   (b) No one has yet woken up

The static universal quantifiers *everyone* and *no one* are in (7) not only interpreted as usual, binding the variable in the predicate, but also as binding the relevant variable in the condition describing the imminent change to the state in which they will be awake.

In conditional contexts, *still* can add significant additional information to the background against which the conditional is to be interpreted. For instance, in (8), the plain conditional without the aspectual adverb would indicate that every situation in which it is raining is a situation in which *D* is also leaving. But the force of *still* in such conditional context is to indicate in addition to the recipient that *D* had a plan to leave, and his plan is to be executed regardless of the possibility that the current situation may be one in which it is raining. In other words, the speaker tells the recipient to cancel any such supposition in his background that rain could prevent *D*'s departure or change his plan to leave.

8. If it is raining, *D* is still leaving

Clearly, the information added by *still* in (8) updates the background of the recipient first with *D*'s plan to leave, and cancels in his own information state the modal conditional that *D* may not be leaving, if it is raining. The final example, showing how subtly the background may be updated by the additional contextual force of *still* in discourse, is presented in (9).

9. (a) Let's have dinner at 8. *D* is still asleep

Assuming that the longer we wait, the more likely it is that *D* wakes up, as sleeping is known to be a transient state, (9) is open to two quite different interpretations, depending on your own subjective beliefs, expectations and norms about dinner time. If you believe that *D* is expected to be at dinner, then (9) tells you that dinner is later than usual, for we are to wait until D wakes up. However, if you believe D is not expected to be at dinner or there may even be a reason to avoid having D wake up during dinner, then (9) tells you that we are to have dinner early, before D has a chance to wake up. Such abductive inferences to be best explanation supplement quite extensive background conditions from all sorts of resources to the interpretation. Misunderstandings may arise, for instance, when the addressee does not know D is a baby, and he himself considers having dinner at 8 late, he might judge D to be rude upon finding him absent from dinner even at that late a time. Or if the recipient considers 8 late to have dinner, he may resent D for delaying dinner by falling asleep, even if he may normally not be at dinner. Resolving such misunderstandings requires sorting out which background conditions have been added to the common ground by communicative acts and which may have entered, possibly subconsciously, solely on the basis of subjective beliefs, personal norms, prejudices or misconceptions.

All these examples of the special dynamic force of the simple aspectual adverb *still* indicate that its semantic rule should be powerful enough to update not only the common ground agents share in communication with new factual information about what was and actually continues to be the case, the future transition to the polar opposite state must also be included, as well as the special effects of intonational meaning or the additional subjective speaker's assessment of the course of events.

2 Aspectual adverbs and intonational meaning

The primary use of aspectual adverbs is clearly to modify a description of a state, indicating it is about to change or has just changed, without actually referring to the event causing this change. Obviously, the event that causes the transition of the negative polarity of not being P to positive P, must be currently going on. The indexical, context dependent temporal adverbs *since* and *until* relate the current reference time r_0 to the time of the polarity transition, binding in effect r_1 to be the first moment at which P before or after the current time r_0, disregarding any later times at which P is the case as irrelevant or at least excluding them from the current context. Correspondingly, in (10) below the semantics of *still* and its three logically related aspectual adverbs is presented in terms of polarity transitions START $(-/+)$ and END $(+/-)$ of static properties and the indexical binding adverbials SINCE and UNTIL, well known in most systems of temporal logic. This clarifies the logical relationships between the four basic aspectual adverbs, showing the compositional interaction between positive and negative polarity, their transitions and these two indexical adverbs. The parameters listed before the bar, $|$, are reference markers, supposedly existentially quantified as in DRS representations and the conditions listed after | present the truth functional content (cf. [van Eijck and Kamp, 1997] for a short introduction to Discourse Representation Theory).

10. Basic aspectual adverbs

a. $[_{IP}x[_{INFL}\ \textit{not yet}\ [_{VP}xP]]] \Rightarrow$
$[r_0, r_1, s, x|P'(s, x, -)\&s \supseteq r_0\&r_1 \supseteq END(P'(s, x, -))\&r_0 < r_1\& UNTIL(r_1, (P'(s, x, -))) \supseteq r_0]$

b. $[_{IP}x[_{INFL}\ \textit{already}\ [_{VP}xP]]] \Rightarrow$
$[r_0, r_1, s, x|P'(s, x, +)\&s \supseteq r_0\&r_1 \supseteq \text{START}(P'(s, x, +))\&r_1 < r_0\& \text{SINCE}(r_1, (P'(s, x, +))) \supseteq r_0]$

c. $[_{IP}x[_{INFL}\ \textit{still}\ [_{VP}xP]]] \Rightarrow$
$[r_0, r_1, s, x|P'(s, x, +)\&s \supseteq r_0\&r_1 \supseteq \text{END}(P'(s, x, +))\&r_0 < r_1\& \text{UNTIL}(r_1, (P'(s, x, +))) \supseteq r_0]$

d. $[_{IP}x[_{INFL}\ \textit{not}[_{VP}xP]\ \textit{anymore}]] \Rightarrow$
$[r_0, r_1, s, x|P'(s, x, -)\&s \supseteq r_0\&r_1 \supseteq \text{START}(P'(s, x, -))\&r_1 < r_0\& \text{SINCE}(r_1, (P'(s, x, -))) \supseteq r_0]$

These four basic aspectual adverbs constitute a logical polarity square in the temporal domain of events, showing the basic logical interaction between the current, past or future reference times, related by *since* and *until*. Limitations of space prevent me from discussing other accounts of the semantics of aspectual adverbs in the literature in any detail here. The interested reader is referred to [Smessaert and ter Meulen, 1994] for a similar account with some minor differences, and our rebuttal of other semantic theories of aspectual adverbs.

3 Intonational meaning of aspectual adverbs

In the examples (4), (5) and (6) above, we saw how aspectual adverbs may be effectively used to express not only factual information about current state and its onset or its termination, but, uttered with marked high pitch prosody, they also convey information about the speaker's attitude regarding the flow of events or its perceived speed. Other languages may express this mix of factual and subjective information differently with aspectual verbs, using, for instance, lexical composition (Dutch) or word order variation (German). Interesting issues of linguistic variability may arise in studying the expressive range of such mixed temporal information, but this paper is limited to very simple cases of English aspectual adverbs. If a speaker feels annoyed or surprised that something is not yet the case, he may of course describe his attitude explicitly stating in a full clause that he is annoyed, surprised or whatever at its not yet being the case. But in English such attitudes may be very effectively indicated with high pitch prosody on aspectual adverbs. Pitch marking of expressions is well known from studies on focus and information structure, where high pitch serves to demarcate new information from what is already assumed, given or otherwise included in the common ground. Along similar lines, the informational purpose of pitch marking aspectual adverbs is to present the subjective content as new, relegating all other supposedly factual information to the background, as if it were already incorporated into the common ground and familiar, hence not at issue in the communication. The speaker uses pitch marked *STILL* when he had expected for one reason or another the described, topical state to have ended earlier and wishes to express his dismay or surprise at its continuing to be the case. For instance, if the speaker says that D is STILL asleep, he must counterfactually have expected D to be awake by now, hence to have ended sleeping or to have woken up in the past. To capture this counterfactual expectation of the speaker in terms of a truth functional operator, a modal operator **ALT** taking as arguments the speaker (sp), the current reference time and a set of conditions, is interpreted as quantifying over speaker ALTernatives to the current course of events, subjectively dependent upon the speaker's epistemic state.

In (11a) the future ($r_0 < r_1$) endpoint r_1 of the continuing current ($s \supseteq r_0$) negative phase of the P-state ($P'(s, x, -)$) should be past ($r_1 < r_0$) according to the speaker's alternative course of current events. Similarly, if a speaker pitch-marks *alREADY*, he indicates that the actual onset of the current positive phase of P took place earlier than he had expected. Again, the pitch marked version *STILL* is a polarity counterpart of *STILL not*, both forward looking towards a earlier alternative polarity transition, and *no LONGER* lexicalizes the pitch marked version of *not anymore* in (2d), looking back to the past transition, considered early. Bold face indicates the primary focus information, even in contexts where the remainder of the content is new to the recipient and hence may be considered secondary focus.

11. The semantics of pitch marked aspectual adverbs.

    a. $[_{IP}x[_{INFL}\ \textit{STILL not}[_{VP}xP]]] \Rightarrow$
    $[r_0, r_1, s, x|P'(s, x, -)\&s \supseteq r_0\&r_1 \supseteq \text{END}(P'(s, x, -))\&r_0 < r_1\&$
    $\text{UNTIL}(r_1, P'(s, x, -)) \supseteq r_0\&\ \mathbf{ALT}(sp, r_0, [r_1 < r_0\&$
    $\text{SINCE}(r_1, P'(s, x, +)) \supseteq r_0])]$

    b. $[_{IP}x[_{INFL}\ \textit{alREADY}[_{VP}xP]]] \Rightarrow$
    $[r_0, r_1, s, x|P'(s, x, +)\&s \supseteq r_0\&r_1 \supseteq \text{START}(P'(s, x, +))\&r_1 < r_0\&$

$\text{SINCE}(r_1, P'(s,x,+)) \supseteq r_0\& \ \mathbf{ALT}(sp, r_0, [r_0 < r_1\&$
$\text{UNTIL}(r_1, P'(e,x,-)) \supseteq r_0]))]$

c. $[_{IP}x[_{INFL}\ \text{STILL}[_{VP}xP]]] \Rightarrow$
$[r_0, r_1, s, x | P'(s,x,+)\&s \supseteq r_0\&r_1 \supseteq \ \text{END}(P'(s,x,+))\&r_0 < r_1\&$
$\text{UNTIL}(r_1, P'(s,x,+)) \supseteq r_0\& \ \mathbf{ALT}(sp, r_0, [r_1 < r_0\&$
$\text{SINCE}(r_1, P'(s,x,+)) \supseteq r_0]))]$

d. $[_{IP}x[_{INFL}\ \textit{no LONGER}[_{VP}xP]]] \Rightarrow$
$[r_0, r_1, s, x | P'(s,x,-)\&s \supseteq r_0\&r_1 \supseteq \ \text{START}\ (P'(s,x,-))\&r_1 < r_0\&$
$\text{SINCE}(r_1, P'(s,x,-)) \supseteq r_0\& \ \mathbf{ALT}(sp, r_0, [r_0 < r_1\&$
$\text{UNTIL}(r_1, P'(e,x,-)) \supseteq r_0]))]$

Assuming the semantics of aspectual adverbs in (10) and the additional information their pitch marked variants express in (11), most of the examples of the first section may be accounted for in application of these rules, though details need to be worked out. What remains a much more challenging puzzle is to account for abductive inferences such as in (9), for the interaction of personal, subjective information with shared information is a complex process, which requires not only more logical analysis, and certainly also psychological study.

4 Summary and conclusion

This paper has addressed some aspects of the semantic behavior of the aspectual adverb still. Its semantic rule requires not only the condition that the state it modifies held in the past and continues to hold at present, it also indicates that the state is soon to end by an imminent polarity reversal. As such this aspectual adverb indicates how to preserve information in a dynamic environment, continuing a past state up to the present. But it also indicates that in the background an event is taking place which will cause a polarity transition in the current state, without actually making reference to this event. If uttered with marked prosody in English, aspectual adverbs such as *still* may add intonation meaning to its ordinary semantic interpretation. In such an enriched interpretation the speaker indicates by marking his utterance with high pitch that he had a counterfactual expectation, hope or perhaps fear, some attitude or other towards the current course of events, evaluating it as slower or faster than his own subjective epistemic state would have it.

In human communication there are modes of expression to indicate what need not change, what is considered background, common ground, while structuring dynamic updates to add new information to the background, before adding new factual information to the focus, considered to be asserted information, subject to acceptance by its recipient before it is considered to be included in the common ground. During the course of communication the information structure constructed by a sequence of updates has the structure in which every layer is contained in another one, as if it were an onion. In allowing semantic representations to revise the content once added by updates, belief revision is restricted in interesting and substantive ways by this structure. Belief revision cannot dig arbitrarily deeply into its layers, but only one layer deeper from any given layer. However, it is always possible to recall older information, bring it back up to foreground by asserting it anew as part of the focus information using perfect tenses. The dynamics of knowledge requires its constant revisability, although such revision is a highly structured and constrained operation. It may well be considered the hallmark rationality to assume that information held to be true to be knowledge, if it is most strongly believed, advocated against contrary information, but also constrained by cognitive resources, effort in understanding, processing time and other such limitations of human cognition.

BIBLIOGRAPHY

[Beaver, 1997] D. Beaver. Presupposition. In [8], pp. 939–1008.

[Kamp and Reyle, 1993] H. Kamp and U. Reyle. *From Discourse to Logic*. Kluwer, Dordrecht, 1993.

[Lascarides and Asher, 1993] A. Lascarides and N. Asher. Temporal Interpretation, Discourse Relations and Commonsense Entailment, *Linguistics and Philosophy*, **16**, 437–493, 1993.

[Lascarides and Asher, 2998] A. Lascarides and N. Asher. Questions in Dialogue, *Linguistics and Philosophy*, **21**, 237–309, 1998.

[ter Meulen, 1995] A. ter Meulen. *Representing Time in Natural Language. The Dynamic Interpretation of Tense and Aspect*. MIT Press, Cambridge, 1995.

[ter Meulen, 2000] A. ter Meulen. Chronoscopes: the dynamic representation of facts and events. In J. Higginbotham *et al.*, eds, *Speaking About Events*, pp. 151–168. Oxford University Press, 2000.

[Smessaert and ter Meulen, 2004] H. Smessaert and A. ter Meulen. Dynamic reasoning with aspectual adverbs. *Linguistics and Philosophy*, **27**, 209–261, 2004.

[van Benthem and ter Meulen, 1997] J. van Benthem and A. ter Meulen, eds. *Handbook of Logic and Language*. Elsevier Science, Amsterdam, & MIT Press, Cambridge, (1997),

[van Eijck and Kamp, 1997] J. van Eijck and H. Kamp. Representing discourse in context. In [8], pp. 179–23.

Sequence-Dominance Grammars

WILFRIED MEYER-VIOL AND RUTH KEMPSON

1 Introduction

In Dynamic Syntax [Kempson *et al.*, 2001],[Meyer-Viol, 1998], henceforth DS, a sentence is interpreted as a sequence of 'instructions packages' to construct a term in a formal representation language, the *logical form* of the sentence in question. Parsing, then, is the process of executing these packages in a left to right order.[1] In this paper we analyze the relations between the syntactic analysis of DS with that of more traditional, generative, frameworks. This analysis, will highlight a fundamental difference between linguistic theories concerning the relation they envisage between *word-order* and *dominance*.

In order to locate DS within the larger class of linguistic theories, we will return to one of the classics of linguistics. In 'Aspects' [Chomsky, 1965] Chomsky describes the task of generative linguistic theories as follows. They must provide for

1. an enumeration of the class $s_1, s_2, ...$ of possible sentences,
2. an enumeration of the class $SD_1, SD_2, ...$ of possible structural descriptions,
3. an enumeration $G_1, G_2, ...$ of possible generative grammars,
4. a specification function F such that $SD_{F(i,j)}$ is the structural description assigned to a sentence s_i by grammar G_j for arbitrary i, j.

Chomsky defines the set of possible sentences as given by a 'universal phonetic theory'. Supposedly these include phonetically possible objects like "*John hit Mary by was*" that are not grammatical sentences. If that is the case, then a grammar will have to assign some kind of 'empty' description

[1] The framework of Dynamic Syntax came into being in a collaboration between Dov Gabbay and Ruth Kempson (ESRC-funded 1990-1992). When the third author of [Kempson *et al.*, 2001] joined the project (EPSRC-funded 1995-1997), the basic intuitions underlying DS were already firmly established.

to some of them to fulfill the demand (4) for arbitrary i, j. We will call the specification function F *defined* on i, j if is does not assign such an empty description. Then, a generative grammar G_j can be identified with its set of possible sentence - structural description pairs as follows

$$G_j = \{(s_i, SD_{F(i,j)}) \mid i \in N, F \text{ is defined on } (i,j)\}.$$

In generative frameworks, the sentence s_i is considered as a *name*, an identifier, and the structural description $SD_{F(i,j)}$ assigns both a sequential and a dominance analysis to the sentence named by s_i.

1.1 Phrase Marker Analysis

The dominance tree involved in a structural description can generally be represented as a tree that may or may not be ordered. Chomsky's analysis does not commit itself either way, although in most generative frameworks structural descriptions are *phrase structure trees*, that is, they are represented by dominance trees the yield of which constitutes the sentence sequence. Figure 1 represents the shape of such theories.

Using phrase markers as structural descriptions a generative grammar G_1 may assign the pairs

1. $(\texttt{the};\texttt{man};\texttt{hits};\texttt{John}, ((\texttt{the},\texttt{man}),(\texttt{hits},\texttt{John}))) \in G_1$,
2. $(\texttt{John};\texttt{the};\texttt{man};\texttt{hits};\epsilon, (\texttt{John}_i, ((\texttt{the},\texttt{man}),(\texttt{hits},\epsilon_i))) \in G_1$.

Note that the word-sequence, occurs at two locations: as the left hand element of a pair and as the yield of the right hand element. This is because in a generative grammar both the sequential order and the structural description are projected as features of a single object, a phrase structure tree. In such a grammar there is no 'essential' relation between sequence and dominance. Phrase structure trees are generally composed of basic trees, either

$w_1, \ldots, w_n :$ $((A((B,C),(D,E)),D),F)$

sequence analysis   dominance analysis

$\nwarrow$ $\nearrow$

PM

Phrase marker

Structural desription of "possible sentence" s

Figure 1. Phrase structure trees as structural descriptions

local trees (a mother and her daughters, in context free grammars) or trees of greater depth (in (L)TAG, for instance). By permuting the precedence order of the daughters in the basic trees of grammar G_1 we get a grammar G_2. Now, every derivation step in G_1 can be matched by a step in G_2 and vice versa. Consequently, G_1 and G_2 cannot be formally distinguished, they are isomorphic, although only G_1 produces the word-order of English. For instance, $((\texttt{the}, \texttt{man}), (\texttt{hits}, \texttt{John})$ and $((\texttt{John}, \texttt{hits}), (\texttt{man}, \texttt{the},))$ have isomorphic dominance structure. Consequently, there is a grammar G_2 isomorphic to G_1 and

$$(\texttt{John}; \texttt{hits}; \texttt{man}; \texttt{the}, ((\texttt{John}, \texttt{hits}), (\texttt{man}, \texttt{the}))) \in G_2.$$

Only one of the isomorphic grammars G_1 and G_2 gives the 'intended' word-order, i.e., that of English. Consequently a generative grammar of English using phrase markers as its analytic tools allows for isomorphic grammars that are not grammars of English. In that sense, such a generative theory does not constitute a *formal* grammar of English: the relation between the sentence sequence and the sentence (dominance) structure is purely conventional.

Now let's consider a theory that suggests a way to establish an essential relation between word order and dominance.

> "When a TAG (Tree Adjoining Grammar) is lexicalized, each operation in that TAG (adjunction or substitution) can be seen as establishing a direct relation between two lexical items (those which anchor the trees being combined). As a direct consequent of this fact, the derivation tree is a dependency tree, since we can identify its nodes with lexemes. But, unlike dependency grammars, a TAG analysis also (at the same time) provides a phrase structure tree."
>
> [Abeille and Rambow, 2000].

Unlike that between the precedence structure of a tree and its dominance structure, the association between a derivation tree and its derived phrase structure tree is 'essential': it is preserved by grammar isomorphies. In a lexicalized framework like LTAG each of the basic trees is associated with a lexeme, so here we have an essential relation between a (dependency) structure of words (the derivation tree) and a dominance structure (the derived tree). As a theory based on phrase structure trees LTAG locates words or lexemes at two locations: as (labels of the) nodes in the derivation trees and as (labels of) the yield of derived trees. The question is: why this double use?

$$\underbrace{\begin{array}{ccc} w_1, \ldots, w_n & : & ((A((B,C),(D,E)),D),F) \\ \uparrow & & \uparrow \\ \text{Sequential, lexical} & & \text{Derived dominance tree} \\ \text{derivation tree} & & \end{array}}_{\text{Structural Description of "possible sentence" } s}$$

Figure 2. Sequence-dominance pairs as structural descriptions

> "The question arises why we should use a phrase structure representation at all and not simply abandon TAG for dependency grammar and the ensuing representational simplification. The reason, [Rambow and Joshi, 1994] suggest is word-order."
>
> ibid.

So, the presence of both a phrase structure tree and a dependency tree is required in order to supply a sequence analysis.

1.2 Sequence-Dominance Analysis

But we can draw a different conclusion from the argument above: once we have an association of the word sequence and a dependency graph of a sentence, we have exhausted the linguistic explanandum and we no longer need the phrase structure tree.

Suppose we have a lexicalized framework with *linear* derivation trees, that is, the class of derivation trees can be associated with a class of word strings. And suppose we take this class as the string language accepted by the grammar. Then we can take dependency graphs as derived trees, i.e., unordered dominance trees, and still have a grammar that satisfies Chomsky's definition. We present an enumeration of the possible sentences, (the actual sentences are the lexical derivation sequences deriving a least one dependency graph), an enumeration of possible structural descriptions , i.e., dependency graphs (the actual ones are derived by at least one derivational sequence), and specification function F, where $SD_{F(i,j)}$ is a dependency tree derived by lexical derivation sequence s_i in grammar G_j. Figure 2 illustrates the shape of such a theory. Arguably, if a grammar generates all sequence dominance pairs for a given language, it satisfies Chomsky's analysis from 'Aspects' and has exhausted the syntactic explanandum with respect to word-order and dominance. Moreover, it establishes an essential relation between sequence and dominance, one that has to be preserved by the isomorphisms of the grammar.

In order to be able to use standard formal techniques we will consider dependency trees in the shape of unordered dominance trees. A local dependency tree is generally thought of as a function immediately dominating its argument. Such a local tree can be turned into a local dominance tree by considering the root node as the value of that function on its arguments now immediately dominating the function and its arguments. This transformation recursively applied to a dependency tree gives a dominance tree with the notion of head daughters (the ones consisting of the functions). For instance, the left example below has the form of a dependency tree $(f(g(x), y)))$, the right example the form of the corresponding dominance tree $(((f, y), (g, x)))$.

$$\texttt{hits}(\texttt{the}(\texttt{man}), \texttt{John}) \quad ((\texttt{hits}, \texttt{John}), (\texttt{the}, \texttt{man}))$$

Using dominance trees the the structural descriptions of 1 and 2 in a sequence-dominance analysis get the following form.

3. $(\texttt{the}; \texttt{man}; \texttt{hits}; \texttt{John}, ((\texttt{hits}, \texttt{John}), (\texttt{the}, \texttt{man}))) \in G_3$,

4. $(\texttt{John}; \texttt{the}; \texttt{man}; \texttt{hits}, ((\texttt{hits}, \texttt{John}), (\texttt{the}, \texttt{man}))) \in G_3$.

In 3 and 4 we see the pairing of a dominance tree with two distinct sequences. As the sequence represents a linear, sequential, derivation of the dominance structure, distinct derivations can generate the same dominance structure: the 'movement' of '`John`' is not reflected on the dominance side. In such a framework movement will be a feature of the (sequential) derivations; unlike in phrase structure frameworks, it is not reflected in the derived tree.

This set-up localizes word-order phenomena on the sequential side. For instance, clauses come in varieties (declaratives, questions, imperatives, but also main clauses, subordinate clauses, relative clauses, etc.) that may systematically differ in word order but, in a dependency analysis, cannot be distinguished w.r.t. dominance structure. For instance, word-order in Dutch subordinate clauses (6) differs systematically from that in main clauses, (5) but this does not affect their dominance structure[2]

5. $(\texttt{the}; \texttt{man}; \texttt{hits}; \texttt{John}, ((\texttt{hits}, \texttt{John}), (\texttt{the}, \texttt{man}))) \in G_4$,

6. $(\texttt{the}; \texttt{man}; \texttt{John}; \texttt{hits}; ((\texttt{hits}, \texttt{John}), (\texttt{the}, \texttt{man}))) \in G_4$.

This brings up the following point. If we assign to 'phonetically possible sentence' s_i sequence-dominance pairs a structural description, then the

[2]Systematically, throughout this paper, we represent Dutch and German examples schematically only, using English words in a purely formulaic way, to focus exclusively on word order differences between the languages.

grammar can distinguish, for instance, questions from declaratives: their sequential sides will differ. The question is then: is this sufficient or do we, moreover, need some marker on the dominance structure to the effect that it is the analysis of a question or a declarative? According solely to Chomsky's requirements above for a linguistic theory, that is, if we consider syntax *autonomous*, then we do not. But if we embed the linguistic theory in a theory of 'how language works', then we tend to position the sequence in the 'outside' world, whereas the dominance structure is located, or represented, in the 'inside' as *the* product, or analysis, of the sequence outside. In that case, we had better have all contributions of the sequence reproduced on the dominance structure, otherwise relevant information would be lost in the analysis.

Theories that use sequence-dominance pairs as structural descriptions fit most comfortably in a purely linguistic framework without assumptions about how we 'process language'. Such theories have an essentially different flavour from lexicalized grammars based on phrase stucture trees, because the derived trees, the target structures of the construction, do not reflect word order in their yield. This obviates the necessity for such complicated tree operations as tree-adjunction in (L)TAG grammars and, as will be seen, it allows us to make do with tree-substitution in context free grammars: to fit in the contribution of a word, only its location in the dominance order has to be determined, its location in the precedence ordering does not have to be accounted for.

1.3 Dynamic Syntax

The grammar that we will develop in this paper is heavily inspired by Dynamic Syntax. In DS, the semantic objects created by parsing a word sequence left-to right are, essentially, unreduced terms of a typed lambda calculus. These are represented as unordered, binary, dominance trees and there is no relation between the yield of such trees and the sequential order of the sentences they analyze.

In this paper we will consider only the syntactic reflection of these semantic objects, that is, syntactic dominance trees. This is straightforward, as the terminal nodes of a semantic tree in DS are mostly in one-to-one correspondence with the words of the sentence, only a small, fixed, set of semantic types is used, and no new types are constructed in a parse (in contrast with, for instance, categorial frameworks). Moreover, each of this small number of types corresponds directly to one of the syntactic categories $S, NP, VP, V, ADJ, ADV, DET$, or N. The *type requirements* of DS will then correspond to *non-terminal symbols* ('requiring' to be rewritten to terminal symbols) and the *pointer* of DS will translate to a more traditional *rewrite location.*

It should be emphasized that we consider DS purely from the perspective of its contributions to word-order and dominance. This leaves out much of what many would consider to be an essential feature of the framework, namely its use of underspecification. Supposedly much of this can be recovered as long as the binding and substitution behaviour can be rendered as constraints on syntactic *indexations* and these indexations allow the use of meta variables. A standard semantics of the syntactic dominance trees, a homomorphic mapping to terms of a typed lambda calculus, for instance, can then interpret these syntactic indexations in the appropriate way. Via this homomorphic mapping the associations can be recovered that DS produces between word sequences and semantic trees.

2 Sequence-Dominance Grammars

We will build our sequence-dominance pairs from scratch on the basis of arbitrary associations between strings of terminal elements and bracketed terms over these elements. A grammar will then consist of a selection of such pairs. In order to organize this selection we will introduce the notion of a categorization, grouping the terminal elements in mutually exclusive (syntactic) categories. We will then work towards local sets of dominance trees, that is sets that can be generated by a context free (tree) grammar and we will investigate sequential derivation strategies generating this local set. Standardly, the basis of such a local set consists of a finite number of local trees (the set can be generated by combining the elements from the basis). But as this does not give us sequential or lexical derivations we will consider bases consisting of tree *paths*. This will lead to general strategy for supplying any local set (with binary branching trees) with a path basis organizing sequential lexical derivations. These derivations then constitute the string language accepted by the grammar.

2.1 Formal Preliminaries

Let T be a **finite** non-empty set. The elements of T we will call **terminal elements**[3]. We define two construction over the set T, the set T^* of **strings over** T and the set $\overline{T}$ of **binary dominance structures** over T.

1. **Strings over** T. Let T^* be the set of all strings over T. The empty string will be represented by 'ϵ', and if $s \in T^*$ and $|s|$ is the length of s, then $s(i)$ gives the i'th element of s. So if $|s| = n$, then $s = s(1); \ldots; s(n)$.

[3]In order to deal with current theories we will assume for the moment that T may contain 'phonologically unrealized', 'empty', elements.

2. **Tree structures over** T. Let $\overline{T}$ be the smallest set such that (1) $T \subseteq \overline{T}$, (2) if $A, B \in \overline{T}$ then $(A, B) \in \overline{T}$. To identify locations in a term A in $\overline{T}$ we define the **index** function i_A of A to be the smallest (partial) function in $(\{0,1\}^*)^{\overline{T}}$ satisfying

   (a) $i_A(A) = \epsilon$

   (b) $i_A((B,C)) = r$ implies $i_A(B) = r0, i_A(C) = r1$

   For $A \in \overline{T}$, we use the notation A_r for the subterm X of A such that $i_A(X) = r$, and we use the notation $A_r(Y)$ for the term A with the subterm A_r replaced by Y.

3. **Connecting strings and trees over** T. For $s \in T^*$ and $A \in \overline{T}$ we set $s \cong_g A$ if g is a one-one-mapping from $\{(i, s) \mid i \leq |s|\}$ to $\{(r, A) \mid A_r \in T\}$ such that $g((i, s)) = (r, A)$ implies $s(i) = A_r$.

For instance, if

$$s = x_4; x_3; x_1; x_2 \in T^*, \quad A = ((x_2, x_4), (x_3, x_1)) \in \overline{T},$$

then $s \cong_g A$, $s(1) = x_4, s(3) = x_1$, $A_{01} = x_4$, and $A_1 = (x_3, x_1)$.
If we assume a precedence ordering on $A \in \overline{T}$ and $s \cong_g A$, then g maps s onto the yield of A preserving that order.

2.2 Pair Grammars

We now consider a grammar G to be a set

$$G \subseteq \{(s, A) \in T^* \times \overline{T} \mid s \cong A\}$$

Defined thus, a grammar determines an arbitrary association between a sequence of terminal elements and a tree structure. The major organizing principles on these pairs we will introduce are **syntactic categories**. Terminal elements are grouped by syntactic categories, and if an association of a sequential structure and a tree structure is accepted by a grammar, then exchanging the terminal elements of the sequence and tree by elements *of the same category* must leave the acceptability of the association unaffected. The syntactic categories also organize the terminal trees. Generally, all terminal elements (in T) have a syntactic category. However, elements of $\overline{T}$, may or may not belong to categories. If they do, they constitute a **dominance analysis**. Within the domain of $\overline{T}$ the syntactic categories reflect also the following invariance: whenever (X, Y) is a tree then so is (X', Y')

where X' (Y') results from X (Y) by replacing some or all trees in X by ones of the same category.[4]

DEFINITION 1 (Categories and Dominance Trees). A **categorization** CAT of a grammar G is a finite set of subsets of $\overline{T}$, i.e., $CAT \subseteq \mathcal{P}(\overline{T})$, such that $X, Y \in CAT$ implies $X = Y$ or $X \cap Y = \emptyset$. For $A \in \bigcup N$ we will use the notation $[A]_{NT}$ to denote the unique $X \in CAT$ such that $A \in X$. If $A, B \in \bigcup CAT$, but $(A, B) \in \overline{T}/\bigcup CAT$, then $[(A, B)]_{NT} = \emptyset$. Then $DT = \bigcup CAT$ is a set of **Dominance Trees** if

1. $T \subseteq DT$
2. if $A, B \in DT$ and $[A_s]_{NT} = [B]_{NT}$, then $A_s(B) \in DT$ and $[A]_{NT} = [A_s(B)]_{NT}$
3. If $(s, A) \in G$ and $s' \cong_g$A', where s', A' are the result of replacing in s, A some or all terminal elements by elements of the same category, then $(s', A') \in G$.

So every word belongs to some category, and descriptions of the same categories are inter-substitutable in a term without affecting its category.

In this general definition no stipulation is made about closure under subtrees. That is $(A, B) \in DT$ does not imply that $A \in DT$ or $B \in DT$, as is correct if we want to include frameworks like TAG which uses basic trees of depth greater then 1. However, for the purpose of this paper we will require that the dominance trees form a local set.

DEFINITION 2 (Context Free Dominance Trees). A set DT of dominance trees is **Context Free** if it satisfies

4. $(A, B) \in DT$ implies $A, B \in DT$

By (4) DT is closed under subtrees.

2.3 Generation of dominance trees

We can now extend our representation of $\overline{T}$-elements to the more familiar labelled bracketing notation of phrase structure trees by replacing $t \in T$, by $({}_{[t]_{NT}}t)$ and (A, B) recursively by $({}_{[(A,B)]_{NT}}A, B)$, giving terms like

$$({}_S({}_{NP}({}_{DET}\texttt{the}), ({}_N\texttt{man})), ({}_{VP}({}_V\texttt{hits}), ({}_{NP}\texttt{John}))).$$

[4]This cannot easily be transferred to the pair perspective: For if we have (X, Y) as above in $(s, A_s(X, Y))$ we need a transformation to $(s', A_s(X', Y'))$ and this cannot be done by replacing words only, for n words in X may be replaced by 1 word in X' and these words need not lie as a contiguous stretch in s.

And to create abstract abstract dominance trees, i.e., **terms**, we have to be able to handle structures like

$$({}_S({}_{NP}({}_{DET}\texttt{the}), N), ({}_{VP}({}_V\texttt{hits}), NP)),$$

in which the categories occur as terminal nodes. We will introduce a set NT (of 'Non-Terminal symbols'), in one-to-one correspondence with elements of CAT as **names for syntactic categories**. The unique element of CAT that $N \in NT$ names will be denoted by CAT_N. Elements of T are considered to name themselves. The set of **terms** is then given by $TERM = \overline{NT \cup T}$, the set $NT \cup T$ closed under pairing. The terms will be interpreted by a function I as subsets of DT as follows.

DEFINITION 3 (Interpretation of terms).

1. If $t \in T$ then $I(t) = \{t\}$,
2. if $N \in NT$ then $I(N) = CAT_N$,
3. for $X, Y \in TERM$, $I((X,Y)) = \{(A,B)) \in DT \mid A \in I(X), B \in I(Y)\}$.

So the interpretation $I(A)$ of a term A is the set of dominance trees to which it can be completed. By abuse of notation we will set for $X \in TERM$: $[X]_{NT} = N$ if $[A]_{NT} = CAT_N$ for some $A \in I(X)$ (so $[NP]_{NT} = NP$, etc.). Note that $I(X)$ is a singleton only if $X \in \overline{T}$ and $I(X) = \emptyset$ if X is some combination involving non-terminals that is not witnessed in DT. For the standard interpretation of the non-terminal symbols $I(S)$ is the set of all sentences, $I(NP)$ the set of all noun phrases, $I((\texttt{the}, N))$ the set of all noun phrases with determiner '$\texttt{the}$', $I(\texttt{John}) = \{\texttt{John}\}, I((\texttt{John}, \texttt{walks})) = \{(\texttt{John}, \texttt{walks})\}$, and

$$I((\texttt{the}, \texttt{the})) = I((DET, DET)) = I((\texttt{man}, NP)) = I((\texttt{John}, \texttt{man})) = \emptyset.$$

The set ADT of **abstract dominance trees** is then given by

$$ADT = \{X \in TERM \mid I(X) \neq \emptyset\}.$$

So $DT \subseteq ADT$ and a pair of non-terminals $(X, Y) \in ADT$ if X and Y can be replaced by dominance trees $A, B \in DT$ respectively, such that $(A, B) \in DT$. Obviously, we would like to find a syntactic characterization of this syntactically defined set. The derivation relation below will deal with that.

The '**generation** ' **relation** $\models$ on terms, with the interpretation "$A \models B$ iff term A generates term B" is defined by

$$A \models B \iff I(B) \subseteq I(A).$$

Generating a term B from a term A means 'zooming in' from a set $I(A)$ $\mathcal{P}DT$ on a subset $I(B)$. So the derivation relation is reflexive transitive and anti-symmetric and $A \models B_s(C)$ and $C \models D$ imply $A \models B_s(D)$.

3 Derivations

The 'zooming in' of generation proceeds through subsets of DT. In order to represent the generation relation by a syntactic **derivation**, or **rewrite** relation, we introduce the notion of a **basis**. A basis is a family of subsets of DT that allows zooming in from S to every element of $I(S)$. In order to reduce $\models$ to the closure in a standard rewrite system we have to consider sets $BAS \subseteq TERM$ that are mapped by I to **finite bases**.

3.1 Finite Bases of Grammars

The categorization NT is **finitely based** if there is a finite set $BAS = \{B^1, \ldots, B^n\} \subseteq ADT$ such that every $A \in DT$ is of the form

$$A = (\ldots(B^{i_1}_{r_1}(B^{i_2}))_{r_2}\ldots)_{r_k}(B^{i_k}))$$

for some $B^{i_1}, \ldots B^{i_k} \in BAS$. That is, every element of DT can be constructed by composing basic trees using tree substitution. We will assume that every basis contains the (finite) set T.
We can then set $A \vdash B$ if B can be created by tree substitution from S using only elements from BAS. That a basis generates all of DT means that the following diagram commutes:

$$\begin{array}{ccc} S & \overset{I}{\mapsto} & I(S) \\ \vdash\downarrow & & \downarrow\supseteq \\ A & \overset{I}{\mapsto} & I(A), \end{array}$$

for S the start position and A a dominance tree in DT, that is, a terminal tree. in other words, $S \vdash A$, S derives A iff $S \models A$, S generates A. This does not mean that $\vdash$ and $\models$ agree on all elements from ADT/DT, but that is not required as we are dealing with a tree grammar: $\vdash$ and $\models$ are strongly equivalent in the standard sense, for they accept the same tree structures.

3.2 Local Bases

We will call a basis BAS for NT **local** if it consist of a finite set of pairs

$$(X, Y) \in ADT, \quad X, Y \in NT$$

together with T

Such a finite basis allows for a syntactic representation of '$\models$'. It determines a finite set CRW of context-free rewrite rules by the stipulation: for $(X, Y) \in BAS, t \in T$,

$$[(X, Y)]_{NT} \Rightarrow (X, Y), \quad [t]_{NT} \Rightarrow t \quad \in CRW,$$

interpreting, for instance $S \Rightarrow (NP, VP)$ and $NP \Rightarrow$ `John`.

The rewrite closure '$\vdash$' of '$\Rightarrow$' then corresponds to the relation of derivability, and we have

$$A \vdash B \iff A \models B.$$

Figure 3 offers a representation of a simple context free grammar. It gives, of course, a too simple analysis of absolutely everything, but it incorporates rudimentary subordinate clauses ($(COMP, S)$), allows for auxiliaries and adjunction of adjectives. We could have added agreement features but have decided to keep the example simple. Intentionally excluded is an analysis of relative clauses. These will be treated in the derivational part, section 6.2. Obviously, the basis can be reformulated as a set of rewrite rules, but this has not been done in order not to confuse matters. We will not use context free rewrite rules to derive the terminal trees, but path rewrite rules. These rules will be seen to derive the set of local trees generated by the local basis. It is important to realize that this is a context free *tree* grammar: it generates unordered trees and the ordering on the pairs in the basis is purely conventional. In the framework we are developing, this local basis is only part of the story. What is missing is a treatment of the *derivations* of the dominance trees. This is where word order and binding phenomena will be located.

As it is our object to isolate sequential derivations rewriting S to dominance trees, we will briefly describe the derivation trees of a tree grammar. Derivation trees represent dependencies between the rules of a grammar as they are used in the derivation of a dominance tree. A derivation tree for a local basis BAS is a tree, each internal node of which is labelled by an element from BAS and each terminal node by an element from T . The structure of the derivation tree reflects the composition ordering of the rules in the generation of a term; siblings represent parallel, 'independent' composition — the order in which the rules are applied does not matter —, whereas a

BAS			*NT*		
(NP, VP)	$\in$	S	`John`	$\in$	NP
(DET, N)	$\in$	NP	`he`	$\in$	NP
(V, NP)	$\in$	VP	`walks`	$\in$	VP
$(COMP, S)$	$\in$	VP	`hits`	$\in$	V
(ADJ, N)	$\in$	N	`the`	$\in$	DET
(AUX, V')	$\in$	V	`man`	$\in$	N
			`tall`	$\in$	ADJ
			`who`	$\in$	QNP
			`whom`	$\in$	QNP
			`hit`	$\in$	V'
			`did`	$\in$	AUX

Figure 3. A basis for a GFG
In this grammar, the set of NT partitions the set T of terminals. The basis of the grammar, BAS, lists pairs of non-terminals that 'generate' DT.

daughter rule is applied to a product of an application of the mother rule. We can impose an arbitrary precedence ordering on the derivation tree and a derivation is then represented by sequence of rewrite rules (elements of $(BAS \cup T)$) respecting both the precedence and the immediate dominance relation.

A grammar G is *lexicalized* if each elements of BAS is uniquely associated with an element of T. In such a grammar, the derivation tree can be seen as a tree each node of which is associated with a terminal element. The relation between a derivation tree (lexicalized or not) and its product, the derived tree, is invariant under replacing terminal elements (terminal rewrite rules) by others of the same category. So it is a 'real' relation in the grammar. That is, it must be preserved under the isomorphisms of the grammar.

Now, it is well-known that a context free grammar, i.e. a local basis, cannot be lexicalized and its derivation trees are definitely not sequential. But local bases for a set of dominance trees are not the only bases available. If we are interested in sequential derivations of DT elements, then we require terms at most one non-terminal of which can be replaced at each stage of the derivation. In frameworks like LTAG or DS, a *pointer* is introduced to select a rewritable non-terminal among all non-terminals, and a pointer *strategy* has to be specified over and above a structural characterization of the tree space.

In this paper we will consider a simpler strategy that stays within the structurally definable class. From the perspective of sequential rewriting, an obvious class of structures to consider is $TERM_1 \subseteq ADT$ consisting of all terms that have at most one non-terminal. These structures have at most one rewritable location, so the question what to rewrite never comes up. Note that $T, NT, DT \subseteq TERM_1$. So our starting points and the intended results of the rewrite derivations belong to this class.

Of course, $TERM_1$ is not closed under context free rewriting, is not the closure of a local basis, because the rewrite rules generally produce pairs of non-terminals, but here is where the path rules come in.

4 Path Rewriting

We will identify a subclass $PATH \subseteq ADT$ in which an algorithm can systematically identify a unique non-terminal to rewrite, if there are non-terminals at all. Because the non-terminal to rewrite can be systematically identified we can define an extended set of terms $PTERM_1$ in which at most one element of NT occurs. Such an $A \in PTERM_1$ will then be mapped by the interpretation function I to a set of abstract dominance trees, called path terms, where the unique non-terminal of A is mapped to the algorithmically identifiable one in elements of $I(A)$.

4.1 Path Terms

In order to define the notion of a **path term** we introduce the notion of c-command [Reinhart, 1981] (see also [Kayne, 1994] for a discussion of the relation between c-command and word order)

For $A \in ADT$ and $s, s' \in Ran(I_A)$, we use the notation

$$A_s \prec_A A_{s'},$$

to express A_s **c-commands** $A_{s'}$ **in term** A, that is A_s **is a sibling of a node that dominates** $A_{s'}$ **in** A.

1. $A_{r1} \prec_{A_r} A_{r0}$ and $A_{r0} \prec_{A_r} A_{r1}$, provided $r0, r1 \in Ran(i_A)$
2. $A_s \prec_{A_q} A_r$ implies $A_s \prec_{A_q} A_{ri}$, $i \in \{0,1\}$, provided $ri \in Ran(i_A)$.
3. $A_s \prec_A A_r$ iff there is some $q \in Ran(I_A)$ such that $A_s \prec_{A_q} A_r$.
4. $A_s \prec_A^M A_r$ iff $A_s \prec_A A_r$ and $A_s, A_r \in M$

So the $\prec_A^{NT \cup T}$ relation is the c-command relation between the the T, NT elements of A. As we are interested in sequential rewriting and rewriting only takes place on elements of NT, the $\prec_A^{NT}$ relation has special significance.

DEFINITION 4 (Path Terms). A **path term** A is an element of ADT such that $\prec_A^{NT}$ and its converse are **linear**, that is, total, irreflexive, transitive and non-branching.

Notice that all elements A of T, NT or DT are path terms, as $\prec_A^{NT} = \emptyset$ and thus satisfies all the above (universal) demands.

Claim I: A is a path term iff there is no $s \in Ran(I_A)$ such that $A_s = (X, Y)$, $X, Y \in NT$,

If $A_s = (X, Y), X, Y \in NT$ then A is not a path term, for $X \prec_A^{NT} Y$ and $Y \prec_A^{NT} X$, and, by transitivity, this would give reflexive points. Conversely, by induction on the structure of A it is straightforward to show that $\prec_A^{NT}$ is linear if there is no s such that $A_s = (X, Y)$.

Claim II: $A_r \prec_A A_s$ iff there some path term A_q and $A_r \prec_{A_q} A_s$.

So all c-command domains can be 'built up' from linear ones. For instance $((\mathtt{hits}, \mathtt{John}), (\mathtt{the}, \mathtt{man})) \in DT$ can be be built up from the path terms

$$(VP, (\mathtt{the}, N)), (V, \mathtt{John}), \mathtt{man}, \mathtt{hits}.$$

A term $A \in ADT/NT$ is a path term iff it is the sequential composition of local trees, that is, every tree composition replaces a non terminal *that has been introduced by the previous composition*, the final composition replacing a non-terminal with a terminal. The following are path terms:

$$(X, (Y, (Z, A))), \quad (X, (B, (Z, A))) \quad A, B \in T.$$

In a path term there is an $\prec_A^{NT}$-maximum[5]. This maximum we will consider its *rewrite location* and we denote this location by A_p, so $A_p = A_s$ if s is this unique $\prec_A^{NT}$-maximum and $A_p = A_\epsilon$ otherwise.

Path terms will be composed at their maxima: if A is a path term, then $A(B)$ will denote $A_p(B)$, that is A with its $\prec_A^{NT}$-maximum replaced by B.

A path term A is **basic** if there are no path terms B, C such that $A = B(C)$. For instance, the left example above is basic, the right example is not (it is of the form $(X, (B, U))((Z, A))$, where $U = [(Z, A)]_{NT}$, and $(X, (B, U))$ and (Z, A) are basic path terms). So a basic path term A is a **path** through an NT-tree ending in a single terminal element which we will denote by $w(a)$.

Let $BPATH$ be the set of all basic path terms such that for all $t \in T$ and $X \in NT$, there are at most a **finite** number of basic paths A such that

[5] and a minimum.

$[A]_{NT} = X$ and $w(A) = t$. This means that we have no recursion on basic paths: there is no basic path A such that $[A_r]_{NT} = [A_s]_{NT}$ for any initial part r of s, where s is an initial part of p.

We define $PATH$ to be the smallest subset of ADT satisfying

1. if $BPATH \subseteq PATH$
2. if $A, B \in PATH$ and $A_p = [B]_{NT}$, then $A(B) \in PATH$

The set $PATH$ contains only terms with a linear c-command relation on its non-terminals, so it is a proper subset of ADT. But note that $T, NT \subseteq PATH$. It is, moreover, easy to show that we can always arrange matters such that $DT \subseteq PATH$. Consequently, the generative capacity of $PATH$ w.r.t DT coincides with that of ADT.

4.2 Path Bases

In order to represent elements of $BPATH$ by terms with at most one non-terminal, we will concentrate on the **choices** involved in a path term. At every non-terminal node a path term selects one of the sibling non-terminals to develop. This choice is grammatically meaningful if it depends only on the categories of the elements involved. So, from the point of view of the path grammar we see the pairs $(X, Y) \in BAS$ such that $[(X, Y)]_{NT} = Z$, as recording two possible continuations of a path at Z. A **path rule** then chooses to continue either down the X branch or the Y branch. This choice is wholly determined by the categories of mothers and daughters. We will denote such choices by terms of the form

$$[^{\downarrow}_{Z}X] \quad [^{\downarrow}_{Z}Y].$$

(recall that $[(X, Y)]_{NT} = Z$.) We do not lose information in these path constructs for any two of the trio $X, Y, [(X, Y)]_{NT}$ determine the remaining one (a consequence of definition 1.2). So, although a path may continue $[(X, Y)]_{NT}; X$, the category of X's sibling is fixed by the path. For instance, if $([DET, N)]_{NT} = NP$, then specifying NP and DET (N) fixes N (DET).

We will also introduce a second construct,

$$[^{*}_{X}Y]$$

strongly motivated by a similar construct in DS, to allow a restricted form of 'movement' in the derivation of path terms. We can interpret this construct as expressing "X dominates Y". As was argued in section 1.2, movement in a pair perspective, with a pairing of a sequential derivation tree and derived dominance tree, is located on the sequential side, so this construct has its effects in the derivation tree and not in the derived tree. Within the confines of this paper we will motivate this construct by its consequences in the examples of section 6.

DEFINITION 5 (Path terms). We extend the set of terms $TERM$ to $PTERM$ by,

1. $T, NT \subseteq PTERM$,
2. if $X, Y \in PTERM$, then $(X, Y) \in PTERM$
3. if $X \in NT, Y \in PTERM$, then $[^{\downarrow}_{X} Y] \in PTERM$
4. if $X \in TERM, Y \in PTERM$, then $[^{*}_{X} Y] \in PTERM$.

The set of **terminal paths** $TPATH$ of $PTERM$ is defined by

1. $DT \subseteq TPATH$,
2. if $X, \in NT, Y \in TPATH$, then $[^{\downarrow}_{X} Y] \in TPATH$,
3. if $X \in TERM, Y \in TPATH$, then $[^{*}_{X} Y] \in TPATH$,
4. if $X, Y \in TPATH$, then $(X, Y) \in TPATH$.

To identify locations in such terms we extend the index function to the partial constructs, $i_A \in (\{0, 1, \downarrow, *\}^*)^{PTERM}$:

$$i_A([^{j}_{B}, C]) = s \text{ implies } i_A(C) = sj,\ j \in \{\downarrow, *\}$$

The new constructions are interpreted by extending the interpretation function (definition 3) $I \in \mathcal{P}(DT)^{PTERM}$ as follows

4. $I([^{\downarrow}_{X} A]) = \{(B, C) \in I(X) \mid B \in I(A), C \in \overline{T \cup NT}\}$.
5. $I([^{*}_{A} B]) = \{C(D) \in DT \mid I(C) \subseteq I(A), D \in I(B)\}$.

Note that $[[^{\downarrow}_{X} A]]_{NT} = X$ and $[[^{*}_{X} Y]]_{NT} = [X]_{NT}$.

Thus,

$I([^{\downarrow}_{S}[^{\downarrow}_{NP}\mathtt{the}]]) = \{[^{\downarrow}_{S}(\mathtt{the}, X)] \mid X \in N\} = \{(Y, (\mathtt{the}, X)) \mid X \in N, Y \in VP\}$.

Consequently, we have the following equivalence between a term in $PTERM$ and a term in $PATH$:

$$I([^{\downarrow}_{S}[^{\downarrow}_{NP}\mathtt{the}]]) = I([^{\downarrow}_{S}(\mathtt{the}, N)]) = I((VP, (\mathtt{the}, N))).$$

DEFINITION 6 (BAS-states). An BAS-**state** σ is an assignment of path constructs to elements of NT such that

1. for all $(X, Y) \in BAS$: $[\downarrow_{[(X,Y)]_{NT}} X] \in \sigma([(X,Y)]_{NT})$ or $[\downarrow_{[(X,Y)]_{NT}} Y] \in \sigma([(X,Y)]_{NT})$,
2. $<_\sigma$ is non-circular

   where $X <^1_\sigma Y$ iff $[\downarrow_X Y] \in \sigma(X)$, $X <^{n+1}_\sigma Y$ iff there is some Z:$X <^n_\sigma Z <^1_\sigma Y$, and $X <_\sigma Y$ if there is some n such that $X <^n_\sigma Y$.

By the first condition, for every pair in the state σ makes a choice of daughter. This guarantees that there are enough path constructs to 'derive' all of DT. By the second condition each sequence of choice is finite, i.e., cannot be continued after a finite number of steps.

Now let $\Sigma(BAS)$ be the set of all BAS-states. These represent families of choices of elements of $(X, Y) \in BAS$ to (X, Y) and each will reflect a word-order strategy.

The states do not affect the set of derived trees; the same terminal structures can be derived in any state. They do, however, affect the derivation order by selecting different paths.

5 The Path Grammar

A path term selects at every node one of the two daughters as a candidate for development, and it is this choice that we will represent. If $(Y, Z) \in BAS$ and Y is the daughter selected by the path term, then we can express this by a **path rewrite rule** of the form

$$X \Rightarrow [\downarrow_X Y],$$

"X can rewritten to X with daughter, $(\downarrow)$, Y". So the set BAS determines the set of path rewrite rules, and a **path rewrite grammar** then consists of a selection of such rules that allows every terminal tree to be reached, that is, there is at least one path rewrite rule for every element from BAS. Every path rule $X \Rightarrow [\downarrow_X Y]$ corresponds uniquely to a an element $(Y, Z) \in BAS$ for some Z, namely the Z such that $[(Y, Z)]_{NT} = X$. And a sequence of path rule applications determines uniquely a sequence of BAS-elements.

Note that if only path rewrite rules are used, then, at every derivation step, the product has at most one non-terminal. Thus path rewriting is sequential and we can define substitution at s in $A_s(Y)$ simply as $A(Y)$, because Y replaces **the** non-terminal in A. When such a single non-terminal is rewritten by a lexical rule, then we have a terminal path, one that cannot be rewritten any more. The unique tree term corresponding to this path is a path term. In such a term A the c-command relation on the non-terminals is linear and thus has a maximal element A_s. This element is always a sibling of some fully rewritten term and it is the non-terminal that will be exposed to the path rewrite system as a new non-terminal to rewrite. The movement is simple: having rewritten for instance $A([^{\downarrow}_{NP}DET])$ to $A([^{\downarrow}_{NP}, \texttt{the}])$, we consider the corresponding term $(DET, N) \in NP$, and allow the move from $A([^{\downarrow}_{NP}, \texttt{the}])$ to $A((\texttt{the}, N))$. This again gives a term with a single non-terminal. This depends on the fact that BAS is a *local* basis. (1) We only have local trees in BAS, a mother with her daughters. If the rules allowed rewrite to trees with a greater depth (as is the case in most tree rewrite grammars), then more than one non-terminal would be exposed by the above rule. (2) the local trees have binary branching. Again, if a mother could have more than two daughters, then more than one non-terminal would be introduced by the rule, hindering sequentiality.

In a path rewrite derivation, at every moment only one non-terminal is available, and the composition strategy allows us to derive all elements of DT. So the strategy does not lose relevant products. We can even retrieve full flexibility in rewrite order, by introducing both possible path rules for every BAS element. Still, all derivations are necessarily sequential, but there will be such sequential derivations for any order of siblings.

The second path construct has the form $[^{*}_{X}Y]$ expressing "term X **dominates** term Y" and we will have, for instance the path rule $S \Rightarrow [^{*}_{S}NP]$, "$S$ can be rewritten to a S dominating an NP term". Once we have path-rewritten the NP to a terminal construction, eg. $[^{*}_{S}(\texttt{the}, \texttt{man})]$, we can start path-rewriting S. If this gives a term C, we may derive $[^{*}_{C}(\texttt{the}, \texttt{man})]$. When now NP is the rewritable non-terminal in C, then $[^{*}_{C}(\texttt{the}, \texttt{man})]$ may be replaced by $C((\texttt{the}, \texttt{man}))$.

5.1 Word order and Clauses

Word-order is determined by BAS-states. A grammar will use a finite set of them, as there are generally systematic word-order differences between clause types like declarative, questions, imperatives. And within these various types we have subordinate clauses and relative clauses, also subject to word-order variation in a variety of languages. Every state influences word order by two facts. The first is the presence of either one or both rules of the form $X \Rightarrow [^{\downarrow}_{X}Z]$ for $Z \in \{Y, U\}, (U, V) \in BAS$. If only one is present, a word-order invariant is fixed. If both are present then both word orders are acceptable. The second is supplied by the $[^{*}_{X}, Y]$ construct which allows 'long distance movement' for selected categories.

BAS-states, that is word-order strategies, seem to be determined at the clause level, formally the starting symbol S. We will consider the start set

$SBAS \subseteq BAS$ by

$$SBAS = \{(X, Y) \in BAS \mid [(X, Y)]_{NT} = S\}.$$

The Basis will then be presented by the disjoint pair

$$(SBAS, RBAS)$$

where $RBAS = BAS/SBAS$ and we let rewrite rules, local trees come only from $RBAS$. This allows us to define a **derivational** notion of locality. In the grammar we no longer have rewrite rules with S on the left hand side. So, having started, there is no recursion on S. Instead we introduce '**derivation rules**' $S \vdash Y$ for $Y \in SBAS$ (for instance, $S \vdash (NP, VP)$). We then extend the rewrite semantics with the following cut rule

$$A \vdash B \;\; B_s \vdash C, C \in DT \;\; \text{implies} \;\; A \vdash B_s(C).$$

We have re-instituted recursion on S by adding a cut rule. This rule only replaces embedded S by a terminal term, one that cannot be rewritten any further. Thus constituents of a category S must be rewritten as a whole (without interference) before they are composed. There are no rewrite products with two half-finished constituents of category S.[6]

5.2 The Formal Grammar

Let $G = (BAS, \Delta)$ consist of a local basis $BAS = (SBAS, RBAS)$ and a set Δ of BAS-states, a **path grammar** $P(G)$ for G is a quadruple

$$P(G) = (S, NT, T, (RW^{\sigma}, SRW^{\sigma}))_{\sigma \in \Delta}$$

where S, T, NT are determined by G and RW^{σ} and SRW^{σ} are finite sets of **path rewrite rules** and **start rules**, respectively, of the following form:

1. $[t]_{NT} \Rightarrow t \in RW^{\sigma}$ for all $t \in T, \sigma \in \Delta$
2. $[(X, Y)]_{NT} \Rightarrow Y \in RW^{\sigma}$ if $Y \in \sigma([(X, Y)]_{NT})$, for all $(X, Y) \in RBAS$
3. $SRW^{\sigma} \subseteq \{[^i_S, Y]^{\sigma'} \mid [^i_S Y] \in \sigma(S), \sigma' \in \Delta, i \in \{\downarrow, *\}\}$

[6]Alternatively, we can remove only some of the $S \Rightarrow A$ rules, but not all. Then only some of the S constituents can be considered domains. For instance, In general S-terms are **tensed domains**, it is established at the level of S and thus a candidate for establishment at the local level (without the cut rule). Then non-tensed domains don't require cut and we may have rewrite rules with LHS S witnessing this.

We will use the notation $\sigma(S) \mapsto \sigma'$ if $[^i_S Y]^{\sigma'} \in SRW^\sigma$ for some i, Y, and we will assume that there is a selected default (main clause) strategy $\rho \in \Delta$.

DEFINITION 7 (Path rewrite semantics). The rewrite semantics of $P(G)$ is formalized by the derivation relation

$$A \mid\!\sim^\sigma_\nu B$$

meaning "term A derives term B using the sequence of words ν and strategy σ". This relation is inductively defined as follows:

For $t \in T, \nu, \mu \in T^*$ and $\epsilon \in T^*$ the empty sequence let:

1. Structural Rules:
   (a) $A \mid\!\sim^\sigma_\epsilon A$, for all $\sigma \in \Delta$
   (b) $S \mid\!\sim^\sigma_\epsilon Y$ if $Y^\sigma \in SRW^\rho$
   (c) if $A \mid\!\sim^\sigma_\nu B$ and $B_p \mid\!\sim^{\sigma'}_\mu, D$, then $A \mid\!\sim^\sigma_{\nu\mu} B(D)$,
   provided $B_p = S$, $D \in DT$, and $\sigma(S) \mapsto \sigma'$.

2. Rewrite Rules:
   (a) if $A \mid\!\sim^\sigma_\nu B$ and $B_p \Rightarrow Y \in RW^\sigma$, then $A \mid\!\sim^\sigma_\nu B(Y)$,
   (b) if $A \mid\!\sim^\sigma_\nu B$ and $B_p \Rightarrow t \in RW^\sigma$, then $A \mid\!\sim^\sigma_{\nu t} B(t)$,

3. Path elimination Rules: If $[(Y, Z)]_{NT} = X$ and $E \in DT$, then
   (a) $A \mid\!\sim^\sigma_\nu B([^\downarrow_X, E])$ implies $A \mid\!\sim^\sigma_\nu B((E, Z))$, provided $[E]_{NT} = Y$,
   (b) $A \mid\!\sim^\sigma_\nu B([^\downarrow_X, E])$ implies $A \mid\!\sim^\sigma_\nu B((Y, E))$, provided $[E]_{NT} = Z$.

4. Movement propagation and Elimination Rules: if $B \in DT$, then
   (a) if $A \mid\sim^\sigma_\nu [^*_X B]$ and $X \mid\!\cong^{\sigma'}_\mu C$, then $A \mid\sim^\sigma_{\nu\mu} [^*_C B]$, provided $\sigma([X]_{NT}) \mapsto \sigma'$,
   where $A \mid\!\cong^\sigma_\mu B$ if $A \mid\!\sim^\sigma_\mu B$, or there is a C such that $A \mid\!\cong_\nu C, C \mid\!\sim^{\sigma'}_\lambda B$ and $\mu = \nu; \lambda$.
   (b) if $A \mid\!\sim^\sigma_\nu [^*_C B]$, then $A \mid\!\sim^\sigma_\nu C(B)$, provided $[B]_{NT} = C_p$

The structural descriptions determined by the grammar are fixed by the set

$$SD(P(G)) = \{(s, A) \in T^* \times \overline{T} \mid \exists\sigma \in \Delta, S \mid\!\sim^\sigma_s A\}.$$

and the string language accepted by the grammar consists of [7]

$$LAN(P(G)) = \{s \in T^* \mid \exists A \in \overline{T}, (s, A) \in SD(P(G))\}.$$

We will set $A \mid\!\sim B$ iff there is a sequence $\mu \in T^*$ and a BAS-state σ such that $A \mid\!\sim^{\sigma}_{\mu} B$.

All derivations on $P(G)$ are sequential, that is the import of the first claim.

Claim III: If $A \in PTERM_1$ and $A \mid\!\sim B$, then $B \in PTERM_1$.

This can be shown by induction on the length of derivations. As has been mentioned, this is a reflection of two features of G: all rewrite rules correspond to trees of depth one and all the local trees are binary. If either were not the case, then the admissible rules would introduce more than one non-terminal. Consequently, starting from a non-terminal (an element of $TERM_1$) every derivation will be wholly sequential as there is at most one non-terminal to rewrite.

Every derivable term without path constructs is a terminal tree. So all ADT terms that are derivable are elements of $NT \cup T \cup DT$.

Every rewrite that starts from a non-terminal and uses only rewrite rules creates finite sequences, that is they represent a *path* through a linear term, e.g.,

$$S \mapsto [^{\downarrow}_{S}NP] \mapsto [^{\downarrow}_{S}[^{\downarrow}_{NP}DET]] \mapsto [^{\downarrow}_{S}[^{\downarrow}_{NP}\mathtt{the}]]$$

Only when a rewrite sequence ends in a terminal can 3a, 3b, be applied making a new non-terminal available to the rewrite rules.

$$[^{\downarrow}_{S}[^{\downarrow}_{NP}\mathtt{the}]] \mapsto [^{\downarrow}_{S}(\mathtt{the}, N)] \mapsto [^{\downarrow}_{S}(\mathtt{the}, \mathtt{man})] \mapsto ((\mathtt{the}, \mathtt{man}), VP).$$

The rules 3a, 3b start at an $\prec^{NT}_{A}$-maximum and work downwards. We can also arrange for them to start at the (unique) $\prec^{NT}_{A}$-minimum. This will be referred to in section 6.3 on cross-serial dependencies. Locality is divided over the rules 1b, 1c. A proper subterm S of a term A cannot be rewritten, by 3a. But we can set up a separate derivation starting from S, and when this results in a completed term, then, by rule 1c, this can replace S in A.

[7] In a more dynamic formulation we consider pairs in $T^* \times PTERM$ and set

$$\langle s, A\rangle \prec \langle s', A'\rangle \iff s = w; s' \;\&\; A \mid\!\sim_w A'.$$

In each $\prec$ transition the first word of s is consumed. Let $\preceq$ be the reflexive transitive closure of $\prec$. In this formulation the **string language** is defined by those $s \in T^*$ such that $(s, S) \preceq (\epsilon, A)$ for some $A \in DT(S)$. That is

$$LANG(P(G)) = \{(s, A) \in T^* \times \overline{T} \mid \langle s, S\rangle \preceq \langle \epsilon, A\rangle\}.$$

BAS			*NT*			*SRW & RW*		
(NP, VP)	$\in$	S	`John`$_x$	$\in$	NP	$\mathbf{S}$	$\mid\!\sim_\epsilon$	S'
(DET, N)	$\in$	NP	`he`$_x$	$\in$	NP	$\mathbf{S}$	$\mid\!\sim_\epsilon$	$[^{*}_{S}\mathbf{NP}]$
(V, NP)	$\in$	VP	`walks`	$\in$	VP	$\mathbf{S}$	$\mid\!\sim_\epsilon$	$[^{*}_{S'}\mathbf{QNP}]$
$(COMP, S)$	$\in$	VP	`hits`	$\in$	V	S	$\mid\!\sim_\epsilon$	$[^{\downarrow}_{S}\mathbf{NP}]$
(ADJ, N)	$\in$	N	`hit`	$\in$	V'	S_x	$\mid\!\sim_\epsilon$	$[^{*}_{S}\mathbf{QNP}_x]$
(AUX, V')	$\in$	V	`the`$_x$	$\in$	DET	S'	$\mid\!\sim_\epsilon$	$[^{\downarrow}_{S}\mathbf{QNP}]$
			`man`	$\in$	N	S'	$\mid\!\sim_\epsilon$	$[^{*}_{S}AUX]$
			`tall`	$\in$	ADJ	$\mathbf{NP}$	$\Rightarrow$	NP
			`who`$_x$	$\in$	QNP	$\mathbf{QNP}$	$\Rightarrow$	QNP
			`whom`$_x$	$\in$	QNP	NP	$\Rightarrow$	$[^{\downarrow}_{NP}, DET]$
			`did`	$\in$	AUX	VP	$\Rightarrow$	$[^{\downarrow}_{VP}, TV]$
						V	$\Rightarrow$	$[^{\downarrow}_{V}AUX]$
						N	$\Rightarrow$	$[^{\downarrow}_{N}, ADJ]$

Figure 4. An Extended GFG

The **path** rules in *RW* assign to each pair in *BAS* an element from that pair. The elements of *SRW* provide starting positions.

In this way, all rewrite derivation steps are localized to a 'domain'.

The movement constructions 4a and 4b respectively propagate a term and incorporate it. In 4a we need the *transitive closure* of the derivation relation in order to allow terms to move into subordinate clauses of subordinate clauses. Due to the lexicalization of the derivations, the definition of the transitive closure, ' $\mid\!\cong$', is harmless. It is a nice feature of a lexicalized framework that we cannot freely generate structure under transitive closure because, without terminal elements, every derivation halts after a finite number of steps. The terminal elements that subscribe the relation constrain, as it were, the reach of the transitive closure.

All rules are 'safe' in the sense that they do not derive terms $A \in DT$ that cannot already be derived from A using context free rewrites. Moreover,

Claim IV: For $A \in DT$, $S \models A \iff S \mid\!\sim A$.

A terminal tree can be derived from the start symbol under $\mid\!\sim$ iff it can be generated by the local basis. So we haven't lost any trees by restricting to purely sequential derivations.

6 Examples

Figure 4 represents an example grammar for the set BAS previously defined in figure 3. For simplicity, we have dropped the reference to BAS-states σ, these we have incorporated by introducing a variety of start terms, each introducing a specific path strategy. This suffices for the examples, but it will not do in general.[8] We have a universal starting position ($\mathbf{S}$), a declarative form (S), an interrogative form S' and a relative clause form S_x. These distinct forms only affect the path rules that apply, that is the rewrite ordering, but in the resulting terminal tree, these distinctions cannot be discovered. For instance, interrogative and declarative sentences may derive the same terminal tree: they differ in derivation order but not in the derived products.

Note that the path basis introduces variable indexations that are absent in BAS: binding is a phenomenon that can only be really defined by including the derivation of the derived tree. There, locality can be dynamically exploited. By SRW, extraction of NP is licensed at the starting position $\mathbf{S}$, extracting of AUX at S', the question form, and of QNP at the relative clause form (see section 6.2). Once extracted, a different starting position is used (so an extraction type can only happen once). Figure 5 displays a simple declarative with and without movement of NPs. Figure 6 shows the derivation of two questions, and figure 7 shows a derivation involving a relative clause. Section 6.1 explains the indexations. The most typical feature of these derivations is that in every line of at most one terminal appears. Indentations in the examples reflect change in starting position, that is, change in word order strategy. All derivations start from $\mathbf{S}$ (this corresponds to the start strategy ρ used in definition 7).

6.1 Coindexing

As derivations proceed by composing path terms, and these are defined relative to the notion of c-command, we can expect binding to be expressible in the path grammar, although it is not expressible in the underlying context free basis. That is, the sequential derivations of a local set allow (free) indexation to be definable. In this paper, we will only give a brief description. In the grammar indices are introduced by terminal elements and propagated by the following rules : if $(Y, Z) \in NP$ then

1. $A \mid\!\sim [{}^{\downarrow}_{NP} B_x]$ implies $A \mid\!\sim (Y, B)_x$, provided $B \in DT, [B]_{NT} = Z$
2. $A \mid\!\sim [{}^{\downarrow}_{NP} B_x]$ implies $A \mid\!\sim (B, Z)_x$, provided $B \in DT, [B]_{NT} = Y$

[8]For instance, Dutch subordinate clauses are verb final, main clause are not. This requires a choice $[(V, NP)]_{NT} \Rightarrow NP$ in relative clauses and a choice $[(V, NP)]_{NT} \Rightarrow V$ in main clauses.

$$
\begin{array}{l}
\mathbf{S} \mid\!\sim_\epsilon [^{\downarrow}_{S} NP] \\
\mathbf{S} \mid\!\sim_\epsilon [^{\downarrow}_{S}[^{\downarrow}_{NP} DET]] \\
\mathbf{S} \mid\!\sim_{\mathtt{t}} [^{\downarrow}_{S}[^{\downarrow}_{NP} \mathtt{the}]] \\
\mathbf{S} \mid\!\sim_{\mathtt{t}} [^{\downarrow}_{S}(\mathtt{the}, N)] \\
\mathbf{S} \mid\!\sim_{\mathtt{t;m}} [^{\downarrow}_{S}(\mathtt{the}, \mathtt{man})] \\
\mathbf{S} \mid\!\sim_{\mathtt{t;m}} (VP, (\mathtt{the}, \mathtt{man})) \\
\mathbf{S} \mid\!\sim_{\mathtt{t;m}} ([^{\downarrow}_{VP} V], (\mathtt{the}, \mathtt{man})) \\
\mathbf{S} \mid\!\sim_{\mathtt{t;m;h}} ([^{\downarrow}_{VP} \mathtt{hits}], (\mathtt{the}, \mathtt{man})) \\
\mathbf{S} \mid\!\sim_{\mathtt{t;m;h}} ((\mathtt{hits}, NP), (\mathtt{the}, \mathtt{man})) \\
\mathbf{S} \mid\!\sim_{\mathtt{t;m;h;J}} ((\mathtt{hits}, \mathtt{John}), (\mathtt{the}, \mathtt{man})) \\
\text{“}\mathtt{the; man; hits; John}\text{”}
\end{array}
\qquad
\begin{array}{l}
\mathbf{S} \mid\!\sim_\epsilon [^{*}_{S} NP] \\
\mathbf{S} \mid\!\sim_{\mathtt{J}} [^{*}_{S} \mathtt{John}] \\
\quad S \mid\!\sim_\epsilon [^{\downarrow}_{S}[^{\downarrow}_{NP} DET]] \\
\quad S \mid\!\sim_{\mathtt{t}} [^{\downarrow}_{S}[^{\downarrow}_{NP} \mathtt{the}]] \\
\quad S \mid\!\sim_{\mathtt{t}} [^{\downarrow}_{S}(\mathtt{the}, N)] \\
\quad S \mid\!\sim_{\mathtt{t;m}} [^{\downarrow}_{S}(\mathtt{the}, \mathtt{man})] \\
\quad S \mid\!\sim_{\mathtt{t;m}} (VP, (\mathtt{the}, \mathtt{man})) \\
\quad S \mid\!\sim_{\mathtt{t;m}} ([^{\downarrow}_{VP} V], (\mathtt{the}, \mathtt{man})) \\
\quad S \mid\!\sim_{\mathtt{t;m;h}} ([^{\downarrow}_{VP} \mathtt{hits}], (\mathtt{the}, \mathtt{man})) \\
\quad S \mid\!\sim_{\mathtt{t;m;h}} ((\mathtt{hits}, NP), (\mathtt{the}, \mathtt{man})) \\
\mathbf{S} \mid\!\sim_{\mathtt{J;t;m;h}} ((\mathtt{hits}, \mathtt{John}), (\mathtt{the}, \mathtt{man})) \\
\text{“}\mathtt{John; the; man; hits}\text{”}
\end{array}
$$

Figure 5. Note that the movement of '`John`' is only reflected on the sequential side

The first two rules propagate the variable introduced by a determiner. In order to formulate locality restrictions we define

$$V(A_r) = \{x \in VAR \mid B_x \prec_A A_r\}, \qquad V(A) =_{df} V(A_p).$$

As A_p is the location at which rewriting can take place in A, $V(A)$ gives the variables of all nodes that c-command the rewrite location in A. When now an NP non-terminal is rewritten by a terminal element, then the assignment of the variable may be constrained by the variables in $V(A)$.
We set:

1. $S \mid\!\sim B$ and $B_p \Rightarrow \mathtt{John}_x \in RW$, then $S \mid\!\sim B(\mathtt{John}_x)$, provided $x \notin VAR(B)$
2. $S \mid\!\sim B$ and $B_p \Rightarrow \mathtt{he}_x \in RW$, then $S \mid\!\sim B(\mathtt{he}_x)$, provided $x \notin V(B)$
3. $S \mid\!\sim B$ and $B_p \Rightarrow \mathtt{himself}_x \in RW$, then $S \mid\!\sim B(\mathtt{himself}_x)$, provided $x \in V(B)$
4. $S \mid\!\sim B$ and $B_p \Rightarrow \mathtt{the}_x \in RW$, then $S \mid\!\sim B(\mathtt{the}_x)$, provided $x \notin V(A)$

This gives the well-known locality conditions for names, pronouns and anaphors.

6.2 Linked Structures

Now we come to a purely derivational construction, **linked trees**, interpreting relative clauses.[9] From the perspective of a path grammar the two

[9] This concept was due to Dov Gabbay in early work with the second author ([Kempson and Gabbay, 1998]), and jointly developed in [Kempson *et al.*, 2001].

$\mathbf{S} \mid\!\sim_{\epsilon} [^{*}_{S} AUX]$
$\mathbf{S} \mid\!\sim_{\mathtt{d}} [^{*}_{S}\mathtt{did}]$
$S \mid\!\sim_{\epsilon} [^{\downarrow}_{S}[^{\downarrow}_{NP} DET]]$
$S \mid\!\sim_{\mathtt{t}} [^{\downarrow}_{S}[^{\downarrow}_{NP}\mathtt{the}]]$
$S \mid\!\sim_{\mathtt{t}} [^{\downarrow}_{S}(\mathtt{the}, N)]$
$S \mid\!\sim_{\mathtt{t;m}} [^{\downarrow}_{S}(\mathtt{the}, \mathtt{man})]$
$S \mid\!\sim_{\mathtt{t;m}} (VP, (\mathtt{the}, \mathtt{man}))$
$S \mid\!\sim_{\mathtt{t;m}} ([^{\downarrow}_{VP} V], (\mathtt{the}, \mathtt{man}))$
$S \mid\!\sim_{\mathtt{t;m}} ([^{\downarrow}_{VP}[^{\downarrow}_{V} AUX]], (\mathtt{the}, \mathtt{man}))$
$S \mid\!\sim_{\mathtt{d;t;m}} ([^{\downarrow}_{VP}[^{\downarrow}_{V}\mathtt{did}]], (\mathtt{the}, \mathtt{man}))$
$\mathbf{S} \mid\!\sim_{\mathtt{d;t;m}} ([^{\downarrow}_{VP}(\mathtt{did}, V')], (\mathtt{the}, \mathtt{man}))$
$\mathbf{S} \mid\!\sim_{\mathtt{d;t;m;h}} ([^{\downarrow}_{VP}(\mathtt{did}, \mathtt{hit})], (\mathtt{the}, \mathtt{man}))$
$\mathbf{S} \mid\!\sim_{\mathtt{d;t;m;h}} (((\mathtt{did}, \mathtt{hit}), NP), (\mathtt{the}, \mathtt{man}))$
$\mathbf{S} \mid\!\sim_{\mathtt{d;t;m;h;J}} (((\mathtt{did}, \mathtt{hit}), \mathtt{John}), (\mathtt{the}, \mathtt{man}))$

$\mathbf{S} \mid\!\sim_{\epsilon} [^{*}_{S'} QNP]$
$\mathbf{S} \mid\!\sim_{\mathtt{w}} [^{*}_{S'}\mathtt{whom}]$
$S' \mid\!\sim_{\epsilon} [^{*}_{S} AUX]$
$S' \mid\!\sim_{\mathtt{d}} [^{*}_{S}\mathtt{did}]$
$S \mid\!\sim_{\epsilon} [^{\downarrow}_{S}[^{\downarrow}_{NP} DET]]$
$S \mid\!\sim_{\mathtt{t}} [^{\downarrow}_{S}[^{\downarrow}_{NP}\mathtt{the}]]$
$S \mid\!\sim_{\mathtt{t}} [^{\downarrow}_{S}(\mathtt{the}, N)]$
$S \mid\!\sim_{\mathtt{t;m}} [^{\downarrow}_{S}(\mathtt{the}, \mathtt{man})]$
$S \mid\!\sim_{\mathtt{t;m}} (VP, (\mathtt{the}, \mathtt{man}))$
$S \mid\!\sim_{\mathtt{t;m}} ([^{\downarrow}_{VP} V], (\mathtt{the}, \mathtt{man}))$
$S \mid\!\sim_{\mathtt{t;m}} ([^{\downarrow}_{VP}[^{\downarrow}_{V} AUX]], (\mathtt{the}, \mathtt{man}))$
$S' \mid\!\sim_{\mathtt{d;t;m}} ([^{\downarrow}_{VP}[^{\downarrow}_{V}\mathtt{did}]], (\mathtt{the}, \mathtt{man}))$
$S' \mid\!\sim_{\mathtt{d;t;m}} ([^{\downarrow}_{VP}(\mathtt{did}, V')], (\mathtt{the}, \mathtt{man}))$
$S' \mid\!\sim_{\mathtt{d;t;m;h}} ([^{\downarrow}_{VP}(\mathtt{did}, \mathtt{hit})], (\mathtt{the}, \mathtt{man}))$
$S' \mid\!\sim_{\mathtt{d;t;m;h}} (((\mathtt{did}, \mathtt{hit}), NP), (\mathtt{the}, \mathtt{man}))$
$\mathbf{S} \mid\!\sim_{\mathtt{w;d;t;m;h}} (((\mathtt{did}, \mathtt{hit}), \mathtt{whom}), (\mathtt{the}, \mathtt{man}))$

Figure 6. Two question forms. Noteworthy is that the first sentence gives the same dominance tree as "the;man;did;hit;John"

features that distinguish relative clauses from subordinate ones are that they are islands with respect to extraction and that they share an index with their head. The first fact precludes the movement construction to enter relative clause representations: relative clauses cannot be subtrees. As there are only trees in DT (all nodes have a unique mother), the structures must be disjoint The second fact requires the two *disjoint* terminal trees to share an index. The following rule accounts for these two facts.

> $A \mid\!\sim_{\mu} C(B_x)$ and $S_x \mid\!\sim_{\nu} D$ implies $A \mid\!\sim_{\mu;\nu} \langle C(B_x), D\rangle$ provided $B, D \in DT$, $[B]_{NT} = NP$.

This introduces a **link construction** $\langle A, B\rangle$ where $B \in DT$, $[B]_{NT} = S$, and A may be an unfinished product. Naturally, $I(\langle X, Y\rangle) = \{\langle A, B\rangle \mid A \in I(X), B \in I(B)$. We add the structural rule:

> $A \mid\!\sim \langle B, C\rangle$ and $B \mid\!\sim D$ implies $A \mid\!\sim \langle D, B\rangle$

to let derivations proceed per coordinate. The trees involved in the link are disjoint, but, by the derivation rule, a linked tree is introduced with an index from the linking tree: the variable associated with the 'head' A_x of the linking clause is transferred into the linked clause S_x. This transferred variable now forces the introduction of $[^{*}QNP_x]$ by the rule

$$S_x \mid\!\sim [^{*}_{S}\mathbf{QNP}_x],$$

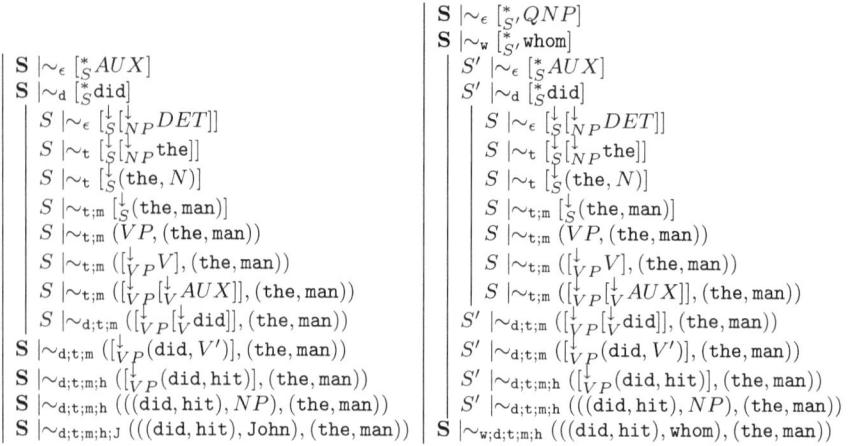

$\mathbf{S} \mid\!\sim_{\epsilon} [^{\downarrow}_{s}NP]$
$\mathbf{S} \mid\!\sim_{\epsilon} [^{\downarrow}_{S}[^{\downarrow}_{NP}DET]]$
$\mathbf{S} \mid\!\sim_{\mathtt{t}} [^{\downarrow}_{S}[^{\downarrow}_{NP}\mathtt{the}_x]]$
$\mathbf{S} \mid\!\sim_{\mathtt{t}} [^{\downarrow}_{S}[^{\downarrow}_{NP}(\mathtt{the}, N)_x]]$
$\mathbf{S} \mid\!\sim_{\mathtt{t;m}} [^{\downarrow}_{S}(\mathtt{the}, \mathtt{man})_x]$
$\quad S_x \mid\!\sim_{\epsilon} [^{*}_{S}QNP_x]$
$\quad S_x \mid\!\sim_{\mathtt{w}} [^{*}_{S}\mathtt{who}_x]$
$\quad\quad S \mid\!\sim_{\epsilon} [^{\downarrow}_{S}NP]$
$\quad S_x \mid\!\sim_{\mathtt{w}} [^{\downarrow}_{S}\mathtt{who}_x]$
$\quad S_x \mid\!\sim_{\mathtt{w}} (\mathtt{who}_x, VP)$
$\quad S_x \mid\!\sim_{\mathtt{w;w}} (\mathtt{who}_x, \mathtt{walks})$
$\mathbf{S} \mid\!\sim_{\mathtt{t;m;w;w}} \langle[^{\downarrow}_{S}(\mathtt{the}, \mathtt{man})_x], (\mathtt{who}_x, \mathtt{walks})\rangle]$
$\mathbf{S} \mid\!\sim_{\mathtt{t;m;w;w}} (\langle(\mathtt{the}, \mathtt{man})_x, VP), (\mathtt{who}_x, \mathtt{walks})\rangle$
$\mathbf{S} \mid\!\sim_{\mathtt{t;m;w;w}} ((\mathtt{the}, \mathtt{man})_x, [^{\downarrow}_{VP}V]), (\mathtt{who}_x, \mathtt{walks})\rangle$
$\mathbf{S} \mid\!\sim_{\mathtt{t;m;w;w;h}} ((\mathtt{the}, \mathtt{man})_x, [^{\downarrow}_{VP}\mathtt{hits}]), (\mathtt{who}_x, \mathtt{walks})\rangle$
$\mathbf{S} \mid\!\sim_{\mathtt{t;m;w;w;h}} ((\mathtt{the}, \mathtt{man})_x, (\mathtt{hits}, NP)), (\mathtt{who}_x, \mathtt{walks})\rangle$
$\mathbf{S} \mid\!\sim_{\mathtt{t;m;w;w;h;J}}, ((\mathtt{the}, \mathtt{man})_x, ((\mathtt{hits}, \mathtt{John})), (\mathtt{who}_x, \mathtt{walks})\rangle$

Figure 7. The derivation induced by the sentence *the man, who walks, hits John* involving a relative clause.

of our toy grammar. Thus the terminal linked tree will have to incorporate a WH-expression from T sharing an index with an element from the linking tree. This process is recursive. [10] Figure 7 gives a sample derivation involving a relative clause. Notice the transfer of the index of the head of the relative clause, $(\mathtt{the}, \mathtt{man})_x$, into the relative clause.

6.3 Cross-serial Dependencies

An interesting test case for path grammars are the so-called cross serial dependencies in German and Dutch. They form the basis for a variety of proofs that the grammatical sentences of Dutch and German cannot be generated by a context free grammar [Huybrechts, 1976], [Groenink, 1997]. In a path grammar over a CFG, on the other hand, matters are reasonably straightforward (the treatment here should be compared with, for instance, [Steedman, 2000]). Consider the sequence - dominance pairs associated to a typical sentence with cross-serial dependencies:

English order deriving "$(\mathtt{that}, (\mathtt{John}, (\mathtt{saw}, (\mathtt{Mary}, (\mathtt{help}, (\mathtt{Fred}, \mathtt{swim}))))))$":

$(\mathtt{that}; \mathtt{John}; \mathtt{saw}; \mathtt{Mary}; \mathtt{help}; \mathtt{Fred}; \mathtt{swim},$

$$(\mathtt{that}, (\mathtt{John}, (\mathtt{saw}, (\mathtt{Mary}, (\mathtt{help}, (\mathtt{Fred}, \mathtt{swim}))))))$$

[10] This fits a semantic analysis of relative clauses as *presuppositions*.

German order deriving the same dominance tree::

$(\texttt{that}; \texttt{John}; \texttt{Mary}; \texttt{Fred}; \texttt{swim}; \texttt{help}; \texttt{saw},$

$$(\texttt{that}, (\texttt{John}, (\texttt{saw}, (\texttt{Mary}, (\texttt{help}, (\texttt{Fred}, \texttt{swim}))))))$$

Dutch order for the dominance tree :

$(\texttt{that}; \texttt{John}; \texttt{Mary}; \texttt{Fred}; \texttt{saw}; \texttt{help}; \texttt{swim},$

$$(\texttt{that}, (\texttt{John}, (\texttt{saw}, (\texttt{Mary}, (\texttt{help}, (\texttt{Fred}, \texttt{swim}))))))$$

The local basis has in all cases $(NP, VP) \in S$ and $(V, S) \in VP$.

Difference in word order comes down to distinct path strategies. This part is immediate. The path choices for English are $[^{\downarrow}_{S}NP]$ and $[^{\downarrow}_{VP}V]$, whereas German and Dutch are specified by the choices $[^{\downarrow}_{S}NP]$, $[^{\downarrow}_{VP}S]$. Without further ado, this gives the German word order. For the Dutch order we are forced to take a closer look at path terms. In a path term A not only is the $\prec^{NT}_{A}$-maximum uniquely defined, but also its $\prec^{NT}_{A}$- minimum. So the non-terminals in a path term can be replaced by paths, top-down or bottom up. The dependency structure of Dutch suggests that this is a parameter that can vary across (maybe also within) languages. Let A be the the term derived by the sequence $\texttt{John}; \texttt{Mary}; \texttt{Fred}$:

$$A = (\texttt{John}, [^{\downarrow}_{VP}(\texttt{Mary}, [^{\downarrow}_{VP}[^{\downarrow}_{S}\texttt{Fred}]])]),$$

and consider the unique path term A' such that $I(A) = I(A')$

$$A' = (\texttt{John}, (V, (\texttt{Mary}, (V, (\texttt{Fred}, V))))).$$

German order we get by working down $\prec^{NT}_{A'}$, using the maxima,

$$(\texttt{John}, [^{\downarrow}_{VP}(\texttt{Mary}, [^{\downarrow}_{VP}(\texttt{Fred}, V)])]).$$

And the Dutch order by working up along the $\prec^{NT}_{A'}$, using the minimum

$$(\texttt{John}, (V, (\texttt{Mary}, [^{\downarrow}_{VP}[^{\downarrow}_{S}\texttt{Fred}]]))).$$

In order to get rules using the $\prec^{NT}_{A}$- minimum, we have to take a closer look at the index function i_A. If $A \in ADT$, then I_A takes values in $\{0,1\}^*$, that is, it assigns strings of 0's and 1's, and if $A \in PTERM$, then it takes values in $\{0, 1, \downarrow, *\}^*$. For $p, q \in \{0, 1, \downarrow, *\}^*$ we set $p < q$ iff $q = p; i, i \in \{0, 1, \downarrow, *\}$ and we let $\leq$ be the reflexive transitive closure of $<$. So for $A \in ADT$, if $r \in Ran(I_A)$ and $q \leq r$, then $q \in Ran(I_A)$, For $A \in DT$, this means that $Ran(I_A)$ is an unordered tree domain. For $A \in PTERM$ we let

$$F(A) = Ran(i^A) \cap \{0,1\}^*.$$

that is, $F(A)$ contains those node identifiers of A that have no $\downarrow$ or $*$ occurrences. Note that $F(A)$ is again closed under sub-sequences. Now consider the variants of the 3a,3b clauses of definition 7. If $[(Y,Z)]_{NT} = X$ and $E \in TPATH$, then

3a'. $A \mid\sim^{\sigma}_{\nu} B([^{\downarrow}_{X}, E])$ implies $A \mid\sim^{\sigma}_{\nu} B((E,Z))$, provided $[E]_{NT} = Y$ and $I(B) = p$ implies $p \in F(B)$,

3b. $A \mid\sim^{\sigma}_{\nu} B([^{\downarrow}_{X}, E])$ implies $A \mid\sim^{\sigma}_{\nu} B((Y,E))$, provided $[E]_{NT} = Z$ and $I(B) = p$ implies $p \in F(B)$.

There are two differences. Firstly, the constraint '$E \in DT$' is replaced by '$E \in TPATH$' (see definition 5 for this notion). Secondly, the rewrite location B_p in B has to be specified by $p \in \{0,1\}^*$. This gives the last expansion above, it prevents the application of the rule to all but the outermost one of a nested sequence of path constructs.

Of course, further analysis is required[11], but, according to us, the simplicity of such a solution speaks strongly in favour of a (derivation sequence - dominance tree) analysis of structural descriptions.

7 Conclusion

It is noteworthy that *any* context free (tree) grammar with binary branching allows for the addition of path strategies and thus sequential, lexical, derivations. Accordingly, any such context free grammar can determine families of string languages by the selection of a set of path bases for that grammar. These string languages exhibit interesting word order phenomena and allow long distance movement of constituents.

Derivational strategies correspond to word order strategies. These seem to be established mainly at the clause level. At the same level we have to define phenomena like agreement, binding and quantifier scope. Without a path derivational component, these phenomena lie mostly beyond the scope of context free grammars. The addition of a path basis to a context free grammar, however, makes them amenable to analysis. The main reason seems to be the intimate connection between paths and c-command domains. This gets us government under c-command, and thus binding domains, almost for free.

The framework of path grammars allows us to locate word order phenomena in the derivational part, independent of the dominance part. This independence can only be really appreciated by considering the dependency

[11] For instance, it seems that the path term itself should control which of the variants of rules 3a,3b should be applied. This should not be left to the grammar.

counterparts of dominance trees. This is where the structural identity that distinguishes a declarative from a question, a main clause from a relative clause, etc, is usually immediately evident.

To transfer an analysis of a linguistic phenomenon, that is formulated in terms of phrase structure trees, to the framework of path grammars, we have to ask ourselves: which features of the analysis are there to account for word-order and which features are due to the reality of dominance? Only if we can tease these apart is a reanalysis in terms of path grammars feasible; and it is the fine structure provided by such grammars that enables us to elegantly model a range of discontinuity phenomena of natural language that are otherwise problematic.

BIBLIOGRAPHY

[Abeille and Rambow, 2000] A. Abeille and O. Rambow. *Tree Adloining Grammars.* Joh Wiley & Sons Ltd, 2000.

[Chomsky, 1965] N. Chomsky. *Aspects of the Theory of Syntax.* MIT Press, 1965.

[Groenink, 1997] A. Groenink. *Surface without Structure.* PhD thesis, University of Utrecht, 1997.

[Huybrechts, 1976] R. Huybrechts. Overlapping dependencies in dutch. In *Utrecht working papers in linguistics*, pages 1:24–65. 1976.

[Kayne, 1994] R. Kayne. *The Antsymmetry of Syntax.* M.I.T press, 1994.

[Kempson and Gabbay, 1998] R. Kempson and D. Gabbay. Crossover: a unified view. *Journal of Linguistics*, 34:73–124, 1998.

[Kempson *et al.*, 2001] Ruth Kempson, Wilfried Meyer-Viol, and Dov Gabbay. *Dynamic Syntax: The Flow of Language Understanding.* Blackwell, 2001.

[Meyer-Viol, 1998] W. Meyer-Viol. Sequential composition of logical forms. In M. Moortgat, editor, *Logical Aspects of Computational Linguistics*, pages 159–178. Springer-Verlag, 1998.

[Rambow and Joshi, 1994] O. Rambow and A. Joshi. A formal look at dependency grammars and phrase-structure grammars, with special consideration of word-order phenomena. In L Wanner, editor, *Current Issues In Meaning-Text Theory.* Pinter, London, 1994.

[Reinhart, 1981] T. Reinhart. Definite np anaphora and c-command domains. *Linguistic Inquiry*, 12:605–635, 1981.

[Steedman, 2000] M. Steedman. *The Syntactic Process.* MIT Press, Cambridge, MA, 2000.

A Hierarchical Analysis of Propositional Temporal Logic Based on Intervals

BEN MOSZKOWSKI

1 Introduction

Following the seminal paper by Pnueli [Pnueli, 1977], temporal logic [Manna and Pnueli, 1981; Kröger, 1987; Emerson, 1990] has become one of the main formalisms used in computer science for reasoning about the dynamic behaviour of systems. In particular, propositional linear-time temporal logic (PTL) and some variants of it have been extensively studied and used. In a relatively recent and significant article, Lichtenstein and Pnueli [Lichtenstein and Pnueli, 2000] give a detailed analysis of PTL which is meant to largely subsume and supercede earlier ones. Indeed, the work appears to have the rather ambitious goal of coming close to offering the last word on the subject and is perhaps best described in the authors' own words:

> The paper summarizes work of over 20 years and is intended to provide a definitive reference to the version of propositional temporal logic used for the specification and verification of reactive systems.

The version of PTL considered by Lichtenstein and Pnueli has discrete time and past time. Both a decision procedure and axiomatic completeness are investigated and a new simplified axiom system is presented. The approach makes use of semantic tableaux and throughout the presentation the treatment of PTL with past-time operators runs in parallel with the future-only version. The authors choose in particular to use tableaux since they offer a basis for uniformly showing axiomatic completeness and also obtaining a practical decision procedure. The extensive material about past time is distinctly marked so that one can optionally delete it to obtain an analysis limited to the future fragment of PTL.

We present a novel framework for investigating PTL which significantly differs from the methods of Lichtenstein and Pnueli and earlier treatments such as [Gabbay *et al.*, 1980; Wolper, 1985; Kröger, 1987; Goldblatt, 1987]. It is used to obtain standard results such as a small model property, a practical decision procedure and axiomatic completeness. However, instead of relying on semantic tableaux, filtration and other previous techniques, our method is based on an interval-oriented analysis of certain kinds of low-level PTL formulas called *transition configurations*. An important feature of this approach is that it provides a natural hierarchical means of reducing full PTL to this subset and also reduces both PTL with the *until* operator and past time to versions without them. Therefore the overwhelming bulk of the analysis only needs to deal PTL with neither *until* nor past time. The low-level formulas also have associated practical decision procedures, including a simple symbolic one based on Binary Decision Diagrams (BDDs) [Bryant, 1986] which we have implemented.

Our approach extensively employs intervals of time which are represented as finite and countably infinite sequences of states and described by formulas in a propositional version of Interval Temporal Logic (ITL) [Moszkowski, 1983b; Moszkowski, 1983a; Halpern *et al.*, 1983; Moszkowski, 1985; Moszkowski, 1986] (see also [ITL, URL]). By using a hierarchical, interval-oriented framework, the approach differs from that of Lichtenstein and Pnueli and previous ones which in general utilise sets of formulas and sequences of such sets (also referred to as *paths*). We instead relate transition configurations to semantically equivalent formulas in propositional ITL. Time intervals facilitate an analysis which naturally relates larger intervals with smaller ones. The process of doing this can be explicitly expressed in ITL in a way not possible within previous frameworks which lack both a formalisation of intervals and logical operators concerning various kinds of sequential composition of intervals.

We have extended the prototype implementation of our PTL decision procedure to support finite-time analysis of an interval-based formalism called *Fusion Logic* (FL) previously used by us in [Moszkowski, 2004a]. The decision procedure for FL exploits a reduction of FL formulas to PTL ones which only works for finite time. A brief introduction to FL is given in §13.4 since FL is a natural extension of our framework for studying PTL and furthermore demonstrates another hierarchy involving both PTL and intervals. We plan in future work to give a more detailed discussion of the decision procedure for FL as well as some other issues concerning FL.

The basic version of PTL used here is described in detail in Sect. 3 but we will now briefly summarise some of the features in order to be able to overview some key aspects of our work. We postpone the treatment of *until* and past time in order to later handle them in a natural hierarchical manner. Both finite and infinite time are permitted, whereas most versions of

PTL deal solely with the latter. One reason for including finite time is to allow us to naturally capture parts of our infinite-time analysis within PTL formulas concerning finite-time subintervals. The only two primitive temporal operators initially considered are $\bigcirc$ (strong next) and $\Diamond$ (eventually) although some others are definable in terms of them (e.g., $\Box$ (henceforth)).

We believe that the interval-based analysis presented here complements existing approaches since it provides a notational way to articulate various issues in model construction which are equally relevant within a more conventional analysis but are normally only considered at the metalevel. This fits nicely with one of the main purposes of a logic which is to provide a notation for explicitly and formally expressing reasoning processes.

Our preliminary work in [Moszkowski, 2004b] contains an earlier description of this material but was limited to showing axiomatic completeness for PTL without past time. In the mean time, we have significantly extended the notation, methods and their scope of application. The structure of presentation has also been refined.

The use of intervals here seems to go well with a growing general awareness even in industry of the desirability for temporal logics which go beyond conventional point-based constructs to also handle behavioural specifications involving intervals of time. As evidence for this we mention the Property Specification Language PSL/Sugar[PSLSugar, URL]. This is a modified version of a language Sugar [Beer *et al.*, 2001] developed at IBM/Haifa. PSL/Sugar has been ratified as an international standard for precisely expressing a hardware system's design properties so that they can then be tested using simulation and model checking. It includes a temporal logic with regular expressions and other operators for sequential composition. In addition, the IEEE Design Automation Standards Committee has recently approved a project to produce a candidate standard for Verisity Ltd.'s [Verisity, URL] *e* language which is intended for testing and verification[1]. A subset of *e* called *temporal e* was influenced in part by ITL [Morley, 1999; Hollander *et al.*, 2001; Verisity Ltd., 2003]. The IEEE Standards Association has assigned the project the number 1647 [IEEE1647, URL].

Structure of Presentation

Let us now summarise the structure of the rest of this paper. Section 2 mentions some related work and gives a comparison with our approach. Section 3 presents the version of PTL we use. Section 4 summarises the propositional version of ITL which we use in the analysis. Section 5 introduces low level PTL formulas called *transition configurations* and relates them to some semantically equivalent propositional ITL formulas which simplify

[1]Verisity has been acquired by Cadence Design Systems [Cadence, URL].

the subsequent analysis. Section 6 proves the existence of small models for transition configurations. Section 7 shows how to relate the satisfiability of the two main kinds of transition configurations with simple interval-oriented tests. Section 8 deals with a practical BDD-based decision procedure for transition configurations. Section 9 concerns axiomatic completeness for an important subset of PTL in which the only temporal operator is *Next* ($\bigcirc$). Section 10 looks at a PTL axiom system and axiomatic completeness for transition configurations. Section 11 presents formulas called *invariants* and *invariant configurations* which together serve as a bridge between the previously mentioned transition configurations and arbitrary PTL formulas. Section 12 discusses how to generalise the previous results to work with arbitrary PTL formulas. Section 13 hierarchically extends our approach to deal with both the temporal operators *until* and past time and also briefly looks at a superset of PTL called *Fusion Logic.* Section 14 concludes with some brief discussion.

2 Background

Temporal logics have become a popular topic of study in theoretical computer science and are also being utilised by industry to locate faults in digital circuit designs, communication protocols and other applications. Issues such as small models, proof systems, axiomatic completeness and decision procedures for PTL (almost always limited to infinite time) have been extensively investigated by Gabbay et al. [Gabbay *et al.*, 1980], Wolper [Wolper, 1985], Kröger [Kröger, 1987], Goldblatt [Goldblatt, 1987], Lichtenstein and Pnueli [Lichtenstein and Pnueli, 2000], Lange and Stirling [Lange and Stirling, 2001], Pucella [Pucella, 2005] (who also considers PTL with finite time) and others. French[French, 2000] elaborates on the presentation by Gabbay et al. [Gabbay *et al.*, 1980].

Vardi and Wolper [Vardi and Wolper, 1986] and Bernholtz, Vardi and Wolper [Bernholtz *et al.*, 1994] describe decisions procedures for various temporal logics based on a reduction to ω-automata. They do not consider axiomatic completeness. Wolper [Wolper, 2001] presents a tutorial on such decision procedure for PTL with infinite time.

Ben-Ari et al. [Ben-Ari *et al.*, 1981; Ben-Ari *et al.*, 1983], Wolper [Wolper, 1981; Wolper, 1983] and Banieqbal and Barringer [Banieqbal and Barringer, 1986] develop closely related proofs of completeness for logics which include PTL as a subset or are branching-time versions of it. The book by Rescher and Urquhart [Rescher and Urquhart, 1971] is an early source of tableau-based completeness proofs for temporal logics. The survey by Emerson [Emerson, 1990] includes material about axiom systems for both linear and branching-time temporal logic.

Fisher [Fisher, 1992; Fisher, 1997] (see also later work by Fisher, Dixon and Peim [Fisher *et al.*, 2001] and Bolotov, Fisher and Dixon [Bolotov *et al.*, 2002]) presents a normal form for PTL called *Separated Normal Form* (SNF) which consists of formulas having the syntax $\Box \bigwedge_i A_i$, where each A_i can be one of the following:

$$\mathbf{start} \supset \bigvee_c l_c \qquad \bigcirc\bigwedge_a k_a \supset \bigcirc\bigvee_d l_d \qquad \bigcirc\bigwedge_b k_b \supset \Diamond l\ .$$

Here each particular k_a, k_b, l, l_c and l_d is a literal (i.e., a proposition variable or its negation). Some versions of SNF permit past-time constructs or have other relatively minor differences. Applications include theorem proving, executable specifications and representing ω-automata. We mention SNF here since it is a PTL normal form which somewhat resembles what we call invariants and formally introduce in Sect. 11.

3 Overview of PTL

This section summarises the basic version of PTL used here. Later on in Sect. 13 we augment PTL with the operator *until* and past time.

3.1 Syntax of PTL

We now describe the syntax of permitted PTL formulas. In what follows, p is any propositional variable and both X and Y denote PTL formulas:

$$p \qquad \mathit{true} \qquad \neg X \qquad X \vee Y \qquad \bigcirc X \text{ (“strong next”)} \qquad \Diamond X \text{ (“eventually”)}\ .$$

We include *true* as a primitive so as to avoid a definition of it which contains some specific variable. This is not strictly necessary. Other conventional logic operators such as *false*, $X \wedge Y$ and $X \supset Y$ (X implies Y) are defined in the usual way. Also, $\Box X$ (“henceforth”) is defined as $\neg\Diamond\neg X$.

3.2 Semantics of PTL

The version of PTL considered here uses discrete, linear time which is represented by intervals each consisting of a sequence of one or more states. More precisely, an interval σ is any finite or infinite sequence of one or more states $\sigma_0, \sigma_1, \ldots$. Each state σ_i in σ maps each propositional variable p, $q, \ldots$ to one of the boolean values *true* and *false*. The value of p in the state σ_i is denoted $\sigma_i(p)$. A finite interval σ has an *interval length* $|\sigma| \geq 0$ which equals the number of states minus 1 and is hence always greater than or equal to 0. We regard the smallest nonzero interval length 1 as a *unit* of (abstract) time. For example, an interval with 6 states has interval length 5 or equivalently 5 time units. These units do not correspond to any particular notion of physical time. The interval length of an infinite interval is taken to be ω. The term *subinterval* refers to any interval obtained from some *contiguous* subsequence of another interval's states.

We call a one-state interval (i.e., interval length 0) an *empty interval.* A two-state interval (i.e., interval length 1) is called a *unit interval.* Both kinds of intervals play an important role in our analysis.

The notation $\sigma \models X$ denotes that the interval σ *satisfies* the PTL formula X. We now give a definition of this using induction on X's syntax:

- Propositional variable: $\sigma \models p$ iff p is true in the initial state σ_0 (i.e., $\sigma_0(p) = \mathit{true}$).
- True: $\sigma \models \mathit{true}$ trivially holds for any σ.
- Negation: $\sigma \models \neg X$ iff $\sigma \not\models X$.
- Disjunction: $\sigma \models X \vee Y$ iff $\sigma \models X$ or $\sigma \models Y$.
- Next: $\sigma \models \bigcirc X$ iff $\sigma' \models X$, where σ contains at least two states and σ' denotes the suffix subinterval $\sigma_1\sigma_2\ldots$ which starts from second state σ_1 in σ.
- Eventually: $\sigma \models \Diamond X$ iff $\sigma' \models X$, for some suffix subinterval σ' of σ (perhaps σ itself).

Table 1 shows a variety of other useful temporal operators which are definable in PTL. It includes operators for testing whether an interval is finite or infinite and whether the interval has exactly one state or two states. Most of the operators only become relevant when finite intervals are permitted. Therefore, readers who are just familiar with conventional PTL and infinite time will have previously encountered only a few of the operators. Figure 1 illustrates the use of some of these operators with formulas and sample intervals. In the figure, the logical values *true* and *false* are respectively abbreviated as "`t`" and "`f`". In what follows, we frequently use $\boxed{\scriptstyle m}$ instead of $\Box$ since we need to test pairs of adjacent states in a interval. The operator $\boxed{\scriptstyle m}$ is better suited for this since it does not "run off the end" when examining finite intervals. The fourth example in Figure 1 serves as an example of this feature. As a consequence, $\boxed{\scriptstyle m}$ is easier to work with in our interval-based analysis as is later shown in Theorem 18.

DEFINITION 1 (Satisfiability and Validity) *For any interval σ and* PTL *formula X, if σ satisfies X (i.e., $\sigma \models X$ holds), then X is said to be* satisfiable, *denoted as $\dashv X$. A formula X satisfied by all intervals is* valid, *denoted as $\models X$.*

Standard derived PTL *operators:*		
$\Box X \stackrel{\text{def}}{\equiv} \neg\Diamond\neg X$		Henceforth
$\Diamond^+ X \stackrel{\text{def}}{\equiv} \bigcirc\Diamond X$		Eventually in strict future
$\Box^+ X \stackrel{\text{def}}{\equiv} \neg\Diamond^+\neg X$		Henceforth in strict future (not used here)
PTL *operators primarily for finite intervals:*		
$\mathit{more} \stackrel{\text{def}}{\equiv} \bigcirc\,\mathit{true}$		More than one state
$\mathit{empty} \stackrel{\text{def}}{\equiv} \neg\mathit{more}$		Only one state (*empty interval*)
$ⓦ X \stackrel{\text{def}}{\equiv} \neg\bigcirc\neg X$		Weak next (same as $\mathit{more} \supset \bigcirc X$)
$\mathit{skip} \stackrel{\text{def}}{\equiv} \bigcirc\,\mathit{empty}$		Exactly two states (*unit interval*)
$X? \stackrel{\text{def}}{\equiv} X \wedge \mathit{empty}$		Empty interval with test
$\$X \stackrel{\text{def}}{\equiv} X \wedge \mathit{skip}$		Unit interval with test
PTL *operators for finite and infinite intervals:*		
$\mathit{finite} \stackrel{\text{def}}{\equiv} \Diamond\,\mathit{empty}$		Finite interval
$\mathit{inf} \stackrel{\text{def}}{\equiv} \neg\mathit{finite}$		Infinite interval
$\mathit{sfin}\ X \stackrel{\text{def}}{\equiv} \Diamond(\mathit{empty} \wedge X)$		Strong test of final state
$\mathit{fin}\ X \stackrel{\text{def}}{\equiv} \Box(\mathit{empty} \supset X)$		Weak test of final state
$\Diamond_m X \stackrel{\text{def}}{\equiv} \Diamond(\mathit{more} \wedge X)$		Sometime before the very end
$\Box_m X \stackrel{\text{def}}{\equiv} \Box(\mathit{more} \supset X)$		Henceforth except perhaps at very end

Table 1. Some definable PTL operators

Formula	Interval
$skip \wedge sfin\ \neg p$	p: t f
$\bigcirc \$(p \supset \bigcirc \neg p) \wedge \neg \$(p \wedge \bigcirc p)$	p: t t f
$\Diamond_{\mathrm{m}}\, p \wedge \neg \Diamond_{\mathrm{m}} \neg p \wedge \Diamond p \wedge \Diamond \neg p$	p: t t t f
$\Box_{\mathrm{m}}(p \supset \bigcirc \neg p) \wedge \neg \Box(p \supset \bigcirc \neg p)$	p: t f t f f t
$\Box_{\mathrm{m}}(p \supset \Diamond \neg p) \wedge sfin\ p$	p: t t t t f t

Figure 1. Some examples of formulas with derived PTL operators

DEFINITION 2 (Tautologies) *A* tautology *is any formula which is a substitution instance of some valid nonmodal propositional formula.*

For example, the formula $\bigcirc X \vee \Diamond Y \supset \Diamond Y$ is a tautology since it is a substitution instance of the valid nonmodal formula $\models p \vee q \supset q$. It is not hard to show that all tautologies are themselves valid since intuitively a tautology is any valid formula which does not require modal reasoning to justify its truth.

Convention for variables denoting individual formulas and sets of formulas: In what follows, the variable w refers to a *state formula*, that is, a formula with no temporal operators. Furthermore, PROP denotes the set of all state formulas. For any finite set of variables V, PROP_V denotes the set of all state formulas only having variables in V. Likewise, the set PTL_V denotes the set of all formulas in PTL only containing variables in V. For example, the formula $p \wedge \Diamond q$ is in $\mathrm{PTL}_{\{p,q\}}$ but not in $\mathrm{PTL}_{\{p\}}$.

3.3 Example of the Hierarchical Process

Our analysis of PTL reduces arbitrary PTL formulas to lower level ones with a much more restricted syntax. The next PTL formula serves as a simple example to motivate some of the notation and conventions later introduced:

$$\Box \Diamond p \wedge \Box \Diamond \neg p \ .$$

This is reducible to the formula $\Box I \wedge w$, where I and w are given below:

$$I: \ (r_1 \equiv \Diamond p) \wedge (r_2 \equiv \Diamond \neg r_1) \wedge (r_3 \equiv \Diamond \neg p) \wedge (r_4 \equiv \Diamond \neg r_3)$$
$$w: \ \neg r_2 \wedge \neg r_4 \ .$$

The auxiliary variables $r_1, \ldots, r_4$ are used to avoid any nesting of temporal operators in I. We can the conjunction I an *invariant* and the conjunction $\Box I \wedge w$ an *invariant configuration.* Both are formally introduced later in Sect. 11. It can be shown that the original formula $\Box \Diamond p \wedge \Box \Diamond \neg p$ is satisfiable iff the invariant configuration $\Box I \wedge w$ is.

When analysing behaviour in finite time, we further transform the invariant configuration $\Box I \wedge w$ to another special kind of conjunction $\Box T \wedge w \wedge$ *finite*, where T and w are as follows:

$$\begin{aligned} T: \quad & (r_1 \equiv (p \vee \bigcirc r_1)) \wedge (r_2 \equiv (\neg r_1 \vee \bigcirc r_2)) \\ & \wedge (r_3 \equiv (\neg p \vee \bigcirc r_3)) \wedge (r_4 \equiv (\neg r_3 \vee \bigcirc r_4)) \\ w: \quad & \neg r_2 \wedge \neg r_4 \ . \end{aligned}$$

We call T a *transition formula* and $\Box T \wedge w$ a *transition configuration* (formally defined in Section 5). Note that T contains no $\Diamond$ constructs. In fact, T is a formula in the important subset of PTL called NL^1 (later defined in Definition 6) in which the only temporal constructs are $\bigcirc$ operators not nested within other $\bigcirc$ operators. It can be shown that the original formula $\Box \Diamond p \wedge \Box \Diamond \neg p$ is satisfiable in finite time iff the transition configuration $\Box T \wedge w \wedge$ *finite* is satisfiable. As we later show in Sect. 5, formulas in NL^1 play a fundamental role in our analysis of transition configurations.

3.4 Notation for Accessing Parts of Conjunctions

From the examples just given it can be seen that we often manipulate formulas which are conjunctions. The next three definitions provide some helpful notation for denoting the number of conjuncts of such a formula and for accessing one or more of them.

DEFINITION 3 (Size of a Conjunction) *For any conjunction C of zero or more conjuncts, let the notation $|C|$ denote the number of C's conjuncts.*

DEFINITION 4 (Indexing of a Conjunction's Conjuncts) *For each $k : 1 \leq k \leq |C|$, we let $C[k]$ denote the k-th conjunct.*

Observe that if a conjunction C has length $|C| = 0$, there are no conjuncts to be indexed.

DEFINITION 5 (Parts of a Conjunction) *Suppose C is a conjunction and k and l are natural numbers such that $1 \le k \le |C|$ and $0 \le l \le |C|$. The notation $C[k : l]$ denotes the conjunction of consecutive conjuncts in C starting with $C[k]$ and finishing with $C[l]$, inclusive, i.e., $C[k] \wedge \cdots \wedge C[l]$ (which contains $l - k + 1$ conjuncts).*

Note that for any conjunction C, the formula $C[1 : 0]$ denotes *true* and $C[1 : |C|]$ is identical to C. Also, for any $k : 1 \le k \le |C|$, both $C[k]$ and $C[k : k]$ refer to the same conjunct.

3.5 Next Logic

We now define an important subset of PTL involving the operator $\bigcirc$:

DEFINITION 6 (Next Logic) *The set of* PTL *formulas in which the only primitive temporal operator is $\bigcirc$ is called* Next-Logic (NL). *The subset of* NL *in which no $\bigcirc$ is nested within another $\bigcirc$ is denoted as* NL^1.

For example, the NL formula $p \wedge \bigcirc q$ is in NL^1, whereas the NL formula $p \wedge \bigcirc(q \vee \bigcirc p)$ is not.

The variables T, T' and T'' denote formulas in NL^1.

Let NL^1_V be the set of all formulas in NL^1 only having variables in V.

An NL^1 formula cannot probe past an interval's second state. Hence, if two nonempty intervals share the same first two states, then the truth value of an NL^1 formula T for both intervals is identical. The next lemma describes this more precisely and is later used in the proof of the important Theorem 18.

LEMMA 7 *Let σ and σ' be two nonempty intervals which share the same first two states (i.e., $\sigma_0 = \sigma'_0$ and $\sigma_1 = \sigma'_1$). Then, for any formula T in NL^1, σ satisfies T iff σ' satisfies T.*

Proof. Induction on T's syntax ensures that it cannot distinguish between σ and σ'. ∎

LEMMA 8 *The following are equivalent for any NL^1 formula T:*

(a) The formula T is satisfiable in some nonempty interval.

(b) The formula skip $\wedge$ T is satisfiable.

Proof. *(a) $\Rightarrow$ (b):* Suppose some nonempty interval σ satisfies the formula T. Now σ contains at least two states. Let σ' denote the subinterval consisting the first two states in σ. Now σ' satisfies the formula *skip*. Furthermore, the formula T is in NL^1. Lemma 7 consequently ensures that the interval σ', like σ, satisfies the formula T because both two intervals share the same first two states. Therefore σ' satisfies the formula $skip \wedge T$.

(b) $\Rightarrow$ *(a)*: If some interval σ satisfies the PTL formula $skip \wedge T$, then σ is clearly nonempty and also satisfies T. ■

4 Propositional Interval Temporal Logic

We now describe the version of quantifier-free propositional ITL (PITL) used here for systematically analysing transition configurations. More on ITL can be found in [Moszkowski, 1983b; Moszkowski, 1983a; Halpern *et al.*, 1983; Moszkowski, 1985; Moszkowski, 1986; Moszkowski, 1994; Moszkowski, 1998; Moszkowski, 2000; Moszkowski, 2004a] (see also [ITL, URL]). The same discrete-time intervals are used as in PTL. In addition, all PTL constructs are permitted as well as two other ones. Hence, any PTL formula is also a PITL formula.

Here is the syntax of PITL's two extra constructs, where A and B are themselves PITL formulas:

$$A; B \ (chop) \qquad A^* \ (chop\text{-}star) \ .$$

The semantics of the other constructs in PITL is as in PTL and is therefore omitted here.

Before defining the semantics of chop and chop-star, we introduce some notation for describing subintervals of an interval σ. For natural numbers i, j with $i \leq j \leq |\sigma|$, let $\sigma_{i:j}$ denotes the subinterval having interval length $j - i$ (i.e., $j - i + 1$ states) and with starting state σ_i and final state σ_j. Furthermore, if σ is an infinite interval, let $\sigma_{i:\omega}$ denote that (infinite) suffix subinterval starting with state σ_i.

The formula $A; B$ is true on σ (i.e., $\sigma \models A; B$) iff one of the following holds:

- For some natural number $i : 0 \leq i \leq |\sigma|$, the interval σ can be divided into two subintervals $\sigma_{0:i}$ and $\sigma_{i:|\sigma|}$ sharing the state σ_i such that both $\sigma_{0:i} \models A$ and $\sigma_{i:|\sigma|} \models B$ hold.
- The interval σ itself has infinite length and $\sigma \models A$ holds.

The formula A^* is true on σ (i.e., $\sigma \models A^*$) iff one of the following holds:

- The interval σ has finite length and there exists some natural number $n \geq 0$ and finite sequence of natural numbers $l_0 \leq l_1 \leq \cdots \leq l_n$ where $l_0 = 0$ and $l_n = |\sigma|$, such that for each $i : 0 \leq i < n$, $\sigma_{l_i:l_{i+1}} \models A$ holds.

$\Diamond\!\!\!\!\!\cdot\, A$ ($\Diamond_i A$)	$\stackrel{\text{def}}{\equiv}$	$A;\, \mathit{true}$	A is true in some initial subinterval
$\boxed{i}\, A$	$\stackrel{\text{def}}{\equiv}$	$\neg \Diamond_i \neg A$	A is true in all initial subintervals
$\Diamond_a A$	$\stackrel{\text{def}}{\equiv}$	$\mathit{finite};\, A;\, \mathit{true}$	A is true in some subinterval
$\boxed{a}\, A$	$\stackrel{\text{def}}{\equiv}$	$\neg \Diamond_a \neg A$	A is true in all subintervals

Table 2. Some useful derived PITL operators

The behaviour of chop-star on empty intervals is a frequent source of confusion and it is therefore important to note that any formula A^* (including *false**) is true on a one-state interval. This is because in the semantics of chop-star for a one-state interval we can always set $n = 0$ and therefore ignore the values of variables in the interval σ.

- The interval σ has infinite length and there exists some $n \geq 0$ and finite sequence of natural numbers $l_0 \leq l_1 \leq \cdots \leq l_n$ where $l_0 = 0$, such that for each $i : 0 \leq i < n$, $\sigma_{l_i:l_{i+1}} \models A$ holds and also $\sigma_{l_n:\omega} \models A$ holds.
- The interval σ has infinite length and there exists some countably infinite strictly ascending sequence of natural numbers $l_0 < l_1 < \cdots$ where $l_0 = 0$, such that for each $i : i \geq 0$, $\sigma_{l_i:l_{i+1}} \models A$ holds.

Figure 2 pictorially illustrates the semantics of *chop* and *chop-star* in both finite and infinite time and also shows some simple PITL formulas together with intervals which satisfy them. For some sample formulas we include in parentheses versions using conventional PTL logic operators which were previously introduced in Sect. 3.

We make use of the following definitions of two straightforward forms of iteration expressible with chop and chop-star:

$$A^+ \stackrel{\text{def}}{\equiv} A; A^* \qquad A^\omega \stackrel{\text{def}}{\equiv} (A \wedge \mathit{finite})^* \wedge \mathit{inf} \ .$$

In addition, for any $n \geq 0$, we define A^n to be the formula *empty* if $n = 0$ and otherwise to be $A; A^{n-1}$. The constructs $A^{\leq n}$ and $A^{<n}$ are defined to be the disjunctions $\bigvee_{k \leq n} A^k$ and $\bigvee_{k<n} A^k$, respectively.

Other derived operators are also possible. Table 2 shows some especially useful ones.

The notions of satisfiability and validity already introduced in Definition 1 for PTL naturally generalise to PITL.

Let PITL_V be the set of all PITL formulas only having variables in V.

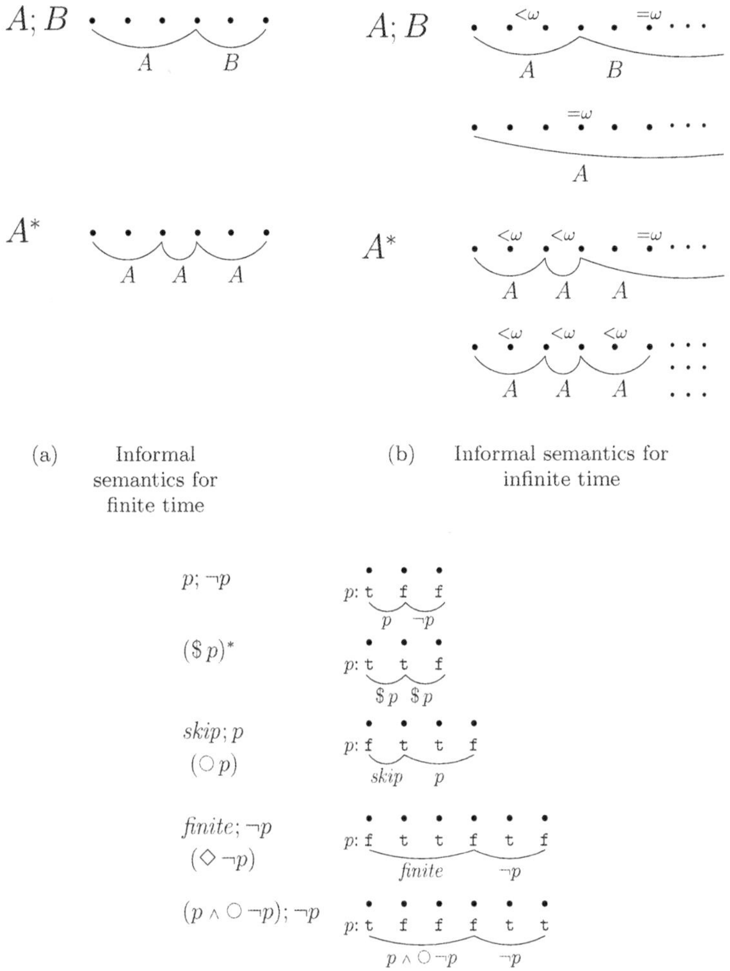

(a) Informal semantics for finite time

(b) Informal semantics for infinite time

(c) Some finite-time examples

Figure 2. Informal PITL semantics and examples

The operators chop and chop-star provide a way to decompose a formula about an interval into formulas concerning subintervals. We now define an operator on intervals which in a sense does the opposite of what chop does. It takes a pair of suitable intervals and sequentially joins them together. This later facilitates the construction of an interval satisfying a suitable chop formula from intervals satisfying the formula's operands.

DEFINITION 9 (Fusion) *Let* σ *and* σ' *be two intervals. The definition of the* fusion *of them, denoted* $\sigma \circ \sigma'$, *has two cases, depending on whether* σ *has finite length or not:*

- *If* σ *has finite length, we require that last state of* σ *equals the first state of* σ'. *The fusion of the* σ *with* σ' *is then the interval obtained by appending the two intervals together so as to include only one copy of the shared state.*
- *Otherwise, the fusion is* σ *itself, no matter what* σ' *is.*

For example, suppose s_1, s_2 and s_3 are states. If σ is the interval $s_1 s_2$ and σ' is the interval $s_2 s_3$, then their fusion $\sigma \circ \sigma'$ equals the three-state interval $s_1 s_2 s_3$, rather than the four-state interval $s_1 s_2 s_2 s_3$ which concatenation yields. Note that when σ has finite length and σ and σ' do not share the relevant state, then their fusion is undefined. If both σ and σ' are finite and compatible, then the interval $\sigma \circ \sigma'$ contains the total sum of states in σ and σ' minus one. Hence the interval length of $\sigma \circ \sigma'$ equals the sum of the interval lengths of σ and σ'. Pratt first defined fusion for describing the semantics of a process logic [Pratt, 1979] and called it *fusion product.*

It is worth comparing chop and fusion. Fusion is a general operation definable for such things as strings (i.e., sequences of letters) or intervals (i.e., sequences of states). As used here, it starts with two suitable intervals and joins them together. In contrast, chop is a logical operator which starts with an overall interval and then tests for the existence of a way to split it into two fusible subintervals. Furthermore, the semantics of the chop operator can be defined using fusion, whereas fusion is for our purposes a semantic concept, not a logical construct.

Here is a lemma relating chop with fusion:

LEMMA 10 *A* PITL *formula* $A; B$ *is satisfiable iff there exist two intervals* σ *and* σ' *such that the fusion of them* $\sigma \circ \sigma'$ *is defined and one of the following is true:*

- *The interval* σ *has finite length, it satisfies* A *and the interval* σ' *satisfies* B.

- *The interval σ has infinite length and it satisfies A.*

This lemma provides a way to reduce the problem of constructing an interval satisfying $A; B$ to that of constructing intervals satisfying A and B.

The next definition introduces a special kind of state formula which is indispensable for interval-based reasoning:

DEFINITION 11 (Atoms and V-Atoms) *An* atom *is any finite conjunction in which each conjunct is some propositional variable or its negation and no two conjuncts share the same variable. The set of all atoms is denoted $Atoms$. The Greek letters α, β and γ denote individual atoms. For any finite set of propositional variables V, let $Atoms_V$ be some set of $2^{|V|}$ logically distinct atoms containing exactly the variables in V. We refer to such atoms as* V-atoms.

For example, we can let $Atoms_{\{p,q\}}$ be the set of the four logically distinct atoms shown below:

$$p \wedge q \qquad p \wedge \neg q \qquad \neg p \wedge q \qquad \neg p \wedge \neg q \ .$$

One simple convention is to assume that the propositional variables in an atom occur from left to right in lexical order. For any finite set of variables V, this immediately leads to a suitable set of $2^{|V|}$ different V-atoms.

The next definition of a notion of canonical states and intervals together with the subsequent Lemma 13 will be extensively utilised to facilitate reasoning about intervals.

DEFINITION 12 (Canonical States and Intervals) *For any finite set of variables V and state s, we say that s is a V-*state *if s assigns each variable* not *in V the value false.*

*Similarly, for any finite set of variables V and interval σ, we say that σ is a V-*interval *if σ's states all assign each variable* not *in V the value false.*

Furthermore, for any set of variables V, we can denote a finite V-state by the unique V-atom which the state satisfies. In addition, a V-interval can be denoted the unique sequence of V-atoms associated with its V-states.

For example, for any V-atoms α and β, the two-atom sequence $\alpha\beta$ denotes a finite V-interval with V-states denoted by α and β, respectively. Hence, $\alpha\beta \models X$ denotes that the two-state V-interval $\alpha\beta$ satisfies the formula X. If X is in PTL_V, then $\alpha\beta \models X$ holds iff the conjunction $\alpha \wedge \bigcirc\beta? \wedge X$ is satisfiable. Furthermore a single V-atom can be regarded as a one-state V-interval. For example, $\alpha \models X$ denotes that the one-state V-interval α satisfies X. For any X in PTL_V, this is the case iff the conjunction

$\alpha \wedge X \wedge empty$ is satisfiable. Similarly, the notation $\alpha\beta\alpha \models X$ denotes that the V-interval $\alpha\beta\alpha$, which has two identical states, satisfies the formula X.

Recall that PITL_V denotes the set of all formulas in PITL only containing variables in V. The next lemma ensures that any satisfiable PITL_V formula is satisfied by some V-interval.

LEMMA 13 *An interval σ satisfies a* PITL_V *formula A iff there exists a V-interval with the same number of states, agrees with σ on the values of the variables in V and moreover satisfies A.*

Proof. Let σ' be the V-interval obtained from σ by setting all variables not in the set V to *false* in each state. The semantics in PITL of A ignores such variables. ■

The following lemma employs V-atoms and the PTL construct *finite* to express a simple sufficient condition which ensures that any two intervals which respectively satisfy the two parts of a chop formula with a particular syntax given in the lemma can be fused together into an interval which satisfies the overall chop formula.

LEMMA 14 *For any V-atom α and* PITL_V *formulas A and B, the following are equivalent:*

(a) The formula $(A \wedge finite); (\alpha \wedge B)$ is satisfiable.

(b) The formulas $A \wedge sfin\ \alpha$ and $\alpha \wedge B$ are satisfiable.

Proof. $(a) \Rightarrow (b)$: If some interval σ satisfies the formula $(A \wedge finite); (\alpha \wedge B)$, then by the semantics of chop there exist two subintervals of σ denoted here as σ' and σ'' such that the subinterval σ' satisfies $A \wedge finite$ and moreover if σ' has finite length, then σ'' satisfies $\alpha \wedge B$. The right subformula *finite* in $A \wedge finite$ ensures that σ' is indeed finite and therefore σ'' does satisfies $\alpha \wedge B$.

$(b) \Rightarrow (a)$: If the two formulas $A \wedge sfin\ \alpha$ and $\alpha \wedge B$ are satisfiable, then by Lemma 13 some V-intervals σ and σ' satisfy them. Now σ is finite due to the subformula $sfin\ \alpha$. Also, the last state of σ and the first state of σ' both equal the V-state denoted by the V-atom α. Hence σ and σ' can be fused and the fusion $\sigma \circ \sigma'$ satisfies the formula $(A \wedge finite); (\alpha \wedge B)$. ■

Here is a lemma similar to Lemma 14 for later use with chop-omega:

LEMMA 15 *For any V-atom α and* PITL_V *formula A, the following are equivalent:*

- *The formula* $(\alpha \wedge A)^{\omega}$ *is satisfiable.*
- *The formula* $\alpha \wedge A \wedge \bigcirc\, \mathit{sfin}\ \alpha$ *is satisfiable.*

Like Lemma 14, Lemma 15 can be proved using V-intervals. In Lemma 15 we use the subformula $\bigcirc\, \mathit{sfin}\ \alpha$ instead of $\mathit{sfin}\ \alpha$ since $\alpha \wedge A \wedge \mathit{sfin}\ \alpha$ might only be satisfiable on empty intervals. The fusion of ω copies of such an empty interval together would itself be an empty interval and clearing not satisfy the formula $(\alpha \wedge A)^{\omega}$ which requires an ω-interval.

5 Transition Configurations

Starting with a finite set of variables V, an NL^1_V formula T and a state formula *init* in PROP_V, we consider small models, a decision procedure and axiomatic completeness for certain low-level formulas referred to here as *transition configurations*. These formulas play a central role in our approach. The analysis of arbitrary PTL formulas can be ultimately reduced to that of transition configurations.

Before actually formally defining transition configurations, we need to introduce the concept of a *conditional liveness formula* which is a specific kind of conjunction necessary for reasoning about liveness properties involving infinite time. The definition therefore makes use of some general notation already introduced in Definitions 3–5 for manipulating conjunctions.

DEFINITION 16 (Conditional Liveness Formulas) *A* conditional liveness formula L *is a conjunction of* $|L|$ *implications* $L[1] \wedge \cdots \wedge L[|L|]$. *Each implication has the form* $w \supset \Diamond_{m}\, w'$, *where* w *and* w' *are two state formulas. For convenience, we let* $\eta_{L[k]}$ *denote the left operand of the* k*-th implication in* L. *Similarly,* $\theta_{L[k]}$ *denotes the operand of the* $\Diamond_{m}$ *formula in the* k*-th implication* $L[k]$*'s right side. Therefore, for each* $k : 1 \leq k \leq |L|$, *the implications* $L[k]$ *and* $\eta_{L[k]} \supset \Diamond_{m}\, \theta_{L[k]}$ *denote the same formula.*

For any V*-atom* α, *any* L *in* PTL_V *and any* $k : 1 \leq k \leq |L|$, *if the formula* $\alpha \wedge \eta_{L[k]}$ *is satisfiable, we say that* α enables L*'s* k*-th implication* $L[k]$.

Here is a sample conditional liveness formula:

$$(1) \quad ((p \vee \neg q) \supset \Diamond_{m} \neg p) \quad \wedge \quad (q \supset \Diamond_{m}(p \equiv \neg q)) \quad \wedge \quad (\mathit{true} \supset \Diamond_{m}(p \supset q))\ .$$

Note that $\Diamond_{m}$ behaves the same as $\Diamond$ on infinite intervals so the effect in formulas concerning infinite time is the same. However, in finite intervals $\Diamond_{m}$, like its dual $\Box_{m}$, ignores the final state. In principle, either $\Diamond_{m}$ or $\Diamond$ can be used in conditional liveness formulas and the choice between them appears to be largely a matter of taste. Nevertheless, we prefer $\Diamond_{m}$ in part because it facilitates an interesting generalisation of both conditional liveness formulas and another kind of formula called an *invariant* which is introduced later in Sect. 11. This generalisation will be mentioned in §13.3.

Here is the definition of transition configurations:

DEFINITION 17 (Transition Configurations) *A* transition configuration *is a formula of the form* $\Box T \wedge X$*, where the formula* T *is in* NL^1_V*, and the* PTL_V *formula* X *has one of the four forms shown below:*

Type of transition configuration	*Syntax of X*
Finite-time	$init \wedge finite$
Infinite-time	$init \wedge \Box \Diamond^+ L$
Final	$w \wedge empty$
Periodic	$\alpha \wedge L \wedge \Box \Diamond^+(\alpha \wedge L)$

Here init is a state formula in PROP_V *which corresponds to some* initial *condition,* w *is some state formula in* PROP_V*,* L *is a conditional liveness formula in* PTL_V *and* α *is a* V*-atom. If init is the formula true, it can be omitted. The same applies with* w*.*

For example, the conjunction $\Box(more \supset (p \equiv \bigcirc p)) \wedge p \wedge finite$ is a finite-time transition configuration which is true exactly for finite intervals in which p is always true.

Note: In the course of analysing transition configurations, we will assume that V, T, *init* and L are fixed.

We will show that finite-time and infinite-time transition configurations are equivalent to certain PITL_V formulas for which we can more readily establish such things as the existence of periodic models, small models, a decision procedure and axiomatic completeness. Table 3 shows the corresponding PITL_V formula for each kind of transition configuration and where the equivalence of the two is proved. Here $\vec{V} \leftarrow \vec{V}$ denotes that the initial value of each variable occurring in the set of variables V equals its final value. It can be expressed as the PTL_V formula $finite \supset \bigwedge_{v \in V}(v \equiv fin\ v)$ and is semantically equivalent to the disjunction $\bigvee_{\alpha \in Atoms_V}(\alpha \wedge fin\ \alpha)$.

In order to perform interval-based analysis on transition configurations, we need to relate $\Box T$ to the PITL formula $(\$T)^*$. Now the PTL formula $\boxdot T$, which is very similar to $\Box T$, was previously defined in Table 1 to be true on an interval iff T is true in all of the interval's nonempty suffix subintervals. It turns out that due to T being in NL^1, the formula $(\$T)^*$ is semantically equivalent to $\boxdot T$. Intuitively, this is because an NL^1 formula cannot probe past the second state of an interval. Therefore, if two nonempty intervals share the same first two states, then the truth value of T for both intervals

Type of transition configuration	PITL_V formula	Where proved
Finite-time	$((\$T)^* \wedge \mathit{init} \wedge \mathit{finite}); (T \wedge \mathit{empty})$	Theorem 24
Infinite-time	$((\$T)^* \wedge \mathit{init} \wedge \mathit{finite});$ $\big((\$T)^* \wedge L \wedge (\vec{V} \leftarrow \vec{V})\big)^{\omega}$	Theorem 35
Final	$T \wedge w \wedge \mathit{empty}$	straightforward
Periodic	$((\$T)^* \wedge \alpha \wedge L)^{\omega}$	Theorem 33

Table 3. Reduction of transition configurations to PITL_V formulas

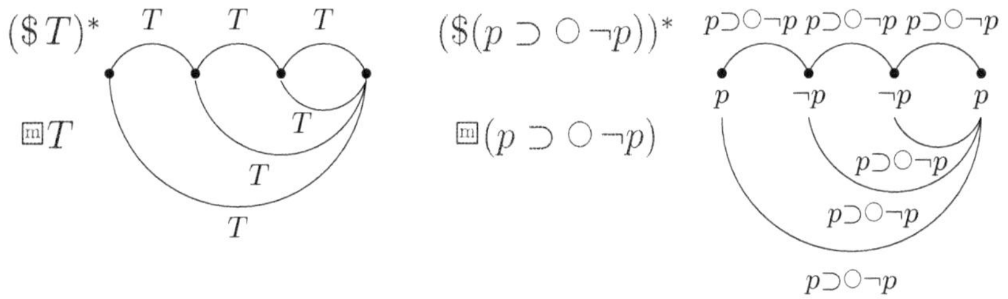

Figure 3. Illustration of equivalence of $(\$T)^*$ and $\boxed{m}\,T$

is identical (recall Lemma 7). Figure 3 illustrates this with two instances of an interval containing 4 states. The second version uses the concrete NL[1] formula $p \supset \bigcirc \neg p$ and shows specific values for the proposition variable p. Both $(\$T)^*$ and $\boxed{m}\,T$ test each pair of adjacent states. The equivalence consequently permits us to express $(\$T)^*$ in PTL_V using $\boxed{m}\,T$. In addition, it is often useful to express $\boxed{m}\,T$ as $(\$T)^*$ because the later turns out to be much more suitable for interval-based reasoning.

The following important theorem formally establishes the semantic equivalence of the PITL formula $(\$T)^*$ and $\boxed{m}\,T$:

THEOREM 18 *The* PITL *formula* $(\$T)^*$ *and the* PTL *formula* $\boxed{m}\,T$ *are semantically equivalent and hence the equivalence* $(\$T)^* \equiv \boxed{m}\,T$ *is valid.*

Proof. Given an interval σ, we can put each two-state (unit) subinterval in one-to-one correspondence with the suffix (nonempty) subinterval which shares the same starting state. Now σ satisfies $(\$T)^*$ iff T is true on all of σ's unit subintervals. Similarly, σ satisfies $\boxed{m}\,T$ iff T is true on all of σ's nonempty suffix subintervals. By the previous Lemma 7 a given unit subinterval satisfies T iff the matching suffix (nonempty) subinterval satisfies T. Consequently, the overall interval satisfies $(\$T)^*$ iff it satisfies $\boxed{m}\,T$. ■

It is not hard to check that on a one-state (empty) interval, $\boxed{\mathrm{m}}\, T$ is trivially true. On a two-state (unit) interval, it is semantically equivalent to the formula T itself.

Also note that the PTL formula $\Box T$ is semantically equivalent to the PTL formula $\boxed{\mathrm{m}}\, T \wedge \mathit{fin}\ T$. This fact and Theorem 18 together establish that $\Box T$ is also semantically equivalent to the PITL formula $(\$T)^* \wedge \mathit{fin}\ T$. Therefore, the $\Box T$ formula in transition configurations can be readily re-expressed in PITL as the conjunction $(\$T)^* \wedge \mathit{fin}\ T$. This will assist our interval-based analysis of transition configurations.

REMARK 19 *We have discussed the important semantic equivalence of* $(\$T)^*$ *and* $\boxed{\mathrm{m}}\, T$ *with quite a few people who themselves have a considerable amount of experience with both* PTL *and* PITL. *Originally we thought that this amounted to a straightforward application of temporal logic. However, to our surprise, these people found the equivalence and its applications to be nontrivial and interesting. For this reason, we have designated the statement of the equivalence of* $(\$T)^*$ *and* $\boxed{\mathrm{m}}\, T$ *to be a theorem (i.e., the previous Theorem 18), rather than merely a lemma.*

Here is a corollary of Theorem 18 for infinite time:

COROLLARY 20 *The* PTL_V *formula* $\Box T$ *and the* PITL_V *formula* $(\$T)^*$ *are semantically equivalent on infinite intervals and hence the implication* $\mathit{inf} \supset \Box T \equiv (\$T)^*$ *is valid.*

Proof. This readily follows from Theorem 18 and the semantic equivalence of $\boxed{\mathrm{m}}\, T$ and $\Box T$ on infinite intervals. ∎

The next two Lemmas 21 and 22 subsequently provide a basis for relating finite-time transition configurations to final ones and also for relating infinite-time transition configurations to periodic ones.

LEMMA 21 *For any* PITL *formula* A*, the next equivalence is valid:*

$$\models \quad \Box T \wedge \Diamond A \ \equiv\ (\$T)^* \wedge \Diamond(\Box T \wedge A) \ .$$

Proof. We first establish the validity of the PTL formula $\Box p \equiv \boxed{\mathrm{m}}\, p \wedge \Diamond \Box p$ which itself leads to the validity of the formula $\Box p \wedge \Diamond q \equiv \boxed{\mathrm{m}}\, p \wedge \Diamond(\Box p \wedge q)$. We then substitute T into p and A into q. Finally, Theorem 18 permits us to replace $\boxed{\mathrm{m}}\, T$ by $(\$T)^*$. ∎

LEMMA 22 *For any state formula* w *and* PITL *formula* A*, the next equivalence is valid:*

$$(2) \quad \Box T \wedge w \wedge \Diamond A \quad \equiv \quad ((\$ T)^* \wedge w \wedge \mathit{finite}); (\Box T \wedge A) \ .$$

Proof. Lemma 21 ensures that $\Box T \wedge \Diamond A$ is semantically equivalent to the conjunction $(\$ T)^* \wedge \Diamond(\Box T \wedge A)$. This is itself semantically equivalent to the next PITL_V formula:

$$((\$ T)^* \wedge \mathit{finite}); ((\$ T)^* \wedge \Box T \wedge A) \ .$$

Now $\Box T$ trivially implies $\boxed{\scriptstyle m} T$ which by Theorem 18 is semantically equivalent to $(\$ T)^*$. This consequently permits us to simplify the subformula $(\$ T)^* \wedge \Box T$ into $\Box T$ to obtain the next valid equivalence:

$$\models \quad \Box T \wedge \Diamond A \quad \equiv \quad ((\$ T)^* \wedge \mathit{finite}); (\Box T \wedge A) \ .$$

Simple temporal reasoning permits us to suitably add the state formula w to each side to obtain the validity of the formula (2). ■

The following Lemma 23 and Theorem 24 concern reducing a finite-time transition configuration to the associated semantically equivalent PITL_V formula in Table 3 which is easier to later analyse:

LEMMA 23 *The following equivalence is valid for finite-time transition configurations and relates them to final configurations:*

$$(3) \quad \models \quad \Box T \wedge \mathit{init} \wedge \mathit{finite} \ \equiv \ ((\$ T)^* \wedge \mathit{init} \wedge \mathit{finite}); (\Box T \wedge \mathit{empty}) \ .$$

Proof. The formula *finite* is defined to be $\Diamond\, \mathit{empty}$. Therefore Lemma 22 ensures the validity of the equivalence (3). ■

Theorem 24 builds on Lemma 23 by reducing a finite-time transition configuration to a chop formula in PITL_V which is even easier to analysis because its righthand operand is in NL^1_V:

THEOREM 24 *The following equivalence is valid for finite-time transition configurations:*

$$\models \quad \Box T \wedge \mathit{init} \wedge \mathit{finite} \ \equiv \ ((\$ T)^* \wedge \mathit{init} \wedge \mathit{finite}); (T \wedge \mathit{empty}) \ .$$

Proof. This readily follows from Lemma 23 and the fact that in an empty interval, the formulas $\Box T$ and T are equivalent. ■

Note that the PITL formula $((\$ T)^* \wedge \mathit{init} \wedge \mathit{finite}); (T \wedge \mathit{empty})$ can also be expressed as the semantically equivalent PITL formulas $\mathit{init}?; ((\$ T)^* \wedge \mathit{finite}); T?$ and $\mathit{init} \wedge (\$ T)^* \wedge \mathit{sfin}\ T$. Each form has its benefits. We prefer

$T \wedge \mathit{empty}$ over the equivalent $T?$ since some readers might get confused upon seeing the operator ? with an operand which is a temporal formula even though this is permitted in PITL.

We now turn to analysing infinite-time transition configurations. The first step involves relating them to periodic transition configurations. The next Lemma 25 does this:

LEMMA 25 *The following equivalence is valid for infinite-time transition configurations:*

$$(4) \quad \Box T \wedge \mathit{init} \wedge \Box\Diamond^+ L$$
$$\equiv \; ((\$T)^* \wedge \mathit{init} \wedge \mathit{finite}); \bigvee_{\alpha\in \mathit{Atoms}_V} (\Box T \wedge \alpha \wedge L \wedge \Box\Diamond^+(\alpha \wedge L)) \; .$$

Proof. Observe that in an infinite interval if L is always eventually true then for at least one of the finite number of V-atoms, the conjunction $\alpha \wedge L$ is also always eventually true. Therefore simple temporal reasoning yields that $\Box\Diamond^+ L$ is semantically equivalent to the disjunction $\bigvee_{\alpha\in \mathit{Atoms}_V} \Box\Diamond^+(\alpha \wedge L)$. The subformula $\Box\Diamond^+(\alpha \wedge L)$ can be re-expressed as $\Diamond(\alpha \wedge L \wedge \Box\Diamond^+(\alpha \wedge L))$ so the next equivalence concerning $\Box\Diamond^+ L$ is valid:

$$(5) \quad \models \quad \Box\Diamond^+ L \;\equiv\; \Diamond \bigvee\nolimits_{\alpha\in \mathit{Atoms}_V} (\alpha \wedge L \wedge \Box\Diamond^+(\alpha \wedge L)) \; .$$

We then use Lemma 22 to establish the equivalence below for some arbitrary V-atom α:

$$\models \quad \Box T \wedge \mathit{init} \wedge \Diamond(\Box T \wedge \alpha \wedge L \wedge \Box\Diamond^+(\alpha \wedge L))$$
$$\equiv \; ((\$T)^* \wedge \mathit{init} \wedge \mathit{finite}); (\Box T \wedge \alpha \wedge L \wedge \Box\Diamond^+(\alpha \wedge L)) \; .$$

Some simple temporal reasoning involving chop and $\bigvee$ yields the next valid equivalence:

$$(6) \quad \models \Box T \wedge \mathit{init} \wedge \Diamond \bigvee\nolimits_{\alpha\in \mathit{Atoms}_V} (\Box T \wedge \alpha \wedge L \wedge \Box\Diamond^+(\alpha \wedge L))$$
$$\equiv \; ((\$T)^* \wedge \mathit{init} \wedge \mathit{finite}); \bigvee\nolimits_{\alpha\in \mathit{Atoms}_V} (\Box T \wedge \alpha \wedge L \wedge \Box\Diamond^+(\alpha \wedge L)).$$

The combination of this and the previously mentioned semantic equivalence (5) establishes the validity of the equivalence (4). ∎

Much of the remainder of the analysis consists of showing how to further reduce a periodic transition configuration $\Box T \wedge \alpha \wedge L \wedge \Box\Diamond^+(\alpha \wedge L)$ to the semantically equivalent formula $((\$T)^* \wedge \alpha \wedge L)^\omega$. We first need to introduce a derived PITL operator which turns out to be useful for analysing periodic behaviour in infinite intervals.

DEFINITION 26 (The Operator $\Diamond_{\mathrm{f}}$) *For any* PITL *formula* A*, let the* PITL *formula* $\Diamond_{\mathrm{f}} A$ *is defined to be* $(A \wedge \mathit{finite}); \mathit{true}$. *Therefore,* $\Diamond_{\mathrm{f}} A$ *true on an interval iff* A *is true on some finite subinterval starting at the beginning of the overall interval.*

Note that $\Diamond_{\mathrm{f}} A$ can also be expressed with the derived operator $\Diamond_{\mathrm{i}}$ (itself previously defined in Table 2) as $\Diamond_{\mathrm{i}}(A \wedge \mathit{finite})$.

It is worthwhile to define a notion of fixpoints of the operator $\Diamond_{\mathrm{f}}$:

DEFINITION 27 (Fixpoints of the Operator $\Diamond_{\mathrm{f}}$) *A* PITL *formula* A *is a fixpoint of* $\Diamond_{\mathrm{f}}$ *iff the equivalence* $A \equiv \Diamond_{\mathrm{f}} A$ *is valid.*

Fixpoints of $\Diamond_{\mathrm{f}}$ are easier to move out of subintervals than are arbitrary formulas. Incidentally, for any PITL formula A, the formula $\Diamond_{\mathrm{f}} A$ is a trivial fixpoint of $\Diamond_{\mathrm{f}}$ since $\Diamond_{\mathrm{f}} A$ and $\Diamond_{\mathrm{f}} \Diamond_{\mathrm{f}} A$ are semantically equivalent. We will shortly show that all conditional liveness formulas are $\Diamond_{\mathrm{f}}$-fixpoints and later use this in the analysis of infinite intervals.

We extensively investigate fixpoints of various temporal operators and their application to compositional reasoning in [Moszkowski, 1994; Moszkowski, 1995; Moszkowski, 1996; Moszkowski, 1998].

The next lemma characterises a broad syntactic class of formulas which are $\Diamond_{\mathrm{f}}$-fixpoints and is easy to check:

LEMMA 28 *Every state formula is a* $\Diamond_{\mathrm{f}}$*-fixpoint. Furthermore, if the* PITL *formulas* A *and* B *are* $\Diamond_{\mathrm{f}}$*-fixpoints, then so are the* PITL *formulas* $A \wedge B$, $A \vee B$, $\bigcirc A$ *and* $\Diamond A$.

LEMMA 29 *Every conditional liveness formula is a* $\Diamond_{\mathrm{f}}$*-fixpoint.*

Proof. A conditional liveness formula is a conjunction of implications each which has the form $w \supset \Diamond_{\mathrm{m}} w'$ for some state formulas w and w'. If we replace $\supset$ and $\Diamond_{\mathrm{m}}$ by their definitions, then the implication reduces to the formula $\neg w \vee \Diamond((\bigcirc\, \mathit{true}) \wedge w')$. Lemma 28 then ensures that this is a $\Diamond_{\mathrm{f}}$-fixpoint. Consequently, the original implication $w \supset \Diamond_{\mathrm{m}} w'$ is one as well. Therefore by Lemma 28, the conjunction of such implications which constitutes a conditional liveness formula is also a $\Diamond_{\mathrm{f}}$-fixpoint. ■

Observe that by Lemmas 28 and 29, the formula $\alpha \wedge L$ is itself a $\Diamond_{\mathrm{f}}$-fixpoint because both α and L are $\Diamond_{\mathrm{f}}$-fixpoints.

Now the formula $\alpha \wedge L \wedge \Box \Diamond^{+}(\alpha \wedge L)$ is itself an instance of the PITL formula $A \wedge \Box \Diamond^{+} A$. We will shortly use Lemma 28 and the next two Lemmas 30 and 31 to prove Theorem 32 which itself shows that if A is a $\Diamond_{\mathrm{f}}$-fixpoint, then the formula $A \wedge \Box \Diamond^{+} A$ can be re-expressed as A^{ω}. This will let us re-express $\alpha \wedge L \wedge \Box \Diamond^{+}(\alpha \wedge L)$ as $(\alpha \wedge L)^{\omega}$.

LEMMA 30 *For any* PITL *formula A which is a $\Diamond\!\!\!\!\mid\;$-fixpoint, the next equivalence is valid:*

$$(A \wedge \mathit{finite}); \mathit{finite} \quad \equiv \quad A \wedge \mathit{finite} \ .$$

LEMMA 31 *For any* PITL *formula A which is a $\Diamond\!\!\!\!\mid\;$-fixpoint, the following implication concerning $A \wedge \Box \Diamond^{+}$ is valid:*

(7) $\models \quad A \wedge \Box \Diamond^{+} A \quad \supset \quad (A \wedge \mathit{more} \wedge \mathit{finite}); (A \wedge \Box \Diamond^{+} A) \ .$

Proof. Let σ be an interval satisfying the antecedent $A \wedge \Box \Diamond^{+} A$. The second conjunct $\Box \Diamond^{+} A$ ensures that σ has infinite time. Also, simple temporal reasoning ensures that the following implication concerning $A \wedge \Box \Diamond^{+} A$ is valid:

(8) $\models \quad A \wedge \Box \Diamond^{+} A \quad \supset \quad \Box \Diamond (A \wedge \Box \Diamond^{+} A) \ .$

Therefore, σ satisfies the formula $\Box \Diamond (A \wedge \Box \Diamond^{+} A)$.

Now A is assumed to be a $\Diamond\!\!\!\!\mid\;$-fixpoint. Also *more* is a $\Diamond\!\!\!\!\mid\;$-fixpoint by Lemma 28 since it is defined to be $\bigcirc\, \mathit{true}$. Therefore Lemma 28 ensures that the conjunction $A \wedge \mathit{more}$ is itself a $\Diamond\!\!\!\!\mid\;$-fixpoint. Consequently, the interval σ also satisfies $\Diamond\!\!\!\!\mid\;(A \wedge \mathit{more})$ which by the definition of $\Diamond\!\!\!\!\mid\;$ is the same as $(A \wedge \mathit{more} \wedge \mathit{finite}); \mathit{true}$. This together with the fact the σ satisfies the formula $\Box \Diamond (A \wedge \Box \Diamond^{+} A)$ and some simple interval-based temporal reasoning yield that the interval σ also satisfies the next formula:

(9) $(A \wedge \mathit{more} \wedge \mathit{finite}); \Diamond (A \wedge \Box \Diamond^{+} A) \ .$

We now re-express the subformula $\Diamond (A \wedge \Box \Diamond^{+} A)$ in (9) using chop and *finite* as $\mathit{finite}; (A \wedge \Box \Diamond^{+} A)$. Therefore σ satisfies the formula shown below:

(10) $(A \wedge \mathit{more} \wedge \mathit{finite}); \mathit{finite}; (A \wedge \Box \Diamond^{+} A) \ .$

Lemma 30 and the fact that $A \wedge \mathit{more}$ is a $\Diamond\!\!\!\!\mid\;$-fixpoint together yield the next valid equivalence:

$$\models \quad (A \wedge \mathit{more} \wedge \mathit{finite}); \mathit{finite} \quad \equiv \quad A \wedge \mathit{more} \wedge \mathit{finite} \ .$$

We then use this to re-express formula (10) as follows:

(11) $(A \wedge \mathit{more} \wedge \mathit{finite}); (A \wedge \Box \Diamond^{+} A) \ .$

Therefore the interval σ also satisfies formula (11). Combining all of the reasoning results in ensuring that the implication (7) is indeed valid. ■

THEOREM 32 *For any* PITL *formula A which is a $\Diamond\!\!\!\!\mid\;$-fixpoint, the next equivalence is valid:*

(12) $\models \quad A \wedge \Box \Diamond^+ A \quad \equiv \quad A^\omega$.

Proof. *Left side implies right side:* Simple temporal reasoning establishes that the implication below is valid for any PITL formulas B and C:

(13) $\models \quad \Box\big(B \supset (C \wedge \mathit{more} \wedge \mathit{finite}); B\big) \quad \supset \quad (B \supset C^\omega)$.

Lemma 31 ensures that the next implication concerning $A \wedge \Box \Diamond^+ A$ is valid:

(14) $\models \quad A \wedge \Box \Diamond^+ A \quad \supset \quad (A \wedge \mathit{more} \wedge \mathit{finite}); (A \wedge \Box \Diamond^+ A)$.

Now the validity of implication (14) readily yields the validity of the next formula:

(15) $\models \quad \Box\big((A \wedge \Box \Diamond^+ A) \supset (A \wedge \mathit{more} \wedge \mathit{finite}); (A \wedge \Box \Diamond^+ A)\big)$.

We can now achieve our main goal of showing the validity of the implication $A \wedge \Box \Diamond^+ A \supset A^\omega$ since the validity of implications (13) and (15) and modus ponens establish this.

Right side implies left side: The following PITL formula is valid by simple temporal reasoning:

$$\models \quad A^\omega \quad \supset \quad \Diamond\!\!\!\!\mid\;\, A \wedge \Diamond^+ A^\omega \ .$$

This together with further simple temporal reasoning including induction over time ensures the validity of the implication $A^\omega \supset \Diamond\!\!\!\!\mid\;\, A \wedge \Box \Diamond^+ \Diamond\!\!\!\!\mid\;\, A$. The assumption that A is a $\Diamond\!\!\!\!\mid\;\,$-fixpoint then yields the desired validity of the implication $A^\omega \supset A \wedge \Box \Diamond^+ A$. ■

The next Theorem 33 relates any periodic transition configuration with its associated PITL_V formula shown in Table 3:

THEOREM 33 *The next equivalence concerning a periodic transition configuration is valid:*

(16) $\models \quad \Box T \wedge \alpha \wedge L \wedge \Box \Diamond^+(\alpha \wedge L) \quad \equiv \quad ((\$T)^* \wedge \alpha \wedge L)^\omega$.

Proof. Lemmas 28 and 29 ensure that the formula $\alpha \wedge L$ is itself a $\Diamond\!\!\!\!\mid\;\,$-fixpoint because both α and L are $\Diamond\!\!\!\!\mid\;\,$-fixpoints. Therefore Theorem 32 yields the validity of the equivalence $\alpha \wedge L \wedge \Box \Diamond^+(\alpha \wedge L) \equiv (\alpha \wedge L)^\omega$. Now we conjoin $\Box T$ to each side of the equivalence. We then use the fact that $\Box T$ and $(\$T)^*$ are semantically equivalent in infinite time (Corollary 20) so the equivalence below is valid:

$$\models \quad \Box T \wedge \alpha \wedge L \wedge \Box \Diamond^+(\alpha \wedge L) \quad \equiv \quad (\$T)^* \wedge (\alpha \wedge L)^\omega \ .$$

Finally, we re-express $(\$T)^* \wedge (\alpha \wedge L)^\omega$ as $((\$T)^* \wedge \alpha \wedge L)^\omega$, thereby obtaining the validity of formula (16). ■

The following Lemma 34 concerning a disjunction of periodic transition configurations is needed to justify our reduction of the satisfiability of a infinite-time transition configuration to the associated PITL_V formula shown in Table 3:

LEMMA 34 *The next equivalence is valid:*

$$(17) \quad \models \quad \bigvee_{\alpha \in Atoms_V} (\Box T \wedge \alpha \wedge L \wedge \Box \Diamond^+(\alpha \wedge L)) \equiv \big((\$T)^* \wedge L \wedge (\vec{V} \leftarrow \vec{V})\big)^{\omega} .$$

Proof. Theorem 33 ensures that the equivalence given below is valid:

$$\models \quad \Box T \wedge \alpha \wedge L \wedge \Box \Diamond^+(\alpha \wedge L) \quad \equiv \quad ((\$T)^* \wedge \alpha \wedge L)^{\omega} .$$

Simple temporal reasoning establishes that the equivalence's righthand operand $((\$T)^* \wedge \alpha \wedge L)^{\omega}$ can then be re-expressed as the formula $\alpha \wedge ((\$T)^* \wedge L \wedge (\vec{V} \leftarrow \vec{V}))^{\omega}$. Some further simple reasoning about the operator $\bigvee$ yields the validity of the equivalence (17). ■

The equivalence of an infinite-time transition configuration with the associated PITL_V formula shown in Table 3 is now established:

THEOREM 35 *The following equivalence is valid for infinite-time transition configurations:*

$$\models \quad \Box T \wedge init \wedge \Box \Diamond^+ L \equiv ((\$T)^* \wedge init \wedge finite); \big((\$T)^* \wedge L \wedge (\vec{V} \leftarrow \vec{V})\big)^{\omega} .$$

Proof. This readily follows from Lemma 25 which relates infinite-time transition configurations to periodic transition configurations and Lemma 34 which re-expresses the disjunction of several periodic transition configurations using chop-omega. ■

The remaining material in this section deals with relating transition configurations involving infinite time to other formulas involving periodicity as well as to formulas about finite time. The connections are interesting in themselves and also later utilised.

The next oLemmas 36 and 37 help later to establish small models, decidability and axiomatic completeness for periodic transition configurations:

LEMMA 36 *For any V-atom α and* PITL_V *formula A, the following are equivalent:*

(a) *The formula* $(\alpha \wedge A)^\omega$ *is satisfiable.*

(b) *The formula* $(\alpha \wedge A)^\omega$ *has a periodic model.*

(c) *The formula* $\alpha \wedge A \wedge \bigcirc\, sfin\ \alpha$ *is satisfiable.*

Proof. (a) $\Rightarrow$ (c): Suppose the interval σ satisfies $(\alpha \wedge A)^\omega$. It then also satisfies the PITL$_V$ formula $(\alpha \wedge A \wedge \bigcirc\, sfin\ \alpha)^\omega$. The semantics of chop-omega then ensures that the subformula $\alpha \wedge A \wedge \bigcirc\, sfin\ \alpha$ is itself satisfiable.

(c) $\Rightarrow$ (b): Suppose the interval σ satisfies $\alpha \wedge A \wedge \bigcirc\, sfin\ \alpha$. As a consequence of α being a V-atom and A being a PITL$_V$ formula together with Lemma 13, we can assume without loss of generality that σ is a V-interval. We then readily fuse ω instances of σ together to obtain a periodic interval satisfying the formula $(\alpha \wedge A)^\omega$.

(b) $\Rightarrow$ (a): Clearly if some periodic interval satisfies $(\alpha \wedge A)^\omega$, then this formula is satisfiable. ■

Lemma 37 shows that any satisfiable periodic transition configuration has a periodic model. Subsequently, Theorem 40 establishes that any satisfiable infinite-time transition configuration has an ultimately periodic model (i.e., an interval with a periodic suffix):

LEMMA 37 *For any V-atom α, the following are equivalent:*

(a) *The periodic transition configuration* $\Box T \wedge \alpha \wedge L \wedge \Box\Diamond^+(\alpha \wedge L)$ *is satisfiable.*

(b) *The periodic transition configuration* $\Box T \wedge \alpha \wedge L \wedge \Box\Diamond^+(\alpha \wedge L)$ *has a periodic model.*

(c) *The formula* $(\$T)^* \wedge \alpha \wedge L \wedge \bigcirc\, sfin\ \alpha$ *is satisfiable.*

Proof. Theorem 33 reduces the periodic transition configuration to the semantically equivalent PITL$_V$ formula $((\$T)^* \wedge \alpha \wedge L)^\omega$. We then utilise Lemma 36. ■

LEMMA 38 *For any V-atom α and* PITL$_V$ *formulas A and B, the following are equivalent:*

(a) *The formula* $(A \wedge finite); (\alpha \wedge B)^\omega$ *is satisfiable.*

(b) *The formula* $(A \wedge finite); (\alpha \wedge B)^\omega$ *has an ultimately periodic model (i.e., an interval with a periodic suffix).*

(c) The formula $(A \wedge \mathit{finite}); (\alpha \wedge B \wedge \bigcirc\, \mathit{sfin}\ \alpha)$ *is satisfiable.*

Proof. *(a)* $\Rightarrow$ *(c):* If the formula $(A \wedge \mathit{finite}); (\alpha \wedge B)^\omega$ is satisfiable then the PITL$_V$ formula $(A \wedge \mathit{finite}); (\alpha \wedge B \wedge \bigcirc\, \mathit{sfin}\ \alpha)^\omega$ is also satisfiable. From this readily follows the satisfiability of the formula $(A \wedge \mathit{finite}); (\alpha \wedge B \wedge \bigcirc\, \mathit{sfin}\ \alpha)$.

(c) $\Rightarrow$ *(b):* If the formula $(A \wedge \mathit{finite}); (\alpha \wedge B \wedge \bigcirc\, \mathit{sfin}\ \alpha)$ is satisfiable then Lemma 14 ensures that the two formulas $A \wedge \mathit{finite} \wedge \mathit{fin}\ \alpha$ and $\alpha \wedge B \wedge \bigcirc\, \mathit{sfin}\ \alpha$ are also satisfiable. Lemma 36 then yields that the formula $(\alpha \wedge B)^\omega$ has a periodic model. Suppose the interval σ satisfies $A \wedge \mathit{finite} \wedge \mathit{fin}\ \alpha$ and the interval σ' is a periodic model of $(\alpha \wedge B)^\omega$. Lemma 13 permits us to assume that σ and σ' are V-intervals. We can fuse σ together with σ' to obtain an ultimately periodic model for $(A \wedge \mathit{finite}); (\alpha \wedge B)^\omega$.

(b) $\Rightarrow$ *(a):* Clearly if some ultimately periodic interval satisfies $(A \wedge \mathit{finite}); (\alpha \wedge B)^\omega$, then this formula is satisfiable. ■

LEMMA 39 *For any* PITL$_V$ *formulas A and B, the following are equivalent:*

(a) The formula $(A \wedge \mathit{finite}); (B \wedge (\vec{V} \leftarrow \vec{V}))^\omega$ *is satisfiable.*

(b) The formula $(A \wedge \mathit{finite}); (B \wedge (\vec{V} \leftarrow \vec{V}))^\omega$ *has an ultimately periodic model.*

(c) The formula $(A \wedge \mathit{finite}); (B \wedge \mathit{more} \wedge \mathit{finite} \wedge (\vec{V} \leftarrow \vec{V}))$ *is satisfiable.*

Proof. This follows from Lemma 38 and simple temporal reasoning involving chop and the operator $\bigvee$. We also make use of the following valid equivalences concerning $\vec{V} \leftarrow \vec{V}$, the formula B and any V-atom α:

$$\models \quad \alpha \wedge B \wedge \bigcirc\, \mathit{sfin}\ \alpha \quad \equiv \quad \alpha \wedge B \wedge \mathit{more} \wedge \mathit{finite} \wedge (\vec{V} \leftarrow \vec{V})$$
$$\models \quad (\alpha \wedge B)^\omega \quad \equiv \quad \alpha \wedge (B \wedge (\vec{V} \leftarrow \vec{V}))^\omega \ .$$

■

THEOREM 40 *The following are equivalent:*

(a) The infinite-time transition configuration $\Box T \wedge \mathit{init} \wedge \Box \Diamond^+ L$ *is satisfiable.*

(b) The infinite-time transition configuration $\Box T \wedge \mathit{init} \wedge \Box \Diamond^+ L$ *has an ultimately periodic model.*

(c) The PITL$_V$ *formula* $\big((\$T)^* \wedge \mathit{init} \wedge \mathit{finite}\big); \big((\$T)^* \wedge L \wedge \mathit{more} \wedge \mathit{finite} \wedge (\vec{V} \leftarrow \vec{V})\big)$ *is satisfiable (in finite time).*

Type of transition configuration	Upper bounds	Where proved
Finite-time	Interval length less than $\lvert Atoms_V \rvert$	Theorem 42
Infinite-time	Initial part $< \lvert Atoms_V \rvert$, Period $\leq (\lvert L \rvert + 1) \cdot \lvert Atoms_V \rvert$	Theorem 49
Final	Interval length is 0	straightforward
Periodic	Period $\leq (\lvert L \rvert + 1) \cdot \lvert Atoms_V \rvert$	Lemma 48

Table 4. Summary of upper bounds of intervals for transition configurations

(d) The PTL_V *formula* $\boxdot T \wedge init \wedge \Diamond(L \wedge finite \wedge more \wedge (\vec{V} \leftarrow \vec{V}))$ *is satisfiable (in finite time).*

Proof. We need to obtain formulas which are in a form suitable for Lemma 39. First of all, Theorem 35 permits us to re-express the infinite-time transition configuration $\Box T \wedge init \wedge \Box\Diamond^+ L$ as the formula $((\$T)^* \wedge init \wedge finite); \big((\$T)^* \wedge L \wedge (\vec{V} \leftarrow \vec{V})\big)^\omega$. Recall that Theorem 18 shows the semantic equivalence of the formulas $\boxdot T$ and $(\$T)^*$. Therefore, simple interval-based temporal reasoning ensures that formulas in (c) and (d) are semantically equivalent. We complete the proof by invoking Lemma 39. ∎

6 Small Models for Transition Configurations

We now turn to giving upper bounds on small models for satisfiable transition configurations. This is later used in Sect. 8 to construct a decision procedure for them. Table 4 summarises the upper bounds for intervals satisfying the various kinds of transition configurations and where the results are proved.

It will be necessary to employ the fact (e.g., in Theorem 42 and Lemma 46) that the formula $\alpha \wedge (\$T)^* \wedge sfin\ \beta$ is satisfiable iff a simple variant of it is satisfiable in an interval of bounded interval length. The following lemma deals with this:

LEMMA 41 *For any* V*-atoms* α *and* β*, the formula* $\alpha \wedge (\$T)^* \wedge sfin\ \beta$ *is satisfiable iff the formula* $\alpha \wedge (\$T)^{<\lvert Atoms_V \rvert} \wedge sfin\ \beta$ *is satisfiable. Hence, if the formula* $\alpha \wedge (\$T)^* \wedge sfin\ \beta$ *is satisfiable iff it is satisfiable in an interval having interval length less than* $\lvert Atoms_V \rvert$.

Proof. Any interval satisfying $\alpha \wedge (\$T)^{<\lvert Atoms_V \rvert} \wedge sfin\ \beta$ can be readily seen to also satisfy $\alpha \wedge (\$T)^* \wedge sfin\ \beta$. Let us now establish the converse by doing a proof by contradiction. Suppose $\alpha \wedge (\$T)^* \wedge sfin\ \beta$ is satisfiable but $\alpha \wedge (\$T)^{<|Atoms_V|} \wedge \mathit{sfin}\ \beta$ is not. Let σ be any interval which has the smallest length of those which satisfy $\alpha \wedge (\$T)^* \wedge \mathit{sfin}\ \beta$. Lemma 13 permits us to assume that σ is a V-interval. Now σ's length is greater than or equal to $|Atoms_V|$ and therefore contains at least $|Atoms_V| + 1$ states. Consequently, some V-state occurs at least twice in σ. Let the V-atom γ denote this state. It follows that σ satisfies the following PITL$_V$ formula:

$$\alpha \wedge \big((\$T)^*; \gamma?; (\$T)^+; \gamma?; (\$T)^*\big) \wedge \mathit{sfin}\ \beta\ .$$

Therefore σ contains two proper subintervals σ' and σ'' which respectively satisfy the PITL$_V$ formulas $\alpha \wedge (\$T)^* \wedge \mathit{sfin}\ \gamma$ and $\gamma \wedge (\$T)^* \wedge \mathit{sfin}\ \beta$. In addition, the last state of σ' is the same as the first one of σ'' so σ' and σ'' can be fused together. The fusion $\sigma' \circ \sigma''$ has length strictly less than that of σ and furthermore, like σ, satisfies the formula $\alpha \wedge (\$T)^* \wedge \mathit{sfin}\ \beta$. But this violates the assumption that σ was amongst the shortest such intervals and yields a contradiction. ■

THEOREM 42 *If a finite-time transition configuration* $\Box T \wedge \mathit{init} \wedge \mathit{finite}$ *is satisfiable, then it is satisfied by some finite interval of length less than* $|Atoms_V|$.

Proof. Theorem 24 ensures that the finite-time transition configuration $\Box T \wedge \mathit{init} \wedge \mathit{finite}$ is semantically equivalent to the formula $((\$T)^* \wedge \mathit{init} \wedge \mathit{finite}); (T \wedge \mathit{empty})$. This is satisfiable iff for some V-atom α, the formula $((\$T)^* \wedge \mathit{init} \wedge \mathit{finite}); (\alpha \wedge T \wedge \mathit{empty})$ is satisfiable. Now Lemma 14 ensures that this itself is satisfiable iff the formulas $(\$T)^* \wedge \mathit{init} \wedge \mathit{sfin}\ \alpha$ and $\alpha \wedge T \wedge \mathit{empty}$ are both satisfiable. By Lemma 41, the first of these is satisfiable iff the formula $(\$T)^{<|Atoms_V|} \wedge \mathit{init} \wedge \mathit{sfin}\ \alpha$ is satisfiable. Lemma 13 permits us to assume without loss of generality that the intervals satisfying the formulas $(\$T)^{<|Atoms_V|} \wedge \mathit{init} \wedge \mathit{sfin}\ \alpha$ and $\alpha \wedge T \wedge \mathit{empty}$ are V-intervals. We then fuse the intervals together to obtain one of interval length less than $|Atoms_V|$ which satisfies $((\$T)^* \wedge \mathit{init} \wedge \mathit{finite}); (T \wedge \mathit{empty})$ and hence also satisfies the semantically equivalent finite-time transition configuration. ■

The next definition is required for analysing infinite-time configurations and makes use of the earlier Definitions 3–5 concerning conjunctions and Definition 16 concerning conditional liveness formulas

DEFINITION 43 (Enabled Liveness Formula) *An* enabled liveness formula *En is a conjunction of* $|En|$ *formulas in which for each* $k : 1 \le k \le |En|$, *the subformula* $En[k]$ *is of the form* $\Diamond\!\!\!\!\!\!\ \ \ w$, *for some state formula* w. *The state formulas* $\theta_{En[1]}, \ldots, \theta_{En[|En|]}$ *denote the* $|En|$ *liveness tests in* En *so that* $En[k]$ *and* $\Diamond\!\!\!\!\!{\scriptscriptstyle\mathrm{m}}\;\theta_{En[k]}$ *refer to the same formula.*

For any V*-atom* α *and conditional liveness formula* L*, we will also define* $En_{L,\alpha}$ *to be the enabled liveness formula containing the* L*'s liveness tests which are enabled by* α *(recall Definition 16). Let* S *be the set of indices of* L*'s implications which are enabled by* α*. Then* $En_{L,\alpha}$ *is the conjunction* $\bigwedge_{j \in S} \Diamond\!\!\!\!\!{\scriptscriptstyle\mathrm{m}}\;\theta_{L[j]}$.

For example, suppose V is the set $\{p, q\}$, α is the V-atom $\neg p \wedge q$ and L is the conditional liveness formula $((p \vee \neg q) \supset \Diamond\!\!\!\!\!{\scriptscriptstyle\mathrm{m}}\;\neg p) \wedge (q \supset \Diamond\!\!\!\!\!{\scriptscriptstyle\mathrm{m}}\;(p \equiv \neg q)) \wedge (\mathit{true} \supset \Diamond\!\!\!\!\!{\scriptscriptstyle\mathrm{m}}\;(p \supset q))$ mentioned earlier as formula (1). Then $En_{L,\alpha}$ is the conjunction $\Diamond\!\!\!\!\!{\scriptscriptstyle\mathrm{m}}\;(p \equiv \neg q) \wedge \Diamond\!\!\!\!\!{\scriptscriptstyle\mathrm{m}}\;(p \supset q)$.

LEMMA 44 *For any* V*-atom* α *and conditional liveness formula* L *in* PTL_V*, the conjunctions* $\alpha \wedge L$ *and* $\alpha \wedge En_{L,\alpha}$ *are semantically equivalent*

Not surprisingly, the hardest part of the proof of existence of small models for infinite-time transition configurations involves finding small models for periodic transition configurations. Recall that Lemma 37 relates the satisfiability of the periodic transition configuration $\Box T \wedge \alpha \wedge L \wedge \Box \Diamond^{+}(\alpha \wedge L)$ to that of the PITL_V formula $(\$T)^* \wedge \alpha \wedge L \wedge \bigcirc\, \mathit{sfin}\ \alpha$. We will use the equivalence of $\alpha \wedge L$ and $\alpha \wedge En_{L,\alpha}$ to assist in the analysis of bounded models of $(\$T)^* \wedge \alpha \wedge L \wedge \bigcirc\, \mathit{sfin}\ \alpha$. These can then be used to obtain a bounded periodic model for the original periodic transition configuration.

LEMMA 45 *For any* V*-atom* α *and conditional liveness formula* L *in* PTL_V*, the following equivalence is valid:*

$$\models \quad (\$T)^* \wedge \alpha \wedge L \wedge \bigcirc\, \mathit{sfin}\ \alpha \quad \equiv \quad (\$T)^* \wedge \alpha \wedge En_{L,\alpha} \wedge \bigcirc\, \mathit{sfin}\ \alpha \ .$$

Proof. This readily follows from the earlier Lemma 44 concerning the semantic equivalence of the formulas $\alpha \wedge L$ and $\alpha \wedge En_{L,\alpha}$. ■

The next Lemma 46 shortens the nonempty, finite model expressed by the formula $(\$T)^* \wedge \alpha \wedge En \wedge \bigcirc\, \mathit{sfin}\ \alpha$ to one having a bounded length by adapting the technique presented earlier in Lemma 41 concerning a bounded model for the formula $(\$T)^* \wedge \alpha \wedge \mathit{sfin}\ \beta$.

LEMMA 46 *For any* V*-atom* α *and enabled liveness formula* En *in* PTL_V*, if the formula* $(\$T)^* \wedge \alpha \wedge En \wedge \bigcirc\, \mathit{sfin}\ \alpha$ *is satisfiable, then it is satisfied by a interval having interval length at most* $(|En| + 1)\,|\mathit{Atoms}_V|$.

Proof. If the formula $(\$T)^* \wedge \alpha \wedge En \wedge \bigcirc sfin\ \alpha$ is satisfiable, then by Lemma 13 there exists some satisfying V-interval. We can fuse $|En| + 1$ copies of this interval together to obtain a V-interval σ which satisfies the formula $\big((\$T)^* \wedge \alpha \wedge En \wedge finite\big)^{|En|+1} \wedge \bigcirc sfin\ \alpha$. It is not hard to check than σ itself satisfies the original formula $(\$T)^* \wedge \alpha \wedge En \wedge \bigcirc sfin\ \alpha$ since each liveness test in En is satisfied somewhere in σ prior to the last state. Furthermore, there exist a sequence of $|En|$ V-atoms $\gamma_1, \ldots, \gamma_{|En|}$ such that for each $j : 1 \leq j \leq |En|$, the state formula $\gamma_j \wedge \theta_{En[j]}$ is satisfied by some state prior to the last one and the V-interval σ satisfies the next formula:

$$\alpha \wedge \big((\$T)^*; \gamma_1?; \ldots; (\$T)^*; \gamma_{|En|}?; (\$T)^+\big) \wedge \bigcirc sfin\ \alpha\ .$$

If a gap between two of the $|En|$ selected states satisfying their respective liveness tests has interval length of at least $|Atoms_V|$, then within the gap, some state occurs twice. Such a gap can then be shortened in the manner of Lemma 41. By means of this we obtain from the V-interval σ another V-interval having bounded length and satisfying the formula below:

$$\alpha \wedge \big((\$T)^{<|Atoms_V|}; \gamma_1?; \ldots; (\$T)^{<|Atoms_V|}; \gamma_{|En|}?; (\$T)^{\leq|Atoms_V|}\big) \\ \wedge \bigcirc sfin\ \alpha\ .$$

The resulting new interval is nonempty and has interval length not exceeding $(|En| + 1)\,|Atoms_V|$. Moreover it still satisfies $(\$T)^* \wedge \alpha \wedge En \wedge \bigcirc sfin\ \alpha$. ∎

LEMMA 47 *If the formula $(\$T)^* \wedge \alpha \wedge L \wedge \bigcirc sfin\ \alpha$ is satisfiable, then it is satisfiable on a finite, nonempty interval with interval length at most $(|L| + 1)\,|Atoms_V|$.*

Proof. From Lemma 46 we have that if the formula $(\$T)^* \wedge \alpha \wedge En_{L,\alpha} \wedge \bigcirc sfin\ \alpha$ is satisfiable, then it is satisfiable on a finite, nonempty interval having interval length at most $(|En_{L,\alpha}| + 1)\,|Atoms_V|$. Lemma 44 ensures that the conjunctions $\alpha \wedge L$ and $\alpha \wedge En_{L,\alpha}$ are semantically equivalent. In addition, we have $|En_{L,\alpha}| \leq |L|$. Therefore, if the formula $(\$T)^* \wedge \alpha \wedge L \wedge \bigcirc sfin\ \alpha$ is satisfiable, then it is satisfiable on a finite, nonempty interval with interval length at most $(|L| + 1)\,|Atoms_V|$. ∎

LEMMA 48 *If the periodic transition configuration $\Box T \wedge \alpha \wedge L \wedge \Box \Diamond^+(\alpha \wedge L)$ is satisfiable, then it is satisfied by a periodic interval with period of interval length at most $(|L| + 1)\,|Atoms_V|$.*

Proof. Lemma 37 ensures that if the periodic transition configuration is satisfiable, then the formula $(\$T)^* \wedge \alpha \wedge L \wedge \bigcirc sfin\ \alpha$ is satisfiable. By

Lemma 47, if this is satisfiable, then it has a satisfying interval having interval length at most $(|L|+1)\,|Atoms_V|$. Lemma 13 permits us to assume without loss of generality that the interval is a V-interval. We can then fuse ω copies of it together to obtain a periodic interval which has a period with interval length at most $(|L|+1)\,|Atoms_V|$ and also satisfies the formula $((\$T)^* \wedge \alpha \wedge L)^\omega$. Theorem 33 establishes that this formula is equivalent to the original periodic transition configuration. ■

THEOREM 49 *If the infinite-time transition configuration* $\Box T \wedge init \wedge \Box\Diamond^+ L$ *is satisfiable, then it is satisfied by an ultimately periodic interval consisting of an initial segment having interval length less than* $|Atoms_V|$ *fused with a periodic interval having a period with interval length of at most* $(|L|+1)\,|Atoms_V|$.

Proof. If some interval satisfies the formula $\Box T \wedge init \wedge \Box\Diamond^+ L$, then Lemma 25 ensures that the interval also satisfies the next semantically equivalent formula:

$$(18)\quad ((\$T)^* \wedge init \wedge finite); \bigvee_{\alpha\in Atoms_V}(\Box T \wedge \alpha \wedge L \wedge \Box\Diamond^+(\alpha \wedge L))\ .$$

Lemma 14 and simple temporal reasoning establish that for some V-atom α the two formulas $(\$T)^* \wedge init \wedge sfin\ \alpha$ and $\Box T \wedge \alpha \wedge L \wedge \Box\Diamond^+(\alpha \wedge L)$ are satisfiable. By Lemma 41, the first formula is satisfiable in some interval σ having interval length less than $|Atoms_V|$. Lemma 48 yields some periodic interval σ' which satisfies the second formula and possesses a period with interval length of at most $(|L|+1)\,|Atoms_V|$. Lemma 13 permits us to assume that σ and σ' are V-intervals. Therefore the last state of σ is the same as the first one of σ' since both states satisfy α. The fusion $\sigma \circ \sigma'$ is itself ultimately periodic and satisfies the formula (18). Hence it also satisfies the semantically equivalent original infinite-time transition configuration $\Box T \wedge init \wedge \Box\Diamond^+ L$ as well. In addition, the interval $\sigma \circ \sigma'$ has an initial segment having interval length less than $|Atoms_V|$ fused with a periodic interval with period of interval length at most $(|L|+1)\,|Atoms_V|$. ■

7 Decomposition of Transition Configurations

We now prove the two Theorems 50 and 53 which respectively relate the satisfiability of finite-time and infinite-time transition configurations with simple interval-oriented tests. These theorems are later used in Sect. 8 as part of the justification of the our PTL decision procedure and in Sect. 10 as part of the completeness proof of an axiom system for PTL.

THEOREM 50 (Decomposition of Finite-Time Transition Configurations) *The following are equivalent:*

(a) *The finite-time configuration* $\Box T \wedge init \wedge finite$ *is satisfiable.*

(b) *For some* V*-atoms* α *and* β*, the three formulas below are satisfiable:*

$$\alpha \wedge init \qquad (\$T)^* \wedge \alpha \wedge sfin\ \beta \qquad T \wedge \beta \wedge empty\ .$$

Proof. Theorem 24 ensures that the finite-time configuration is semantically equivalent to the next PITL formula:

$$((\$T)^* \wedge init \wedge finite); (T \wedge empty)\ .$$

Now simple interval-based reasoning guarantees that this is satisfiable iff for some V-atoms α and β, the next formula is satisfiable:

$$((\$T)^* \wedge \alpha \wedge init \wedge finite); (T \wedge \beta \wedge empty)\ .$$

Lemma 14 ensures that this is itself satisfiable iff the next two formulas are:

$$(\$T)^* \wedge \alpha \wedge init \wedge sfin\ \beta \qquad T \wedge \beta \wedge empty\ .$$

Finally, simple temporal reasoning ensures that the first of these is itself is satisfiable iff the following two formulas are satisfiable:

$$\alpha \wedge init \qquad (\$T)^* \wedge \alpha \wedge sfin\ \beta\ .$$

■

We now turn to decomposing an infinite-time transition configuration:

LEMMA 51 *The infinite-time transition configuration* $\Box T \wedge init \wedge \Box \Diamond^+ L$ *is satisfiable iff for some* V*-atoms* α *and* β*, the following formulas are satisfiable:*

(19) $(\$T)^* \wedge \alpha \wedge init \wedge sfin\ \beta \qquad (\$T)^* \wedge \beta \wedge En_{L,\beta} \wedge \bigcirc\, sfin\ \beta\ .$

Proof. Theorem 40 ensures that the infinite-time configuration is satisfiable iff the next PITL_V formula is satisfiable:

$$((\$T)^* \wedge init \wedge finite); \big((\$T)^* \wedge L \wedge more \wedge finite \wedge (\vec{V} \leftarrow \vec{V})\big)\ .$$

Simple interval-based temporal reasoning ensures that this itself is satisfiable iff for some V-atoms α and β, next formula is satisfiable:

$$(20)\quad \big((\$T)^* \wedge \alpha \wedge \mathit{init} \wedge \mathit{finite}\big);\big((\$T)^* \wedge \beta \wedge L \wedge \bigcirc\, \mathit{sfin}\ \beta\big)\ .$$

Now Lemma 44 guarantees the semantic equivalence of the conjunctions $\beta \wedge L$ and $\beta \wedge En_{L,\beta}$. We therefore can replace L by $En_{L,\beta}$ in formula (20). Finally, Lemma 14 yields that the resulting formula is itself satisfiable iff the two formulas in (19) are satisfiable. ■

The next lemma concerning enabled liveness formulas is shortly used in Theorem 53 to analyse the satisfiability of infinite-time configurations:

LEMMA 52 *For any V-atom α and enabled liveness formula En, the following are equivalent:*

(a) The formula $(\$T)^* \wedge \alpha \wedge En \wedge \bigcirc\, \mathit{sfin}\ \alpha$ is satisfiable.

(b) For some $|En|$ V-atoms $\gamma_1, \ldots, \gamma_{|En|}$ (not necessarily distinct), the following are all satisfiable:

$$(\$T)^* \wedge \alpha \wedge \bigcirc\, \mathit{sfin}\ \alpha$$

$$\textit{for each } \gamma_i\text{:}\quad (\$T)^* \wedge \alpha \wedge \mathit{sfin}\ \gamma_i \qquad \gamma_i \wedge \theta_{En_{L,\alpha}[i]} \qquad (\$T)^* \wedge \gamma_i \wedge \mathit{sfin}\ \alpha\ .$$

Proof. Induction on the length of En and simple interval-based reasoning can be used to demonstrate that the formula $(\$T)^* \wedge \alpha \wedge En \wedge \bigcirc\, \mathit{sfin}\ \alpha$ is satisfiable iff the formula $(\$T)^* \wedge \alpha \wedge \bigcirc\, \mathit{sfin}\ \alpha$ is satisfiable and also for some V-atoms $\gamma_1, \ldots, \gamma_{|En|}$, for each γ_i the following formula is satisfiable:

$$(21)\quad (\$T)^* \wedge \alpha \wedge \Diamond(\gamma_i \wedge \theta_{En[i]}) \wedge \mathit{sfin}\ \alpha\ .$$

This guarantees that for each liveness test $\theta_{En[i]}$ in En, the V-atom α can reach some V-atom γ_i which satisfies $\theta_{En[i]}$ and this V-atom γ_i itself can reach back to α. We can re-express (21) as the semantically equivalent formula below:

$$\big((\$T)^* \wedge \alpha \wedge \mathit{finite}\big);\big((\$T)^* \wedge \gamma_i \wedge \theta_{En[i]} \wedge \mathit{sfin}\ \alpha\big)\ .$$

Lemma 14 ensures that this is satisfiable iff the next two formulas are:

$$(\$T)^* \wedge \alpha \wedge \mathit{sfin}\ \gamma_i \qquad (\$T)^* \wedge \gamma_i \wedge \theta_{En[i]} \wedge \mathit{sfin}\ \alpha\ .$$

The second one is satisfiable iff the two formulas shown below are satisfiable:

$$\gamma_i \wedge \theta_{En[i]} \qquad (\$T)^* \wedge \gamma_i \wedge \mathit{sfin}\ \alpha\ .$$

■

THEOREM 53 (Decomposition of Infinite-Time Transition Configurations) *The following are equivalent:*

(a) The infinite-time configuration $\Box T \wedge init \wedge \Box \Diamond^{+} L$ *is satisfiable.*

(b) For some V-atoms α, β *and* $\gamma_1, \ldots, \gamma_{|En_{L,\beta}|}$ *(not necessarily distinct), the following are all satisfiable:*

$$\alpha \wedge init \qquad (\$T)^* \wedge \alpha \wedge sfin\ \beta \qquad (\$T)^* \wedge \beta \wedge \bigcirc\, sfin\ \beta$$
$$\textit{for each } \gamma_i\!: \quad (\$T)^* \wedge \beta \wedge sfin\ \gamma_i \quad \gamma_i \wedge \theta_{En_{L,\beta}[i]} \quad (\$T)^* \wedge \gamma_i \wedge sfin\ \beta\ .$$

Proof. Lemma 51 establishes that the infinite-time configuration $\Box T \wedge init \wedge \Box \Diamond^{+} L$ is satisfiable iff there exist some V-atoms α and β for which the next two formulas are satisfiable:

$$(22)\quad (\$T)^* \wedge \alpha \wedge init \wedge sfin\ \beta \qquad (\$T)^* \wedge \beta \wedge En_{L,\beta} \wedge \bigcirc\, sfin\ \beta\ .$$

Now simple temporal reasoning ensures that the first of these is itself is satisfiable iff the following two formulas are satisfiable:

$$\alpha \wedge init \qquad (\$T)^* \wedge \alpha \wedge sfin\ \beta\ .$$

Furthermore, Lemma 52 guarantees that the second formula in (22) is satisfiable iff the formula $(\$T)^* \wedge \beta \wedge \bigcirc\, sfin\ \beta$ is satisfiable and furthermore for some V-atoms $\gamma_1, \ldots, \gamma_{|En_{L,\beta}|}$ (not necessarily distinct), the following are all satisfiable for each γ_i:

$$(\$T)^* \wedge \beta \wedge sfin\ \gamma_i \qquad \gamma_i \wedge \theta_{En_{L,\beta}[i]} \qquad (\$T)^* \wedge \gamma_i \wedge sfin\ \beta\ .$$

■

8 A Decision Procedure

We now describe decision procedures for finite-time and infinite-time transition configurations based on Binary Decision Diagrams (BDDs) [Bryant, 1986; Bryant, 1992] which provide an efficient basis for performing many computational tasks involving reductions to reasoning about formulas in propositional logic. We had little difficultly implementing the algorithms using the popular Colorado University Decision Diagram Package (CUDD) [CUDD, URL] developed by Somenzi. Our prototype tool consists of a front-end coded in the CLISP [CLISP, URL] implementation of Common Lisp [ANSI, 1999] as well as a back-end coded in Perl [Perl, URL]. The back-end employs a Perl-oriented interface to CUDD written by Somenzi and called PerlDD [PerlDD, URL]. The front-end accepts arbitrary PTL formulas and converts them to transition configurations using methods later described in Sections 11 and 12. The transition configurations are then passed to the back-end which analyses them using BDDs. In this section we describe the basis for performing this analysis.

The remainder of this section assumes that the reader already has some familiarity with BDDs.

Our algorithm for finite-time transition configurations adapts methods for *symbolic state space traversal* described by Coudert, Berthet and Madre [Coudert *et al.*, 1989b; Coudert *et al.*, 1989a; Coudert *et al.*, 1990] (see also Kropf [Kropf, 1999; Clarke *et al.*, 2000]) for use with BDD-based representations of formulas in propositional logic. It simultaneously greatly benefits from closely related methods first employed by McMillan in symbolic model checking [McMillan, 1993; Burch *et al.*, 1992; Clarke *et al.*, 2000] which also include the automatic generation of counterexamples for unsatisfiable formulas and, similarly, witnesses for satisfiable ones. Recall that Theorem 50 shows that the finite-time transition configuration $\Box T \wedge \mathit{init} \wedge \mathit{finite}$ is satisfiable iff for some V-atoms α and β, the next three formulas are satisfiable:

$$\alpha \wedge \mathit{init} \qquad (\$T)^* \wedge \alpha \wedge \mathit{sfin}\ \beta \qquad T \wedge \beta \wedge \mathit{empty}\ .$$

We can readily search for suitable V-atoms using BDDs. Three BDDs Γ_1, Γ_2 and Γ_3 are initially constructed. In what follows, please recall the notion $\dashv X$ introduced in Definition 1 to denote that the formula X is satisfiable. We first describe the roles of the BDDs Γ_1, Γ_2 and Γ_3 before actually constructing them:

- The BDD Γ_1 represents the state formula *init* and hence the set of V-atoms satisfying *init* (i.e., the set $\{\alpha \in \mathit{Atoms}_V : \alpha \models \mathit{init}\}$). This is the same as the set $\{\alpha \in \mathit{Atoms}_V : \dashv \alpha \wedge \mathit{init}\}$.

- The second BDD Γ_2 captures all pairs of V-atoms corresponding to unit (i.e., two-state) intervals satisfying T. In other words, it corresponds to the set $\{\langle\alpha, \beta\rangle \in \mathit{Atoms}_V^2 : \alpha\beta \models T\}$. This is the same as the set $\{\langle\alpha, \beta\rangle \in \mathit{Atoms}_V^2 : \dashv T \wedge \alpha \wedge \mathit{skip} \wedge \mathit{sfin}\ \beta\}$.

- The third BDD Γ_3 captures the behaviour of T in an empty interval. Therefore Γ_3 represents the set of all V-atoms satisfying the formula $T \wedge \mathit{empty}$ (i.e., the set $\{\alpha \in \mathit{Atoms}_V : \alpha \models T\}$). This is the same as the set $\{\alpha \in \mathit{Atoms}_V : \dashv T \wedge \alpha \wedge \mathit{empty}\}$

In the course of manipulating the BDDs we make use of two sets of variables. They include the original ones (e.g., p, r_1, ..., r_4) as well as primed versions (e.g., p', r'_1, ..., r'_4). For convenience, we often do not distinguish between a BDD and the propositional logic formula it represents.

Let V and V' respectively denote the two sets of variables. We now construct the BDDs Γ_1, Γ_3 and Γ_2 as follows:

- Let Γ_1 be the formula *init*.
- Obtain Γ_2 from the formula T by replacing all variables in the scope of any $\bigcirc$ constructs by corresponding ones in V' and then deleting all $\bigcirc$ operators (but not the associated operands) to obtain a formula in conventional propositional logic. We refer to this process of constructing Γ_2 from T by the term *flattening*.
- Obtain Γ_3 from the formula T by replacing each $\bigcirc$ construct by *false*.

The BDDs Γ_1 and Γ_3 both only can contain variables in V whereas Γ_2 can contain variables in V and V'.

Suppose T and *init* are the following formulas mentioned earlier in §3.3:

$$T\colon\ (r_1 \equiv (p \vee \bigcirc r_1)) \wedge (r_2 \equiv (\neg r_1 \vee \bigcirc r_2)) \qquad \mathit{init}\colon\ \neg r_2 \wedge \neg r_4\ .$$
$$\wedge\, (r_3 \equiv (\neg p \vee \bigcirc r_3)) \wedge (r_4 \equiv (\neg r_3 \vee \bigcirc r_4))$$

Here are the associated Γ_1, Γ_2 and Γ_3 for these T and *init*:

$$\Gamma_1\colon\ \neg r_2 \wedge \neg r_4$$
$$\Gamma_2\colon\ (r_1 \equiv (p \vee r_1')) \wedge (r_2 \equiv (\neg r_1 \vee r_2'))$$
$$\wedge\, (r_3 \equiv (\neg p \vee r_3')) \wedge (r_4 \equiv (\neg r_3 \vee r_4'))$$
$$\Gamma_3\colon\ (r_1 \equiv (p \vee \mathit{false})) \wedge (r_2 \equiv (\neg r_1 \vee \mathit{false}))$$
$$\wedge\, (r_3 \equiv (\neg p \vee \mathit{false})) \wedge (r_4 \equiv (\neg r_3 \vee \mathit{false}))\ .$$

The connection between the BDDs for Γ_1 and Γ_3 and the previously mentioned sets of V-atoms they are meant to capture is straightforward. In order to justify the less intuitive relationship between the construction for Γ_2 and the earlier associated set of pairs of V-atoms, we present a lemma relating Γ_2 with T.

LEMMA 54 *For any V-atoms α and β, the following are equivalent:*

(a) The formula $T \wedge \alpha \wedge \mathit{skip} \wedge \mathit{sfin}\ \beta$ is satisfiable (i.e., $\alpha\beta \models T$).

(b) The propositional logic formula $\Gamma_2 \wedge \alpha \wedge \beta_V^{V'}$ is satisfiable.

Proof. $(a) \Rightarrow (b)$: Suppose the formula $T \wedge \alpha \wedge \mathit{skip} \wedge \mathit{sfin}\ \beta$ is satisfiable. Then the flattening of T into Γ_2 readily yields that the formula $\Gamma_2 \wedge \alpha \wedge \beta_V^{V'}$ is satisfiable.

(b) $\Rightarrow$ *(a)*: If the propositional logic formula $\Gamma_2 \wedge \alpha \wedge \beta_V^{V'}$ is satisfiable, then the flattening of $\bigcirc$ constructs in Γ_2 readily yields that the NL[1] formula $T \wedge \alpha \wedge \bigcirc\beta$ is satisfiable. Clearly any interval satisfying it has at least two states. Hence by the earlier Lemma 8 the formula $skip \wedge T \wedge \alpha \wedge \bigcirc\beta$ is satisfiable. Simple temporal reasoning then ensures that the semantically equivalent formula $T \wedge \alpha \wedge skip \wedge sfin\ \beta$ is also satisfiable. ∎

We use Γ_2 to together with the first BDD Γ_1 to iteratively calculate a sequence of BDDs $\Delta_0, \ldots, \Delta_k, \ldots$ so that for any k, Δ_k describes all V-atoms which can be reached from one which satisfies *init* in exactly k steps. In order words, Δ_k represents the following set:

$$\{\beta \in Atoms_V : \text{for some } \alpha \in Atoms_V, \dashv (\$T)^k \wedge \alpha \wedge init \wedge sfin\ \beta\} \ .$$

We set Δ_0 to be Γ_1. Therefore, every variable in Δ_0 is in V. Each Δ_{k+1} is calculated to be semantically equivalent to the next quantified propositional logic formula in which renaming ensures that all free variables are in V:

$$(23) \quad \left(\exists V.\, (\Delta_k \wedge \Gamma_2)\right)_{V'}^{V} \ .$$

Due to the final renaming, the sole variables left in the BDD Δ_{k+1} itself are elements of V. The only BDD operations required to calculate Δ_{k+1} from (23) are logical-and, existential quantification (which actually yields a BDD representing a semantically equivalent quantifier-free formula) and renaming which are all standard ones.

REMARK 55 *Within the CUDD system, the entire calculation to obtain* $\exists V.\, (\Delta_k \wedge \Gamma_2)$ *can even be done by a single CUDD operation tailored to handle this specific kind of common BDD manipulation. Furthermore, the renaming of variables in* V' *to those in* V *is actually achieved by taking the BDD obtained for* $\exists V.\, (\Delta_k \wedge \Gamma_2)$ *and then performing a single CUDD operation which yields another BDD in which the variables in* V *are swapped with the corresponding ones in* V'.

For any given Δ_k which has been calculated, we next determine the logical-and of Γ_3 and Δ_k and then proceed as follows:

1. If the logical-and is not false, then there is some V-atom β satisfying $T \wedge empty$ which can be reached in k steps from a V-atom α satisfying *init*. Therefore the next three formulas are all satisfiable:

   $$\alpha \wedge init \qquad (\$T)^k \wedge \alpha \wedge sfin\ \beta \qquad T \wedge \beta \wedge empty \ .$$

   Now the second formula ensures the satisfiability of the formula $(\$T)^* \wedge \alpha \wedge sfin\ \beta$. Therefore Theorem 50 can be invoked to obtain the satisfiability of the original finite-time transition configuration $\Box T \wedge$ *init* $\wedge$ *finite*. We therefore do not need to calculate any further Δ_k's.

2. Otherwise, the logical-and is false so we must continue to iterate.

During the iteration process, we maintain a BDD representing the set of all V-atoms so far reachable from one satisfying *init*. This BDD corresponds to the formula $\bigvee_{0 \le i \le k} \Delta_i$ which equals the next set:

$$\{\beta \in Atoms_V : \text{for some } \alpha \in Atoms_V, \dashv (\$T)^{\le k} \wedge \alpha \wedge init \wedge sfin\ \beta\} \ .$$

If no such β exists which also satisfies $T \wedge$ *empty*, the BDD eventually converges to a value corresponding to the set of all V-atoms reachable from V-atoms which satisfy *init*. The following set denotes this:

$$\{\beta \in Atoms_V : \ \text{for some } \alpha \in Atoms_V, \ \dashv (\$T)^* \wedge \alpha \wedge init \wedge sfin\ \beta\} \ .$$

We then terminate the algorithm with a report that the original transition configuration $\Box T \wedge$ *init* $\wedge$ *finite* is unsatisfiable. Even though Lemma 41 bounds the number of iterations, in some cases convergence takes too long. This necessitates a preset iteration limit or a facility for manual intervention in order to force premature termination of the loop.

If for some n, the algorithm succeeds after n iterations and determines that the transition configuration is satisfiable, then a sample V-interval having $n+1$ states and which satisfies the formula can be calculated. This involves standard BDD methods for constructing such examples and is done by working backward through the BDDs $\Delta_n, \Delta_{n-1}, \ldots \Delta_0$ to find a suitable sequence of $n+1$ V-atoms to serve as a V-interval satisfying the transition configuration. The algorithm can be also readily adapted to only determine values for a subset of the variables in V.

8.1 Dealing with Infinite Time

For testing an infinite-time transition configuration $\Box T \wedge$ *init* $\wedge \Box \Diamond^+ L$, we can make use of Theorem 40 which guarantees that this formula is satisfiable iff the next PTL_V formula is satisfiable:

$$\boxed{m}\, T \wedge init \wedge \Diamond(L \wedge finite \wedge more \wedge (\vec{V} \leftarrow \vec{V})) \ .$$

The previously described satisfiability algorithm for finite-time can therefore be utilised. However, we must first transform this second formula to some suitable finite-time transition configuration using techniques later described in Sect. 12 for reducing arbitrary PTL formulas to finite-time transition configurations. Alternatively, more sophisticated algorithms using Theorem 53 can be employed to directly analyse the infinite-time transition configuration using BDD-based techniques. Space does not permit more details here.

Axioms:	
N1 (K).	$\vdash \textcircled{w}(X \supset X') \supset \textcircled{w}X \supset \textcircled{w}X'$
N2 (D_c).	$\vdash \bigcirc X \supset \textcircled{w}X$
Inference rules:	
NR1.	If X is a tautology, then $\vdash X$
NR2 (MP).	If $\vdash X \supset X'$ and $\vdash X$, then $\vdash X'$
NR3 (RN).	If $\vdash X$, then $\vdash \textcircled{w}X$

Table 5. Complete axiom system for NL (Modal system $K{+}D_c$)

9 Axiom System for NL

In preparation for the proof of axiomatic completeness for PTL, we now consider an axiom system for NL. The axiomatic completeness of NL later plays a major role in the completeness proof for PTL.

Within this section, the variables X, X', X_0 and X_0' denote NL *formulas.*

Table 5 contains a complete axiom system for NL adapted from the modal logic $K{+}D_c$. Here $\textcircled{w}$ ("weak next"), previously defined in Table 1 to be a derived operator, is instead regarded as a primitive construct. We can consider $\bigcirc X$ to be an abbreviation for $\neg \textcircled{w} \neg X$. Hughes and Cresswell [Hughes and Cresswell, 1996, Problem 6.8 on p. 123 with solution on p. 379] briefly discuss how to show deductive completeness of the logic $K{+}D_c$.

Table 6 contains a complete axiom system for NL in which $\bigcirc$, rather than $\textcircled{w}$, is the primitive operator. Consequently, $\textcircled{w}$ is derived in the manner already shown in Table 1. The axiom system is essentially one of several M-based axiomatisations of normal systems of modal logic covered by Chellas [Chellas, 1980] with the addition of the axiom D_c. This second axiom system appears preferable for our purposes since our definition of PTL also takes $\bigcirc$ to be primitive. We therefore use this axiom system here although the methods employed can be easily adapted to the first NL axiom system.

DEFINITION 56 (Theoremhood and Consistency for NL) *If some* NL *formula X is deducible from the axiom system, we call it an* NL theorem *and denote this theoremhood as $\vdash_{\mathrm{NL}} X$. We define X to be* NL-consistent *if $\neg X$ is* not *an* NL *theorem, i.e., $\nvdash_{\mathrm{NL}} \neg X$.*

Below are some representative lemmas about satisfiability and consistency of NL formulas. They are subsequently used in the completeness proof for the NL axiom system in Table 6.

Axioms:	
N1′ (N◇).	$\vdash \neg \bigcirc \mathit{false}$
N2′ (C◇).	$\vdash \bigcirc(X \vee X') \supset \bigcirc X \vee \bigcirc X'$
N3′ (D_c).	$\vdash \bigcirc X \supset ⓦ X$
Inference rules:	
NR1′.	If X is a tautology, then $\vdash X$
NR2′ (MP).	If $\vdash X \supset X'$ and $\vdash X$, then $\vdash X'$
NR3′ (RM◇).	If $\vdash X \supset X'$, then $\vdash \bigcirc X \supset \bigcirc X'$

Table 6. Alternative complete axiom system for NL

LEMMA 57 *For any state formula w and* NL *formula X, if w is satisfiable, then the* NL *conjunction $w \wedge \neg \bigcirc X$ is satisfied by some one-state interval.*

LEMMA 58 *For any state formula w and* NL *formula X, if both w and X are satisfiable, then so is the formula $w \wedge \bigcirc X$.*
In such as case, if X itself is satisfied by an interval having at most n states, then $w \wedge \bigcirc X$ is satisfied by an interval having at most $n+1$ states,

LEMMA 59 *For any* NL *formula X, if $\bigcirc X$ is* NL*-consistent, then so X.*

For any NL formulas X and X', the following are deducible as NL theorems and shortly used to simplify formulas:

(24) $\vdash_{\mathrm{NL}} \bigcirc(X \wedge X') \equiv \bigcirc X \wedge \bigcirc X'$
(25) $\vdash_{\mathrm{NL}} \bigcirc(X \wedge \neg X') \equiv \bigcirc X \wedge \neg \bigcirc X'$
(26) $\vdash_{\mathrm{NL}} \neg \bigcirc(X \vee X') \equiv \neg \bigcirc X \wedge \neg \bigcirc X'$.

Axiomatic completeness is usually defined to mean that every valid formula is deducible as a theorem. However, we will make use of the following variant way of expressing completeness:

LEMMA 60 (Alternative Notion of Completeness) *A logic's axiom system is complete iff each consistent formula is satisfiable.*

THEOREM 61 (Completeness of Alternative NL Axiom System) *The* NL *axiom system in Table 6 is complete.*

Proof. The proof involves the kind of consistency-based reasoning found later in the paper. Using Lemma 60, we show that any NL formula X_0 which is NL-consistent (i.e., $\nvdash_{\mathrm{NL}} \neg X_0$) has a satisfying finite interval. Let

Axioms:		Inference rules:	
T1.	$\vdash \Box(X \supset Y) \supset \Box X \supset \Box Y$	R1.	If X is a tautology, then $\vdash X$
T2.	$\vdash \bigcirc X \supset ⓦ X$	R2.	If $\vdash X \supset Y$ and $\vdash X$, then $\vdash Y$
T3.	$\vdash \bigcirc(X \supset Y) \supset \bigcirc X \supset \bigcirc Y$	R3.	If $\vdash X$, then $\vdash \Box X$
T4.	$\vdash \Box X \supset X \wedge ⓦ \Box X$		
T5.	$\vdash \Box(X \supset ⓦ X) \supset X \supset \Box X$		

Table 7. Modified version of Pnueli's complete PTL axiom system DX

n be the next-height of X_0, i.e., the maximum nesting of $\bigcirc$s in X_0. We do induction on n to show that X_0 is satisfied by some interval with at most $n+1$ states. For $n > 0$, we regard the temporal constructs in X_0 which are not nested in other temporal constructs as being primitive. Then conventional propositional reasoning yields a deducibly equivalent formula X_0' in disjunctive normal form. As least one disjunct is consistent. Lemmas 57–59 and equivalences (24)–(26) are invoked to obtain a satisfying interval. ■

10 Axiomatic Completeness for Transition Configurations

We now turn to describing a PTL axiom system with which axiomatic completeness can be shown for transition configurations discussed.

The PTL axiom system used here is shown in Table 7 and is adapted from another similar PTL axiom system DX proposed by Pnueli [Pnueli, 1977]. Gabbay et al. [Gabbay *et al.*, 1980] showed that DX is complete. Pnueli's original system uses strong versions of $\Diamond$ and $\Box$ (which we denote as $\Diamond^+$ and $\Box^+$, respectively) which do not examine the current state. In addition, Pnueli's system only deals with infinite time. However, Gabbay et al. [Gabbay *et al.*, 1980] also include a variant system called $\mathrm{D^0X}$ based on the conventional $\Diamond$ and $\Box$ operators which examine the current state. The version presented here does this as well and furthermore permits both finite and infinite time.

DEFINITION 62 (Theoremhood and Consistency for PTL) *If the* PTL *formula X is deducible from the axiom system, we call it a* PTL theorem *and denote this theoremhood as $\vdash X$. We define X to be* consistent *if $\neg X$ is* not *a theorem, i.e., $\nvdash \neg X$.*

In the course of proving completeness for PTL we make use of a definition of completeness for sets of formulas such as sets of transitions configurations:

Lemma	Summary
65	If $\dashv \boxed{m}\, T \wedge \alpha \wedge \bigcirc \beta$, then $\Dashv T \wedge \alpha \wedge \mathit{skip} \wedge \mathit{sfin}\ \beta$
67	If $\dashv \boxed{m}\, T \wedge \alpha \wedge \Diamond \beta$, then $\Dashv (\$T)^{*} \wedge \alpha \wedge \mathit{sfin}\ \beta$
68	If $\dashv \boxed{m}\, T \wedge \alpha \wedge \Diamond^{+} \beta$, then $\Dashv (\$T)^{*} \wedge \alpha \wedge \bigcirc \mathit{sfin}\ \beta$

Table 8. Summary of some basic lemmas for consistency and satisfiability

DEFINITION 63 (Completeness for a Set of Formulas) *An axiom system is said to be complete for a set of formulas $\{X_1, \ldots, X_n\}$ if the consistency of any X_i implies that X_i is also satisfiable.*

Now the Alternative Notion of Completeness (Lemma 60) can also be readily adapted to sets of formulas. Indeed, our goal in the rest of this section is to show that any consistent transition configuration is also satisfiable.

The next lemma permits us to utilise within PTL the axiomatic completeness of the NL proof system:

THEOREM 64 (Completeness for NL formulas in PTL) *The* PTL *axiom system is complete for the set of* NL *formulas.*

Proof. Theorem 61 establishes the completeness of the alternative NL axiom system in Table 6. We then show that any NL theorem is also a PTL theorem. This can be done by demonstrating that all axioms and inferences rules in the NL axiom system are derivable from PTL ones. ■

10.1 Some Basic Lemmas for Completeness

In this subsection, we deal with another part of the completeness proof. We utilise ways to go from certain specific kinds of consistent formulas involving reachability to intervals in order to later construct models for consistent transition configurations in §10.2. Table 8 summarises the basic lemmas proved here. Within the table, we use the notation $\Dashv X$ already introduced in Definition 1 to denote that the formula X is satisfiable and $\dashv X$ to denote that X is consistent.

LEMMA 65 *For any V-atoms α and β, if the formula $\boxed{m}\, T \wedge \alpha \wedge \bigcirc \beta$ is consistent, then the formula $T \wedge \alpha \wedge \mathit{skip} \wedge \mathit{sfin}\ \beta$ is satisfiable.*

Proof. From the consistency of the formula $\boxed{m}\, T \wedge \alpha \wedge \bigcirc \beta$ and simple temporal reasoning, we obtain the consistency of the NL^1_V formula $T \wedge \alpha \wedge \bigcirc \beta$. Theorem 64 concerning axiomatic completeness for NL formulas in the PTL axiom system then ensures that this is satisfiable. Clearly any interval satisfying it has at least two states. Hence by the earlier Lemma 8 the formula $\mathit{skip} \wedge T \wedge \alpha \wedge \bigcirc\beta$ is also satisfiable. Consequently, simple temporal reasoning yields that the semantically equivalent formula $T \wedge \alpha \wedge \mathit{skip} \wedge \mathit{sfin}\ \beta$ is satisfiable as well. ■

For any V-atom α, within the next two lemmas we let S_α denote the subset of Atoms_V containing exactly every V-atom γ for which the following formula, which concerns reachability from α, is satisfiable:

$$(\$T)^* \wedge \alpha \wedge \mathit{sfin}\ \gamma \ .$$

Here is a more formal definition of S_α:

$$S_\alpha \quad \stackrel{\text{def}}{=} \quad \{\gamma \in \mathit{Atoms}_V : \ \Dashv (\$T)^* \wedge \alpha \wedge \mathit{sfin}\ \gamma\} \ .$$

LEMMA 66 *For any V-atom α, the following formula is a* PTL *theorem:*

$$(27) \vdash \quad \boxed{\mathrm{m}}\, T \wedge \alpha \supset \Box \bigvee_{\gamma \in S_\alpha} \gamma \ .$$

Proof. The following formulas are valid and in NL[1]. Hence, they are theorems by the completeness of the PTL axiom system for NL[1] formulas (Theorem 64):

$$\vdash \quad \alpha \supset \bigvee_{\gamma \in S_\alpha} \gamma \qquad\qquad \vdash \quad \mathit{more} \wedge T \wedge \bigvee_{\gamma \in S_\alpha} \gamma \quad \supset \quad \bigcirc \bigvee_{\gamma \in S_\alpha} \gamma \ .$$

From these and simple temporal reasoning we can readily deduce our goal (27). ■

LEMMA 67 *For any V-atoms α and β, if the formula $\boxed{\mathrm{m}}\, T \wedge \alpha \wedge \Diamond\beta$ is consistent, then the formula $(\$T)^* \wedge \alpha \wedge \mathit{sfin}\ \beta$ is satisfiable.*

Proof. Suppose on the contrary that $(\$T)^* \wedge \alpha \wedge \mathit{sfin}\ \beta$ is unsatisfiable. Now α is in the set S_α, whereas β is not. Hence, the following formula concerning β not being in S_α is valid and thus a propositional tautology:

$$(28) \vdash \quad \bigvee_{\gamma \in S_\alpha} \gamma \quad \supset \quad \neg\beta \ .$$

Furthermore, the previous Lemma 66 ensures that the next implication is a PTL theorem:

$$(29) \vdash \quad \boxed{\mathrm{m}}\, T \wedge \alpha \supset \Box \bigvee_{\gamma \in S_\alpha} \gamma \ .$$

The two implications (28) and (29) together with some simple temporal reasoning let us deduce that α can never reach β:

$$\vdash \quad \boxed{\scriptstyle m}\, T \wedge \alpha \supset \Box \neg \beta \; .$$

From this and the general equivalence $\vdash \Box \neg \beta \equiv \neg \Diamond \beta$ we can deduce the following PTL theorem:

$$\vdash \quad \boxed{\scriptstyle m}\, T \wedge \alpha \supset \neg \Diamond \beta \; .$$

Therefore, the formula $\boxed{\scriptstyle m}\, T \wedge \alpha \wedge \Diamond \beta$ is inconsistent. This contradicts the lemma's assumption. ■

LEMMA 68 *For any V-atoms α and β, if the formula $\boxed{\scriptstyle m}\, T \wedge \alpha \wedge \Diamond^{+} \beta$ is consistent, then the formula $(\$T)^{*} \wedge \alpha \wedge \bigcirc\, \mathit{sfin}\ \beta$ is satisfiable.*

Proof. From the consistency of the formula $\boxed{\scriptstyle m}\, T \wedge \alpha \wedge \Diamond^{+} \beta$, we readily deduce for some V-atom γ the consistency of the two PTL_V formulas below:

$$\boxed{\scriptstyle m}\, T \wedge \alpha \wedge \Diamond \gamma \qquad \boxed{\scriptstyle m}\, T \wedge \gamma \wedge \bigcirc \beta \; .$$

The consistency of the first formula $\boxed{\scriptstyle m}\, T \wedge \alpha \wedge \Diamond \gamma$ and Lemma 68 yield that the formula $(\$T)^{*} \wedge \alpha \wedge \bigcirc\, \mathit{sfin}\ \gamma$ is satisfiable. Lemma 65 and the second formula $\boxed{\scriptstyle m}\, T \wedge \gamma \wedge \bigcirc \beta$ then guarantee that the formula $T \wedge \gamma \wedge \mathit{skip} \wedge \mathit{sfin}\ \beta$ is satisfiable. Lemma 14 then yields that the next formula is satisfiable:

$$\big((\$T)^{*} \wedge \alpha \wedge \mathit{more} \wedge \mathit{finite}\big); \big(T \wedge \gamma \wedge \mathit{skip} \wedge \mathit{sfin}\ \beta\big) \; .$$

From this and some further simple interval-based reasoning we can establish our goal, namely, that the formula $(\$T)^{*} \wedge \alpha \wedge \bigcirc\, \mathit{sfin}\ \beta$ is satisfiable. ■

10.2 Completeness for Transition Configurations

We now apply the material presented in the previous §10.1 to ultimately establish completeness for finite- and infinite-time transition configurations. Here is a summary of the completeness theorems for them:

Type of transition	Where proved
Finite-time	Theorem 69
Infinite-time	Theorem 70

The remaining two kinds of transition configurations are subordinate to these. For the sake of brevity, we do not consider them here.

THEOREM 69 *Completeness holds for any finite-time transition configuration $\Box T \wedge \mathit{init} \wedge \mathit{finite}$.*

Proof. From the consistency of the finite-time transition configuration $\Box T \wedge init \wedge finite$ and simple temporal reasoning we can demonstrate that for some V-atoms α and β, the next formula is consistent:

$$\boxed{m}\, T \wedge \alpha \wedge init \wedge sfin\,(T \wedge \beta) \ .$$

From this and further simple temporal reasoning it is readily follows that the following formulas are all consistent:

$$\alpha \wedge init \qquad \boxed{m}\, T \wedge \alpha \wedge \Diamond \beta \qquad T \wedge \beta \wedge empty \ .$$

The first of these is itself satisfiable since any consistent formula in PROP is satisfiable. The second one and Lemma 67 yields that the PITL formula $(\$T)^* \wedge \alpha \wedge sfin\ \beta$ is satisfiable. The third formula $T \wedge \beta \wedge empty$ is in NL^1 and hence by Theorem 64 satisfiable. Hence the following formulas are all satisfiable:

$$\alpha \wedge init \qquad (\$T)^* \wedge \alpha \wedge sfin\ \beta \qquad T \wedge \beta \wedge empty \ .$$

This and Theorem 50 then yield the satisfiability of the finite-time transition configuration $\Box T \wedge init \wedge finite$. ■

THEOREM 70 *Completeness holds for any infinite-time transition configuration* $\Box T \wedge init \wedge \Box \Diamond^+ L$.

Proof. From the consistency of the infinite-time transition configuration $\Box T \wedge init \wedge \Box \Diamond^+ L$ and simple temporal reasoning we can demonstrate that for some V-atoms α and β, the next formula is consistent:

(30) $\boxed{m}\, T \wedge \alpha \wedge init \wedge \Box \Diamond^+(\beta \wedge L) \ .$

Lemma 44 ensures that the formulas $\beta \wedge L$ and $\beta \wedge En_{L,\beta}$ are semantically equivalent. The proof of this only requires simple propositional reasoning not involving the temporal operators in L. Hence the next equivalence is readily deducible as a PTL theorem using substitution into a propositional tautology (see Definition 2 and PTL inference rule R1 in Table 7):

(31) $\vdash \quad \beta \wedge L \ \equiv \ \beta \wedge En_{L,\beta} \ .$

From the consistency of formula (30) and the deducibility of formula (31), we can show the consistency of the next formula:

$$\boxed{m}\, T \wedge \alpha \wedge init \wedge \Box \Diamond^+(\beta \wedge En_{L,\beta}) \ .$$

This and simple temporal reasoning then together yield the consistency of the following formulas involving some additional V-atoms $\gamma_1, \ldots, \gamma_{|En_{L,\beta}|}$ (not necessarily distinct):

$$\begin{array}{llll} \alpha \wedge \mathit{init} & \boxed{\mathrm{m}}\, T \wedge \alpha \wedge \Diamond \beta & \boxed{\mathrm{m}}\, T \wedge \beta \wedge \Diamond^{+} \beta & \\ \text{for each } \gamma_i\text{:} & \boxed{\mathrm{m}}\, T \wedge \beta \wedge \Diamond \gamma_i & \gamma_i \wedge \theta_{En_{L,\beta}[i]} & \boxed{\mathrm{m}}\, T \wedge \gamma_i \wedge \Diamond \beta \ . \end{array}$$

The consistency of the propositional formulas $\alpha \wedge \mathit{init}$ and $\gamma_i \wedge \theta_{En_{L,\beta}[i]}$ for each V-atom γ_i ensures they are satisfiable. Lemma 67 is then applied to the remaining consistent formulas, except for $\boxed{\mathrm{m}}\, T \wedge \beta \wedge \Diamond^{+} \beta$ which requires Lemma 68. The combined result is that the following formulas are all satisfiable:

$$\begin{array}{llll} \alpha \wedge \mathit{init} & (\$T)^{*} \wedge \alpha \wedge \mathit{sfin}\ \beta & (\$T)^{*} \wedge \beta \wedge \bigcirc\, \mathit{sfin}\ \beta & \\ \text{for each } \gamma_i\text{:} & (\$T)^{*} \wedge \beta \wedge \mathit{sfin}\ \gamma_i & \gamma_i \wedge \theta_{En_{L,\beta}[i]} & (\$T)^{*} \wedge \gamma_i \wedge \mathit{sfin}\ \beta \ . \end{array}$$

Hence by Theorem 53, the original consistent infinite-time transition configuration is indeed satisfiable. ■

11 Invariants and Related Formulas

We now introduce the concepts of invariants and invariant configurations which act as a middle level between transition configurations and full PTL. Satisfiability, existence of small models, decidability and axiomatic completeness for invariant configurations can be readily related to the analysis of transition configurations. Furthermore, it is not hard to reduce arbitrary PTL formulas to invariant configurations.

The analysis of invariant configurations and arbitrary PTL formulas does not require any further interval-based reasoning or PITL.

DEFINITION 71 (Invariant) *An* invariant *is any finite conjunction of zero or more equivalences in which each equivalence's left side is a distinct propositional variable and each equivalence's right side is one of the following:*

- *Some* PTL *formula of the form $\Diamond w$, where w is itself a propositional formula containing no temporal constructs.*
- *Some* NL^1 *formula.*

The variables occurring on the left sides of equivalences are called *dependent variables* and any other variables are called *independent variables.* The right sides are called *dependent formulas* and each equivalence is itself called a *dependency.* Hence for a given invariant I, it follows that $|I|$ denotes the number of dependencies in I. Also, for any $k : 1 \leq k \leq |I|$, $I[k]$ denote the k-th dependency in I. Each dependency containing $\Diamond$ is referred to as a *$\Diamond$-dependency.* Observe that a dependent variable can be referenced in any dependent formula including the one associated with it.

Below is a sample invariant referred to as I_1:

$$I_1: \quad (r_1 \equiv \Diamond(p \wedge \neg q)) \ \wedge \ (r_2 \equiv (r_1 \wedge \bigcirc r_2)) \ .$$

Here $|I_1|$ equals 2, the first dependency $I[1]$ is the equivalence $r_1 \equiv \Diamond(p \wedge \neg q)$ and the second dependency $I[2]$ is the equivalence $r_2 \equiv (r_1 \wedge \bigcirc r_2)$.

Note that an invariant is not necessarily satisfiable as in $r_1 \equiv \neg r_1$.

We can view an invariant I as being any conjunction having the form $\bigwedge_{k:1\leq k\leq |I|}(u_k \equiv \phi_k)$ so that u_k is the k-th dependent propositional variable and ϕ_k is the k-th dependent formula in I. Observe that for any $k : 1 \leq k \leq |I|$, the conjunct $I[k]$ has the form $u_k \equiv \phi_k$ and I itself can be expressed as $\bigwedge_{k:1\leq k\leq |I|} I[k]$.

Starting with an invariant I and a state formula w containing only variables occurring in I, we analyse certain low-level formulas referred to here as *invariant configurations.*

DEFINITION 72 (Invariant Configurations) *An* invariant configuration *is a formula of the form* $\Box I \wedge X$ *where the* PTL *formula* X *only has variables occurring in* I *and is in one of three categories shown below:*

Type of invariant configuration	*Syntax of X*
Basic	w
Finite-time	$w \wedge$ *finite*
Infinite-time	$w \wedge$ *inf*

Here w *is a state formula.*

For example, the conjunction $\Box I_1 \wedge r_2$ is a basic invariant configuration which is true for intervals which are infinite, have r_1 and r_2 always true and p and $\neg q$ both always eventually true.

Note: In order to somewhat simplify the notation in the reduction of invariant configurations to transition configurations, we assume without loss of generality that all of an invariant's conjuncts containing $\Diamond$ precede any others. It is not hard to re-order an arbitrary invariant's conjuncts to obtain a semantically equivalent one which obeys this requirement.

We now associate with an invariant I a transition formula T_I and a conditional liveness formula L_I. They serve to expeditiously reduce invariant configurations to transition configurations previously analysed in earlier sections. Definition 73 below describes T_I. The subsequent Definition 75 describes the form of L_I.

DEFINITION 73 (Transition Formula for an Invariant) *For any invariant I, the associated transition formula T_I is an* NL^1 *formula which captures I's transitional behaviour between pairs of adjacent states. It is defined to be the conjunction of equivalences in which each dependency in I containing $\Diamond$ and hence of the form $r \equiv \Diamond w$, for some propositional variable r and state formula w, is replaced by the $\Diamond$-free equivalence $r \equiv (w \vee \bigcirc r)$.*

Observe that the transition formula T_I is in NL^1 and is also a well-formed invariant. Also, for any $k : 1 \leq k \leq |I|$, if the dependency $I[k]$ does not contain $\Diamond$, then it and T_I's corresponding dependency $T_I[k]$ are identical.

Here is the transition formula T_{I_1} associated with I_1:

$$T_{I_1}: \quad \big(r_1 \equiv ((p \wedge \neg q) \vee \bigcirc r_1)\big) \;\wedge\; \big(r_2 \equiv (r_1 \wedge \bigcirc r_2)\big) \ .$$

Let us now introduce some simple notation needed for reasoning about liveness and $\Diamond$-dependencies. This will be used in the definition of an invariant's associated conditional liveness formula.

DEFINITION 74 (Liveness Tests of an Invariant) *For any invariant I having n $\Diamond$-dependencies, define n different* liveness tests $\theta_{I[1]}, \ldots, \theta_{I[n]}$ *so that for each $k : 1 \leq k \leq n$, the k-th dependency in I is expressible as $u_k \equiv \Diamond \theta_{I[k]}$.*

For instance, the sample invariant I_1 has a single liveness test $\theta_{I[1]}$ which denotes the formula $p \wedge \neg q$. Note that each $\theta_{I[k]}$ is always a state formula. If an invariant I has n $\Diamond$-dependencies, then for each $k : 1 \leq k \leq n$, T_I's dependency $T_I[k]$ identical to the equivalence $u_k \equiv (\theta_{I[k]} \vee \bigcirc u_k)$.

Given an invariant I, we now associate a specific conditional liveness formula L_I with it:

DEFINITION 75 (Conditional Liveness Formula of an Invariant) *The conditional liveness formula L_I of an invariant I which has n $\Diamond$-dependencies is itself a conjunction of n implications. For each $k : 1 \leq k \leq n$, the k-th implication is obtained by simply replacing the outermost equivalence operator in I's k-th $\Diamond$-dependency by the implication operator and using $\Diamond^{\mathrm{m}}$ instead of $\Diamond$. Therefore, for each $k : 1 \leq k \leq n$, the dependency $I[k]$ has the form $u_k \equiv \Diamond \theta_{I[k]}$ and the implication $L_I[k]$ has the form $u_k \supset \Diamond^{\mathrm{m}} \theta_{I[k]}$.*

The definition of I's conditional liveness formula L_I intentionally ignores any NL^1 dependencies in I since T_I already adequately deals with them. As a result, L_I can contain fewer conjuncts than I and T_I. Below is the conditional liveness formula L_{I_1} associated with invariant I_1:

$$L_{I_1}: \quad (r_1 \supset \Diamond^{\mathrm{m}}(p \wedge \neg q)) \ .$$

It is not hard to see that, unlike an invariant's transition formula, the invariant's associated conditional liveness formula is not a well-formed invariant.

11.1 Reduction of Basic Invariant Configurations

We now consider the relationship between a basic invariant configuration and its associated finite-time and infinite-time invariant configurations. This permits us to focus the remaining analysis on the two later kinds of invariant configurations.

LEMMA 76 *A basic invariant configuration* $\Box I \wedge w$ *is satisfiable iff at least one of its associated finite-time and infinite-time invariant configurations is satisfiable.*

Proof. This follows from the validity of the formula *finite* $\vee$ *inf* and simple propositional reasoning. ∎

Each finite-time and infinite-time invariant configuration has a corresponding semantically equivalent transition configuration of the same kind as is now shown:

	Invariant configuration	Transition configuration	Where proved
Finite time	$\Box I \wedge w \wedge \mathit{finite}$	$\Box T_I \wedge w \wedge \mathit{finite}$	Theorem 78
Infinite time	$\Box I \wedge w \wedge \mathit{inf}$	$\Box T_I \wedge w \wedge \Box \Diamond^+ L_I$	Theorem 85

Observe that the reductions from the two types of the invariant configurations to the corresponding transition configurations do not introduce any extra variables. In what follows we prove that a finite-time invariant configuration is semantically equivalent to its associated finite-time transition configuration and similarly a infinite-time invariant configuration is semantically equivalent to its associated infinite-time transition configuration.

LEMMA 77 *The formulas* $\Box I$ *and* $\Box T_I$ *are semantically equivalent on finite intervals. In other words, the following implication is valid:*

$$\models \quad \mathit{finite} \quad \supset \quad \Box I \equiv \Box T_I \ .$$

Proof. We can represent $\Box I$ as the conjunction $\bigwedge_{k:1 \le k \le |I|} \Box I[k]$ and similarly represent $\Box T_I$ as the conjunction $\bigwedge_{k:1 \le k \le |I|} \Box T_I[k]$. For any $k : 1 \le k \le |I|$, if $I[k]$ is in NL^1 then $T_I[k]$ is identical to it and hence $\Box I[k]$ and $\Box T_I[k]$ are identical. Otherwise, $\Box I[k]$ can be seen as a substitution instance of the PTL formula $\Box(p \equiv \Diamond q)$ containing two propositional variables p and q. Now $\Box T_I[k]$ therefore corresponds to the formula $\Box(p \equiv (q \vee \bigcirc p))$. Simple temporal reasoning can then be used to show that each of these implies the other in any finite interval. ∎

Let us note that the validity for finite time of the relevant equivalence $\Box(p \equiv \Diamond q) \equiv \Box(p \equiv (q \vee \bigcirc p))$ can even be readily checked by a computer implementation of a decision procedure for PTL with finite time.

THEOREM 78 *A finite-time invariant configuration is semanticallyequivalent to its associated finite-time transition configuration.*

Proof. This readily follows from Lemma 77 and propositional reasoning. ■

Unfortunately, the equivalent $\Box I \equiv \Box T_I$ can fail to be valid for infinite time if I contains $\Diamond$-dependencies. This is because the formulas $\Box(p \equiv \Diamond q)$ and $\Box(p \equiv (q \vee \bigcirc p))$ are not semantically equivalent on infinite-time intervals since the second formula does not necessarily imply the first one. For example, in an infinite interval where p is always true and q is always false, the second formula $\Box(p \equiv (q \vee \bigcirc p))$ is true but the first one $\Box(p \equiv \Diamond q)$ is false. Therefore, in infinite time, if I contains $\Diamond$-dependencies, then $\Box I$ is not necessarily equivalent to $\Box T_I$ since $\Box T_I$ does not fully capture I's $\Diamond$-dependencies. However, the next lemma holds even for infinite time:

LEMMA 79 *The* PTL *implication* $\Box I \supset \Box T_I$ *is valid.*

Proof. As before, we represent $\Box I$ as the conjunction $\bigwedge_{k:1 \leq k \leq |I|} \Box I[k]$ and similarly represent $\Box T_I$ as the conjunction $\bigwedge_{k:1 \leq k \leq |I|} \Box T_I[k]$. The lemma's proof then reduces to showing that for each $k : 1 \leq k \leq |I|$, the implication $\Box I[k] \supset \Box T_I[k]$ is valid. If $I[k]$ is an NL^1 formula, than $T_I[k]$ is identical to it and therefore the formula $\Box I[k] \supset \Box T_I[k]$ is trivially valid. Otherwise, we make use of the valid PTL formula $\Box(p \equiv \Diamond q) \supset \Box(p \equiv (q \vee \bigcirc p))$. ■

We just noted in Lemma 79 that the formula $\Box I \supset \Box T_I$ is valid for both finite and infinite time. However if I contains $\Diamond$-dependencies, then the converse implication $\Box T_I \supset \Box I$ is not necessarily valid for infinite time because the implication $\Box(p \equiv (q \vee \bigcirc p)) \supset \Box(p \equiv \Diamond q)$ can fail to be true. However, the following weaker implication is valid for infinite time and turns out to be useful:

(32) $\models \quad \Box(p \equiv (q \vee \bigcirc p)) \quad \supset \quad \Box(\Diamond q \supset p)$.

Here we use the formula $\Diamond q \supset p$ instead of the stronger equivalence $p \equiv \Diamond q$. Let us now replace p by u_k and q by $\theta_{I[k]}$. The resulting implication's antecedent is $\Box T_I[k]$ and the consequent is $\Box(\Diamond \theta_{I[k]} \supset u_k)$. Clearly the consequent is weaker that $\Box I[k]$ (i.e., $\Box(u_k \equiv \Diamond \theta_{I[k]})$).

It readily follows from L_I's definition together with the preceding discussion about the relationship between $\Box I$ and $\Box T_I$ in infinite time that the equivalence $\Box I \equiv (\Box T_I \wedge \Box L_I)$ is valid for infinite intervals as is the semantically equivalent, more concise one $\Box I \equiv \Box(T_I \wedge L_I)$. Consequently, we have the following lemma:

LEMMA 80 *The formula* $inf \supset \big(\Box I \equiv \Box(T_I \wedge L_I)\big)$ *is valid.*

Proof. Only the $\Diamond$-dependencies in I and T_I differ. Hence, the proof mainly involves checking that the following similar equivalence holds for each $\Diamond$-dependency $I[k]$ for $k : 1 \leq k \leq |L_I|$:

$$\models \quad inf \supset \big(\Box I[k] \equiv \Box(T_I[k] \wedge L_I[k])\big) \ .$$

This is a substitution of the following valid PTL formula:

$$\models \quad inf \supset \Big(\Box(p \equiv \Diamond q) \ \equiv\ \Box\big((p \equiv (q \vee \bigcirc p)) \ \wedge\ (p \supset \Diamond q)\big)\Big) \ .$$

The proof of validity uses the valid implication (32). ■

The next few lemmas lead up to Theorem 85 which establishes the semantic equivalence of an infinite-time invariant configuration and its associated infinite-time transition configuration:

LEMMA 81 *The implication* $T_I \wedge \bigcirc L_I \supset L_I$ *is valid.*

Proof. Ensuring validity reduces to showing that for each $k : 1 \leq k \leq |L_I|$, the implication $T_I[k] \wedge \bigcirc L_I[k] \supset L_I[k]$ is valid. Now $I[k]$ is a $\Diamond$-dependency. Therefore $T_I[k]$ denotes $u_k \equiv \Diamond \theta_{I[k]}$ and $L_I[k]$ denotes $u_k \supset \Diamond_{\mathrm{m}} \theta_{I[k]}$. We make use of the valid PTL formula now given which captures the essence of the implication $T_I[k] \wedge \bigcirc L_I[k] \supset L_I[k]$:

$$\models \quad (p \equiv (q \vee \bigcirc p)) \ \wedge\ \bigcirc(p \supset \Diamond_{\mathrm{m}} q) \quad \supset \quad (p \supset \Diamond_{\mathrm{m}} q) \ .$$

By substituting u_k into p and $\theta_{I[k]}$ into q, we obtain the validity of the formula $T_I[k] \wedge \bigcirc L_I[k] \supset L_I[k]$. ■

LEMMA 82 *The implication* $\Box(T_I \wedge \Diamond^+ L_I) \supset \Box L_I$ *is valid.*

Proof. We use simple temporal reasoning to ensure the validity of the following PTL implication:

$$\models \quad \Box((p \wedge \bigcirc q) \supset q) \quad \supset \quad \big(\Box(p \wedge \Diamond^+ q) \supset \Box q\big) \ .$$

Substitution of T_I and L_I into p and q, respectively, together with Lemma 81 (to obtain $\models \Box(T_I \wedge \bigcirc L_I \supset L_I)$) and modus ponens yields our goal. ■

LEMMA 83 *The equivalence* $\Box(T_I \wedge L_I) \equiv \Box(T_I \wedge \Diamond^+ L_I)$ *is valid.*

Proof. We use simple temporal reasoning to ensure the validity of the following PTL implication:

$$\models \quad \big(\Box(p \wedge \Diamond^+ q) \supset \Box q\big) \quad \supset \quad \big(\Box p \supset (\Box \Diamond^+ q \equiv \Box q)\big) \ .$$

The substitution of T_I and L_I into p and q, respectively, together with Lemma 82 and modus ponens yields the validity of the next formula:

$$\models \quad \Box T_I \quad \supset \quad (\Box \Diamond^+ L_I \equiv \Box L_I) \ .$$

This and some simple temporal reasoning ensures our goal, namely, the validity of the equivalence $\Box(T_I \wedge L_I) \equiv \Box(T_I \wedge \Diamond^+ L_I)$. ■

LEMMA 84 *The formula* $\mathit{inf} \supset \big(\Box I \equiv \Box(T_I \wedge \Diamond^+ L_I)\big)$ *is valid.*

Proof. We use Lemmas 80 and 83 and propositional reasoning. ■

THEOREM 85 *An infinite-time invariant configuration is semantically equivalent to its associated infinite-time transition configuration.*

Proof. This readily follows from Lemma 84 and simple temporal reasoning. ■

The soundness of the reductions to the associated transition configurations ensures that we can use the decision procedure described in Sect. 8.

11.2 Bounded Models for Invariant Configurations

Let V be the set of variables occurring in the invariant I. The theorem given below gives the small model property for basic invariant configurations:

THEOREM 86 *A basic invariant configuration* $\Box I \wedge w$ *is satisfiable iff it is satisfied by some some finite interval with interval length less than* $|\mathit{Atoms}_V|$ *or by an infinite, ultimately periodic one consisting of an initial segment with interval length at most* $|\mathit{Atoms}_V|$ *fused with a remaining infinite periodic part with a period having interval length at most* $(|L_I| + 1)\,|\mathit{Atoms}_V|$.

Proof. Suppose $\Box I \wedge w$ is satisfiable. We will consider the two cases of finite and infinite intervals separately:

Case for finite intervals: Theorem 78 ensures that the finite-time invariant configuration $\Box I \wedge w \wedge \mathit{finite}$ and its associated finite-time transition configuration $\Box T_I \wedge w \wedge \mathit{finite}$ are semantically equivalent. The construction of T_I ensures that any variable occurring in it is a member of the set

V. Lemma 42 therefore establishes that if the conjunction $\Box T_I \wedge w \wedge \mathit{finite}$ is satisfiable, then a satisfying interval exists having less interval length than $|\mathit{Atoms}_V|$. This interval consequently also satisfies the basic invariant configuration $\Box I \wedge w$.

Case for infinite intervals: Theorem 85 ensures that the infinite-time invariant configuration $\Box I \wedge w \wedge \mathit{inf}$ and its associated infinite-time transition configuration $\Box T_I \wedge w \wedge \Box \Diamond^+ L_I$ are semantically equivalent. From Lemma 49 we have that this second formula is satisfied by an infinite interval consisting of an initial segment having interval length less than $|\mathit{Atoms}_V|$ fused with a periodic interval with period having interval length at most $(|L_I| + 1)\,|\mathit{Atoms}_V|$. The overall ultimately periodic interval therefore also satisfies the formula $\Box I \wedge w$. ■

11.3 Axiomatic Completeness for Basic Invariant Configurations

THEOREM 87 *Completeness holds for basic invariant configurations.*

Proof. Subsection 11.1 describes how to construct a transition configuration from a finite-time or infinite-time invariant configuration. The various valid formulas involved there can be deduced as PTL theorems. This and the previously shown axiomatic completeness for finite-time or infinite-time transition configurations proved in Theorems 69 and 70 ensure that any consistent finite-time or infinite-time invariant configuration is satisfiable. Now if a basic invariant configuration is consistent, then at least one of the associated finite-time or infinite-time invariant configurations is as well. Any interval satisfying it can also be used for the basic invariant configuration. This demonstrates the desired axiomatic completeness. ■

12 Dealing with Arbitrary PTL Formulas

So far we have only looked at bounded models and axiomatic completeness for certain kinds of PTL formulas. For an arbitrary PTL formula X, it is straightforward to construct an invariant I linearly bounded by the size of X and containing a finite number of dependent variables $u_1, u_2, \ldots, u_{|I|}$ not themselves occurring in X so as to mimic the semantics of X in the sense that X is satisfiable iff $\Box I \wedge u_{|I|}$ is satisfiable and in addition the implication $\Box I \supset (u_{|I|} \equiv X)$ is valid.

One possible translation will be detailed shortly. Before describing it, we need to discuss a convention for systematically renaming an invariant's dependent variables. Normally, the first dependent variable in an invariant I constructed here from an PTL formula is r_1 and the last is $r_{|I|}$. However, we inductively construct the invariants by combining smaller invariants into

X	$\mathcal{H}(X)$
T	$r_1 \equiv T$, for any NL^1 formula T.
$\neg Y$	$\mathcal{H}(Y) \wedge (r_{\lvert\mathcal{H}(Y)\rvert+1} \equiv \neg r_{\lvert\mathcal{H}(Y)\rvert})$
$Y \vee Y'$	$\mathcal{H}(Y) \wedge \mathcal{H}(Y') \uparrow m \wedge (r_{m+n+1} \equiv r_m \vee r_{m+n})$, where $m = \lvert\mathcal{H}(Y)\rvert$ and $n = \lvert\mathcal{H}(Y')\rvert$.
$\Diamond Y$	$\mathcal{H}(Y) \wedge (r_{\lvert\mathcal{H}(Y)\rvert+1} \equiv \Diamond r_{\lvert\mathcal{H}(Y)\rvert})$

Table 9. Definition of $\mathcal{H}(X)$

larger ones and often must alter the indices of the dependent variables to avoid clashes. A operator on formulas to suitably do this is now defined:

DEFINITION 88 (Shifting of Subscripts in Invariants) *For any invariant I, the operation $I \uparrow k$ is defined to be the invariant obtained by replacing $u_1, \ldots, u_{\lvert I\rvert}$ by $r_{1+k}, \ldots, r_{\lvert I\rvert+k}$, i.e., $I^{r_{1+k},\ldots,r_{\lvert I\rvert+k}}_{u_1,\ldots,u_{\lvert I\rvert}}$.*

It is not hard to see that if I's dependent variables are themselves the distinct variables $r_1, \ldots, r_{\lvert I\rvert}$, then $I \uparrow k$ shifts the subscripts of them so that each r_j becomes r_{j+k}. Therefore, the first dependent variable becomes r_{1+k} instead of r_1, the second becomes r_{2+k} and so forth. In other words, $I \uparrow k$ denotes the same formula as the conjunction $\bigwedge_{1 \le j \le \lvert I\rvert}(r_{j+k} \equiv (\phi_j)^{r_{1+k},\ldots,r_{\lvert I\rvert+k}}_{u_1,\ldots,u_{\lvert I\rvert}})$.

Without loss of generality, let X be an PTL formula which does not contain any of the variables $r_1, r_2, \ldots$. Table 9 contains the definition of a function $\mathcal{H}(X)$ which translates X into an invariant containing some of the variables r_1, r_2, ... as dependent variables. In order to reduce the number of dependent variables, the first case is used whenever the formula is in NL^1 even if one of the next two cases for negation and logical-or is applicable.

Table 10 contains a sample PTL formula X_0, an equivalent formula X_0' having no logical-ands, implications or $\Box$ constructs, and the invariant $\mathcal{H}(X_0')$ and the initial condition $r_{\lvert\mathcal{H}(X_0')\rvert}$. We also include a version of X_0' which shows how the dependencies correspond to the subformulas in X_0'.

It is straightforward to utilise more sophisticated methods which construct invariants directly from formulas with other logical operators such as logical-and and $\Box$. In addition, it is not hard to systematically produce invariants containing a lot fewer dependencies then the ones generated by $\mathcal{H}$. In fact, our prototype implementation of the decision procedure described in Sect. 8 makes use of such techniques and others as well. Here is an invariant and initial formula produced by the decision procedure directly

X_0:	$\Box\big((p \supset \bigcirc\Diamond q) \wedge \Diamond(\neg p \vee \bigcirc q)\big)$
X'_0:	$\neg\Diamond\neg\Big(\neg(\neg(\neg p \vee \bigcirc\Diamond q) \vee \neg\Diamond(\neg p \vee \bigcirc q))\Big)$
X'_0 with dependent variables shown:	$\overbrace{\neg \underbrace{\Diamond \overbrace{\neg \Big(\underbrace{\neg \big(\overbrace{\underbrace{\neg \overbrace{(\underbrace{\neg p}_{r_1} \vee \underbrace{\bigcirc \overbrace{\Diamond \underbrace{q}_{r_2}}^{r_3}}_{r_4})}^{r_5}}_{r_6} \vee \underbrace{\neg \overbrace{\Diamond \underbrace{(\neg p \vee \bigcirc q)}_{r_7}}^{r_8}}_{r_9}}^{r_{10}} \big)}_{r_{11}} \Big)}^{r_{12}}}_{r_{13}}}^{r_{14}}$
$\mathcal{H}(X'_0)$:	$(r_1 \equiv \neg p) \wedge (r_2 \equiv q) \wedge (r_3 \equiv \Diamond r_2) \wedge (r_4 \equiv \bigcirc r_3)$ $\wedge\ (r_5 \equiv (r_1 \vee r_4)) \wedge (r_6 \equiv \neg r_5) \wedge (r_7 \equiv (\neg p \vee \bigcirc q))$ $\wedge\ (r_8 \equiv \Diamond r_7) \wedge (r_9 \equiv \neg r_8) \wedge (r_{10} \equiv (r_6 \vee r_9))$ $\wedge\ (r_{11} \equiv \neg r_{10}) \wedge (r_{12} \equiv \neg r_{11}) \wedge (r_{13} \equiv \Diamond r_{12})$ $\wedge\ (r_{14} \equiv \neg r_{13})$
$r_{\lvert\mathcal{H}(X'_0)\rvert}$:	r_{14}

Table 10. Example of invariant obtained by applying $\mathcal{H}$ to a PTL formula

from the formula X_0:

$$\begin{array}{ll} X_0\colon & \Box((p \supset \bigcirc\Diamond q) \wedge \Diamond(\neg p \vee \bigcirc q)) \\ I'\colon & (r_1 \equiv \Diamond q) \wedge (r_2 \equiv \bigcirc r_1) \wedge (r_3 \equiv \bigcirc q) \wedge (r_4 \equiv \Diamond(\neg p \vee r_3)) \\ & \quad \wedge\ (r_5 \equiv \Diamond \neg((p \supset r_2) \wedge r_4)) \\ \mathit{init}'\colon & \neg r_5 \end{array}$$

We omit further details.

It is easy to check that $\mathcal{H}(X)$ contains at most one dependent variable for each variable and operator in X so the total number of dependent variables in $\mathcal{H}(X)$ is bounded by X's size and indeed the size of $\mathcal{H}(X)$ is linearly bounded by X's size. It is also easy to check by doing induction on X's syntactic structure that X is satisfiable iff the basic invariant configuration $\Box\,\mathcal{H}(X) \wedge r_{|\mathcal{H}(X)|}$ is satisfiable. Furthermore, the implication $\Box\,\mathcal{H}(X) \supset (r_{|\mathcal{H}(X)|} \equiv X)$ can be shown to be valid. Consequently, $\Box\,\mathcal{H}(X) \wedge r_{|\mathcal{H}(X)|}$ is used to represent X's behaviour (modulo the dependent variables which act as auxiliary ones). The bounded model for the invariant configuration (see Theorem 86) satisfies X as well. The decision procedure described in Sect. 8 can be utilised to check the satisfiability of arbitrary PTL formulas by reducing them first to basic invariant configurations and then testing the associated finite-time and infinite-time transition configurations (see §11.1). Axiomatic completeness for X readily reduces to that for the invariant configuration $\Box\,\mathcal{H}(X) \wedge r_{|\mathcal{H}(X)|}$.

13 Some Additional Features

This section describes a number of extensions to our approach. They include the temporal operator *until* and past-time constructs and also a subset of PITL called *Fusion Logic* (FL) which includes constructs of the sort found in Propositional Dynamic Logic (PDL). In addition, the liveness tests found in conditional liveness formulas and invariants can be generalised to be of the form $\circledast T$, where T is an NL^1 formula, rather than just a state formula. We will consider each of these issues in turn. For the sake of brevity, the presentation is briefer and less formal than in the previous sections.

13.1 The Operator *until*

The operator *until* is a binary operator with the syntax $X\,\mathcal{U}\,Y$, where X and Y are PTL formulas. Recall from Sect. 4 that for any interval σ and natural number k which does not exceed σ's interval length, $\sigma_{k:|\sigma|}$ denotes the suffix subinterval obtained by deleting the first k states from σ. Here is the semantics of *until*:

$$\sigma \models X \mathcal{U} Y \quad \text{iff}$$
$$\text{for some } k \leq |\sigma|,\ \sigma_{k:|\sigma|} \models Y \text{ and for all } j : 0 \leq j < k, \sigma_{j:|\sigma|} \models X \ .$$

Observe that the operator $\Diamond$ can be expressed in terms of *until* since $\Diamond X$ is semantically equivalent to the formula *true until* X.

We can alter the definition of invariants by replacing $\Diamond$-dependencies with dependencies of the form $r \equiv (w \mathcal{U} w')$, where w and w' are state formulas. If the j-th dependency $I[j]$ of an invariant I is such a dependency (called an *until-dependency*), then the corresponding conjunction $T_I[j]$ in I's transition formula T_I has the form $r \equiv (w' \vee (w \wedge \bigcirc r))$. The associated conjunction $L_I[j]$ in L_I is $r \supset \Diamond\!\!\!\!\!\!\bigcirc\, w'$. It is not hard to modify the material in Sect. 11 to ensure that finite-time and infinite-time invariant configurations remain semantically equivalent to the associated transition configurations.

13.2 Past Time

Let us now consider PTL with a bounded past. The syntax is modified to include the two additional primitive operators $\ominus X$ (read *previous* X) and $\diamondminus X$ (read *once* X). The set of PTL formulas including past-time constructs is denoted as PTL^{P}. The semantics of a PTL formula X is now expressed as $(\sigma, k) \models X$ where k is any natural number not exceeding $|\sigma|$. For example, the semantics of $\ominus$ and $\diamondminus$ are as follows:

$$(\sigma, k) \models \ominus X \quad \text{iff} \quad k > 0 \text{ and } (\sigma, k-1) \models X$$
$$(\sigma, k) \models \diamondminus X \quad \text{iff} \quad \text{for some } j : 0 \leq j \leq k,\ \ (\sigma, j) \models X \ .$$

We define the operator $\boxminus X$ (read *so-far* X) as $\neg \diamondminus \neg X$ and the operator $\circleddash X$ (read *weak previous* X) as $\neg \ominus \neg X$. The operator *first* is defined to be $\neg \ominus$ *true* and tests for the first state of an interval. An past-time version of *until* called *since* can also be included but we omit the details.

A PTL^{P} formula X is defined to satisfiable iff $(\sigma, k) \models X$ holds for some pair (σ, k) with $k \leq |\sigma|$. The formula X is valid iff $(\sigma, k) \models X$ holds for every pair (σ, k) with $k \leq |\sigma|$. Note that these straightforward definitions of satisfiability and validity correspond to the so-called *floating framework* of PTL with past time. However, Manna and Pnueli propose another interesting approach called the *anchored framework* [Manna and Pnueli, 1989] (also discussed in [Lichtenstein and Pnueli, 2000]) which they argue is superior. In this framework, satisfiability and validity only examine pairs of the form $(\sigma, 0)$. There exist ways to go between the two conventions but we will not delve into this here and instead simply assume the more traditional floating interpretation.

We now define an analogue of the set of formulas NL:

DEFINITION 89 (Previous Logic) *The set of* PTL *formulas in which the only primitive temporal operator is* $\ominus$ *is called* Previous Logic (PrevL). *The subset of* PrevL *with no* $\ominus$ *nested in another* $\ominus$ *is denoted as* PrevL^1.

We let the variables Z and Z' denote formulas in PrevL^1. Also, PrevL^1_V denotes the set of all formulas in PrevL^1 only having variables in V.

The following definitions extend the notation of transition configurations to deal with past time:

DEFINITION 90 (Past-Time Transition Configurations) *A* past-time transition configuration *is any formula of the form* $\boxminus\Box(T \wedge Z) \wedge X$, *where* T *is in* NL^1_V, Z *is in* PrevL^1_V, *and the formula* X *is in* PTL_V *and is in one of the two categories shown below:*

Type of configuration	*Syntax of X*
Finite-time	$w \wedge \mathit{finite}$
Infinite-time	$w \wedge \Box\Diamond^+ L$

Here w *is a state formula in* PROP_V *and* L *is a conditional liveness formula in* PTL_V.

The formula $\boxminus\Box(T \wedge Z)$ contains both $\boxminus$ and $\Box$ to ensure that both T and Z are true everywhere in the interval.

The analysis of a finite-time or infinite-time past-time transition configurations can be easily reduced to reasoning in PTL without past time. Let us demonstrate this by first examining how to test the satisfiability of a finite-time past-time transition configuration $\boxminus\Box(T \wedge Z) \wedge w \wedge \mathit{finite}$. This involves finding an interval σ and natural number $k \leq |\sigma|$, such that $(\sigma, k) \models \boxminus\Box(T \wedge Z) \wedge w \wedge \mathit{finite}$ holds. Note that this past-time transition configuration is satisfiable iff the following formula, which shifts reasoning back to an interval's starting state, is satisfiable:

$$(33) \quad \diamondminus\big(\Box(T \wedge Z) \wedge \mathit{first} \wedge \Diamond w \wedge \mathit{finite}\big) \ .$$

Here we can dispense with the operator $\boxminus$ since $\boxminus\Box$ and $\Box$ have the same semantics at the starting state.

Now for any $\mathrm{PTL}^{\mathrm{P}}$ formula X, the formula $\diamondminus X$ is satisfiable iff X is satisfiable. Hence, the formula (33) is satisfiable iff its subformula $\Box(T \wedge Z) \wedge \mathit{first} \wedge \Diamond w \wedge \mathit{finite}$ is satisfiable. Let us now define the NL^1_V formula T' by replacing each $\ominus$ construct in Z by its operand and by taking each state formula in Z which does not occur in $\ominus$ and enclosing it in $\bigcirc$. For example, if Z is the formula $p \vee \ominus(q \wedge r)$, then T' is $(\bigcirc p) \vee (q \wedge r)$. Furthermore, let w' be the state formula in PROP_V obtained from Z by replacing each $\ominus$ construct by *false*. In our example, w' is $p \vee \mathit{false}$. It can be readily checked that the following formula relating Z and T' is true at any interval's initial state: $\Box Z \equiv \boxed{\scriptstyle m}\, T' \wedge w'$. Therefore, the original finite-time past-time transition configuration is satisfiable iff the following formula in PTL without past time is satisfiable:

(34) $\Box(T \wedge (\mathit{more} \supset T')) \wedge w' \wedge \Diamond w \wedge \mathit{finite}$.

This is still not a well-formed finite-time transition configuration due to the presence of the formula $\Diamond w$. However, $\Diamond w$ can be reduced by introducing a new propositional variable r as shown in the next formula:

(35) $\Box\big(T \wedge (\mathit{more} \supset T') \wedge (r \equiv (w \vee \bigcirc r))\big) \wedge w' \wedge r \wedge \mathit{finite}$.

The reduction of the original past-time transition configuration $\boxminus\Box(T \wedge Z) \wedge w \wedge \mathit{finite}$ to the finite-time transition configuration (35) systematically relates all aspects of the analysis of the past-time transition configuration to the purely future-only reasoning presented earlier. This includes bounded models, decision procedures and axiomatic completeness.

An alternative way to reduce the PTL formula (34) involves interval-based reasoning. We first re-expressing the formula in PTL as the next semantically equivalent conjunction:

(36) $\boxed{\scriptstyle m}(T \wedge (\mathit{more} \supset T')) \wedge w' \wedge \Diamond w \wedge \mathit{sfin}\ T$.

This makes use of the valid PTL equivalence $(\Box X \wedge \mathit{finite}) \equiv (\boxed{\scriptstyle m} X \wedge \mathit{sfin}\ X)$, for any PTL formula X. However, in our case we can omit the subformula $\mathit{more} \supset T'$ in the *sfin* construct since the operator *more* ensures that it is trivially true in the associated empty interval. Let T'' denote the subformula $T \wedge (\mathit{more} \supset T')$. Theorem 18 ensures the semantic equivalence of $\boxed{\scriptstyle m}\, T''$ and $(\$\, T'')^*$. Now the formula (36) can in turn be itself re-expressed as the following chop-formula:

(37) $((T'')^* \wedge w' \wedge \mathit{finite});\ ((T'')^* \wedge w \wedge \mathit{sfin}\ T)$.

Let w'' denote a state formula obtained by replacing every $\bigcirc$ construct in T by *false*. Consequently, w'' is true exactly in states for which $T \wedge \mathit{empty}$ is true. It follows that we can test for satisfiability of formula (37) by adapting the symbolic methods mentioned in Sect. 8 to solve for V-atoms α, β and γ for which the following formulas are satisfiable:

$$\alpha \wedge w \qquad (\$\, T'')^* \wedge \alpha \wedge \mathit{sfin}\ \beta \qquad \beta \wedge w' \qquad (\$\, T'')^* \wedge \beta \wedge \mathit{sfin}\ \gamma \qquad \gamma \wedge w'' \ .$$

Further details are omitted here.

The treatment for a infinite-time past-time transition configuration is nearly identical to that for a finite-time one since the assumption of a bounded past still applies and avoids the need for a past-time conditional liveness formula. First of all, we replace the subformula *finite* by $\Box \Diamond^{+} L$.

$$\Box(T \wedge T') \wedge w' \wedge \Diamond w \wedge \Box \Diamond^{+} L \ .$$

The use of infinite time ensures we can omit the instance of *more* found in the finite-time formulas (34) and (35) since T and $more \supset T$ are semantically equivalent on an infinite interval. The formula $\Diamond w$ is itself reduced by introducing a new propositional variable r and conjoining a new implication to L to obtain the well-formed infinite-time transition configuration below:

$$\Box(T \wedge T') \wedge w' \wedge r \wedge \Box \Diamond^{+}(L \wedge (r \supset \overset{fin}{\Diamond} w)) \ .$$

So far we have only considered finite- and infinite-time transition configurations. Invariants (and hence also invariant configurations) can be extended to support past-time reasoning by adding two new kinds of dependencies. The first has the form $u \equiv Z$ and the second has the form $u \equiv \diamondminus w$. The use of $\diamondminus$ does not involve I's conditional liveness formula L_I due to the assumption of a bounded past. The definitions of invariant configurations remain the same and the reduction of them to past-time transition configurations is straightforward since no dependency contains both future- and past-time temporal constructs. Furthermore, dependencies containing the temporal operator *since* (a conventional past-time analogue of the operator *until*) are not much harder to handle than $\diamondminus$-dependencies. The reduction of an arbitrary $\mathrm{PTL}^{\mathrm{P}}$ formula to an invariant with past time is also straightforward.

13.3 Generalised Conditional Liveness Formulas and Invariants

Conditional liveness formulas and invariants require that any operand of $\overset{fin}{\Diamond}$ and $\Diamond$, respectively, is a state formula. We can slightly relax this requirement and permit arbitrary formulas in NL^1. This makes invariants more succinct since a formula such as *sfin* w can now be expressed using only one dependency such as $u_k \equiv \Diamond(empty \wedge w)$ instead of requiring two. The formula $\Box \Diamond^{+} w$ can be expressed with the invariant $u_k \equiv \Diamond(w \wedge \bigcirc u_k)$. The overall analysis of such invariants only differs slightly from that for the basic version of invariants. Invariants with *until*-dependencies (see §13.1) can be analogously generalised to permit *until*-dependencies of the form $u_k \equiv (T \,\mathcal{U}\, T')$, where both T and T' are in NL^1.

13.4 Fusion Logic

Regular expressions are a standard notation for representing regular languages. However, within PITL, it is more appropriate to use languages based on the fusion operator rather than conventional concatenation. This involves a variation of regular expressions called here *fusion expressions.* We now define a PITL-based representation of them which is in fact a special subset of PITL formulas. This subset will then provide the basis for a generalisation of PTL called *Fusion Logic* (FL) which is also itself a subset of PITL. We originally used Fusion Logic in [Moszkowski, 2004a] as a kind of intermediate logic when we reduced the problem of showing axiomatic completeness of Propositional Interval Temporal Logic (PITL) with finite time to showing axiomatic completeness for PTL. Fusion Logic is closely related to Propositional Dynamic Logic (PDL) [Fischer and Ladner, 1977; Fischer and Ladner, 1979; Kozen and Tiuryn, 1990; Harel, 1984; Harel *et al.*, 2000; Harel *et al.*, 2002]. A major reason for discussing Fusion Logic here is because it is not hard to extend our decision procedure for PTL with finite time to also handle more expressive interval-oriented FL formulas by simply reducing FL formulas to lower level PTL formulas of the kinds already discussed. This demonstrates another link between PTL and intervals and has practical applications.

DEFINITION 91 (Fusion Expression Formulas) *The set of* fusion expression formulas, *denoted* FE, *consists of* PITL *formulas with the syntax given below, where w is a state formula, T is in* NL^1 *and E and F themselves denote* FE *formulas:*

$$w? \qquad E \vee F \qquad \$\, T \qquad E; F \qquad E^* \ .$$

The syntax of FE *formulas is like that of programs in Propositional Dynamic Logic without rich tests. However* FE *has a semantics based on sequences of states rather than binary relations.*

For any set of variables V, let FE_V *denote the set of* FE *formulas containing only variables in V.*

Unlike letters in conventional regular expressions, any nonmodal formula can be used in $w?$. For example, *false*? is permitted even though it is unsatisfiable. Consider the following FE formula:

$$\big((\$ \bigcirc p); (q?)\big) \vee (\$ \neg q)^*.$$

This is true on an interval if either the interval has exactly two states and p and q are both true in the second state or it has some arbitrary number of states, say k, with q false in each of the first $k-1$ states.

REMARK 92 (Expressing concatenation) *It is important to note that the conventional concatenation of two* FE *formulas E and F can be achieved through the use of the* FE *formula* $E; (\$\, true); F$. *Here* $\$\, true$ *is itself an*

FE *formula which is an alternative way to express the* PTL *operator skip.*
This temporal operation on E and F is sometimes called "chomp", *since it is a slight variation of chop. Hence, in the context of temporal logic,* FE *formulas can largely subsume regular expressions although there are slightly different conventions for such things as empty words. We omit the details.*

We now present the sublogic of PITL called here Fusion Logic. In essence, Fusion Logic augments conventional PTL with the fusion expression formulas already introduced.

DEFINITION 93 (Fusion Logic) *Here is the syntax of* FL *where p is any propositional variable, E is any* FE *formula and X and Y are themselves formulas in* FL*:*

$$p \qquad \neg X \qquad X \vee Y \qquad \bigcirc X \qquad \Diamond X \qquad \langle E \rangle X.$$

We define the new construct $\langle E \rangle X$ (called "FL-chop"*) and its dual $[E]X$ (called* "FL-yields"*) using the primitive* PITL *constructs chop and $\neg$:*

$$\langle E \rangle X \stackrel{\mathrm{def}}{\equiv} E;X \qquad [E]X \stackrel{\mathrm{def}}{\equiv} \neg\langle E \rangle \neg X.$$

Within an FL formula, $\bigcirc$, $\Diamond$ and FL-chop are treated as primitive constructs. Unlike PITL, FL limits the left sides of chop to being FE formulas.

In [Moszkowski, 2004a], we described an earlier version of FL having *skip* as a primitive FE formula instead of $\$T$. As we noted earlier in Remark 92, the PTL formula *skip* can be expressed in FE as $\$\,true$. The two versions of FL can readily be shown to be equally expressive since $\$T$ can be replaced with a semantically equivalent disjunction of formulas by using of ?, *skip* and chop. For example, the FE formula $\$(p \supset \bigcirc q)$ is semantically equivalent to the FE formula $((\neg p)?; skip) \vee (skip; q?)$. In practice, the version described here is much more natural and succinct.

Henriksen and Thiagarajan [Henriksen and Thiagarajan, 1997; Henriksen and Thiagarajan, 1999] investigate a formalism related to Wolper's ETL [Wolper, 1981; Wolper, 1983] and called *Dynamic Linear Time Temporal Logic* which combines PTL and PDL in a linear-time framework with infinite time. It is similar to our Fusion Logic and uses multiple atomic programs instead of the FE operators ? and \$.

REMARK 94 *The temporal operators $\bigcirc$ and $\Diamond$ which are primitives in* FL *can actually be expressed as instances of* FL-*chop if finite time is assumed:*

$$\models \bigcirc X \equiv \langle \$\,true \rangle X \qquad \models \Diamond X \equiv \langle (\$\,true)^* \rangle X.$$

In spite of FL being a proper subset of PITL, they have the same expressiveness. This is discussed in [Moszkowski, 2004a], where a hierarchical reduction of FL formulas to PTL formulas is also given but is limited to dealing with finite-time intervals. This reduction provides the basis of a decision procedure for FL with finite-time. We plan to describe in future work a hierarchical reduction to transition configurations (also restricted to finite-time). Such transition configurations can then be tested with the decision procedure described in Sect. 8. Like the first reduction in [Moszkowski, 2004a], this reduction can also be used for proving the completeness of an axiom system for FL with finite time.

14 Discussion

We conclude with a look at some issues connected with PTL and FL.

As noted earlier, a number of PTL decision procedures are tableau-based algorithms. These include ones described by Wolper [Wolper, 1985], Emerson [Emerson, 1990] and Lichtenstein and Pnueli [Lichtenstein and Pnueli, 2000]. It would be interesting to investigate whether a tableau-based algorithm could be naturally obtained from some kind of modified version of our analysis of transition configurations.

The BDD-based techniques described in Sect. 8 can be adapted to check in real time that an executing system is not violating assertions expressed in PTL or FL as it runs. Whether FL in particular is useful for this in practice is unclear. In addition, it would appear that the reachability analysis necessary for our approach to work can, as with Bounded Model Checking (BMC) [Clarke *et al.*, 2001], employ SAT-based techniques for PTL and FL instead of BDDs. However, such a SAT-based approach, unlike the BDD-based one, normally cannot exhaustively test for unsatisfiability because in BMC there is no notion corresponding to convergence of BDDs to the set of all atoms reachable from some starting one. Rather BMC works by employing SAT to find at most a single solution not exceeding some predetermined maximum bounded length which for practical reasons is generally much less than the worst-case bounds derived from formula syntax. If a solution is not found, this is typically not by itself sufficient to exclude the existence of larger satisfying intervals.

We have used versions of invariants, transition formulas and conditional liveness formulas to analyse Propositional Dynamic Logic (PDL) without the need for Fischer-Ladner closures. Indeed, this was the original motivation for conditional liveness formulas. However, at present the benefits and novelty of utilising our approach for PDL are less compelling than for PTL.

Acknowledgements

We thank Antonio Cau, Jordan Dimitrov, Rodolfo Gómez and Helge Janicke for comments on versions of this work. In the course of discussions, Howard Bowman, Shmuel Katz, Maciej Koutny and Simon Thompson also made helpful suggestions leading to improvements in the presentation of the material. We are especially grateful to Hussein Zedan for his patience and encouragement during the time this research was undertaken.

BIBLIOGRAPHY

[ANSI, 1999] ANSI. Common Lisp: Standard ANSI INCITS 226-1994 (R1999) (formerly ANSI X3.226-1994 (R1999)). URL: `http://www.ansi.org`, 1999.

[Banieqbal and Barringer, 1986] Behnam Banieqbal and Howard Barringer. A study of an extended temporal logic and a temporal fixed point calculus. Technical Report UMCS-86-10-2, Dept. of Computer Science, University of Manchester, England, October 1986. revised June 1987.

[Beer *et al.*, 2001] Ilan Beer, Shoham Ben-David, et al. The temporal logic Sugar. In Gérard Berry, Hubert Comon, and Alain Finkel, editors, *13th Conference on Computer-Aided Verification (CAV01), Paris, France, 18–22 July 2001*, volume 2102 of *LNCS*, pages 363–367, Berlin, 2001. Springer-Verlag.

[Ben-Ari *et al.*, 1981] Mordechai Ben-Ari, Zohar Manna, and Amir Pnueli. The temporal logic of branching time. In *Eighth ACM Symposium on Principles of Programming Languages*, pages 164–176. ACM, JAN 1981.

[Ben-Ari *et al.*, 1983] Mordechai Ben-Ari, Zohar Manna, and Amir Pnueli. The temporal logic of branching time. *Acta Informatica*, 20(3):207–226, 1983.

[Bernholtz *et al.*, 1994] Orna Bernholtz, Moshe Y. Vardi, and Pierre Wolper. An automata-theoretic approach to branching-time model checking. In *Computer Aided Verification, Proc. 6th Int'l. Workshop*, volume 818 of *LNCS*, pages 142–155, Stanford, California, June 1994. Springer-Verlag.

[Bolotov *et al.*, 2002] Alexander Bolotov, Michael Fisher, and Clare Dixon. On the relationship between ω-automata and temporal logic normal forms. *Journal of Logic and Computation*, 12(4):561–581, August 2002. Available as `http://www3.oup.co.uk/logcom/hdb/Volume_12/Issue_04/pdf/120561.pdf`.

[Bryant, 1986] Randal E. Bryant. Graph-based algorithms for Boolean function manipulation. *IEEE Transactions on Computers*, C-35(8), 1986.

[Bryant, 1992] Randal E. Bryant. Symbolic Boolean manipulation with ordered binary-decision diagrams. *ACM Computing Surveys*, 24(3):293–318, September 1992.

[Burch *et al.*, 1992] J. R. Burch, E. M. Clarke, K. L. McMillan, D. L. Dill, and L. J. Hwang. Symbolic model checking: 10^{20} states and beyond. *Inf. and Comp.*, 98(2):142–170, June 1992.

[Cadence, URL] Cadence Design Systems. `http://www.cadence.com/`, URL.

[Chellas, 1980] Brian F. Chellas. *Modal Logic: An Introduction.* Cambridge University Press, Cambridge, England, 1980.

[Clarke *et al.*, 2000] Edmund M. Clarke, Orna Grumberg, and Doron A. Peled. *Model Checking.* MIT Press, Cambridge, Massachusetts, 2000.

[Clarke *et al.*, 2001] E. Clarke, A. Biere, R. Raimi, and Y. Zhu. Bounded model checking using satisfiability solving. *Formal Methods in System Design*, 19(1), July 2001.

[CLISP, URL] CLISP: An ANSI Common Lisp implementation. `http://clisp.cons.org`, URL.

[Coudert *et al.*, 1989a] Olivier Coudert, Christian Berthet, and Jean Christophe Madre. Verification of sequential machines using boolean functional vectors. In L. Claesen, editor, *Proc. IFIP International Workshop on Applied Formal Methods for Correct VLSI Design*, pages 111–128, Leuven, Belgium, November 1989.

[Coudert *et al.*, 1989b] Olivier Coudert, Christian Berthet, and Jean Christophe Madre. Verification of synchronous sequential machines based on symbolic execution. In Joseph Sifakis, editor, *Automatic Verification Methods for Finite State Systems, International Workshop, Grenoble, France, June 12-14, 1989, Proceedings*, volume 407 of *Lecture Notes in Computer Science*, pages 365–373. Springer, 1989.

[Coudert *et al.*, 1990] Olivier Coudert, Christian Berthet, and Jean Christophe Madre. A unified framework for the formal verification of sequential circuits. In *Proc. IEEE International Conf. on Computer Aided Design*, pages 126–129, November 1990.

[CUDD, URL] Colorado University Decision Diagram Package (CUDD). Available at `http://vlsi.colorado.edu/~fabio`, URL.

[Emerson, 1990] E. Allen Emerson. Temporal and modal logic. In Jan van Leeuwen, editor, *Handbook of Theoretical Computer Science*, volume B: Formal Models and Semantics, chapter 16, pages 995–1072. Elsevier/MIT Press, Amsterdam, 1990.

[Fischer and Ladner, 1977] Michael J. Fischer and Richard E. Ladner. Propositional modal logic of programs (extended abstract). In *Conference Record of the Ninth Annual ACM Symposium on Theory of Computing*, pages 286–294, Boulder, Colorado, 2–4 May 1977.

[Fischer and Ladner, 1979] Michael J. Fischer and Richard E. Ladner. Propositional dynamic logic of regular programs. *Journal of Computer and System Sciences*, 18(2):194–211, April 1979.

[Fisher *et al.*, 2001] Michael Fisher, Clare Dixon, and Martin Peim. Clausal temporal resolution. *ACM Transactions on Computational Logic*, 2(1):12–56, January 2001.

[Fisher, 1992] Michael Fisher. A normal form for first-order temporal formulae. In Deepak Kapur, editor, *Automated Deduction - CADE-11, 11th International Conference on Automated Deduction, Saratoga Springs, NY, USA, June 15-18, 1992, Proceedings*, volume 607 of *LNCS*, pages 370–384. Springer-Verlag, 1992.

[Fisher, 1997] Michael Fisher. A normal form for temporal logic and its application in theorem-proving and execution. *Journal of Logic and Computation*, 7(4):429–456, August 1997.

[French, 2000] Tim French. A proof of the completeness of PLTL. Available as `http://www.cs.uwa.edu.au/~tim/papers/pltlcomp.ps`, 2000.

[Gabbay *et al.*, 1980] D. Gabbay, A. Pnueli, S. Shelah, and J. Stavi. On the temporal analysis of fairness. In *Seventh Annual ACM Symposium on Principles of Programming Languages*, pages 163–173, 1980.

[Goldblatt, 1987] R. Goldblatt. *Logics of Time and Computation*, volume 7 of *CSLI Lecture Notes*. CLSI/SRI International, 333 Ravenswood Av., Menlo Park, CA 94025, 1987.

[Halpern *et al.*, 1983] J. Halpern, Z. Manna, and B. Moszkowski. A hardware semantics based on temporal intervals. In J. Diaz, editor, *Proceedings of the 10-th International Colloquium on Automata, Languages and Programming*, volume 154 of *LNCS*, pages 278–291, Berlin, 1983. Springer-Verlag.

[Harel *et al.*, 2000] David Harel, Dexter Kozen, and Jerzy Tiuryn. *Dynamic Logic*. MIT Press, Cambridge, Massachusetts, 2000.

[Harel *et al.*, 2002] David Harel, Dexter Kozen, and Jerzy Tiuryn. Dynamic logic. In Dov Gabbay and Franz Guenthner, editors, *Handbook of Philosophical Logic*, volume 4, pages 99–217. Kluwer Academic Publishers, Dordrecht, 2nd edition edition, 2002.

[Harel, 1984] David Harel. Dynamic logic. In Dov Gabbay and Franz Guenthner, editors, *Handbook of Philosophical Logic*, volume II, pages 497–604. Reidel Publishing Company, Dordrecht, 1984.

[Henriksen and Thiagarajan, 1997] Jesper G. Henriksen and P. S. Thiagarajan. Dynamic linear time temporal logic. Technical Report RS-97-8, BRICS, Department of Computer Science, University of Aarhus, Aarhus, Denmark, April 1997. Available at `http://www.brics.dk/RS/97/8/`.

[Henriksen and Thiagarajan, 1999] Jesper G. Henriksen and P. S. Thiagarajan. Dynamic linear time temporal logic. *Annals of Pure and Applied Logic*, 96(1-3):187–207, 1999.

[Hollander *et al.*, 2001] Yoav Hollander, Matthew Morley, and Amos Noy. The *e* language: A fresh separation of concerns. In *Technology of Object-Oriented Languages and Systems (Proceedings of 38th Int'l. TOOLS Conference, TOOLS Europe 2001)*, pages 41–50. IEEE Computer Society Press, March 2001. All authors at Verisity, Ltd., Rosh-Ha-Ain, Israel. Presented at TOOLS Europe 2001, 38th International TOOLS Conference, Zürich Technopark, Zürich, Switzerland, March 12–14, 2001.

[Hughes and Cresswell, 1996] George E. Hughes and Max J. Cresswell. *A New Introduction to Modal Logic*. Routledge, London, 1996.

[IEEE1647, URL] IEEE Candidate Standard 1647. Produced by the **e** Functional Verification Language Working Group. `http://www.ieee1647.org/`, URL.

[ITL, URL] Interval Temporal Logic (ITL) homepage. URL: `http://www.cse.dmu.ac.uk/~cau/itlhomepage/itlhomepage.html`, URL.

[Kozen and Tiuryn, 1990] Dexter Kozen and Jerzy Tiuryn. Logics of programs. In Jan van Leeuwen, editor, *Handbook of Theoretical Computer Science*, volume B, pages 789–840. Elsevier Science Publishers, Amsterdam, 1990.

[Kröger, 1987] F. Kröger. *Temporal Logic of Programs*, volume 8 of *EATCS Monographs on Theoretical Computer Science*. Springer-Verlag, 1987.

[Kropf, 1999] Thomas Kropf. *Introduction to Formal Hardware Verification*. Springer-Verlag, Heidelberg, Germany, 1999.

[Lange and Stirling, 2001] Martin Lange and Colin Stirling. Focus games for satisfiability and completeness of temporal logic. In *Proc. 16th Annual IEEE Symp. on Logic in Computer Science, LICS'01*, pages 357–365, Boston, MA, USA, June 2001. IEEE Computer Society Press.

[Lichtenstein and Pnueli, 2000] Orna Lichtenstein and Amir Pnueli. Propositional temporal logics: Decidability and completeness. *Logic Journal of the IGPL*, 8(1):55–85, 2000. Available at `http://www3.oup.co.uk/igpl/Volume_08/Issue_01/#Lichtenstein`.

[Manna and Pnueli, 1981] Z. Manna and A. Pnueli. Verification of concurrent programs: the temporal framework. In Robert S. Boyer and J. Strother Moore, editors, *The Correctness Problem in Computer Science*, pages 215–273, New York, 1981. Academic Press.

[Manna and Pnueli, 1989] Z. Manna and A. Pnueli. The anchored version of the temporal framework. In J. W. De Bakker, Willem-Paul de Roever, and Grzegorz Rozenberg, editors, *Linear Time, Branching Time, and Partial Order in Logics and Models for Concurrency (REX Workshop 1988)*, volume 354 of *LNCS*, pages 201–284. Springer-Verlag, 1989.

[McMillan, 1993] Kenneth L. McMillan. *Symbolic model checking*. Kluwer Academic Publishers, Boston, Mass., 1993.

[Morley, 1999] Matthew J. Morley. Semantics of temporal *e*. In T. F. Melham and F. G. Moller, editors, Banff'99 *Higher Order Workshop: Formal Methods in Computation, Ullapool, Scotland, 9–11 Sept. 1999*, pages 138–142. University of Glasgow, Department of Computing Science Technical Report, 1999.

[Moszkowski, 1983a] B. Moszkowski. *Reasoning about Digital Circuits*. PhD thesis, Department of Computer Science, Stanford University, June 1983. Technical report STAN–CS–83–970.

[Moszkowski, 1983b] B. Moszkowski. A temporal logic for multi-level reasoning about hardware. In *Proceedings of the 6-th International Symposium on Computer Hardware Description Languages*, pages 79–90, Pittsburgh, Pennsylvania, May 1983. North-Holland Pub. Co.

[Moszkowski, 1985] B. Moszkowski. A temporal logic for multilevel reasoning about hardware. *Computer*, 18:10–19, 1985.

[Moszkowski, 1986] B. Moszkowski. *Executing Temporal Logic Programs*. Cambridge University Press, Cambridge, England, 1986.

[Moszkowski, 1994] Ben Moszkowski. Some very compositional temporal properties. In E.-R. Olderog, editor, *Programming Concepts, Methods and Calculi*, volume A-56 of *IFIP Transactions*, pages 307–326. IFIP, Elsevier Science B.V. (North–Holland), 1994.

[Moszkowski, 1995] Ben Moszkowski. Compositional reasoning about projected and infinite time. In *Proceedings of the First IEEE Int'l Conf. on Engineering of Complex Computer Systems (ICECCS'95)*, pages 238–245. IEEE Computer Society Press, 1995.

[Moszkowski, 1996] Ben Moszkowski. Using temporal fixpoints to compositionally reason about liveness. In He Jifeng, John Cooke, and Peter Wallis, editors, *BCS-FACS 7th Refinement Workshop*, electronic Workshops in Computing, London, 1996. BCS-FACS, Springer-Verlag and British Computer Society.

[Moszkowski, 1998] Ben Moszkowski. Compositional reasoning using Interval Temporal Logic and Tempura. In Willem-Paul de Roever, Hans Langmaack, and Amir Pnueli, editors, *Compositionality: The Significant Difference*, volume 1536 of *LNCS*, pages 439–464, Berlin, 1998. Springer-Verlag.

[Moszkowski, 2000] Ben Moszkowski. An automata-theoretic completeness proof for Interval Temporal Logic (extended abstract). In Ugo Montanari, José Rolim, and Emo Welzl, editors, *Proceedings of the 27th International Colloquium on Automata, Languages and Programming (ICALP 2000)*, volume 1853 of *LNCS*, pages 223–234, Geneva, Switzerland, July 2000. Springer-Verlag.

[Moszkowski, 2004a] Ben Moszkowski. A hierarchical completeness proof for Propositional Interval Temporal Logic with finite time. *Journal of Applied Non-Classical Logics*, 14(1–2):55–104, 2004. Special issue on Interval Temporal Logics and Duration Calculi. V. Goranko and A. Montanari guest eds.

[Moszkowski, 2004b] Ben Moszkowski. A hierarchical completeness proof for propositional temporal logic. In Nachum Dershowitz, editor, *Verification: Theory and Practice: Essays Dedicated to Zohar Manna on the Occasion of His 64th Birthday*, volume 2772 of *LNCS*, pages 480–523. Springer-Verlag, Heidelberg, 2004.

[PerlDD, URL] PerlDD: Perl extensions to CUDD [CUDD, URL]. Available at `http://vlsi.colorado.edu/~fabio`, URL.

[Perl, URL] The Perl programming language. `http://www.perl.org`, URL.

[Pnueli, 1977] A. Pnueli. The temporal logic of programs. In *Proceedings of the 18th Symposium on the Foundation of Computer Science*, pages 46–57. ACM, 1977.

[Pratt, 1979] V. R. Pratt. Process logic. In *Sixth Annual ACM Symposium on Principles of Programming Languages*, pages 93–100, 1979.

[PSLSugar, URL] PSL/Sugar Consortium. `http://www.pslsugar.org`, URL.

[Pucella, 2005] Riccardo Pucella. Logic column 11: The finite and the infinite in temporal logic. *SIGACT News*, 36(1):86–99, 2005. Available at Computing Research Repository (CoRR): `http://arxiv.org/abs/cs.LO/0502031`.

[Rescher and Urquhart, 1971] N. Rescher and A. Urquhart. *Temporal Logic*. Springer-Verlag, New York, 1971.

[Vardi and Wolper, 1986] Moshe Y. Vardi and Pierre Wolper. Automata-theoretic techniques for modal logics of programs. *Journal of Computer and System Sciences*, 32(2):183–221, April 1986.

[Verisity Ltd., 2003] Verisity Ltd. Semantics of temporal *e*. Revised version of Morley [Morley, 1999]. Available from website of IEEE candidate standard 1647 as `http://www.ieee1647.org/downloads/temporale_denotational.pdf`, December 2003.

[Verisity, URL] Verisity Ltd. (acquired by Cadence Design Systems [Cadence, URL] in 2005). `http://www.cadence.com/verisity/`, URL.

[Wolper, 1981] Pierre Wolper. Temporal logic can be more expressive. In *Proc. 22nd Annual Symposium on Foundations of Computer Science (FOCS)*, pages 340–348, Nashville, Tennessee, October 1981. IEEE Computer Society.

[Wolper, 1983] Pierre Wolper. Temporal logic can be more expressive. *Information and Control*, 56(1-2):72–99, 1983.

[Wolper, 1985] Pierre Wolper. The tableau method for temporal logic: An overview. *Logique et Analyse*, 110–111:119–136, 1985.

[Wolper, 2001] Pierre Wolper. Constructing automata from temporal logic formulas: A tutorial. In *Lectures on Formal Methods in Performance Analysis (First EEF/Euro Summer School on Trends in Computer Science)*, volume 2090 of *LNCS*, pages 261–277. Springer-Verlag, July 2001.

Nesting Patterns in Fibred Logics of Context

ROLF NOSSUM

Dedicated to Dov Gabbay on his 60th birthday.

In recent years, many logical systems for contextual reasoning with the $ist(c, \lambda)$ modality have been proposed. In the formula $ist(c, \lambda)$, c is a context and λ a proposition, and the formula is taken as true whenever λ is true in context c.

When the truth of a formula is evaluated, the evaluation takes place in some context. This also applies to ist formulas. In the notation of [McCarthy and Buvač, 1994]

$$b : ist(c, \lambda)$$

is taken as the assertion that the proposition λ is true in context c, itself asserted in another context b. This invites inquiry into how the assertion of $b : ist(c, \lambda)$ relates to the truth of the formula $ist(b, ist(c, \lambda))$.

Some systems of context logic have deduction rules for moving in and out of context, often resembling the rule of necessitation and the inverse rule of necessitation in modal logic, and conforming to the following general patterns:

$$(1) \quad \text{Enter: } \frac{\vdash u : ist(v, \lambda)}{\vdash w : \lambda} \qquad \text{Exit: } \frac{\vdash x : \lambda}{\vdash y : ist(z, \lambda)}$$

Depending on the interplay between u, v, w, resp. x, y, z, in (1), logical systems with different characteristics are obtained.

For example, [Buvač, 1996] has these rules:

$$(2) \quad \text{Enter: } \frac{\vdash x : ist(c, \lambda)}{\vdash c : \lambda} \qquad \text{Exit: } \frac{\vdash c : \lambda}{\vdash x : ist(c, \lambda)}$$

while [Buvač *et al.*, 1995] have the following ones:

$$(3) \quad \text{Enter: } \frac{\vdash x : ist(c, \lambda)}{\vdash xc : \lambda} \qquad \text{Exit: } \frac{\vdash xc : \lambda}{\vdash x : ist(c, \lambda)}$$

We start by describing the algebraic context logic of [Nossum, 2001; Nossum, 2002], which has the systems of [Buvač *et al.*, 1995; Buvač, 1996] as special cases. Then we describe the self fibred context logic of [Gabbay, 1999], which specializes to the quantificational context logic of [Buvač, 1996] by ad hoc restrictions on its semantical structure. Finally we demonstrate how the algebraic properties of [Nossum, 2002] correspond to certain semantical conditions in the self fibred context logic of [Gabbay, 1999].

1 Algebraically structured contexts

The logic of [Nossum, 2001] subsumes several other logics of *ist* in the literature by varying an algebraic component. Consider any context constructor $\oplus$ which combines a previously accumulated context x with the context c being entered into:

(4) Enter: $\dfrac{\vdash x : ist(c, \lambda)}{\vdash x \oplus c : \lambda}$ Exit: $\dfrac{\vdash x \oplus c : \lambda}{\vdash x : ist(c, \lambda)}$

Along with these deduction rules the algebraic properties of the context constructor $\oplus$ are specified by equations. For example, one might have a purely associative constructor:

(5) $(u \oplus v) \oplus w = u \oplus (v \oplus w)$

or an idempotent one:

(6) $u \oplus u = u$

or a commutative one:

(7) $u \oplus v = v \oplus u$

or even what has come to be known as a *flat* one, corresponding to a much stronger condition on $\oplus$:

(8) $u \oplus v = v$

or a combination of these and/or other algebraic properties. Associativity corresponds to maintaining a sequence of contexts entered into and not exited from, i.e. a stack-like reasoning discipline, and idempotence means it is redundant to enter a context you are already in. Commutativity means that all pairs of contexts are independent, while flatness corresponds to rejecting any accumulation of contextual information. The algebra on context terms allows to vary the reasoning discipline by changing the algebraic equations.

The language of this logic is a sorted first order predicate language $\mathcal{L}$ with identity, augmented by the special modality $ist(\ldots, \ldots)$. The sorts are C for contexts and T for other objects of discourse.

DEFINITION 1 ($\mathcal{L}$, the set of well-formed formulas).

$$\mathcal{L} ::= P \mid \neg\mathcal{L} \mid \mathcal{L} \rightarrow \mathcal{L} \mid \forall V.\mathcal{L} \mid ist(C, \mathcal{L})$$

where P is a set of atomic predicates on sorted terms, including the identity predicate for each sort, V is a set of sorted variables, and C is a set of context names.

A formula $\lambda \in \mathcal{L}$, asserted in a context $c \in C$:

$$c : \lambda$$

is interpreted in a structure called a *rigid interpretation*, to be defined next.

Let the set C of contexts and the set T of objects of discourse be given a priori. These are nonempty and no more than countable, and shall stay fixed throughout. Let the constants (names) of each sort of the language be rigid designators, i.e. let them correspond 1-1 to elements of C resp T, so that we may identify the sets of constants of sort C with C itself, and correspondingly with T.

DEFINITION 2 (Rigid interpretations). A rigid interpretation is a first-order interpretation of the language, in which:

- the domain for objects of discourse is T
- the domains for contexts is $C^{\oplus}_{=}$, the qoutient of the set of $\oplus$ terms under the equivalence relation imposed by the algebraic equations
- each constant of sort T is interpreted according to the 1-1 correspondence mentioned above
- each constant of sort C is interpreted as its own equivalence class modulo the algebraic equations
- the $\oplus$ symbol is interpreted homomorphically, i.e. $x \oplus y$ is interpreted as the set of terms $\widehat{x} \oplus \widehat{y}$ such that $\widehat{x}$ is in the set interpreting x and $\widehat{y}$ is in the set interpreting y.
- the identity predicate for each sort is interpreted as the corresponding identity relation

DEFINITION 3 (x-continuants). For $x \in C^{\oplus}$, an x-continuant is any term $((x \oplus c_1) \oplus c_2) \ldots \oplus c_m$ where $m \geq 0$ and $c_i \in C, 1 \leq i \leq m$.

Note that x itself is an x-continuant, with $m = 0$. The parenthetical structure is a part of the definition, so the following are examples of x-continuants

$$x \oplus c, \qquad (x \oplus c) \oplus d$$

but

$$x \oplus (c \oplus d)$$

is not an x-continuant.

It may well be that different x-continuants are equal by force of the algebraic equations, for example

$$(((x \oplus c_1) \oplus c_2) \ldots \oplus c_m) = (((x \oplus d_1) \oplus d_2) \ldots \oplus d_n)$$

but this does not in general imply that

$$((c_1 \oplus c_2) \ldots \oplus c_m) = ((d_1 \oplus d_2) \ldots \oplus d_n)$$

unless additional information about $\oplus$ is available.

DEFINITION 4 (x-bundles). For $x \in C^{\oplus}$, the set of x-bundles is the quotient of the set of x-continuants under the equivalence relation imposed by the algebraic equations.

DEFINITION 5 (x-models). An x-model is a function from x-bundles to sets of rigid interpretations.

$M(y)$ denotes the set of rigid interpretations that M associates with the x-bundle containing y.

We can now define truth and falsity of formulas asserted in context. Let $x \in C^{\oplus}, \lambda \in \mathcal{L}$, and let M be an x-model:

$$M \models x : \lambda \text{ iff } M, I \models x : \lambda \text{ for all } I \in M(x)$$

where for $I \in M(x)$

$$\begin{array}{lcll} M, I \models & x : p & \text{iff } I \text{ interprets } p \text{ as true} & (9) \\ M, I \models & x : \neg\lambda & \text{iff } M, I \not\models x : \lambda & \\ M, I \models & x : \lambda \rightarrow \gamma & \text{iff } M, I \models x : \lambda \ \text{ implies } M, I \models x : \gamma & \\ M, I \models & x : \forall v.\lambda(v) & \text{iff } M, I \models x : \lambda(t) \ \text{ for all } t \text{ of correct sort} & \\ M, I \models & x : ist(c, \lambda) & \text{iff } M, J \models x \oplus c : \lambda \ \text{ for all } J \in M(x \oplus c) & \end{array}$$

The table on page 445 gives axiom schemata and rules of deduction which are sound and complete with respect to the semantical framework. Actually, the Enter rule is redundant, being entailed by the rest of the system, but is included in the table for symmetry. For the proofs, consult [Nossum, 2001].

Table 1. **Axiom schemata and deduction rules, Section 1**

Rules for changing context:

$$\text{Enter: } \frac{\vdash x : ist(c, \lambda)}{\vdash x \oplus c : \lambda} \qquad \text{Exit: } \frac{\vdash x \oplus c : \lambda}{\vdash x : ist(c, \lambda)}$$

Equational properties:

Reflexivity: $\vdash x : y = y$

Congruence: $\vdash x : y = z \to (\lambda(y) \to \lambda(z))$

Algebraic equations: $\vdash x : y_i = z_i \qquad y_i, z_i \in C^{\oplus} \qquad 1 \leq i \leq N$

Propositional properties:

PL: $\vdash x : \lambda$ whenever λ is an instance of a propositional tautology

$$\text{MP: } \frac{\vdash x : \lambda \quad \vdash x : \lambda \to \chi}{\vdash x : \chi}$$

Modal properties:

K: $\vdash x : ist(c, \lambda \to \chi) \to (ist(c, \lambda) \to ist(c, \chi))$

Δ : $\vdash x : \neg ist(c, ist(d, \lambda)) \to ist(c, \neg ist(d, \lambda))$

Nesting: $\vdash x : ((x \oplus c_1) \ldots \oplus c_m) = ((x \oplus d_1) \ldots \oplus d_n) \to$
$(ist(c_1, \ldots, ist(c_m, \lambda)) \to ist(d_1, \ldots, ist(d_n, \lambda)))$

Quantificational properties:

UI: $\vdash x : \forall v.\lambda(v) \to \lambda(t)$ $\qquad$ UG: $\dfrac{\vdash x : \xi \to \lambda(t)}{\vdash x : \xi \to \forall v.\lambda(v)}$

where v is not free in ξ

BF: $\vdash x : \forall v.ist(c, \lambda(v)) \to ist(c, \forall v.\lambda(v))$

where c is not the variable v

2 Special cases

The algebraic framework embedded in the table on page 445 admits arbitrary algebras of context combination. When focusing on a particular case, say taking $\oplus$ to be set union, for instance, it is customary to take the equational properties of $\oplus$ for granted, rather than stating them explicitely in the axiomatics.

For many cases, a considerably simplified axiomatic presentation is possible, notably the class of associative $\oplus$ operators.

When $\oplus$ is associative, the algebraic equations in the table on page 445 can be listed as

(10) $x : (u \oplus v) \oplus w = u \oplus (v \oplus w)$

plus additional equations

(11) $x : u_{i1} \oplus \ldots \oplus u_{im_i} = v_{i1} \oplus \ldots \oplus v_{in_i} \quad i = 2 \ldots N$

where parentheses are disposed of because of associativity. If the intention were that $\oplus$ is set union for instance, these would be axioms of commutativity and idempotency, and it would be customary to take those properties for granted rather than listing them explicitely in the axiomatics, and to replace $\oplus$ with the more familiar $\cup$ symbol in other parts of the system.

In [Nossum, 2002], this representational device is made precise for all associative $\oplus$ systems where the algebra has no function symbol besides $\oplus$. This class is called *AFG systems* (associative finite ground systems). It is shown that by taking associativity for granted, and replacing each ground axiom schema of the form

(12) $x : c_1 \oplus \ldots \oplus c_m = d_1 \oplus \ldots \oplus d_n$

with a corresponding axiom schema

(13) $x : ist(c_1, \ldots, ist(c_m, \lambda)) = ist(d_1, \ldots, ist(d_n, \lambda))$

the axiom of nesting can be removed.

Varying the algebraic equations in this way immediately yields some of the logics of *ist* in the literature as special cases, as well as suggesting some new ones:

The propositional logic of context of Buvač, Buvač, and Mason in [Buvač *et al.*, 1995] has purely associative context composition, i.e. a composite context is represented as the sequence of its constituent parts. There are therefore no equations of the form (11) to worry about, and since their system is propositional without equality, the equational and quantificational schemas on page 445 are irrelevant. The ensuing system is therefore

$$\text{Enter: } \frac{\vdash x : ist(c,\lambda)}{\vdash xc : \lambda} \qquad \text{Exit: } \frac{\vdash xc : \lambda}{\vdash x : ist(c,\lambda)}$$

PL: $\vdash x : \lambda$ whenever λ is an instance of a propositional tautology

$$\text{MP: } \frac{\vdash x : \lambda \quad \vdash x : \lambda \rightarrow \chi}{\vdash x : \chi}$$

K: $\vdash x : ist(c, \lambda \rightarrow \chi) \rightarrow (ist(c,\lambda) \rightarrow ist(c,\chi))$

Δ : $\vdash x : \neg ist(c, ist(d,\lambda)) \rightarrow ist(c, \neg ist(d,\lambda))$

which is seen to correspond exactly to the one in [Buvač *et al.*, 1995]. As noted previously, the Enter rule is entailed by the rest of the system.

In a follow-up paper [Buvač, 1996], Buvač develops a quantificational logic of *ist*, with equality, for 'flat' contexts, where entering and exiting context is according to the following rules:

$$\text{Enter: } \frac{\vdash x : ist(c,\lambda)}{\vdash c : \lambda} \qquad \text{Exit: } \frac{\vdash c : \lambda}{\vdash x : ist(c,\lambda)}$$

These rules correspond to a liberal import and export regime on results derived within particular contexts. Flat contexts obey this drastic equation

(14) $x : u \oplus v = v$

which becomes the following axiom schema of the form (13):

(15) $x : ist(c, ist(d,\lambda)) \leftrightarrow ist(d,\lambda)$

Thus we have a quantificational logic of flat contexts with equality:

$$\text{Enter: } \frac{\vdash x : ist(c, \lambda)}{\vdash c : \lambda} \qquad \text{Exit: } \frac{\vdash c : \lambda}{\vdash x : ist(c, \lambda)}$$

$$\text{Refl: } \quad \vdash x : y = y$$

$$\text{Congr: } \quad \vdash x : y = z \to (\lambda(y) \to \lambda(z))$$

$$\text{Flat: } \quad \vdash x : ist(c, ist(d, \lambda)) \leftrightarrow ist(d, \lambda)$$

PL: $\vdash x : \lambda$ whenever λ is an instance of a propositional tautology

$$\text{MP: } \frac{\vdash x : \lambda \quad \vdash x : \lambda \to \chi}{\vdash x : \chi}$$

$$\text{K: } \quad \vdash x : ist(c, \lambda \to \chi) \to (ist(c, \lambda) \to ist(c, \chi))$$

$$\Delta : \quad \vdash x : \neg ist(c, ist(d, \lambda)) \to ist(c, \neg ist(d, \lambda))$$

$$\text{UI: } \quad \vdash x : \forall v.\lambda(v) \to \lambda(t) \qquad \text{UG: } \frac{\vdash x : \xi \to \lambda(t)}{\vdash x : \xi \to \forall v.\lambda(v)}$$

where v is not free in ξ

$$\text{BF: } \quad \vdash x : \forall v.ist(c, \lambda(v)) \to ist(c, \forall v.\lambda(v))$$

where c is not the variable v

which corresponds exactly to the one given in [Buvač, 1996]. Also the semantical frameworks coincide, since (14) forces all congruence classes to be singletons.

We can, however, have a quantificational logic of *ist* without committing to flat contexts, for example taking context combination as commutative or commutative+idempotent, getting multisets, respectively sets, of contexts. These can be cast as AFG systems, cfr [Nossum, 2002], yielding these axiom schemas:

$$\text{Comm: } \quad \vdash x : ist(c, ist(d, \lambda)) \leftrightarrow ist(d, ist(c, \lambda)) \tag{16}$$

$$\text{Idem: } \quad \vdash x : ist(c, ist(c, \lambda)) \leftrightarrow ist(c, \lambda) \tag{17}$$

We have, in short, a palette of options, to be combined according to need. Any AFG algebra can be combined with equality and/or quantification, to obtain a logic of *ist* with the corresponding properties.

For example, in applications with equality and quantification, and where microcontexts combine according to the union of sets, the model structure consists of mappings from x-bundles (constructed according to set equality) to sets of rigid interpretations, and the following axioms and rules are sound and complete with respect to that class of models:

Enter: $\dfrac{\vdash x : ist(c, \lambda)}{\vdash x \cup \{c\} : \lambda}$ Exit: $\dfrac{\vdash x \cup \{c\} : \lambda}{\vdash x : ist(c, \lambda)}$

Refl: $\vdash x : y = y$

Congr: $\vdash x : y = z \to (\lambda(y) \to \lambda(z))$

Comm: $\vdash x : ist(c, ist(d, \lambda)) \leftrightarrow ist(d, ist(c, \lambda))$

Idem: $\vdash x : ist(c, ist(c, \lambda)) \leftrightarrow ist(c, \lambda)$

PL: $\vdash x : \lambda$ whenever λ is an instance of a propositional tautology

MP: $\dfrac{\vdash x : \lambda \quad \vdash x : \lambda \to \chi}{\vdash x : \chi}$

K: $\vdash x : ist(c, \lambda \to \chi) \to (ist(c, \lambda) \to ist(c, \chi))$

Δ : $\vdash x : \neg ist(c, ist(d, \lambda)) \to ist(c, \neg ist(d, \lambda))$

UI: $\vdash x : \forall v.\lambda(v) \to \lambda(t)$ UG: $\dfrac{\vdash x : \xi \to \lambda(t)}{\vdash x : \xi \to \forall v.\lambda(v)}$

where v is not free in ξ

BF: $\vdash x : \forall v.ist(c, \lambda(v)) \to ist(c, \forall v.\lambda(v))$

where c is not the variable v

3 Construction of context logic by self fibring

Gabbay [Gabbay, 1999] defines the ist-language through a recursive construction, starting with a free predicate logic containing a binary predicate $ist(x, y)$. Note that so far both coordinates of ist are terms, not formulas. Now in a manœvre called *self fibring*, the language is folded into itself by allowing formulas in the y coordinate of ist. Gabbay provides a semantical framework capable of handling fibred formulas, and obtains an axiomatic presentation of the resulting logic.

DEFINITION 6 (Models for fibred languages). The models for fibred languages are of the form $\Sigma = (S, D, a, F, h, g)$, satisfying certain conditions:

- S is a set of labels naming classical models.
- D is a domain common to all $s \in S$.
- $a \in S$ is a designated classical model (the "actual world").
- F is a fibring function associating with each $X \subseteq D$ and $t \in S$ a set of labels $F(X, t) \subseteq S$
- h is an interpretation associating for each $t \in S$ and each **m**-place predicate P a subset $h(t, P) \subseteq D^m$

- g is a rigid assigment giving for each variable or constant x of the language an element $g(x) \in D$.

Satisfaction of $ist(c, \phi)$ at a point t is defined as follows:

$$t \models ist(c, \phi) \quad \text{iff} \quad s \models \phi, \text{ for all } s \in \mathcal{M}_c^t \tag{18}$$

$$\text{where} \quad \mathcal{M}_c^t = F(X_{t,c}, t) \tag{19}$$

$$\text{and} \quad X_{t,c} = \{y \mid t \models ist(c, y)\} \tag{20}$$

In order to align this evaluation at points ($t \models ist(c, \phi)$) with evaluation at contexts as in Buvač *et al.* ($t : ist(c, \phi)$), Gabbay makes a further simplification of his semantical structure by setting

$$S = D$$

thus identifying points of evaluation with contexts ([Gabbay, 1999] p.196).

The following Hilbert system with free quantifiers (x) is sound and complete with respect to the resulting semantical framework.

1. Axioms and rules of free predicate logic
   - All substitution instances of truth functional tautologies
   - $\dfrac{\vdash \phi, \vdash \phi \rightarrow \psi}{\vdash \psi}$
   - $(x)(\phi(x) \wedge \psi(x)) \rightarrow (x)\phi(x) \wedge (x)\psi(x)$
   - $(x)\phi(x) \wedge (x)(\phi(x) \rightarrow \psi(x)) \rightarrow (x)\psi(x)$
   - $\dfrac{\vdash \phi \rightarrow \psi(x)}{\vdash \phi \rightarrow (x)\psi(x)}$ where x is not free in ϕ
   - $(x)(y)\phi \rightarrow (y)(x)\phi$
   - $(u)(\psi \wedge (x)\phi(x) \rightarrow \phi(u))$.
2. Modal rules K
   - $\dfrac{\vdash \wedge\phi_i \rightarrow \phi}{\vdash ist(c, \wedge\phi_i) \rightarrow ist(c, \phi)}$
   - $(x)\ ist(c, \phi(x)) \rightarrow ist(c, (x)\ \phi(x))$.

For the proof, cfr [Gabbay, 1999] theorem 7.6 . Gabbay proceeds from here to introduce certain restrictions on the semantical framework, which transforms the above system into Buvač's quantificational logic of context

[Buvač, 1996]. In particular, certain constraints on F and h are shown ([Gabbay, 1999], theorem 10.28) to correspond to flatness of context nesting:

$$ist(c, ist(d, \phi)) \leftrightarrow ist(d, \phi)$$

which when added to the above axioms and rules yields a close variant of Buvač's system.

4 Semantical correspondences in self fibred context logic

Let us show that arbitrary AFG patterns of context nesting, not only flatness, correspond to certain restrictions on the semantical structure of Gabbay's fibred logic of context.

By analogy with [Nossum, 2002], we'll translate each associative equation of the form

(21) $c_1 \oplus \ldots \oplus c_m = d_1 \oplus \ldots \oplus d_n$

where $c_1, \ldots, c_m, d_1, \ldots, d_n$ are context names, to a corresponding axiom schema

(22) $ist(c_1, \ldots, ist(c_m, \lambda)) \leftrightarrow ist(d_1, \ldots, ist(d_n, \lambda))$

and give conditions for the satisfaction of the latter.

Consider initially the evaluation of

$$t \models ist(c, ist(d, \phi))$$

which develops into

$$r \models \phi \text{ for all } r \in \mathcal{M}_d^s \text{ such that } s \in \mathcal{M}_c^t$$

where $\mathcal{M}$ is as defined in (19). This motivates us to extend the definition of $\mathcal{M}$ to composite contexts, composed from ground contexts with the $\oplus$ operator, which is assumed associative:

DEFINITION 7 ($\mathcal{M}^t_{c_1 \oplus \ldots \oplus c_m}$).

$$\begin{aligned}\mathcal{M}^t_{c_1 \oplus \ldots \oplus c_m} &= \mathcal{M}^t_c \text{ if } m = 1\\ &= \{v \mid \text{ there exists } u \text{ such that } u \in \mathcal{M}^t_{c_1} \text{ and } v \in \mathcal{M}^u_{c_2 \oplus \ldots \oplus c_m}\}\\ &\quad \text{if } m > 1\end{aligned}$$

and it is straightforward to verify the following

THEOREM 8. *The axiom schema*

$$ist(c_1, \ldots, ist(c_m, \lambda)) \leftrightarrow ist(d_1, \ldots, ist(d_n, \lambda))$$

is satisfied in those structures where for all $t, c_1, \ldots, c_m, d_1, \ldots, d_n$

$$\mathcal{M}^t_{c_1 \oplus \ldots \oplus c_m} = \mathcal{M}^t_{d_1 \oplus \ldots \oplus d_n}$$

From this, we can read off semantical conditions corresponding to

(23) Flatness: $\mathcal{M}^t_{c \oplus d} = \mathcal{M}^t_d$ for all t, c, d

(24) Idempotence: $\mathcal{M}^t_{c \oplus c} = \mathcal{M}^t_c$ for all t, c

(25) Commutativity: $\mathcal{M}^t_{c \oplus d} = \mathcal{M}^t_{d \oplus c}$ for all t, c, d

and every other AFG pattern of context nesting.

BIBLIOGRAPHY

[Buvač *et al.*, 1995] Saša Buvač, Vanja Buvač, and Ian A. Mason. Metamathematics of context. *Fundamenta Informaticae*, 23(3), 1995.

[Buvač, 1996] Saša Buvač. Quantificational logic of context. In *Proceedings of the Thirteenth National Conference on Artificial Intelligence*, 1996.

[Gabbay, 1999] Dov Gabbay. *Fibring Logics.* Oxford Logic Guides 38; 475pp. Clarendon Press, Oxford, 1999.

[McCarthy and Buvač, 1994] John McCarthy and Saša Buvač. Formalizing Context (Expanded Notes). Technical Note STAN-CS-TN-94-13, Stanford University, 1994.

[Nossum, 2001] Rolf Nossum. A uniform quantificational logic for algebraic notions of context. Skriftserien 82, Agder University College, N-4604 Kristiansand, Dec 2001. ISBN 82-7117-448-7.

[Nossum, 2002] Rolf Nossum. Propositional logic for ground semigroups of context. *Logic Journal of the IGPL*, 10(3):273–297, 2002.

Modelling Periodic Temporal Notions by Labelled Partitionings — The PartLib Library

HANS JÜRGEN OHLBACH

1 Motivation and Introduction

In 1998 I published with Dov Gabbay a very first paper about a system we called *calendar logic* [Ohlbach and Gabbay, 1998; Ohlbach, 2000]. It was a theoretical formalisation of calendar systems and operations on calendar systems. In the meantime a growing community of computer scientists is working at modelling calendar systems and more abstract temporal notions. A reason why this work is becoming more important is the globalisation, and with this the need to make software familiar with all the different calendar systems and temporal notions used at our planet. In particular the Semantic Web initiative [Berners-Lee *et al.*, 1999] generates a need for detailed modelling of temporal notions. As one of the activities in the EU network of excellence REWERSE (http://www.rewerse.net) I am developing the ideas of these early papers with Dov further into a support system for representing and manipulating all kinds of temporal notions. This chapter describes the component of the system that deals with periodic temporal notions.

The basic time units of calendar systems, years, months, weeks, days etc. are the prototypes of periodic temporal notions. Because time is one of the most important parameters of our life, the representation of temporal notions, and in particular periodic temporal notions, is necessary in many computer applications. There have been quite intensive studies of periodic temporal notions from various points of view. One can distinguish at least three approaches.

First of all, there is the important work of Dershowitz and Reingold [Dershowitz and Reingold, 1997] who analysed existing calendar systems and came up with algorithms for converting date information from one system to another. These algorithms are the basis for the implementation of concrete calendar systems in computer programs.

On a more abstract level there is all the work about the mathematical representation of periodic temporal notions as *time granularities*, or similar kind of mathematical objects. A good overview is given in the book of Bettini, Jajoda and Wang [Bettini *et al.*, 2000]. This work is particularly motivated by the need to represent time in temporal databases. A selection of papers about the abundant work in this area is [Bettini and R.D.Sibi, 2000; Ning *et al.*, 2002; Kline *et al.*, 1999; Soo and Snodgrass, 1992; Leban *et al.*, 1986; Niezette and Stevenne, 1993; Dyreson *et al.*, 2000; Bettini *et al.*, 1998; Egidi and Terenziani, 2004; Bettini *et al.*, 2004; Goralwalla *et al.*, 2001; Bry *et al.*, 2004]. Since time granularities are the most important objects in this area, we introduce them already at this early place in the chapter. A time granularity is usually defined as a mapping of a subset of the integers to sets of intervals in the time domain, the *granules*. This mapping must have certain properties in order to count as time granularity. Another way to explain time granularities is: a *granule* is a, possibly non-convex finite subinterval of the time domain. A *time granularity* is a sequence of such granules. One can require that this sequence is consecutive, i.e. the rightmost time point of a granule n comes before the leftmost time point of the granule $n + 1$. Sometimes, however, overlapping granules are also considered [Egidi and Terenziani, 2004]. The simplest time granularities are in fact partitionings of the time domain. All basic time units, years, months etc., are of this type. The granules consist of one single interval, and there are no gaps between them. Granules consisting of one single interval only, but with gaps between them, can, for example, be used to model 'weekend'. The time spans between the weekends are the gaps between the granules. Granules consisting of several intervals are useful to model notions like 'my working day', where there is a lunch break which should not count as part of 'my working day'. Overlapping granules might be used to model, for example, the union of 'my working day' and 'my wife's working day'. The 'time granularity community' has developed ways for constructing time granularities, usually as algebraic operations on previously constructed time granularities. Conversion operations between different granularities have been defined. Relations between different time granularities have been developed, and applications, mainly in the area of temporal databases, have been considered.

An even further abstraction is possible by axiomatising temporal notions in an expressive enough logic, for example in first order predicate logic. The SOL time theory (SOL for Structured Temporal Object) of Diana Cuckierman with a first order formalisation of time loops is a prominent example for this approach [Cukierman, 2003; Cukierman and Delgrande, 1998; Cukierman and Delgrande, 2004].

This chapter presents an alternative to the granularities approach. We represent periodic temporal notions as partitionings of the real numbers, which is the simplest form of granularities. To compensate for this very weak structure, we introduce names (labels) for the partitions. The labels carry information about the meaning of the partitions. As we shall see, this separation of structure and meaning has a number of algorithmic advantages. A built-in label is 'gap'. It can be used to denote a partition which logically does not belong to a given partitioning. For example, the time between two subsequent school holidays in a school holiday partitioning can be labelled 'gap'. The label 'gap' allows one to simulate the granules of time granularities while separating the algorithms into the ones dealing with the partitionings and the other ones dealing with the labels and the granules.

The CTTN System

The work on partitionings for modelling periodic temporal notions is part of the CTTN-project. The CTTN system (Computational Treatment of Temporal Notions) [Ohlbach, 2005a] is a system for understanding, representing and manipulating complex temporal notions from everyday life. The basic modules in the CTTN system are the FuTI library for representing and manipulating fuzzy time intervals [Ohlbach, 2005b], the PartLib library for representing and manipulating periodic temporal notions, different calendar systems, and finally the GeTS language [Ohlbach, 2005c]. The GeTS language (GeoTemporal Specifications) is a typed functional programming language with a lot of built-in data types and operations for manipulating temporal notions. The GeTS language has in particular access to the FuTI library for dealing with crisp and fuzzy time intervals, and to the PartLib library for dealing with periodic temporal notions. A simple example for a definition in GeTS is

$$tomorrow = partition(now(), day, 1, 1).$$

$now()$ yields the current moment in time, measured in seconds. day refers to the day partitioning from the currently activated calendar, $partition(\ldots, day, 1, 1)$ creates the interval that corresponds to the day partition of the next day. The algorithms for computing the boundaries of this interval are presented in this document.

REMARK 1. It is important to notice that the ideas and techniques presented in this chapter are only one piece in a bigger mosaic. The labelled partitionings together with the operations described in this chapter are only one of several components in the GeTS language, which is the main tool for representing and working with temporal notions. Not all operations which are in principle possible with partitionings are therefore realized in the PartLib library itself, but on another level of the CTTN–system. In particular, there is no direct support for logical inferencing with the partitionings in the PartLib library. ■

Guidelines

The guidelines for the particular approach presented in this chapter, and realized in the PartLib library were:

1. The reality should be taken serious:

This means that all phenomena in real calendar systems and realistic periodic temporal notions should be taken into account. The consequences of this can be illustrated with the following definition of *month* taken from [Ning *et al.*, 2002]: The authors defined `day` first, then

$$\texttt{pseudomonth} = Alter^{12}_{11,-1}(day, Alter^{12}_{9,-1}(day, Alter^{12}_{6,-1}(day, Alter^{12}_{4,-1}(day, Alter^{12}_{2,-3}(day, Group_{31}(day))))))$$

and finally

$$\texttt{month} = Alter^{12\cdot 400}_{2+12\cdot 399,1}(day, Alter^{12\cdot 100}_{2+12\cdot 99,-1}(day, Alter^{12\cdot 4}_{2+12\cdot 3,1}(day, \texttt{pseudomonth}))).$$

The last definition takes leap days into account. *Alter* is the *alternating tick* operator and *Group* is the grouping operator. It is not necessary here to understand these operators. The point is that in a user friendly implementation of these operators an evaluator for arithmetic expressions like $2 + 12 \cdot 3$ is needed. This should be be no problem as long as the expressions are simple enough. For more complex temporal notions, however, the expressions become also more complex, and eventually a full size programming language is needed here. For example, for modelling ecclesiastical calendars, one needs to calculate the Easter date, and the algorithm for this is too complex to be expressed as a simple arithmetic formula.

The consequence for the PartLib library was to introduce a partitioning type *algorithmic partitionings*, which is specified by providing concrete algorithms in a concrete programming language (C++ in this case). Nevertheless, this is an exception. The general guideline is 'as algorithmic as necessary, as symbolic as possible'. The algorithmic partitionings can be used to define what some authors call *basic calendars* [Leban *et al.*, 1986; Egidi and Terenziani, 2004].

2. Separation of structure and meaning:

An infinite sequence of non-overlapping granules is in principle also a partitioning of the time domain if the gaps between the granules and within the granules are considered as part of the partitioning. Therefore one can turn a partitioning into a granularity by labelling certain partitions as gaps, and labelling the partitions which should belong to a granule with a common name. This has the advantage that the algorithms can be separated into a part which deals with the structure of the partitions, and a part which deals with the labels. Moreover, the labels can be used for other purposes. For example, the labels of a bus timetable can be the bus identifiers, and these can be keys for a bus database.

Notice that the notion of a 'label' in this chapter is different to the notion of a 'label' in the literature about granularities. Labels in this chapter are *names*, i.e. strings like 'Monday', 'Tuesday' etc. The 'labels' in the literature about granularities correspond to *coordinates* in this chapter.

PartLib provides no means for representing overlapping granularities as a single object. They must be represented as two separate partitionings.

3. Compact data structures and efficient algorithms:
The partitionings must be represented with finite data structures which support a number of particular algorithms. This excludes certain problematic operators for constructing new partitionings (granularities) from existing ones. An example is the *union* operator. To understand this, consider a representation of, say, 'Tom's working day' and 'Jane's working day'. If Tom's working day is every day from 8 am until 5 pm, and 'Jane's working day' is every day from 9 am until 6 pm, then the union operation on these two partitionings is unproblematic. The resulting partitioning is easily representable in a compact way. If, however, Tom's job is to watch the moon in an observatory, then his working day may need to follow the moon phases, which can be represented with an algorithmic partitioning. The union of the two partitionings 'Tom's working day' and 'Jane's working day' is now a really complicated object, and not easily represented. Therefore we exclude operations like union, intersection etc. in PartLib itself. Instead we provide such operations in the GeTS language. What is easy to realize is an operation which cuts 'Tom's working days' and 'Jane's working day' out of a given *finite* interval and then applies the set operation to the two intervals. Therefore 'Tom's and Jane's working day' would not be represented as a partitioning, but as a function that takes a time interval I and returns the subintervals of I which corresponds to the union of Tom's working days and Jane's working days in I.

4. Intuitive specification of user defined partitionings:
Most basic time units of calendar systems have a non-trivial algorithmic component. In the CTTN–system they are therefore realized as built-in partitionings. Many others, however, are application specific or user defined. Therefore various authors have come up with algebraic operations for constructing new partitionings (granularities) from given ones [Egidi and Terenziani, 2004]. The art is to find a basic set of operations which

allows one to define new partitionings in a way which is intuitive to the user, and which provides good data structures for the algorithms. This set should be as small as possible in order to reduce the burden to develop the corresponding algorithms. On the other hand, it should be powerful and expressive enough that most real world examples for periodic temporal notions can be specified.

Besides the algorithmic partitionings, PartLib provides two more basic types of specifications: 'duration partitionings' and 'folded partitionings'. Duration partitionings are specified by an anchor time and a sequence of 'durations'. A *duration* is something like '1 month + 3 day', where 'month' and 'day' represent previously defined partitionings. For example, I could define 'my weekend' as a *duration partitioning* with anchor time 2004/7/23, 4 pm (Friday July, 23rd, 2004, 4 pm) and durations: ('8 hour + 2 day', '4 day + 16 hour'). The first interval would be labelled 'weekend', and the second interval would be labelled 'gap'.

Notice that this specification is different to ('56 hour', '112 hour'). The difference is that when standard time changes to daylight savings time then the day is only 23 hours long, and when daylight savings time changes to standard time then the day is 25 hours long. A proper representation of 'day' as an algorithmic partitioning can take this into account, such that '8 hour + 2 day' would be the correct time shift even in this case. The specification of 'weekend' with a duration of '56 hour', however, would become wrong during the daylight savings time period.

PartLib provides two specialisations of duration partitionings. They allow for faster algorithms, and they are more intuitive in certain cases. The first specialisation, the *regular partitionings* covers the case that the durations are all of the same kind 'n P' where P is always the same partitioning. For example, 'semester' could be defined this way. The anchor time is the start of, say, the winter semester. The durations are ('6 month', '6 month'). The first partition would be labelled 'winter semester', and the second partition would be labelled 'summer semester'. The algorithms for this simple kind of duration partitioning are more efficient than in the general case.

Another specialisation are *date partitionings*. In this version the partitions are specified by concrete dates. In many countries, for example, people used to count the years from the beginning of the reigns of their emperors, and these are concrete dates. At a first glance, this specifies a partitioning of only a finite part of the time domain. To take this into account many of the algorithms would need to check whether the time points under consideration are in the valid part of the time line where the partitioning is specified, or not. With a simple trick, however, one can turn this finite partitioning into an infinite partitioning, and thus avoid these special cases. The trick is to turn the difference between two consecutive dates into durations. For example, the two dates 2004/5/10 and 2006/8/15 can be turned into a duration '2 year + 3 month + 5 day'. This way a date partitioning is turned into a duration partitioning. The finite part of the date partitioning is then automatically extrapolated into the infinite future and past. PartLib provides means to define boundaries for the partitionings, but these boundaries are not checked by the algorithms. It is up to the application of PartLib to check the boundaries.

Duration partitionings are the second basic type of partitionings. The third type are *folded partitionings*. Consider a bus timetable, which changes from season to season. The best way to specify this, would be to specify the seasons first, and for each season to specify the particular bus timetable. The 'folded partitioning' specification operation takes as input a *frame partitioning*, for example the seasons, and a sequence of *folded partitionings*, for example the four different bus timetables. It maps the folded partitionings automatically to the right frame partition, such that from the outside the whole thing looks like an ordinary partitioning.

5. Support for certain key operations with partitionings:
A very natural operation is to measure the distance between two time points in terms of a given time unit. 'The distance between t_1 and t_2 is 3.5 weeks', could, for example, be a useful information. Measuring the distance between two time points in terms of partitions of fixed length, for example seconds, is no point. It becomes more difficult if the time units have varying lengths. 'The distance between t_1 and t_2 is 3.5 months' is a nontrivial statement, because it depends on the location of t_1 and t_2 on the time line.

PartLib provides two *length* functions. The first one measure the distance between two time points in partitions of a given partitioning, and the second one measures the distance in granules. 'The distance between t_1 and t_2 is 1 working day', for example, is a possible outcome, even if 'working day' is defined not as a partitioning, but as a granule with a gap in it (for lunch time).

The second very natural operation is a shift operation, also in terms of partitions or granules. For example, one can ask a PartLib method to shift a time point t by, say 3.5 months, or 3.5 working days into the future. Since the lengths of the partitions and granules may vary, the concrete amount, t is shifted, depends on the location of t on the time line. It turns out that the notion of a time shift by some partitions is ambiguous. There are at least two different ways to do this, with more or less intuitive results. This problem is discussed in detail in Section 4.

After a brief review of PartLib's time domain we present the formal definitions of the basic concepts and then discuss the specification mechanisms and the operations on partitionings. The interface to the PartLib implementation is presented in the appendix.

Time Measurements and the Semantics of Computer Time

The backbone of our time representation is a reference time line, measured in seconds. The relation between the artificial counting of seconds in the computer and the real flow of time on our planet is determined by the physics of time measurement. Before the adoption of the UTC standard (Coordinated Universal Time) in 1972, a second was just the 86400th fraction of a day, measured between two subsequent zeniths of the sun. Since the rotation of the earth is not perfectly stable over the year, and, moreover, slows down from year to year, these seconds corresponded to varying time intervals. After the adoption of the UTC standard, a second corresponds to exactly 9.192.631.770 cycles of the light emitted when an electron jumps between the two lowest hyperfine levels of the Cesium 133 atom (measured and coordinated by the 'Bureau International des Poids et Mesures' in Paris, URL: http://www.bipm.fr/). The synchronisation with the rotation of the earth is achieved by inserting a leap second almost every year by the International Earth Rotation Service (URL: http://hpiers.obspm.fr/). Therefore the seconds in our modelling of partitionings for the time before 1972 correspond to a fraction of the day. For the time after 1972 they correspond to the atomic seconds of the TAI standard (Temps Atomique International).

In this chapter it is assumed that there is a *global reference time GRT*, measured in seconds or some fraction of a second. *GRT* is actually isomorphic to the real numbers, but the number 0 corresponds to a particular point in time. As it is common in Unix systems, the origin of the reference time in our examples is the beginning of the year 1970 at the 0-meridian.

REMARK 2. Since the real numbers $\mathbb{R}$ are used as the time axis, we speak of the *earliest* or *leftmost* number, time point or interval if we mean the one closest to $-\infty$. We speak of the *latest* or *rightmost* number, time point or interval if we mean the one closest to $+\infty$. ■

A Top Level View of the System

Partitionings are infinite structures. For specifying and implementing infinite structures, however, it is necessary to find finite representations. The algorithms can, of course, only work on these finite representations. There may, however, be different finite representations and hence different algorithms for the same task. The different finite representations are caused by the different ways the structures are specified. From a software engineering point of view, we have therefore two sides of the system, in this case the

PartLib library. At one side there are the different specification types for partitionings. Certain algorithms, for example the mapping between partition boundaries and partition numbers, rely on the concrete specification type. At the other side there is the application interface to the library. The API hides the fact that there are different types of partitionings and provides a uniform access to them. The API is realized in PartLib as an abstract class `Partitioning`. As many algorithms as possible are defined for the abstract class. They may, however, use algorithms which are specific for the specification types.

The next section describes the aspects of abstract partitionings, which is essentially the application interface. The concrete specification types are then introduced in Section 8.

2 Partitionings

A partitioning of the real numbers $\mathbb{R}$ may be for example $(..., [-100, 0[, [0, 100[, [100, 101[, [101, 500[, ...)$. The intervals in the partitionings considered in this chapter need not be of the same length (because time units like years are not of the same length either). The intervals can, however, be enumerated by integers (their *coordinates*). For example, we could have the following enumeration

$$\begin{array}{cccccc} ... & [-100\ 0[& [0\ 100[& [100\ 101[& [101\ 500[& ... \\ ... & -1 & 0 & 1 & 2 & ... \end{array}$$

It is not by chance that half open intervals are used in this example. Since the partitions in a partitioning do not overlap, one cannot use closed intervals because the endpoints of the closed intervals would be in two different partitions. Open intervals can not be used either because then the infima and suprema of the intervals would not be in any partition at all. Therefore only half open intervals can be used, either of the type $[a, b[$, or of the type $]a, b]$. In most cases there is no preference for either of the two types, but both types should not be used together. In this chapter we therefore use the first type $[a, b[$.

Since all time measurements are done in discrete units (seconds, ticks of a clock, hyperfine transitions in a Cesium atom etc.) it makes sense to take integers as boundaries of the partitions. In the examples they represent seconds. Any other fraction of a second is possible as well. Multiples of seconds are not possible without losing precision because the leap seconds are ignored in this case.

The formal definition for partitionings of $\mathbb{R}$ which is used in this chapter is:

DEFINITION 3 (Partitioning). A partitioning P of the real numbers $\mathbb{R}$ is a sequence

$$\ldots [t_{-1}, t_0[, [t_0, t_1[, [t_1, t_2[, \ldots$$

of non-empty half open intervals in $\mathbb{R}$ with integer boundaries. ■

Some useful notations for partitionings are defined:

DEFINITION 4 (Notations). Let P be a partitioning.

1. For a partition $p = [s, t[$ in P let $p_{[} \stackrel{\text{def}}{=} s$ be the left boundary of p and let $p_{]} \stackrel{\text{def}}{=} t$ be the right boundary of p.

2. For a time point t and a partitioning P, let t^P be the partition in P which contains t.

■

A sequence of finite partitions of the real numbers is in fact isomorphic to the integers. This can be exploited to give the partitions *addresses* or *coordinates*. The coordinates are very useful for navigating through sequences of partitions. Therefore we introduce *coordinate mappings*. In principle, there are many different coordinate mappings for a given partitioning, but for the intended application of the partitioning concept described in this chapter, one single coordinate mapping is sufficient. Therefore this unique coordinate mapping becomes an integral component of a partitioning.

DEFINITION 5 (Coordinate Mapping). A *coordinate mapping* of a partitioning P is a bijective mapping between the intervals in P and the integers. Since we usually use one single coordinate mapping for a partitioning P, we can just use P itself to indicate the mapping.

Therefore let p^P be the *coordinate* of the partition p in P.

For a coordinate i let i^P be the partition which corresponds to i.

For a time point t let $P.pc(t) \stackrel{\text{def}}{=} (t^P)^P$ be the coordinate of the partition containing t. (*pc* stands for 'partition coordinate').
Let $P.sopT(t) \stackrel{\text{def}}{=} t^P{}_{[}$ be the start of the partition containing t.
Let $P.eopT(t) \stackrel{\text{def}}{=} t^P{}_{[}$ be the end of the partition containing t.

For a coordinate i let $P.sopC(i) \stackrel{\text{def}}{=} i^P{}_{[}$ be the start of the partition with coordinate i.
Let $P.eopC(i) \stackrel{\text{def}}{=} i^P{}_{[}$ be the end of the partition with coordinate i.

For a time point t we define

$$P.lopT(t) \stackrel{\text{def}}{=} P.eopT(t) - P.sopT(t)$$

as the length of the partition containing t.

For a coordinate i we define

$$P.lopC(i) \stackrel{\text{def}}{=} P.lopT(P.sopC(i))$$

as the length of the partition with coordinate i. ■

The two pictures below illustrate the transitions between time points, coordinates and partitions:

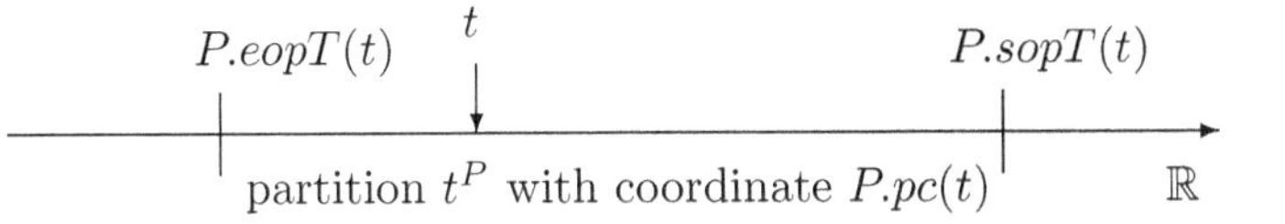

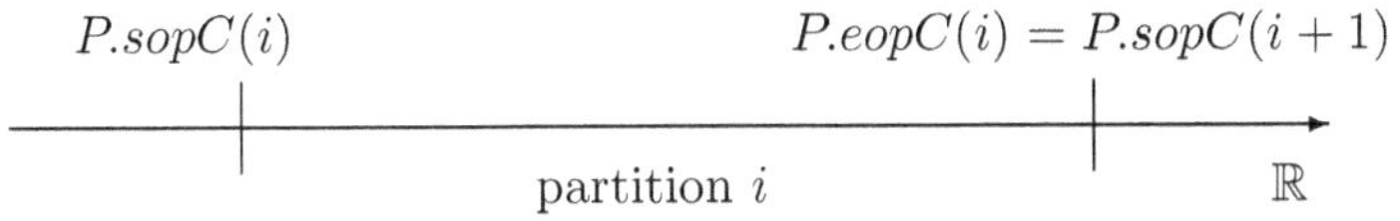

EXAMPLE 6 (Seconds and Minutes). The partitioning for seconds consists of the sequence of intervals $\ldots[-i,-i{+}1[\ldots[0,1[\ldots[k,k{+}1[\ldots$. The interval $[0,1[$ represents the first second in January 1st 1970, and its coordinate is 0.

The partitioning for minutes is

			1970/1/1 ↓			
$ref.time:$	...	$[-60,0[$	$[0,60[$	$[60,120[$	$[120,180[$	...
$coordinate:$	...	-1	0	1	2	,...

■

Remarks:

1. Partitions are not explictly represented in PartLib, only time points and coordinates. The functions which map time points to coordinates and back are therefore the important ones. Thus, the formulations of the algorithms below do not refer to partitions, but use these functions.

2. We use the notation $P.pc(t)$ for the function that maps a time point t to the coordinate of its partition in the partitioning P. Alternative notations would be $pc(P,t)$ or $pc_P(t)$ or $pc^P(t)$. The notation $P.pc(t)$ comes from object oriented programming. P is an object (instance of a class), and pc is a method in this class. This notation has two advantages. First of all, it is a bridge to the actual implementation where the program code looks just so. Secondly, it emphasises the special role of P as the context for the pc

function as well as a number of other functions. Therefore we shall use the dot notation '*P.*' for most of the functions which depend on partitionings and other objects. If it is clear from the context, which object is meant, we may omit this object, and just use the function name.

2.1 Length of Intervals in Partitions

It is very common to measure the length of intervals or the distance between time points in terms of time units. Examples are 'The train A arrives in the station 5 minutes before the train B leaves it'. 'Tomorrow I go on a adventure trip and will be back in 3 months time'.

A very useful function is therefore $P.length(t_1, t_2)$, which measures the length of the distance between t_1 and t_2 in terms of partitions of the partitioning P. For example, $month.length(t_1, t_2)$ measures the length of $[t_1, t_2[$ in months. Since partitions may have different lengths, this is a nontrivial operation.

The idea for the method can be illustrated with the following picture

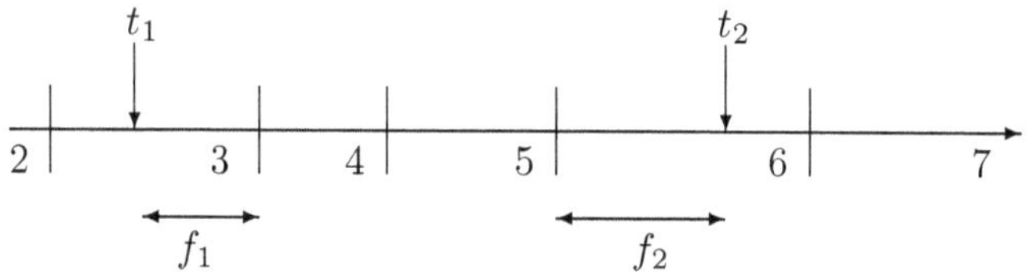

The distance between t_1 and t_2 is the sum of the relative length of f_1, measured as a fraction of the length of partition 3, plus the relative length of f_2, measured as a fraction of the length of partition 6, plus the number of partitions in between.

DEFINITION 7 (Length in Partitions). Let P be a partitioning.

For two time points t_1 and t_2 with $t_1 \leq t_2$ we define

$$P.lengthP(t_1, t_2) \stackrel{\text{def}}{=} \begin{cases} \dfrac{t_2 - t_1}{P.lopT(t_1)} & \text{if } P.pc(t_1) = P.pc(t_2) \\[2ex] P.pc(t_2) - P.pc(t_1) - 1 + & \\ \dfrac{P.eopT(t_1) - t_1}{P.lopT(t_1)} + \dfrac{t_2 - P.sopT(t_2)}{P.lopT(t_2)} & \text{otherwise} \end{cases}$$

If $t_2 < t_1$ then $P.lengthP(t_1, t_2) \stackrel{\text{def}}{=} -P.lengthP(t_2, t_1)$. ∎

$P.lengthP(t_1, t_2)$ is continuous. That means if t_1 is kept fixed and t_2 is moved, or the other way round, then $P.lengthP(t_1, t_2)$ makes no jumps. It is, however, not differentiable at the points where t_2 crosses the boundaries of neighbouring partitions with different length.

$P.lengthP(t_1, t_2)$ can be used to measure the absolute length of the interval $[t_1, t_2[$ if P is the partitioning for seconds or smaller time units. If P

is the partitioning for minutes we can get the effect that an interval of 60 seconds length is smaller than one minute. This is the case for those minutes which contain leap seconds. Similar things happen for the coarser time units. We may get $day.length(t_1, t_2) < 1$ even if $hour.length(t_1, t_2) = 24$. This happens when daylight savings time is disabled just during the interval $[t_1, t_2[$ and the day is 25 hours long.

DEFINITION 8 (modulo, remainder, $\lfloor ... \rfloor$ and $| \dots |$).
The mod and remainder functions are used to map integers to non-negative indices $0, \dots, n-1$. Therefore we need versions where the resulting values are between 0 and $n-1$, even for negative numbers. mod and remainder are defined for positive numbers as usual. For the negative numbers there are two different possibilities. We need the version where the resulting value is positive.

That means, $k \bmod n$ is chosen such that for example $4 \bmod 3 = 1$ and $-4 \bmod 3 = 2$.

$m/n = k$ remainder l is chosen such that for example $4/3 = 1$ remainder 1 and $-4/3 = -2$ remainder 2.

Let $\lfloor m \rfloor$ be the integer part of m such that $\lfloor 3.5 \rfloor = 3$ and $\lfloor -3.5 \rfloor = -3$.

For an interval s let $|s|$ be the length of the interval. ■

2.2 Labels

For many periodic temporal notions there are standard names for the partitions. For example, days are named 'Monday', 'Tuesday' etc., months are named 'January', 'February' etc., seasons are named 'winter', 'spring' etc. If these names are attached as *labels* at the partitions, temporal notions like 'next summer' etc. can be modelled in an elegant way.

DEFINITION 9 (Labelled Partitionings). A *Labelling* L is a finite sequence of labels (strings) $l_0, \dots, l_{n-1}$.

A labelling $L = l_0, \dots, l_{n-1}$ is turned into a *labelling function* $L(i)$ for a coordinate i:

$$L(i) \stackrel{\text{def}}{=} l_{i \ mod \ n}$$

In the sequel we shall identify the labelling L with the labelling function $L(i)$. ■

A labelling L can now be very easily attached to a partitioning: the partition with coordinate i gets label $L(i)$.

EXAMPLE 10 (The Labelling of Days). The origin of the reference time is again January 1st 1970. This was a Thursday. Therefore we choose as labelling for the day partitioning

$$L \stackrel{\text{def}}{=} Th, Fr, Sa, Su, Mo, Tu, We.$$

The following correspondences are obtained:

ref.time :	...	$[-86400, 0[$	$[0, 86400[$	$[86400, 172800[$	...
coordinate :	...	-1	0	1	...
label :	...	We	Th	Fr	...

This means, for example, $L(-1) = We$, i.e. December 31 1969 was a Wednesday. ■

2.3 Granules

The label 'gap' is a reserved keyword. It can be used to denote gaps between semantically related partitions. For example, if the partitions represent school holidays then the periods between the school holidays can be named 'gap'. Gaps, together with the possibility to use the same label at different positions in a labelling makes it possible to define *granules*, with the same effect as in the original definitions of granularities. As an example, consider again the definition of 'working day' as a period of time between 8 o'clock am and 5 o'clock pm, interrupted by a lunch break between 1 o'clock pm and 2 o'clock pm. This could be defined in two stages. First, we define a partitioning with an anchor time at some particular Monday 8 o'clock am, and durations '5 hour', '1 hour', '3 hour', '16 hour'. The '1 hour' period is the lunch break and the '16 hour' period is the time between two 'working days'. A suitable labelling is 'working_day, gap, working_day, gap'.

A *generator for granules* is characterised by a maximally long subsequence of a labelling such that the non-gap labels in the generator are the same. 'working_day, gap, working_day' in the above labelling is a generator for a granule, and this is the only one in this example.

REMARK 11. The definition of a *granule* in this chapter includes internal gaps. A granule that corresponds to the generator 'working_day, gap, working_day' includes therefore the 'gap partition (the lunch time). A granule therefore corresponds always to a *convex* interval. Since a granule is generated by a labelling, the information about which part of the granule is a gap, is always available. Therefore it is technically simpler to define it this way, and to let the algorithms exploit information about internal gaps. ■

If a labelling is attached at a partitioning, we get granules as subsets of the time line. The next picture illustrates this for the 'working_day' example. 'wd' stands for 'working_day'.

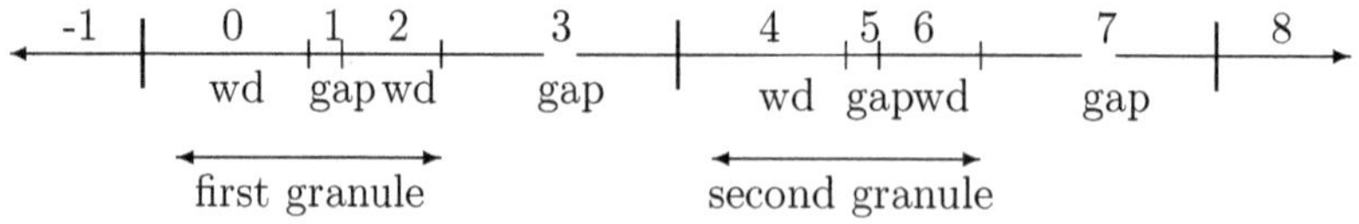

Only the partitions 0-2, 4-6 etc. represent granules. The partitions 2-4 are not a granule, although they are also labelled 'working_day, gap, working_day'. This is prevented because the definition of granule is based on the initial labelling, in this case 'working_day, gap, working_day, gap'.

DEFINITION 12 (Granule). Let $L = l_0, \ldots, l_{n-1}$ be a labelling.

1. A *generator for granules* of the labelling L is a maximal subsequence $G = l_k \ldots l_m$ of L such that
   (i) $l_k \neq gap$ and $l_m \neq gap$, and
   (ii) all non-gap labels in G are the same.

2. A granule of a labelling function $L(\ldots)$ (Def. 9) is a sequence $k \ldots m$ of coordinates, such that $l_{k \bmod n}, \ldots l_{m \bmod n}$ is a generator for granules of the labelling L.

3. A granule of a labelled partitioning P with labelling L is an interval $k^P \cup \ldots \cup m^P$, such that $k \ldots m$ is a granule of the labelling function $L(\ldots)$.

■

A labelling like 'Monday, Tuesday, ...' without gaps and with the labels all being different has granules labelled with a single label each. A labelling may of course also have several non-trivial granules. To see this, let us take our 'working_day' example a bit further. The underlying partitioning partitions also the weekends. If we want to specify that 'working_days' are only from Monday till Friday, we could do this with the following labelling: 'wdMo, gap, wdMo, gap, wdTu, gap, wdTu, gap, wdWe, gap, wdWe, gap, wdTh, gap, wdTh, gap, wdFr, gap, wdFr, gap, gap, gap, gap, gap, gap, gap, gap, gap'. This labelling has five different granules 'wdMo, gap, wdMo', 'wdTu, gap, wdTu', 'wdWe, gap, wdWe', 'wdTh, gap, wdTh', 'wdFr, gap, wdFr', one for each day. The last 8 gaps exclude the weekend.
A partitioning without an explicit labelling defines of course also granules: each partition is a granule consisting just of this single partition.

We define a function *closestGranule* which returns for a partition coordinate or a time point the coordinates of the closest granule. The version *closestGranuleC*$(i, direction)$ takes a coordinate as argument, and the version *closestGranuleT*$(t, direction)$ takes a time point as argument. *direction* is only relevant for time points between granules. It can be either `past`, `future`, or `closest`. `past` causes that these time points are mapped to the previous granule. `future` causes that these time points are mapped to the next granule in the future, and finally `closest` causes that the closest granule is chosen, with a preference for `future` when the time point lies exactly in the middle between two granules. The *closestGranule* function is used in particular for the function *lengthG* (Def. 14) which measures intervals in terms of granules.

DEFINITION 13 (Closest Granule). Let L be a labelling of a partitioning P, i a coordinate and t a time point. *direction* is a keyword which can be `past`, `future`, or `closest`.

The function $L.closestGranuleC(i, direction)$ returns a pair (k, m) of coordinates, such that
(i) $k \ldots m$ is a granule of the labelling function $L(\ldots)$ (Def. 12), and
(ii) either i is within a granule, i.e. $k \leq i \leq m$, or i is not within a granule and
(ii,a) $direction$ = `past` and $k \ldots m$ is the closest granule before i, or
(ii,b) $direction$ = `future` and $k \ldots m$ is the closest granule after i, or
(ii,c) $direction$ = `closest` and i is closer to the granule $k \ldots m$ than to any other granule. That means precisely, if (a, b) are the boundaries of another granule, and $b \leq i \leq k$ then $k-i \leq i-b$, and if $m \leq i \leq a$ then $a-i \leq i-m$.

The function $L.closestGranuleT(t)$ returns a pair (k, m) of coordinates, such that
(i) (k, m) is a granule of the labelling function $L(\ldots)$, and
(ii) either t is within a granule, t is not within a granule, and
(ii,a) $direction$ = `past` and $k \ldots m$ is the closest granule before t, or
(ii,b) $direction$ = `future` and $k \ldots m$ is the closest granule after t, or
(ii,c) $direction$ = `closest` and t is closer to the granule $k \ldots m$ than to any other granule. That means precisely, if (a, b) are the boundaries of another granule, and $P.eopC(b) \leq t \leq P.sopC(k)$ then $P.sopC(k) - t \leq t - P.eopC(b)$, and if $P.eopC(m) \leq t \leq P.sopC(a)$ then $P.sopC(a) - t \leq t - P.eopC(m)$. ∎

Length of intervals in granules:
Once the notion of 'working_day' is defined by means of granules, it is quite natural to measure intervals in terms of the length of a 'working_day'. For example, you may want to say that 'these four hours are half a working day'. This is not much of a problem if the granules all have the same length. If not, it depends on the position of the interval. 'Half a working day' may mean something different if I measure it on a Monday where I work 8 hours, or on a Friday, where I work, say, only 4 hours. Even worse, what if the interval lies on a weekend, and therefore outside any granules? It may still make sense to say that 'these four hours are half a working day', if I consider to shift the interval into a working day.

The function $lengthG(t_1, t_2, direction)$ measures the distance between t_1 and t_2 in terms of granules. It deals with the problem that part of the interval $[t_1, t_2[$ or even the whole interval may lie outside any granule. The basic idea is to split the interval $[t_1, t_2[$ according to the given partitioning, determine for each subinterval the closest granule (Def. 13), and measure the subinterval in terms of this closest granule. The arrows in the next picture show which granules are used to measure the different parts of the interval. *direction* = `closest` is chosen in this example.

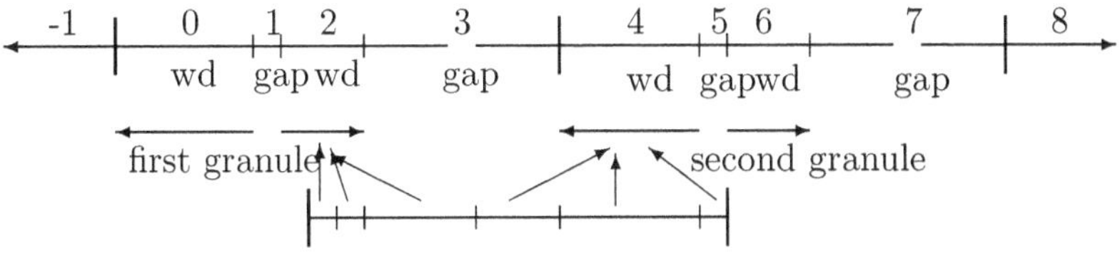

DEFINITION 14 (Length in Granules). Let P be a partitioning which is labelled with a labelling L. The function $P.lengthG(t_1, t_2, direction)$ measures the distance between t_1 and t_2 in terms of granules. *direction* is again one of the keywords `past`, `future`, or `closest`. The function works as follows:

Step 1: The interval $[t_1, t_2[$ is partitioned into subintervals $s_1, \ldots$ such that the subinterval boundaries are aligned with the partition boundaries of P.

Step 2: The middle point t_i for each subinterval s_i is computed, and the coordinates (l, m) of the granule closest to t_i is determined by the *closestGranuleT* function (Def. 13). Let l_i be the length of the non-gap partitions in this granule.

Step 3: The relative lengths $|s_i|/l_i$ are added together to give the result of $P.lengthG(t_1, t_2, direction)$. ■

2.4 Relations Between Partitionings

PartLib supports four different relations between partitionings.

DEFINITION 15 (Relations Between Partitionings). Let P and Q be two partitionings. We define the following four relations between P and Q.

1. 'P has shorter partitions than Q' if the largest partition of P is shorter than the shortest partition of Q (P is finer grained than Q.)

2. 'P has shorter granules than Q' if the largest granule of P is shorter than the shortest granule of Q.

3. 'P includes the partitions of Q' if each partition of P is a subset of a partition of Q.

4. 'P includes the granules of Q' if each granule of P is a subset of a granule of Q. ■

These four relations are easy to define, but difficult to compute because for algorithmic partitionings it is in general not possible to compute the minimal or maximal partition length. Therefore PartLib uses an approximation. It generates random time points and computes the partition and granule length for these random points and certain points in their neighbourhood. How many points in the neighbourhood are to be checked is controlled by the 'repetition' parameter. For example, for the partitioning 'month' one would check 12 subsequent months for each random time point. This way, the 'local structure' of the partitionings is checked completely, whereas only finite samples check the 'global structure', in the 'month' example, the leap years.

The relations 'includes the partitions of' and 'includes the granules of' are also checked with finitely many randomly generated time points and their neighbourhoods.

3 Date Formats

In the subsequent section various notions are introduced in a recursive way: date formats, shifts, durations, and different types of partitionings.

We start with the definition of date formats. Date formats and dates in PartLib are data structures, not date strings. Concrete date strings depend on calendar systems and conventions or standards. They are dealt with in the interface to the CTTN system, not in the PartLib library.

DEFINITION 16 (Date Format). A *date format* DF is a sequence $P_0/\ldots/P_k$ of partitionings.

A *date* in a date format DF is a sequence $d_0/\ldots/d_n$ of integers with $n \leq k$. ■

In principle, the date formats can consist of arbitrary partitionings. In most calendar systems there are, however, a few particular date formats. The Gregorian calendar, for example, has the two date formats year/month/day/hour/minute/second (where the names stand for the corresponding partitions), and year/week/day/hour/minute/second.

There are, however, two big differences between common date formats and the particular interpretation of the numbers in the dates we need in this chapter. Consider the date format year/month/day/hour/minute/second. The first difference is the interpretation of the year. The number 30, for example, for the 'year' part in the date format is the coordinate of the year. If the year 1970 has coordinate 0 then '30' is the coordinate of the year 2000. The next difference has to do with the way we count months, weeks and days. Usually these are counted from 1. That means, January is month 1, the first week in a year is week 1, and Monday is day 1. In contrast to this, we count hours, minutes and seconds from 0. The first hour in a day is hour 0, the first minute in an hour is minute 0 etc. Our interpretation of date formats like the above is that months, days etc. denote shifts, instead of absolute values. In this interpretation the date 30/1 denotes a shift of 1 month from the beginning of the year with coordinate 30 (i.e. the year 2000). This is the beginning of February. Thus, all time units are counted from 0.

By interpreting the numbers as shifts, there is no problem to deal with arbitrary big numbers, and even with negative numbers. The date 30/200/-50, for example, in the date format year/month/day/hour/minute/second denotes a time point t which is obtained from the beginning s of the year 2000 (30 years after 1970), by shifting s 200 months into the future, and from there 50 days into the past.

With these ideas in mind we can define a function *date*, which turns a time point into a date of the given format.

DEFINITION 17 (*date*). Let $DF = P_0/\dots/P_k$ be a date format, and t a time point.
Let $d_0/\dots/d_k \stackrel{\text{def}}{=} DF.date(t)$ where
$d_0 \stackrel{\text{def}}{=} P_0.pc(t)$ and
$t_i \stackrel{\text{def}}{=} P_i.sopT(P_{i-1}.sopT(t))$ and
$d_i \stackrel{\text{def}}{=} \lfloor P_i.lengthP(t_i, t) \rfloor$ for $i = 1 \dots k$. ∎

d_0 is the absolute coordinate of the P_0 partition containing t. t_1 is the starting point of the P_1 partition containing t. The d_i are then calculated as P_i increments compared to t_i where t_i is the beginning of the p_i partition containing t.

For example, in the date format year/week/day/... d_0 is the coordinate of the year containing t. t_1 is the beginning of the first week overlapping with the year d_0. d_1 is then the integer part of the number of weeks between t_1 and t. t_2 is the beginning of the first day in the week containing t etc.

The *date* function is exact only if the last partitioning P_k corresponds to the integers of the time axis, usually seconds, or fractions of a second.

It is also possible to turn a date $d = d_0/\dots/d_n$ in a date format $DF = P_0/\dots/P_k$ into a time point. The function $TimePoint(d)$ gives the dates a precise semantics.

DEFINITION 18 (Time Point $DF.TimePoint(d)$). Let $DF = P_0/\dots/P_k$ be a date format and let $d = d_0/\dots/d_n$ be a date in this format. The corresponding time point is defined:

$DF.TimePoint(d) \stackrel{\text{def}}{=} t_n$ where
$t_0 \stackrel{\text{def}}{=} P_0.sopC(d_0)$ and
$t_i \stackrel{\text{def}}{=} P_i.sopC(P_i.pc(t_{i-1}) + d_i)$ for $i = 1 \ldots n$. ■

In the date format year/week/day/..., for example, t_0 is the beginning of the year with coordinate d_0. $week.pc(t_0)$ is the coordinate of the week containing t_0. $week.pc(t_0) + d_1$ is the coordinate of the week which is d_1 weeks into the year. $t_1 = week.sopC(week.pc(t_0) + d_1)$ is the beginning of this week. t_2 is then the beginning of the day which is d_2 days into this week etc. Finally, t_k is the coordinate of the second denoted by the given date.

By induction on the length of a date, one can prove that turning a time point t into a date and the date back into a time point then we end up at the same time point t.

PROPOSITION 19. *For a date format DF whose last partitioning corresponds to the integers in the time axis, and a time point t:*

$$DF.TimePoint(DF.date(t)) = t.$$

■

The other direction, $DF.date(DF.TimePoint(d)) = d$. need not hold, because there may be quite different dates which represent the same time point. For example, February 1st 2000 may be represented by 30/1/0 or by 30/0/31.

4 Shift Functions

Notions like 'in two weeks time' or 'three years from now' etc. denote time shifts. They can be realized by a function which maps a time point t to a time point t' such that $t' - t$ is just the required distance of 'two weeks' or 'three years' etc. Shift functions are of particular importance to the PartLib library because some of the specification mechanisms for partitionings require to shift an anchor point by certain durations. Therefore this needs to be explained in detail.

4.1 Length Oriented Shift Function

We define a function $P.shiftPL(t, m)$ which shifts a time point t to a time point $t' = P.shiftPL(t, m)$ such that $P.lengthP(t' - t) = m$. ('shiftPL' stands for 'shift Partitions Length oriented', in contrast to 'shiftPD', which stands for 'shift Partitions Date oriented').

EXAMPLE 20 (for shiftPL). The algorithm for this function can be best understood by the following example:

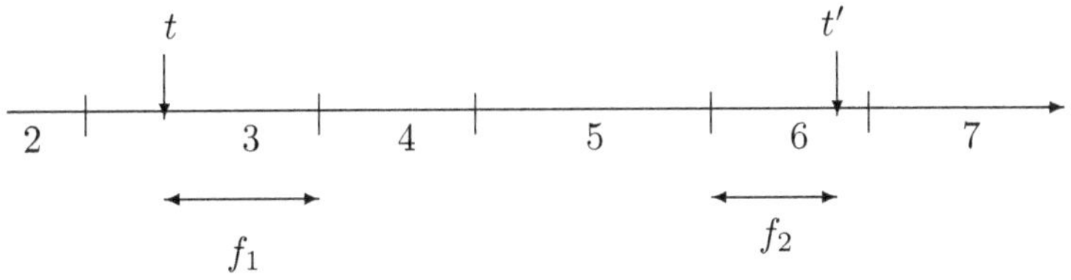

Suppose we want to shift the time point t by 3.5 partitions. First, the relative distance f_1 between t and the end of the partition containing t is measured. Suppose it is 0.75. That means from the end of the partition we need to move forward still 2.75 partitions. We can move forward 2 partitions by just adding the 2 to the coordinate 4. We end up at the start of partition 6. From there we need to move forward 0.75 partitions, which is just 75% of the length of partition 6. ■

DEFINITION 21 (shiftPL). Let P be a partitioning, t a time point, and m a real number.
If $m \geq 0$ then

$$shiftPL(t) \stackrel{\text{def}}{=} \begin{cases} t + m * lopT(t) & \text{if } t + m \cdot lopT(t) \leq eopT(t) \\ sopC(pc(t) + \lfloor m \rfloor) + (m - \lfloor m \rfloor) \cdot lopC(pc(t) + \lfloor m \rfloor) & \\ \qquad \text{if } sopT(t) = t & \\ sopC(pc(t) + \lfloor m' \rfloor + 1) & \\ + (m - m') \cdot lopC(pc(t) + \lfloor m' \rfloor + 1) & \text{otherwise} \end{cases}$$

where $m' \stackrel{\text{def}}{=} m - (eopT(t) - t)/lopT(t)$.
If $m < 0$ then

$$shiftPL(t) \stackrel{\text{def}}{=} \begin{cases} t + m * lopT(t) & \text{if } sopT(t) \leq t + m \cdot lopT(t) \\ sopC(pc(t) + \lfloor m \rfloor) + (m - \lfloor m \rfloor) \cdot lopC(pc(t) + \lfloor m \rfloor - 1) & \\ \qquad \text{if } sopT(t) = t & \\ sopC(pc(t) + \lfloor m' \rfloor) & \\ +(m - m') \cdot lopC(pc(t) + \lfloor m' \rfloor - 1) & \text{otherwise} \end{cases}$$

where $m' \stackrel{\text{def}}{=} m + (t - sopT(t))/lopT(t)$.

■

It is not so difficult to see that the shiftPL function shifts a time point t by m partitions to a time point t' such that the distance $t' - t$ is just m.

PROPOSITION 22 (Soundness of shiftPL). *For any time point t:*

$$shiftPL(t, m) - t = lengthP(t, shiftPL(t, m)) = m$$

Furthermore

$$shiftPL(shiftPL(t, m_1), m_2) = shiftPL(t, m_1 + m_2)$$ ■

The proof is technical and gives no new insight. It is therefore omitted.

Unfortunately the shiftPL function does not always give intuitive results. Suppose the time point t is noon at March, 15th, and we want to shift t by

1 month. March has 31 days. Therefore the distance to the end of March is exactly 0.5 months. Thus, we need to move exactly 0.5 times the length of April into April. April has 30 days. 0.5 times its length is exactly 14 days. Thus, we end up at midnight April, 14th.

This is not what one would usually expect. We would expect to shift t to the same time of the day as we started with. With the above definition of shiftPL this happens only by chance, or when the partitions have the same length.

4.2 Date Oriented Shift Function

PartLib provides another shift function, $shiftPD$ which avoids the above problems and gives more intuitive results. The idea is to do the calculations not on the level of reference time points, but on the level of dates. If, for example, t represents 2004/1/15, then 'in one month time' usually means 2004/2/15. That means the reference time must be turned into a date, the date must be manipulated, and then the manipulated date is turned back into a reference time. This is quite straight forward if the partitioning represents a basic time unit of a calendar system (year, month, week, day etc.), and this calendar system has a date format where the time unit occurs. In the Gregorian calendar this is the case, even for the time unit 'weeks'. 'In two weeks time' requires to turn the reference time into a date format which uses weeks. The corresponding date format uses the counting of weeks in the year (ISO 8601). For example, 2004/42/1 means Tuesday[1] in week 42 in the year 2004. In two weeks time would then be 2004/45/1.

The next problem is to deal with fractional shifts. How can one implement, say, 'in 3.5 months time'? The idea is as follows: suppose the date format is year/month/day/hour/minute/second, and the reference time corresponds to, say, 34/1/20/10/5/1. First we make a shift by three months and we end up at 34/4/20/10/5/1. This is a day in May. From the date format we take the information that the next finer grained time unit is 'day'. May has 31 days. $0.5 * 31 = 15.5$. Therefore we need to shift the date first by 15 days, and we end up at 34/4/34/10/5/1. There is still a remaining shift of half a day. The next finer grained time unit is hour. One day has 24 hours. $0.5 * 24 = 12$. Thus, the last date is shifted by 12 hours, and the final date is now 34/4/34/22/5/1. This is turned back into a reference time.

This version of 'shift' gives more intuitive results. The drawback is that $shiftPD(t, m) - t = lengthP(t, shiftPD(t, m)) = m$ is usually no longer true. $shiftPD$ has in fact not much to do with $lengthP$.

[1] According to ISO 8601, the first day in a week is Monday. In the standard notation this is day number 1. Since we count days from 0, Monday is day 0 and Tuesday is day 1.

The concrete definition of *shiftPD* depends on the partitioning type. Therefore, we give them in the corresponding sections below (see for example Def. 34).

4.3 Shift by Granules

A statement like 'we must move this task by three working days' refers to a shift of time points which is measured in granules. Since granules are multiples of partitions, it is not necessary to have a *shiftGranules* function. A simpler idea is to turn a number m of granules into a corresponding number n of partitions and then to apply the date oriented or length oriented shift function for partitions. Therefore PartLib provides only a function $granules2Partitions(t, m)$. It turns the number m of granules into a number n of partitions such that a shift of the time point t by m granules corresponds to a shift by n partitions. m can be any positive or negative integer or fractional number.

The algorithm for *granules2Partitions* has to distinguish different cases:

case 1: the time point t is within a granule;

> **case 1a:** t is within a non-gap partition of the granule;
>
> **case 1b:** t is within a gap partition of the granule;

case 2: t is in a gap partition between two granules. This case can be reduced to case 1a by inverting the role of gaps and granules. That means the granules together with the intra–granule gaps, become gaps, and the gaps between the granules become granules.

These cases are quite trivial if the structure of the granules is always the same. An example where the granules have a different structure may be a working day consisting of a day shift (ds) and a night shift (ns). As indicated in the figure below the day shift consists of two non-gap blocks (morning and afternoon) separated by a lunch break. The night shift consists of three non-gap blocks separated by two breaks.

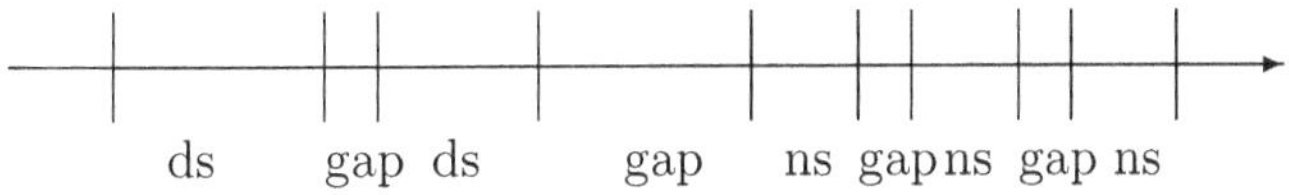

Let us illustrate the difficulties with a still quite simple example. If the time point t is, say, in the afternoon block of the day shift, where should 'one granule further' be? It should be in the night shift, but not in one of the breaks of the night shift. The function *granules2Partitions* tries to identify in this case a corresponding non-gap block in the night shift by measuring the relative distance of the non-gap block containing t to the left boundary of the granule, in this case $2/2$ (t is in non-gap block 2 of two non-gap blocks). Since the night shift contains three non-gap blocks, $2/2 * 3 = 3$, i.e. the corresponding non-gap block in the night shift is the third block. This is a heuristic. Experience has to show whether there are better ones.

We present the cases 1a and 1b of the algorithm for *granules2Partitions* by means of typical examples. Only the case $m > 0$ is explained in detail. The case $m < 0$ is analogous.

EXAMPLE 23 (for case 1a of *granules2Partitions*:). Suppose we want to shift the time point t in the picture below by $m = 1.5$ granules. t is within a non-gap partition of a granule and $m > 0$. We determine the number of partitions to be shifted in 5 steps.

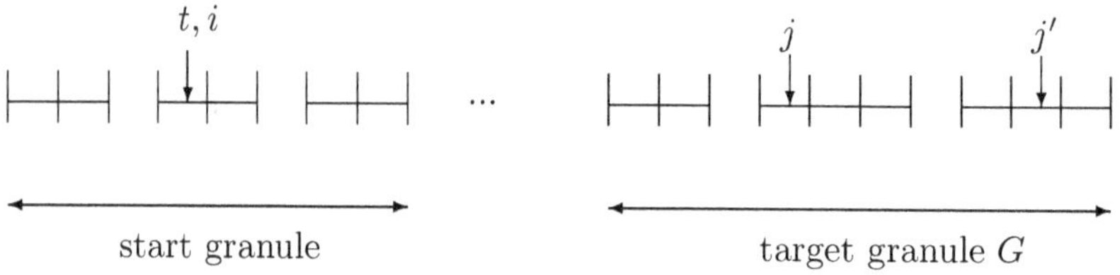

Let $i \stackrel{\text{def}}{=} P.pc(t)$ be the coordinate of the partition containing t. Let b be the relative distance of the non-gap block containing t. In the example t is in block 2 of three blocks. Therefore $b = 2/3$. Let d be the relative distance of partition i within the non-gap block containing t. In the example t is in partition 1 of a 2-partition non-gap block. Therefore $d = 1/2$.

Step 1: we determine the target granule G into which the time point is to be shifted by moving from the start granule containing t $\lfloor m \rfloor$ granules forward. In this example, it is just one granule further.

Step 2: we determine the target block B in the target granule G as $b' = b * b_G$ where b_G is the number of non-gap blocks in G. The target granule in the example has 3 non-gap blocks. Therefore $b' = 2/3 * 3 = 2$, i.e. the second block is the target non-gap block in G.

Step 3: we determine the target partition in block B by $d' = d * d_G$ where d_G is the number of non-gap partitions in the target block. In the example $d' = 1/2 * 3 = 1.5 \sim 2$, i.e. the second partition in block B is the target partition. Let j be the coordinate of this partition.

Step 4: Let $m \stackrel{\text{def}}{=} m - \lfloor m \rfloor$. If $m' = 0$ then $j - i$ is the result of *granules2Partitions*. If $m' > 0$ let $e \stackrel{\text{def}}{=} m' * |G|$ where $|G|$ is the number of non-gap partitions in G. In the example $e = 0.5 * 8 = 4$. Let j' be the partition obtained by moving from j e non-gap partitions forward. If j' is within the target granule G, then $j' - i$ is the result of *granules2Partitions*.

Step 5: If j' is beyond the target granule G, let f be the relative distance between the partition j and the end of the granule G, counting only non-gap partitions. In the example $f = 5/8$. Since j' is beyond the target granule G, $f > m'$. Let G' be the next granule after G. We must move the target partition into G'. Now let j'' be the coordinate of the partition obtained by moving $\lfloor (f - m') * |G'| \rfloor$ non-gap partitions forward from the start of granule G'. $j'' - i$ is now the result of *granules2Partitions*. ■

EXAMPLE 24 (for case 1b of *granules2Partitions*:). Suppose we want to shift the time point t in the picture below by $m = 1.5$ granules. t is within a gap partition of a granule and $m > 0$. We determine the number of partitions to be shifted in 5 steps.

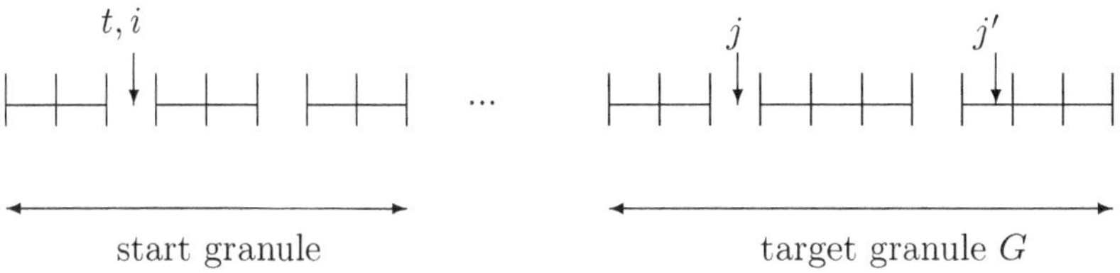

Let $i \stackrel{\text{def}}{=} P.pc(t)$ be the coordinate of the partition containing t. Let b be the relative distance of the gap block containing t. In the example t is in block 1 of two blocks. Therefore $b = 1/2$. Let d be the relative distance of partition i within the block containing t. In the example t is in partition 1 of a 1-partition gap block. Therefore $d = 1$.

Step 1: we determine the target granule G into which the time point is to be shifted by moving from the start granule containing t $\lfloor m \rfloor$ granules forward. In this example, it is just one granule further. If G contains no gaps at all, we proceed as in case 1a (Example 23), while ignoring that there are gaps in the start granule.

Step 2: we determine the target block B in the target granule G as $b' = b * b_G$ where b_G is the number of gap blocks in G. The target granule in the example has 2 non-gap blocks. Therefore $b' = 1/2 * 2 = 1$, i.e. block 1 is the target gap block in G.

Step 3: we determine the target partition in block B by $d' = d * d_G$ where d_G is the number of gap partitions in the target block. In the example $d' = 1 * 1 = 1$, i.e. the first partition in block B is the target partition. Let j' be the coordinate of this partition.

Step 4: Let $m' \stackrel{\text{def}}{=} m - \lfloor m \rfloor$. If $m' = 0$ then $j - i$ is the result of *granules2Partitions*. If $m' > 0$ we proceed as in case 1a, but ignore that there are gaps in the granules. Let $e \stackrel{\text{def}}{=} m' * |G|$ where $|G|$ is the number of partitions in G. In the example $e = 0.5 * 10 = 5$. Let j' be the partition obtained by moving from j e partitions forward. If j' is within the target granule G, then $j' - i$ is the result of *granules2Partitions*.

Step 5: If j' is beyond the target granule G, let f be the relative distance between the partition j and the end of the granule G, counting all partitions. In the example $f = 7/10$. Since j' is beyond the target granule G, $f > m'$. Let G' be the next granule after G. We must move the target partition into G'. Now let j'' be the coordinate of the partition obtained by moving $\lfloor (f - m') * |G'| \rfloor$ partitions forward from the start of granule G'. $j'' - i$ is now the result of *granules2Partitions*. ■

5 Durations

The partitionings are the mathematical model of periodic time units, such as years, months etc. This offers the possibility to define *durations*. A duration may for example be '3 months + 2 weeks'. Months and weeks are represented as partitionings, and 3 and 2 denote the number of partitions in these partitionings. The numbers need not be integers, but can be arbitrary real numbers.

A duration can be interpreted as the length of an interval. The numbers should not be negative in this case. A duration, however, can also be interpreted as a time shift. In this interpretation negative numbers make perfectly sense. $d = (-2\ week), (-3\ month)$, for example, denotes a backward shift of 2 weeks followed by a backward shift of 3 months.

DEFINITION 25 (Duration). A *duration* $d = (d_0, P_0), \ldots, (d_k, P_k)$ is a list of pairs where the d_i are real numbers and the P_i are partitionings.

If a duration is interpreted as a shift of a time point, it may be necessary to turn the shift around, in the backwards direction. Therefore the inverse of a duration is defined:

$-d \stackrel{\text{def}}{=} (-d_k, P_k), \ldots, (-d_0, P_0)$ ■

For example, if $d = (3\ month), (2\ week)$ then $-d = (-2\ week), (-3\ month)$.

In Def. 34 below a `shift` function for algorithmic partitionings is introduced. It shifts a time point by a number of partitions in a given partitioning. Any such shift function can be lifted to operate on durations.

DEFINITION 26 (`shift` for Durations). Given a function $P.shift(t, m)$, which shifts a time point t by m partitions of the partitioning P, we define a corresponding `shift` function for durations:

$$D.shift(t) \stackrel{\text{def}}{=} P_k.shift(\ldots P_1.shift(P_0.shift(t, d_0), d_1) \ldots, d_k)$$

where t is a time point and $D = (d_0, P_0), \ldots, (d_k, P_k)$ is a duration. ■

For example, if $D = (3\ month), (2\ week)$ then

$$D.shift(t) = week.shift(month.shift(t, 3), 2),$$

i.e. t is shifted by 3 months first, and the resulting time point is then shifted by 2 weeks.

6 Calendar Systems

Calendar systems are essentially certain sets of partitionings. The Gregorian calendar, for example, can be modelled by defining suitable partitionings for years, months, weeks, days, hours, minutes and seconds. Since the CTTN–system has a global reference time, and there is always a mapping between the coordinate system of a partitioning and the global reference time, it is possible to navigate between different granularities of the same calendar system, and between different granularities of different calendar systems.

There is, however, more information associated with calendar systems. Date strings, for example, must be parsed in a calendar specific way. The fact, that we count months and weeks from number 1, but hours, minutes and seconds from number 0, must be encoded in a calendar specific parser for date strings.

If longer historical periods in a particular region of the world are to be covered, it may even be necessary to combine two or more calendar systems. In the western world it is the Julian and the Gregorian calendar system which have been used over the past 2000 years. Therefore the real calendar system is not either the Julian or the Gregorian system, but a combination of both.

By these and a few other reasons, 'calendar system' is not a concept in the PartLib library, but in a separate component of the CTTN–system. Therefore calendar systems are not further mentioned in this chapter.

7 Global and Local Reference Time

The global reference time GRT corresponds directly to UTC time. With the introduction of a *local reference time* LRT for each partitioning it is possible to deal with leap seconds and time zones. The purpose is that the algorithms for the different partitions can just use the local reference time, and do not need to deal with leap seconds and time zones.

The transition from GRT to LRT is done in two steps. The first step deals with the leap seconds and the second step deals with time zones. The transition from LRT to GRT goes the other way round.

A correction function for leap seconds is defined first. The function $lsG(t)$ defined below ('G' for 'global') computes the accumulated leap seconds until the global reference time point t. The function $lsL(t)$ ('L' for 'local') also computes the accumulated leap seconds, but until the 'local' reference time point t. $lsG(t)$ is used for the transition from GRT to LRT, whereas $lsL(t)$ is used for the other direction. Unfortunately, it is computationally difficult to derive one version from the other. It is much more efficient when both functions are generated from a table of leap second corrections. (see http://www.ptb.de/de/org/4/43/432/ssec.htm).

DEFINITION 27 (Correction Function for Leap Seconds). If t is a time point in the global reference time then $lsG(t)$ computes the accumulated number of leap seconds until t.

If t is a time point in a reference time where the leap second corrections have already been done then $lsL(t)$ computes the accumulated number of leap seconds until t. ■

EXAMPLE 28 (Correction Function for Leap Seconds). The first 10 leap seconds were introduced for the last minute in 1971. The reference time for the regular end of this minute is 63072000. Therefore $lsG(t) = 0$ for all $t \leq 63072000$ and $lsG(63072000 + n) = n$ for $0 \leq n \leq 10$.

$lsG(t)$ remains constant with value 10 from $t = 63072010$ until $t = 94694410$. $lsG(94694411) = 11$, because another leap second was introduced in the last minute of 1972.

$lsL(t) = 0$ for all $t \leq 63072000$ as well, but $lsL(63072001) = 10$.
$lsL(94694400) = 10$ and $lsL(94694401) = 11$ etc. ■

Time zones are characterised by an offset between the local time and the time at the 0-meridian. For example, if at the 0-meridian it is 0 o'clock then the offset to the time zone of Germany is 1 hour, i.e. in Germany it is already 1 am. The time zone offset is in this case +3600 seconds, and the local reference time is 3600 seconds ahead of the global reference time.

DEFINITION 29 (Transition between GRT and LRT). Given the correction functions lsG and lsL for leap seconds (Def. 27) and a time zone offset tzo, we define

$$LRT(t) \stackrel{\text{def}}{=} t + tzo - lsG(t).$$

for a global reference time t. $LRT(t)$ computes the local reference time from the global reference time.

The function

$$GRT(t) \stackrel{\text{def}}{=} t - tzo + lsL(t)$$

computes the global reference time from the local reference time. ■

8 Specification of Partitionings

A partitioning is usually an infinite sequence of intervals, and these intervals need not be of the same length. Therefore it is not possible to specify such a partitioning by just enumerating its partitions.

Three basically different ways to specify partitionings are presented, and it is shown how a specification corresponds to a partitioning and an associated coordinate mapping. Since partitions themselves are not explicitly represented in PartLib, but only time points and coordinates, it is sufficient to give for each type of specification of a partitioning P corresponding definitions of the 'start of partition' function $sopC(i)$ and the 'partition coordinate' function $pc(t)$. $sopC(i)$ maps a coordinate i to the starting point of the corresponding partition and $pc(t)$ maps a time point t to the coordinate of the partition containing t.

The 'start of partition' function determines the partitioning completely because

$$p_i \stackrel{\text{def}}{=} [sopC(i), sopC(i+1)[.$$

The 'partition coordinate' function $pc(t)$ is then derivable:

$$(1) \quad P.pc(t) \stackrel{\text{def}}{=} \min_i (P.sopC(i) \leq t < P.sopC(i)).$$

This way, however, $pc(t)$ can only be computed through search, which is extremely inefficient. Therefore we give a more efficient definition of $pc(t)$ in each case. In most cases the algorithm for $pc(t)$ tries at first a good guess i' for the coordinate, and then searches in the neighbourhood of i' until the condition (1) is satisfied.

8.1 Algorithmic Partitionings

The first type of partitionings is mainly used for modelling the basic time units of calendar systems, years, months etc. The specification consists of an average length of the partitions, a correction function and an offset against time point 0.

DEFINITION 30 (Specification of Algorithmic Partitionings).
Algorithmic partitionings are specified by the components (avl, po, cf, DF) where

1. avl is the average length of a partition, given in the finest time unit;

2. po is an offset for the partition with coordinate 0, also given in the finest time unit,

3. $cf(i)$ is a correction function, and

4. DF is a date format. The date format is needed for the $shiftPD$ function.

■

The correction function $cf(i)$ computes for a partition with coordinate i the difference between the reference time of the beginning of the partition with coordinate i, and the estimated beginning $i \cdot avl$.

DEFINITION 31 (The Function *sopC* for Algorithmic Partitionings).
The algorithmic partitioning P which is specified by the components (avl, po, cf, DF) (Def. 30) has the following 'start of partition' function:

$$P.sopC(i) \stackrel{\text{def}}{=} GRT(i \cdot avl + cf(i) + po).$$ ■

Partitionings whose partitions have constant length in the local reference time only need a correction function that returns the constant 0. This is the case for seconds, minutes, and hours. It is no longer the case for days if daylight saving time regulations are taken into account.

EXAMPLE 32 (Basic Time Units for the Gregorian Calendar).
The specification of the basic time units as algorithmic partitionings for the Gregorian Calendar are:

second: average length: 1, offset: 0, correction function: $\lambda(n)0$.

minute: average length: 60, offset: 0, correction function: $\lambda(n)0$.

hour: average length: 3600, offset: 0, correction function: $\lambda(n)0$.

day: average length: 86400, offset: 0, correction function: $-3600 \cdot h$ if the day i is during the daylight saving time period, 0 otherwise.
The number h is usually 1 (for 1 hour). Exceptions are, for example, the year 1947 in Germany, where in the night of 1947/5/11 the clock was set forward a second time by 1 hour such that the offset against standard time was 2 hours.

week: average length: 604800, offset -259200, correction function: again, this function has to return an offset of $-3600 \cdot h$ for the weeks during the daylight saving time periods.

month: average length: 2592000 (30 days), offset 0, correction function: this function has to deal with the different length of the months and the daylight saving time regulations.

year: average length: 31536000 (365 days), offset 0, correction function: this function has to deal with leap years only. The effects of daylight saving time regulations are averaged out over the year. ■

The 'partition coordinate' function $pc(t)$ maps a reference time point t to the coordinate of the partition containing t. For algorithmic partitionings this function is more complicated than $sopC(i)$ because it needs to use the correction function $cf(i)$, which takes a coordinate as input, and this is the coordinate which is yet to be computed. Therefore the basic idea for the algorithm is to use a fixed point iteration which calls $sopC(i)$ for guessed coordinates until the resulting time point matches the given time point. The algorithm is described rather informally, but the key steps should become clear.

DEFINITION 33 (The Function $pc(t)$ for Algorithmic Partitionings). Let t be a local reference time point for the given partitioning $P = (avl, po, cf, DF)$. The algorithm for $pc(i)$ starts with a first guess $i \stackrel{\text{def}}{=} t/avl$ for the coordinate of the partition containing t. Since this guess is wrong in general, there is a first iteration which brings i closer to the correct solution:

Starting with an initial value for i', a fixed point iteration is performed until i' falls under a certain threshold[2]: Let $r \stackrel{\text{def}}{=} sopC(i)$. If $r \geq t$ let $r' \stackrel{\text{def}}{=} sopC(i-1)$ and compute $i' \stackrel{\text{def}}{=} (r-t)/(r-r')$ to get a better estimate $i \stackrel{\text{def}}{=} i - i'$ for the correct coordinate. If $r < t$, i is increased in a similar way[3].

The second phase of the algorithm is simpler: the correct coordinate is searched by just decreasing or increasing i by 1, until $sopC(i) \leq t < sopC(i+1)$ holds. The result of the function $pc(t)$ is then the coordinate i for which this condition holds. ■

During the first phase, the algorithm performs big jumps to get very close to the correct solution. In the second phase it does the fine tuning by searching in the neighbourhood of the coordinate which was computed in the first phase. This phase guarantees that the result is correct, i.e. it satisfies the condition (1) for $pc(t)$. The algorithm converges in very few (usually $<$ 10) steps to the correct solution even if the average length of the partitions is quite different to their individual length.

We can now define the date oriented shift function for algorithmic partitionings. The idea of it was already explained in Section 4.2.

DEFINITION 34 (Date Oriented `shiftPD` for Algorithmic Partitionings). The function $P.shiftPD(t, m)$ where t is a GRT time point, m is a real number and P is an algorithmic partitioning (sec. 8.1) with date format $DF = P_0, \ldots$ performs the following steps:

1. Let $d = d_1/\ldots/d_k \stackrel{\text{def}}{=} DF.date(t)$ (Def. 17);

2. i, d and m are now modified destructively:
   $while(m \neq 0 \text{ and } i \leq k)$

[2] The threshold in the implementation is 3. The initial value for i' can be any number greater than the threshold. $i' = 10$ is fine.

[3] This version of the fixed point iteration is slightly simplified. It can happen that i' oscillates around the correct i. If this happens, the iteration is immediately stopped.

(a) $d_i \stackrel{\text{def}}{=} d_i + \lfloor m \rfloor$

(b) $if(i < k)$
$t \stackrel{\text{def}}{=} DF.TimePoint(d)$ (Def. 18)
$m \stackrel{\text{def}}{=} (m - \lfloor m \rfloor) \cdot P_{i+1}.lengthP(P_i.sopT(t), P_i.eopT(t))$
$i \stackrel{\text{def}}{=} i + 1.$

3. the result of $P.shiftPD(t, m)$ is now $DF.TimePoint(d)$. ■

Although the shiftPD function gives intuitive results in most cases, it has a number of drawbacks. One of them was already mentioned: shiftPD has not much to do with the lengthP function. Measuring the shifted distance with the lengthP function does usually not give the expected results.

Another drawback is that shifting a time point t first by m_1 partitions, and then by m_2 partitions may not be the same as shifting t by $m_1 + m_2$ partitions. This holds only if m_1 and m_2 are integers. A counter example for the case that m_1 and m_2 are factional values is:

EXAMPLE 35 (Counterexample). Suppose we want to shift the time point 0 twice by 1.5 months. The date for 0 is 0/0/0. Since February 1970 has 28 days, a shift by 1.5 months ends up at 0/1/14. Another shift by 1 month yields 0/2/14. This is in March. March has 31 days. Therefore a shift by 0.5 months means a shift by 15.5 days. We end up at 0/2/29/12. This is different to the result of a direct shift by 3 months: 0/3/0.

Nevertheless, the shiftPD is usually preferable. A striking example which illustrates the difference between shiftPD and shiftPL is such a simple thing as a shift by 1 day. If it is 5 pm, a shift by 1 day should end up at 5 pm next day. This is the case with the shiftPD function, even if during the night, standard time has been changed to daylight savings time. In contrast, the shiftPL function yields a very odd result in this case.

Other periodic temporal notions which can be modelled by algorithmic partitionings are, for example, sunrises and sunsets, moon phases, the church year which starts with Easter, etc. The specification for the western version of Easter would be: average length: 31536000 (1 year), offset: 7516800 (1970/3/29) and a correction function which actually computes the precise date of Easter (see for example [Dershowitz and Reingold, 1997]). The specification of all these partitionings must be accompanied by an appropriate `shiftPD` function.

The specification of the algorithmic partitionings requires the correction function $cf(i)$, and this is a piece of code. Therefore algorithmic partitionings are usually hard coded in the application program. This is different for the remaining partitioning types. They can be specified purely symbolically.

Therefore one can read their specification from a file or a database at run time. This makes the system very flexible.

8.2 Duration Partitionings

Duration partitionings are specified by an anchor time and a sequence of 'durations'. For example, I could define 'my weekend' as a *duration partitioning* with anchor time 2004/7/23, 4 pm (Friday July, 23rd, 2004, 4 pm) and durations: ('8 hour + 2 day', '4 day + 16 hour'). The first interval would be labelled 'weekend', and the second interval would be labelled 'gap'.

DEFINITION 36 (Specification of a Duration Partitioning).
A duration partitioning is specified by the tuple $(t_A, (D_0 \dots D_{n-1}), shift)$ where

1. t_A is the anchor time point (in the global reference time);
2. $D_0 \dots D_{n-1}$ is a list of durations (Def. 25);
3. $shift$ is a shift function for durations (Def. 26). ∎

The coordinates for a duration partitioning are such that the first partition after the anchor time point has coordinate 0. The next picture illustrates the situation.

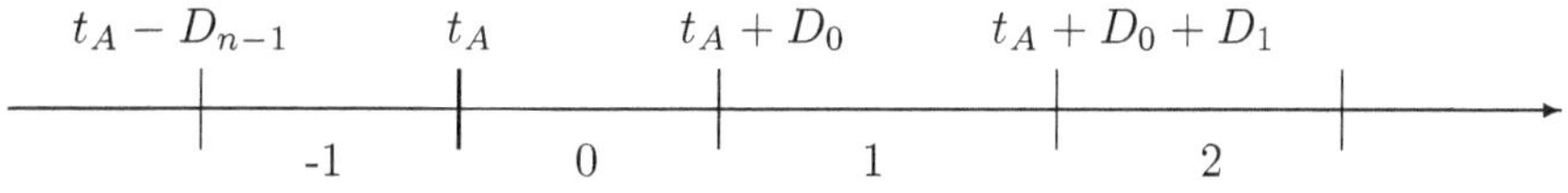

The durations in the specification of a duration partitioning can be very irregular. Therefore there is not much of a chance to realize a 'start of partition' function *sopC* other than by just looping i times over $D_0 \dots D_{n-1}$.

DEFINITION 37 (The Function *sopC* for Duration Partitionings).
The duration partitioning P which is specified by the data $(t_A, (D_0 \dots D_{n-1}), shift)$ (Def. 36) has the following 'start of partition' function:

$$sopC(i) \stackrel{\text{def}}{=} t_i$$

where t_i is determined by shifting the anchor time point t_A i times:
Let $t_0 \stackrel{\text{def}}{=} t_A$.
if($i \geq 0$): for $j = 1, \dots, i$: $t_j \stackrel{\text{def}}{=} D_{(j-1) \bmod n}.shift(t_{j-1})$. (Def. 26)
if($i < 0$): for $j = 1, \dots, -i$: $t_j \stackrel{\text{def}}{=} -D_{(n-j) \bmod n}.shift(t_{j-1})$.
∎

Notice that the shift function for durations uses the shiftPD function (Sec. 4.1). Because two shifts by m_1 and m_2 partitions are not necessarily the same as one shift by $m_1 + m_2$ partitions (Ex. 35) , this has an effect on the meaning of duration partitionings. A duration partitioning with a duration '1.5 month + 1.5 month' is not the same as a duration partitioning with a duration '3 month'.

Because there is no further assumption about the durations in the specification of duration partitionings, there is not much chance to optimise the 'partition coordinate' function pc either. Therefore the definition (1) is taken as algorithm for pc.

The `shiftPD` function cannot be optimised either. It also loops over the duration $D_0 \dots D_{n-1}$ and calls the shift function for durations as often as necessary.

DEFINITION 38 (`shiftPD` for Duration Partitionings).
Let $P = (t_A, (D_0 \dots D_{n-1}))$ be a duration partitioning. The $P.shiftPD(t, m)$ function for a time point t and a real number m performs the following steps:

Let $(k \text{ remainder } l) \stackrel{\text{def}}{=} m/n$.

If $m \geq 0$:

if $(m \geq 1)$: for $i = 0, \dots, \lfloor m \rfloor - 1$ let $t \stackrel{\text{def}}{=} D_{i \bmod n}.shift(t)$

let $t \stackrel{\text{def}}{=} t + (m - \lfloor m \rfloor) \cdot (D_{\lfloor m \rfloor \bmod n}).shift(t) - t)$.

If $m < 0$:

for $i = 0, \dots, -\lfloor m \rfloor - 1$ let $t \stackrel{\text{def}}{=} -D_{n-1-(i \bmod n)}.shift(t)$

let $t \stackrel{\text{def}}{=} t + (m - \lfloor m \rfloor) \cdot (t - D_{(\lfloor m \rfloor \bmod n)}.shift(t)$.

The result of $P.shift(t, m)$ is now the modified t. ∎

8.3 Regular Partitionings

A special case of a duration partitioning is when all durations are of the form 'n P' and the partitioning P is the same in all durations. For this case there are more efficient ways than looping over lists of durations. Therefore, and because many partitionings are of this type, it makes sense to treat them in a special way.

A typical example is the notion of a semester at a university. In the Munich case, the dates could be: anchor time: October 2000. The shifts are: 6 months (with label 'winter semester') and 6 months (with label 'summer semester'). This defines a partitioning with partition 0 starting at the anchor time, and then extending into the past and the future. The partition with coordinate 0 in this example is the winter semester 2000/2001.

DEFINITION 39 (Specification of Regular Partitionings). A regular partitioning is specified by the triple $(t_A, U, (s_0 \dots s_{n-1}), shift)$ where

1. t_A is the anchor time point (in the global reference time)
2. U is a partitioning, the time unit for the shifts;
3. $s_0 \dots s_{n-1}$ is a list of real numbers, the shifts; $shift$ is a shift function for partitionings. ■

The partitions of regular partitionings are obtained by shifting the anchor point t_A first by s_0 time units U, and then by $s_0 + s_1$ time units U etc.

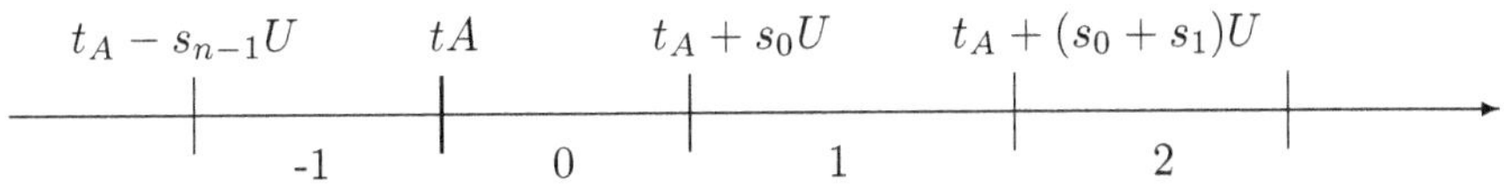

This picture illustrates a subtle, but important difference to duration partitionings. The partition boundaries for duration partitions are computed by successively applying the shift function for the corresponding durations. The shift function for durations uses internally the date oriented shiftPD function, for which $shift(shift(t, m_1), m_2) \neq shift(t, m_1 + m_2)$ may be the case.

In contrast to this the partition boundaries for regular partitions are computed by first adding the shifts together, and then shifting the anchor time in one single step. Both versions yield the same results for integer shifts. This is not guaranteed if the shifts are fractional.

DEFINITION 40 (The Function *sopC* for Regular Partitionings). The regular partitioning P which is specified by the data $(t_A, U, (s_0 \dots s_{n-1}), shift)$ (Def. 39) has the following 'start of partition' function:

$$sopC(i) \stackrel{\text{def}}{=} U.shift(t_A, s(i))$$

where $s(i) \stackrel{\text{def}}{=} k \cdot \Sigma_{j=0}^{n-1} s_j + \begin{cases} \Sigma_{j=0}^{l} s_j & \text{if } i \geq 0 \\ \Sigma_{j=0}^{l-1} s_j & \text{otherwise} \end{cases}$

and $(k \text{ remainder } l) \stackrel{\text{def}}{=} i/n$.

■

EXAMPLE 41. Let the anchor date be 2000/1, and let the shifts be 3,4,5 months.

$sopC(5)$ is calculated as follows:
$5/3 = 1$ remainder 2.
$i' = 1 \cdot (3 + 4 + 5) + (3 + 4) = 19$.
This means 2000/1 is to be shifted by 19 month, and we end up at the beginning of 2001/7.

$sopC(-5)$ is calculated as follows:
$-5/3 = -2$ remainder 1.
$i' = -2 \cdot (3+4+5)+3 = -21$.
This means 2000/1 is to be shifted by -21 month, and we end up at the beginning of 1998/3. ∎

The 'partition coordinate' function pc turns the reference time into a coordinate i of the time unit U and then uses the difference between i and the coordinate of the anchor time to compute the number of shifts which are necessary to get from the anchor time to the reference time. This is the coordinate of the partition containing t.

DEFINITION 42 (The Function pc for Regular Partitionings).
Let $(t_A, U, (s_0 \dots s_{n-1}), shift)$ be the specification of a regular partitioning.
Let t be a local reference time point.
Let $i \stackrel{\text{def}}{=} U.pc(t) - U.pc(t_A)$ and $(k \text{ remainder } l) \stackrel{\text{def}}{=} i/\Sigma_{j=0}^{n-1} s_j$.

Let $P.pc(t) \stackrel{\text{def}}{=} k \cdot n + \max_{i \geq 0}((\Sigma_{j=0}^{i} s_j) \leq l)$. ∎

EXAMPLE 43. Let the anchor date be 2000/1, and let the shifts be 3.5,4.5,5.5 months.

Let t be such that $i \stackrel{\text{def}}{=} U.pc(t) - U.pc(t_A) = 21$.
k *remainder* $l = i/(3.5+4.5+5.5) = 1$ *remainder* 7.5.
$P.pc(t) = 1 \cdot 3 + 1 = 4$. This is the partition between month 17 and 21.5.

Now let t be such that $i \stackrel{\text{def}}{=} U.pc(t) - U.pc(t_A) = -21$.
k *remainder* $l = -21/13.5 = -2$ remainder 8.
$P.pc(t) = -2 \cdot 3 + 2 = -4$.
This is the partition between month -23.5 and -19. ∎

shiftPD:
The `shiftPD` function for regular partitionings is explained informally with the following example: Suppose we have a specification of *trimesters* in the following way: Anchor date: 2000/10, trimesters: 3,4,5 months.

We want to shift a time point t by 8.5 trimesters. The following steps are necessary

1. $8/3 = 2$ *remainder* 2. That means, first a shift of 2 full cycles of 3+4+5 = 12 months is performed, and we end up at a time point t_1.
2. The index i' of the partition containing t_1 is computed, suppose it is 1, i.e. t_1 lies in the first trimester. In order to get two trimesters further, a shift of 3 + 4 = 7 months is necessary, and we end up at time point t_2 which is in the third trimester.
3. The third trimester has 5 month. $5 \cdot 0.5 = 2.5$, i.e. we shift t_2 by another 2.5 month.

DEFINITION 44 (`shiftPD` for Regular Partitionings).
Let $P = (t_A, U, (s_0, \ldots, s_{n-1}), shift)$ be a regular partitioning (Def. 39). The function $P.shiftPD(t, m)$ performs the following steps:

1. Let (k remainder l) $\stackrel{\text{def}}{=} \lfloor m \rfloor / n$
   Let $t_1 \stackrel{\text{def}}{=} U.shift(t, k \cdot \Sigma_{j=0}^{n-1} s_j)$.
2. Let $i' \stackrel{\text{def}}{=} t_1^P$.
   Let $t_2 \stackrel{\text{def}}{=} U.shift(t, \Sigma_{j=0}^{l-1} s_{(i'+j) \bmod n})$.
3. Return $U.shift(t_2, (m - \lfloor m \rfloor) \cdot s_{(i'+l) \bmod n})$.

■

Further examples for periodic temporal notions which can be encoded as regular partitionings are decades, centuries, the British financial year which starts April 1st, the dates of a particular lecture, which is every week at the same time etc.

8.4 Date Partitionings

Date Partitionings are specified by providing the boundaries of the partitions as concrete dates.

An example could be the dates of the Time conferences: 1994/5/4 Time94 1994/5/5 gap 1995/4/26 Time95 1995/4/27 gap 1996/5/19 Time96 1996/5/21 gap 1997/5/10 Time97 1997/5/12 gap 1998/5/16 Time98 1998/5/18 gap 1999/5/1 Time99 1999/5/3 gap 2000/7/7 Time00 2000/7/10 gap 2001/6/14 Time01 2001/6/17 gap 2002/7/7 Time02 2002/7/10 gap 2003/7/8 Time03 2003/7/11 gap 2004/7/1 Time04 2004/7/4.

Another example could be the seasons: 2000/3/21 spring 2000/6/21 summer 2000/9/23 autumn 2000/12/21 winter 2001/3/21.

In the introduction I explained the trick how to turn these finitely many dates into an infinite partitioning: the differences between two subsequent dates are turned into durations. The durations are then used to extrapolate the partitioning into the infinity.

The seasons example above shows that this makes really sense. The time difference between the dates can be expressed as durations '3 month' for spring, '3 month + 2 day' for summer, '3 month - 2 day' for autumn and '3 month' for winter. The extrapolation of this now yields the seasons for the whole time line. This works even for the leap years, in which winter is one day longer. This is covered by the duration '3 month' for winter, which is one day longer in leap years.

DEFINITION 45 (Specification of Date Partitionings). A date partitioning is specified as a list dates $d_0, \ldots, d_n$ in a date format DF. ■

The dates can very easily be turned into durations: Suppose the date format is $DF \stackrel{\text{def}}{=} P_0, \ldots$. The difference between two dates $d_i = d_{i,0}/d_{i,1}/\ldots$ and $d_{i+1} = d_{i+1,0}/d_{i+1,1}/\ldots$ yields the following duration: $(d_{i+1,0} - d_{i,0}, P_0)$, $(d_{i+1,1} - d_{i,1}, P_1), \ldots$, and this is sufficient to specify a duration partitioning. The anchor time is given by $DF.timePoint(d_0)$.

Notice that the coordinate calculation for duration partitionings which applies the shift function for durations, and this uses the date oriented shift function for partitionings, undos the above subtractions. Therefore it reconstructs exactly the original dates.

8.5 Folded Partitionings

Another basic type of partitionings are *folded partitionings*. We explained already the bus timetable example. The bus timetable changes from season to season. The best way to specify this, would be to specify the seasons first, and for each season to specify the particular bus timetable. The 'folded partitioning' specification operation takes as input a *frame partitioning*, for example the seasons, and a sequence of *component partitionings*, for example the four different bus timetables. It maps the component partitionings automatically to the right frame partition, such that from the outside the whole thing looks like an ordinary partitioning.

DEFINITION 46 (Folded Partitioning). A folded partitioning P is specified by a *frame partitioning* F and a finite list $P_0, \ldots, P_{n-1}$ of *component partitionings*. The component partitionings must meet the following condition: The partitions of the component partitionings must be aligned with the partitions of the frame partition. That means, for $i = 0, \ldots, n-1$, the start of each frame partition must be the start of a component partition in P_i, and the end of each frame partition must be the end of a component partition in P_i;

A folded partitioning is called *constant* iff each frame partition with coordinate i contains the same number of component partitions as the frame partition with coordinate i mod n. ■

The condition is in principle not necessary, but if it is not met, it complicates the algorithms enormously. It excludes, for example, that 'week' can be used as component partitioning when the frame partition is 'year'. 'month', however, can be used because years start and end with a month.

The figure below gives an idea how the coordinates for the folded partitioning are obtained from the coordinates of the frame partitioning and the component partitionings.

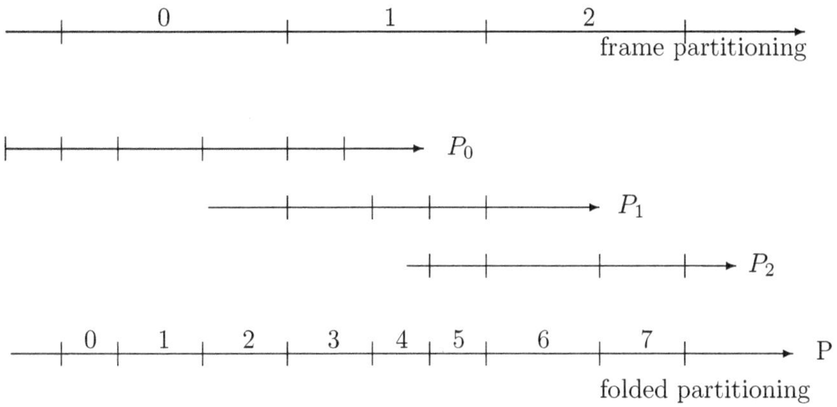

The algorithms for the 'partition coordinate' function *pc* and for the 'start of partition' function *sopC* are much more efficient if the partitioning is constant, i.e. the number of component partitions per frame partition of each component partitioning P_i does not change over time. In this case it is possible to compute the total number of component partitions up to the beginning of a given frame partition by a few multiplications and additions. If the partitioning is not constant, one must iterate through all frame partitions and add the corresponding numbers together.

The bus timetable example is typical for a folded partitioning which is *not* constant. This is due to the leap years. Since the winter is one day longer in a leap year, the bus timetable in this winter has more partitions than in the other winters. An example for a constant folded partitioning could be the definition of working hours with shifts. In the first week I may have a morning shift, in the second week an evening shift, and in the third week a night shift. The number of partitions labelled 'working hour' and 'gap' would not change in this case.

The 'partition coordinate' function *pc* needs an auxiliary function *partitionsUpTo*, which computes for a frame coordinate the number of partitions up to the beginning of this frame partition. More precisely: for a positive frame coordinate *frco*, *partitionsUpTo* computes the number of component partitions from frame partition 0 (inclusive) up to the beginning of frame partition with coordinate *frco*, and for negative frame coordinates, it computes the *negated* number of component partitions from frame partition -1 (inclusive) down to and including the frame partition with coordinate *frco*.

We define this function for constant folded partitionings and for non-constant folded partitionings separately.

DEFINITION 47 (*partitionsUpTo* for constant folded partitionings). Let P

be a constant folded partitioning with component partitionings $P_0 \ldots, P_{n-1}$.

We define the function $partitionsUpTo(frco)$ where $frco$ is a coordinate in the frame partitioning as follows:

For $i = 0, \ldots, n-1$ let $partitions(i)$ be the number of component partitions per frame partition P_i. Since the partitioning is constant, it is sufficient to compute these numbers for the first n frame partitions.

Let $cycle \stackrel{\text{def}}{=} \Sigma_{i=0}^{n-1} partitions(i)$.

Let $frco/n = blocks\ remainder\ rest$.

Now we have:

$$partitionsUpTo(frco) \stackrel{\text{def}}{=} blocks * cycle + \Sigma_{i=0}^{rest-1} partitions(i).$$ ■

DEFINITION 48 ($partitionsUpTo$ for normal folded partitionings). Let P be a folded partitioning with frame partitioning F and component partitionings $P_0 \ldots, P_{n-1}$.

We define the function $partitionsUpTo(frco)$ where $frco$ is a coordinate in the frame partitioning as follows:

If $frco \geq 0$ let $from \stackrel{\text{def}}{=} 0$ and $to \stackrel{\text{def}}{=} frco - 1$.
If $frco < 0$ let $from \stackrel{\text{def}}{=} frco + 1$ and $to \stackrel{\text{def}}{=} -1$.

Let $n \stackrel{\text{def}}{=} \Sigma_{i=from}^{to} P_{i \bmod n}.pc(F.eopC(i)) - P_{i \bmod n}.pc(F.sopC(i))$.

If $(frco \geq 0)$ let $partitionsUpTo(frco) \stackrel{\text{def}}{=} n$.
If $(frco < 0)$ let $partitionsUpTo(frco) \stackrel{\text{def}}{=} -n$. ■

Now we can define the 'partition coordinate' function $pc(t)$. It uses the $partitionsUpTo$ function to get the number of partitions up to the frame partition containing t, and then computes the remaining partitions locally with the corresponding component partitioning.

DEFINITION 49 (The Function *pc* for Folded Partitionings). Let P be a folded partitioning with frame partitioning F and component partitionings $P_0, \ldots, P_{n-1}$. Let t be a global reference time point.

Let $frco \stackrel{\text{def}}{=} F.pc(t)$ be the frame coordinate for time point t.

We have:

$$pc(t) \stackrel{\text{def}}{=} partitionsUpTo(frco) + P_{frco \bmod n}.pc(t) - P_{frco \bmod n}.pc(F.sopC(frco))$$

■

The 'start of partition' function *sopC* needs an auxiliary function $frameCoordinate(co) = (frco, rist)$, which computes for a given coordinate of a partition (i) the coordinate of the frame partition which contains this partition and (ii) the coordinate of the first component partition within this frame partition. There are again different definitions for constant folded partitionings and for normal folded partitionings.

DEFINITION 50 (*frameCoordinate* for Constant Folded Partitionings). Let P be a constant folded partitioning with component partitionings $P_0 \ldots, P_{n-1}$.

We define the function $frameCoordinate(co)$ where co is a P-coordinate as follows:

For $i = 0, \ldots, n-1$ let $partitions(i)$ be the number of component partitions per frame partition P_i.

Let $cycle \stackrel{\text{def}}{=} \Sigma_{i=0}^{n-1} partitions(i)$.

Let $co/cycle = blocks\ remainder\ rest$

Let $frco \stackrel{\text{def}}{=} blocks \cdot n$ be the coordinate of the first frame partition p_0 in the partitions $p_0, \ldots, p_{n-1}$ which correspond to $P_0, \ldots, P_{n-1}$, and where one of the p_i contains the component partition with coordinate co.

Let $i \stackrel{\text{def}}{=} max_i(\Sigma_{k=0}^{i} partitions(k) \leq rest)$ be the offset from $frco$, such that $frco + i$ is the coordinate of the frame partition containing the component partition with coordinate co.

This way we get

$frameCoordinate(co) \stackrel{\text{def}}{=} (frco + i, blocks \cdot cycle + \Sigma_{k=0}^{i} partitions(k))$. ■

DEFINITION 51 (*frameCoordinate* for Normal Folded Partitionings). Let P be a constant folded partitioning with frame partitioning F and component partitionings $P_0 \ldots, P_{n-1}$.

We define the function $frameCoordinate(co)$, where co is a P-coordinate as follows:

If $co \geq 0$ let

$$frco \stackrel{\text{def}}{=} max_i(\Sigma_{k=0}^{i} P_{k \bmod n}.pc(F.eopC(k)) - P_{k \bmod n}.pc(F.sopC(k)) \leq co)$$

$$first \stackrel{\text{def}}{=} \Sigma_{k=0}^{frco} P_{k \bmod n}.pc(F.eopC(k)) - P_{k \bmod n}.pc(F.sopC(k)) \leq co)$$

If $co < 0$ let

$$frco \stackrel{\text{def}}{=} -max_i(\Sigma_{k=1}^{i} P_{-k \bmod n}.pc(F.eopC(-k)) - P_{-k \bmod n}.pc(F.sopC(-k)) \leq co)$$

$$first \stackrel{\text{def}}{=} \Sigma_{k=1}^{-frco} P_{-k \bmod n}.pc(F.eopC(-k)) - P_{-k \bmod n}.pc(F.sopC(-k)) \leq co)$$

This way we get

$frameCoordinate(co) \stackrel{\text{def}}{=} (frco, first)$. ■

The 'start of partition' function *sopC* is now defined as follows:

DEFINITION 52 (The Function *sopC* for Folded Partitionings).
Let P be a folded partitioning with frame partitioning F and component partitionings $P_0, \ldots, P_{n-1}$. Let co be a coordinate of P.

Let $(frco, first) \stackrel{\text{def}}{=} frameCoordinate(co)$

Let $border = F.sopC(frco)$ be the left boundary of the frame partition containing the component partition with coordinate co.
Now we define

$sopC(co) \stackrel{\text{def}}{=} P_{frco \bmod n}.sopC(P_{frco \bmod n}.pc(border) + (co - first))$ ■

All operations on folded partitionings are straightforward if the coordinates involved remain within a single frame partition. The corresponding operation on the corresponding component partition can be used in this case. If the coordinates, however, cross the border of a frame partition then a special treatment is necessary. We explain this for the $shiftPD(t, m)$ function. All other functions are similar.

The $shiftPD(t, m)$ function for folded partitionings is straightforward if the shifted time point still remains in the same frame partition. In this case it is sufficient to use the $shiftPD$ function of the corresponding component partitioning. If this shift crosses the border of the frame partition then the relative distance to this border is subtracted from m, and then $shiftPD$ is called recursively for the border time point and the reduced m-value.

DEFINITION 53 (`shiftPD` for Folded Partitionings). Let P be a folded partitioning with frame partitioning F and component partitionings $P_0, \ldots,$ P_{n-1}. Let t be a time point and m a real number.

$P.shiftPD(t, m)$ performs the following steps:

1. Let $co \stackrel{\text{def}}{=} F.pc(t)$ be the coordinate of the frame partition containing t.

2. Let $C \stackrel{\text{def}}{=} P_{co \bmod n}$ the component partitioning assigned to the partition containing t.

3. Let $s \stackrel{\text{def}}{=} C.shiftPD(t, m)$ be the shifted time point.

4. If s is still in the same frame partition as t, return s.

5. If s is not the in the same frame partition, let m' be the relative distance between t and the end t' (if $m > 0$) or the start t' (if $m < 0$) of the frame partition containing t.

6. Call $P.shiftPD(t', m - m')$ recursively.

∎

9 Summary

The basic ideas of the PartLib library for representing periodic temporal notions are explained in this chapter. We use partitionings of the real numbers as basic mathematical structures. The partitionings can be labelled with symbolic names. The labels can be used for various purposes, in particular for defining *granules*, i.e. clusters of partitions which belong together semantically.

The approach proposed in this chapter is a mix of algorithmic and symbolic specifications. The basic time units are realized as algorithmic partitionings where the details of the calendar system is hard-coded in the correction functions. All other periodic temporal notions are specified symbolically as regular, duration or folded partitionings. This seems to be a good compromise between the efficiency of compiled code for the difficult parts of calendar systems, and the flexibility of symbolic specifications for application specific parts.

The PartLib library is an open source C++ system. The details of its interface are explained in [Ohlbach, 2005d].

BIBLIOGRAPHY

[Berners-Lee *et al.*, 1999] T. Berners-Lee, M. Fischetti, and M. Dertouzos. *Weaving the Web: The Original Design and Ultimate Destiny of the World Wide Web.* Harper, San Francisco, September 1999. ISBN: 0062515861.

[Bettini and R.D.Sibi, 2000] C. Bettini and R.D.Sibi. Symbolic representation of user-defined time granularities. *Annals of Mathematics and Artificial Intelligence*, 30:53–92, 2000. Kluwer Academic Publishers.

[Bettini *et al.*, 1998] Claudio Bettini, Curtis E. Dyreson, William S. Evans, Richard T. Snodgrass, and X. Sean Wang. *Temporal Databases, Research and Practice*, volume 1399 of *LNCS*, chapter A Glossary of Time Granularity Concepts, pages 406–413. Springer Verlag, 1998.

[Bettini *et al.*, 2000] Claudio Bettini, Sushil Jajodia, and Sean X. Wang. *Time Granularities in Databases, Data Mining and Temporal Reasoning.* Springer Verlag, 2000.

[Bettini *et al.*, 2004] Claudio Bettini, Sergio Mascetti, and X. Sean Wang. Mapping calendar expressions into periodical granularities. In C. Combi and G. Ligozat, editors, *Proc. of the 11th International Symposium on Temporal Representation and Reasoning*, pages 87–95, Los Alamitos, California, 2004. IEEE.

[Bry *et al.*, 2004] François Bry, Frank-André Rieß, and Stephanie Spranger. CaTTS: Calendar Types and Constraints for Web Applications. research report, PMS-FB-2004-24 PMS-FB-2004-24, Institute for Informatics, University of Munich, 2004.

[Cukierman and Delgrande, 1998] Diana R. Cukierman and James P. Delgrande. Expressing time intervals and repetition within a formalization of calendars. *Computational Intelligence*, 14(4):563–597, 1998.

[Cukierman and Delgrande, 2004] Diana R. Cukierman and James P. Delgrande. The SOL time theory: A formalization of structured temporal objects and repetition. In C. Combi and G. Ligozat, editors, *Proc. of the 11th International Symposium on Temporal Representation and Reasoning*, pages 71–34, Los Alamitos, California, 2004. IEEE.

[Cukierman, 2003] Diana R. Cukierman. *A Formalization of Structured Temporal Objects and Repetition*. PhD thesis, Simon Franser University, Vancouver, Canada, 2003.

[Dershowitz and Reingold, 1997] Nachum Dershowitz and Edward M. Reingold. *Calendrical Calculations*. Cambridge University Press, 1997.

[Dyreson *et al.*, 2000] Curtis E. Dyreson, Wikkima S. Evans, Hing Lin, and Richard T. Snodgrass. Efficiently supporting temporal granularities. *IEEE Transactions on Knowledge and Data Engineering*, 12(4):568–587, 2000.

[Egidi and Terenziani, 2004] Lavinia Egidi and Paolo Terenziani. A lattice of classes of user-defined symbolic periodicities. In C. Combi and G. Ligozat, editors, *Proc. of the 11th International Symposium on Temporal Representation and Reasoning*, pages 13–20, Los Alamitos, California, 2004. IEEE.

[Goralwalla *et al.*, 2001] I.A. Goralwalla, Y. Leontiev, M.T. Ozsu, D. Szafron, and C. Combi. Temporal granularity: Completing the picture. *Journal of Intelligent Information Systems*, 16(1):41–63, 2001.

[Kline *et al.*, 1999] Nick Kline, Jie Li, and Richard Snodgrass. Specifying multiple calendars, calendric systems and field tables and functions in timeadt. Technical Report TR-41, Time Center Report, May 1999.

[Leban *et al.*, 1986] B. Leban, D. Mcdonald, and D.Foster. A representation for collections of temporal intervals. In *Proc. of the American National Conference on Artificial Intelligence (AAAI)*, pages 367–371. Morgan Kaufmann, Los Altos, CA, 1986.

[Niezette and Stevenne, 1993] M. Niezette and J. Stevenne. An efficient symbolic representation of periodic time. In *Proc. of the first International Conference on Information and Knowledge Management*, volume 752 of *Lecture Notes in Computer Science*, pages 161–169. Springer Verlag, 1993.

[Ning *et al.*, 2002] Peng Ning, X. Sean Wang, and Sushil Jajodia. An algebraic representation of calendars. *Annals of Mathematics and Artificial Intelligenc*, 36(1-2):5–38, September 2002. Kluwer Academic Publishers.

[Ohlbach and Gabbay, 1998] Hans Jürgen Ohlbach and Dov Gabbay. Calendar logic. *Journal of Applied Non-Classical Logics*, 8(4), 1998.

[Ohlbach, 2000] Hans Jürgen Ohlbach. Calendar logic. In I. Hodkinson D.M. Gabbay and M. Reynolds, editors, *Temporal Logic: Mathematical Foundations and Computational Aspec ts*, pages 489–586. Oxford University Press, 2000.

[Ohlbach, 2005a] Hans Jürgen Ohlbach. Computational treatment of temporal notions – the CTTN-system. Research Report PMS-FB-2005-30, Inst. für Informatik, LFE PMS, University of Munich, June 2005. URL: http://www.pms.ifi.lmu.de/publikationen/#PMS-FB-2005-30.

[Ohlbach, 2005b] Hans Jürgen Ohlbach. Fuzzy time intervals – the FuTI-library. Research Report PMS-FB-2005-26, Inst. für Informatik, LFE PMS, University of Munich, June 2005. URL: http://www.pms.ifi.lmu.de/publikationen/#PMS-FB-2005-26.

[Ohlbach, 2005c] Hans Jürgen Ohlbach. GeTS – a specification language for geo-temporal notions. Research Report PMS-FB-2005-29, Inst. für Informatik, LFE PMS, University of Munich, June 2005. URL: http://www.pms.ifi.lmu.de/publikationen/#PMS-FB-2005-29.

[Ohlbach, 2005d] Hans Jürgen Ohlbach. Modelling periodic temporal notions by labelled partitionings of the real numbers – the PartLib library. Research Report PMS-FB-2005-28, Inst. für Informatik, LFE PMS, University of Munich, June 2005. URL: http://www.pms.ifi.lmu.de/publikationen/#PMS-FB-2005-28.

[Soo and Snodgrass, 1992] Michael D. Soo and Richard T. Snodgrass. Mixed calendar query language support for temporal constants. Technical Report TR 92-07, Dept. of Computer Science, Univ. of Arizona, February 1992.

A New Basic Set of Transformations between Proofs

Anjolina G. de Oliveira and Ruy J. G. B. de Queiroz

1 Introduction

Deductive systems based on the so-called Curry–Howard isomorphism [Howard, 1980] have an interesting feature: termination and confluence (Church–Rosser property) theorems can be proved by reductions on the terms of the functional calculus. Exploring this important characteristic, we have proved these theorems for the *Labelled Natural Deduction – LND* [de Queiroz and Gabbay, 1999] via a term rewriting system (*TRS*) constructed from the *LND*-terms of the functional calculus [de Oliveira and de Queiroz, 1994]. The *LND* system is an instance of Gabbay's *Labelled Deductive Systems – LDS* [Gabbay, 1996], which is based on a generalization of the functional interpretation of logical connectives [Gabbay and de Queiroz, 1992; de Queiroz *et al.*, to appear] (i.e. Curry–Howard isomorphism).

Proving the *termination* and *confluence* properties for the *TRS* associated to the *LND* system (*LND-TRS*) [de Oliveira and de Queiroz, 1994], we have in fact proved the properties of termination and confluence for the *LND* system, respectively. The *termination* property guarantees the existence of a normal form of the *LND*-terms, while the *confluence* property its uniqueness.

The significance of applying this technique in the proof of the normalisation theorems lies in the presentation of a simple and computational method, which allowed the discovery of a new basic set of transformations between proofs, which we called "ι (iota)-reductions" [de Oliveira, 1995; de Oliveira and de Queiroz, 1997]. With this result, we obtained a confluent system which contains the η-reductions. Traditionally, the η-reductions have not been given an adequate status, as rightly pointed out by Girard in [Girard *et al.*, 1989, p. 16], when he defines the *primary* equations, which correspond to the β-equations and the *secondary* equations, which are the η-equations. Girard [Girard *et al.*, 1989] says that the system given by these equations is consistent and decidable, however he notes the following:

"Although this result holds for the whole set of equations, one only ever considers the first three. It is a consequence of the *Church–Rosser property* and the *normalisation theorem* (...)"

The first three equations, referred to by Girard, are the *primary* ones, i.e. β-equations.

η-rules: expansion versus reduction. Traditionally, the λ-calculus η rules are seen as *reduction* rules such as $\lambda x.(Mx) = M$ (provided that x is not free in M). On the other hand, in Prawitz' style natural deduction the corresponding η rules are taken to be *expansion* rules. As it is argued in [de Queiroz and Gabbay, 1999; de Queiroz *et al.*, to appear], there is, however a sense in which η rules may be seen as *expansion* rules, rather than as *reduction* rules. In a Natural Deduction proof system there is another way of making 'redundant' steps that one can make, apart from the (β-like) '*introduction* followed by *elimination*'. It is the exact inverse of this: an *elimination* step is followed by an *introduction* step. As it turns out, the convertibility relation will be revealing another aspect of the 'propositions-as-types' paradigm, i.e. that there are redundant steps which from the point of view of the definition/presentation of propositions/types are saying that given any arbitrary proof/element from the proposition/type, it must be of the form given by the *introduction* rules. In other words, take an arbitrary inhabitant of the type; it must be equal to the result of applying the steps '*elimination* followed by an *introduction*' to itself. Formally:

$$\frac{c : A \wedge B}{\langle \mathtt{FST}(c), \mathtt{SND}(c)\rangle = c : A \wedge B} \qquad \frac{c : A \rightarrow B}{\lambda x.\mathtt{APP}(c, x) = c : A \rightarrow B}$$

$$\frac{c : A \vee B}{\mathtt{CASE}(c, \upsilon a.\mathtt{inl}(a), \upsilon b.\mathtt{inr}(b)) = c : A \vee B}$$

In the typed λ-calculus literature, this 'inductive' convertibility relation has been referred to as simply 'η'-convertibility.

In order to use techniques of term-rewriting we remain within the λ-calculus tradition by looking at η rule as a *reduction* rule. By doing this, we have been able to prove that the interaction between the so-called *permutative* reductions of Prawitz with the rules for the so-called Skolem-type connectives ($\vee$, $\exists$, $\dot{=}$), leads to non-confluence. A different approach is taken by Ghani [Ghani, 1995], and Altenkirch et al. [Altenkirch *et al.*, 2001], where η is taken to be an expansion rule.

Knuth–Bendix completion: uncovering non-confluence. Applying the so-called *completion procedure*, proposed by Knuth and Bendix in [Knuth and Bendix, 1970], to *LND-TRS*, the following term, which causes non-confluence in the system, is produced (i.e. a divergent critical pair is generated):

$$w(\mathtt{CASE}(c, \upsilon a_1.\mathtt{inl}(a_1), \upsilon a_2.\mathtt{inr}(a_2)))$$

This term rewrites in two different ways[1]:

1. $\leadsto_\eta w(c)$ **2.** $\leadsto_\zeta \texttt{CASE}(c, \upsilon a_1.w(\texttt{inl}(a_1)), \upsilon a_2.w(\texttt{inr}(a_2)))$

The method of Knuth–Bendix says that when a terminating system is not confluent it may be possible to add rules in such a way that the resulting system becomes confluent. Thus applying this procedure to *LND-TRS*, a new rule is added to the system:

$$\texttt{CASE}(c, \upsilon a_1.w(\texttt{inl}(a_1)), \upsilon a_2, w(\texttt{inr}(a_2))) \leadsto_\iota w(c)$$

Since terms represent proof-constructions in the *LND* system, this rule defines a new transformation between proofs:

ι-reduction-$\vee$

$$\dfrac{c : A_1 \vee A_2 \quad \dfrac{\dfrac{[a_1 : A_1]}{\texttt{inl}(a_1) : A_1 \vee A_2} \vee\text{-}intr}{w(\texttt{inl}(a_1)) : W}\mathsf{r} \quad \dfrac{\dfrac{[a_2 : A_2]}{\texttt{inr}(a_2) : A_1 \vee A_2} \vee\text{-}intr}{w(\texttt{inr}(a_2)) : W}\mathsf{r}}{\texttt{CASE}(c, \upsilon a_1.w(\texttt{inl}(a_1)), \upsilon a_2.w(\texttt{inr}(a_2))) : W} \vee\text{-}elim \quad \leadsto_\iota$$

$$\dfrac{c : A_1 \vee A_2}{w(c) : W}\mathsf{r}$$

Similarly, the ι reduction for the existential quantifier is defined [de Oliveira, 1995; de Oliveira and de Queiroz, 1997], since, similarly to $\vee$, the quantifier $\exists$ is a "Skolem-type" connective (i.e. in the *elimination* inference for this type of connective is necessary to open local assumptions):[2]

ι-reduction-$\exists$

$$\dfrac{c : \exists x^D.P(x) \quad \dfrac{\dfrac{[t : D] \quad [g(t) : P(t)]}{\varepsilon y.(g(y), t) : \exists y^D.P(y)} \exists\text{-}intr}{w(\varepsilon y.(g(y), t)) : W}\mathsf{r}}{\texttt{INST}(c, \sigma g.\sigma t.w(\varepsilon y.(g(y), t))) : W} \exists\text{-}elim \quad \leadsto_\iota \quad \dfrac{c : \exists x^D.P(x)}{w(c) : W}\mathsf{r}$$

(where 'ε' is an abstractor).

With this result, we believe that we have answered the question as to why the η-reductions are not considered in the proofs of the normalisation theorems (*confluence* requires ι-reductions). However, by applying a computational and well-defined method, the completion procedure, it seems that this problem of the non-confluence caused by η-reductions is solved.

2 Proof transformations in labelled deduction

The functional interpretation of logical connectives, the so-called Curry-Howard interpretation [Curry, 1934; Howard, 1980], provides an adequate framework for the establishment of various calculi of logical inference. Being an 'enriched' system of natural deduction, in the sense that terms representing proof-constructions (the *labels*) are carried alongside formulas, it

[1]See Appendix B for the definition of the ζ-reductions.

[2]For more details on the treatment of the existential quantifier in labelled natural deduction see [de Queiroz and Gabbay, 1993].

constitutes an apparatus of great usefulness in the formulation of logical calculi in an operational manner. By uncovering a certain harmony between a functional calculus on the labels and a logical calculus on the formulas, it proves to be instrumental in giving mathematical foundations for systems of logic presentation designed to handle meta-level features at the object-level via a labelling mechanism, such as, e.g. D. Gabbay's [Gabbay, 1996] *Labelled Deductive Systems*.

By using labels one can keep track of proof steps. This feature of labelled natural deduction systems helps to recover the 'local' control virtually lost in plain natural deduction systems. The 'global' character of the so-called *improper* inference rules (i.e. those rules who do not only manipulate premises and conclusions, but also involve assumptions, such as, e.g., $\rightarrow$-*introduction*, $\vee$-*elimination*, $\exists$-*elimination*, etc.) is made 'local' by turning the 'discharge functions' into appropriate disciplines for variable-binding via the device of 'abstractors'. As a matter of fact, the use of abstractors to make arbitrary names lose their identity is not a new device: it was already used by Frege in his *Grundgesetze I* [Frege, 1893]. (Think of the step from 'assuming $x : A$ and arriving at $b(x) : B$' to 'assert $\lambda x.b(x) : A \rightarrow B$' as making the arbitrary name 'x' lose the identity it had in $b(x)$ as a name. The binding transforms the name into a mere place-marker for arguments in substitutions.) From this perspective, critical remarks of the sort "Natural Deduction uses global rules (...) which apply to whole deductions, in contrast to rules like elimination rule for $\rightarrow$, which apply to formulas" [Girard, 1989, page 35] will have much less impact than they have w.r.t. 'plain' natural deduction systems. In this sense, *LND* makes it possible to look at η rules as rules of reduction, rather than expansions, and normalisation may be proved via term rewriting. It turns out that the use of term rewriting (and Knuth–Bendix completion) brought about the need for a new kind of rewriting rule, and, thus, a new proof transformation rule. the reductions here called "ι"-rules. Together with the ζ rules [de Queiroz and Gabbay, 1999], they offer a kind of explanation for the non-interference between the main branches and the secondary branches in 'Skolem-type' logical connectives, such as $\vee$, $\exists$ and $\doteq$.[3]

The relevance of the proposal hereby presented is due to basically two reasons:

[3]In [de Queiroz and Gabbay, 1994; de Queiroz and Gabbay, 1999] the study of natural deduction for equality yields an equational fragment of *Labelled Natural Deduction*, with proof rules framed as:

$\doteq$-*introduction*

$$\frac{a =_s b : D}{s(a,b) : \doteq_D (a,b)}$$

1. well-known results for Natural Deduction are proved via a computationally meaningful method: by proving the properties of termination and confluence for the *TRS* associated to *LND*, we prove the properties of termination and confluence of the proof system.

2. The discovery of a new basic set of transformation between proofs: the ι-reductions.

We shall first present the syntax of the *TRS* associated to *LND*, by means of an equational system with ordered sorts. Next, the *TRS* defined in [de Oliveira and de Queiroz, 1994] is presented, and the proofs of termination and confluence are given.

(See Appendix B for the definition of the proof rules of *LND*.)

3 Equivalences between proofs in *LND*

Here we shall present the notion of equivalence between proofs in *LND*, based in the formalism adopted in equational logic with ordered sorts. Such a formalism suit our purposes here basically for two reasons:

- the terms of *LND* have sorts (i.e. the formulas);

- via equational logic with ordered sorts, the correct domain of the operators can be specified. For example, the operator `FST` (which is a *destructor*) cannot have as an argument an operator such as `inl` or `inr`: the terms `FST`(`inl`(x)) and `FST`(`inr`(x)) are not *LND*-terms, as it shall be seen in this section.

DEFINITION 1 (signature). The signature of our LND Σ_{LND} is the set formed by the definitions of subsorts and operators of LND (ol). The subsorts are grouped in the set S_{LND}:

$$S_{LND} = \{T_\wedge,\ T_\vee,\ T_\rightarrow,\ T_\exists,\ T_\forall,\ T_{\doteq},\ T_{atomic},\ T_\sigma,\ T_\theta,\ T_\upsilon,\ T_{wff},\ D,\ D'\}$$

≐-reduction

$$\dfrac{\dfrac{a =_s b : D}{s(a,b) :\doteq_D (a,b)}\ \doteq\text{-}intr \qquad \begin{array}{c}[a =_t b : D]\\ d(t) : C\end{array}}{\mathtt{REWR}(s(a,b), \theta t.d(t)) : C}\ \doteq\text{-}elim \quad \rightsquigarrow_\beta \quad \begin{array}{c}[a =_s b : D]\\ d(s/t) : C\end{array}$$

≐-induction

$$\dfrac{e :\doteq_D (a,b) \qquad \dfrac{[a =_t b : D]}{t(a,b) :\doteq_D (a,b)}\ \doteq\text{-}intr}{\mathtt{REWR}(e, \theta t.t(a,b)) :\doteq_D (a,b)}\ \doteq\text{-}elim \quad \rightsquigarrow_\eta \quad e :\doteq_D (a,b)$$

where 'θ' is an abstractor which binds the occurrences of the (new) variable 't' introduced with the local assumption '$[a =_t b : D]$' as a kind of 'Skolem'-type constant denoting the (presumed) 'reason' why 'a' was assumed to be equal to 'b'.

The properties of confluence and termination for the equational fragment are proved in [de Oliveira and de Queiroz, 1999].

. Subsort declaration:

- $T_\wedge < T_{wff}$
- $T_\vee < T_{wff}$
- $T_\rightarrow < T_{wff}$
- $T_\exists < T_{wff}$
- $T_\forall < T_{wff}$
- $T_{\doteq} < T_{wff}$
- $T_{atomic} < T_{wff}$
- $T_\sigma < T_{wff}$
- $T_\theta < T_{wff}$
- $T_\upsilon < T_{wff}$
- $D' < D$

. Operators:

- $\mathtt{FST} : T_\wedge \rightarrow T_{wff}$, $\mathtt{SND} : T_\wedge \rightarrow T_{wff}$, $\langle\,\rangle : T_{wff} \times T_{wff} \rightarrow T_\wedge$
- $\mathtt{APP} : T_\rightarrow \times T_{wff} \rightarrow T_{wff}$, $\lambda : T_{wff} \times T_{wff} \rightarrow T_\rightarrow$
  The λ-abstractor is a special operator. It constructs terms of the form $\lambda x.f(x)$, whose sort is $T_\rightarrow$. The first argument of λ must be a variable of sort T_{wff} and the second argument, also of sort T_{wff}, may or may not contain a variable of the first argument.
- $\sigma : T_{wff} \times T_{wff} \rightarrow T_\sigma$, $\theta : T_{wff} \times T_{wff} \rightarrow T_\theta$, $\upsilon : T_{wff} \times T_{wff} \rightarrow T_\upsilon$
  Those three operators are similar to operator λ. They construct terms of the form $\sigma x.f(x)$, $\theta x.f(x)$ and $\upsilon x.f(x)$, respectively.
- $\mathtt{CASE} : T_\vee \times T_\upsilon \times T_\upsilon \rightarrow T_{wff}$, $\mathtt{inl} : T_{wff} \rightarrow T_\vee$, $\mathtt{inr} : T_{wff} \rightarrow T_\vee$
- $\mathtt{INST} : T_\exists \times T_\sigma \rightarrow T_{wff}$, $\varepsilon : T_{wff} \times D \rightarrow T_\exists$
  The ε-abstractor is also a special operator, since it constructs terms of the form $\varepsilon x.(f(x), a)$ of sort $T_\exists$.
- $\mathtt{EXTR} : T_\forall \times D \rightarrow T_{wff}$, $\Lambda : T_{wff} \times T_{wff} \rightarrow T_\forall$
  The Λ-abstractor is similar to λ, except for the fact that it constructs terms of the form $\Lambda x.f(x)$ of sort $T_\forall$.
- $\mathtt{REWR} : T_{\doteq} \times T_\theta \rightarrow T_{wff}$, $r : D \times D \rightarrow T_{\doteq}$, $r : T_{wff} \times T_{wff} \rightarrow T_{\doteq}$

According to the *principle of inversion*, one has that the type of a term in the functional side, is determined by the main connective of the formula in the logical calculus. Thus, the sorts $T_\wedge$, $T_\vee$, $T_\rightarrow$, $T_{\dot{=}}$, $T_\forall$ and $T_\exists$ are associated to the logical connectives $(\wedge, \vee, \rightarrow, \dot{=}, \forall, \exists)$. The sort T_{wff} refers to any well-formed formula When a term has an atomic formula as its type, it has the sort T_{atomic}. Since in *LND* there exist two kinds of information in the so-called meta-level – terms which represent proofs and terms which represent objects from a certain domain (see, for example, the rule of $\forall$-introduction) –, the terms may also have a certain domain D as its type and possibly a subdomain D' of D.

A subset of operators of *LND* can be grouped according to the type of proof that they represent.

DEFINITION 2 (‘`constructor`’ and ‘`DESTRUCTOR`’)). A subset of operators of *LND* determines two kinds of signature: $\Sigma_{constructor}$ and $\Sigma_{DESTRUCTOR}$. The set $\Sigma_{constructor}$ is constituted by the declaration of the operators which form the *canonical* proofs corresponding subsort declarations, whereas the set $\Sigma_{DESTRUCTOR}$ is formed by the declaration of the elimination operators together with the corresponding subsort declarations, which form the *non-canonical* proofs.

Operators of $\Sigma_{constructor} = \{\ \lambda\ , \langle\ \rangle,\ \texttt{inl},\ \texttt{inr},\ \Lambda,\ \varepsilon,\ r\}$ (where r is any of the rewriting ‘reasons’ $r \in \{\beta, \eta, \zeta, \iota\}$ as defined by the proof-rules of propositional equality [de Queiroz and Gabbay, 1994])

Operators of $\Sigma_{DESTRUCTOR} = \{\ \texttt{APP}, \texttt{FST}, \texttt{SND}, \texttt{CASE}, \texttt{INST}, \texttt{REWR}\ \}$

The set of terms of *LND* is denoted by $T_{\Sigma-LND}$. The terms of *LND* are defined as follows:

DEFINITION 3 (*LND*-terms). An *LND*-term of type s is defined as:

- Variables and constants, the so-called atomic labels (al), of sort $s \in S_{LND}$ are *LND*-terms. So, if $al \in \Sigma_{LND}$, then $al \in T_{\Sigma-LND}$ with arity$(al) = 0$.

- $ol(u_1, \ldots, u_n)$ is an *LND*-term, given that there exists a declaration of ol:

  $s_1, \ldots, s_n \rightarrow s \in \Sigma_{LND}$, such that the sort t of $ol(u_1, \ldots, u_n)$ is $\leq_\Sigma s$, u_i is an *LND*-term of sort s_i for $i = 1, \ldots n$ with arity$(ol) = n$.

Since terms represent proof constructions in *LND*, the transformations between proofs form an equational system defined as follows.

DEFINITION 4 (equational system for *LND*). The equational system of *LND* (E_{LND}) is made of definitional equalities which reflect the proof transformations, and for this reason the equational axioms are classified into four groups:

The group of β-equations is the following:

$\mathtt{FST}(\langle a_1, a_2\rangle) =_\beta a_1$
$\mathtt{SND}(\langle a_1, a_2\rangle) =_\beta a_2$
$\mathtt{CASE}(\mathtt{inl}(a_1), \upsilon s_1.d(s_1), \upsilon s_2.e(s_2)) =_\beta d(a_1/s_1)$
$\mathtt{CASE}(\mathtt{inr}(a_2), \upsilon s_1.d(s_1), \upsilon s_2.e(s_2)) =_\beta e(a_2/s_2)$
$\mathtt{APP}(\lambda x.b(x), a) =_\beta b(a/x)$
$\mathtt{EXTR}(\Lambda x.f(x), a) =_\beta f(a/x)$
$\mathtt{INST}(\varepsilon x(f(x), a), \sigma g.\sigma t.d(g,t)) =_\beta d(f/g, a/t)$
$\mathtt{REWR}(s(a,b), \theta t.d(t)) =_\beta d(s/t)$

The η-equations are:

$\langle \mathtt{FST}(c), \mathtt{SND}(c)\rangle =_\eta c$
$\mathtt{CASE}(c, \upsilon a_1.\mathtt{inl}(a_1), \upsilon a_2.\mathtt{inr}(a_2)) =_\eta c$
$\lambda x.\mathtt{APP}(c, x) =_\eta c$ (where c does not depend on x)
$\Lambda x.\mathtt{EXTR}(c, x) =_\eta c$ (where x does not occur free in c).
$\mathtt{INST}(c, \sigma g.\sigma t.\varepsilon y(g(y), t)) =_\eta c$
$\mathtt{REWR}(c, \theta t.t(a,b)) =_\eta c$

The ζ-equations are:

$w(\mathtt{CASE}(p, \upsilon s_1.d(s_1), \upsilon s_2.e(s_2))) =_\zeta \mathtt{CASE}(p, \upsilon s_1.w(d(s_1)), \upsilon s_2.w(e(s_2)))$
$w(\mathtt{INST}(c, \sigma g.\sigma t.d(g,t))) =_\zeta \mathtt{INST}(c, \sigma g.\sigma t.w(d(g,t)))$
$w(\mathtt{REWR}(e, \theta t.d(t))) =_\zeta \mathtt{REWR}(e, \theta t.w(d(t)))$

The ι-equations are (cf. Section 4.5):

$\mathtt{CASE}(c, \upsilon a_1.w(\mathtt{inl}(a_1)), \upsilon a_2.w(\mathtt{inr}(a_2))) =_\iota w(c)$
$\mathtt{INST}(c, \sigma g.\sigma t.w(\varepsilon y.(g(y), t))) =_\iota w(c)$
$\mathtt{REWR}(c, \sigma t.w(t(a,b))) =_\iota w(c)$

The pair $\langle \Sigma_{LND}, E_{LND}\rangle$ defines a specification of the set of all equivalences between proofs in *LND*.

An equational specification $\langle \Sigma, E\rangle$, where E is a set of equations between Σ-terms, defines a class of Σ-algebras A, such that $A \models E$. Birkhoff [Birkhoff, 1935] proved completeness for the equational logics, thus making sure that for all terms u, v belonging to a set $T_\Sigma(X)$, $\langle \Sigma, E\rangle \vdash u = v \Leftrightarrow \langle \Sigma, E\rangle \models u = v$. Given an equational specification $\langle \Sigma, E\rangle$, the validity problem (or the uniform word problem) for $\langle \Sigma, E\rangle$ is easily solved, in case it is possible to construct a canonical *TRS* equivalent to $\langle \Sigma, E\rangle$. After constructing a canonical *TRS*, to decide whether $E \vdash u = v$, it suffices to reduce u and v to their corresponding normal forms u' and v' and compare them. However, in general, the main results of the theory of standard rewriting systems do not naturally extend to *TRS*'s with ordered sorts. In order to get all the results, it is necessary that the *TRS* has the *sort decreasing* property.

Next section presents the *TRS*, whose syntax was defined by means of the specification $\langle \Sigma_{LND}, E_{LND} \rangle$, and proves the main properties of the system: *sort decreasing*, *termination* and *confluence*.

4 The term rewriting system for *LND*

We shall present the rewriting system associated to *LND* (*LND-TRS*), and prove *sort decreasing*, *termination* and *confluence*.

4.1 Defining the *LND-TRS*

LND-TRS computes the normal form of proofs in *LND*. Thus, the rules of such a system are based on the transformations between proofs (β, η and ζ) defined in [de Queiroz and Gabbay, 1999], and a new set of proof transformations arise, namely, the ι rules, uncovered during the construction of the proof of confluence of the system as originally defined with only β, η and ζ. This way, the equations of the *LND* equational system, defined in the previous section, are oriented according to these transformations. *LND-TRS* is defined as follows:

DEFINITION 5 (*LND-TRS*). *LND-TRS* is a term rewriting system with ordered sorts which computes the normal form of proofs in *LND*. The rules are the following:

1. $\mathtt{FST}(\langle a_1, a_2 \rangle) \rightsquigarrow_\beta a_1$
2. $\mathtt{SND}(\langle a_1, a_2 \rangle) \rightsquigarrow_\beta a_2$
3. $\mathtt{CASE}(\mathtt{inl}(a_1), \upsilon s_1.d(s_1), \upsilon s_2.e(s_2)) \rightsquigarrow_\beta d(a_1/s_1)$
4. $\mathtt{CASE}(\mathtt{inr}(a_2), \upsilon s_1.d(s_1), \upsilon s_2.e(s_2)) \rightsquigarrow_\beta e(a_2/s_2)$
5. $\mathtt{APP}(\lambda x.b(x), a) \rightsquigarrow_\beta b(a/x)$
6. $\mathtt{EXTR}(\Lambda x.f(x), a) \rightsquigarrow_\beta f(a/x)$
7. $\mathtt{INST}(\varepsilon x(f(x), a), \sigma g.\sigma t.d(g, t)) \rightsquigarrow_\beta d(f/g, a/t)$
8. $\mathtt{REWR}(s(a, b), \theta t.d(t)) \rightsquigarrow_\beta d(s/t)$
9. $\langle \mathtt{FST}(c), \mathtt{SND}(c) \rangle \rightsquigarrow_\eta c$
10. $\mathtt{CASE}(c, \upsilon a_1.\mathtt{inl}(a_1), \upsilon a_2.\mathtt{inr}(a_2)) \rightsquigarrow_\eta c$
11. $\lambda x.\mathtt{APP}(c, x) \rightsquigarrow_\eta c$ (where c does not depend on x)

12. $\Lambda x.\mathtt{EXTR}(c, x) \rightsquigarrow_\eta c$ (where x does not occur free in c).

13. $\mathtt{INST}(c, \sigma g.\sigma t.\varepsilon y(g(y), t)) \rightsquigarrow_\eta c$

14. $\mathtt{REWR}(c, \theta t.t(a, b)) \rightsquigarrow_\eta c$

15. $w(\mathtt{CASE}(p, \upsilon s_1.d(s_1), \upsilon s_2.e(s_2))) \rightsquigarrow_\zeta \mathtt{CASE}(p, \upsilon s_1.w(d((s_1)),$
$\upsilon s_2.w(e(s_2)))$

16. $w(\mathtt{INST}(c, \sigma g.\sigma t.d(g, t))) \rightsquigarrow_\zeta \mathtt{INST}(c, \sigma g.\sigma t.w(d(g, t)))$

17. $w(\mathtt{REWR}(e, \theta t.d(t))) \rightsquigarrow_\zeta \mathtt{REWR}(e, \theta t.w(d(t)))$

18. $\mathtt{CASE}(c, \upsilon a_1.w(\mathtt{inl}(a_1)), \upsilon a_2.w(\mathtt{inr}(a_2))) \rightsquigarrow_\iota w(c)$

19. $\mathtt{INST}(c, \sigma g.\sigma t.w(\varepsilon y.(g(y), t))) \rightsquigarrow_\iota w(c)$

20. $\mathtt{REWR}(c, \sigma t.w(t(a, b))) \rightsquigarrow_\iota w(c)$

The rules of *LND-TRS* come with a subscript which informs the type of proof transformation (β, η, ζ ou ι) which the rule is supposed to represent. This way, one has the notion of β-redex, η-redex and ζ-redex, which characterizes the type of reduction:

DEFINITION 6 (β (η, ζ, ι)-redex). A β-redex is a redex belonging to an *LND*-term which is reducible according to a β rule. Similarly for η-redex and ζ-redex.

DEFINITION 7 (normal *LND*-term). An *LND*-term is in the normal form if it does not contain any β-, η-, ζ- or ι-redex.

4.2 The *sort decreasing* property

In order to make sure that the results for standard *TRS*'s get naturally extended to *TRS*'s with ordered sorts, it is necessary for the *TRS* to be *sort decreasing*. In the case of *LND* this property is trivially verified, since all rewriting rules representing proof transformations keep the sort of the rewritten terms. For example:

$$\mathtt{FST}(\langle a_1, a_2 \rangle) : A \quad \rightsquigarrow_\beta \quad a_1 : A$$

Thus, the *LND-TRS* is *sort decreasing*.

4.3 Defining an order

When comparing the lefthand side and the righthand side of the reductions β, η and ι, one may check that the righthand side is a syntactically simpler term. By defining an order which guarantees the property of termination for a system with only those three sets of equations is a rather trivial task: it would suffice to use a complexity measure on terms, based, for example, on the number of operators. However, this does not apply to the ζ transformations. The set of ζ transformations is characterized as being a set of permutative reductions. A permutative reduction is one such that both sides of the equality contain the same symbols, such as the associative and commutative laws. According the observation by Peterson and Stickel in [Peterson and Stickel, 1981], the permutative reductions represent an additional difficulty in determining the confluence and termination properties. For this reason, one has to be more careful with those kinds of reduction.

For proving the termination property for *LND-TRS* we might adopt a methodology used for cases in which subsets of the rules of the rewriting system have common features, giving rise to various subsystems of rewriting. In this situation, the termination property is proved separately for each subsystem. For the *LND-TRS*, this technique would be useful, since the system could be split into two 'modules': one subsystem formed by the rules of β, η and ι reduction and the other one by the rules of ζ reduction. However, the termination property is not modular for the general case: it is necessary that the subsystems of rewriting possess certain properties [Klop, 1990]. Furthermore, the subsystems of rewriting must be disjoint, i.e., the function symbols and the constant symbols need to be different. In case the alphabet is not distinct, one takes renamed copies of the subsystems [Klop, 1990]. This way, it is preferable to use an order which suits the whole system.

The proof method adopted here does not use such device of splitting the *TRS* into modules, since the chosen order may be applied to the system as a whole. Due to the difficulty presented by the so-called 'permutative' we shall first take a careful look at the set of ζ reductions, and then establish an adequate order for the whole system.

Analyzing the ζ-reductions

The set of ζ reductions, as previously presented, does not illustrate all possible ζ transformations between proofs. The operator w may have arity 1, 2 or 3, so that the set of ζ reductions ζ is extended in the following way:

- $w(\texttt{CASE}(p, \upsilon s_1.d(s_1), \upsilon s_2.e(s_2))) \rightsquigarrow_\zeta \texttt{CASE}(p, \upsilon s_1.w(d(s_1), \upsilon s_2.w(e(s_2)))$
- $w(\texttt{CASE}(p, \upsilon s_1.d(s_1), \upsilon s_2.e(s_2)), u) \rightsquigarrow_\zeta$

$$\texttt{CASE}(p, \upsilon s_1.w(d(s_1), u), \upsilon s_2.w(e(s_2), u))$$

- $w(\texttt{CASE}(p, \upsilon s_1.d(s_1), \upsilon s_2.e(s_2)), u_1, u_2) \rightsquigarrow_\zeta$
  $\texttt{CASE}(p, \upsilon s_1.w(d(s_1), u_1, u_2), \upsilon s_2.w(e(s_2), u_1, u_2))$
- $w(\texttt{INST}(c, \sigma g.\sigma t.d(g,t))) \rightsquigarrow_\zeta \texttt{INST}(c, \sigma g.\sigma t.w(d(g,t)))$
- $w(\texttt{INST}(c, \sigma g.\sigma t.d(g,t)), u) \rightsquigarrow_\zeta \texttt{INST}(c, \sigma g.\sigma t.w(d(g,t), u))$
- $w(\texttt{INST}(c, \sigma g.\sigma t.d(g,t)), u_1, u_2) \rightsquigarrow_\zeta \texttt{INST}(c, \sigma g.\sigma t.w(d(g,t), u_1, u_2))$
- $w(\texttt{REWR}(e, \theta t.d(t))) \rightsquigarrow_\zeta \texttt{REWR}(e, \theta t.w(d(t)))$
- $w(\texttt{REWR}(e, \theta t.d(t)), u) \rightsquigarrow_\zeta \texttt{REWR}(e, \theta t.w(d(t), u))$
- $w(\texttt{REWR}(e, \theta t.d(t)), u_1, u_2) \rightsquigarrow_\zeta \texttt{REWR}(e, \theta t.w(d(t), u_1, u_2))$

Among this set of reductions, the first three rules, which concern permutation of CASE, represent an additional difficulty in determining the order, since, besides the fact that they are supposed to be a commutative reduction, the size of its righthand side is greater than its lefthand side, and the operator w is duplicated.

This leads to the use of recursive path ordering.

DEFINITION 8 (recursive path ordering). Let $>$ be a partial order over the set of operators Σ. The recursive path ordering $>^*$ over the set $T_\Sigma(X)$ of terms over Σ is defined recursively as follows:

$$s = f(s_1, \ldots, s_m) >^* g(t_1, \ldots, t_n) = t$$

if and only if one of the following conditions holds:

1. $f = g$ and $\{s_1, \ldots, s_m\} \gg^* \{t_1, \ldots, t_n\}$
2. $f > g$ and $\{s\} \gg^* \{t_1, \ldots, t_n\}$
3. $f \not\geq g$ and $\{s_1, \ldots, s_m\} \gg^* \quad ou \quad = \quad \{t\}$

where $\gg^*$ is the extension of $>^*$ for multisets.

By using a precedence order in which $w > \texttt{CASE}$, when proving that

$$w(\texttt{CASE}(p, \upsilon s_1.d(s_1), \upsilon s_2.e(s_2))) >^* \texttt{CASE}(p, \upsilon s_1.w(d(s_1), \upsilon s_2.w(e(s_2)))$$

the following situations would have to be analyzed:

- If $w > \texttt{CASE}$ one has to show that

$$\{w(\texttt{CASE}(p, \upsilon s_1.d(s_1), \upsilon s_2.e(s_2)))\} \gg^* \{p, \upsilon s_1.w(d(s_1)), \upsilon s_2.w(e(s_2))\} :$$

  - $w(\texttt{CASE}(p, \upsilon s_1.d(s_1), \upsilon s_2.e(s_2))) >^* p$ by the subterm property.
  - $w(\texttt{CASE}(p, \upsilon s_1.d(s_1), \upsilon s_2.e(s_2))) >^* \upsilon s_1.w(d(s_1))$. In this case, one has a λ-term (i.e. $\upsilon s_1.w(d(s_1))$)[4]; every time a λ-term is compared with some other term, the function from which the variable is being abstracted will be used in the comparison. This way, the term $w(d(s_1))$ will be used in the comparison. Since $w = w$, one has to prove that
    $\{\texttt{CASE}(p, \upsilon s_1.d(s_1), \upsilon s_2.e(s_2)))\} \gg^* \{d(s_1)\}$.
    This relation is true by the subterm property.

  Thus, when $w > \texttt{CASE}$ the recursive path ordering is rather adequate.

- If $w = \texttt{CASE}$, to prove that

$$\texttt{CASE}(\texttt{CASE}(p, \upsilon s_1.d(s_1), \upsilon s_2.e(s_2)), u_1, u_2)) >^* \\ \texttt{CASE}(p, \upsilon s_1.\texttt{CASE}(d(s_1), u_1, u_2), \upsilon s_2.\texttt{CASE}(e(s_2), u_1 u_2))$$

  it is necessary to show that

$$\{\texttt{CASE}(p, \upsilon s_1.d(s_1), \upsilon s_2.e(s_2)), u_1, u_2\} \gg^* \\ \{p, \upsilon s_1.\texttt{CASE}(d(s_1), u_1, u_2), \upsilon s_2.\texttt{CASE}(e(s_2), u_1, u_2)\}.$$

  It is not possible to guarantee that this relation holds since, although by the subterm property one has that $\texttt{CASE}(p, \upsilon s_1.d(s_1), \upsilon s_2.e(s_2)) >^* p$, by this same property the following relations also hold:

  - $u_1 <^* \upsilon s_1.\texttt{CASE}(d(s_1), u_1, u_2)$
  - $u_1 <^* \upsilon s_2.\texttt{CASE}(e(s_2), u_1, u_2)$
  - $u_2 <^* \upsilon s_1.\texttt{CASE}(d(s_1), u_1, u_2)$
  - $u_2 <^* \upsilon s_2.\texttt{CASE}(e(s_2), u_1, u_2)$

  This relation could still be guaranteed if $\texttt{CASE}(p, \upsilon s_1.d(s_1), \upsilon s_2.e(s_2))$ were $>^*$ than $\upsilon s_1.\texttt{CASE}(d(s_1), u_1, u_2)$ and $\upsilon s_2.\texttt{CASE}(e(s_2), u_1, u_2)$:

  - When comparing $\texttt{CASE}(p, \upsilon s_1.d(s_1), \upsilon s_2.e(s_2))$ and $\upsilon s_1.\texttt{CASE}(d(s_1), u_1, u_2)$ one has that $\texttt{CASE} = \texttt{CASE}$, thus it is necessary to prove that

[4] Every term with abstractors λ, θ, υ and σ is here called simply λ-term.

$\{p, \upsilon s_1.d(s_1), \upsilon s_2.e(s_2)\} \gg^* \{d(s_1), u_1, u_2\}$.
It is not possible to guarantee that this relation holds without previous knowledge about p , u_1 and u_2.

- When comparing $\texttt{CASE}(p, \upsilon s_1.d(s_1), \upsilon s_2.e(s_2))$ and $\upsilon s_2.\texttt{CASE}(e(s_2), u_1, u_2)$ a situation similar to the previous case occurs.

From this analysis, the use of recursive path ordering was disconsidered. The same situation occurs with the other permutative equations. In fact, this is the same problem that occurs when trying to apply this ordering to rules which deal with the law of associativity (e.g. $+(+(a,b),c) \rightarrow +(a,+(b,c))$. Plaisted [Plaisted, 1994] had already noticed that to use recursive path ordering in rules of this kind it is necessary to resort to other mechanisms. Thus, it is not difficult to notice that in all permutative reductions the first argument of the term on the righthand side of the reduction is smaller than the first argument of the term on the lefthand side of the reduction. So, one must use an order in which the first argument of the terms of the equation could be compared. It is exactly this "trick" which Dershowitz proposes in [Dershowitz, 1982]. He proposes a recursive path ordering in which an operand may be used as operator:

> "One way of extending the recursive path ordering is to allow some function of a term $f(t_1, \ldots, t_n)$ to serve the role of the operator f. For example, we can consider the kth operand t_k to be the operator, and compare two terms by first recursively comparing their kth operands. This yields a simplification ordering for the same reasons that the original definition does."

By means of the example below this recursive path ordering is illustrated [Dershowitz, 1982]:

EXAMPLE 9. The recursive path ordering given in this example uses an operand as operator. Let us take the *TRS* $R = \{if(if(a,b,c),d,e) \rightarrow if(a, if(b,d,e), if(c,d,e))\}$. The conditional expression "$if(a,b,c)$" represents an "$if - then - else$": "$if\ \ a\ \ then\ \ b\ \ else\ \ c$". This system "normalises" conditional expressions by removing nested if's from the condition "a".

To prove that this system is terminating, the condition is considered as an operator. The condition statement "$if(a,b,c)$" on the lefthand side of the rule is greater (via the subterm property) than the condition "a" on the righthand side. Thus, it is necessary to show that the lefthand side is greater than both the operands at the righthand side, "$if(b,d,e)$" and "$if(c,d,e)$": $if(if(a,b,c),d,e) > if(b,d,e)$ and $if(if(a,b,c),d,e) >$

$if(c, d, e)$, since $if(a, b, c)$ is greater than the operators "b" and "c" (subterm property). It suffices then to prove that the lefthand side is greater than the operands "d" and "e", which is easily proved, once again via the subterm property.

This example fits perfectly within the first three ζ reductions of the set here defined. So, by taking the first argument as an operator and the remaining arguments as operands, the recursive path ordering, adopted in the above example, applies not only to the ζ reductions, but also to the remaining set of reductions in the system. Next, we prove termination using this order.

4.4 Proving the termination property

As it is well-known, in order to prove the property of termination for a *TRS*, after one adopts a recursive path ordering $>^*$, it suffices to show that for every rule $e \to d$ of the system, $e >^* d$. This way, the proof of termination for *LND-TRS* will be done by showing that the lefthand side of every rule in the system is greater than its righthand side.

The proof shown here is based on, apart from the properties of the recursive path ordering, the lemma of embedded homeomorphic relation, given by Dershowitz [Dershowitz, 1979]:

LEMMA 10. . *Let s and t be terms in $T_\Sigma(X)$. If $s \trianglelefteq t$, then $s \leq t$ in any simplification order $>$ over $T_\Sigma(X)$.*

This lemma establish that if a term s is syntactically simpler than a term t, then $t > s$ in the simplification order. Therefore, $t >^* s$ in the recursive path ordering, which is a simplification ordering.

Proof for the subset of β reductions:

1. (a) $\texttt{FST}(\langle a_1, a_2\rangle) \rightsquigarrow_\beta a_1$

   (b) $\texttt{SND}(\langle a_1, a_2\rangle) \rightsquigarrow_\beta a_2$

   - One has that $\texttt{FST}(\langle a_1, a_2\rangle) >^* a_1$ and $\texttt{SND}(\langle a_1, a_2\rangle) >^* a_2$ by the subterm property.

2. (a) $\texttt{CASE}(\texttt{inl}(a_1), \upsilon s_1.d(s_1), \upsilon s_2.e(s_2)) \rightsquigarrow_\beta d(a_1/s_1)$

   (b) $\texttt{CASE}(\texttt{inr}(a_2), \upsilon s_1.d(s_1), \upsilon s_2.e(s_2)) \rightsquigarrow_\beta e(a_2/s_2)$

   - $\texttt{inl}(a_1) >^* a_1$ and $\texttt{inr}(a_2) >^* a_2$, by the subterm property.

   - $\{\texttt{CASE}(\texttt{inl}(a_1), \upsilon s_1.d(s_1), \upsilon s_2.e(s_2))\} \gg^* \emptyset$ and $\{\texttt{CASE}(\texttt{inr}(a_2), \upsilon s_1.d(s_1), \upsilon s_2.e(s_2))\} \gg^* \emptyset$ vacuously.

3. $\mathtt{APP}(\lambda x.b(x), a) \rightsquigarrow_\beta b(a/x)$

    - Comparing $\lambda x.b(x)$ and a:
      Here we have a λ-term. In this case, we compare the function that is being abstracted from (i.e. $b(x)$):
        - $b(x)$ and a: the value of x is a, so, by the subterm property we have that $b(a) >^* a$; and, $\{\mathtt{APP}(\lambda x.b(x), a)\} \gg^* \emptyset$.

4. $\mathtt{EXTR}(\Lambda x.f(x), a) \rightsquigarrow_\beta f(a/x)$

    - Comparing $\Lambda x.f(x)$ and a:
      Again, here we have a λ-term. In this case, we compare the function which is being abstracted from (i.e. $f(x)$):
        - $f(x)$ and a: the value of x is a, so $f(a) >^* a$ by the subterm property; and, $\{\mathtt{EXTR}(\Lambda x.f(x), a)\} \gg^* \emptyset$.

5. $\mathtt{INST}(\varepsilon x(f(x), a), \sigma g.\sigma t.d(g,t)) \rightsquigarrow_\beta d(f/g, a/t)$

    - $\varepsilon x(f(x), a) >^* f$ by the lemma of embedded homeomorphic relation.

    - $\{\mathtt{INST}(\varepsilon x(f(x), a), \sigma g.\sigma t.d(g,t))\} \gg^* \{a\}$ by the subterm property.

6. $\mathtt{REWR}(s(a,b), \theta t.d(t)) \rightsquigarrow_\beta d(s/t)$

    - $s(a,b) >^* s$ by the lemma of embedded homeomorphic relation.

    - $\{\mathtt{REWR}(s(a,b), \theta t.d(t))\} \gg^* \emptyset$ vacuously.

■

Proof for the subset of η reductions:

1. $\langle \mathtt{FST}(c), \mathtt{SND}(c)\rangle \rightsquigarrow_\eta c$

    - $\langle \mathtt{FST}(c), \mathtt{SND}(c)\rangle >^* c$ by the subterm property.

2. $\mathtt{CASE}(c, \upsilon a_1.\mathtt{inl}(a_1), \upsilon a_2.\mathtt{inr}(a_2)) \rightsquigarrow_\eta c$

    - $\mathtt{CASE}(c, \upsilon a_1.\mathtt{inl}(a_1), \upsilon a_2.\mathtt{inr}(a_2)) >^* c$ by the subterm property.

3. $\lambda x.\mathtt{APP}(c, x) \rightsquigarrow_\eta c$

    - $\lambda x.\mathtt{APP}(c, x) >^* c$ by the subterm property.

4. $\Lambda t.\mathtt{EXTR}(c,t) \rightsquigarrow_\eta c$

   - $\Lambda t.\mathtt{EXTR}(c,t) >^* c$ by the subterm property.

5. $\mathtt{INST}(c,\sigma g.\sigma t.\varepsilon y(g(y),t)) \rightsquigarrow_\eta c$

   - $\mathtt{INST}(c,\sigma g.\sigma t.\varepsilon y(g(y),t)) >^* c$ by the subterm property.

6. $\mathtt{REWR}(c,\theta t.t(a,b)) \rightsquigarrow_\eta c$

   - $\mathtt{REWR}(c,\theta t.t(a,b)) >^* c$ by the subterm property.

■

Proof for the subset of ζ reductions:

1. $w(\mathtt{CASE}(p,\upsilon s_1.d(s_1),\upsilon s_2.e(s_2))) \rightsquigarrow_\zeta$;
   $\mathtt{CASE}(p,\upsilon s_1.w(d((s_1)),\upsilon s_2.w(e(s_2)))$

   - $w(\mathtt{CASE}(p,\upsilon s_1.d(s_1),\upsilon s_2.e(s_2))) >^* p$ by the subterm property.
   - $\{w(\mathtt{CASE}(p,\upsilon s_1.d(s_1),\upsilon s_2.e(s_2)))\} \gg^* \{\upsilon s_1.w(d((s_1)),$ $\upsilon s_2.w(e(s_2))\}$:
     - $w(\mathtt{CASE}(p,\upsilon s_1.d(s_1),\upsilon s_2.e(s_2))) >^* \upsilon s_1.w(d(s_1))$ since $\mathtt{CASE}(p,\upsilon s_1.d(s_1),\upsilon s_2.e(s_2)) >^* d(s_1)$ by the subterm property.
     - $w(\mathtt{CASE}(p,\upsilon s_1.d(s_1),\upsilon s_2.e(s_2))) >^* \upsilon s_2.w(e(s_2))$ since $\mathtt{CASE}(p,\upsilon s_1.d(s_1),\upsilon s_2.e(s_2)) >^* e(s_2)$ by the subterm property.

2. $w(\mathtt{INST}(c,\sigma g.\sigma t.d(g,t))) \rightsquigarrow_\zeta \mathtt{INST}(c,\sigma g.\sigma t.w(d(g,t)))$

   - $\mathtt{INST}(c,\sigma g.\sigma t.d(g,t)) >^* c$ by the subterm property.

   - $\{w(\mathtt{INST}(c,\sigma g.\sigma t.d(g,t)))\} \gg^*$
     $\{\sigma g.\sigma t.w(d(g,t))\}$ pois $\mathtt{INST}(c,\sigma g.\sigma t.d(g,t)) > d(g,t)$ by the subterm property.

3. $w(\mathtt{REWR}(e,\theta t.d(t))) \rightsquigarrow_\zeta \mathtt{REWR}(e,\theta t.w(d(t)))$

   - $\mathtt{REWR}(e,\theta t.d(t))) >^* e$ by the subterm property.
   - $\{w(\mathtt{REWR}(e,\theta t.d(t)))\} \gg^* \{\theta t.w(d(t))\}$ pois $\mathtt{REWR}(e,\theta t.d(t)) >^*$ $d(t)$ by the subterm property.

The proof for the rules in which w has arity greater than 1 is done analogously, since the additional arguments are not placed as first operator, and they are added in the same way, be it on the lefthand side or the righthand side of the rules.

The proof for the set of ι reductions is shown in Appendix A.

From the proof of termination of *LND-TRS*, the following result follows:

THEOREM 11 (existence). *Every derivation in LND converts to a normal form.* ■

4.5 Proving the confluence property

Once we have proved the termination property for the *LND-TRS*, it suffices to apply the superposition algorithm to check if there exist divergent critical pairs.

The superposition algorithm tests each rule of the system with a view towards verifying if there are divergent critical pairs. In case there is any divergent critical pair, one needs to give it an orientation as illustrated by the Knuth–Bendix completion procedure.

The proof of confluence had as a result the incorporation of three new rules to *LND-TRS*:

18. $\texttt{CASE}(c, va_1.w(\texttt{inl}(a_1)), va_2, w(\texttt{inr}(a_2))) \rightsquigarrow_\iota w(c)$
19. $\texttt{INST}(c, \sigma g.\sigma t.w(\varepsilon y.(g(y), t))) \rightsquigarrow_\iota w(c)$
20. $\texttt{REWR}(c, \sigma t.w(t(a, b))) \rightsquigarrow_\iota w(c)$

This new set of rules defines new basic transformations between proofs, which do not seem to have appeared in the literature.

ι-$\vee$-reduction

$$\frac{c : A_1 \vee A_2 \qquad \dfrac{\dfrac{[a_1 : A_1]}{\texttt{inl}(a_1) : A_1 \vee A_2}\,\vee\text{-}intr}{w(\texttt{inl}(a_1)) : W}\,\mathsf{r} \qquad \dfrac{\dfrac{[a_2 : A_2]}{\texttt{inr}(a_2) : A_1 \vee A_2}\,\vee\text{-}intr}{w(\texttt{inr}(a_2)) : W}\,\mathsf{r}}{\texttt{CASE}(c, va_1.w(\texttt{inl}(a_1)), va_2.w(\texttt{inr}(a_2))) : W}\,\vee\text{-}elim \quad \rightsquigarrow_\iota \quad \frac{c : A_1 \vee A_2}{w(c) : W}$$

ι-$\exists$-reduction

$$\frac{c : \exists x^D.P(x) \qquad \dfrac{\dfrac{[t : D] \quad [g(t) : P(t)]}{\varepsilon y.(g(y), t) : \exists y^D.P(y)}\,\exists\text{-}intr}{w(\varepsilon y.(g(y), t)) : W}\,\mathsf{r}}{\texttt{INST}(c, \sigma g.\sigma t.w(\varepsilon y.(g(y), t))) : W}\,\exists\text{-}elim \quad \rightsquigarrow_\iota \quad \frac{c : \exists x^D.P(x)}{w(c) : W}\,\mathsf{r}$$

ι-$\doteq$-reduction

$$\dfrac{c :\doteq_D (a,b) \qquad \dfrac{\dfrac{[a =_t b : D]}{t(a,b) :\doteq_D (a,b)}\doteq\text{-}intr}{w(t(a,b)) : W}\mathsf{r}}{\mathtt{REWR}(c, \acute{\theta} t.w(t(a,b))) : W}\doteq\text{-}elim \quad \rightsquigarrow_\iota \quad \dfrac{c :\doteq_D (a,b)}{w(c) : W}\mathsf{r}$$

These transformations, together with the ζ transformations, offer a formal justification for the non-interference of the other branches with the main branch in the *elimination* rules of $\vee$, $\exists$ and $\doteq$: the rule r (an interference in the form of an *introduction*) may be applied to each secondary branches or in the main branch. The example below shows the application of such transformations:

EXAMPLE 12.

$$\dfrac{c : A_1 \vee A_2 \qquad \dfrac{\dfrac{[a_1 : A_1]}{\mathtt{inl}(a_1) : A_1 \vee A_2} \quad b : B}{\langle \mathtt{inl}(a_1), b\rangle : (A_1 \vee A_2) \wedge B} \qquad \dfrac{\dfrac{[a_2 : A_2]}{\mathtt{inr}(a_2) : A_1 \vee A_2} \quad b : B}{\langle \mathtt{inr}(a_2), b\rangle : (A_1 \vee A_2) \wedge B}}{\mathtt{CASE}(c, \acute{v}a_1.\langle \mathtt{inl}(a_1), b\rangle, \acute{v}a_2.\langle \mathtt{inr}(a_2), b\rangle) : (A_1 \vee A_2) \wedge B}$$

$$\rightsquigarrow_\iota \quad \dfrac{c : A_1 \vee A_2 \quad b : B}{\langle c, b\rangle : (A_1 \vee A_2) \wedge B}$$

THEOREM 13 (uniqueness). *Every derivation in LND converts to a unique normal form.*

5 Final remarks

We have defined a *TRS* associated to *LND* (*LND-TRS*), which computes the normal form of proofs in *LND*. The proof of termination for *LND-TRS* had as a consequence the existence of a normal form for *LND* terms, whereas the proof of confluence led to the uniqueness of normal form. The termination of *LND-TRS* was proved via a special kind of recursive path ordering proposed by Dershowitz [Dershowitz, 1979]. Confluence was established via the mechanism of superposition among the rules of the system.

The results presented here seem to be relevant not only to the *LND* system, but seems to be of general interest since they offer a possible answer to the question as to why normalisation proofs for natural deduction systems do not take into account the so-called η-rules as *reduction* rules. We believe that the uncovering of the need for a new set of reduction rules (i.e. the 'ι' rules) in order to obtain confluence brings in new information into the discussion.

BIBLIOGRAPHY

[Altenkirch *et al.*, 2001] T. Altenkirch, P. Dybjer, M. Hofmann and Ph. Scott. Normalization by evaluation for typed lambda calculus with coproducts. In *16th Annual IEEE Symposium on Logic in Computer Science*, 2001, pages 303–310.

[Birkhoff, 1935] G. Birkhoff. On the structure of abstract algebras. In *Proceedings of the Cambridge Philosophical Society* **31**:433–452, 1935.

[Curry, 1934] H. B. Curry. Functionality in Combinatory Logic. In *Proceedings of the National Academy of Sciences of USA* **20**:584–590, 1934.

[Dershowitz, 1979] N. Dershowitz. A note on simplification orderings. *Inf. Proc. Lett.* **5**(9):212–215, 1979.

[Dershowitz, 1982] N. Dershowitz. Ordering for term-rewriting systems. *Theoretical Computer Science*, **17**:279–301, 1982.

[Fenstad,] J. E. Fenstad, editor.1971 *Proceedings of the Second Scandinavian Logic Symposium*, volume 63 of *Studies in Logic and The Foundations of Mathematics*. North-Holland, Amsterdam, viii+405pp, 1971. Proceedings of the Symposium held in Oslo, June 18–20 1970.

[Frege, 1893] G. Frege. *Grundgesetze der Arithmetik. Begriffsschriftlich abgeleitet. I.* Verlag von Hermann Pohle, Jena, 1893. Reprinted in volume 32 of *Olms Paperbacks*, Georg Olms Verlagsbuchhandlung, Hildesheim, 1966, XXXII+254pp. Partial English translation in [Furth, 1964].

[Furth, 1964] M. Furth, editor. *The Basic Laws of Arithmetic. Exposition of the System.* University of California Press, Berkeley and Los Angeles, lxiv+143pp, 1964. Partial English translation of Gottlob Frege's *Grundgesetze der Arithmetik.*

[Gabbay, 1996] D. M. Gabbay. *Labelled Deductive Systems, Volume I - Foundations.* Oxford University Press, 1996.

[Gabbay and de Queiroz, 1992] D. M. Gabbay and R. J. G. B. de Queiroz. Extending the Curry-Howard interpretation to linear, relevant and other resource logics. *The Journal of Symbolic Logic*, **57**(4):1319–1365, December 1992.

[Ghani, 1995] N. Ghani. Beta-Eta Equality for Coproducts. In Proceedings of *Typed Lambda-Calculus and Applications 1995*, Volume 902 in Lecture Notes in Computer Science, pages 171–185. Springer-Verlag 1995.

[Girard, 1989] J.-Y. Girard. Towards a geometry of interaction. In *Categories in Computer science and Logic*, volume 92 of *C ontemporary Mathematics*, pages 69–108. AMS Publications, 1989.

[Girard *et al.*, 1989] J. Y. Girard, Y. Lafont, and P. Taylor. *Proofs and Types.* Cambridge University Press, 1989.

[Howard, 1980] W. A. Howard. The formulae-as-types notion of construction. In J. R. Seldin and J.R. Hindley, editors, *To H. B. Curry: Essays on Combinatory Logic Lambda Calculus and Formalism.* Academic Press, 1980. xxv+606pp.

[Klop, 1990] J. W. Klop. Term rewriting systems. In Abramsky, S., Gabbay, D., and Maibaum, T., editors, *Handbook of Logic in Computer Science*, volume 1, chapter 6. Oxford University Press, 1990.

[Knuth and Bendix, 1970] D. E. Knuth and P. B. Bendix. Simple word problems in universal algebras. In J. Leech, editor, *Computational Problems in Abstract Algebra*, pages 263–297. Pergamon Press, 1970.

[Martin-Löf, 1975] P. Martin-Löf. About Models for Intuitionistic Type Theories and the Notion of Definitional Equality. In S. Kanger, editor, *Proceedings of the Third Scandinavian Logic Symposium*, Series *Studies in Logic and The Foundations of Mathematics*, pages 81–109, Amsterdam, 1975. North-Holland. Symposium held in 1973.

[Martin-Löf, 1984] P. Martin-Löf. *Intuitionistic Type Theory.* Series Studies in Proof Theory. Bibliopolis Naples, iv+91pp., 1984. Notes by Giovanni Sambin of a series of lectures given in Padova, June 1980.

[Peterson and Stickel, 1981] G. E. Peterson and M. E. Stickel. Complete sets of reductions for some equational theories. *JACM*, **28**(2):233–264, April 1981.

[Plaisted, 1994] D. A. Plaisted. Equational reasoning and term rewriting systems. In D. Gabbay, C. Hogger, and J. A. Robinson, editors, *Handbook of Logic in Artificial Intelligence and Logic Programming*, volume 1, pages 273–364. Oxford University Press, 1994.

[de Oliveira, 1995] A. G. de Oliveira. Proof Transformations for Labelled Natural Deduction via Term Rewriting. (In Portuguese). Master's thesis, Depto. de Informática, Universidade Federal de Pernambuco, C.P. 7851, Recife, PE 50732-970, Brasil, April 1995.

[de Oliveira and de Queiroz, 1997] A. G. de Oliveira and R. J. G. B de Queiroz. A New Basic Set of Transformations between Proofs (abstract). *The Bulletin of Symbolic Logic* **3**(1):124–126, March 1997.

[de Oliveira and de Queiroz, 1994] A. G. de Oliveira and R. J. G. B de Queiroz. Term rewriting systems with *LDS*. *Proc. of Brazilian Symposium on Artificial Intelligence, SBIA'94*, 1994, pp. 425–439.

[de Oliveira and de Queiroz, 1999] A. G. de Oliveira and R. J. G. B de Queiroz. A Normalization Procedure for the Equational Fragment of Labelled Natural Deduction. *Logic Journal of the Interest Group in Pure and Applied Logics*, **7**(2):173–215, Oxford Univ Press, 1999.

[Poigné, 1992] A. Poigné. Basic Category Theory. In S. Abramsky, D. Gabbay, and T. Maibaum, editors, *Handbook of Logic in Computer Science. Vol. I.* Oxford University Press, Oxford, 1992.

[Prawitz, 1971] D. Prawitz. Ideas and Results in Proof Theory, 1971. In [Fenstad,], pages 235–307.

[de Queiroz and Gabbay, 1994] R. J. G. B. de Queiroz and D. M. Gabbay. Equality in Labelled Deductive Systems and the functional interpretation of propositional equality. In *Proceedings of the 9th Amsterdam Colloquium*, P. Dekker and M. Stockhof (eds.), ILLC/Department of Philosophy, University of Amsterdam, pp. 547–566, 1994.

[de Queiroz and Gabbay, 1993] R. J. G. B. de Queiroz and D. M. Gabbay. The functional interpretation of the existential quantifier. *Bulletin of the Interest Group in Pure and Applied Logics* **3**(2–3):243–290, 1995. (Presented at *Logic Colloquium '91*, Uppsala, August 9–16 1991. Abstract in *JSL* **58**(2):753–754, 1993.)

[de Queiroz and Gabbay, 1999] R. J. G. B. de Queiroz and D. M. Gabbay. Labelled natural deduction. In *Logic, Language and Reasoning*, H.J. Ohlbach and U. Reyle (eds.), Volume 5 of *Trends in Logic* series, Kluwer Academic Publishers, 1999, pp. 173–250.

[de Queiroz *et al.*, to appear] R. J. G. B. de Queiroz, D. M. Gabbay and A. G. de Oliveira. *The Functional Interpretation of Non-Classical Logics*, Imperial College Press, World Scientific, to appear.

[Troelstra and van Dalen, 1988] A. S. Troelstra and D. van Dalen. *Constructivism in Mathematics: An Introduction. Vol. II*, volume 123 of *Studies in Logic and The Foundations of Mathematics.* North-Holland, Amsterdam, xvii+535pp, 1988.

Appendix A: Proof of Confluence of *LND-TRS*

Here comes the proof of confluence of *LND-TRS* via the application of the Knuth–Bendix procedure. (Only the cases involving the ι rules will be shown. The full list of cases is given in [de Oliveira, 1995].) The result of such a completion procedure is the addition of the following three new rules to the initial rewriting system:

18. $\mathtt{CASE}(c, \upsilon a_1.w(\mathtt{inl}(a_1)), \upsilon a_2, w(\mathtt{inr}(a_2))) \rightsquigarrow_\iota w(c)$
19. $\mathtt{INST}(c, \sigma g.\sigma t.w(\varepsilon y.(g(y), t))) \rightsquigarrow_\iota w(c)$
20. $\mathtt{REWR}(c, \sigma t.w(t(a, b))) \rightsquigarrow_\iota w(c)$

With the addition of those three new rules, *LND-TRS* becomes confluent.

The rules of the initial *LND-TRS* are as follows:

1. $\texttt{FST}(\langle a_1, a_2\rangle) \rightsquigarrow_\beta a_1$
2. $\texttt{SND}(\langle a_1, a_2\rangle) \rightsquigarrow_\beta a_2$
3. $\texttt{CASE}(\texttt{inl}(a_1), \upsilon s_1.d(s_1), \upsilon s_2.e(s_2)) \rightsquigarrow_\beta d(a_1/s_1)$
4. $\texttt{CASE}(\texttt{inr}(a_2), \upsilon s_1.d(s_1), \upsilon s_2.e(s_2)) \rightsquigarrow_\beta e(a_2/s_2)$
5. $\texttt{APP}(\lambda x.b(x), a) \rightsquigarrow_\beta b(a/x)$
6. $\texttt{EXTR}(\Lambda x.f(x), a) \rightsquigarrow_\beta f(a/x)$
7. $\texttt{INST}(\varepsilon x.(f(x), a), \sigma g.\sigma t.d(g,t)) \rightsquigarrow_\beta d(f/g, a/t)$
8. $\texttt{REWR}(s(a,b), \theta t.d(t)) \rightsquigarrow_\beta d(s/t)$
9. $\langle\texttt{FST}(c), \texttt{SND}(c)\rangle \rightsquigarrow_\eta c$
10. $\texttt{CASE}(c, \upsilon a_1.\texttt{inl}(a_1), \upsilon a_2.\texttt{inr}(a_2)) \rightsquigarrow_\eta c$
11. $\lambda x.\texttt{APP}(c, x) \rightsquigarrow_\eta c$ (where c does not depend on x).
12. $\Lambda x.\texttt{EXTR}(c, x) \rightsquigarrow_\eta c$ (where x does not occur free in c).
13. $\texttt{INST}(c, \sigma g.\sigma t.\varepsilon y(g(y), t)) \rightsquigarrow_\eta c$
14. $\texttt{REWR}(c, \theta t.t(a,b)) \rightsquigarrow_\eta c$
15. $w(\texttt{CASE}(p, \upsilon s_1.d(s_1), \upsilon s_2.e(s_2))) \rightsquigarrow_\zeta \texttt{CASE}(p, \upsilon s_1.w(d(s_1)), \upsilon s_2.w(e(s_2)))$
16. $w(\texttt{INST}(c, \sigma g.\sigma t.d(g,t))) \rightsquigarrow_\zeta \texttt{INST}(c, \sigma g.\sigma t.w(d(g,t)))$
17. $w(\texttt{REWR}(e, \theta t.d(t))) \rightsquigarrow_\zeta \texttt{REWR}(e, \theta t.w(d(t)))$

The analysis of superpositions for the key cases is shown below.

- Analyzing rules 15 and 10:

  $$w(\texttt{CASE}(c, \upsilon a_1.\texttt{inl}(a_1), \upsilon a_2.\texttt{inr}(a_2)))$$

  Rewriting sequence 1:

  $\rightsquigarrow_\eta w(c)$

  Rewriting sequence 2:

$\rightsquigarrow_\zeta$ $\texttt{CASE}(c, \upsilon a_1.w(\texttt{inl}(a_1)), \upsilon a_2.w(\texttt{inr}(a_2)))$

In this case there was no confluence, therefore by the Knuth–Bendix procedure, a new rule must be added according to the order previously chosen:

- $\texttt{CASE}(c, \upsilon a_1.w(\texttt{inl}(a_1)), \upsilon a_2.w(\texttt{inr}(a_2))) >^* w(c)$ since $c = c$ and $\{w(\texttt{inl}(a_1)), w(\texttt{inr}(a_2))\} \gg^* \emptyset$ vacuously.

Thus, the following rule, called "ι", is added to the system:

18. $\texttt{CASE}(c, \upsilon a_1.w(inl(a_1)), \upsilon a_2, w(inr(a_2))) \rightsquigarrow_\iota w(c)$

- Analyzing rules 16 and 13:

$$w(\texttt{INST}(c, \sigma g.\sigma t.\varepsilon y.(g(y), t)))$$

Rewriting sequence 1:

$\rightsquigarrow_\eta w(c)$

Rewriting sequence 2:

$\rightsquigarrow_\zeta \texttt{INST}(c, \sigma g.\sigma t.w(\varepsilon y.(g(y), t)))$

This case is similar to the one involving rules 15 and 10.

- $\texttt{INST}(c, \sigma g.\sigma t.w(\varepsilon y.(g(y), t))) >^* w(c)$, since $c = c$ and $\{w(\varepsilon y.(g(y), t))\} \gg^* \emptyset$

Thus, the following new rule is added to the rewriting system:

19. $\texttt{INST}(c, \sigma g.\sigma t.w(\varepsilon y.(g(y), t))) \rightsquigarrow_\iota w(c)$

- Analyzing rules 17 and 14:

$$w(c, \texttt{REWR}(c, \theta t.t(a, b))$$

Rewriting sequence 1:

$\rightsquigarrow_\eta w(c)$

Rewriting sequence 2:

$\rightsquigarrow_\zeta \texttt{REWR}(c, \theta t.w(t(a, b)))$

No confluence, thus, similarly to the case involving 15 and 10, as well as 16 and 13, a new rule must be added to the rewriting system, respecting the previously chosen order:

– $\texttt{REWR}(c, \sigma t.w(t(a,b))) >^* w(c)$, since $c = c$ and $\{w(t(a,b))\} \gg^* \emptyset$ vacuously.

Thus, the following new rule is added to the system:

20. $\texttt{REWR}(c, \sigma t.w(t(a,b))) \rightsquigarrow_\iota w(c)$ ■

Appendix B: Proof rules of *LND*

∧-introduction

$$\frac{a_1 : A_1 \qquad a_2 : A_2}{\langle a_1, a_2 \rangle : A_1 \wedge A_2}$$

∨-introduction

$$\frac{a_1 : A_1}{\texttt{inl}(a_1) : A_1 \vee A_2} \qquad \frac{a_2 : A_2}{\texttt{inr}(a_2) : A_1 \vee A_2}$$

→-introduction

$$\frac{\begin{array}{c}[x : A]\\ b(x) : B\end{array}}{\lambda x.b(x) : A \to B}$$

∀-introduction

$$\frac{\begin{array}{c}[x : D]\\ f(x) : P(x)\end{array}}{\Lambda x.f(x) : \forall x^D.P(x)}$$

∃-introduction

$$\frac{a : D \qquad f(a) : P(a)}{\varepsilon x.(f(x), a) : \exists x^D.P(x)}$$

$\dot{=}$-introduction

$$\frac{a =_s b : D}{s(a,b) : \dot{=}_D (a,b)}$$

β-Type reductions

∧-β-reduction

$$\frac{\dfrac{a_1 : A_1 \qquad a_2 : A_2}{\langle a_1, a_2 \rangle : A_1 \wedge A_2} \wedge\textit{-intr}}{\texttt{FST}(\langle a_1, a_2 \rangle) : A_1} \wedge\textit{-elim} \qquad \rightsquigarrow_\beta \qquad a_1 : A_1$$

$$\frac{\dfrac{a_1 : A_1 \qquad a_2 : A_2}{\langle a_1, a_2 \rangle : A_1 \wedge A_2} \wedge\textit{-intr}}{\texttt{SND}(\langle a_1, a_2 \rangle) : A_2} \wedge\textit{-elim} \qquad \rightsquigarrow_\beta \qquad a_2 : A_2$$

∨-β-reduction

$$\dfrac{\dfrac{a_1 : A_1}{\mathtt{inl}(a_1) : A_1 \vee A_2}\vee\textit{-intr} \qquad \begin{array}{c}[s_1 : A_1]\\ d(s_1) : C\end{array} \quad \begin{array}{c}[s_2 : A_2]\\ e(s_2) : C\end{array}}{\mathtt{CASE}(\mathtt{inl}(a_1), \upsilon s_1.d(s_1), \upsilon s_2.e(s_2)) : C}\vee\textit{-elim} \quad \rightsquigarrow_\beta \quad \begin{array}{c}[a_1 : A_1]\\ d(a_1/s_1) : C\end{array}$$

$$\dfrac{\dfrac{a_2 : A_2}{\mathtt{inr}(a_2) : A_1 \vee A_2}\vee\textit{-intr} \qquad \begin{array}{c}[s_1 : A_1]\\ d(s_1) : C\end{array} \quad \begin{array}{c}[s_2 : A_2]\\ e(s_2) : C\end{array}}{\mathtt{CASE}(\mathtt{inr}(a_2), \upsilon s_1.d(s_1), \upsilon s_2.e(s_2)) : C}\vee\textit{-elim} \quad \rightsquigarrow_\beta \quad \begin{array}{c}[a_2 : A_2]\\ e(a_2/s_2) : C\end{array}$$

→-β-reduction

$$\dfrac{a : A \qquad \dfrac{\begin{array}{c}[x : A]\\ b(x) : B\end{array}}{\lambda x.b(x) : A \to B}\to\textit{-intr}}{\mathtt{APP}(\lambda x.b(x), a) : B}\to\textit{-elim} \quad \rightsquigarrow_\beta \quad \begin{array}{c}[a : A]\\ b(a/x) : B\end{array}$$

∀-β-reduction

$$\dfrac{a : D \qquad \dfrac{\begin{array}{c}[x : D]\\ f(x) : P(x)\end{array}}{\Lambda x.f(x) : \forall x^D.P(x)}\forall\textit{-intr}}{\mathtt{EXTR}(\Lambda x.f(x), a) : P(a)}\forall\textit{-elim} \qquad \begin{array}{c}[a : D]\\ f(a/x) : P(a)\end{array}$$

∃-β-reduction

$$\dfrac{\dfrac{a : D \quad f(a) : P(a)}{\varepsilon x.(f(x), a) : \exists x^D.P(x)}\exists\textit{-intr} \qquad \begin{array}{c}[t : D, g(t) : P(t)]\\ d(g, t) : C\end{array}}{\mathtt{INST}(\varepsilon x.(f(x), a), \sigma g.\sigma t.d(g, t)) : C}\exists\textit{-elim} \rightsquigarrow_\beta \begin{array}{c}[a : D, f(a) : P(a)]\\ d(f/g, a/t) : C\end{array}$$

≐-β-reduction

$$\dfrac{\dfrac{a =_s b : D}{s(a, b) : \doteq_D (a, b)}\doteq\textit{-intr} \qquad \begin{array}{c}[a =_t b : D]\\ d(t) : C\end{array}}{\mathtt{REWR}(s(a, b), \theta t.d(t)) : C}\doteq\textit{-elim} \quad \rightsquigarrow_\beta \quad \begin{array}{c}[a =_s b : D]\\ d(s/t) : C\end{array}$$

η-Type reductions

∧-η-reduction

$$\dfrac{\dfrac{c : A_1 \wedge A_2}{\mathtt{FST}(c) : A_1}\wedge\textit{-elim} \qquad \dfrac{c : A_1 \wedge A_2}{\mathtt{SND}(c) : A_2}\wedge\textit{-elim}}{\langle \mathtt{FST}(c), \mathtt{SND}(c)\rangle : A_1 \wedge A_2}\wedge\textit{-intr} \quad \rightsquigarrow_\eta \quad c : A_1 \wedge A_2$$

∨-η-reduction

$$\frac{c : A_1 \vee A_2 \quad \dfrac{[a_1 : A_1]}{\mathtt{inl}(a_1) : A_1 \vee A_2}\vee\text{-}intr \quad \dfrac{[a_2 : A_2]}{\mathtt{inr}(a_2) : A_1 \vee A_2}\vee\text{-}intr}{\mathtt{CASE}(c, \acute{v}a_1.\mathtt{inl}(a_1), \acute{v}a_2.\mathtt{inr}(a_2)) : A_1 \vee A_2}\vee\text{-}elim \rightsquigarrow_\eta$$

$$c : A_1 \vee A_2$$

→-η-reduction

$$\frac{\dfrac{[x : A] \qquad c : A \to B}{\mathtt{APP}(c, x) : B} \to\text{-}elim}{\lambda x.\mathtt{APP}(c, x) : A \to B} \to\text{-}intr \qquad \rightsquigarrow_\eta \qquad c : A \to B$$

where c does not depend on x.
∀-η-reduction

$$\frac{\dfrac{[t : D] \qquad c : \forall x^D.P(x)}{\mathtt{EXTR}(c, t) : P(t)}\forall\text{-}elim}{\Lambda t.\mathtt{EXTR}(c, t) : \forall t^D.P(t)}\forall\text{-}intr \qquad \rightsquigarrow_\eta \qquad c : \forall x^D.P(x)$$

where x does not occur free in c.
∃-η-reduction

$$\frac{c : \exists x^D.P(x) \quad \dfrac{[t : D] \qquad [g(t) : P(t)]}{\varepsilon y.(g(y), t) : \exists y^D.P(y)}\exists\text{-}intr}{\mathtt{INST}(c, \sigma g.\sigma t.\varepsilon y.(g(y), t)) : \exists y^D.P(y)}\exists\text{-}elim \quad \rightsquigarrow_\eta \quad c : \exists x^D.P(x)$$

≐-η-reduction

$$\frac{e :\doteq_D (a, b) \qquad \dfrac{[a =_t b : D]}{t(a, b) :\doteq_D (a, b)} \doteq\text{-}intr}{\mathtt{REWR}(e, \theta t.t(a, b)) :\doteq_D (a, b)} \doteq\text{-}elim \qquad \rightsquigarrow_\eta \qquad e :\doteq_D (a, b)$$

ζ-Type reductions

For the connectives that make use of 'Skolem'-type procedures of opening local branches with new assumptions, locally introducing new names and making them 'disappear' (or lose their identity via an abstraction) just before coming out of the local context or scope, there is another way of transforming proofs, which goes hand-in-hand with the properties of 'value-range' terms resulting from abstractions.

In natural deduction terminology, these proof transformations are called 'permutative' reductions. Here we give them a direction, so they are no longer permutative.

∨-ζ-reduction

$$\dfrac{\dfrac{p : A_1 \vee A_2 \quad \begin{matrix}[s_1 : A_1]\\ d(s_1) : C\end{matrix} \quad \begin{matrix}[s_2 : A_2]\\ e(s_2) : C\end{matrix}}{\mathtt{CASE}(p, \acute{v}s_1.d(s_1), \acute{v}s_2.e(s_2)) : C}}{w(\mathtt{CASE}(p, \acute{v}s_1.d(s_1), \acute{v}s_2.e(s_2))) : W}\,\mathsf{r} \quad \rightsquigarrow_\zeta$$

$$\dfrac{p : A_1 \vee A_2 \quad \dfrac{\begin{matrix}[s_1 : A_1]\\ d(s_1) : C\end{matrix}}{w(d(s_1)) : W}\,\mathsf{r} \quad \dfrac{\begin{matrix}[s_2 : A_2]\\ e(s_2) : C\end{matrix}}{w(e(s_2)) : W}\,\mathsf{r}}{\mathtt{CASE}(p, \acute{v}s_1.w(d(s_1)), \acute{v}s_2.w(e(s_2))) : W}$$

∃-ζ-reduction

$$\dfrac{\dfrac{e : \exists x^D.P(x) \quad \begin{matrix}[t : D, g(t) : P(t)]\\ d(g,t) : C\end{matrix}}{\mathtt{INST}(e, \sigma g.\sigma t.d(g,t)) : C}}{w(\mathtt{INST}(e, \sigma g.\sigma t.d(g,t))) : W}\,\mathsf{r} \quad \rightsquigarrow_\zeta$$

$$\dfrac{e : \exists x^D.P(x) \quad \dfrac{\begin{matrix}[t : D, g(t) : P(t)]\\ d(g,t) : C\end{matrix}}{w(d(g,t)) : W}\,\mathsf{r}}{\mathtt{INST}(e, \sigma g.\sigma t.w(d(g,t))) : W}$$

≐-ζ-reduction

$$\dfrac{\dfrac{e : \doteq_D (a,b) \quad \begin{matrix}[a =_t b : D]\\ d(t) : C\end{matrix}}{\mathtt{REWR}(e, \theta t.d(t)) : C}}{w(\mathtt{REWR}(e, \theta t.d(t))) : W}\,\mathsf{r} \quad \rightsquigarrow_\zeta \quad \dfrac{e : \doteq_D (a,b) \quad \dfrac{\begin{matrix}[a =_t b : D]\\ d(t) : C\end{matrix}}{w(d(t)) : W}\,\mathsf{r}}{\mathtt{REWR}(e, \theta t.w(d(t))) : W}$$

In all cases the inference rule 'r' may not involve neither the discharge of assumptions nor the binding of a variable.

ζ-Equality

Now, if the functional calculus on the labels is to match the logical calculus on the formulas, we must have the following ζ-equality (read 'zeta'-equality) between terms:

$$w(\mathtt{CASE}(p, \acute{v}s_1.d(s_1), \acute{v}s_2.e(s_2)), u) =_\zeta \mathtt{CASE}(p, \acute{v}s_1.w(d(s_1), u), \acute{v}s_2.w(e(s_2), u))^{5}$$

[5] When defining 'Linearised sum', Girard *et al.* [Girard *et al.*, 1989] give the following equation as the term-equality counterpart to the permutative reduction:
"Finally, the commuting conversions are of the form

$$\mathsf{E}(\delta\ x.u\ y.v\ t) \rightsquigarrow \delta\ x.(\mathsf{E}u)\ y.(\mathsf{E}v)\ t$$

for disjunction,

$$w(\texttt{INST}(e, \sigma g.\sigma t.d(g,t)), u) =_\zeta \texttt{INST}(e, \sigma g.\sigma t.w(d(g,t), u))$$

for the existential quantifier, and

$$w(\texttt{REWR}(e, \theta t.d(t))) =_\zeta \texttt{REWR}(e, \theta t.w(d(t)))$$

for the propositional equality symbol.

Note that in the cases of '$\vee$', '$\exists$' and '$\doteq$' the operator 'w' could be 'pushed inside' the value-range abstraction terms.

In terms of the proof theory, these reductions are stating that the newly opened branches must be independent from the main branch. And, indeed, notice that in the proof-trees above, the step coming after the *elimination* of the connective concerned ($\vee$-, $\exists$- and $\doteq$- *elimination*) is taken to be as general as possible, provided that it does not affect the dependencies on the main branch (i.e. '$p : A_1 \vee A_2$', '$e : \exists x^D.P(x)$', '$e : \doteq_D (a,b)$', respectively). (E.g. any deduction step involving discharge of assumptions may disturb the dependencies.) Those reductions will then uncover β-type redundancies which may be hidden by an $\vee$-, $\exists$- *elimination* rule. Perhaps for this reason, in the literature it is common to restrict that particular step to a deduction to an *elimination* rule where the formula 'C' is to be its major premise.[6]

---

where E is an elimination." [Girard *et al.*, 1989, page 103]

Note the restriction on the step corresponding to the operator 'E' (which corresponds to our 'w'): it has to be an elimination.

In our ζ-equality the operator 'w' does not have to be an eliminatory operator, but it only needs to be such that it preserves the dependencies of the term coming from the main branch, i.e. the step must preserve the free variables on which our 'p' depends. In other words, w cannot be an abstraction over free variables of p.

Our generalised ζ-equality also finds parallels in the recent literature on equational counterparts to commutative diagrams of category theory. For example, in the definition of *binary sums* given by A. Poigné [Poigné, 1992], the counterpart of our ζ-equality for disjunction appears as:

$$h \circ case(f,g) = case(h \circ f, h \circ g)$$

where '$\circ$' is the basic operation of composition. Note that the function 'h' can be pushed inside the '*case*'-term, similarly to our ζ-equality where the 'w' can be pushed inside the υ-abstraction terms of our CASE-expression.

[6]When commenting on the requirements of permutative reductions, Prawitz remarks: "It has been remarked by Martin-Löf that it is only necessary to require in the $\vee$E- and $\exists$E-reductions that the lowest occurrence of C is the major premiss of an elimination. A reduction of this kind can then always be carried out and we can sharpen the requirements as to the normal form accordingly." [Prawitz, 1971, pages 253ff]

And, indeed, for his proof of the Strong Validity Lemma (p. 295) Prawitz needs the condition on the permutative reductions that the step after the $\vee$-*elimination* (resp. $\exists$-*elimination*) be also an *elimination* inference.

No restriction to an *elimination* step is mentioned in [Martin-Löf, 1975]. Rather, it is

The restriction to the case when the step is an *elimination* rule seems to be connected with the idea that the conversions are brought in to help recover the so-called *subformula property*.[7] We would prefer to see the rôle of those rules of proof transformation as that of guaranteeing a 'pact of non-interference' between the main branch and those new branches created by the elimination rules of 'Skolem-type' connectives ($\vee$, $\exists$, $\doteq$).

Thus, in the more general case, it seems as though the restriction (to the case where the formula C is a major premise of an *elimination* inference) is unnecessary. And this is because we can have the following conversion using an *introduction* inference instead:

$$\cfrac{\cfrac{p : A_1 \vee A_2 \qquad \begin{matrix}[s_1 : A_1]\\ d(s_1) : C\end{matrix} \qquad \begin{matrix}[s_2 : A_2]\\ e(s_2) : C\end{matrix}}{\texttt{CASE}(p, \upsilon s_1.d(s_1), \upsilon s_2.e(s_2)) : C}}{\texttt{inl}(\texttt{CASE}(p, \upsilon s_1.d(s_1), \upsilon s_2.e(s_2))) : C \vee U}(*) \qquad \rightsquigarrow_\zeta$$

$$\cfrac{p : A_1 \vee A_2 \qquad \cfrac{\begin{matrix}[s_1 : A_1]\\ d(s_1) : C\end{matrix}}{\texttt{inl}(d(s_1)) : C \vee U} \qquad \cfrac{\begin{matrix}[s_2 : A_2]\\ e(s_2) : C\end{matrix}}{\texttt{inl}(e(s_2)) : C \vee U}}{\texttt{CASE}(p, \upsilon s_1.\texttt{inl}(d(s_1)), \upsilon s_2.\texttt{inl}(e(s_2))) : C \vee U}$$

One can readily notice that the $\vee$-*introduction* step marked '(*)' does not affect the dependencies (i.e. does not involve any assumption discharge), so the constructor '`inl`' can be pushed inside the υ-abstraction terms. The same holds if, instead of $\vee$-*introduction*, one performs an $\wedge$-*introduction* as in:

$$\cfrac{\cfrac{p : A_1 \vee A_2 \qquad \begin{matrix}[s_1 : A_1]\\ d(s_1) : C\end{matrix} \qquad \begin{matrix}[s_2 : A_2]\\ e(s_2) : C\end{matrix}}{\texttt{CASE}(p, \upsilon s_1.d(s_1), \upsilon s_2.e(s_2)) : C} \qquad u : U}{\langle \texttt{CASE}(p, \upsilon s_1.d(s_1), \upsilon s_2.e(s_2)), u\rangle : C \wedge U} \qquad \rightsquigarrow_\zeta$$

$$\cfrac{p : A_1 \vee A_2 \qquad \cfrac{\begin{matrix}[s_1 : A_1]\\ d(s_1) : C\end{matrix} \qquad u : U}{\langle d(s_1), u\rangle : C \wedge U} \qquad \cfrac{\begin{matrix}[s_2 : A_2]\\ e(s_2) : C\end{matrix} \qquad u : U}{\langle e(s_2), u\rangle : C \wedge U}}{\texttt{CASE}(p, \upsilon s_1.\langle d(s_1), u\rangle, \upsilon s_2.\langle e(s_2), u\rangle) : C \wedge U}$$

---

required that the dependencies be preserved:

"(...) the *permutative* rules for $\vee$ and $\exists$, (...) provided the inference from C to D neither binds any free variable nor discharges any assumption in the derivation of $A \vee B$ and $(\exists x)B[x]$, respectively." [Martin-Löf, 1975, pages 100f]

Cf. also other standard texts in the literature where the restriction is unnecessarily imposed: Troelstra and van Dalen's [Troelstra and van Dalen, 1988, pages 534ff] and Girard *et al.*'s [Girard *et al.*, 1989] definitions of *permutative conversions* have the requirement that the step following the $\vee$- ($\exists$-)*elimination* be an 'E-rule' (*Elimination* rule).

[7] Cf. [Girard *et al.*, 1989; Troelstra and van Dalen, 1988].

and, clearly:

$$\langle \texttt{CASE}(p, \upsilon s_1.d(s_1), \upsilon s_2.e(s_2)), u\rangle =_\zeta \qquad \texttt{CASE}(p, \upsilon s_1.\langle d(s_1), u\rangle, \upsilon s_2.\langle e(s_2), u\rangle)$$

given that the pairing operation can be pushed inside the υ-abstraction terms without disturbing the dependencies. One can readily see that the $\wedge$-*introduction* is harmless with respect to the dependencies. (Note that the same observation applies to ζ-reduction of $\exists$, and, as we shall see later on, to the reduction of $\doteq$.)

Should we Send Him to Prison? Paradoxes of Aggregation and Belief Merging

GABRIELLA PIGOZZI

1 Introduction

Social choice theory (Arrow 1963, Arrow *et al.* 2002, Sen 1979) studies the aggregation of individual preferences in order to select the collected preferred alternatives. Way back in 1770, the Marquis de Condorcet proposed a method for the aggregation of preferences which led to the first aggregation problem: the voting paradox. Given a set of individual preferences, we compare each of the alternatives in pairs. For each pair we determine the winner by majority voting, and the final social ordering is obtained by a combination of all partial results. The paradoxical result is that the pairwise majority rule can lead to intransitivity, in contradiction to the preference postulates for linear order.

More recently, social choice theorists have become interested in other aggregation problems like the Ostrogorski paradox (Kelly 1989) and the doctrinal paradox (or discursive dilemma). [1] It has been shown that the discursive dilemma is a generalization of the paradox of voting (List and Pettit 2002). Bezembinder and van Acker (1985) have also studied the relation between the Ostrogorski paradox and the paradox of voting and concluded that in every instance of the first there is an underlying paradox of voting. However, unlike the voting paradox, these two dilemmas arise when the members of a group have to make a judgment (in the form of yes or no) on specific propositions rather than express preferences among candidates. As in the Condorcet paradox, the collective outcome is (in some respect) inconsistent, despite of the individual inputs being consistent. The result is inconsistent with regard to the adopted voting rule in the Ostrogorski paradox, and it is logically inconsistent in the case of the doctrinal paradox. In this paper I will argue that, despite of the differences, these two

[1] For a comprehensive bibliography of the rapidly growing body of literature on the doctrinal paradox, see List 2005.

problems share a similar structure. The conclusion will be twofold: on the one hand, the similarities among the paradoxes support the claim that these problems should be tackled using the same aggregation procedure; on the other hand, applying the same aggregation procedure to these paradoxes will help clarifying the strength and weakness of the aggregation method itself.

In a previous work (Pigozzi 2004) I have shown that the doctrinal paradox dissolves when we apply an operator defined in artificial intelligence in order to merge knowledge bases. In this paper I will apply the same merging operator to the Ostrogorski paradox, and show that this paradoxical outcome is also avoided. The discussion will be informal and conceptual rather than technical. The formal definition of the merging operator can be found in Konieczny and Pino-Pérez (1998, 2002).

In Section 2 I will introduce the doctrinal paradox and summarize the possibility result achieved when we apply one specific merging operator to the doctrinal paradox (Pigozzi 2004). In Section 3 I will explain in which sense the Ostrogorski paradox shares a similar structure with the discursive dilemma, and how we can avoid counterintuitive results by applying a merging operator to this paradox as well. In Section 4 I will outline an alternative approach to deal with aggregation problems like the two investigated in the present paper.

2 The doctrinal paradox

The doctrinal paradox can emerge when the members of a group have to make a judgment (in the form of yes or no) on several logically interconnected propositions, and the individually logically consistent judgments need to be combined into a collective decision. For example, consider a set of premises and a conclusion in which the latter is logically equivalent to the former. When majority voting is applied to some propositions (the premises) it may give a different outcome than majority voting applied to another set of propositions (the conclusion). This phenomenon did first draw the attention of researchers in law (Kornhauser 1992, Kornhauser and Sager 1986, 1993), who illustrated the paradox with the following example. Suppose that three judges have to decide whether a defendant is liable for breaching a contract. According to the legal doctrine, a person is liable of a certain action X (this is the proposition R) if and only if the defendant performed the action X (P) and had contractual obligation not to do X (Q). Now assume that each judge makes a consistent judgment over the propositions P, Q and R, as the following table shows:

	P	Q	R	$(P \wedge Q) \leftrightarrow R$
Judge 1	Yes	Yes	Yes	Yes
Judge 2	Yes	No	No	Yes
Judge 3	No	Yes	No	Yes
Majority	Yes	Yes	No	Yes

Each individual consistently assigns a truth value to each proposition P, Q and R (saying yes to R if and only if both P and Q are believed to be true). However, if simple majority voting is applied only to the premises (P, Q) of the argument (this procedure is called *premise-based procedure*), the result is that there is a majority that believes both P and Q to be true (and, therefore, because of $(P \wedge Q) \leftrightarrow R$, that majority is held to believe that R is also true). At the same time, if majority voting is applied only to the conclusion R (*conclusion-based procedure*), the majority of the group believes that R is false, which conflicts with the aggregation of the premises. The paradox lies precisely in the fact that the two procedures may lead to contradictory results (one accepting and the other rejecting R), depending on whether the majority is taken on the individual judgments of P and Q, or whether the majority is calculated on the individual votes of R. The question is then whether a collective outcome exists in these cases, and if it does, what it is like.[2]

A possibility result (Pigozzi 2004) is obtained when a merging operator, originally introduced in theoretical computer science to merge several finite sets of information, is imported and applied to the problem of judgment aggregation. The justification for this move is that the theory of information merging and group decision-making share a similar difficulty, viz. the definition of operators that produce collective knowledge from individual knowledge bases, and operators that produce a collective decision from individual decisions.

The discursive dilemma and the Ostrogorski paradox disappear as soon as we recognize that the propositions voted by the group via majority do not necessarily define a unique, consistent and collective outcome. It is indeed necessary to exclude the inconsistent collective judgments from the set of the possible results, and ties on several consistent and collective outcomes

[2]The question as to whether a group (once certain conditions are satisfied) is likely to reach the correct decision, is justified by the Condorcet Jury Theorem. The theorem applies when any two pairs of alternatives are given, each member of a group has a probability greater than 0.5 to vote for the right choice and the group votes through a majority rule. Under specific additional (independence) assumptions, the group's probability of choosing the right alternative increases with the size of the group itself and approaches 1 in the limit. List and Goodin (2001) have generalized the Condorcet Jury Theorem to multiple propositions. See also Bovens and Hartmann (2003), Section 3.6.

must be allowed to occur. An outcome in the new aggregation procedure is a consistent assignment to each proposition under judgment (i.e. a consistent truth value assignment to a conclusion *and* to the premises supporting that conclusion in doctrinal paradox terms). We do not anymore have two separate aggregations on the premises and on the conclusion, but an operator that takes the *whole* sequence of truth assignments on premises and conclusion for each individual and assigns a consistent collective truth value to those propositions.

I now explain how a belief merging operator works, avoiding any formal definitions or technicalities, which can be found in the literature on belief merging, as well as the set of axioms that belief merging operators satisfy (Konieczny, and Pino-Pérez 1998, 2002). A knowledge (or belief) base K_i is a finite set of propositional formulas representing the explicit beliefs or information of the base (or agent) i. A merging operator combines various and possibly conflicting knowledge bases. The merged base is a set that *consistently* integrates parts of the knowledge from all the initial bases.

In order to define a merging operator we need to specify how the belief bases are combined into a collective one. To stay closer to the original situations the paradoxes deal with, we consider only equally reliable belief bases. When the bases to merge are mutually consistent, the result can be easily constructed: it is the union of the knowledge bases. Things turn out to be more interesting when the belief bases are in conflict with each other. The outcome of a merging process is a base that integrates parts of the knowledge from (possibly all) the initial bases. In addition, the final knowledge base must satisfy some integrity constraints IC. The IC express additional conditions on which all the individuals agree (in the three-member court example, the IC would be $(P \wedge Q) \leftrightarrow R$).

A belief set $E = \{K_1, K_2, \ldots, K_n\}$ is a finite collection of belief bases K_i each representing an individual judgment set. The aggregation rule $\mathcal{F}$ is a function that assigns a belief base (which corresponds to a consistent collective judgment set in judgment aggregation terms) to IC and to E.

Several types of merging operators have already been proposed in the literature. The one used here is intended to reflect the view of the majority, by maximizing the level of total agreement among the agents. This is also the operator that behaves most similarly to the propositionwise majority voting adopted in the discursive dilemma and in the Ostrogorski paradox. It is a model-based operator ('model' here is to be intended in the usual classical truth functional way), where the models of the merged bases are models of the set of IC, which are preferred according to some distance measure.[3]

[3]Model-based merging operators have been discussed in Konieczny and Pino-Pérez

The merging operator so defined aims at minimizing the total distance between all the individual bases and each possible outcome. The distance measure Konieczny and Pino-Pérez use is the widely known Dalal's distance (Dalal 1988a, 1988b) between two interpretations. According to this measure, the distance between two interpretations is equal to the number of the propositional variables in which two interpretations differ. In this way, an ordering over the interpretations is induced. For example, $w = \{1, 0, 0\}$ and $w' = \{0, 1, 0\}$ are two possible interpretations of $((P \wedge Q) \leftrightarrow R)$, and the distance between w and w' is 2 (as they assign a different truth value to P and to Q, while they agree on R being false).

We are now ready to apply the belief merging (intuitively) introduced to the three-member court problem. The set E is $\{K_1, K_2, K_3\}$, where K_i is the belief base of each judge, and $IC = \{((P \wedge Q) \leftrightarrow R)\}$. The merging procedure takes each individual judgment as a belief base. Each individual makes a judgment over the atomic propositions P, Q and R that satisfies the integrity constraint. This is because the doctrinal paradox assumes each individual to be rational, though this assumption could be relaxed in the merging framework. We can therefore write:

$K_1 = \{P, Q, R\}$
$K_2 = \{P, \neg Q, \neg R\}$
$K_3 = \{\neg P, Q, \neg R\}$

The interpretations for each belief base are the following:

$Mod(K_1) = \{(1, 1, 1)\}$
$Mod(K_2) = \{(1, 0, 0)\}$
$Mod(K_3) = \{(0, 1, 0)\}$

The table below shows the result of the IC majority fusion operator on $E = \{K_1, K_2, K_3\}$. The first three columns are all the possible interpretations for the propositional variables P, Q and R. The rows with a shaded background correspond to the interpretations excluded by IC. The numbers in the columns of K_1, K_2 and K_3 are the Dalal's distances of each K_i from the respective interpretation. Finally, in the last column is $d_{\Sigma}(I, E)$, the numbers expressing the distance between the corresponding interpretation (I) on the three propositions and each belief base in E.

(1998), Liberatore and Schaerf (2000), Lin and Mendelzon (1996, 1999), Revesz (1997).

P	Q	R	K_1	K_2	K_3	$d_\Sigma(I,E)$
1	1	1	0	2	2	4
1	1	0	1	1	1	3
1	0	1	1	3	1	5
1	0	0	2	2	0	4
0	1	1	1	1	3	5
0	1	0	2	0	2	4
0	0	1	2	2	2	6
0	0	0	3	1	1	5

Because $\mathcal{F}$ is an IC merging operator, the collective outcomes are chosen among the interpretations that are *not* excluded by IC. Only $(1,1,1)$, $(1,0,0)$, $(0,1,0)$ and $(0,0,0)$ are the available candidates for the collective judgments. Moreover, $\mathcal{F}$ is a majority operator, and so the interpretations associated with the total minimum distance value are selected. Thus, no paradox arises using this merging operator. $Mod(\mathcal{F}) = \{(1,1,1),(1,0,0),(0,1,0)\}$, which is tantamount to saying that the collective outcome is a tie: $\{K_1 \vee K_2 \vee K_3\}$. We do not have enough information to select a unique collective judgment. Hence, since the judges cannot say that the defendant is liable beyond any reasonable doubt, they should release him.

We now turn to the Ostrogorski paradox and show how this shares a similar structure with the doctrinal paradox, thus justifying the application of belief merging operator also to this problem.

3 One more paradox resolved

In the Ostrogorski paradox there are two parties (right (R) and left (L)) and three issues (for example, economic, environmental, international). Each party has a position on each topic and each individual casts a vote (R or L) on it depending on whether the Right Party or the Left Party represents her opinion on that issue. Suppose there are five voters and that they vote as shown in the table below:

	Issue 1	Issue 2	Issue 3	Party supported
Voter 1	R	L	L	L
Voter 2	L	R	L	L
Voter 3	L	L	R	L
Voter 4	R	R	R	R
Voter 5	R	R	R	R
Majority	R	R	R	L

If each voter votes for the party with which she agrees on a majority of issues, the Left Party wins. However, the Right Party represents the views of the majority of the voters on every issue.

The argumentative structure of the Ostrogorski paradox is similar to the premises-conclusion structure of the discursive dilemma. Here an individual votes for the Right Party (Left Party) if and only if she agrees with the Right Party (Left Party) on at least two of the three issues. As we have seen, in the doctrinal paradox a judge would send the defendant to prison if and only if she believes that the defendant did a certain action X and had contractual obligation not to perform X. Not sending a defendant to prison (R false) could be supported by different reasons (it is enough that one of the two propositions, P or Q, is believed to be false). The consequence of this is that, when we aggregate only on the conclusions, we may not know which reasons supported the final decision. Similarly, in the Ostrogorski paradox the reasons to vote for the Right Party (Left Party) could be one among several judgment sets on the three issues (R,R,R), (R,R,L), (R,L,R), etc. (respectively (L,L,L), (L,L,R), (L,R,L), etc. for the Left Party). Again, when we apply simple majority voting on the party we do not know precisely on which issues that party gained the majority of the votes.

So much about the similarities between the two paradoxes. The only dissimilarity between them is that in the Ostrogorski paradox the issues and the parties are not logically connected (as it happens between premises and conclusion in the doctrinal paradox). It is rather a majority rule on the issues that determines which party an individual should vote for: the individual votes for the party she is in agreement with on the majority of the issues. Nevertheless, if we concur in representing a judgment set as a sequence of Yes and No, 0 and 1, or L and R, we can apply the majority merging operator to the Ostrogorski paradox as well. Again, the IC selects the permissible social outcomes, i.e. those outcomes where a party is elected only if it collected the majority on the issues.

It is worth noticing that, when we characterize a judgment set on premises and conclusions or on issues and parties as a sequence of 0 and 1, we imply that the propositions in the judgment set are treated as propositions of the same sort. For this reason, the difference between those propositions that are 'premises' ('issues') and those that are 'conclusions' ('party') vanishes. But the diversity between kinds of propositions was already lost in the propositionwise majority aggregation in the original formulation of the two paradoxes. For example, when we apply the majority rule to each premise separately, we forget how these propositions were related. If we look at the doctrinal paradox example, we see that there was a majority for P (judges 1 and 2) and *another* majority for Q (judges 1 and 3). When we count how much support P and Q got, we combine these two different majorities into a majority for $(P \wedge Q)$, when in fact two different majorities supported the atomic propositions. The logical connectives are reintroduced only between the propositions that received the highest degree of support. It can be discussed whether aggregating premises and conclusions, issues and parties, taking them to be propositions of the same sort is correct. In Section 4 I will suggest that, if we really intend to grant that these propositions are of a different type, and we therefore want to maintain the argumentative structure of the doctrinal paradox and the Ostrogorski paradox, it would be worth to refer to the literature on floating conclusions. A floating conclusion is a conclusion that is supported by different (and potentially conflicting) arguments. It is not at all clear whether a floating conclusion should always be accepted or not, or if the answer to this question is rather context-dependent.

I will now show how the Ostrogorski paradox is avoided when we apply the majority merging operator introduced in the previous section. Let us call the Right Party 1 and the Left Party 0. Each voter casts her vote on *Issue* 1, *Issue* 2, *Issue* 3 and *Party* that satisfies the integrity constraint (as assumed in the Ostrogorski problem, each voter is rational). The set of IC is the set of allowed judgment sets:

$$IC = \{(1,1,1,1),(0,0,0,0),(1,1,0,1),(1,0,1,1),(0,1,1,1),(0,0,1,0),\\(0,1,0,0),(1,0,0,0)\}$$

The five voters are:
$K_1 = \{R, L, L, L\}$
$K_2 = \{L, R, L, L\}$
$K_3 = \{L, L, R, L\}$
$K_4 = \{R, R, R, R\}$
$K_5 = \{R, R, R, R\}$

The interpretations for each belief base are the following:
$Mod(K_1) = \{(1,0,0,0)\}$

$Mod(K_2) = \{(0,1,0,0)\}$
$Mod(K_3) = \{(0,0,1,0)\}$
$Mod(K_4) = \{(1,1,1,1)\}$
$Mod(K_5) = \{(1,1,1,1)\}$

When we apply the majority belief merging we obtain the result below:

	K_1	K_2	K_3	K_4	K_5	$d_\Sigma(I,E)$
1111	3	3	3	0	0	**9**
1101	2	2	4	1	1	10
1011	2	4	2	1	1	10
1000	0	2	2	3	3	10
0111	4	2	2	1	1	10
0100	2	0	2	3	3	10
0010	2	2	0	3	3	10
0000	1	1	1	4	4	11

Whereas we obtained a tie in the doctrinal paradox, we now find a single optimal solution in the Ostrogorski case. The closest model to each member of the group turns out to be $\{1,1,1,1\}$, which also coincides to the judgment set of the last two voters. As in the doctrinal paradox, the inconsistent outcome resulting from majority voting on the single issues and parties $\{1,1,1,0\}$ is excluded by IC.

4 Floating conclusions

As maintained in Section 3, both belief merging and propositionwise majority voting treat the propositions under judgment in an even handed manner. However, the two paradoxes here considered have an argumentative structure such that the judgments on some propositions are constrained by the judgments on other propositions. One advantage of the belief merging framework is that it provides an interpretation to each proposition in the judgment set. In the case of the discursive dilemma, the belief merging outcome provides a collective conclusion together with the reasons for that decision. Similarly, in the case of the Ostrogorski paradox, we are given not only the party that received the majority of the votes, but also the set of issues that supported this decision.

The problem of finding a unique reason for a decision likewise arises in other areas, for example in the theory of defeasible inheritance nets (Horty *et al.* 1990, Horty 1994). There, Makinson and Schlechta (1991) first studied the *floating conclusion* phenomenon, which became then relevant also in argument systems. A conclusion is floating when it can be supported by two different and potentially incompatible arguments. Floating conclusions can be traced in the discursive dilemma and in the Ostrogorski paradox. In the first paradox we had that the conclusion (R) not to put a person in jail could be because of P is believed to be false but Q is true, or vice versa, or both P and Q are believed to be false. Similarly, in the Ostrogorski paradox a party can win thanks to different combinations of issues.

Intuitively, the theory of defeasible inheritance nets deals with the problem that a knowledge base can be associated with multiple extensions (an extension represents some total set of arguments). Traditionally, two solutions to this problem have been proposed. The *credulous* approach consists in accepting the set of conclusions endorsed by an arbitrary one of the argument extensions. The *skeptical* approach considers the intersection of the extensions. Makinson and Schlechta observed that there are two ways to perform a skeptical solution. We can either intersect the arguments, or we can intersect the conclusions. These two skeptical approaches can yield different results, since a statement may be supported in each extension, but only by different and possibly conflicting arguments (floating conclusion). It is not at all clear whether we should always endorse a floating conclusion or not.

An example that shows that the two skeptical approaches lead to two different conclusions is the Horty's inheritance example. Suppose that John wants to buy a very expensive yacht but he does not have all the money. His parents are very sick and about to die within a month. He has a brother and sister, both reliable. His brother tells him that their mother will leave half a million dollars to him but their father will give that amount to John. On the other hand, his sister tells him that their mother will leave half a million dollars to John and their father will leave that amount to her. Horty concludes that if he were to intersect the arguments, he would not be justified in concluding that he is about to inherit half a million dollars. But if he were to follow the second skeptical strategy (intersecting the conclusions), he would be justified in drawing the conclusion that he is about to inherit all that money, and so he could place the deposit for the yacht. Indeed, both his brother's and his sister's arguments, though contradicting each other, support the conclusion that John will inherit half a million dollars either from his father or from his mother.

When we count (as we do in the aggregation paradoxes) how many people in a group voted for a certain conclusion, we ignore the reasons supporting that decision. The literature on floating conclusions shows that splitting a set of arguments into reasons and conclusions, and 'aggregating' them separately, can lead to opposite consequences. I believe that the research done in the defeasible inheritance nets can throw new light on the area of judgment aggregation. I plan to investigate this relation in a future paper.

5 Conclusions

There are several aggregation dilemmas in social choice theory, and this paper aimed at showing that two of them (the doctrinal paradox and the Ostrogorski paradox) share a similar structure. In both these paradoxes, members of a group have to make a judgment on several propositions, and these judgments are constrained by some rule. In the doctrinal paradox it is a logical rule; the propositions are logically connected and the individual as well as the collective judgments are assumed to be logically consistent. The truth values assigned to the premises entails the conclusion being true or false. In the Ostrogorski paradox the individual and collective choices on the issues determine which party that person and the group support. In both paradoxes there are judgments on some propositions (premises in the case of the doctrinal paradox, issues in the Ostrogorski paradox) that dictate the individual and collective judgments on other propositions (conclusions and parties, respectively).

The difficulty with the aggregation problems is that the set of propositions on which most group members agree is not guaranteed to be a candidate for the collective decision because the set can fail to satisfy consistency even though each individual consistently expressed her judgments. One way to avoid this is to impose some integrity constraints on the collective outcome, not only on the individual judgments.

In a previous work I have claimed that judgment aggregation and belief merging are related in an interesting way and that more exchange between these two areas of research is definitely fruitful. Following the research done by Konieczny and Pino-Pérez, a majoritarian IC merging operator for belief bases has been applied to the doctrinal paradox, and this proved to be a solution for the dilemma. The value of the fusion procedure rests upon the exclusion of inconsistent sets of judgments from the set of the candidates apt to become collective judgments, and in the definition of a preference order on the remaining candidates.

Relying on the similarities between the doctrinal paradox and the Ostrogorski paradox, the same merging operator has been applied to the latter dilemma in the present paper. The result is that also this paradox is avoided and a unique best social outcome is selected.

In the last section I have sketched an alternative approach to the judgment aggregation problems. This makes use of some of the discussions on floating conclusions in the area of defeasible inheritance nets. The discursive dilemma and the Ostrogorski paradox appear because the aggregations on the premises (issues) and on the conclusion (party) go in two different directions. On the one hand, the premises that receive the highest degree of support cannot consistently provide reasons for the most popular conclusion (party). On the other hand the social selected conclusion cannot be inferred by the premises that received the majority of the votes. The need for a unique set of reasons for a certain decision is shared by the multiple extensions problem in the defeasible inheritance nets. There, Makinson and Schlechta have baptized 'floating conclusions' the phenomenon that a conclusion is supported by some arguments contained in every extension but there exists no argument in all the extensions that supports that conclusion. I believe that the interplay between judgment aggregation and defeasible inheritance nets is worth being investigated. I leave this to future research plans.

BIBLIOGRAPHY

[Arrow, 1963] K. J. Arrow. *Social Choice and Individual Values*, Wiley, New York, second edition, 1963.

[Arrow *et al.*, 2002] K. J. Arrow, A.K. Sen, and K. Suzumura, eds. *Handbook of Social Choice and Welfare*, Vol.1, Elsevier, 2002.

[Bezembinder and van Acker, 1985] T. Bezembinder and P. Van Acker. "The Ostrogorski paradox and its relation to nontransitive choice", *Journal of Mathematical Sociology*, 11, 1985.

[Bovens and Hartmann, 2003] L. Bovens and S. Hartmann. *Bayesian Epistemology*, Oxford: Oxford University Press, 2003.

[Dalal, 1988a] M. Dalal. "Updates in propositional databases", technical report, Rutgers University, 1988.

[Dalal, 1988b] M. Dalal. "Investigations into a theory of knowledge base revision: Preliminary report", in *Proceedings of the 7th National Conference of the American Association for Artificial Intelligence*, Saint Paul, Minn., 475–479, 1988.

[Horty, 1994] J. F. Horty. "Some direct theories of nonmonotonic inheritance". In *Handbook of Logic in Artificial Intelligence and Logic Programming, Vol. 3: Nonmonotonic Reasoning and Uncertain Reasoning*, D. Gabbay, C. Hogger, and J. Robinson (eds.), Oxford University Press, 111–187, 1994.

[Horty, 2002] J. F. Horty. "Skepticism and floating conclusions", *Artificial Intelligence*, 135, 55–72, 2002.

[Horty *et al.*, 1990] J. F. Horty, R. H. Thomason and D. S. Touretzky. "A skeptical theory of inheritance in nonmonotonic semantic networks", *Artificial Intelligence*, 42, 311–348, 1990.

[Kelly, 1989] J. S. Kelly. "The Ostrogorski's paradox", *Social Choice and Welfare*, 6, 71–76, 1989.

[Konieczny, 1999] S. Konieczny. *Sur la Logique du Changement: Révision et Fusion de Bases de Connaissance*, Ph.D. dissertation, University of Lille I, France, 1999.

[Konieczny and Pino-Pérez, 1998] S. Konieczny and R. Pino-Pérez. "On the logic of merging", in *Proceedings of KR'98*, Morgan Kaufmann, 488–498, 1998.

[Konieczny and Pino-Pérez, 2002] Konieczny, S., and R. Pino-Pérez (2002) "Merging information under constraints: a logical framework", *Journal of Logic and Computation*, 12(5), 773–808.

[Kornhauser, 1992] L. A. Kornhauser. "Modelling collegial courts. II. Legal doctrine", *Journal of Law, Economics and Organization*, 8, 441–470, 1992.

[Kornhauser and Sager, 1986] L. A. Kornhauser and L.G. Sager. "Unpacking the court", *Yale Law Journal* 82, 1986.

[Kornhauser and Sager, 1993] L. A. Kornhauser and L. G. Sager. "The one and the many: Adjudication in collegial courts", *California Law Review* 81(1), 1–59, 1993.

[Liberatore and Schaerf, 1998] P. Liberatore and M. Schaerf. "Arbitration (or how to merge knowledge bases", in *ProIEEE Transactions on Knowledge and Data Engineering*, 10, 1998.

[Lin and Mendelzon, 1996] J. Lin and A. Mendelzon. "Merging databases under constraints", *International Journal of Cooperative Information Systems*, 7, 55–76, 1996.

[Lin and Mendelzon, 1999] J. Lin and A. Mendelzon. "Knowledge base merging by majority", in *Dynamic Worlds: From the Frame Problem to Knowledge Management*, R. Pareschi and B. Fronhoefer, eds., Kluwer, 1999.

[List, 2005] C. List. Judgment Aggregationa Bibliography on the Discursive Dilemma, the Doctrinal Paradox and Decisions on Multiple Propositions, 2005. `http://personal.lse.ac.uk/LIST/doctrinalparadox.htm`

[List and Goodin, 2001] C. List and and R. E. Goodin. "Epistemic democracy: Generalizing the Condorcet Jury Theorem", *Journal of Political Philosophy*, 9(3), 277–306, 2001.

[List and Pettit, 2002] C. List and P. Pettit. "Aggregating sets of judgments: An impossibility result", *Economics and Philosophy*, 18, 89–110, 2002.

[Makinson and Schlechta, 1991] D. Makinson and K. Schlechta. "Floating conclusions and zombie paths: two deep difficulties in the 'directly skeptical' approach to defeasible inheritance nets", *Artificial Intelligence*, 48, 199–209, 1991.

[Pigozzi, 2004] G. Pigozzi. "Collective decision-making without paradoxes: A fusion approach". Working paper. King's College London, 2004. `http://www.dcs.kcl.ac.uk/staff/pigozzi/publications.html`

[Revesz, 1997] P. Revesz. "On the semantics of arbitration", *International Journal of Algebra and Computation*, 7, 133–160, 1977.

[Sen, 1979] A. Sen. *Collective Choice and Social Welfare*, Elsevier, 1979.

Verification of Procedural Programs

AMIR PNUELI

1 Introduction

In this paper we review Floyd's methods [Floyd, 1967] for verifying weak and strong correctness of sequential programs, and then extend these methods to deal with programs with (possibly recursive) procedures.

The renewed interest in these old-fashioned topics arises from the recent emergence of predicate and ranking abstractions. These abstractions yield boolean programs with procedures, which need to be analyzed and verified.

Historically, formal verification started with the study of sequential programs. *Floyd*'s seminal paper [Floyd, 1967] defined the problem, and outlined the main principles of its solution: using *invariants* for proving partial correctness, and well-founded ranking functions to establish termination. These basic principles underly all subsequent developments in formal verification, including their extensions to *reactive and parallel verification*, methods of *simulation* and *abstraction*, and verification of *functional programs*. Recently, there has been a revival of interest in sequential verification, through encouragement of the use of *assertions* within programs, and intense activity in *program analysis*.

In most expositions of verification methods for sequential programs, such as [Manna, 1974], [Apt and Olderog, 1991], and [Francez, 1992], the application of Floyd's methods is typically restricted to the treatment of programs without procedures. As soon as procedures are introduced into the language, the standard recommendation is to switch into Hoare's logic [Hoare, 1969].

The treatment of procedures using Floyd's approach, which is presented in this paper, has been first developed by the author in lecture notes around 1976, but has never been written before in an article. On the other hand, it served as an inspiration to the paper [Sharir and Pnueli, 1981] which employs a similar approach for the inter-procedural data-flow analysis of programs.

1.1 The Verification Framework

- A *programming language* $\mathcal{P}$. Can be compiled and executed on conventional computing systems.
- A *specification language* $\mathcal{S}$. A higher level non-procedural language which offers a natural vehicle for humans to represent requirements and specification of computing tasks.

Given a verification framework, there are several questions one could ask about relationship between objects in these two languages:

- The *Synthesis Problem:* Given a specification $S \in \mathcal{S}$, construct a program $P \in \mathcal{P}$ which satisfies the specification.
- The *Analysis Problem:* Given a program $P \in \mathcal{P}$, find its corresponding description $S \in \mathcal{S}$.
- The *Verification Problem:* Given a specification $S \in \mathcal{S}$ and a program $P \in \mathcal{P}$, check whether they are compatible, i.e. whether P *satisfies* S.
- The *Debugging Problem:* Given a specification $S \in \mathcal{S}$ and a program $P \in \mathcal{P}$ known not to satisfy S, find a program $P' \in \mathcal{P}$ "close" to P, i.e., transform P into P', such that P' satisfies S.
- The *Optimization Problem:* Given a specification $S \in \mathcal{S}$ and a program $P \in \mathcal{P}$ satisfying S. Among all programs P' "close" to P and satisfying S, find the "best" program (i.e. maximizing some performance metric).

A central notion which appears in all of these questions is that of a program $P \in \mathcal{P}$ *satisfying* a specification $S \in \mathcal{S}$. For that reason, we should study the verification problem first.

In general, all of these problems are difficult, undecidable, and at best, intractable. However, if $\mathcal{S}$ and $\mathcal{P}$ are close enough, they may admit algorithmic solutions. For example, *compilation* can be viewed as a special case of *synthesis*.

The rest of the paper is organized as follows: In Section 2 we present our (graphical) transition-graph notation for the presentation of programs (still without procedures), the operational semantics of programs, specification of programs, and the notions of partial correctness, successful (fault-free) executions, and convergence (termination) of programs. Section 3 introduces the method of inductive assertions for verifying partial correctness of transition-graph programs. In Section 4 we explore some of the properties of the inductive assertion method, and establish its soundness and completeness. Section 5 presents methods for proving success and fault-freedom of programs, which are extensions of the inductive assertion method. In Section 6 we introduce Floyd's method for proving termination of programs which are based on well-founded ranking. Finally, Section 7 augments the transition-graph language with (possibly recursive) procedures and presents appropriate extensions of the inductive assertion and well-founded ranking methods to deal with procedural programs.

2 Programs, Computations, and Specifications

Our programming language will be based on *transition graphs*. We assume a set of typed program variables V.
A *transition graph* is a labeled directed graph such that:

- All nodes are labeled by *locations* ℓ_i.
- There is one *initial node*, usually labeled by ℓ_0, and having no incoming edges.
- There is one *terminal node*, labeled ℓ_t with no outgoing edges.
- Nodes are connected by directed edges labeled by an instruction of the form

  $$c \rightarrow [\vec{y} := \vec{e}]$$

  where c is a boolean expression over V, $\vec{y} \subseteq V$ is a list of variables, and $\vec{e}$ is a list of expressions over V. In cases the assignment part is empty, we can abbreviate the label to a pure condition $c?$.
- Every node is on a path from ℓ_0 to ℓ_t.

EXAMPLE 1 (Integer Square Root Program)
Program INT-SQRT *presented in Fig. 1 computes in* y_1 *the integer square root of the input variable* $x \geq 0$.

2.1 States and Computations

For simplicity, we assume that all program variables range over the same domain D. For example, for program INT-SQRT, D is the domain of integers. We denote by $d = (d_1, \ldots, d_n)$ a sequence of D-values, which represent an *interpretation* (i.e., an assignment of values) of the program variables V.
A *state* of program P is a pair $\langle \ell, d \rangle$ consisting of a label ℓ and a data-interpretation d. A *computation* of program P is a maximal sequence

$$\sigma : \quad \langle \ell^0, d^0 \rangle, \langle \ell^1, d^1 \rangle, \ldots, \langle \ell_k, d^k \rangle \ \ldots,$$

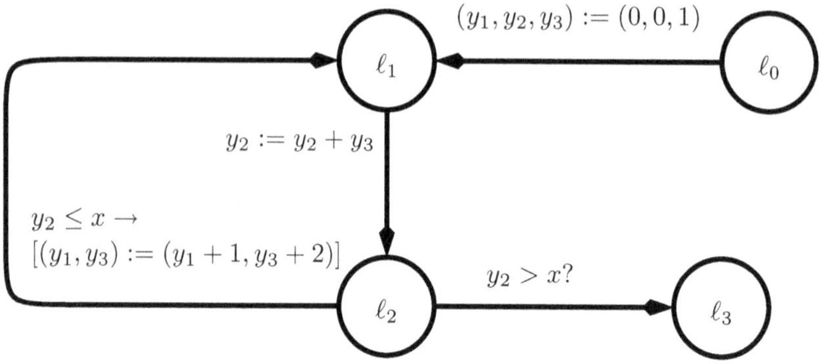

Figure 1. Integer Square Root Program

such that

- $\ell^0 = \ell_0$.
- For each $i = 0, 1, \ldots$, there exists an edge connecting ℓ^i to ℓ^{i+1} and labeled by the instruction $c \rightarrow [\vec{y} := \vec{e}]$, such that $d^i \models c$ and $d^{i+1} = d^i$ **with** $\vec{y} := \vec{e}(d^i)$.

We denote by $\mathcal{C}omp(P, d)$ the set of computations of program P starting at data-state d.

An Example of a Computation

Reconsider program INT-SQRT of Fig. 1. Following is a computation generated for $x = 5$:

$$\begin{array}{llll} \langle \ell_0; (-,-,-) \rangle, & & & \\ \langle \ell_1; (0,0,1) \rangle, & \langle \ell_2; (0,1,1) \rangle, & \langle \ell_1; (1,1,3) \rangle, & \langle \ell_2; (1,4,3) \rangle, \\ \langle \ell_1; (2,4,5) \rangle, & \langle \ell_2; (2,9,5) \rangle, & \langle \ell_3; (2,9,5) \rangle & \end{array}$$

2.2 Results of Computations

Let σ be computation. We define the *result* of the computation σ, denoted $val(\sigma)$, according to the following cases:

- If the computation is finite, and the last state is $\langle \ell_t; d \rangle$, then $val(\sigma) = d$. We refer to such a computation as a *terminating computation*.
- If the computation is finite, and the last state is $\langle \ell; d \rangle$ for some $\ell \neq \ell_t$, we say that the computation *fails* and write $val(\sigma) = fail$. This is possible if all guards on edges departing from location ℓ are false on d. In particular if there are no edges departing from ℓ.

- If the computation is infinite, we say that the computation *diverges*, and write $val(\sigma) = \perp$.

For a program P and initial data-state d, we define the *meaning* of the program P as a function:

$$M(P, d) = \{val(\sigma) \mid \sigma \in \mathcal{C}omp(P, d)\}$$

It is customary to write $M(P, d)$ as $M[P](d)$ to emphasize that M is a mapping which, for each program P yields a function $M[P]$ of the type:

$$M[P] : D^n \mapsto 2^{D^n \cup \{fail, \perp\}}$$

Specifications

A specification for a sequential program is given by a pair (φ, ψ) of first-order formulas, where

- The *pre-condition* φ imposes constraints on the initial data state by which proper computations could start.
- The *post-condition* ψ specifies the properties the terminal data state of a proper computation should satisfy.

For example, a specification for program INT-SQRT can be given by the pair

$$(x \geq 0, \qquad y_1^2 \leq x < (y_1 + 1)^2)$$

According to this specification, on initiation x should have a non-negative value while, on termination y_1 should be such that its square does not exceed x, but the square of $y_1 + 1$ should exceed x.

A computation whose initial state satisfies φ is called a *φ-computation.*

2.3 Correctness Statements

Given a specification (φ, ψ), we can formulate several notions of correctness.

- *Partial Correctness.* Program P is *partially correct* with respect to the specification (φ, ψ) if every terminating φ-computation ends in a ψ-state, i.e.

$$\varphi(d^0) \wedge d \in M[P](d^0) \quad \rightarrow \quad \psi(d)$$

- *Success.* A program is *successful under* φ (φ-successful) if there are no failing φ-computations. That is,

$$\varphi(d^0) \quad \rightarrow \quad fail \notin M[P](d^0)$$

- *Convergence.* A program is *convergent under* φ (φ-convergent, φ-terminating) if there are no divergent φ-computations. That is,

$$\varphi(d^0) \quad \rightarrow \quad \perp \notin M[P](d^0)$$

- *Total Correctness.* Program P is *totally correct* with respect to (φ, ψ) if it is partially correct, successful, and convergent under (φ, ψ).

3 Proving Partial Correctness by the Method of Inductive Assertions

We now present a proof method for proving partial correctness of a program. This proof method is called the method of *inductive assertions* [Floyd, 1967].

Step 1: Identifying a Cut-point Set
A *cut-point set* is a subset of locations $\mathcal{C} \subseteq \mathcal{L}$ such that $\ell_0, \ell_t \in \mathcal{C}$ and every cycle in the program's graph contains at least one cut-point (a member of $\mathcal{C}$).
For example, for program INT-SQRT of Fig. 1, we can choose the cut-point set $\mathcal{C} = \{\ell_0, \ell_2, \ell_3\}$, as shown in Fig. 2.

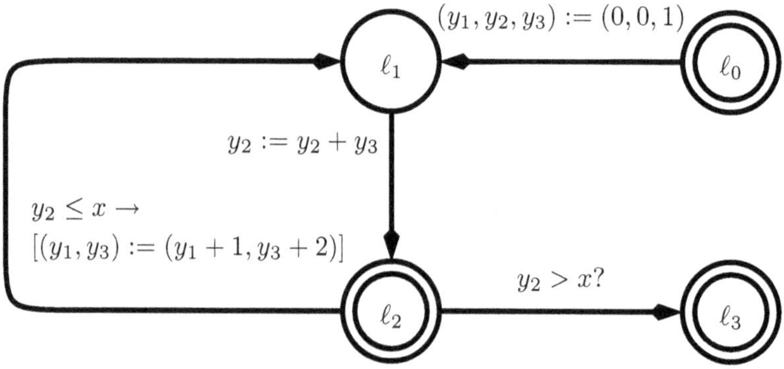

Figure 2. Cut points selected in program INT-SQRT

Step 2: Verification Paths A *verification path* is a path from one cut-point to another cut-point, which does not pass through any other cut-point. For example, in program INT-SQRT of Fig. 2, we have 3 verification paths given by

$$\begin{array}{lcl} \pi_{02} & : & \ell_0, \ell_1, \ell_2 \\ \pi_{22} & : & \ell_2, \ell_1, \ell_2 \\ \pi_{23} & : & \ell_2, \ell_3 \end{array}$$

Summary Guarded Commands

Consider a verification path π where, for simplicity, all assignments are made to the full set of program variables V.

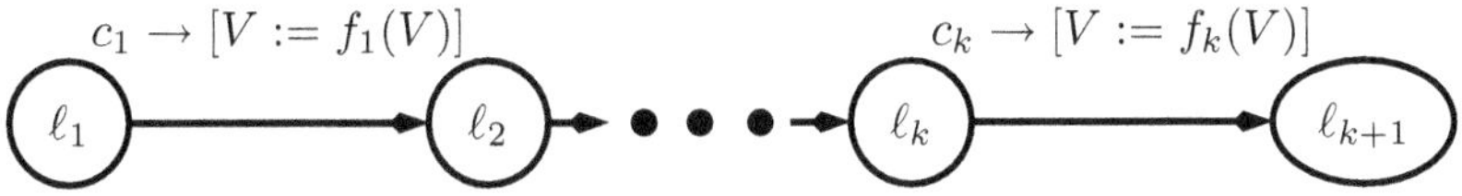

For such a path we can compute a *traversal condition* c_π and a *data transformation* f_π. Condition c_π when satisfied at ℓ_1 guarantees that it is possible to traverse the path π. The transformation f_π specifies the values of V at the end of an execution of π as a function of the values of V in the beginning of such execution. They are respective given by:

$$\begin{array}{lcl} c_\pi & : & c_1(V) \wedge c_2(f_1(V)) \wedge \cdots \wedge c_k(f_{k-1}(\cdots f_1(V)\cdots)) \\ f_\pi & : & f_k(f_{k-1}(\cdots f_2(f_1(V))\cdots)) \end{array}$$

Given these constructs we can summarize the effect of executing the path π by the *summary guarded command* $G_\pi : c_\pi \to [V := f_\pi(V)]$.

Application to INT-SQRT

Apply this procedure to program INT-SQRT, as presented in Fig. 2. The summary guarded commands for the 3 verification paths are given by:

$$\begin{array}{lcl} G_{02} & : & (y_1, y_2, y_3) := (0, 1, 1) \\ G_{22} & : & y_2 \leq x \to [(y_1, y_2, y_3) := (y_1 + 1, y_2 + y_3 + 2, y_3 + 2)] \\ G_{23} & : & y_2 > x \to [(y_1, y_2, y_3) := (y_1, y_2, y_3)] \end{array}$$

Once we derive these summary guarded commands, it is possible to construct the *reduced version* of the original program, as presented in Fig. 3.

This reduced program is *weakly equivalent* to the original program in the sense that it preserves all successful terminating computations and all divergent computations. However, it may lose some failing computations of the original program.

Step 3: Devise an Assertion Network

With each cut-point $\ell_i \in \mathcal{C}$ associate an assertion φ_i (first-order formula) over V.

For example, for program INT-SQRT of Fig. 3, we can form the following assertion network:

$$\begin{array}{lcl} \varphi_0 & : & x \geq 0 \\ \varphi_2 & : & y_1^2 \leq x \ \wedge \ y_2 = (y_1 + 1)^2 \ \wedge \ y_3 = 2y_1 + 1 \\ \varphi_3 & : & y_1^2 \leq x < (y_1 + 1)^2 \end{array}$$

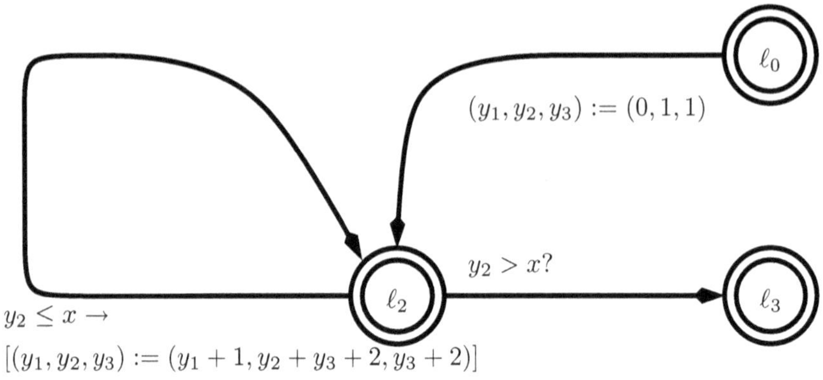

Figure 3. The reduced version of program INT-SQRT

Step 4: Form Verification Conditions

For each verification path π connecting cut-point ℓ_i to cut-point ℓ_j, we form the *verification condition*

$$V_\pi : \quad \varphi_i(V) \wedge c_\pi \quad \rightarrow \quad \varphi_j(f_\pi(V))$$

For example, for program INT-SQRT of Fig. 3, and the assertion network

$$\begin{array}{lcl} \varphi_0 & : & x \geq 0 \\ \varphi_2 & : & y_1^2 \leq x \ \wedge \ y_2 = (y_1+1)^2 \ \wedge \ y_3 = 2y_1 + 1 \\ \varphi_3 & : & y_1^2 \leq x < (y_1+1)^2 \end{array}$$

we obtain the following set of verification conditions:

$$\begin{array}{lcl} V_{02} & : & x \geq 0 \ \rightarrow \ 0^2 \leq x \ \wedge \ 1 = (0+1)^2 \ \wedge \ 1 = 2 \cdot 0 + 1 \\ V_{22} & : & y_1^2 \leq x \ \wedge \ y_2 = (y_1+1)^2 \ \wedge \ y_3 = 2y_1 + 1 \ \wedge \ y_2 \leq x \quad \rightarrow \\ \multicolumn{3}{l}{(y_1+1)^2 \leq x \ \wedge \ y_2 + y_3 + 2 = ((y_1+1)+1)^2 \ \wedge \ y_3 + 2 = 2(y_1+1)+1} \\ V_{23} & : & y_1^2 \leq x \ \wedge \ y_2 = (y_1+1)^2 \ \wedge \ y_3 = 2y_1 + 1 \ \wedge \ y_2 > x \quad \rightarrow \\ & & \hfill y_1^2 \leq x < (y_1+1)^2 \end{array}$$

3.1 Inductive and Invariant Networks

An assertion network $\mathcal{N} = \{\varphi_0, \ldots, \varphi_t\}$ for a program P is said to be *inductive* if all the verification conditions V_π for all verification paths π in P are valid.

Network $\mathcal{N}$ is said to be *invariant* if for every execution state $\langle \ell_i, d \rangle$ occurring in a φ_0-computation, where $\ell_i \in \mathcal{C}$, $d \models \varphi_i$. That is, on every visit of a φ_0-computation at a cut-point ℓ_i the visiting data state satisfies the corresponding assertion φ_i associated with ℓ_i.

CLAIM 2 *Every inductive network is invariant.*

Proof Let $\mathcal{N} = \{\varphi_0, \ldots, \varphi_t\}$ be an inductive network. Let

$$\sigma : \quad \langle \ell_{i_0}, d^0 \rangle \xrightarrow{\pi_0} \langle \ell_{i_1}, d^1 \rangle \xrightarrow{\pi_1} \cdots \xrightarrow{\pi_{k-1}} \langle \ell_{i_k}, d^k \rangle \xrightarrow{\pi_k} \cdots$$

be a φ_0-computation where we explicitly display the sequence of cut-points $\ell_0 = \ell_{i_0}, \ell_{i_1}, \ldots$ visited by σ and the verification paths $\pi_0, \pi_1, \ldots$ connecting them.

We will prove by induction on $j = 0, 1, \ldots$ that $d^j \models \varphi_{i_j}$. For $j = 0$, we consider the cut-point $\ell_{i_0} = \ell_0$. Since σ is a φ_0-computation, we have that $d^0 \models \varphi_0$.

Assume now that $d^j \models \varphi_{i_j}$. We will show that $d^{j+1} \models \varphi_{i_{j+1}}$. Since σ proceeded from ℓ_{i_j} to $\ell_{i_{j+1}}$ through verification path π_j, we know that $d^j \models c_{\pi_j}$ and $d^{j+1} = f_{\pi_j}(d^j)$. We also have that the verification condition

$$V_{\pi_j} : \quad \varphi_{i_j}(d^j) \ \wedge \ c_{\pi_j}(d^j) \quad \rightarrow \quad \varphi_{i_{j+1}}(f_{\pi_j}(d^j))$$

holds. Since both $\varphi_{i_j}(d^j)$ and $c_{\pi_j}(d^j)$ are true, we conclude that the right-hand side $\varphi_{i_{j+1}}(f_{\pi_j}(d^j)) = \varphi_{i_{j+1}}(d^{j+1})$ is also true. It follows that $d^{j+1} \models \varphi_{i_{j+1}}$. ■

Consequences

From Claim 2 we conclude:

COROLLARY 3 *If $\mathcal{N} = \{\varphi_0, \ldots, \varphi_t\}$ is an inductive network, then program P is partially correct with respect to the specification (φ_0, φ_t).*

Let (p, q) be a specification. We say that the network $\mathcal{N} = \{\varphi_0, \ldots, \varphi_t\}$ *entails* the specification (p, q) if the following two implications are valid:

$$p \ \rightarrow \ \varphi_0 \qquad\qquad \varphi_t \ \rightarrow \ q$$

COROLLARY 4 *If $\mathcal{N} = \{\varphi_0, \ldots, \varphi_t\}$ is an inductive network which entails the specification (p, q), then program P is partially correct with respect to (p, q).*

This leads to the final formulation of the inductive assertion proof method.

> In order to prove that program P is partially correct w.r.t specification (p, q), find an assertion network $\mathcal{N} = \{\varphi_0, \ldots, \varphi_t\}$ and prove that $\mathcal{N}$ is inductive and that it entails the specification (p, q).

4 Properties of the Inductive Assertions Method

In this section, we will explore the properties of the inductive assertions method and prove its completeness.

4.1 Dependence on the Cut-Set

Interested in exploring the limitations of the *inductive assertion* method, we first check whether the method is sensitive to the choice of the cut-set $\mathcal{C}$. A special case is that of a *full cut-set* $\mathcal{C} = \mathcal{L}$ in which the cut-set includes all the locations in the program.
The following claim shows that any inductive assertions which is not full, can be extended to a bigger inductive network.

CLAIM 5 (Inductive networks can be extended) *Let* $\mathcal{N} = \langle \mathcal{C}, \{\varphi_\ell \mid \ell \in \mathcal{C}\}\rangle$ *be an inductive assertion network, and* $\widetilde{\ell} \notin \mathcal{C}$ *a location not in* $\mathcal{C}$. *There exists an inductive assertion network over the extended cut-set* $\widetilde{\mathcal{C}} = \mathcal{C} \cup \{\widetilde{\ell}\}$ *which agrees with* $\mathcal{N}$ *on the assertions* φ_ℓ *for all* $\ell \in \mathcal{C}$.

Proof. The extended network has the form $\widetilde{\mathcal{N}} = \langle \widetilde{\mathcal{C}}, \{\psi_\ell \mid \ell \in \widetilde{\mathcal{C}}\}\rangle$, where $\widetilde{\mathcal{C}} = \mathcal{C} \cup \{\widetilde{\ell}\}$. For all $\ell \in \mathcal{C}$, we take $\psi_\ell = \varphi_\ell$. We will present two different constructions for the computation of $\psi_{\widetilde{\ell}}$, one based on backwards propagation while the other is based on forward propagation. ∎

4.2 Backwards Propagation

Let $\Pi_{\widetilde{\ell},\mathcal{C}}$ be the set of paths connecting $\widetilde{\ell}$ to a location in $\mathcal{C}$ without passing through any other cut-point. For each path $\pi \in \Pi_{\widetilde{\ell},\mathcal{C}}$, let $dest(\pi)$, c_π, and f_π denote, respectively, the cut-point at the end of path π, the summary traversal condition, and data transformation associated with π. We define the *pre-condition* for $\mathcal{N}$ at $\widetilde{\ell}$ which is given by:

$$pre(\widetilde{\ell}, \mathcal{N}) : \bigwedge_{\pi \in \Pi_{\widetilde{\ell},\mathcal{C}}} \Big(c_\pi(V) \quad \rightarrow \quad \varphi_{dest(\pi)}(f_\pi(V))\Big)$$

Formula $pre(\widetilde{\ell}, \mathcal{N})$ is the condition at $\widetilde{\ell}$ which guarantees that if execution continues to reach a location $\ell \in \mathcal{C}$, then it will reach it with a data state satisfying φ_ℓ.

We will now show that, under the assumption that $\mathcal{N}$ is inductive, the extended network obtained by taking $\psi_{\widetilde{\ell}} = pre(\widetilde{\ell}, \mathcal{N})$ is also inductive. Clearly we only have to consider new verification paths, i.e. verification paths which appear in $\widetilde{\mathcal{N}}$ but did not exist in $\mathcal{N}$. There are two such classes of paths, which we consider separately.

Backwards Propagation: **Paths from $\widetilde{\ell}$ to $\ell \in \mathcal{C}$**

Let π be a $\widetilde{\mathcal{N}}$-path connecting $\widetilde{\ell}$ to some cut-point $\ell_2 \in \mathcal{C}$. The verification condition for such a path is given by:

$$\psi_{\widetilde{\ell}}(V) \land c_\pi(V) \quad \rightarrow \quad \varphi_{\ell_2}(f_\pi(V))$$

which is equivalent to

$$\psi_{\widetilde{\ell}}(V) \quad \rightarrow \quad (c_\pi(V) \rightarrow \varphi_{\ell_2}(f_\pi(V)))$$

As π connects $\widetilde{\ell}$ to a location in $\mathcal{C}$, this path is one of the members of the set $\Pi_{\widetilde{\ell},\mathcal{C}}$ over which the conjunction defining $\psi_{\widetilde{\ell}} = pre(\widetilde{\ell}, \mathcal{N})$ is taken. Therefore, $c_\pi(V) \rightarrow \varphi_{\ell_2}(f_\pi(V))$ is one of these conjuncts and is implied by $\psi_{\widetilde{\ell}}(V)$.

Backwards Propagation: **Paths from $\ell \in \mathcal{C}$ to $\widetilde{\ell}$**

Next, consider a path π_1 connecting some location $\ell_1 \in \mathcal{C}$ to $\widetilde{\ell}$ and not passing through any other cut-point.

Let $\pi_2 \in \Pi_{\widetilde{\ell},\mathcal{C}}$ be an arbitrary verification path connecting $\widetilde{\ell}$ to some location $\ell_2 = dest(\pi_2)$. Let $\pi = \pi_1 \circ \pi_2$ be the path obtained by following π_1 first and then continuing along π_2 (fusion of π_1 and π_2). Path π was a verification path for the network $\mathcal{N}$. Since $\mathcal{N}$ was inductive, we know that the verification condition

$$(1) \quad \varphi_1(V) \land c_\pi(V) \quad \rightarrow \quad \varphi_2(f_\pi(V))$$

is valid. It is not difficult to relate the traversal condition and data transformation of π to these of its constituents. These are given by

$$(2) \quad c_\pi(V) = c_{\pi_1}(V) \land c_{\pi_2}(f_{\pi_1}(V)) \qquad \text{and} \qquad f_\pi(V) = f_{\pi_2}(f_{\pi_1}(V))$$

With these relation, Formula 1 can be rewritten as

$$\varphi_1(V) \land c_{\pi_1}(V) \quad \rightarrow \quad (c_{\pi_2}(f_{\pi_1}(V)) \rightarrow \varphi_2(f_{\pi_2}(f_{\pi_1}(V))))$$

or, equivalently, as

$$\varphi_1(V) \land c_{\pi_1}(V) \quad \rightarrow \quad \textbf{let } \overline{V} = f_{\pi_1}(V) \textbf{ in } \left(c_{\pi_2}(\overline{V}) \rightarrow \varphi_2(f_{\pi_2}(\overline{V}))\right)$$

Taking the conjunction of this implication over all paths $\pi_2 \in \Pi_{\widetilde{\ell},\mathcal{C}}$, we obtain

$$\varphi_1(V) \land c_{\pi_1}(V) \quad \rightarrow \quad \textbf{let } \overline{V} = f_{\pi_1}(V) \textbf{ in } \bigwedge_{\pi_2 \in \Pi_{\widetilde{\ell},\mathcal{C}}} \left(c_{\pi_2}(\overline{V}) \rightarrow \varphi_2(f_{\pi_2}(\overline{V}))\right)$$

or, equivalently,

$$\varphi_1(V) \land c_{\pi_1}(V) \quad \rightarrow \quad \psi_{\widetilde{\ell}}(f_{\pi_1}(V))$$

4.3 Forward Propagation

An alternate definition of $\psi_{\widetilde{\ell}}$ is based on the consideration of paths from $\mathcal{C}$ to $\widetilde{\ell}$.

Let $\Pi_{\mathcal{C},\widetilde{\ell}}$ be the set of paths connecting a location in $\mathcal{C}$ to $\widetilde{\ell}$ without passing through any other cut-point. For each path $\pi \in \Pi_{\mathcal{C},\widetilde{\ell}}$, let $srce(\pi)$, c_π, and f_π denote, respectively, the cut-point at the beginning of path π, the summary traversal condition, and data transformation associated with π. We define the *post-condition* for $\mathcal{N}$ at $\widetilde{\ell}$ which is given by:

$$post(\mathcal{N},\widetilde{\ell}): \quad \exists V_0 \bigvee_{\pi\in\Pi_{\mathcal{C},\widetilde{\ell}}} \left(\varphi_{srce(\pi)}(V_0) \ \wedge\ c_\pi(V_0) \ \wedge\ V = f_\pi(V_0)\right)$$

Formula $post(\mathcal{N},\widetilde{\ell})$ characterize the states which can be reached at location $\widetilde{\ell}$ by an execution whose previous visit to a location $\ell \in \mathcal{C}$ was with a data state satisfying φ_ℓ.

We will now show that, under the assumption that $\mathcal{N}$ is inductive, the extended network obtained by taking $\psi_{\widetilde{\ell}} = post(\mathcal{N},\widetilde{\ell})$ is also inductive. Clearly we only have to consider new verification paths, i.e. verification paths which appear in $\widetilde{\mathcal{N}}$ but did not exist in $\mathcal{N}$. As before, there are two such classes of paths, which we consider separately.

Forward Propagation: **Paths from $\ell \in \mathcal{C}$ to $\widetilde{\ell}$**

Let π be a $\widetilde{\mathcal{N}}$-path connecting a location $\ell_1 \in \mathcal{C}$ to $\widetilde{\ell}$. The verification condition for such a path is given by:

$$\varphi_{\ell_1}(V_0) \ \wedge\ c_\pi(V_0) \quad \rightarrow \quad \psi_{\widetilde{\ell}}(f_\pi(V_0))$$

which is validity-equivalent to

$$\varphi_{srce(\pi)}(V_0) \ \wedge\ c_\pi(V_0) \ \wedge\ V = f_\pi(V_0) \quad \rightarrow \quad \psi_{\widetilde{\ell}}(V)$$

As π connects a location in $\mathcal{C}$ to $\widetilde{\ell}$, this path is one of the members of the set $\Pi_{\mathcal{C},\widetilde{\ell}}$ over which the disjunction defining $\psi_{\widetilde{\ell}} = post(\mathcal{N},\widetilde{\ell})$ is taken. Therefore, $\varphi_{srce(\pi)}(V_0) \ \wedge\ c_\pi(V_0) \ \wedge\ V = f_\pi(V_0)$ is one of these disjuncts and hence implies $\psi_{\widetilde{\ell}}(V)$.

Forward Propagation: **Paths from $\widetilde{\ell}$ to $\ell \in \mathcal{C}$**

Next, consider a path π_2 connecting $\widetilde{\ell}$ to some location $\ell_2 \in \mathcal{C}$ and not passing through any other cut-point.

Let $\pi_1 \in \Pi_{\widetilde{\ell},\mathcal{C}}$ be an arbitrary verification path connecting some location $\ell_1 = srce(\pi_1)$ to $\widetilde{\ell}$. Let $\pi = \pi_1 \circ \pi_2$ be the fusion of paths π_1 and π_2. Path π

was a verification path for the network $\mathcal{N}$. Since $\mathcal{N}$ was inductive, we know that the verification condition

$$(3) \quad \varphi_1(V) \wedge c_\pi(V) \quad \rightarrow \quad \varphi_2(f_\pi(V))$$

is valid. Using the relations between the traversal condition and data transformation of π to these of its constituents, Formula 3 can be rewritten as

$$\varphi_1(V) \wedge c_{\pi_1}(V) \quad \rightarrow \quad (c_{\pi_2}(f_{\pi_1}(V)) \rightarrow \varphi_2(f_{\pi_2}(f_{\pi_1}(V))))$$

which is validity-equivalent to

$$(\exists V_0 : \varphi_1(V_0) \wedge c_{\pi_1}(V_0) \wedge V = f_{\pi_1}(V_0)) \quad \rightarrow \quad (c_{\pi_2}(V) \rightarrow \varphi_2(f_{\pi_2}(V)))$$

Taking the conjunction of this implication over all paths $\pi_1 \in \Pi_{\mathcal{C},\widetilde{\ell}}$, we obtain

$$\left[\bigvee_{\pi_1 \in \Pi_{\mathcal{C},\widetilde{\ell}}} \exists V_0 : \varphi_1(V_0) \wedge c_{\pi_1}(V_0) \wedge V = f_{\pi_1}(V_0)\right] \quad \rightarrow \quad (c_{\pi_2}(V) \rightarrow \varphi_2(f_{\pi_2}(V)))$$

or, equivalently,

$$\psi_{\widetilde{\ell}}(V) \wedge c_{\pi_2}(V) \quad \rightarrow \quad \varphi_2(f_{\pi_2}(V))$$

▟

4.4 Removing Cut-Points

In the previous discussion we have shown that it is always possible to add more cut-points to an inductive network, while maintaining inductiveness. We will now show that it is also possible to *remove* cut-points, provided the remaining set is still a cut-set.

CLAIM 6 *Let $\mathcal{N} = \langle \mathcal{C}, \{\varphi_\ell \mid \ell \in \mathcal{C}\}\rangle$ be an inductive network. Let $\widetilde{\ell} \in \mathcal{C}$ be a location in $\mathcal{C}$ such that $\overline{\mathcal{C}} = \mathcal{C} - \{\widetilde{\ell}\}$ is a cut-set. Then the network $\overline{\mathcal{N}} = \langle \overline{\mathcal{C}}, \{\varphi_\ell \mid \ell \in \overline{\mathcal{C}}\}\rangle$, obtained by removing $\widetilde{\ell}$ and $\varphi_{\widetilde{\ell}}$ from $\mathcal{N}$, is also inductive.*

Proof. We only need to consider "new" verification paths, i.e. paths which exist in $\overline{\mathcal{N}}$ but not in $\mathcal{N}$. Such a path π connecting $\ell_1 \in \overline{\mathcal{C}}$ to $\ell_2 \in \overline{\mathcal{C}}$ must be the fusion $\pi = \pi_1 \circ \pi_2$ of two paths, where path π_1 connects ℓ_1 to $\widetilde{\ell}$, while π_2 connects $\widetilde{\ell}$ to ℓ_2.

Since both π_1 and π_2 are verification paths in the inductive network $\mathcal{N}$, we know that the following implications are valid:

$$\begin{array}{lcl} \varphi_{\ell_1}(V_1) \wedge c_{\pi_1}(V_1) & \rightarrow & \varphi_{\tilde{\ell}}\ (f_{\pi_1}(V_1)) \\ \varphi_{\tilde{\ell}}\ (V_2) \wedge c_{\pi_2}(V_2) & \rightarrow & \varphi_{\ell_2}(f_{\pi_2}(V_2)) \end{array}$$

Substituting $f_{\pi_1}(V_1)$ for V_2 in the second implication, and combining the two together yields

$$\varphi_{\ell_1}(V) \wedge c_{\pi_1}(V) \wedge c_{\pi_2}(f_{\pi_1}(V)) \quad \rightarrow \quad \varphi_{\ell_2}(f_{\pi_2}(f_{\pi_1}(V)))$$

which, using the relations in Formula 2, can be rewritten as

$$\varphi_{\ell_1}(V) \wedge c_{\pi}(V) \quad \rightarrow \quad \varphi_{\ell_2}(f_{\pi}(V))$$

■

4.5 Method is Independent of the Choice of the Cut-Set

The preceding discussions can be summarized by the statement that the success or failure of an application of the inductive assertion method is independent of the particular choice of the cut-set $\mathcal{C}$.

Technically, this can be summarized by the following corollary:

COROLLARY 7 *Let $\langle p, q\rangle$ be a specification for program P, and let $\mathcal{C}_1$ and $\mathcal{C}_2$ be two cut-sets. Then, there exists an inductive network $\mathcal{N}_1$ based on $\mathcal{C}_1$ and entailing $\langle p, q\rangle$ iff there exists an inductive network $\mathcal{N}_2$ based on $\mathcal{C}_2$ and entailing $\langle p, q\rangle$.*

The proof of this statement can be obtained by starting with a $\mathcal{C}_1$-based inductive network $\mathcal{N}_1$ and incrementally adding missing locations, using Claim 5, until we obtain a full network. Then we start removing locations which do not belong to $\mathcal{C}_2$, relying on Claim 6, until we obtain a network $\mathcal{N}_2$ based on $\mathcal{C}_2$. Since both ℓ_0 and ℓ_t belong to both $\mathcal{N}_1$ and $\mathcal{N}_2$, their associated assertions are preserved throughout the entire process. Therefore, if $\mathcal{N}_1$ entails $\langle p, q\rangle$ then so does $\mathcal{N}_2$.

4.6 Completeness of the Method

An important question which arises whenever a proof method is introduced is that of *completeness*. Namely, is it the case that, whenever a program is partially correct w.r.t a specification $\langle p, q\rangle$, this fact can be proven, using the *inductive assertion* method?

In our case, the answer is positive, and we will prepare the necessary constructs for proving this fact.

Let ℓ_i and ℓ_j be two locations in the program which are connected by a direct edge, labeled by the guarded command $c_{ij} \rightarrow [V := f_{ij}(V)]$. We define the formula

$$\rho_{ij}(V_1, V_2): \quad c_{ij}(V_1) \ \wedge \ V_2 = f_{ij}(V_1)$$

Obviously, ρ_{ij} is satisfied by the two data states $V_1 = d_1$ and $V_2 = d_2$ iff there exists a computation step $\langle \ell_i, d_1 \rangle \rightarrow \langle \ell_j, d_2 \rangle$ leading from the execution state $\langle \ell_i, d_1 \rangle$ to the execution state $\langle \ell_j, d_2 \rangle$.

Let E denote the set of direct *edges* in the program, i.e. set of pairs (ℓ_i, ℓ_j), such that there is a direct edge from ℓ_i to ℓ_j. Assume that we consider a fixed specification $\langle p, q \rangle$ for program P. For simplicity, we assume first that the program has a single data variable (V) which ranges over domain D.

The Minimal Predicate at Location ℓ

For each location ℓ in the program, we define the *minimal predicate* $M_\ell(V)$. It is intended that a data state d satisfies M_ℓ iff

> There exists a p-computation reaching the execution state $\langle \ell, d \rangle$.

This can be formalized by the following extended predicate logic formula:

$$M_\ell(V): \quad \left(\begin{array}{l} \exists k \geq 0 : \exists loc : [0..k] \mapsto [0..t], A : [0..k] \mapsto D : \\ \quad loc[0] = 0 \ \wedge \ p(A[0]) \ \wedge \ \ell_{loc[k]} = \ell \ \wedge \ V = A[k] \ \wedge \\ \quad \forall r : [0..k) : \bigvee\limits_{(i,j) \in E} loc[r] = i \ \wedge \ loc[r{+}1] = j \ \wedge \\ \quad \rho_{ij}(A[r], A[r{+}1]) \end{array} \right)$$

The formula states the existence of two arrays of size $k+1$ for some $k \geq 0$. The array $loc[0..k]$ encodes the indices of the locations traversed during the computation from ℓ_0 to ℓ, while the array $A[0..k]$ encodes the sequence of data states encountered during this computation. The conjunction $loc[0] = 0 \wedge p(A[0])$ ensures that the execution state $\langle \ell_{loc[0]}, A[0] \rangle$ is an initial state. The conjunction $\ell_{loc[k]} = \ell \ \wedge \ V = A[k]$ ensure that the last execution state encoded by these two arrays is of the form $\langle \ell, d \rangle$ where d equals the current value of V. The conjunction under the $\forall r$ quantification guarantees that the sequence evolve according to the rules of the program P and is, therefore, a computation of P.

The Claim of Completeness

CLAIM 8 (Completeness of the *inductive assertions* method) *Let P be a program which is partially correct w.r.t the specification $\langle p, q \rangle$. Then there exists an inductive assertion network which entails $\langle p, q \rangle$.*

Proof. For the cut-set we take $\mathcal{C} = \mathcal{L}$, i.e. a full cut-set. As the assertion associated with location ℓ we take the minimal predicate M_ℓ. It remains to show that this assertion network is inductive and that it entails $\langle p, q \rangle$.
To show inductiveness, let ℓ_i and ℓ_j be two locations which are connected by a direct edge. We have to show that if a data state d satisfies $\varphi_i \wedge c_{ij}$, where $\varphi_i = M_i$, then $f_{ij}(d)$ satisfies $\varphi_j = M_j$. By definition of M_i, $d \models M_i$ implies that there exists a p-computation segment $\sigma_i : s_0, \ldots, \langle \ell_i, d \rangle$ reaching location ℓ_i with data state d. Since $d \models c_{ij}$, we can extend σ_i by one more step to obtain the p-computation segment $\sigma_j : s_0, \ldots, \langle \ell_i, d \rangle, \langle \ell_j, f_{ij}(d) \rangle$, i.e., a computation segment reaching location ℓ_j with data state $f_{ij}(d)$. By the definition of M_j, it follows that $f_{ij}(d) \models M_j$.

Next, we have to show that the defined network entails the specification $\langle p, q \rangle$. Since location ℓ_0 has no incoming edges, the only computation segments reaching ℓ_0 must be singleton sequences of the form $\langle \ell_0, d \rangle$, where $d \models p$. It follows that the minimal predicate M_{ℓ_0} equals p, and therefore is trivially implied by p.

For the terminal location, we have to show that $M_{\ell_t} \to q$. Let d be a data state satisfying M_{ℓ_t}. This implies that d is a possible final result of a terminating p-computation. Since Claim 8 assumes that P is partially correct w.r.t $\langle p, q \rangle$, d must satisfy q. It follows that M_{ℓ_t} implies q. ■

What we Have Not Proven

It is not very difficult to remove the restriction that V consists of a single data variable. Assume, instead, that $V = \{y_1, \ldots, y_m\}$. Then, the only difference is that the array A will have to be a two-dimensional array of the form $A[1..m, 0..k]$.
Also, allowing the formula for M_ℓ to quantify over arrays, is not a very significant deviation from conventional first-order logic. If the data domain D is the naturals or integers or, for that matter, any other recursive data structure, we can use Gödel encoding, in order to reduce arrays of any dimension and sequences to natural numbers.

On the other hand, the reader should notice that the only thing we proved is the existence of an assertion network whose verification conditions are *valid*. Nowhere did we claim that these conditions are provable in any formal system. Rather, all the available undecidability and incompleteness results for first-order logic imply that that there does not exist a single formal system in which these verification conditions can always be proven.

Furthermore, the "construction" of inductive network as outlined in the completeness proof, should not be interpreted to mean that such construction is useful for any actual application. The main reason why this construction is not useful is that its soundness relies on the a priori assumption that the program is partially correct w.r.t $\langle p, q\rangle$. If we are already ensured of this fact, there is no remaining motivation for applying the inductive assertion method.

5 Verifying Success and Absence of Faults

We consider now methods by which we can verify the property of *success*, i.e. *deadlock-freedom*. As in the case of partial correctness, the method starts by identifying a cut-set $\mathcal{C}$ and an associated *assertion network*.
First, consider the case that the cut-set is *full*, i.e. contains all locations in the program.
Consider a node $\ell \in \mathcal{L}$ in the program. Let $c_1, \ldots, c_k$ be the guards on all edges departing from node ℓ. We define the *exit condition* for ℓ to be

$$E_\ell : \quad c_1 \vee \cdots \vee c_k$$

The following claim summarizes the first version of a rule for proving success.

CLAIM 9 *In order to prove that program P is p-successful (i.e., no p-computation ever deadlocks), it is sufficient to find a full network $\mathcal{N} : \{\varphi_\ell \mid \ell \in \mathcal{L}\}$, satisfying the following requirements:*

1. *The network $\mathcal{N}$ is inductive.*
2. $p \quad \rightarrow \quad \varphi_{\ell_0}$
3. $\varphi_\ell \quad \rightarrow \quad E_\ell \qquad$ *for every $\ell \in \mathcal{L}$*

Proof. Let $\sigma : \langle \ell_0, d_0\rangle, \ldots, \langle \ell, d_m\rangle$ be a p-computation segment reaching location ℓ. By premise *2*, σ is also a φ_{ℓ_0}-computation. By Claim 2 and premise *1*, $d_k \models E_\ell$. By premise *3*, $d_m \models E_\ell$ which implies that at least one of the edges departing from location ℓ is enabled. Thus, σ cannot deadlock at ℓ. ∎

5.1 Extensions to Non-Full Networks

We now extend the method to apply also in the case that the network $\mathcal{N}$ is not full. Consider a partial network $\mathcal{N} : \langle \mathcal{C}, \{\varphi_\ell \mid \ell \in \mathcal{C}\}\rangle$.
Let $\widetilde{\ell} \notin \mathcal{C}$ be a location not in $\mathcal{C}$. As previously introduced, let $\Pi_{\mathcal{C},\widetilde{\ell}}$ be the set of paths connecting a location in $\mathcal{C}$ to $\widetilde{\ell}$ without passing through any other cut-point. For each path $\pi \in \Pi_{\mathcal{C},\widetilde{\ell}}$, let $srce(\pi)$, c_π, and f_π denote, respectively, the cut-point at the beginning of path π, the summary traversal condition, and data transformation associated with π.
The following claim summarizes the general rule for proving success, using an arbitrary network.

CLAIM 10 *In order to prove that program P is p-successful (i.e., no p-computation ever deadlocks), it is sufficient to find a network $\mathcal{N} : \langle \mathcal{C}, \{\varphi_\ell \mid \ell \in \mathcal{C}\}$, satisfying the following requirements:*

1. *The network $\mathcal{N}$ is inductive.*
2. $p \quad \rightarrow \quad \varphi_{\ell_0}$
3. $\varphi_\ell \quad \rightarrow \quad E_\ell$ *for every* $\ell \in \mathcal{C}$
4. $\varphi_{srce(\pi)}(V) \wedge c_\pi(V) \rightarrow E_{\widetilde{\ell}}(f_\pi(V))$ *for every* $\widetilde{\ell} \notin \mathcal{C}$ *and path* $\pi \in \Pi_{\mathcal{C},\widetilde{\ell}}$

Proof. Assume that there exists a network $\mathcal{N} : \langle \mathcal{C}, \{\varphi_\ell \mid \ell \in \mathcal{C}\}$ which satisfies the four requirements of Claim 10 but program P is not p-successful.

In that case, there exists a p-computation segment σ reaching some location $\widetilde{\ell}$ with data state d such that $d \not\models E_{\widetilde{\ell}}$. We consider two cases:

Case $\widetilde{\ell} \in \mathcal{C}$
Since $\mathcal{N}$ is inductive, data state d must satisfy $\varphi_{\widetilde{\ell}}$. However, this contradicts requirement *3* of the claim and the assumption $d \not\models E_{\widetilde{\ell}}$. Therefore, this case is impossible.

Case $\widetilde{\ell} \notin \mathcal{C}$
In this case we consider the last cut-point location visited by the execution σ. Assume that this is location ℓ, and since then σ followed the path $\pi \in \Pi_{\mathcal{C},\widetilde{\ell}}$ on its way to $\widetilde{\ell}$. Let d_0 be the data state with which σ last visited location ℓ. Since σ followed the path π, we know that $d_0 \models c_\pi$ and $d = f_\pi(d_0)$. Due to the inductiveness of $\mathcal{N}$, we also know that $d_0 \models \varphi_\ell$. Substituting these facts in requirement *4* of the claim, we conclude that $d \models E_{\widetilde{\ell}}$ which contradicts the assumption $d \not\models E_{\widetilde{\ell}}$.
We thus conclude that programs P is p-successful. ■

5.2 *Example:* Binary Search

Program BINARY-SEARCH , presented in Fig. 4, is expected to identify the precise range in which an input number x falls. We are given a sorted array of numbers $a[1..n]$ such that $a[1] < x < a[n]$ and $x \neq a[i]$ for all $i \in [1..n]$. The specification is given by $\langle p, q\rangle$, where

$$\begin{array}{ll} p: & n > 1 \ \wedge\ (a[1] < x < a[n]) \ \wedge\ \mathit{sorted}(a,1,n) \ \wedge\ \forall i : [1..n] : x \neq a[i] \\ q: & (1 \leq L < n) \ \wedge\ (a[L] < x < a[L+1]) \end{array}$$

In order to prove success for this program, we consider locations ℓ_1 and ℓ_2. Their exit conditions are respectively given by

$$E_{\ell_1} : H - L \geq 1 \qquad \text{and} \qquad E_{\ell_2} : x \neq a[m]$$

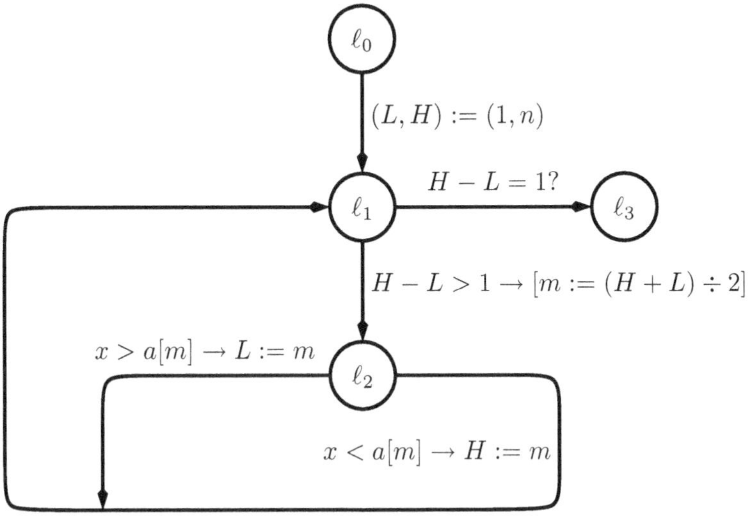

Figure 4. Program BINARY-SEARCH

As the cut-set, we take $\mathcal{C} = \{\ell_0, \ell_1, \ell_3\}$. The assertion network is given by

$$\begin{array}{ll} \varphi_0: & p \\ \varphi_1: & (1 \leq L < H \leq n) \ \wedge \ \forall i : [1..n] : x \neq a[i] \\ \varphi_3: & \mathit{True} \end{array}$$

Inductiveness of the network follows from the following property:

$$H - L > 1 \quad \rightarrow \quad L < (H + L) \div 2 < H$$

Absence of deadlock at locations ℓ_1 and ℓ_2 follows, respectively, from requirements *3* and *4* as follows:

$$\textit{3} \text{ for } \ell_1: \quad \underbrace{\cdots \wedge L < H \wedge \cdots}_{\varphi_1} \quad \rightarrow \quad \underbrace{H - L \geq 1}_{E_{\ell_1}}$$

$$\textit{4} \text{ for } \ell_2: \quad \underbrace{(1 \leq L < H \leq n) \ \wedge \ \forall i : [1..n] : x \neq a[i]}_{\varphi_1} \ \wedge \ \underbrace{H - L > 1}_{c_\pi} \quad \rightarrow \quad \underbrace{x \neq a[(H + L) \div 2]}_{E_{\ell_2}(f_\pi(V))}$$

5.3 Freedom from Faults

The correctness criterion and proof method for deadlock absence can be extended to guarantee *absence of faults*. Examples of faults during execution are: accessing a variable which has not been assigned a value, an array access with an out-of-range subscript, division by 0, arithmetic overflow, extracting the square root of a negative argument, etc.
An execution which does not generate any faults is called a *fault-free execution*. A program whose p-computations are all fault-free is called *p-fault-free*. For an assignment $\alpha : V := f(V)$, it is possible to formulate a *safety condition* Γ_α which guarantees fault freedom in the execution of α. For example, the safety condition for the assignment $\alpha : A[i] := sqrt(A[j]/k)$ is given by

$$\Gamma_\alpha : \quad i \in range(A) \ \land \ j \in range(A) \ \land \ k \neq 0 \ \land \ A[j]/k \geq 0$$

For a guarded command $\gamma : c \rightarrow [V := f(V)]$, we define the safety condition as $\Gamma_\gamma = \Gamma_c \ \land \ (c \rightarrow \Gamma_{V:=f(V)})$.
Let ℓ be a location in a program, and let $\gamma_1, \ldots, \gamma_k$ be the guarded commands labeling the edges departing from ℓ. Then, we define the safety condition for ℓ to be the conjunction $\Gamma_\ell : \quad \Gamma_{\gamma_1} \ \land \ \cdots \ \land \ \Gamma_{\gamma_k}$.
To prove that a program is p-fault free, we can use the proof method of Claim 10 where we replace the exit condition E_ℓ by the safety condition Γ_ℓ.

5.4 *Example:* Binary Search

Reconsider program BINARY-SEARCH of Fig. 4. The only non-trivial safety condition is $\Gamma_{\ell_2} : 1 \leq m \leq n$. To prove p-fault-freedom for this program, we take the same assertion network as before. The verification condition for location ℓ_2 is given by

$$\underbrace{(1 \leq L < H \leq n) \ \land \ \cdots}_{\varphi_1} \ \land \ \underbrace{H - L > 1}_{c_\pi} \quad \rightarrow \quad \underbrace{1 \leq (H + L) \div 2 \leq n}_{\Gamma_{\ell_2}(f_\pi(V))}$$

which is obviously valid.

6 Proving Termination

All proofs of termination (convergence) rely on the construction of *well founded ranking functions*.
We define a *well-founded domain* to be a pair $(\mathcal{A}, \succ)$ consisting of a domain $\mathcal{A}$ and an ordering relation $\succ$ over $\mathcal{A}$ such that there does not exist an infinitely descending sequence

$$a_0 \succ a_1 \succ \cdots$$

of $\mathcal{A}$-elements.

For example, the natural numbers with the $>$ ordering forms a well-founded domain, denoted $(\mathbb{N}, >)$. When there is no danger of confusion, we refer to the well-founded domain $(\mathcal{A}, \succ)$, simply as $\mathcal{A}$. For elements $a, b \in \mathcal{A}$, we write $a \succeq b$ if either $a \succ b$ or $a = b$.

6.1 Composite Well-Founded Domains

Given two well-founded domains $(\mathcal{A}_1, \succ_1)$ and $(\mathcal{A}_2, \succ_2)$, we introduce two ways to construct a composite well-founded domain.

The *cross product* $\mathcal{A}_1 \times \mathcal{A}_2$ is the well-founded domain $(\mathcal{A}, \succ)$, where $\mathcal{A} = \mathcal{A}_1 \times \mathcal{A}_2$ and

$$(a_1, a_2) \succ (b_1, b_2) \iff (a_1 \succ_1 b_1 \wedge a_2 \succeq_2 b_2) \vee (a_1 \succeq_1 b_1 \wedge a_2 \succ_2 b_2)$$

The *lexicographic product* $\mathcal{A}_1 \times_{lex} \mathcal{A}_2$ is the well-founded domain $(\mathcal{A}, \succ)$, where $\mathcal{A} = \mathcal{A}_1 \times \mathcal{A}_2$ and

$$(a_1, a_2) \succ_{lex} (b_1, b_2) \iff (a_1 \succ_1 b_1) \vee (a_1 = b_1 \wedge a_2 \succ_2 b_2)$$

CLAIM 11 *If both $(\mathcal{A}_1, \succ_1)$ and $(\mathcal{A}_2, \succ_2)$ are well-founded, then so are $\mathcal{A}_1 \times \mathcal{A}_2$ and $\mathcal{A}_1 \times_{lex} \mathcal{A}_2$.*

Proof. It is sufficient to show that $\mathcal{A}_1 \times_{lex} \mathcal{A}_2$ is well-founded.

Assume to the contrary, that there exists an infinitely descending sequence

$$(a_1, b_1) \succ_{lex} (a_2, b_2) \succ_{lex} \cdots$$

From the definition of $\succ_{lex}$ it follows that the sequence of first pair members satisfies $a_1 \succeq_1 a_2 \succeq_1 \cdots$. Since $\mathcal{A}_1$ is well founded, it follows that there exists some position k such that $a_k = a_{k+1} = \cdots$. Therefore, the sequence $b_k \succ_2 b_{k+1} \succ_2 \cdots$ must be infinitely descending, contradicting the well-foundedness of $\mathcal{A}_2$. ■

6.2 The Well-Founded Ranking Functions Method for Proving Termination of Programs

The following claim outlines and prove the soundness of the *ranking function* method for verifying convergence.

CLAIM 12 (Ranking Functions Method) *Let P be a program with a pre-condition specification p. Let $\mathcal{N} : \langle \mathcal{C}, \{\varphi_\ell \mid \ell \in \mathcal{C}\}\rangle$ be an assertion network, $(\mathcal{A}, \succ)$ be a well-founded domain, and $\{\delta_\ell \mid \ell \in \mathcal{C}\}$ be a network of ranking functions, each mapping states into elements of $\mathcal{A}$. If the following requirements are satisfied:*

1. *The network $\mathcal{N}$ is inductive.*

2. $p \rightarrow \varphi_{\ell_0}$.

3. *For every verification path π connecting location ℓ_i to location ℓ_j, the condition* $\varphi_i(V) \wedge c_\pi(V) \rightarrow \delta_i(V) \succ \delta_j(f_\pi(V))$ *is valid.*

then program P is p-convergent.

Proof. Assume to the contrary, that all requirements of the claim are met, yet there exists a divergent p-computation of the form

$$\sigma : \langle \ell_{i_0}, d_0 \rangle \xrightarrow{\pi_0} \langle \ell_{i_1}, d_1 \rangle \xrightarrow{\pi_1} \cdots \xrightarrow{\pi_{k-1}} \langle \ell_{i_k}, d_k \rangle \xrightarrow{\pi_k} \cdots$$

where we explicitly display the sequence of cut-points $\ell_0 = \ell_{i_0}, \ell_{i_1}, \ldots$ visited by σ and the verification paths $\pi_0, \pi_1, \ldots$ connecting them.
We can show by induction that the infinite sequence

$$\delta_{\ell_{i_0}}(d_0) \succ \delta_{\ell_{i_1}}(d_1) \succ \cdots \succ \delta_{\ell_{i_k}}(d_k) \succ \cdots$$

is an infinite descending chain of well-founded elements, which is impossible. ∎

6.3 Termination Networks

We refer to a network $\mathcal{N} = \langle \mathcal{C}, \{\varphi_\ell, \delta_\ell \mid \ell \in \mathcal{C}\} \rangle$, which assigns an assertion φ_ℓ and a ranking function δ_ℓ to each cut-point $\ell \in \mathcal{C}$ as a *termination network*. A termination network which satisfies requirements *1* and *3* of Claim 12 is called a valid network.

6.4 *Example:* Integer Square Root

Reconsider program INT-SQRT with the specification pre-condition $p : x \geq 0$.

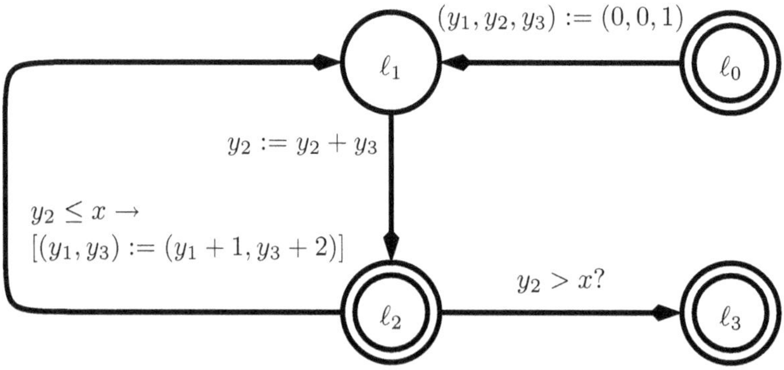

As a well-founded domain we choose $(\mathbb{N}, >)$. For the cut-set, we choose $\{\ell_0, \ell_2, \ell_3\}$. As assertion and ranking networks, we choose

ℓ_i	φ_{ℓ_i}	δ_{ℓ_i}
ℓ_0	$x \geq 0$	$x + 2$
ℓ_2	$y_3 > 0 \ \wedge\ y_2 \leq x + y_3 + 2$	$\lvert x + y_3 + 3 - y_2 \rvert$
ℓ_3	1	0

Examination of this example shows that checking descent along the paths $\ell_0 \to \ell_1 \to \ell_2$ and $\ell_2 \to \ell_3$ is completely unnecessary since such paths can appear at most once in any computation.

6.5 An Improved Method

To make use of the above observation, we introduce the notion of an acyclic decomposition.

An *acyclic decomposition* of a program P is a sequence of subsets of the nodes of P, $K_1, \ldots, K_k$ such that $K_1, \ldots, K_k$ is a partition of all the locations of P and, if there is an edge from a location $\ell_i \in K_i$ to a location $\ell_j \in K_j$ then ,necessarily $i \leq j$. We refer to $K_1, \ldots, K_k$ as the *components* of the decomposition. A trivial case of a decomposition is that of taking $K_1 = P$ yielding a decomposition with a single component.

The other extreme is when we decompose the graph of P into its *maximal strongly connected components MSCC*'s.

For a given acyclic decomposition, we say that a verification path is an *intra-component path* if it is fully contained in a single component of the decomposition.

CLAIM 13 (Improved Version) *In the application of the ranking function method, it is sufficient to require that condition 3 of Claim 12 holds for intra-component verification paths.*

Proof. Assume that we successfully managed to apply the improved method using the well-founded domain $(\mathcal{A}, \succ)$ with ranking functions $\{\delta_\ell \mid \ell \in \mathcal{C}\}$. We extend the well-founded domain into $\widetilde{\mathcal{A}} = [1..k] \times_{lex} \mathcal{A}$. As the extended ranking function, we take $\widetilde{\delta_\ell}(V) = (k+1-i, \delta_\ell(V))$ for every location $\ell \in \mathcal{C}$ which belongs to component K_i in the acyclic decomposition. ■

6.6 *Examples:* The GCD Program

We will illustrate the method on several examples. Consider first program GCD , presented in Fig. 5, for computing the gcd of two positive integers.

As the well-founded domain we choose $(\mathbb{N}, >)$. Decomposing the program into its MSCC yields the decomposition $\{\ell_0\}, \{\ell_1\}, \{\ell_2\}$. It follows that

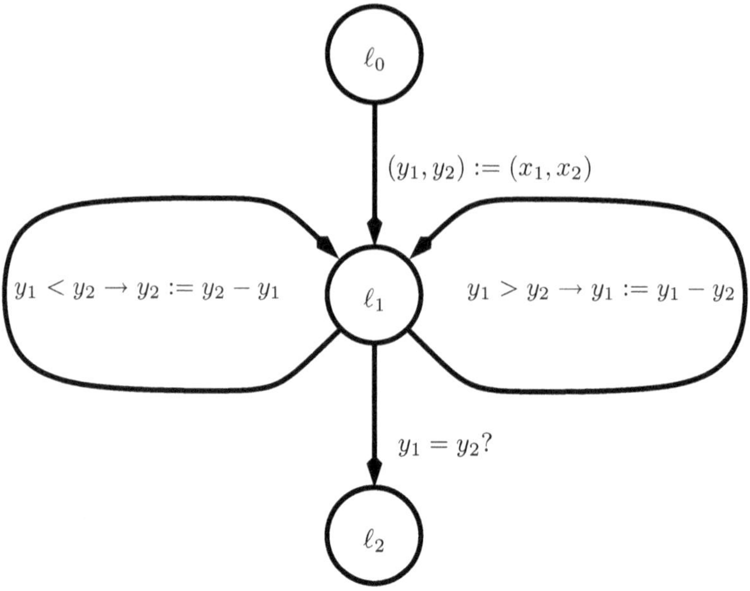

Figure 5. Program GCD

the only intra-component verification paths are $\ell_1 \rightarrow$ *left* $\rightarrow \ell_1$ and $\ell_1 \rightarrow$ *right* $\rightarrow \ell_1$. Consequently, we propose the following assertion and ranking-function networks:

ℓ_i	φ_{ℓ_i}	δ_{ℓ_i}
ℓ_0	$x_1 > 0 \wedge x_2 > 0$	0
ℓ_1	$y_1 > 0 \wedge y_2 > 0$	$\lvert y_1 + y_2 \rvert$
ℓ_2	1	0

For example, the descent condition for the path $\pi : \ell_1 \rightarrow$ *right* $\rightarrow \ell_1$ is given by

$$\underbrace{y_1 > 0 \wedge y_2 > 0}_{\varphi_1} \wedge \underbrace{y_1 > y_2}_{c_\pi} \quad \rightarrow \quad |y_1 + y_2| > |(y_1 - y_2) + y_2|$$

which is obviously valid.

6.7 Additional Example: *Integer Division*

As additional example, consider program INT-DIV , presented in Fig. 6. As the cut-set we choose $\{\ell_0, \ell_1, \ell_3, \ell_4\}$. An *MSCC* decomposition yields

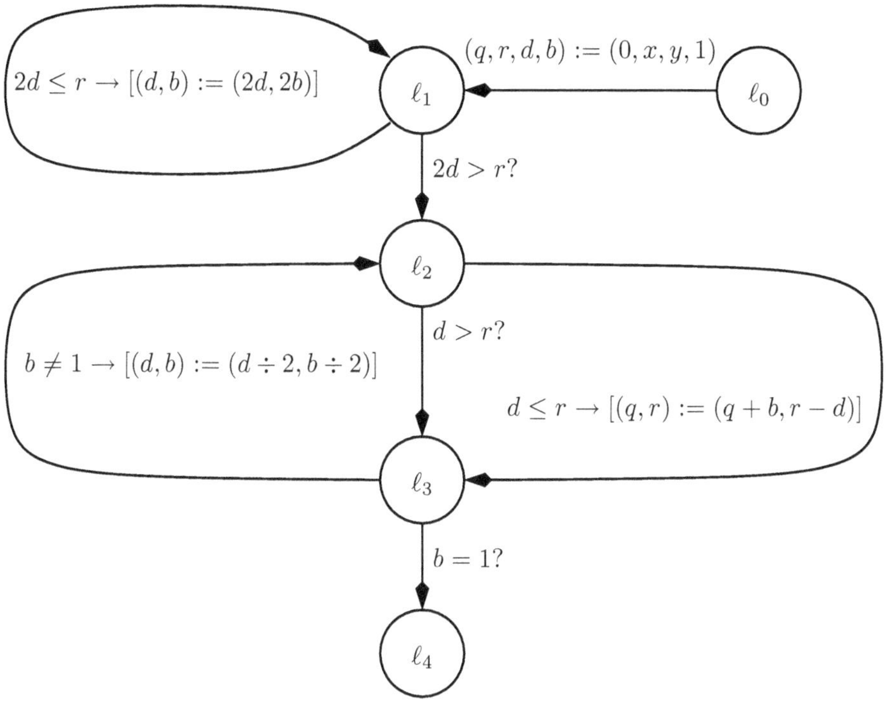

Figure 6. Program INT-DIV

$\{\ell_0\}, \{\ell_1\}, \{\ell_2, \ell_3\}, \{\ell_4\}$. Consequently, we take the termination network to be

ℓ_i	φ_i	δ_i
ℓ_0	$x \geq 0 \ \wedge \ y > 0$	0
ℓ_1	$b > 0 \ \wedge \ d > 0$	$\|r - d\|$
ℓ_3	$b > 0$	$\|b\|$
ℓ_4	1	0

For example, the descent condition for the path $\ell_1 \to \ell_1$ is

$$b > 0 \ \wedge \ d > 0 \ \wedge \ 2d \leq r \quad \to \quad |r - d| > |r - 2d|$$

6.8 *Example:* Bubble Sort

Next, consider program BUBBLE-SORT, presented in Fig. 7.

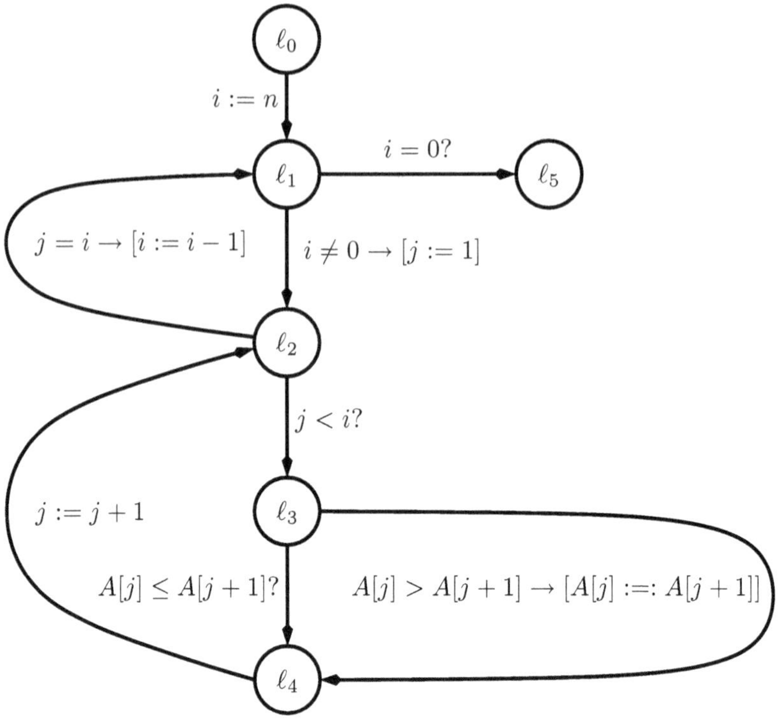

Figure 7. Program BUBBLE-SORT

As the well-founded domain we take $\mathbb{N} \times_{lex} \mathbb{N}$. The termination network is given by:

ℓ_i	φ_{ℓ_i}	δ_{ℓ_i}
ℓ_0	$n > 0$	0
ℓ_2	$1 \leq j \leq i \leq n$	$(\|i\|, \|i - j\|)$
ℓ_5	1	0

Obviously, on the path $\ell_2 \rightarrow \ell_1 \rightarrow \ell_2$, the rank $|i|$ decreases while $|i - j|$ may possibly increase. On both paths of the form $\ell_2 \rightarrow \ell_3 \rightarrow \ell_4 \rightarrow \ell_2$ rank $|i - j|$ decreases while $|i|$ stays the same.

6.9 Dependence on the Cut-Set

As for the inductive assertion method, we proceed to show that the proof method of well-founded ranking is independent of the particular cut-set selected for the proof.

CLAIM 14 (Termination networks can be extended)
Let $\mathcal{N} = \langle \mathcal{C}, \{\varphi_\ell, \delta_\ell \mid \ell \in \mathcal{C}\}\rangle$ be a p-valid termination network, and $\widetilde{\ell} \notin \mathcal{C}$ a location not in $\mathcal{C}$. There exists a p-valid termination network over the extended cut-set $\widetilde{\mathcal{C}} = \mathcal{C} \cup \{\widetilde{\ell}\}$ which agrees with $\mathcal{N}$ on the assertions φ_ℓ for all $\ell \in \mathcal{C}$.

Proof. Assume that the well-founded domain used for the network $\mathcal{N}$ is $(\mathcal{A}, \succ)$. The extended network has the form $\widetilde{\mathcal{N}} = \langle \widetilde{\mathcal{C}}, \{\psi_\ell, \widetilde{\delta}_\ell \mid \ell \in \widetilde{\mathcal{C}}\}\rangle$, where $\widetilde{\mathcal{C}} = \mathcal{C} \cup \{\widetilde{\ell}\}$. For all $\ell \in \mathcal{C}$, we take $\psi_\ell = \varphi_\ell$. For $\widetilde{\ell}$, we take $\psi_{\widetilde{\ell}} = pre(\widetilde{\ell}, \mathcal{N})$ as defined in the proof of Claim 5.

As the well-founded domain, we take $\mathcal{A} \times_{lex} \{0,1\} \cup \{\perp\}$, where $\perp$ is a special element, considered to be smaller than any element in $\mathcal{A} \times_{lex} \{0,1\}$. For all $\ell \in \mathcal{C}$, we take $\widetilde{\delta}_\ell = (\delta_\ell, 0)$. For $\widetilde{\ell}$, we take

$$\widetilde{\delta}_{\widetilde{\ell}}(V) = \max_{\pi \in \Pi_{\widetilde{\ell},\mathcal{C}}} \left(\textbf{if } c_\pi(V) \textbf{ then } \delta_{dest(\pi)}(f_\pi(V), 1) \textbf{ else } \perp\right)$$

Thus, if there exists at least one path π connecting $\widetilde{\ell}$ to some $\ell \in \mathcal{C}$ such that V satisfies c_π, then we take $\widetilde{\delta}_{\widetilde{\ell}}(V)$ to be the maximum of $\delta_{dest(\pi)}(f_\pi(V), 1)$ over all such paths. On the other hand, if for all such paths π $V \not\models c_\pi$, then $\widetilde{\delta}_{\widetilde{\ell}}(V) = \perp$.

Invoking the proof of Claim 5, we establish that the assertion network $\{\psi_\ell \mid \ell \in \widetilde{\mathcal{C}}\}$ is inductive. It remains to show that the descent condition *3* holds for every verification path π in the cut-set $\widetilde{\mathcal{C}}$. Again, it is sufficient to consider only "new" verification paths, i.e., paths which either depart or arrive to the new cut-point $\widetilde{\ell}$.

Paths departing from $\widetilde{\ell}$
Let π_2 be a "new" verification path leading from $\widetilde{\ell}$ to $\ell_2 \in \mathcal{C}$, and let V be a data state such that $V \models c_\pi$. Since $\pi_2 \in \Pi_{\widetilde{\ell},\mathcal{C}}$ and $V \models c_\pi$, π is one of the paths over which the maximum is taken in the definition of $\widetilde{\delta}_{\widetilde{\ell}}(V)$. It follows that $\widetilde{\delta}_{\widetilde{\ell}}(V) \succeq (\delta_{\ell_2}(f_{\pi_2}(V)), 1)$, which implies $\widetilde{\delta}_{\widetilde{\ell}}(V) \succ (\delta_{\ell_2}(f_{\pi_2}(V)), 0) = \widetilde{\delta}_{\ell_2}(f_{\pi_2}(V))$. We can therefore conclude the descent condition for path π_2 which requires

$$\textit{3.} \qquad \psi_{\widetilde{\ell}}(V) \wedge c_{\pi_2}(V) \quad \rightarrow \quad \widetilde{\delta}_{\widetilde{\ell}}(V) \succ \widetilde{\delta}_{\ell_2}(f_{\pi_2}(V))$$

Paths Arriving to $\widetilde{\ell}$
Let π_1 be a "new" verification path leading from $\ell_1 \in \mathcal{C}$ to $\widetilde{\ell}$, and let V_1 be a data state satisfying $\psi_{\ell_1} \wedge c_{\pi_1}$. Let $V = f_{\pi_1}(V_1)$ be the data state obtained at the end of path π_1, when execution reaches $\widetilde{\ell}$. We consider two cases:

- In the first case, there is no path π_2 connecting $\widetilde{\ell}$ to $\mathcal{C}$ such that $V \models c_{\pi_2}$. In this case $\widetilde{\delta}_{\widetilde{\ell}}(V) = \perp$ and there is obviously a descent between $\widetilde{\delta}_{\ell_1}(V_1) = (\delta_{\ell_1}(V_1), 0)$ and $\widetilde{\delta}_{\widetilde{\ell}}(V) = \widetilde{\delta}_{\widetilde{\ell}}(f_{\pi_1}(V_1)) = \perp$.
- In the other case, there exist one or more paths π_2 connecting $\widetilde{\ell}$ to $\mathcal{C}$ such that $V \models c_{\pi_2}$. Let us pick the path π_2 such that $\widetilde{\delta}_{\widetilde{\ell}}(V) = (\delta_{\ell_2}(f_{\pi_2}(V)), 1)$. Assume that π_2 connects $\widetilde{\ell}$ to $\ell_2 \in \mathcal{C}$. The combined path $\pi = \pi_1 \circ \pi_2$ is one of the "old" verification paths and leads from ℓ_1 to ℓ_2. Since we assumed that the termination network $\mathcal{N}$ satisfied all the descent conditions, it satisfies in particular

$$\varphi_{\ell_1}(V_1) \wedge c_\pi(V_1) \quad \rightarrow \quad \delta_{\ell_1}(V_1) \succ \delta_{\ell_2}(f_\pi(V_1))$$

Recall that the traversal condition and data transformation of a composed path are related to these of its constituents by

$$c_\pi(V_1) = c_{\pi_1}(V_1) \wedge c_{\pi_2}(f_{\pi_1}(V_1)) \qquad \text{and} \qquad f_\pi(V_1) = f_{\pi_2}(f_{\pi_1}(V_1))$$

Using these relations and the facts that $\psi_{\ell_1} = \varphi_{\ell_1}$, $V_1 \models \psi_{\ell_1} \wedge c_{\pi_1}$, $V = f_{\pi_1}(V_1)$, $V \models c_{\pi_2}$, we can conclude that $\delta_{\ell_1}(V_1) \succ \delta_{\ell_2}(f_\pi(V_1)) = \delta_{\ell_2}(f_{\pi_2}(V))$. It follows that

$$\widetilde{\delta}_{\ell_1}(V_1) \;=\; (\delta_{\ell_1}(V_1), 0) \;\succ\; (\delta_{\ell_2}(f_{\pi_2}(V)), 1) \;=\; \widetilde{\delta}_{\widetilde{\ell}}(V) \;=\; \widetilde{\delta}_{\widetilde{\ell}}(f_{\pi_1}(V))$$

establishing the descent from ℓ_1 to $\widetilde{\ell}$. ■

6.10 Removing Cut-Points

In the previous discussion we have shown that it is always possible to add more cut-points to a valid termination network, while maintaining validity. We will now show that it is also possible to *remove* cut-points, provided the remaining set is still a cut-set.

CLAIM 15 *Let $\mathcal{N} = \langle \mathcal{C}, \{\varphi_\ell, \delta_\ell \mid \ell \in \mathcal{C}\}\rangle$ be a valid termination network. Let $\widetilde{\ell} \in \mathcal{C}$ be a location in $\mathcal{C}$ such that $\overline{\mathcal{C}} = \mathcal{C} - \{\widetilde{\ell}\}$ is a cut-set. Then the network $\overline{\mathcal{N}} = \langle \overline{\mathcal{C}}, \{\varphi_\ell, \delta_\ell \mid \ell \in \overline{\mathcal{C}}\}\rangle$, obtained by removing $\widetilde{\ell}$ and $\varphi_{\widetilde{\ell}}$ from $\mathcal{N}$, is also valid.*

Proof. Inductiveness of the reduced assertion network follows from the arguments of Claim 6. It remains to show that the reduced ranking functions maintain the descent conditions over all verification paths.

We only need to consider "new" verification paths, i.e. paths which exist in $\overline{\mathcal{N}}$ but not in $\mathcal{N}$. Such a path π connecting $\ell_1 \in \overline{\mathcal{C}}$ to $\ell_2 \in \overline{\mathcal{C}}$ must be the fusion $\pi = \pi_1 \circ \pi_2$ of two paths, where path π_1 connects ℓ_1 to $\widetilde{\ell}$, while π_2 connects $\widetilde{\ell}$ to ℓ_2. Let V_1 be a data state which satisfies $\varphi_1(V_1)$ and $c_\pi(V_1) = c_{\pi_1}(V_1) \wedge c_{\pi_2}(f_{\pi_1}(V_1))$. Let $V = f_{\pi_1}(V_1)$ be the data state arising at $\widetilde{\ell}$ while traversing the path π. Due to the $\mathcal{N}$-valid verification condition $\varphi_{\ell_1}(V_1) \wedge c_{\pi_1}(V_1) \rightarrow \varphi_{\widetilde{\ell}}(f_{\pi_1}(V_1))$ and the $\mathcal{N}$-valid descent condition $\varphi_{\ell_1}(V_1) \wedge c_{\pi_1}(V_1) \rightarrow \delta_{\ell_1}(V_1) \succ \delta_{\widetilde{\ell}}(f_{\pi_1}(V_1))$, we can conclude that $\varphi_{\widetilde{\ell}}(V) = 1$ and $\delta_{\ell_1}(V_1) \succ \delta_{\widetilde{\ell}}(V)$. As $c_{\pi_2}(V) = c_{\pi_2}(f_{\pi_1}(V_1))$ is implied by $c_\pi(V_1)$, we can use the $\mathcal{N}$-valid descent condition $\varphi_{\widetilde{\ell}}(V) \wedge c_{\pi_2}(V) \rightarrow \delta_{\widetilde{\ell}}(V) \succ \delta_{\ell_2}(f_{\pi_2}(V))$, to conclude $\delta_{\widetilde{\ell}}(V) \succ \delta_{\ell_2}(f_{\pi_2}(V)) = \delta_{\ell_2}(f_\pi(V_1))$. Together with $\delta_{\ell_1}(V_1) \succ \delta_{\widetilde{\ell}}(V)$, this implies

$$\delta_{\ell_1}(V_1) \quad \succ \quad \delta_{\ell_2}(f_\pi(V_1))$$

as required. ■

6.11 Method is Independent of the Choice of the Cut-Set

The preceding discussions can be summarized by the statement that the success or failure of an application of the ranking functions method is independent of the particular choice of the cut-set $\mathcal{C}$.

Technically, this can be summarized by the following corollary:

COROLLARY 16 *Let p be a pre-condition specification for program P, and let $\mathcal{C}_1$ and $\mathcal{C}_2$ be two cut-sets. Then, there exists a valid termination network $\mathcal{N}_1$ based on $\mathcal{C}_1$ and entailing p iff there exists a valid termination network $\mathcal{N}_2$ based on $\mathcal{C}_2$ and entailing p.*

The proof of this statement can be obtained by starting with a $\mathcal{C}_1$-based valid termination network $\mathcal{N}_1$ and incrementally adding missing locations, using Claim 14, until we obtain a full network. Then we start removing locations which do not belong to $\mathcal{C}_2$, relying on Claim 15, until we obtain a network $\mathcal{N}_2$ based on $\mathcal{C}_2$. Since ℓ_0 belongs to both $\mathcal{N}_1$ and $\mathcal{N}_2$, its associated assertion is preserved throughout the entire process. Therefore, if $\mathcal{N}_1$ entails p then so does $\mathcal{N}_2$.

6.12 König's Lemma

In preparation for proving the completeness of the ranking function method, we present a useful lemma due to König [Kleene, 2002]. A tree is said to be of *finite degree* if each node has only finitely many direct descendants. A tree is called *infinite* if it contains infinitely many nodes.

LEMMA 17 (König) *An infinite tree of finite degree contains an infinite path.*

Proof. Let T be an infinite tree of finite degree. Let n_0 be the root of the tree and $n_1, \ldots, n_k$ its immediate descendants. We represent this configuration in Fig. 8.

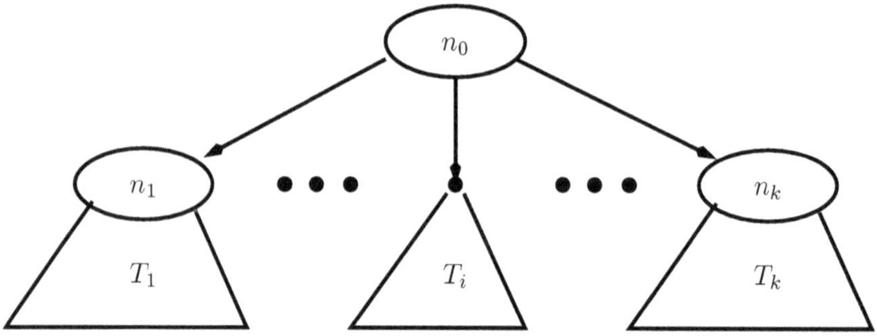

Figure 8. An infinite tree of Finite Degree

Each of the descendants $n_1, \ldots, n_k$ is the root of a subtree. Since T is infinite, at least one of the subtrees $T_1, \ldots, T_k$ must be infinite. Assume it is T_i. Consider $n_{i1}, \ldots, n_{im}$, the immediate descendants of n_i. They, in turn, are the roots of a set of subtrees, at least one of which must be infinite. Continuing in this manner, we trace an infinite path within the tree T where, at each stage, we consider a root of an infinite subtree. It follows that T contains an infinite path. ■

6.13 Completeness of the Method

Finally, we state and prove the completeness of the ranking functions method for verifying p-termination of programs.

CLAIM 18 (Completeness of the *ranking functions* method) *Let P be a program which is p-terminating. Then there exists a valid termination network which entails p.*

Proof. Let $\mathcal{C}$ be an arbitrary cut-set. As the assertion associated with location ℓ we take the minimal predicate M_ℓ, as defined in the proof of Claim 8. As shown in that proof, this assertion network is inductive and entails p.
As the well-founded domain, we take $(\mathbb{N}, >)$. As the ranking function δ_ℓ, we take the function $L_\ell(V)$ whose value for $V = d$ is defined by

> *If $d \not\models \varphi_\ell$ then $L_\ell(d) = 0$. Otherwise, $L_\ell(d)$ equals the length of the longest computation segment, initiated at location ℓ with data state $V = d$.*

We have to show that $L_\ell(d)$ is always defined. The only possibility for $L_\ell(d)$ to be undefined is that $d \models \varphi_\ell$ but the length of d-computation segments departing from ℓ is unbounded. In this case, we construct a tree T as follows:

The root of T is the execution state $\langle \ell, d \rangle$. As immediate descendants of $\langle \ell, d \rangle$ we list the execution states $\langle \ell_1, d_1 \rangle, \cdots, \langle \ell_k, d_k \rangle$ which can be obtained by a single computation step from $\langle \ell, d \rangle$. As direct descendants of these second generation nodes, we list all executions states which can be obtained by a single computation step from these states. The resulting tree T is presented in Fig. 9.

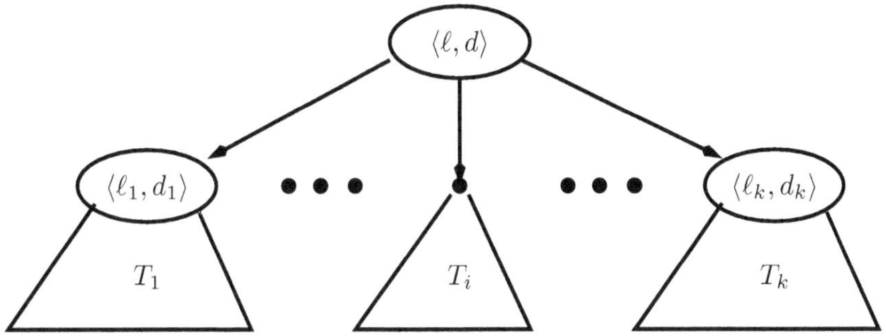

Figure 9. The constructed tree T

The tree T is obviously of finite degree, since there are only finitely many edges departing from each location in the program graph. According to *König*'s lemma, either the tree is finite, or it has an infinite path.

An infinite path in this tree implies the existence of an infinite computation segment π^∞ whose first state is $\langle \ell, d \rangle$. Since $d \models \varphi_\ell = M_\ell$ there exists a p-computation segment of the form $\pi = \langle \ell_0, d_0 \rangle, \ldots, \langle \ell, d \rangle$. Obviously, the concatenation $\pi \circ \pi^\infty$ is a divergent p-computation of program P which, according to the hypothesis of the claim, is impossible.

We thus conclude that the tree rooted at $\langle \ell_0, d \rangle$ is finite, which implies that there exists a longest computation segment departing from $\langle \ell_0, d \rangle$. This shows that the function $L_\ell(V)$ is always defined.

Proof Concluded

It remains to show that the ranking functions $L_\ell(V)$ satisfy the descent conditions. Consider a verification path π leading from ℓ_1 to ℓ_2. Assume that $\varphi_{\ell_1}(d_1) = c_\pi(d_1) = 1$. We have to show that $\delta_{\ell_1}(d_1) > \delta_{\ell_2}(f_\pi(d_1))$.

Let $d_2 = f_\pi(d_1)$. Since $\varphi_{\ell_1}(d_1) = c_\pi(d_1) = 1$ and the assertion network is inductive, we can conclude that $\varphi_{\ell_2}(d_2) = 1$. Therefore, $L_{\ell_2}(d_2)$ is the length of some computation segment $\sigma = \langle \ell_2, d_2 \rangle, \ldots$. Similarly, $L_{\ell_1}(d_1)$ is the length of the longest computation segment departing from $\langle \ell_1, d_1 \rangle$. Since $\pi \circ \sigma$ is one of the computation segments departing from $\langle \ell_1, d_1 \rangle$, we have that $L_{\ell_1}(d_1) \geq |\pi| + |\sigma| = |\pi| + L_{\ell_2}(d_2)$. Since $|\pi| > 0$, we conclude that $L_{\ell_1}(d_1) > L_{\ell_2}(d_2)$. ■

6.14 Additional Examples: An Efficient GCD Program

We conclude our discussion of methods for proving termination by presenting two more examples. The first is an efficient program for computing the *gcd* of two positive integers, using subtractions and left and right shifts (multiplication and division by 2). Program EFFICIENT-GCD is presented in Fig. 10.

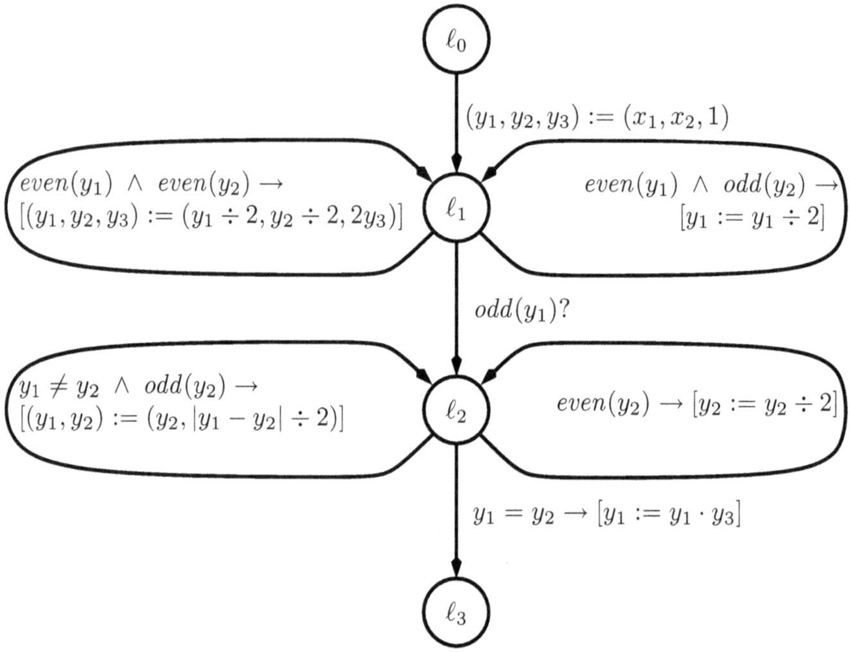

Figure 10. Program EFFICIENT-GCD

As our assertion network, we take

$$\begin{aligned} \varphi_0 &: \quad x_1 > 0 \wedge x_2 > 0 \\ \varphi_1 &: \quad y_1 > 0 \wedge y_2 > 0 \\ \varphi_2 &: \quad y_1 > 0 \wedge y_2 > 0 \wedge odd(y_1) \end{aligned}$$

In order to prove partial correctness we could just add the conjunct $\gcd(x_1, x_2) = \gcd(y_1, y_2) \cdot y_3$ to φ_1 and φ_2. As ranking function for ℓ_1, we can take

$$\delta_1 : \quad y_1$$

To choose a ranking function at ℓ_2 we try a liner combination of the form $ay_1 + by_2$. This linear combination must be descending on all paths from ℓ_2 to itself. This leads to the following inequalities:

$$\begin{array}{rcl} ay_1 + by_2 & > & ay_1 + b\frac{y_2}{2} \\ ay_1 + by_2 & > & ay_2 + b\frac{|y_1 - y_2|}{2} \end{array} \quad \sim \quad \begin{array}{rcl} 2ay_1 + 2by_2 & > & 2ay_1 + by_2 \\ 2ay_1 + 2by_2 & > & 2ay_2 + b|y_1 - y_2| \end{array}$$

The first inequality can be simplified to $b > 0$. The second inequality can be expanded into two cases, depending on whether $y_1 \geq y_2$ or $y_1 < y_2$. This give rise to the following two inequalities:

$$\begin{array}{rcl} 2ay_1 + 2by_2 & > & 2ay_2 + b(y_1 - y_2) \\ 2ay_1 + 2by_2 & > & 2ay_2 + b(y_2 - y_1) \end{array}$$

Collecting the coefficients of y_1 and y_2, we get

$$\begin{array}{lcl} y_1(2a - b) + y_2(3b - 2a) & > & 0 \\ y_1(2a + b) + y_2(b - 2a) & > & 0 \end{array}$$

Since these inequalities should hold for arbitrary positive values of y_1 and y_2 we must have all coefficients nonnegative, and their sum positive. Thus, we have to find a and b satisfying

$$\begin{array}{lll} b > 0 & 2a - b \geq 0 & 3b - 2a \geq 0 \\ & 2a + b \geq 0 & b - 2a \geq 0 \end{array}$$

A possible solution is $a = 1$, $b = 2$, which leads to the ranking function

$$\delta_2 : \quad y_1 + 2 \cdot y_2$$

guaranteed to descend on all self edges departing from ℓ_2.

6.15 Structured GCD

Consider program STRUCTURED-GCD (presented in Fig. 11) which also computes the *gcd* of two positive integers.

As the assertion network we can take

$$\begin{array}{ll} \varphi_0 : & x_1 > 0 \ \wedge \ x_2 > 0 \\ \varphi_2 = \varphi_3 : & y_1 > 0 \ \wedge \ y_2 > 0 \end{array}$$

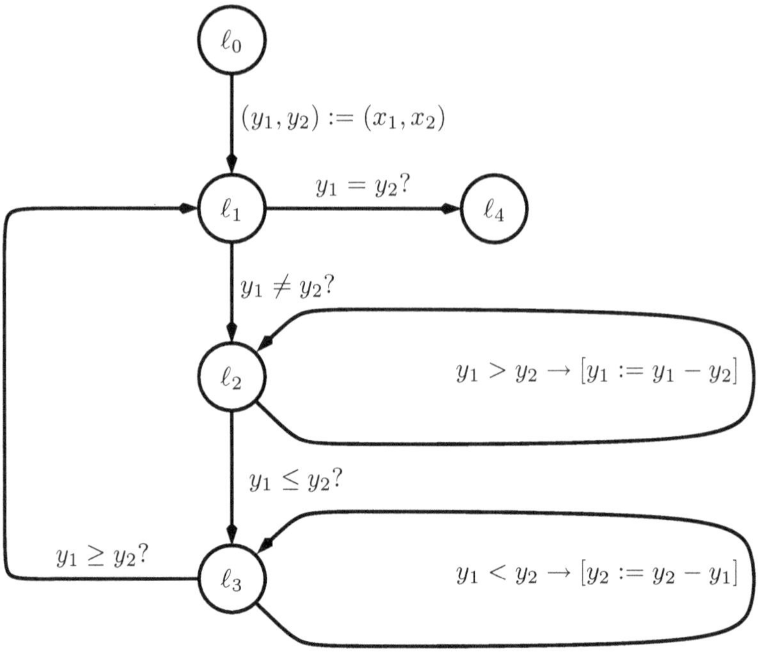

Figure 11. Program STRUCTURED-GCD

It is obvious that the function $\delta : y_1 + y_2$ should be a component of the ranking at both ℓ_2 and ℓ_3. However, δ by itself is insufficient, because it dose not decrease on the verification path $\ell_2 \rightarrow \ell_3$. Therefore, we will be looking for ranking functions of the form $\delta_i : (y_1 + y_2, \eta_i)$ for $i = 2, 3$.

This characteristic example has the feature that it is easy to identify a *primary ranking function* (i.e., $y_1 + y_2$) which is known to decrease on every loop. The question is how to augment it by the secondary rank $\eta_i(V)$ such that there will be a decrease on every intra-component verification path.

A heuristics that often works is to let η_i be the maximal number of cut-points which are encountered on a path departing from ℓ_i until we reach an edge which causes the primary rank to decrease or until we reach the terminal location ℓ_t. For example, η_2 can be defined as follows, distinguishing between three cases:

$y_1 > y_2$	0	The first edge departing from ℓ_2 decrements $y_1 + y_2$
$y_1 < y_2$	1	The path departing from ℓ_2 is $\ell_2 \to \ell_1 \to decrement(y_1 + y_2)$
$y_1 = y_2$	2	The path departing from ℓ_2 is $\ell_2 \to \ell_1 \to \ell_4$

Consequently, we can take

$$\delta_2(y_1, y_2) = (y_1 + y_2,\ \textbf{if } y_1 > y_2 \textbf{ then } 0 \textbf{ else-if } y_1 < y_2 \textbf{ then } 1 \textbf{ else } 2)$$

Similarly, we can take

$$\delta_3(y_1, y_2) = (y_1 + y_2,\ \textbf{if } y_1 > y_2 \textbf{ then } 1 \textbf{ else-if } y_1 < y_2 \textbf{ then } 0 \textbf{ else } 1)$$

A more compact representation of similar ranking functions can be given by

$$\begin{array}{ll} \delta_2: & (y_1 + y_2,\ 2(y1 \leq y_2)) \\ \delta_3: & (y_1 + y_2,\ \quad y_1 \geq y_2) \end{array}$$

7 Procedural Programs

We will now extend our treatment of programs to the consideration of programs with procedures. A program P in the extended language consists of $m{+}1$ *modules*: $P_0, P_1, \ldots, P_m$, where P_0 is the *main* module, and $P_1, \ldots, P_m$ are *procedures* which may be called from P_0 or from other procedures.

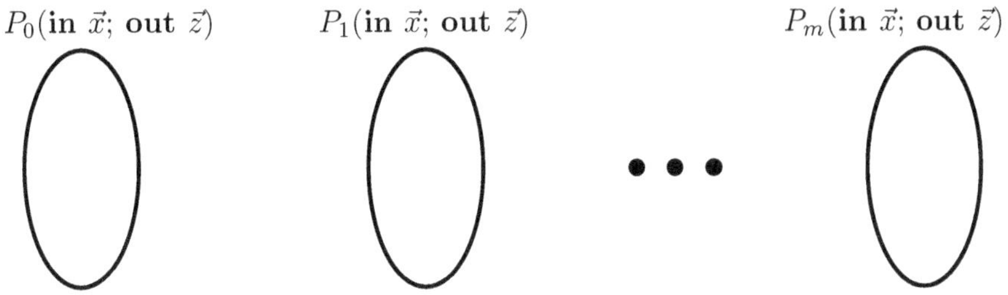

Each module P_i is presented as a flow-graph with its own set of locations $\mathcal{L}_i = \{\ell_0^i, \ell_1^i, \ldots, \ell_t^i\}$. It must have ℓ_0^i as its only entry point, ℓ_t^i as its only exit, and every other location must be on a path from ℓ_0^i to ℓ_t^i.

The variables of each module P_i are partitioned into $\vec{y} = (\vec{x}; \vec{u}; \vec{z})$. We refer to $\vec{x}$,$\vec{y}$, and $\vec{z}$ as the *input*, *working*, and *output* variables, respectively. A module cannot modify its own input variables.
Edges in the graph are labeled by an instruction which must be one of

- An *assignment* $c(\vec{y}) \to [\vec{v} := f(\vec{y})]$, where the left-hand side variables $\vec{v} \subseteq \{\vec{u}, \vec{z}\}$ may not include any member of $\vec{x}$.

- A *procedure call* $c(\vec{y}) \rightarrow P_j(\vec{e}; \vec{v})$, where $\vec{e}$ is a list of expressions over $\vec{y}$, and $\vec{v} \subseteq \{\vec{u}, \vec{z}\}$ is a list of distinct variables not including any member of $\vec{x}$. We refer to $\vec{e}$ and $\vec{y}$ as the *actual arguments* of the call.

7.1 Computations of Procedural Programs

A *ξ-computation* of module P_i is a sequence of states and their labeled transitions:

$$\sigma : \langle \ell_0^i; (\xi, \vec{\bot}, \vec{\bot}) \rangle \xrightarrow{\lambda_1} \langle \ell^1; \vec{d_1} \rangle \xrightarrow{\lambda_2} \langle \ell^2; \vec{d_2} \rangle \cdots$$

The values $\vec{\bot}$ denote uninitialized values. Labels in the transitions are either names of edges in the program or the special label *return*. Each transition $\langle \ell; \vec{d} \rangle \xrightarrow{\lambda} \langle \ell'; \vec{d'} \rangle$ in a computation must be justified by one of the following cases:

Assignment There exists an edge e in the program P (not necessarily in P_i)

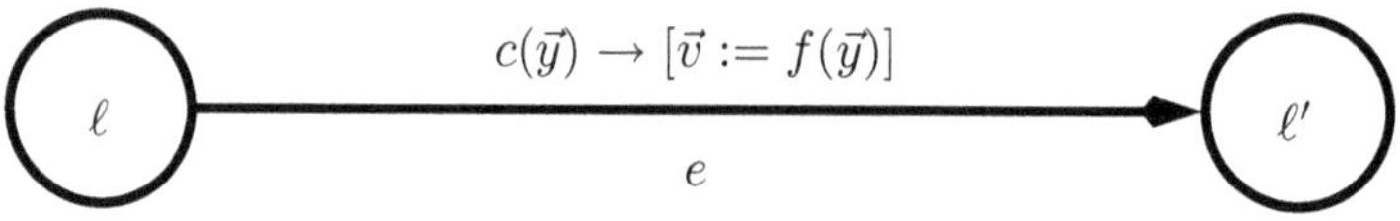

such that $\lambda = e$, $\vec{d} \models c$, and $\vec{d'} = (\vec{d}$ **with** $\vec{v} = f(\vec{d}))$, i.e. $\vec{d'}$ is obtained from $\vec{d}$ by replacing the values corresponding to the variables $\vec{v}$ by $f(\vec{d})$.

Procedure Call There exists an edge e in the program P

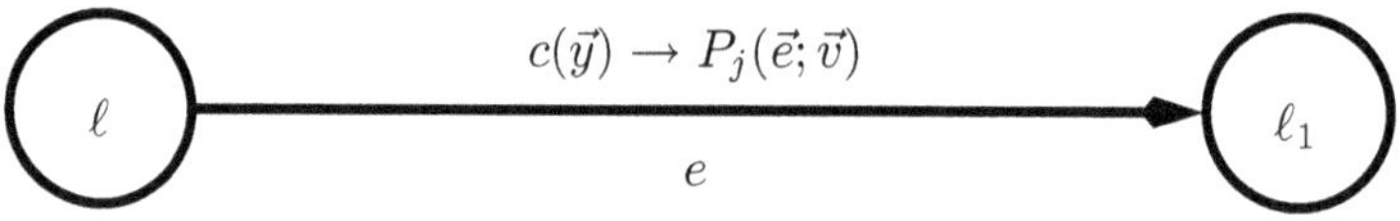

such that $\lambda = e$, $\vec{d} \models c$, $\ell' = \ell_0^j$, and $\vec{d'} = (\vec{e}(\vec{d}); \vec{\bot}; \vec{\bot})$. Thus, $\ell' = \ell_0^j$ is the first location in the called procedure P_j, and $\vec{d'}$ are the initial values on entry to P_j. We assume that the working and result variables in P_j are uninitialized.

7.2 Computations Continued: Procedure Return

Finally we consider a transition $\langle \ell; (\xi; \eta; \zeta) \rangle \overset{return}{\longrightarrow} \langle \ell'; \vec{d'} \rangle$. To justify such a transition, there must exists a procedure P_j (the procedure from which we return), such that $\ell = \ell_t^j$ (the terminal location of P_j), and we should be able to identify a suffix of the current computation of the form

$$\langle \ell_1; \vec{d_1} \rangle \xrightarrow{e_1} \underbrace{\langle \ell_0^j; (\xi; \vec{\bot}; \vec{\bot}) \rangle \xrightarrow{e_2} \cdots \xrightarrow{e_k} \langle \ell; (\xi; \eta; \zeta) \rangle}_{\sigma_1}$$

such that the segment σ_1 is *balanced* (has an equal number of *call*'s and *return*'s), e_1 is a call edge of the form

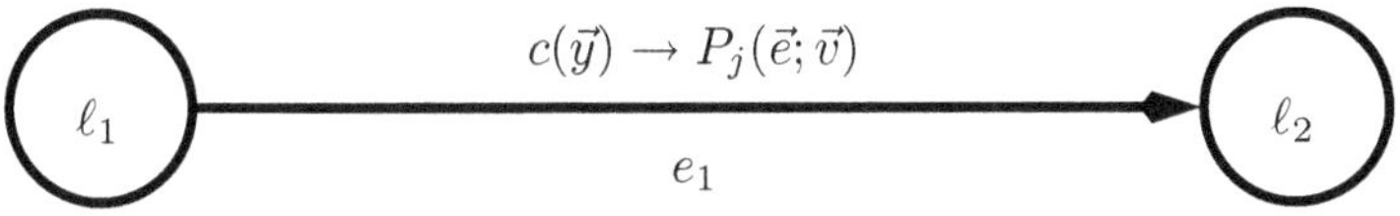

$\ell' = \ell_2$, and $\vec{d'} = (\vec{d_1}$ **with** $\vec{v} = \zeta)$.

7.3 Results of Computations

Given a computation σ, it is possible to assign to each execution state in σ a *depth* which is a natural number equal to $\#(call) - \#(return)$ from the beginning of the computation up to the current state. A computation is called *maximal* if it cannot be extended any further.
Maximal computations fall into one of the possible categories:

- *Terminating computations* – The computation σ is finite, and its last state is $\langle \ell_t^i; (\xi; \eta; \zeta) \rangle$. We define

  $$val(\sigma) = \zeta$$

- *Failing Computations* – The computation σ is finite, but its last location is not a terminal location of any procedure. This is a case of a deadlock and we write

  $$val(\sigma) = fail$$

- *Divergent computations* – σ is infinite. Define

  $$val(\sigma) = \perp$$

For a module P_i, we define the meaning of P_i to be the set of all possible outcomes.

$$\mathcal{M}[P_i](\xi) = \mathcal{M}(P_i; \xi) = \{val(\sigma) \mid \sigma \text{ is a maximal } \xi\text{-computation of } P_i\}$$

For the entire program P, we define

$$\mathcal{M}[P] = \mathcal{M}[P_0]$$

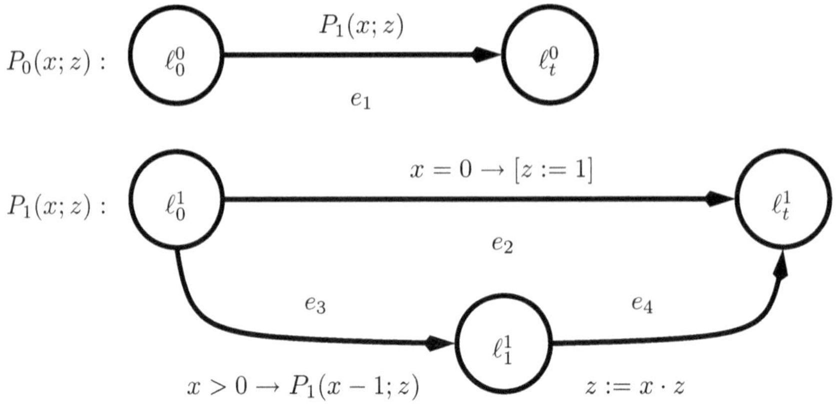

Figure 12. Program FACTORIAL

7.4 Example: Factorial

In Fig. 12 we present program FACTORIAL which computes the factorial of a natural number.

Following is a computation of this program for input $x = 3$:

$$\begin{array}{l}
\langle \ell_0^0;\ (3,\bot)\rangle \xrightarrow{e_1} \\
\quad \langle \ell_0^1;\ (3,\bot)\rangle \xrightarrow{e_3} \\
\quad\quad \langle \ell_0^1;\ (2,\bot)\rangle \xrightarrow{e_3} \\
\quad\quad\quad \langle \ell_0^1;\ (1,\bot)\rangle \xrightarrow{e_3} \\
\quad\quad\quad\quad \langle \ell_0^1;\ (0,\bot)\rangle \xrightarrow{e_2} \langle \ell_t^1;\ (0,1)\rangle \xrightarrow{return} \\
\quad\quad\quad \langle \ell_1^1;\ (1,1)\rangle \xrightarrow{e_4} \langle \ell_t^1;\ (1,1)\rangle \xrightarrow{return} \\
\quad\quad \langle \ell_1^1;\ (2,1)\rangle \xrightarrow{e_4} \langle \ell_t^1;\ (2,2)\rangle \xrightarrow{return} \\
\quad \langle \ell_1^1;\ (3,2)\rangle \xrightarrow{e_4} \langle \ell_t^1;\ (3,6)\rangle \xrightarrow{return} \\
\langle \ell_t^0;\ (3,6)\rangle
\end{array}$$

Consequently, $\mathcal{M}(\mathit{factorial}, 3) = 6$.

7.5 Proving Partial Correctness

We extend the inductive assertion method to deal with *procedural programs*. A *cut-set* $\mathcal{C}$ is a set of locations in $\mathcal{L} = \mathcal{L}_0 \cup \cdots \cup \mathcal{L}_m$ such that:

1. Every loop in each P_i, $i = 0, \ldots, m$ contains at least one location of $\mathcal{C}$.

2. For every $i = 0, \ldots, m$, both ℓ_0^i and ℓ_t^i belong to $\mathcal{C}$.

2. For every edge $\ell_i \xrightarrow{e} \ell_j$ labeled by a procedure call, both ℓ_i and ℓ_j belong to $\mathcal{C}$.

An *assertion network* associates an assertion $\varphi_i^j(\vec{y})$ with each location ℓ_i^j. For each module P_k, we denote φ_0^k by p_k and require that $p_k = p_k(\vec{x})$ depends only on the input variables of the module. Similarly, we denote φ_t^k by q_k and require that $q_k = q_k(\vec{x}; \vec{z})$ depends only on the input and output variables of the module.

The *input predicate* $p_k(\vec{x})$ imposes constraints on the input variables we expect on entry to module P_k. The *output predicate* $q_k(\vec{x}; \vec{z})$ specifies the relation between the output results and the input values.

The Verification Conditions

We consider two types of verification conditions.

Let π be a verification path leading from location ℓ_i to location ℓ_j such that all edges in π are labeled by guarded assignment instructions. We refer to such a path as an *assignment path*. As usual, let c_π denote the traversal condition for π, and let $\vec{y} := f_\pi(\vec{y})$ summarize the data transformation effected by the execution of the path. With such a path we associate the following verification condition:

$$V_\pi : \quad \varphi_i(\vec{y}) \wedge c_\pi(\vec{y}) \quad \rightarrow \quad \varphi_j(f_\pi(\vec{y}))$$

The other type of verification condition is associated with a *procedure call*. Consider an edge of the following form:

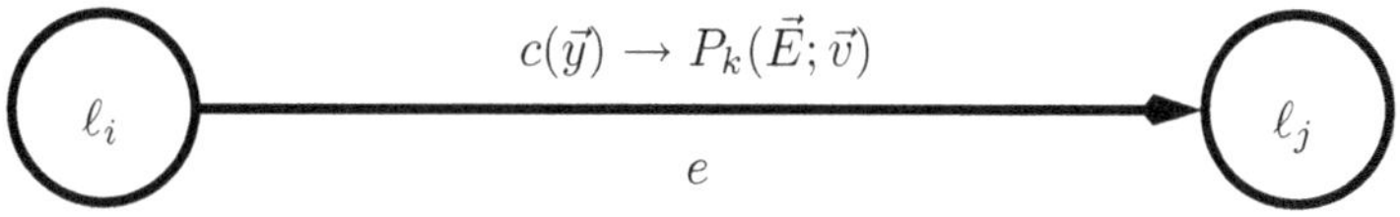

With the (length one) verification path e, we associate the following two verification conditions:

$$\begin{array}{llcl} V_{entry}: & \varphi_i(\vec{y}) \wedge c(\vec{y}) & \rightarrow & p_k(\vec{E}(\vec{y})) \\ V_{exit}: & \varphi_i(\vec{y}) \wedge c(\vec{y}) \wedge q_k(\vec{E}(\vec{y}); \vec{z}') & \rightarrow & \varphi_j(\vec{y})[\vec{v} \mapsto \vec{z}'] \end{array}$$

where $\varphi_j(\vec{y})[\vec{v} \mapsto \vec{z}']$ is obtained from $\varphi_j(\vec{y})$ by replacing variables in $\vec{v}$ by corresponding variables in $\vec{z}'$.

7.6 Soundness of the Method

An assertion network which satisfies all the verification conditions is called an *inductive network*. An assertion network is defined to be *p-invariant* if every p-computation σ which reaches location $\ell \in \mathcal{C}$ with data state $\vec{y} = \vec{d}$ satisfies $\vec{d} \models \varphi_\ell$.

CLAIM 19 *An inductive assertion network whose assertion at ℓ_0^0 is p_0 is a p_0-invariant network.*

The claim can be proved by induction on the number of cut-points which the computation σ visits.

COROLLARY 20 *If the network $\mathcal{N}$ is inductive for program P, then P is partially correct w.r.t the specification $\langle p_0, q_0 \rangle$. Furthermore, if $\mathcal{N}$ entails the specification $\langle p, q \rangle$, then P is partially correct w.r.t $\langle p, q \rangle$.*

7.7 Example: Factorial

Reconsider program FACTORIAL of Fig. 12. We will prove that this program is partially correct w.r.t the specification

$$p : x \geq 0 \qquad\qquad q : z = x!$$

As the cut-set we take all locations. The proposed assertion network is given by

$$\begin{array}{ll} p_0 = p_1 : & x \geq 0 \\ q_0 = q_1 : & z = x! \\ \varphi_1^1 : & x > 0 \ \wedge \ z = (x-1)! \end{array}$$

This gives rise to the following set of valid verification conditions:

$$\begin{array}{lll}
V_{e_1}^{entry} : x \geq 0 & \rightarrow & x \geq 0 \\
V_{e_1}^{exit} : \ x \geq 0 \ \wedge \ \underbrace{z' = x!}_{q_1(x,z')} & \rightarrow & \underbrace{z' = x!}_{q_0[z \mapsto z']} \\
V_{e_2} : \ x \geq 0 \ \wedge \ x = 0 & \rightarrow & \underbrace{1 = x!}_{q_0(f_{e_1}(x;z))} \\
V_{e_3}^{entry} : x \geq 0 \ \wedge \ x > 0 & \rightarrow & \underbrace{x - 1 \geq 0}_{p_1(x-1)} \\
V_{e_3}^{exit} : \ x \geq 0 \wedge x > 0 \wedge \underbrace{x > 0 \wedge z' = (x-1)!}_{q_1(x-1,z')} & \rightarrow & \underbrace{x > 0 \wedge z' = (x-1)!}_{\varphi_1^1(x,z')} \\
V_{e_4} : \ x > 0 \ \wedge \ z = (x-1)! & \rightarrow & \underbrace{x \cdot z = x!}_{q_1(x,x \cdot z)}
\end{array}$$

7.8 Example: Fibonacci

As another example, consider program FIBONACCI (presented in Fig. 13) which. for input $x \geq 0$, computes the x'th element of the *Fibonacci* series:

$$2,\ 1,\ 3,\ 4,\ 7,\ 11,\ \ldots$$

where $a_0 = 2$, $a_1 = 1$, and $a_{i+2} = a_{i+1} + a_i$ for $i \geq 0$.

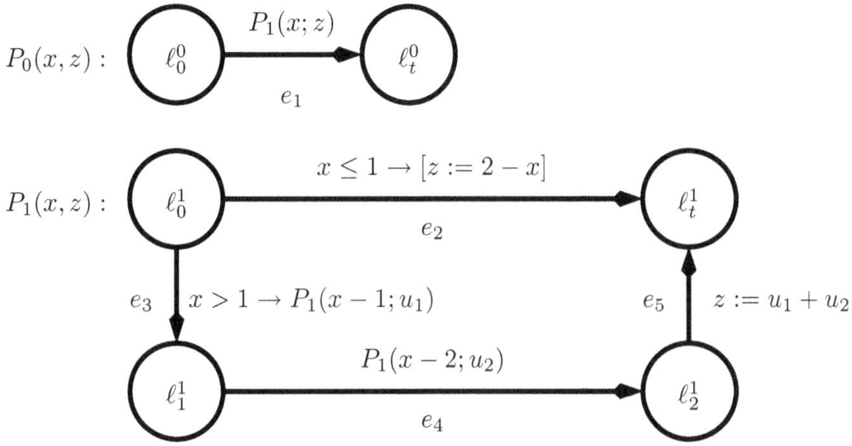

Figure 13. Program FIBONACCI

As the specification, we take

$$p : x \geq 0 \qquad\qquad q : z = \alpha^x + \beta^x$$

where α and β are the two roots of quadratic equation $u^2 - u - 1 = 0$. As the cut-set we take all locations. The proposed assertion network is given by

$$\begin{array}{ll}
p_0 = p_1 : & x \geq 0 \\
q_0 = q_1 : & z = \alpha^x + \beta^x \\
\varphi_1^1 : & x \geq 2 \ \wedge\ u_1 = \alpha^{x-1} + \beta^{x-1} \\
\varphi_2^1 : & x \geq 2 \ \wedge\ u_1 = \alpha^{x-1} + \beta^{x-1} \ \wedge\ u_2 = \alpha^{x-2} + \beta^{x-2}
\end{array}$$

Verification Conditions for Fibonacci

The preceding choices give rise to the following set of valid verification conditions:

$$
\begin{array}{lll}
V_{e_1}^{entry}: & x \geq 0 & \rightarrow \quad x \geq 0 \\
V_{e_1}^{exit}: & x \geq 0 \ \wedge \ z' = \alpha^x + \beta^x & \rightarrow \quad z' = \alpha^x + \beta^x \\
V_{e_2}: & x \geq 0 \ \wedge \ x \leq 1 & \rightarrow \quad 2 - x = \alpha^x + \beta^x \\
V_{e_3}^{entry}: & x \geq 0 \ \wedge \ x \geq 2 & \rightarrow \quad x - 1 \geq 0 \\
V_{e_3}^{exit}: & x \geq 0 \ \wedge \ x \geq 2 \ \wedge \ u_1 = \alpha^{x-1} + \beta^{x-1} & \rightarrow \\
 & \multicolumn{2}{r}{x \geq 2 \ \wedge \ u_1 = \alpha^{x-1} + \beta^{x-1}}
\end{array}
$$

$$
\begin{array}{ll}
V_{e_4}^{entry}: & x \geq 2 \ \wedge \ u_1 = \alpha^{x-1} + \beta^{x-1} \quad \rightarrow \quad x - 2 \geq 0 \\
V_{e_4}^{exit}: & \underbrace{x \geq 2 \ \wedge \ u_1 = \alpha^{x-1} + \beta^{x-1}}_{\varphi_1^1} \ \wedge \ \underbrace{u_2 = \alpha^{x-2} + \beta^{x-2}}_{q_1(x-2,u_2)} \quad \rightarrow \\
 & \qquad \underbrace{x \geq 2 \ \wedge \ u_1 = \alpha^{x-1} + \beta^{x-1} \ \wedge \ u_2 = \alpha^{x-2} + \beta^{x-2}}_{\varphi_2^1}
\end{array}
$$

$$
\begin{array}{ll}
V_{e_5}: & x \geq 2 \ \wedge \ u_1 = \alpha^{x-1} + \beta^{x-1} \ \wedge \ u_2 = \alpha^{x-2} + \beta^{x-2} \quad \rightarrow \\
 & \qquad \underbrace{u_1 + u_2 = \alpha^x + \beta^x}_{q_1(x,u_1+u_2)}
\end{array}
$$

7.9 Proving Success (Deadlock Absence) of Procedural Programs

As in the case of non-procedural programs, we define for each location $\ell \in \mathcal{L}$, its *exit condition*

$$E_\ell : \quad c_1 \ \vee \ \cdots \ \vee \ c_k$$

where $c_1, \ldots, c_k$ are the guards on all edges departing from node ℓ. For a cut-set $\mathcal{C}$ and location $\widetilde{\ell} \notin \mathcal{C}$ we denote by $\Pi_{\mathcal{C},\widetilde{\ell}}$ the set of paths connecting a location in $\mathcal{C}$ to $\widetilde{\ell}$ without passing through any other cut-point. For each path $\pi \in \Pi_{\mathcal{C},\widetilde{\ell}}$, let *srce*$(\pi)$, c_π, and f_π denote, respectively, the cut-point at the beginning of path π, the summary traversal condition, and data transformation associated with π.

The following claim summarizes the general rule for proving success.

CLAIM 21 *In order to prove that program P is p-successful (i.e., no p-computation ever deadlocks), it is sufficient to find a network $\mathcal{N} : \langle \mathcal{C}, \{\varphi_\ell \mid \ell \in \mathcal{C}\}$, satisfying the following requirements:*

1. *The network* $\mathcal{N}$ *is inductive.*
2. $p \quad \rightarrow \quad p_0$
3. $\varphi_\ell \quad \rightarrow \quad E_\ell$ *for every* $\ell \in \mathcal{C}$
4. $\varphi_{srce(\pi)}(V) \wedge c_\pi(V) \quad \rightarrow \quad E_{\widetilde{\ell}}(f_\pi(V))$
for every $\widetilde{\ell} \notin \mathcal{C}$ *and path* $\pi \in \Pi_{\mathcal{C},\widetilde{\ell}}$

7.10 Proving Termination of Procedural Programs

The method starts by constructing an inductive assertion network $\mathcal{N}$. We then choose a well-founded domain $(\mathcal{A}, \succ)$. With each cut-point $\ell \in \mathcal{C}$ we associate a *ranking function* δ_ℓ. Ranking functions associated with procedure entry points ℓ_0^k depend only on $\vec{x}$, while ranking functions associated with terminal locations ℓ_t^k will depend only on $(\vec{x}, \vec{z})$.
We then form *descent conditions* as follows:
- For an assignment path π connecting location ℓ_i to ℓ_j, we form the following descent condition:

$$D_\pi: \quad \varphi_i(\vec{y}) \wedge c_\pi(\vec{y}) \quad \rightarrow \quad \delta_i(\vec{y}) \succ \delta_j(f_\pi(\vec{y}))$$

- For a call edge of the form:

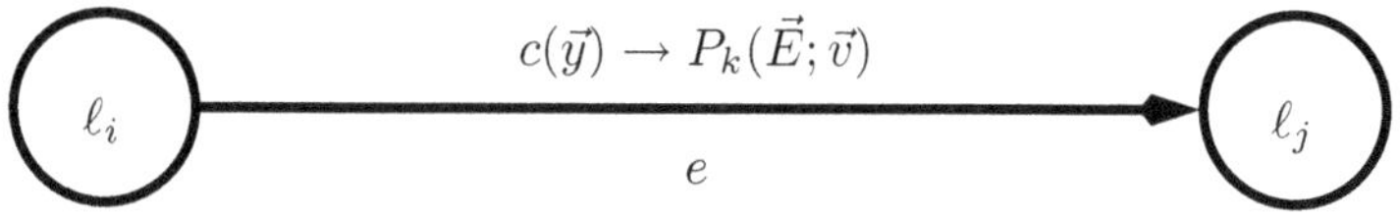

we form the following two descent conditions:

$$D_e^{in}: \quad \varphi_i(\vec{y}) \wedge c(\vec{y}) \quad \rightarrow \quad \delta(\vec{y}) \succ \delta_0^k(\vec{E}(\vec{y}))$$
$$D_e^{out}: \quad \varphi_i(\vec{y}) \wedge c(\vec{y}) \wedge q_k(\vec{E}(\vec{y}); \vec{z}') \quad \rightarrow \quad \delta_i(\vec{y}) \succ \delta_j(\vec{y})[\vec{v} \mapsto \vec{z}']$$

The Proof Method

The proof method is summarized in the following claim:

CLAIM 22 (Verifying Termination of Procedural Programs)
Let P *be a procedural program with a pre-condition specification* p. *Let* $\mathcal{N} : \langle \mathcal{C}, \{\varphi_\ell \mid \ell \in \mathcal{C}\} \rangle$ *be an assertion network,* $(\mathcal{A}, \succ)$ *be a well-founded domain, and* $\{\delta_\ell \mid \ell \in \mathcal{C}\}$ *be a network of ranking functions, each mapping states into elements of* $\mathcal{A}$. *If the following requirements are satisfied:*

1. *The network* $\mathcal{N}$ *is inductive.*
2. $p \quad \rightarrow \quad p_0$
3. *The 3 types of descent conditions are valid for all generated verification paths.*

then program P *is* p*-convergent.*

7.11 Example: Factorial

Reconsider the program FACTORIAL of Fig. 12. As a first step, let us prove the termination of procedure P_1 under the pre-condition $x \geq 0$. As the well-founded domain we choose $\mathbb{N} \times_{lex} [0..2]$. As the termination network we choose:

ℓ_i	φ_{ℓ_i}	δ_{ℓ_i}
ℓ_0^1	$x \geq 0$	$(\lvert x\rvert, 2)$
ℓ_1^1	$x > 0$	$(\lvert x\rvert, 1)$
ℓ_t^1	1	$(\lvert x\rvert, 0)$

which leads to the following descent conditions:

$$\begin{array}{llcl} D_{e_2}: & x \geq 0 \ \wedge\ x = 0 & \rightarrow & (|x|, 2) \succ (|x|, 0) \\ D_{e_3}^{in}: & x \geq 0 \ \wedge\ x > 0 & \rightarrow & (|x|, 2) \succ (|x-1|, 2) \\ D_{e_3}^{out}: & x \geq 0 \ \wedge\ x > 0 & \rightarrow & (|x|, 2) \succ (|x|, 1) \\ D_{e_4}: & x > 0 & \rightarrow & (|x|, 1) \succ (|x|, 0) \end{array}$$

7.12 Verifying Complete Programs

Note that if we wish to verify termination of procedure P_0, we need to use the following more cumbersome ranking network:

ℓ_i	φ_{ℓ_i}	δ_{ℓ_i}
ℓ_0^0	$x \geq 0$	$(\lvert x\rvert, 3)$
ℓ_t^0	1	$(\lvert x\rvert, 0)$
ℓ_0^1	$x \geq 0$	$(\lvert x\rvert, 2)$
ℓ_1^1	$x > 0$	$(\lvert x\rvert, 1)$
ℓ_t^1	1	$(\lvert x\rvert, 0)$

which includes also ranking for the locations of P_0.

7.13 Acyclic Decomposition of Call Graphs

The task of verifying entire program can be somewhat simplified by considering the *call graph* of a procedural program. This is a graph which contains a node for each procedure P_i, $i = 0, \ldots, k$, and a directed edge connecting node P_i to P_j if procedure P_i contains a call to procedure P_j. For example, the call graph for program *factorial* is given by:

Assume that we have constructed the call graph for program P and computed an *acyclic decomposition* $K_1, \ldots, K_k$ of the procedures, such that if procedure $P_i \in K_i$ calls procedure $P_j \in K_j$ then, necessarily $i \leq j$. We refer to $K_1, \ldots, K_k$ as *procedure clusters*, and to an edge on which P_i calls P_j where both procedures belong to the same cluster an *intra-cluster* edge. Using this terminology, we have the following:

CLAIM 23 (Improved Version) *In the application of the ranking function method, it is sufficient to require that conditions D_e^{in} and D_e^{out} of Claim 22 hold for intra-cluster edges e.*

7.14 Example: Ackerman's Function

As our last example, we consider *Ackerman*'s function which can be defined by the following recursive definition:

$$
\begin{array}{lcl}
A(0, x_2) & = & x_2 + 1 \\
A(x_1 + 1, 0) & = & A(x_1, 1) \\
A(x_1 + 1, x_2 + 1) & = & A(x_1, A(x_1 + 1, x_2))
\end{array}
$$

A procedural program for computing $A(x_1, x_2)$ is given by program ACKERMANN in Fig. 14.

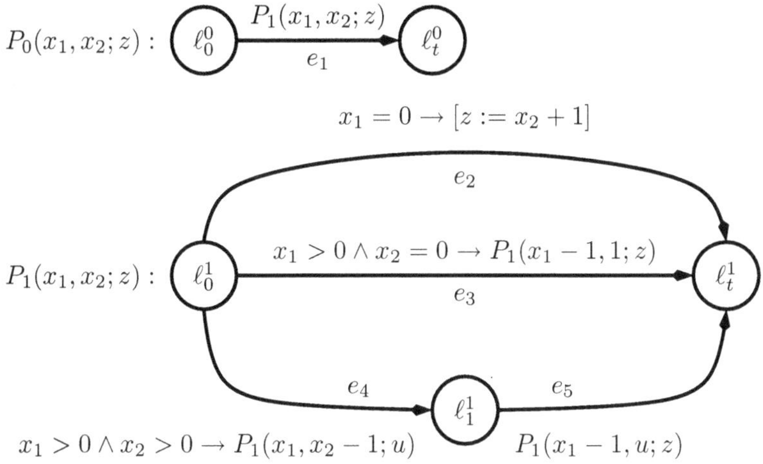

Figure 14. Program ACKERMANN

The termination network is given by:

ℓ_i	φ_{ℓ_i}	δ_{ℓ_i}
ℓ_0^1	$x_1 \geq 0 \ \wedge \ x_2 \geq 0$	$(\lvert x_1\rvert, \lvert x_2\rvert, 2)$
ℓ_1^1	$x_1 > 0 \ \wedge \ x_2 > 0$	$(\lvert x_1\rvert, \lvert x_2\rvert, 1)$
ℓ_t^1	1	$(\lvert x_1\rvert, \lvert x_2\rvert, 0)$

which gives rise to the following descent conditions:

$$
\begin{array}{lll}
D_{e_2}: & x_1 \geq 0 \wedge x_2 \geq 0 \wedge x_1 = 0 & \rightarrow (|x_1|, |x_2|, 2) \succ (|x_1|, |x_2|, 0) \\
D_{e_3}^{in}: & x_1 \geq 0 \wedge x_2 \geq 0 \wedge x_1 > 0 \wedge x_2 = 0 & \rightarrow (|x_1|, |x_2|, 2) \succ (|x_1-1|, 1, 2) \\
D_{e_3}^{out}: & x_1 \geq 0 \wedge x_2 \geq 0 \wedge x_1 > 0 \wedge x_2 = 0 & \rightarrow (|x_1|, |x_2|, 2) \succ (|x_1|, |x_2|, 0) \\
D_{e_4}^{in}: & x_1 \geq 0 \wedge x_2 \geq 0 \wedge x_1 > 0 \wedge x_2 > 0 & \rightarrow (|x_1|, |x_2|, 2) \succ (|x_1|, |x_2-1|, 2) \\
D_{e_4}^{out}: & x_1 \geq 0 \wedge x_2 \geq 0 \wedge x_1 > 0 \wedge x_2 > 0 & \rightarrow (|x_1|, |x_2|, 2) \succ (|x_1|, |x_2|, 1) \\
D_{e_5}^{in}: & x_1 > 0 \wedge x_2 > 0 & \rightarrow (|x_1|, |x_2|, 2) \succ (|x_1-1|, |u|, 2) \\
D_{e_5}^{in}: & x_1 > 0 \wedge x_2 > 0 & \rightarrow (|x_1|, |x_2|, 1) \succ (|x_1|, |x_2|, 0)
\end{array}
$$

8 Conclusions and Discussions

In this paper we presented an extension of Floyd's methods to deal with programs that have procedures. The key idea on which this extension is based is that it is sufficient to maintain for each procedure the initial values of all the local variables (including the input parameters) as they were on the most recent entry into the procedure. These initial values are used as parameters for all the intermediate assertions within the procedure.

The soundness of the method relies on the restrictions that the parameter transfer mechanism is that of "value-result". That is, values of the input parameters $\vec{x}$ are copied into local variables on entry, and the values of the result parameters $\vec{z}$ are copied into the actual arguments on exit. An implied constraint is that there should be no aliasing among the input actual arguments or among the result actual arguments. However, there may be an aliasing between an input and a result actual argument.

Acknowledgment

This paper is dedicated to Prof. Dov Gabbay, a personal friend and an extremely diverse and creative scientist whom I have continuously admired from the first day of our meeting. Dov's view of logic as a living universal discipline which has a bearing on every branch of science as well as every aspect of our daily life is a source of inspiration and imitation to all of his followers and disciples among whom I am happy to be counted.

This research has been supported in part by NSF grant CCR-0205571, ONR grant N00014-99-1-0131, and Israel Science Foundation grant 106/02-1.

BIBLIOGRAPHY

[Apt and Olderog, 1991] K.R. Apt and E.-R. Olderog. *Verification of Sequential and Concurrent Programs.* Springer-Verlag, 1991.

[Floyd, 1967] R.W. Floyd. Assigning meanings to programs. *Proc. Symposia in Applied Mathematics*, 19:19–32, 1967.

[Francez, 1992] N. Francez. *Program Verification.* Addison-Wesley, 1992.

[Hoare, 1969] C.A.R Hoare. An axiomatic basis for computer programming. *Comm. ACM*, 12(10):576–580, 1969.

[Kleene, 2002] S.C. Kleene. *Mathematical Logic.* Dover, New York, 2002.

[Manna, 1974] Z. Manna. *Mathematical Theory of Computation.* McGraw-Hill, 1974.

[Sharir and Pnueli, 1981] M. Sharir and A. Pnueli. Two approaches to inter-procedural data-flow analysis. In Jones and Muchnik, editors, *Program Flow Analysis: Theory and Applications.* Prentice-Hall, 1981.

Iterated Revision and Automatic Similarity Generation

Odinaldo Rodrigues

1 Introduction

In the original AGM formulation, revision is a function $\circ$ that takes a belief set K and a sentence A as inputs and returns as a result a revised belief set $K \circ A$. Because the two arguments of $\circ$ are of different types — the *infinite set* K and the *finite formula* A — in this particular setting they cannot be interchanged. However, it is well known that Katsuno and Mendelzon [Katsuno and Mendelzon, 1992] provided a characterization of the AGM postulates for belief sets that are finitely representable as a *formula* K. In this scenario, K, A and $K \circ A$ are all formulae and therefore can be applied in the revision function in any of its arguments.[1]

On the other hand, when reasoning about a sequence of belief revision operations $\Delta = \delta_1 \circ \delta_2 \circ \ldots \circ \delta_k$ one usually thinks of the process as being forward in time, i.e., $(\ldots(\delta_1 \circ \delta_2) \circ \ldots) \circ \delta_k$. Since the success postulate requires that the revising sentence is accepted in the revised belief set, AGM imposes an implicit priority *expectation* of each δ_i in the sequence over its predecessor δ_{i-1} $(i > 1)$. However, since AGM is coherentist, at each successive revision step $K_1 = \delta_1 \circ \delta_2$, $K_2 = K_1 \circ \delta_3$, …, the relative priorities between δ_1, δ_2, …, etc., are *forgotten*. Formalisms that propose rationality guidelines for iterated revision try to recover the lost memory by explicitly recording the sequence of revisions [Lehmann, 1995] and/or by making a distinction between the *belief state* of an agent and his/her *epistemic state* [Darwiche and Pearl, 1994; Boutilier, 1996]. In this setting, revision now takes into account not only the belief state of an agent and the new information, but also how the agent *arrived* at that particular belief state. Semantical characterizations of the iteration process stipulate conditions on how revisions should react with respect to previous inputs and can be done via similarity orderings between worlds for a given epistemic state.

[1] Furthermore, the semantical counterpart of the representation is based on the notion of models and that of a *faithful assignment*, which indeed does not rely on the syntactic nature of K or A.

For the finite case mentioned before, it is possible to apply $\circ$ in any order and in [Gabbay and Rodrigues, 1997], we argued that $\circ$ is in general non-associative. It turns out that there are advantages to applying $\circ$ backwards in a sequence Δ, i.e., $\delta_1 \circ (\delta_2 \circ \ldots (\delta_{k-1} \circ \delta_k) \ldots)$. In particular, it allows for a refinement of the notion of similarity between worlds for a given sequence or revisions Δ; paves the way for the investigation of rational ways of automatically adjusting the similarity ordering after revisions are performed and reconciliates coherence with memory.

In this work, we analyse some of the aspects involved in this paradigm and its relation to formalizations of iterated revision found in the literature, including Darwiche and Pearl's approach [Darwiche and Pearl, 1996]; Lehmann's postulates for iterated revision [Lehmann, 1995] and Boutilier's MC-revision models [Boutilier, 1996].

2 Preliminaries

In [Katsuno and Mendelzon, 1992], Katsuno and Mendelzon provided a reformulation of the AGM postulates for belief revision for belief sets representable by finite sets of propositional formulae, that we call here *belief bases.* Since a belief base is a finite set, it can be associated with the conjunction of its formulae. We reproduce Katsuno and Mendelzon's reformulation below, where K, A and B are formulae of a propositional language $\mathcal{L}$ constructed from a finite set of symbols $\mathcal{P}$. $K \circ A$ represents the belief base obtained from revising K by A.

AGM postulates for belief revision rewritten for finite belief bases

(R1) $K \circ A$ implies A
(R2) If $K \wedge A$ is satisfiable, then $K \circ A \equiv K \wedge A$
(R3) If A is satisfiable, then $K \circ A$ is also satisfiable
(R4) If $K_1 \equiv K_2$ and $A \equiv B$, then $K_1 \circ A \equiv K_2 \circ B$
(R5) $(K \circ A) \wedge B$ implies $K \circ (A \wedge B)$
(R6) If $(K \circ A) \wedge B$ is satisfiable, then $K \circ (A \wedge B)$ implies $(K \circ A) \wedge B$

Valuations of $\mathcal{L}$ are constructed by assigning $\{\top, \bot\}$ to each element of $\mathcal{P}$ and extending the assignment to complex formulae in the usual way. $\mathcal{I}$ is used to denote the set of all valuations and the set $\text{mod}(K)$ — the *models of* K — is defined as $\{I \in \mathcal{I} \mid I \Vdash K\}$.

In the same article, Katsuno and Mendelzon provided a semantical characterisation of all revision operators satisfying the AGM postulates. The general idea is reminiscent of Grove's systems of spheres [Grove, 1988] and uses a system of pre-orders to compare similarities between worlds with respect to a given belief set. The first step in the characterisation is the definition of minimum requirements the pre-orders have to satisfy. For valuations I and I' and a belief set K, $I \leq_K I'$ denotes the fact that valuation I is at least as similar to K as I' is. As usual, $I <_K I'$ denotes $I \leq_K I'$ and $I' \not\leq_K I$.

DEFINITION 1 (Faithful assignment for belief revision). A *faithful assignment for belief revision* is a function mapping each propositional formula K to a *pre-order* $\leq_K$ on $\mathcal{I}$, such that

(F1) If $I, I' \in \mathrm{mod}(K)$, then $I <_K I'$ does not hold.

(F2) If $I \in \mathrm{mod}(K)$ and $I' \notin \mathrm{mod}(K)$, then $I <_K I'$ holds.

(F3) If $K \leftrightarrow K'$, then $\leq_K = \leq_{K'}$.

A faithful assignment is called *total* if its associated pre-orders are total. What Definition 1 effectively says is that, provided K is satisfiable, *i)* any two distinct models of K are either incomparably or equivalently similar to K; *ii)* models of K are strictly more similar to K than any non-model of K and *iii)* pre-orders assigned to logically equivalent sentences are identical. Note that no constraints on the granularity of $\leq_K$ for non-models of K are imposed.

NOTATION 2. Let $\leq$ be a pre-order on a set S, and $M \subseteq S$. The expression $\min_{\leq}(M)$ will denote the set $\{m \in M \mid \neg\exists m' \in M \text{ such that } m' < m\}$.

The following theorem [Katsuno and Mendelzon, 1992] establishes the correspondence between faithful assignments and revision operators satisfying the AGM postulates:

THEOREM 3. *A revision operator* $\circ$ *satisfies postulates* (R1)–(R6) *if and only if there exists a* total *faithful assignment* $\leq_K$ *for each formula* K*, such that* $\mathrm{mod}(K \circ A) = \min_{\leq_K}(\mathrm{mod}(A))$.

Since the assignments associated with operators verifying the AGM postulates are in fact total, $\min_{\leq_K}(\mathcal{I})$ forms an equivalence class consisting of the valuations in $\mathrm{mod}(K)$ (by (F1) and (F2)).

Katsuno and Mendelzon's characterisation is perhaps the most intuitive way to understand the revision process. In order to verify (R1), the models of $K \circ A$ must be included in the models of A (if any). Intuitively, one expects these to be those which preserve as much as possible of the informational content of K. The measurement of similarity with respect to K is given by the ordering $\leq_K$. The minimum requirements $\leq_K$ must fulfill in order for the associated revision operator to verify the information preservation requirements given by (R1)–(R6) are specified in Definition 1 and hence the minimal elements in mod(A) are exactly the models of A which *best* preserve the informational content of K (with respect to $\leq_K$).

Notice that if K is consistent with A, it is easy to see by Definition 1 that $\min_{\leq_K}(\text{mod}(A)) = \text{mod}(K) \cap \text{mod}(A)$. According to Theorem 3, these are the models of $K \circ A$ — meeting exactly the requirements imposed by (R2). Also notice that K is considered as a whole in the similarity measurement $\leq_K$. This reflects well the coherentist view adopted by AGM.

2.1 Iterating Revision Operations

One of the main criticisms against the AGM formulation is its lack of guidance with respect to the iteration fo the revision operation. The only postulates related to iteration are (R5) and (R6), but they simply constrain the interaction between revisions and *expansions*:

(R5) $(K \circ A) \wedge B$ implies $K \circ (A \wedge B)$
(R6) If $(K \circ A) \wedge B$ is satisfiable, then $K \circ (A \wedge B)$ implies $(K \circ A) \wedge B$

A number of extensions to the original AGM formulation have been proposed to rationalise the behaviour of the iteration process. One of the most discussed was proposed by Darwiche and Pearl [Darwiche and Pearl, 1994]. In that work, they point out that a distinction between belief states and *epistemic states* is essential for the successful modelling of the iterative process, although a reformulation of the AGM postulates in order to reflect this only appeared later [Darwiche and Pearl, 1996; Darwiche and Pearl, 1997]. Essentially, an epistemic state is a somehwhat richer structure from which a belief state is derived. An immediate consequence of this change in paradigm is that equivalent belief states may be derived from different epistemic states. This departure from pure coherentism is supported by a number of authors [Friedman and Halpern, 1996; Lehmann, 1995; Rodrigues, 2003; Ryan, 1992; Nebel, 1992b].

The starting point is again the assumption of a finite belief base. In Katsuno and Mendelzon's terms, the epistemic state would be represented by the belief base, whereas the belief state would be associated with the set of logical consequences of this base. Since an epistemic state carries more information than a belief set, revisions must take into account differences arising from the distinction. Postulates (R1)–(R6) were reformulated with this in mind and motivated by the observation that they were incompatible with the new proposed postulates (C1)–(C4) [Freund and Lehmann, 1994].

In the following presentation, Ψ will be used to represent an epistemic state and bel(Ψ) to represent the belief set obtained from Ψ. However, we follow Darwiche and Pearl and use Ψ instead of bel(Ψ) where the context is clear in order to lighten the notation. By this we mean that, for instance in (R*4) below, it is the epistemic states Ψ_1 and Ψ_2 that are meant in the first half of the postulate, but the belief sets bel($\Psi_1 \circ A_1$) and bel($\Psi_2 \circ A_2$) in the second one.

Darwiche and Pearl's postulates for belief revision of epistemic states

(R*1) $\Psi \circ A$ implies A
(R*2) If $\Psi \wedge A$ is satisfiable, then $\Psi \circ A \equiv \Psi \wedge A$
(R*3) If A is satisfiable, then $\Psi \circ A$ is also satisfiable
(R*4) If $\Psi_1 = \Psi_2$ and $A_1 \equiv A_2$, then $\Psi_1 \circ A_1 \equiv \Psi_2 \circ A_2$
(R*5) $(\Psi \circ A) \wedge B$ implies $\Psi \circ (A \wedge B)$
(R*6) If $(\Psi \circ A) \wedge B$ is satisfiable, then $\Psi \circ (A \wedge B)$ implies $(\Psi \circ A) \wedge B$

The above presentation is essentially the same as Katsuno and Mendelzon's, except for (R*4) which is strictly weaker than (R4). In (R*4), the condition for the equivalence of the resulting belief sets is that the original epistemic states are *identical*, instead of equivalent as in (R4). This is reflected immediately in the semantical characterisaton of revision operators satisfying (R*1)–(R*6) given in [Darwiche and Pearl, 1997].

To simplify notation we will simply use mod(Ψ) to denote the set of models of an epistemic state Ψ (when we really mean mod(bel(Ψ))). In the formulation below, $I \preceq_\Psi I'$ represents the fact that I is at least as good a state of affairs as I' is with respect to the epistemic state Ψ.

DEFINITION 4 (Faithful assignment for revision of epistemic states). A *faithful assignment for belief revision of epistemic states* is a function mapping each epistemic state Ψ to a total pre-order $\preceq_\Psi$ on $\mathcal{I}$, such that

1. If $I, I' \in \text{mod}(\Psi)$, then $I \equiv_\Psi I'$.
2. If $I \in \text{mod}(\Psi)$ and $I' \notin \text{mod}(\Psi)$, then $I \prec_\Psi I'$ holds.
3. If $\Psi = \Delta$, then $\preceq_\Psi = \preceq_\Delta$.

Obviously, $I \in \text{mod}(\Psi)$ if and only if $I \Vdash \text{bel}(\Psi)$. Note that the last item requires the two epistemic states to be identical for their epistemic orderings to be identical. In practice, bel(Ψ) could be identical to bel(Γ) without Ψ and Γ being necessarily the same. Darwiche and Pearl proved the following theorem that is the counterpart of Theorem 3 for epistemic states:

THEOREM 5. *A revision operator $\circ$ satisfies postulates* (R*1)—(R*6) *if there exists a faithful assignment that maps each epistemic state Ψ to a total pre-order $\preceq_\Psi$, such that* $\text{mod}(\Psi \circ A) = \min_{\preceq_\Psi}(\text{mod}(A))$.

(R*1)–(R*6) were then augmented with postulates that deal specifically with the iteration of the revision process.

(C1) If $A \models B$, then $(\Psi \circ B) \circ A \equiv \Psi \circ A$
(C2) If $A \models \neg B$, then $(\Psi \circ B) \circ A \equiv \Psi \circ A$
(C3) If $\Psi \circ A \models B$, then $(\Psi \circ B) \circ A \models B$
(C4) If $\Psi \circ A \not\models \neg B$, then $(\Psi \circ B) \circ A \not\models \neg B$

If Ψ is the current epistemic state and A is used to revise Ψ, then the ordering $\preceq_\Psi$ must relate to $\preceq_{\Psi \circ A}$ appropriately in order for (C1)–(C4) to hold. The following representation theorem by Darwiche and Pearl [Darwiche and Pearl, 1997] dictates how faithful assignments of two consecutive epistemic states must relate.

THEOREM 6. *If a given revision operator* $\circ$ *satisfies postulates* (R*1)—(R*6), *then* $\circ$ *satisfies* (C1)—(C4) *if and only if the operator and its corresponding faithful assignment satisfy:*

(CR1) *If* $I, I' \in \text{mod}(A)$, *then* $I \preceq_\Psi I'$ *iff* $I \preceq_{\Psi \circ A} I'$
(CR2) *If* $I, I' \in \text{mod}(\neg A)$, *then* $I \preceq_\Psi I'$ *iff* $I \preceq_{\Psi \circ A} I'$
(CR3) *If* $I \in \text{mod}(A)$ *and* $I' \in \text{mod}(\neg A)$, *then* $I \prec_\Psi I'$ *implies* $I \prec_{\Psi \circ A} I'$
(CR4) *If* $I \in \text{mod}(A)$ *and* $I' \in \text{mod}(\neg A)$, *then* $I \preceq_\Psi I'$ *implies* $I \preceq_{\Psi \circ A} I'$

(CR1) states that the relationship between models of a sentence A should not change with respect to an epistemic state before and after a revision by A. (CR2) is related to (C2) and it could be said that the motivation here is reminiscent of the original conditions for (F1)–(F3) for faithful assignments for belief revision seen at the beginning of this section. They do not impose any constraints on the behaviour of the similarity ordering with respect to non-models of a sentence. However, it is arguable whether the fact that two valuations I and I' do not satisfy a sentence A is sufficient to consider them equally ranked after an epistemic state Ψ is revised by A. (CR3) and (CR4) can be seen in the following way: (CR4) "preserves" the preference in favour of a model of a sentence A against a non-model of A after a revision by A, whereas (CR3) "preserves" the non-preference of non-models of A against models of A.

3 Formalization of the Problem

In this section, we present the general setting in which we will discuss the iteration of the revision process. For simplicity, we consider epistemic states that arise from a finite sequence of input sentences $\Delta = [\delta_1, \delta_2, \ldots, \delta_k]$, where for all $j > i$, δ_j is received after δ_i. In [Gabbay and Rodrigues, 1997], we called this a *prioritised database* (PDB) and required that each δ_i was a sentence in disjunctive normal form (DNF). That requirement was due to the fact that our revision operator takes advantage of properties of a formula in DNF [Gabbay and Rodrigues, 1996]. In this paper, we will be considering *any* revision operation $\circ$ satisfying the AGM postulates, and thus that restriction is lifted. Other terms have been used in the literature to define such sequences. For instance, they were called *linear prioritised belief bases* by Nebel [Nebel, 1991b] and *epistemic states* by Lehmann [Lehmann, 1995].

Now suppose that the belief set associated with a PDB Δ is obtained by successively applying $\circ$ to the sentences in Δ. Provided that $\circ$ is not associative, there will be therefore two natural ways of interpreting the sequence of revisions in Δ: either by considering the operation left associative or by considering it right associative. We will use $^*\Delta$ (read *left delta*) to denote the left associative interpretation of the sequence of revisions in Δ and Δ^* to denote the right associative one. Formally,

DEFINITION 7. Let $\Delta = [\varphi_1, \varphi_2, \ldots, \varphi_k]$ be a PDB.

$$^*\Delta = \begin{cases} \varphi_k & \Rightarrow \text{ if } k = 1 \\ ((\varphi_1 \circ \varphi_2) \circ \ldots) \circ \varphi_k & \Rightarrow \text{ if } k > 1 \end{cases}$$

$$\Delta^* = \begin{cases} \varphi_k & \Rightarrow \text{ if } k = 1 \\ \varphi_1 \circ (\ldots \circ (\varphi_{k-1} \circ \varphi_k)) & \Rightarrow \text{ if } k > 1 \end{cases}$$

In [Gabbay and Rodrigues, 1997], we argued that the right associative interpretation was the most interesting one, because the inevitable re-application of the revision steps, although expensive, provided an opportunity to revisit previous decisions. Our next goal is to analyse properties of similarity orderings associated with a sequence Δ when $\circ$ is applied to the elements of Δ using the right associative interpretation.

4 Properties of right associative revision sequences

In this section, we provide a semantical characterization of sequences of revisions performed using the right associative interpretation of the operation. That is, for a sequence of inputs $\Delta = [\delta_1, \ldots, \delta_k]$, we consider the sequence of revisions $\delta_1 \circ (\ldots \circ (\delta_{k-1} \circ \delta_k))$, where $\circ$ is *any* revision operator satisfying the AGM postulates (finite formulation).

We start by defining an ordering $\preceq_\Delta$ that will help us to analyse how valuations of $\mathcal{L}$ compare with each other with respect to the epistemic state Δ. As before $I \leq_\Delta I'$ denotes the fact that I is at least as similar to Δ as I' is.

DEFINITION 8. Let $\Delta = [\delta_1, \delta_2, \ldots, \delta_k]$ be a PDB and consider the belief set Δ^* where the revision operator $\circ$ satisfies the AGM postulates. Let $\leq_{\delta_i}$

be the faithful assignment for the operation for each sentence δ_i in Δ as in Definition 1, and take $i, j \in \{1, \ldots, k\}$.

$$I \preceq_\Delta I' \text{ iff for all } i,\ I' <_{\delta_i} I \text{ implies } \exists j > i \text{ such that } I <_{\delta_j} I'.$$

The ordering above was motivated by the observations about associativity of $\circ$ made in [Gabbay and Rodrigues, 1997] and first appeared in [Rodrigues, 1998], where a comprehensive account of its properties can be found.

Note that if Δ is ε, that is, the empty sequence, then $I \equiv_\Delta I'$, for all $I, I' \in \mathcal{I}$ (vacuously). The greater the index of the sentence, the more recent the information it represents is. Thus, what the definition above says is that the failure of an valuation I to be at least as good at satisfying a sentence received at time i as another valuation I' can only be compensated by $I's$ being better than I' at satisfying a sentence at some later time j. Similarity with respect to the sentences at each point is determined by the faithful assignment of the operator used in the sequence of revisions.

Of course, if there is just one sentence in the sequence Δ, we expect $\preceq_\Delta$ to behave exactly as the faithful assignment for that sentence. Thus,

PROPOSITION 9. *Suppose $\Delta = [\gamma]$, for some sentence γ.*

$$I \preceq_\Delta I' \text{ iff } I \leq_\gamma I'$$

Proof. ($\Rightarrow$) Suppose $I \preceq_\Delta I'$. By Definition 8, for all i, $I' <_{\delta_i} I$ implies $\exists j > i$ such that $I <_{\delta_j} I'$. There is just one sentence in Δ. Thus, $\neg\exists j > 1 \therefore I' \not<_\gamma I$. Since $\leq_\gamma$ is total, $I \leq_\gamma I'$.
($\Leftarrow$) Suppose $I \leq_\gamma I'$, but $I \not\preceq_\Delta I'$. By Definition 8, if $I \not\preceq_\Delta I'$, then there exists i such that $I' <_{\delta_i} I$ and $\neg\exists j > i$ such that $I <_{\delta_j} I'$. Since there is just one sentence in Δ, then $\delta_{i=1} = \gamma$. If $I' <_\gamma I$, then $I \not\leq_\gamma I'$, a contradiction. ■

In general, we expect $\preceq_\Delta$ to behave well. That is, for it to be at least reflexive and transitive.

PROPOSITION 10. *$\preceq_\Delta$ is a pre-order.*

Proof. Reflexivity follows directly from reflexivity of each $\leq_{\delta_i}$. As for transivity, suppose $I \preceq_\Delta I'$ and $I' \preceq_\Delta O$. We shall show that $I \preceq_\Delta O$.

Suppose $I \not\leq_{\delta_i} O$, for some i, we have to show that $\exists j > i$ such that $I <_{\delta_j} O$.

i) If $I \leq_{\delta_i} I'$, then either $I' \leq_{\delta_i} O$ or $O <_{\delta_i} I'$ (since $\leq_{\delta_i}$ is total). In the first case, $I \leq_{\delta_i} O$, because $\leq_{\delta_i}$ is transitive, a contradiction. If $O <_{\delta_i} I'$, then from $I' \preceq_\Delta O$, $\exists x > i$ such that $I' <_{\delta_x} O$. If $I \leq_{\delta_x} I'$, then we are done, because $I <_{\delta_x} O$, so we just take $j = x$. Otherwise, $I' <_{\delta_x} I$, and then $\exists y > x$ such that $I <_{\delta_y} I'$. But $y > x > i$, and hence $I' \leq_{\delta_y} O$ and then $I <_{\delta_y} O$. Set $j = y$ in this case.

ii) If $I \not\leq_{\delta_i} I'$, then $I' <_{\delta_i} I$. Since $I \preceq_\Delta I'$, take $x > i$ such that $I <_{\delta_x} I'$ and for all $k \geq x$, $I \leq_{\delta_k} I'$ (such x exists by Proposition 11). If $I' \leq_{\delta_x} O$, then $I <_{\delta_x} O$, and thus we just need to set $j = x$. Otherwise, $O <_{\delta_x} I'$, and since $I' \preceq_\Delta O$, $\exists y > x$ such that $I' <_{\delta_y} O$. But $y > x > i$. Thus, $I \leq_{\delta_y} I'$, and hence $I <_{\delta_y} O$. We just set $j = y$ in this case and the proof is finished.

■

Since the sequence Δ is finite, there must be a point in time from which the tie between two valuations is settled.[2]

PROPOSITION 11. *$I \preceq_\Delta I'$ iff for all i, $I' <_{\delta_i} I$ implies $\exists j > i$ such that $I <_{\delta_j} I'$ and $\forall k \geq j$, $I \leq_{\delta_k} I'$.*

Proof. ($\Leftarrow$) Straightforward. ($\Rightarrow$) Suppose $I \preceq_\Delta I'$ and that $I' <_{\delta_i} I$, for some i. Take the maximum j such that $j > i$ and $I <_{\delta_j} I'$ holds. The existence of such j is guaranteed by Definition 8 and by the fact that we are considering finite lists of sentences[3]. For any $k \geq j$, it is not the case that $I' <_{\delta_k} I$, because then it would be the case that $I \not\preceq_\Delta I'$. Since the orderings $\leq_{\delta_{i's}}$ are all total, it follows that $I \leq_{\delta_k} I'$, for all $k \geq j$. ■

PROPOSITION 12. *$\preceq_\Delta$ is total.*

Proof. This is easy to show. For suppose $I \not\preceq_\Delta I'$. By Definition 8, there exists a point x such that $I' <_{\delta_x} I$ and for all $y > x$, it is not the case that $I <_{\delta_y} I'$. Since all $\leq_{\delta_y}$ are total, it follows that $I' \leq_{\delta_y} I$. This implies that $I' \preceq_\Delta I$, by Proposition 11. Clearly, $\neg\exists i \geq x$ such that $I <_{\delta_i} I'$, and if there exists $i < x$ such that $I <_{\delta_i} I'$, then all we have to do is to set $j = y$. ■

DEFINITION 13. For each PDB Δ, $\varkappa$ is a function defined as follows:

$$\varkappa(\Delta) = \{\preceq_\Delta, \mathcal{I}\}.$$

where $\preceq_\Delta$ conforms to Definition 8.

THEOREM 14. *$\varkappa$ is a faithful assignment for revision of epistemic states.*

[2] In reality, we only need that the sequence is not associated with an infinite ascending chain.

[3] Or orderings with a maximum

Proof. We have to prove that $\varkappa$ satisfies the conditions stated in Definition 4. By Proposition 10, for each epistemic state Δ, $\preceq_\Delta$ is a pre-order. By Proposition 12, $\preceq_\Delta$ is also total. If Δ is empty or $\Delta = [\gamma]$, for some sentence γ, then the three conditions follow immediately from Definition 1 and from Definition 9. Thus, suppose $\Delta = [\delta_1, \ldots, \delta_k]$ for some $k > 1$. As before, we will use the expression mod (Δ) to represent the set of models of the belief state obtained from the epistemic state Δ. Since we consider the right associative employment of the operator $\circ$, this amounts to considering $\text{mod}(\Delta) = \text{mod}(\delta_1 \circ (\ldots \circ (\delta_{k-1} \circ \delta_k) \ldots))$. The three conditions and the respective proofs are listed below.

1. $I, I' \in \text{mod}(\Delta)$ implies $I \equiv_\Delta I'$.

   Suppose $I \in \text{mod}(\Delta)$, $I' \in \text{mod}(\Delta)$, but $I \prec_\Delta I'$. If $I \in \text{mod}(\Delta)$, then $I \in \min_{\leq_{\delta_1}}(\min_{\leq_{\delta_2}}(\ldots \text{mod}\,(\delta_k) \ldots))$. On the other hand, if $I \prec_\Delta I'$, then there exists i such that $I' <_{\delta_i} I$ and $\neg\exists j > i$ such that $I <_{\delta_j} I'$. If $i = k$, then by Definition 1, $I \notin \text{mod}(\delta_k)$ and hence $I \notin \text{mod}(\Delta)$, a contradiction. If $i < k$, then from $I' \in \text{mod}(\Delta)$, it follows that $I \notin \min_{\leq_{\delta_i}}(\ldots (\text{mod}(\delta_k) \ldots))$ and thus $I \notin \text{mod}(\Delta)$, again a contradiction.

2. $I \in \text{mod}(\Delta)$ and $I' \notin \text{mod}(\Delta)$ implies $I \prec_\Delta I'$.

   Suppose $I \in \text{mod}(\Delta)$, $I' \notin \text{mod}(\Delta)$, but $I' \preceq_\Delta I$. If $I' \notin \text{mod}(\Delta)$, then $I' \notin \min_{\leq_{\delta_1}}(\min_{\leq_{\delta_2}}(\ldots \text{mod}\,(\delta_k) \ldots))$. If $I' \notin \text{mod}(\delta_k)$, then $I <_{\delta_k} I'$, and thus $I' \not\preceq_\Delta I$, a contradiction. Therefore, take the greatest index i, such that $I' \notin \min_{\leq_{\delta_i}}(\min_{\leq_{\delta_{i+1}}}(\ldots \text{mod}\,(\delta_k) \ldots))$, but $I' \in \min_{\leq_{\delta_{i+1}}}(\ldots \text{mod}\,(\delta_k) \ldots)$ (if $i = k-1$, then $I' \in \text{mod}(\delta_k)$). Obviously, $I \in \min_{\leq_{\delta_{i+1}}}(\ldots \text{mod}\,(\delta_k) \ldots)$. It follows that $I <_{\delta_i} I'$, and $\neg\exists j > i$ such that $I' <_{\delta_j} I$, and hence $I' \not\preceq_\Delta I$, a contradiction.

3. $\Delta = \Gamma$ implies $\preceq_\Delta = \preceq_\Gamma$.

   Immediate.

■

NOTATION 15. $\Delta; \beta$ will be used to denote the sequence obtained by appending β to the right end of the sequence Δ, i.e., $\Delta :: [\beta]$.

THEOREM 16. *Let Δ be a PDB and consider the revision of Δ by a sentence β given by $(\Delta; \beta)^*$. This revision scheme satisfies postulates* (R*1)–(R*6).

Proof. All we have to show is that for any sequence Δ, $\text{mod}((\Delta; \beta)^*) = \min_{\preceq_\Delta}(\text{mod}(\beta))$.

We start with the limiting cases. If $\Delta = \varepsilon$, then $\Delta;\beta = [\beta]$, and hence $\mathrm{mod}([\beta]^*) = \mathrm{mod}(\beta)$. It is easy to see that $\min_{\preceq_\varepsilon}(\mathrm{mod}(\beta)) = \mathrm{mod}(\beta)$. It is also straightforward to prove that the theorem holds when $\mathrm{mod}(\beta)$ is empty.

Now suppose $\Delta = [\delta_1, \ldots, \delta_k]$, where $k \geq 1$. We want to show that that $\mathrm{mod}((\Delta;\beta)^*) = \mathrm{mod}(\delta_1 \circ (\ldots (\delta_k \circ \beta))) = \min_{\preceq_\Delta}(\mathrm{mod}(\beta))$.
($\subseteq$) Suppose $I \in \mathrm{mod}(\Delta \circ \beta)$, but $I \notin \min_{\preceq_\Delta}(\mathrm{mod}(\beta))$. If $I \in \mathrm{mod}(\Delta \circ \beta)$, then $I \in \mathrm{mod}(\beta)$. It follows that $\exists J \in \mathrm{mod}(\beta)$ such that $J \prec_\Delta I$. By Proposition 11, $\exists x$ such that $J <_{\delta_x} I$ and $\forall y > x$, $J \leq_{\delta_y} I$. On the other hand, if $I \in \mathrm{mod}(\Delta \circ \beta)$, then

$$I \in \min_{\leq_{\delta_1}}(\ldots(\min_{\leq_{\delta_x}}(\min_{\leq_{\delta_{x+1}}}(\ldots \min_{\leq_{\delta_k}}(\mathrm{mod}(\beta)))))).$$

We know that $J <_{\delta_x} I$. Notice that if $I \in \min_{\leq_{\delta_{x+1}}}(\ldots(\min_{\leq_{\delta_k}}(\mathrm{mod}(\beta))))$, so does J. This is guaranteed by the fact that $\forall y > x$, $J \leq_{\delta_y} I$. But this is a contradiction, because $J <_{\delta_x} I$ and therefore $I \notin \min_{\delta_x}(\min_{\delta_{x+1}}(\ldots \min_{\delta_k}(\mathrm{mod}(\beta))))$.
($\supseteq$) Suppose $I \in \min_{\preceq_\Delta}(\mathrm{mod}(\beta))$, but $I \notin \mathrm{mod}(\Delta \circ \beta)$. It follows that $I \notin \min_{\leq_{\delta_1}}(\ldots(\min_{\leq_{\delta_x}}(\min_{\leq_{\delta_{x+1}}}(\ldots \min_{\leq_{\delta_k}}(\mathrm{mod}(\beta))))))$.

Obviously, $I \in \mathrm{mod}(\beta)$, so let x be the greatest natural number for which $I \notin \min_{\leq_{\delta_x}}(\min_{\leq_{\delta_{x+1}}}(\ldots(\min_{\leq_{\delta_k}}(\mathrm{mod}(\beta)))))$. It follows that $\exists J \in \min_{\leq_{\delta_{x+1}}}(\ldots(\min_{\leq_{\delta_k}}(\mathrm{mod}(\beta))))$ such that $J <_{\delta_x} I$ (if $x = k$, then the expression on the right hand side of $\in$ is just $\mathrm{mod}(\beta)$). But $I \in \min_{\preceq_\Delta}(\mathrm{mod}(\beta))$, therefore by Definition 8 $\exists y > x$ such that $I <_{\delta_y} J$. If $x = k$, this is a contradiction. If $x < k$, then this contradicts $J \in \min_{\leq_{\delta_{x+1}}}(\ldots(\min_{\leq_{\delta_k}}(\mathrm{mod}(\beta))))$. ∎

We now discuss more specifically how the ordering $\preceq$ evolves as new sentences are added to a PDB. The evolution itself depends on the particular characteristics of the faithful assignment for the revision operator. For illustration purposes, we pick $\leq_\delta$ as determined by the distance function used by Dalal and others [Dalal, 1988; Gabbay and Rodrigues, 1997], in which $I \leq_\delta J$ if and only if the minimum number of disagreements of truth-values of propositional letters between I and any model of δ is at most the minimum number of disagreements between J and any model of δ.

Suppose the initial PDB is composed solely by the sentence p. The only ordering in this case is $\leq_p$. Therefore, we have the ordering below. For simplicity, we consider $\mathcal{L}$ over $[p, q, r]$ only. A valuation I is represented as a sequence of binary digits PQR where $P = 1$ iff $I \Vdash p$, and $P = 0$ otherwise; $Q = 1$ iff $I \Vdash q$, and $Q = 0$ otherwise and $R = 1$ iff $I \Vdash r$, and $R = 0$ otherwise. A valuation appearing lower in the graph is preferred to a valuation appearing above it (e.g., $100 \prec 000$ in the graph below) — valuations at the same level are equivalent module $\preceq$.

$$\begin{array}{c} 000, 001, 010, 011 \\ \uparrow \\ 100, 101, 110, 111 \end{array}$$

There is not much information in the PDB when only p is present and the ordering above reflects that; it only makes a distinction between the valuations that satisfy p and those which do not. This is required by (F2) of Definition 1. The class with the models of p must be as represented above because of (F1) and the fact the faithful assignments must be total. The models of $\neg p$ are are not constrained in any way by Definition 1.

By adding the sentence $p \rightarrow q$ to the PDB above, the ordering for the PDB changes to reflect the new priorities:

$$\begin{array}{c} 100, 101 \\ \uparrow \\ 000, 001, 010, 011 \\ \uparrow \\ 110, 111 \end{array}$$

Now we have three classes of valuations: in the minimal (preferred) one we have 110 and 111, which are exactly the only two models of p and $p \rightarrow q$. According to this similarity ordering, the next best alternatives to these valuations are contained in the next class — 000, 001, 010, 011. These are exactly the valuations that failing to satisfy p, at least satisfy the most important sentence in the PDB ($p \rightarrow q$). Finally, the last level contains the valuations that do not satisfy $p \rightarrow q$, but at least satisfy p.

The addition of r to the PDB results in the following rearrangement of the similarity ordering above:

$$\begin{array}{c} 100 \\ \uparrow \\ 000, 010 \\ \uparrow \\ 110 \\ \uparrow \\ 101 \\ \uparrow \\ 001, 011 \\ \uparrow \\ 111 \end{array}$$

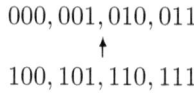

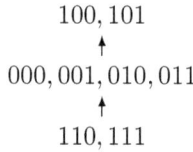

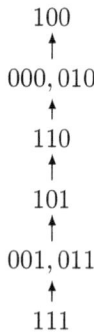

The minimal elements of this ordering are exactly the models of the three sentences. The next class contains the valuations that satisfy the two sentences with highest priority in the PDB, followed by the class with the valuations that satisfy the sentence with highest priority and the one with least priority.

The other four classes of the ordering in the top half follow a similar reasoning. They contain the valuations that fail to satisfy r. That not being possible, the next best thing is to satisfy the other two sentences. This is represented by valuation 110. The next two valuations satisfy the second sentence and the least preferred one satisfies only the least important sentence.

Perhaps a little more interesting is to see how the ordering above is obtained. First, order the valuations according to the most recent information received: r. This will result in a number of equivalence classes totally ranked amongst themselves (since each faithful assignment is total). Next, order each of the classes according to the faithful assignment of the sentence immediately before it in the PDB: $p \to q$. Finally, order the resulting classes according to the next sentence p (see Figure 1).

$$
\mathcal{I} \quad \xrightarrow{\leq_r} \quad
\begin{array}{c}
000, 010, \\
100, 110 \\
\uparrow \\
001, 011, \\
101, 111
\end{array}
\quad \xrightarrow{\leq_{p \to q}} \quad
\begin{array}{c}
100 \\
\uparrow \\
000, 010, 110 \\
\uparrow \\
101 \\
\uparrow \\
001, 011, 111
\end{array}
\quad \xrightarrow{\leq_p} \quad
\begin{array}{c}
100 \\
\uparrow \\
000, 010 \\
\uparrow \\
110 \\
\uparrow \\
101 \\
\uparrow \\
001, 011 \\
\uparrow \\
111
\end{array}
$$

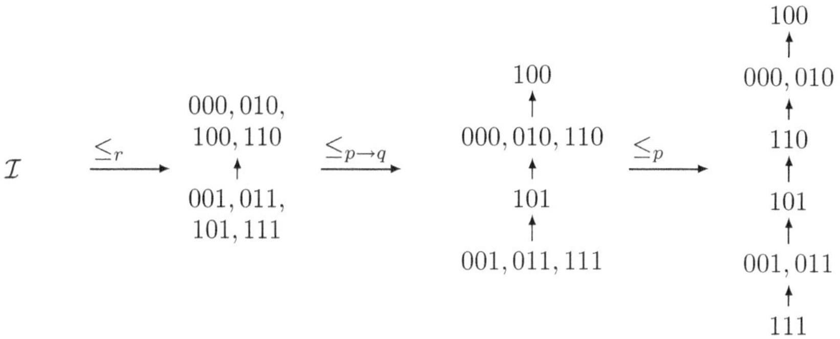

Figure 1. Embedded orderings in PDBs.

More formally, we can show the relation between the ordering of a given PDB Δ and the ordering of Δ revised by a sentence β, in the following way:

PROPOSITION 17. *Let $\Delta = [\delta_1, \ldots, \delta_k]$ be a PDB, β a sentence, $\preceq_\Delta$ the faithful assignment for the epistemic state Δ and $\leq_\beta$ the faithful assignment for the sentence β.*

$$I \preceq_{\Delta \circ \beta} I' \text{ iff } I' \leq_\beta I \text{ implies } I \leq_\beta I' \text{ and } I \preceq_\Delta I'$$

where $\preceq_{\Delta\circ\beta}$ *is the faithful assignment for the PDB* $[\delta_1, \ldots, \delta_k, \beta]$.

Proof. ($\Rightarrow$) Suppose $I \preceq_{\Delta\circ\beta} I'$ and $I' \leq_\beta I$. By Definition 8, for $0 \leq i \leq k+1$, $I' <_{\delta_i} I$ implies $\exists j$, $i > j \leq k+1$ such that $I <_{\delta_j} I'$. Clearly, $I \not<_{\delta_{k+1}} I'$, since $I' \leq_\beta I$. Therefore, $I \leq_\beta I'$ and if $I' <_{\delta_i} I$ for $0 \leq i \leq k$, then $\exists j$, $i < j \leq k$ such that $I <_{\delta_j} I'$. By Definition 8 again, $I \preceq_\Delta I'$.
($\Leftarrow$) Suppose $I' \leq_\beta I$, but $I \not\preceq_{\Delta\circ\beta} I'$. If $I \not\preceq_{\Delta\circ\beta} I'$, then there exists i, $0 \leq i \leq k+1$ such that $I' <_{\delta_i} I$ and $\neg\exists j > i$ such that $I <_{\delta_j} I'$. This is a contradiction, for if $i = k+1$, then $I \not\leq_\beta I'$ and if $i \leq k$, then $I \not\preceq_\Delta I'$. On the other hand, if $I' \not\leq_\beta I$, then $I <_\beta I'$, and $I \preceq_{\Delta\circ\beta} I'$ follows trivially. ■

5 Relations with Darwiche and Pearl's Postulates

The characterization given in Proposition 17 above allows us to relate revisions achieved by PDBs with Darwiche and Pearl's formalization in the following way:

THEOREM 18. *The revision method achieved by PDBs satisfies postulates* (C1), (C3) *and* (C4).

Proof. The proof is obtained via Darwiche and Pearl's semantic characterization of postulates (C1), (C3) and (C4), namely conditions (CR1), (CR3) and (CR4), and Proposition 17 above.

(CR1) If $I, I' \in \text{mod}(\beta)$, then $I \preceq_\Delta I'$ iff $I \preceq_{\Delta\circ\beta} I'$

Suppose $I, I' \in \text{mod}(\beta)$. By Definition 1, $I, I' \in \text{mod}(\beta)$ implies $I \leq_\beta I'$ and $I' \leq_\beta I$.
($\Rightarrow$) Suppose $I \preceq_\Delta I'$. By Proposition 17, from $I' \leq_\beta B$, $I \leq_\beta$ and $I \preceq_\Delta I'$, then $I \preceq_{\Delta\circ\beta} I'$.
($\Leftarrow$) Suppose $I \preceq_{\Delta\circ\beta} I'$. By Proposition 17, since $I' \leq_\beta I$, then $I \leq_\beta I'$ and $I \preceq_\Delta I'$.

(CR3) If $I \in \text{mod}(\beta)$ and $I' \in \text{mod}(\neg\beta)$, then $I \prec_\Delta I'$ implies $I \prec_{\Delta\circ\beta} I'$
(CR4) If $I \in \text{mod}(\beta)$ and $I' \in \text{mod}(\neg\beta)$, then $I \preceq_\Delta I'$ implies $I \preceq_{\Delta\circ\beta} I'$

For the two requirements above, we prove a stronger result:

(CR$\star$) If $I \in \text{mod}(\beta)$ and $I' \in \text{mod}(\neg\beta)$, then $I \prec_{\Delta\circ\beta} I'$

Suppose $I \in \text{mod}(\beta)$ and $I' \in \text{mod}(\neg\beta)$. It follows that $I' \notin \text{mod}(\beta)$. By Definition 1, $I <_\beta I'$, and hence by Definition 8, $I \prec_{\Delta\circ\beta} I'$. ■

However, postulated (C2) is not satisfied. (C2) is one of the most controversial postulates of the Darwiche and Pearl's series. It says that if an agent learns B first and then is given some information that contradicts this evidence, say A, then he/she should completely ignore the *whole* content of the information in B. (C2) was shown to be incompatible with the original AGM postulates by Freund and Lehmann in [Freund and Lehmann, 1994]. However, if one considers the reformulation for epistemic states given by Darwiche and Pearl, they become indeed compatible. (C2) (and other similar versions of it) seems to undermine the principle of minimal change. It does not seem reasonable to discard completely the information B simply because of the arrival of the contradictory information A. Suppose B is $p \wedge q$ and A is $\neg p \vee \neg q$. Since $\neg p \vee \neg q \models \neg(p \wedge q)$, (C2) applies, but the question remains of whether the agent should completely discard the information conveyed by $p \wedge q$ in face of $\neg p \vee \neg q$. After all, the belief in $p \leftrightarrow \neg q$ is compatible with $\neg p \vee \neg q$ and manages to keep some of the informational content of $p \wedge q$. However, compliance with (C2) requires the agent to completely ignore the fact that he/she ever believed in $p \wedge q$. The example by which Darwiche and Pearl justify the plausibility of (C2) [Darwiche and Pearl, 1997, page 12] is arguable and we discuss it at length next.

EXAMPLE 19. An agent's current epistemic state includes the sentence smart $\wedge$ rich, representing the belief that lady X is smart and she is also rich. The epistemic state is then revised by $\neg$smart and, subsequently, by smart. Since smart $\wedge$ rich is followed by some information that contradicts this observation, namely, $\neg$smart, (C2) applies and requires that the resulting epistemic state is equivalent to the initial state revised by the second observation only. In other words, the intermediate observation should be disregarded:

$$((\text{smart} \wedge \text{rich}) \circ \neg\text{smart}) \circ \text{smart} \equiv (\text{smart} \wedge \text{rich}) \circ \text{smart}$$

The reason why (C2) seems reasonable in this example is because nothing of the informational content of the first revising sentence ($\neg$smart) can be kept after the second revision (by $\neg$smart) is performed. After all, what could be consistently kept from $\neg$smart in the face of smart?

However, consider the following modified scenario. Start with an initially empty epistemic state which is to be revised by smart $\wedge$ rich and then by $\neg$smart:

$$(\top \circ (\text{smart} \wedge \text{rich})) \circ \neg\text{smart} \stackrel{?}{\equiv} \top \circ \neg\text{smart}$$

The postulate applies here too, because $\neg$smart $\models \neg$(smart $\wedge$ rich). If one adopts the principle of minimal change, it seems counterintuitive to give up the belief in "rich" just in face of "$\neg$smart". It might have been the case that the observation with respect to lady X's being smart was wrong, but this does not necessarily mean that the belief in her being rich was inaccurate too. The acceptance of (C2) requires this though.

Reanalysing (CR2), we can restate our point of view differently:

(CR2) If $I, I' \in \text{mod}(\neg A)$, then $I \preceq_\Psi I'$ iff $I \preceq_{\Psi \circ A} I'$.

The fact that both I and I' do not satisfy A is not sufficient to consider them equally good at satisfying A. Therefore, the ordering $\leq_A$ might change the way I and I' relate in $\preceq_{\Psi \circ A}$ with respect to the way they used to relate in $\preceq_\Psi$.

Regarding (C3) and (C4), we have proved that the following condition holds:

(CR$\star$) If $I \in \text{mod}(\beta)$ and $I' \in \text{mod}(\neg\beta)$, then $I \prec_{\Delta \circ \beta} I'$.

(CR$\star$) is stronger than each of the corresponding conditions on the faithful assigmnent orderings (CR3) and (CR4):

(CR3) If $I \in \text{mod}(\beta)$ and $I' \in \text{mod}(\neg\beta)$, then $I \prec_\Delta I'$ implies $I \prec_{\Delta \circ \beta} I'$
(CR4) If $I \in \text{mod}(\beta)$ and $I' \in \text{mod}(\neg\beta)$, then $I \preceq_\Delta I'$ implies $I \preceq_{\Delta \circ \beta} I'$

The motivation for (CR$\star$) comes from the following observation: when Δ is revised by β, the new top priority is to satisfy β (if possible at all). If I satisfies β, but I' does not, it should not matter how I and I' related before with respect to $\preceq_\Delta$. In the new ordering $\preceq_{\Delta \circ \beta}$, I should be preferred to I'.

6 PDBs and Minimal Change of Conditional Beliefs

In [Boutilier, 1996], Boutilier proposed a formalism that allowed iteration of the revision process by defining how a similarity ordering for the initial belief set K should evolve during successive revision operations. As with Darwiche and Pearl, Boutilier also distinguishes between belief states and epistemic states. He calls the former *objective belief sets* whereas the latter is implicitly defined by a combination of a belief set K and an associated *revision model*. A revision model is a structure used to determine how revisions are performed for a given belief set K. The details are presented below. As before, we consider a language $\mathcal{L}$ over a set of propositional variables $\mathcal{P}$.

DEFINITION 20 (Revision model). Given a belief set K, a *revision model* for K is a tuple $M = \langle W, \leq, s \rangle$, where W is a set of worlds; $\leq$ is a total pre-order on W and $s : \mathcal{P} \longrightarrow 2^W$ is a valuation function.

A valuation s identifies those worlds in W in which a given propositional variable holds and is extended to complex formulae as usual. $M \Vdash_w A$ denotes the fact that $w \in s(A)$ in M. The collection of all such worlds is denoted $|A|$. For worlds $v, w \in W$, $v \leq w$ is interpreted as world v being at least as *plausible* a state of affairs as world w is. Since $\leq$ is total, it induces a total order on equivalence classes in W and can be seen as a ranking relation (much in the same way as in Grove's systems of spheres or as in Katsuno and Mendelzon's faithful assignments).

Revision models are required to verify a number of rationality conditions. One such condition is that $|K|$ is constituted exactly by the minimum equivalence class in W. In other words,

$$w \leq v \text{ for all } v \in W \text{ iff } M \Vdash_w K$$

It is also required that for all $p \in \mathcal{P}$, $s(p) \neq \emptyset$. In addition, a well foundedness condition on $\leq$ is imposed. For full details, please check [Boutilier, 1996].

Conditional assertions of the kind $A \Box\!\!\rightarrow B$ are evaluated in a model M in the following way:

DEFINITION 21 (Evaluation of conditionals in a revision model). Let M be a revision model and A and B formulae of propositional logic.

$$M \Vdash A \Box\!\!\rightarrow B \text{ iff } \min(M, A) \subseteq |B|$$

where
$\min(M, A) = \{w \in W \mid M \Vdash_w A, \text{ and} M \Vdash_v A \text{ implies } m \leq v \text{ for all } v \in W\}$[4]

This is used to define a revision function $\circ_b$ for a particular revision model M:

DEFINITION 22. Let K be a belief set; M a revision model for K and A and B formulae of propositional logic.

$$K \circ_b A = \{B \mid M \Vdash A \Box\!\!\rightarrow B\}$$

Note that $\circ_b$ is defined in terms of unnested conditionals, since A and B are restricted to pure propositional logic formulae only. In [Boutilier, 1994], the following correspondences were proved:

THEOREM 23. *If M is a K-revision model and $\circ_b$ the revision function determined by M according to Definition 22, then $\circ_b$ verifies the AGM postulates for belief revision.*

THEOREM 24. *Let $\circ$ be any revision function satisfying the AGM postulates. For any belief set K, there is a K-revision model M such that $K \circ A = K \circ_b A$.*

Note that if A is contradictory, then $\min(M, A) = \emptyset$. Therefore, $M \Vdash A \Box\!\!\rightarrow B$ for all B and hence $K \circ_b A = K_\perp$, as expected.[5]

[4] This is similar to $\min_{\leq}(\text{mod}(A))$, except that worlds are used instead of valuations.
[5] $K_\perp$ is the inconsistent belief set.

Note that the revision function $\circ_b$ is entirely defined by a revision model M. In fact, it is easy to show the correspondence between properties of a revision model and those of a faithful assignment.

Boutilier argued that revision could not be iterated directly because M is given for K and $K \circ_b A$ is an entirely different structure from K. However, he showed how a revision model M could be updated for $K \circ_b A$. Basically, his idea was to impose minimal conditions on how $\leq$ should change:

DEFINITION 25 (Minimal conditional revision operator). Let $M = \langle W, \leq, s\rangle$ be a revision model and A a sentence of propositional logic. The *minimal conditional revision operator* (MC-revision operator) $*$ maps M into $M^*_A = \langle W, \leq', v\rangle$ such that *i)* if $M \in \min(M, A)$, then $M \leq' N$ for all $N \in \mathcal{I}$ and $N \leq' M$ if and only if $N \in \min(M, A)$; and *ii)* if $M, N \notin \min(M, A)$, then $M \leq' N$ if and only if $M \leq N$.

In other words, the worlds in $\min(M, A)$ become the preferred class of $\leq'$ and the relationship between all other worlds remain unchanged. This rearrangement of $\leq$ only works well if revisions are *kind*, i.e., forward compatible in Boutilier's terminology.

DEFINITION 26 (Forward compatibility). Let M be a revision model determining a MC-revision function $*$. The revision sequence $A_1, \ldots, A_n$ is *forward compatible* with respect to $*$ (or model M), if and only if $\neg A_{i+1} \notin (\ldots((K \circ_b A_1) \circ_b A_2) \ldots) \circ_b A_i$ for each $1 \leq i < n$.

One could argue that revision sequences that are always forward compatible are not interesting, since they can be accomplished by expansions only. Boutilier has noticed that:

THEOREM 27. *If* $A_1, \ldots, A_n$ *is forward compatible for* K, *then* $((K \circ_b A_1) \circ_b A_2 \ldots) \circ_b A_n = K \circ_b (A_1 \wedge \ldots \wedge A_n)$.

Unfortunately, when an incompatible sentence comes in, the model forgets all revisions occurring up to the last sentence in the sequence that is compatible with the new information:

THEOREM 28 (Forgetfulness). *Let* $A_1, \ldots, A_{n+1}$ *be an incompatible sequence such that* $A_1, \ldots, A_n$ *is compatible. Let* k *be the maximal element of* $\{i \leq n \mid \neg A_{n+1} \notin ((K \circ_b A_1) \circ_b A_2 \ldots) \circ_b A_i\}$. *It follows that* $((K \circ_b A_1) \circ_b A_2 \ldots) \circ_b A_{n+1} = K \circ_b (A_1 \wedge \ldots \wedge A_k \wedge A_{n+1})$.

In this sense, the revision scheme obtained by associating $\circ$ to the right is more forgiving, as can be seen in the next example:

EXAMPLE 29. Consider the PDB $\Delta = [p, q \wedge r, \neg q]$ and suppose the revision model for p is $\langle W, \leq_p, v\rangle$, where $M \leq_p$. The MC-revision operator will update $\leq_p$ successively as follows.

$$
\begin{array}{ccccc}
\leq_{[p]} & \longrightarrow & \leq_{[p,q\wedge r]} & \longrightarrow & \leq_{[p,q\wedge r,\neg q]} \\
 & & & & \\
000,001,010,011 & & 000,001,010,011 & & 000,001,010,011 \\
\uparrow & & \uparrow & & \uparrow \\
100,101,110,111 & & 100,101,110 & & 110 \\
 & & \uparrow & & \uparrow \\
 & & 111 & & 111 \\
 & & & & \uparrow \\
 & & & & 100,101
\end{array}
$$

As a result, $(p \circ_b (q \wedge r)) \circ_{\neg q} = \mathrm{Cn}(p \wedge \neg q)$ (the first revision is forgotten).

$$
\begin{array}{ccccc}
\leq_{[p]} & \longrightarrow & \leq_{[p,q\wedge r]} & \longrightarrow & \leq_{[p,q\wedge r,\neg q]} \\
 & & & & \\
000,001,010,011 & & 000 & & 010 \\
\uparrow & & \uparrow & & \uparrow \\
100,101,110,111 & & 100 & & 110 \\
 & & \uparrow & & \uparrow \\
 & & 001,010 & & 011 \\
 & & \uparrow & & \uparrow \\
 & & 101,110 & & 111 \\
 & & \uparrow & & \uparrow \\
 & & 011 & & 000 \\
 & & \uparrow & & \uparrow \\
 & & 111 & & 100 \\
 & & & & \uparrow \\
 & & & & 001 \\
 & & & & \uparrow \\
 & & & & 101
\end{array}
$$

and $[p, q \wedge r, \neg q]^* = \mathrm{Cn}(p \wedge \neg q \wedge r)$, in which at least the information r of $q \wedge r$ is retained.

7 PDBs and Lehmann's Belief Revision, Revised

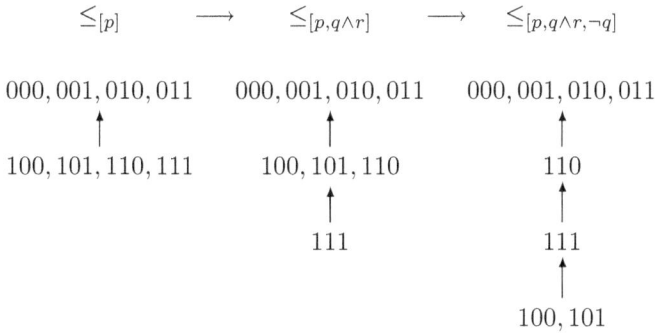

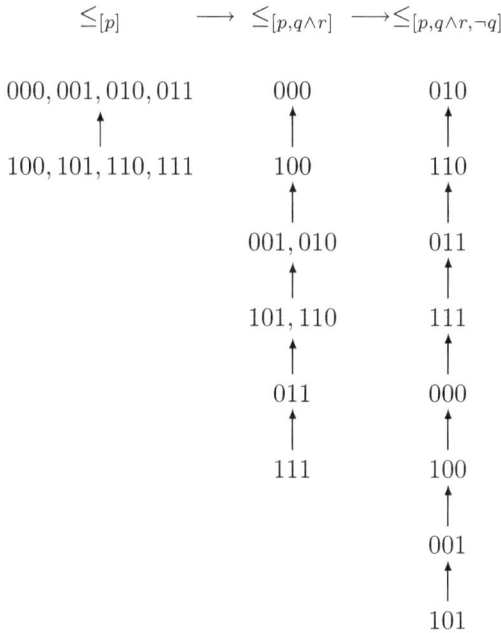

Lehmann also argued in [Lehmann, 1995] that the AGM postulates in their original belief set interpretation are incompatible with some desired properties of the iteration process (including postulate (C1) seen in Section 2.1)

and advocates a revision of the complete framework. This led to the definition of the new postulates (I1)–(I7) presented below.

In his framework, he considers belief states resulting from a finite sequence of revisions by non-contradictory formulae. Epistemic states can thus be seen as before as sequences of formulae, and the corresponding belief states are obtained by application of a particular revision procedure which is represented by $[\ ]$. For instance, $[\alpha]$ represents the belief set obtained from the initial belief set revised by α. Subsequent revisions are represented by appending formulae to the end of the list: $[\alpha \cdot \beta]$ is the belief set obtained by revising $[\alpha]$ by β, and so forth. Lehmann's postulates are listed below:

Lehmann's postulates for iterated belief revision

(I1)	$[\alpha]$ is a consistent theory
(I2)	$\alpha \in [\sigma \cdot \alpha]$
(I3)	If $\beta \in [\sigma \cdot \alpha]$, then $\alpha \to \beta \in [\sigma]$
(I4)	If $\alpha \in [\sigma]$, then $[\sigma \cdot \gamma] = [\sigma \cdot \alpha \cdot \gamma]$
(I5)	If $\beta \models \alpha$, then $[\sigma \cdot \alpha \cdot \beta \cdot \gamma] = [\sigma \cdot \beta \cdot \gamma]$
(I6)	If $\neg\beta \notin [\sigma \cdot \alpha]$, then $[\sigma \cdot \alpha \cdot \beta \cdot \gamma] = [\sigma \cdot \alpha \cdot \alpha \wedge \beta \cdot \gamma]$
(I7)	$[\sigma \cdot \neg\beta \cdot \beta] \subseteq \mathrm{Cn}([\sigma], \beta)$

The only postulates that actually add new properties to the original AGM presentation are (I5) and (I7). The other five postulates are reformulations of the AGM ones for the new framework. It is worth emphasizing though, that as for Darwiche and Pearl, the epistemic state plays a fundamental role in the postulates. The condition below, which is a consequence of (R4), does not follow from (I1)–(I7):

(IS) If $[\sigma] = [\gamma]$, then $[\sigma \cdot \alpha] = [\gamma \cdot \alpha]$

(I5) is related to Darwiche and Pearl's (C1). It says that if α is a logical consequence of β, then revising by α and then revising by β is the same as revising by β only.

(I7) is a weaker version of Darwiche and Pearl's (C2). It asserts that beliefs acquired from the revision by $\neg\beta$ should not be retained if the belief state is immediately revised by β. Notice that the precondition for (C2) is strictly weaker than that for (I7), whereas its postcondition is in turn at least as strong.

A negative result of this framework is that any revision procedure satisfying postulates (I1)–(I7) becomes trivial for an arbitrarily long sequence of revisions. The trivial revision procedure is the one which assigns $\mathrm{Cn}(A)$ to $[\sigma \cdot A]$, whenever $\neg A \in [\sigma]$, and $\mathrm{Cn}([\sigma]; A)$, otherwise.

We do not provide a full comparison between properties of revisions whose ordering are related to a right associative interpretation of the operation and Lehmann's postulates, but we put forward some questions.

We have seen that one of the negative results of Lehmann's approach is that for any long enough sequence of revisions, the revision procedure becomes trivial. This obviously goes against the principle of minimal change. Lehmann suggests that, because postulates (I1)–(I7) are closely related to AGM's original postulates, no reasonable revision satisfying these postulates can comply with the principle of minimal change. It seems to us that the problem lies in postulate (I7) below.

(I7) $[\sigma \cdot \neg\beta \cdot \beta] \subseteq \mathrm{Cn}([\sigma], \beta)$

(I7) is a weaker version of Darwiche and Pearl's (C2), which we have also argued to be counterintuitive with respect to minimal change. The revision procedure defined by PDBs, even though not complying with (I1)–(I7), does comply with the original AGM postulates and the widely accepted (C1),(C3) and (C4). More importantly, the longer the sequence of revisions, the more structured the ordering for the PDB becomes and less trivial the whole revision process is.

8 Conclusions and Discussion

In this paper we revisited the problem of iterated revision by proposing a new way to automatically re-generate the similarity ordering of an epistemic state after a revision is performed. We drew some comparisons with a number of other formalisms for iterated revision found in the literature.

Boutilier's idea of re-organising the initial similarity ordering was a good step towards automatic re-calculation of the similarity ordering for an epistemic *state*. However, it suffers from some drawbacks. In particular, it is too dependent on the first revision model $M = \langle W, \leq, v \rangle$ for the initial knowledge base K: iterated revisions of K by δ_1, δ_2, ... refine $\leq$ in so far as prioritising *some* models of δ_1, δ_2, etc., successively on top of $\leq$, but only according to the initial $\leq$. An immediate consequence of this is that as soon as a sentence δ_i which is inconsistent with another sentence δ_j appearing earlier in the sequence comes in, all revision steps between δ_j and and δ_i are ignored (see Theorem 28).

Our idea for the automatic re-generation of the similarity ordering stems from looking at a sequence of revision operations and seeing the operator as a right associative one. We have shown that this interpretation has a number of desired properties.

One drawback with the current right associative interpretation is that whenever a new revision needs to be performed *all* previous revisions have to be re-applied. Since revision operations are known to be computationally expensive [Nebel, 1991a; Nebel, 1992a; Nebel, 1998; Gärdenfors and Rott, 1995; Eiter and Gottlob, 1992], this may have serious implications. However, it may be cheaper to calculate the result of the revision *semantically*, i.e., via its models. The idea is as follows. The similarity orderings for each sentence in a sequence $\Delta = [\delta_1, \ldots, \delta_k]$ have to be total (a faithful assignment requirement). When a new sentence δ_{k+1} comes in, one simply refines each equivalence class in $\leq_{\delta_{k+1}}$ so that the valuations in the classes are ordered exactly as they used to be in $\leq_\Delta$. The new ordering can then be used to calculate the models of $\Delta; \delta_{k+1}$.

It seems to us that any practical belief revision reasoning system will have to rely somehow or revisiting previous decisions of acceptances (or not) of evidence received. If one considers the analysis of whether to incorporate the evidence at the meta-level, this attitude will not be in opposition to the principle of the primacy of the update, since this principle requires the new evidence to be accepted after a revision is performed, but in fact does not require the revision itself *to be* performed.

We believe that a combination of the technique for refining an epistemic state's similarity ordering proposed here and some mechanism to weight in the plausibility of new evidence so that the revision process may be declined in special cases would be an interesting exercise.

Acknowledgement

Thank you Dov for your continual inspiration, thirst for knowledge and vision.

BIBLIOGRAPHY

[Boutilier, 1994] C. Boutilier. Unifying default reasoning and belief revision in a modal framework. *Artificial Intelligence*, 68:33–85, 1994.

[Boutilier, 1996] C. Boutilier. Iterated revision and minimal change of conditional beliefs. *Journal of Philosophical Logic*, 25:262–305, 1996.

[Dalal, 1988] Mukesh Dalal. Investigations into a theory of knowledge base revision: Preliminary report. In Paul Rosenbloom and Peter Szolovits, editors, *Proceedings of the Seventh National Conference on Artificial Intelligence*, volume 2, pages 475–479, Menlo Park, California, 1988. AAAI Press.

[Darwiche and Pearl, 1994] Adnan Darwiche and Judea Pearl. On the logic of iterated belief revision. In Ronald Fagin, editor, *Proceedings of the 5th International Conference on Principles of Knowledge Representation and Reasoning*, pages 5–23. Morgan Kaufmann, Pacific Grove, CA, March 1994.

[Darwiche and Pearl, 1996] Adnan Darwiche and Judea Pearl. On the logic of iterated belief revision. Technical Report R-202, Cognitive Science Laboratory, Computer Science Department, University of California, Los Angeles, CA 90024, November 1996.

[Darwiche and Pearl, 1997] Adnan Darwiche and Judea Pearl. On the logic of iterated belief revision. *Artificial Intelligence*, 89:1–29, 1997.

[Eiter and Gottlob, 1992] T. Eiter and G. Gottlob. On the complexity of propositional knowledge base revision, updates and counterfactuals. In *ACM Symposium on Principles of Database Systems*, 1992.

[Freund and Lehmann, 1994] Michael Freund and Daniel Lehmann. Belief revision and rational inference. Technical Report TR 94-16, The Leibniz Center for Research in Computer Science, Institute of Computer Science, Hebrew University, July 1994.

[Friedman and Halpern, 1996] Nir Friedman and Joseph Y. Halpern. Belief revision: A critique. In *Proceedings of the 5th International Conference on Principles of Knowledge Representation and Reasoning*, pages 421–431, Cambridge, Massachusetts, 1996.

[Gabbay and Rodrigues, 1996] Dov Gabbay and Odinaldo Rodrigues. A methodology for iterated theory change. In Dov M. Gabbay and Hans Jürgen Ohlbach, editors, *Practical Reasoning - First International Conference on Formal and Applied Practical Reasoning, FAPR'96*, Lecture Notes in Artificial Intelligence. Springer Verlag, 1996.

[Gabbay and Rodrigues, 1997] Dov Gabbay and Odinaldo Rodrigues. Structured belief bases: a practical approach to prioritised base revision. In Dov M. Gabbay, Rudolf Kruse, Andreas Nonnengart, and Hans Jürgen Ohlbach, editors, *Proceedings of First Internation Joint Conference on Qualitative and Quantitative Practical Reasoning*, pages 267–281. Springer-Verlag, June 1997.

[Gärdenfors and Rott, 1995] P. Gärdenfors and Hans Rott. Belief revision. In C. J. Hogger Dov Gabbay and J. A. Robinson, editors, *Handbook of Logic in Artificial Intelligence and Logic Programming*, volume 4, pages 35–132. Oxford University Press, 1995.

[Grove, 1988] Adam Grove. Two modellings for theory change. *Journal of Philosophical Logic*, 17:157–170, 1988.

[Katsuno and Mendelzon, 1992] H. Katsuno and A. O. Mendelzon. On the difference between updating a knowledge base and revising it. In P. Gärdenfors, editor, *Belief Revision*, pages 183–203. Cambridge University Press, 1992.

[Lehmann, 1995] Daniel Lehmann. Belief revision, revised. In *Proceedings of the 14th International Joint Conference of Artificial Intelligence (IJCAI-95)*, pages 1534–1540, 1995.

[Nebel, 1991a] B. Nebel. Belief revision and default reasoning: syntax-based approaches. In *Proc. Second International Conference on Principles of Knowledge Representation and Reasoning (KR '91)*, pages 417–428. Morgan Kaufmann, San Francisco, CA, 1991.

[Nebel, 1991b] Bernhard Nebel. Belief revision and default reasoning: Syntax-based approaches. In J. Allen, R. Fikes, and E. Sandewall, editors, *Proceedings of the Second International Conference on Principles of Knowledge Representation and Reasoning*, pages 417–428. Morgan Kaufmann, 1991.

[Nebel, 1992a] B. Nebel. Syntax-based approaches to belief revision. In Peter Gärdenfors, editor, *Belief Revision*, number 29 in Cambridge Tracts in Theoretical Computer Science, pages 52–88. Cambridge University Press, 1992.

[Nebel, 1992b] Bernhard Nebel. Syntax based approaches to belief revision. *Belief Revision*, pages 52–88, 1992.

[Nebel, 1998] B. Nebel. How hard is it to revise a belief base. In D. Dubois and H. Prade, editors, *Handbook of Defeasible Reasoning and Uncertainty Management Systems*, volume 3: Belief Change, pages 77–145. Kluwer Academic Publishers, Dordrecht, The Netherlands, 1998.

[Rodrigues, 1998] Odinaldo Rodrigues. *A methodology for iterated information change.* PhD thesis, Department of Computing, Imperial College, January 1998.

[Rodrigues, 2003] Odinaldo Rodrigues. Structured clusters: A framework to reason with contradictory interests. *Journal of Logic and Computation*, 13(1):69–97, 2003.

[Ryan, 1992] Mark Ryan. *Ordered Presentation of Theories - Default Reasoning and Belief Revision.* PhD thesis, Department of Computing, Imperial College, U.K., 1992.

Inference in Temporal Next-Time Logic

VLADIMIR V. RYBAKOV

I dedicate this paper to Dov Gabbay. His bright talent and energy have enabled him to fill many areas of logic with pioneering and important results.

1 Introduction, Motivation

Logical inference is fundamental in mathematical logic. It can be viewed in many ways: classical proof theory, inferences in non-classical logics, formalizations in algebraic logic. A branch in this research is the study of most strong kinds of logical inference in propositional logic — so called admissible inferences, admissible inference rules, which have been introduced by Lorenzen (1955, [Lorenzen, 1955]). Until recently the majority of known results have been concerned with transitive modal logics or superintuitionistic logics (cf. [Mints, 1976; Ghilardi, 1999; Friedman, 1975; Iemhoff, 2001; Rybakov, 1997; Rybakov, 2001; Rybakov *et al.*, 2000a; Rybakov and Fedorishin, 2000; Rybakov *et al.*, 2000b]). Other logics, e.g. intransitive modal logics, multi-modal logics or temporal logics are more resistant to positive results within the technique, e.g. to find algorithms recognizing admissible inferences. In this paper we make an attempt to find an approach to temporal logics.

In present time, temporal logic is an active issue; it has found numerous applications in Artificial Intelligence and Computer Science. Tense Logic was introduced by Arthur Prior (1957, 1967, cf. [Prior, 1967]) as a result of an interest in the relationship between tense and modality. Enumerating applications of temporal logic to AI and CS, temporal logic is a natural logic for hardware verification (cf.[Cyrluk and Narendran, 1994]); this logic has a significant role in the formal specification and verification of concurrent and distributed systems (cf. Pnueli [1977]). It has a numerous applications to safety, liveness and fairness (cf. Manna, Pnueli [1992], Emerson [1990]). A good reference point is the book Advances in Temporal Logic ([Barringer *et al.*, 1999]) describing modern applications. For instance, model checking is also an active area for temporal logic (cf. [McMillan, 1993; van der Hoek and Wooldridge, 2002]). So, in sum, temporal logic is an active research field in many aspects. Since initial symbolic investigations in temporal logic, using Kripke/Hintikka semantics (cf. Segerberg, [1970]) and developing algebraic mathematical models (cf. [van Benthem, 1991]), this area evolved as a solid branch of non-classical logic.

Evidently, inference rules are basic instruments in classical proof theory and, in particular, proof theory of temporal logic. But inference rules of non-classical logic may also be involved for description of subtle properties of models which are problematic to be expressed merely by formulas. D. Gabbay was the pioneer of this approach. In (Gabbay, [1981]) he suggested the *irreflexivity rule*:

$$(ir) := \frac{\neg(p \to \Diamond p) \to \varphi}{\varphi}$$

(where p does not occur in the formula φ).

The rule (ir) is actually saying that any element of a model, where φ is not valid, should be irreflexive; it was implemented in [Gabbay, 1981] for the proof of the completeness theorem. Based on this ideas, axiomatizations of various classes of multi-modal frames by Sahlqvist formulas and special derivation rules were constructed by Venema [1993].

The aim of our paper is to investige logical inference in the temporal Next-Time logic **NXT**. We demonstrate with examples the distinction between evident and admissible inferences of **NXT**. In order to develop an enhanced technique to describe admissible inferences, we extend instruments constructed before for the study of inference rules in modal and intuitionistic logics (cf. Rybakov [2003]–[1997]). We start by a description of admissible inferences given in a semantic manner — via truth-values of inferences in special infinite constructive models. Special suggested normal reduced forms for arbitrary inference rules form a basis for general techniques to handle admissible inferences. The main result of the paper is the final Theorem 27 stating that the logic NXT is decidable w.r.t. admissible inferences. In the paper we use term *inference* but not *inference rule* in order to emphasize that we are primarily interested in descriptions of immediate logical consequence but not in the construction of derivations.

2 Temporal Next-Time Logic, Basic Facts

We start by giving general definitions, notation and facts concerning the Next-time temporal logic. The *language of the Next-time temporal logic* is the standard one for Boolean propositional logic enriched by two additional temporal logical operations: $\Box$ and $\bigcirc$. So, it has a countable set of propositional letters: $p_1, p_2, \dots$, Boolean logical operations $\wedge, \vee, \neg, \to$, and the above-mentioned unary operations $\Box$ and $\bigcirc$. The formation rules for well-formed formulas (in the sequel, wffs) are standard, and, for the new operations, for any wff A, $\Box A$ and $\bigcirc A$ are wffs. $\Box A$ has the following meaning: *the formula A will always be true*, and $\bigcirc A$ can be read: *the formula A will be true at the next time point (tomorrow)*. The logical operation $\Diamond$ is the compound operation: $\Diamond A := \neg\Box\neg A$. Taking into account the chosen language and the meaning of temporal operations, we restrict the general term *temporal logic* as follows.

DEFINITION 1. *A temporal logic L is any set of wffs possessing the properties: (i) L is closed w.r.t. substitutions and inference rules: $x, x \to y/y$, $x/\Box x$ and $x/\bigcirc x$; (ii) L contains all tautologies of the Boolean propositional logic, all axioms of the modal logic K for $\Box$ and $\bigcirc$ as modalities, and the axiom $\Box p_1 \to \bigcirc p_1$.*

A formula C is said to be *valid* in a logic L if $C \in L$. Kripke semantics for wffs of temporal logics in the chosen language consists of bi-frames (frames, for short in the sequel). A frame is a triple $\mathcal{F} := \langle F, R, Next\rangle$, where F is a set, R is a binary reflexive, transitive and quasi-linear relation in F i.e.

$$\forall a, x, y \in F(aRx \& aRy \Rightarrow xRy \vee yRx)$$

and $Next$ is a unary function: $Next : F \to F$ such that $\forall a \in F$, $aRNext(a)$. The element $Next(a)$ is said to be the next element for a.

A *valuation* V of some propositional letters in a frame $\mathcal{F} := \langle F, R, Next\rangle$ is a mapping of these letters in the set of all subsets of F: for any such letter p, $V(p) \subseteq F$. Being given a valuation V of all letters from a formula A in a frame $\mathcal{F}$, V can be extended to all subformulas of A and A itself in the standard way: (i) for any $a \in F$, p is valid at a (abbreviation $a \Vdash_V p$) iff $a \in V(p)$, (ii) the usual steps for Boolean operations, and (iii)

$$a \Vdash_V \Box C \Leftrightarrow (\forall b \in F[aRb \Rightarrow b \Vdash_V C]), \quad a \Vdash_V \bigcirc C \Leftrightarrow Next(a) \Vdash_V C.$$

In accordance with the definition above for $V(p)$ and $\Vdash_V$, for any formula A, $V(A) := \{c \mid c \in F, c \Vdash_V A\}$. A *model* is a frame with a given valuation. A formula C *is valid* in a frame $\mathcal{F}$ w.r.t. a given valuation V (or in the model $\langle \mathcal{F}, V\rangle$) iff for any $a \in \mathcal{F}$, $a \Vdash_V C$. A formula C is *valid in a frame* $\mathcal{F}$ (notation is $\mathcal{F} \Vdash C$), iff C is valid in $\mathcal{F}$ w.r.t. any valuation V of its propositional letters. For any frame $\mathcal{F}$, $L(\mathcal{F})$ is the set of all formulas valid in $\mathcal{F}$. It is easy to see that any $L(\mathcal{F})$ is a temporal logic. Indeed all conditions from Definition 1 can be directly easily verified. For a class of frames $\mathcal{K}$, $L(\mathcal{K}) := \bigcap\{L(\mathcal{F}) \mid \mathcal{F} \in \mathcal{K}\}$ is *the temporal logic generated by* $\mathcal{K}$. For a logic L, a frame $\mathcal{F}$ is L-*frame* iff $L \subseteq L(\mathcal{F})$. All definitions above just specify the ones for multi-modal logic for the case of temporal logic with operation Next-time (cf. among modern literature D. Gabbay *et al.*, [2003]).

Let the frame $\mathcal{N} := \langle N, \leq, Next \rangle$ be the set of all natural numbers with the usual linear ordering $\leq$, where $Next(n) := n + 1$. The following is a natural model of the time-flow: any natural number n can be viewed as a certain time-point; $n + 1$ is the next time-point for n, and if $n \leq m$ holds then m is a future time-point for n.

DEFINITION 2. *The temporal Next-Time logic* **NXT** *is the set of all formulas valid in the frame* $\mathcal{N}$, *i.e.* $\mathbf{NXT} := L(\mathcal{N})$.

In what follows, we will use standard notation, definitions and results from modal logics, they will work because our **NXT** is a particular case of bi-modal logic. To describe the logic **NXT** in terms of finite frames we need the frames $F(n, k)$ described below.

Consider an interval of natural numbers $[1, n]$ with standard relation $\leq$ and the partial unary function $Next$, where $Next(m) := m+1$ for all $m < n$. That is $Next$ is not defined only for n.

Take the frame $\bigcirc_k$ which is the k-element frame $\langle \{a_1, \ldots, a_k\}, R, Next \rangle$, where $Next(a_k) = a_1$ and $Next(a_j) = a_{j+1}$ for $j < k$, and $a_j R a_i$ for all elements a_i and a_j.

The frame $F(n, k) := [1, n] \Rightarrow \bigcirc_k$ is the sequential concatenation of $[1, n]$ and $\bigcirc_k$. More precisely, this frame has as the base set the disjoint union of the base set of $\bigcirc_k$ and the set $[1, n]$, $Next(n) := a_1$, and, for all other elements b of $F(n, k)$, $Next(b)$ has the same value as it has in $[1, n]$ or $\bigcirc_k$. The relation R in $F(n, k)$ is the union of $\leq$ from $[1, n]$, R from $\bigcirc_k$ and the set of pairs $\{(x, y) \mid x \in [1, n], y \in \bigcirc_k\}$. We call a_1 the *entry point* in the loop $\bigcirc_k$. The frame $F(0, k)$ is the frame $F(n, k)$ with omitted $[1, n]$. A geometrical sketch of the frame $F(n, k)$ is given below at the picture Fig. 1.

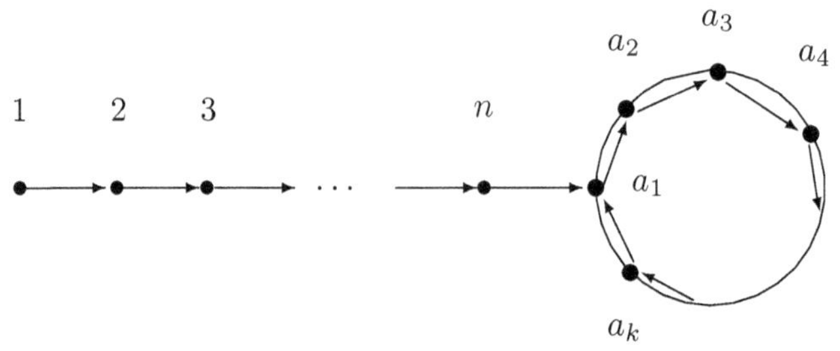

Figure 1. A draft of the frame $F(n, k)$

It is easy to see that if a formula A is valid in the frame $\mathcal{N}$ then A must be valid also in all frames $F(n, k)$ (because $F(n, k)$ is a p-morphic image of

$\mathcal{N}$ w.r.t both $Next$ and $\leq$ in $\mathcal{N}$). So, $L(\mathcal{N}) \subseteq L(F(n,k))$.

For a given temporal logic L, L is said to have the *finite model property* iff $L = L(\mathcal{K})$ from some class $\mathcal{K}$ of finite frames. In this case $\mathcal{K}$ is called a *class generating* L. As a simple illustration, using standard technique we can prove

LEMMA 3. The logic **NXT** has the finite model property. The class consisting of all frames $F(n,k)$ is a generating class for **NXT**.

Proof. Indeed, assume $\langle N, \leq, Next\rangle \not\Vdash_S \gamma$ for some valuation S. Then there is an $a \in N$, where $a \not\Vdash_S \gamma$. Let $Sub(\gamma)$ be the set of all subformulas of γ. For any $X \subseteq Sub(\gamma)$ we set:

$$Desc(X) := \bigwedge_{\varphi \in X} \varphi \wedge \bigwedge_{\varphi \in [Sub(\gamma) \setminus X]} \neg\varphi.$$

Consider the set $Rpr(\gamma) := \{Desc(X) \mid X \subseteq Sub(\gamma)\}$. Let $min(\gamma)$ be the $\leq$-minimal number a from N where the set $Sub(\gamma)_\diamond(a) := \{\beta \mid \beta \in Rpr(\gamma), a \Vdash_S \diamond\beta\}$ is smallest among all possible ones. The existence is evident because $\forall n_1, n_2 \in N,\ n_1 \leq n_2 \Rightarrow Sub(\gamma)_\diamond(n_2) \subseteq Sub(\gamma)_\diamond(n_1)$. For any subset $X \subseteq Sub(\gamma)$, take the $\leq$-smallest element e_X in the set

$$Val(X) := \{c \mid c \geq Next(min(\gamma)), c \Vdash_S Desc(X)\},$$

if this set is nonempty. Take the $\leq$-maximal element m among all elements e_X. For any number $c \in N$, we set $Sub_\gamma(c) := \{\beta \mid \beta \in Sub(\gamma), c \Vdash_S \beta\}$.

Consider elements $min(\gamma)$ and $Next(min(\gamma))$, and take $\leq$-minimal element $b \in N$, where $b > m$ and $Sub_\gamma(b) = Sub_\gamma(min(\gamma))$. The existence of such element is guaranteed by our choice of $min(\gamma)$.

Now we transform the model $\langle \mathcal{N}, S\rangle$, by deleting all numbers c, where $c > b$ and by taking

$$Next(b) := Next(min(\gamma)),$$

where $Next$ remains the same for other elements. The relation $\leq$ again remains to be the same on numbers from $[1, min(\gamma)]$, and we set xRy for all $x, y > min(\gamma)$, for $x \in [1, min(\gamma)]$ and $y \geq min(\gamma)$, and for all x, y where $x \leq y$. We assume the valuation S is just transferred to the new frame from our original one. The resulting model $\mathcal{M}$ is based on a frame of the type $F(n,k)$. It can be easily verified by conventional induction on length of formulas from $Sub(\gamma)$ that the truth-values of such formulas at corresponding elements of the original model $\langle \mathcal{N}, S\rangle$ and the model $\mathcal{M}$

coincide. So, $\mathcal{M}$ will refute γ. We briefly expose steps of this induction below. Only computations of truth-values for formulas of kind $\Diamond\beta$ and $\bigcirc\beta$ are nontrivial. For $\bigcirc\beta$ only the case

$$(\mathcal{N}, b) \Vdash_S \bigcirc \beta \Leftrightarrow (\mathcal{M}, b) \Vdash_S \bigcirc \beta$$

is nontrivial. Assume first $(\mathcal{N}, b) \Vdash_S \bigcirc \beta$. Then we have $\bigcirc\beta \in Sub_\gamma(b) = = Sub_\gamma(min(\gamma))$, so $(\mathcal{N}, min(\gamma)) \Vdash_S \bigcirc \beta$, and $(\mathcal{N}, Next(min(\gamma))) \Vdash_S \beta$. Hence by inductive hypothesis $(\mathcal{M}, Next(min(\gamma))) \Vdash_S \beta$, and we conclude $(\mathcal{M}, b) \Vdash_S \bigcirc \beta$.

Conversely, let $(\mathcal{M}, b) \Vdash_S \bigcirc \beta$. Then $(\mathcal{M}, Next(min(\gamma)) \Vdash_S \beta$, and by inductive hypothesis it follows that $(\mathcal{N}, Next(min(\gamma)) \Vdash_S \beta$, and also $(\mathcal{N}, min(\gamma) \Vdash_S \bigcirc \beta$, consequently $\bigcirc\beta \in Sub_\gamma(min(\gamma)) = Sub_\gamma(b)$. Hence $(\mathcal{N}, b) \Vdash_S \bigcirc \beta$ as required.

To verify the inductive step for $\Diamond\beta$ assume $(\mathcal{N}, x) \Vdash_S \Diamond\beta$. Then there is y, where $x \leq y \in \mathcal{N}$, $(\mathcal{N}, y) \Vdash_S \beta$. If $y \in \mathcal{M}$ and xRy, then $(\mathcal{M}, y) \Vdash_S \beta$ and $(\mathcal{M}, x) \Vdash_S \Diamond\beta$. The case $y \in \mathcal{M}$ but not xRy is impossible. So, assume $y \notin \mathcal{M}$. Then $min(\gamma) < b < y$ and by the choice of elements e_X there is $e_{Sub_\gamma(y)} \in \mathcal{M}, e_{Sub_\gamma(y)} \geq Next(min(\gamma))$, and

$$(\mathcal{N}, e_{Sub_\gamma(y)}) \Vdash_S Desc(Sub_\gamma(y)).$$

In particular, $(\mathcal{N}, e_{Sub_\gamma(y)}) \Vdash_S \beta$. Then by inductive assumption we conclude $(\mathcal{M}, e_{Sub_\gamma(y)}) \Vdash_S \beta$ and $xRe_{Sub_\gamma(y)}$, hence $(\mathcal{M}, x) \Vdash_S \Diamond\beta$.

Conversely, assume that $(\mathcal{M}, x) \Vdash_S \Diamond\beta$. Then there is y, where xRy and $(\mathcal{M}, y) \Vdash_S \beta$. Next, by inductive assumption $(\mathcal{N}, y) \Vdash_S \beta$. If $x \leq y$ then $(\mathcal{N}, x) \Vdash_S \Diamond\beta$ as required. If $y < x$ (but in $\mathcal{M}$, xRy) then $min(\gamma) < y$. Since $(\mathcal{N}, y) \Vdash_S Desc(Sub_\gamma(y))$ and $(\mathcal{N}, y) \Vdash_S \Diamond Desc(Sub_\gamma(y))$, and because of the choice of $min(\gamma)$, we conclude $(\mathcal{N}, x) \Vdash_S \Diamond Desc(Sub_\gamma(y))$ and consequently $(\mathcal{N}, x) \Vdash_S \Diamond\beta$ as required. ■

3 Admissible Inference, Preliminary Facts

The general view on the logical inference problem is as follows: given with a set of assumptions $A_1, \ldots, A_n$, what are *correct logical consequences* B from these assumptions? As a mechanism to analyze logical consequence we propose *inferences*, logical sequents defined below. Writing any given formula in the displayed form $A(p_1,\dots,p_n)$ signifies that this formula contains propositional letters $p_1,\dots,p_n$; for formulas $C_1,\dots,C_n$, $A(C_1,\dots,C_n)$ is the formula constructed from $A(p_1,\dots,p_n)$ by substitution formulas C_i in place of letters p_i. For a collection $A_1(x_1,\dots,x_n)$, ... , $A_m(x_1,\dots,x_n)$, $B(x_1,\dots,x_n)$ of formulas, the expression

$$\mathbf{inf} := \frac{A_1(x_1,\dots,x_n),\dots,A_m(x_1,\dots,x_n)}{B(x_1,\dots,x_n)},$$

(or, for short, $inf := A_1(x_1,\dots,x_n),\dots,A_m(x_1,\dots,x_n)/B(x_1,\dots,x_n)$) is called *inference*. Formulas $A_1(x_1,\dots,x_n)$, ... , $A_m(x_1,\dots,x_n)$ are *premises* (or assumptions) and $B(x_1,\dots,x_n)$ is the conclusion of **inf**. The informal meaning of **inf** is: $B(x_1,...,x_n)$ is a logical consequence of formulas $A_1(x_1,...,x_n)$, ... , $A_m(x_1,...,x_n)$. Note that any formula A can be viewed as the inference $\top/A$ with always true premise. One way of thinking about logical consequence is as follows. The correct inferences are those when we can derive the conclusion B from the premises by axioms and inference rules of the given logic. A reasonable question is: are only such inferences correct? For instance, how could we reason about correct logical inferences if we do not know the axiomatic systems for a given logic (say, if this logic is generated by its semantics). To approach these questions we start from

DEFINITION 4. *For an inference* **inf** $:= A_1,\dots,A_n/B$, *we say that* **inf** *is evident in a logic* L *iff the formula* $\Box[\bigwedge_{1\le i\le n} A_i]\wedge \bigwedge_{1\le i\le n} A_i \to B$ *is valid in* L.

LEMMA 5. *The inference* $\bigcirc x/\Box x$ *is not evident in* **NXT**.

Proof. To refute the formula $\Box \bigcirc x \wedge \bigcirc x \to \Box x$ in $\mathcal{N}$ it is sufficient to take the valuation V: $V(x) := \{m \mid m > 1\}$. So this formula is invalid in **NXT**. ■

However, the inference $\bigcirc x/\Box x$ can be viewed as a correct logical inference for **NXT**. Indeed, consider the notion of admissible inference suggested by Lorenzen (1955, [Lorenzen, 1955]).

DEFINITION 6. *An inference* **inf** $:= A(x_1,\dots,x_n)$, ... , $A_m(x_1,\dots,x_n)/$ $B(x_1,\dots,x_n)$ *is admissible in a logic* L *iff, for any formulas* $C_1,\dots,C_n$,

$$[\forall i \in \{1,\dots,m\} A_i(C_1,\dots,C_n) \in L] \Longrightarrow B(C_1,\dots,C_n) \in L.]$$

In other terms, an inference **inf** is admissible in L iff L is closed w.r.t. **inf**. Therefore, admissible inferences are correct logical consequences, actually they are strongest logical inferences compatible with the given logic L (considering L as the set of all formulas valid in L).

LEMMA 7. *The inference* $\bigcirc x/\Box x$ *is admissible in* **NXT**.

Proof. Assume $\bigcirc x/\Box x$ is not admissible in **NXT**. Then there is a formula C such that $\bigcirc C \in$ **NXT** but $\Box C \notin$ **NXT**. Because **NXT** is generated by frames $F(n,k)$ (Lemma 3), there is an $F(n,k)$ refuting $\Box C$ by a valuation V. Then there is a frame $F(m,k)$ $(m \leq n)$ refuting C by a valuation V in its $\leq$-smallest element 1, or C is refuted in the frame $F(0,k)$ itself. In the first case, take the frame $F(m+1,k)$ extending the frame $F(m,n)$ by joining new $\leq$-smallest element with the valuation V transferred from $F(m,k)$. Then, for $a = Next(1)$, in the frame $F(m+1,k)$, $a \not\Vdash_V C$ holds because $1 \not\Vdash_V C$ in the frame $F(m,k)$. Therefore, in the frame $F(m+1,k)$, $1 \not\Vdash_V \bigcirc C$ holds, which contradicts $\bigcirc C \in$ **NXT**. The second case can be considered similarly. ■

A logic L is said to be *structurally complete* if any admissible in L inference is evident in L. Using Lemmas 5, 7, we have

PROPOSITION 8. *The logic* **NXT** *is structurally incomplete.*

EXAMPLE 9 (Inferences).

$$\frac{\bigcirc\bigcirc x \rightarrow \bigcirc\bigcirc y}{\bigcirc x \rightarrow \bigcirc y}, \qquad \frac{\bigcirc^k x}{\bigcirc x}, \text{ for } k > 1.$$

are also admissible but not evident in **NXT**. These examples are very simple, however there are admissible inferences which cannot be described by simple standard techniques similar to the one used above. To develop more powerful instruments, we need some particular infinite constructive models for **NXT**. To represent the structure of these models we have to prepare some technical results.

Consider the frame $F(0,k)$ described above. This frame is simply time-cluster: i.e. any element is future-time point for all others. Consider a valuation V of a finite set of propositional letters in this frame. For any element $c \in F(0,k)$, $Var_V(c)$ is the set of all propositional letters valid at c w.r.t. the valuation V.

DEFINITION 10. *Valuation* V *is* perfect *if for any two distinct elements* $a, b \in F(0,k)$ *the following two sequences differ:*

$$Seq(a) := \{Var_V(a), Var_V(Next(a)), Var_V(Next^2(a)), \ldots,$$

$$Var_V(Next^{k-1}(a))\}$$

$$Seq(b) := \{Var_V(b), Var_V(Next(b)), Var_V(Next^2(b)), \ldots,$$

$Var_V(Next^{k-1}(b))\}$.

For any formula A, $deg_{\Box}(A)$ is the modal degree A w.r.t. $\Box$, i.e. the maximal number of nested occurrences of operations $\Box$; respectively, $deg_{\bigcirc}(A)$ is the modal degree of A w.r.t. $\bigcirc$, and $deg_{Mod}(A)$ is the modal degree of A w.r.t. both modalities $\bigcirc$ and $\Box$, i.e. the maximal number of nested occurrences of any type modalities. Consider a frame $F(0,k)$ with a given valuation V.

LEMMA 11. The following holds:

(i) If V is perfect then, for any pair of elements $c, d \in F(0,k)$, there is a formula A with $deg_{\Box}(A) = 0$ and $deg_{\bigcirc}(A) \leq k$, which distinguishes c and d: $c \Vdash_V A$ and $d \nVdash_V A$.

(ii) If $Var_V(a) = Var_V(Next(a))$ for all $a \in F(0,k)$ then all elements of $F(0,k)$ are non-distinguishable: for any formula A and any $a \in F(0,k)$ $a \Vdash_V A \Leftrightarrow Next(a) \Vdash_V A$. If V is imperfect then there are at least two distinct elements which are non-distinguishable by formulas.

(iii) If V is perfect then any element $e \in F(0,k)$ is definable in $F(0,k)$ by a formula A_e with $deg_{\Box}(A_e) = 0$ and $deg_{\bigcirc}(A_e) \leq k$: $\forall b \in F(0,k)$ $(b \Vdash_V A_e$ iff $b = e)$.

Proof. The proof is a straightforward, conventional verification. ■

We need some information concerning special definable sets in the frames $F(n,k) := [1,n] \Rightarrow \bigcirc_k$ for arbitrary n and k. Let V be a valuation in $F(n,k)$ of a finite set of propositional letters.

LEMMA 12. *The following holds:*

(i) If $\forall a \in \bigcirc_k$ $Var_V(a) = Var_V(Next(a))$ and $Var_V(n) = Var_V(a_k)$ then for any formula A, $a_k \Vdash_V A \Leftrightarrow n \Vdash_V A$. If $\forall a \in \bigcirc_k$ $Var_V(a) = Var_V(Next(a))$ but $Var_V(n) \neq Var_V(a_k)$ then there is a formula A such that $n \Vdash_V A$, $b \nVdash_V A$ for all worlds $b \in \bigcirc_k$, $deg_{\bigcirc}(A) \leq k+1$, $deg_{\Box}(A) \leq 1$ and $deg_{Mod}(A) \leq k+2$.

(ii) Suppose V is perfect in $F(0,k)$. If $Var_V(n) = Var_V(a_k)$ then, for any formula A, $n \Vdash_V A \Leftrightarrow a_k \Vdash_V A$. If $Var_V(n) \neq Var_V(a_k)$ then, for some formula B where $deg_{\bigcirc}(B) \leq 2k+2$, $deg_{\Box}(B) \leq 1$ and $deg_{Mod}(B) \leq 2k+3$, $n \Vdash_V B$ and $a \nVdash_V B$ for all $a \in \bigcirc_k$.

(iii) *If* V *is perfect on* $F(0,k)$ *and* $Var_V(n) \neq Var_V(a_k)$ *then for any* m, $1 \leq m \leq n$, *there is a formula* φ_m *such that* $\forall x \in F(n,k)$, $x \Vdash_V \varphi_m \Leftrightarrow x = m$, $deg_{\bigcirc}(\varphi_m) \leq 2k+n+2$, $deg_{\Box}(\varphi_m) \leq 1$ *and* $deg_{Mod}(\varphi_m) \leq 2k+n+3$.

(iv) *If* V *is perfect on* $F(0,k)$ *and* $Var_V(n) \neq Var_V(a_k)$ *then any element* b *of the frame* $F(n,k)$ *is definable: there is a formula* A_b, *where* $deg_{Mod}(A_b) \leq 2k+n+4$, *which is valid only on the element* b.

Proof. The proof is a long conventional tedious verification moving from the statement (i) to the statement (iv) and using previously proven steps and statements from Lemma 11. ∎

Construction of models $\mathcal{M}(k, \mathbf{NXT})$

Using the two previous lemmas we introduce the so-called k-characterizing constructive models for the logic **NXT**. Given a set $p_1, \ldots, p_k$ of propositional letters, given a frame $F(0,n) := \bigcirc_n$ with a perfect valuation V of letters $p_1, \ldots, p_k$, recall that for any model $\mathcal{M}$ with a valuation V and any $a \in \mathcal{M}$, $Var_V(a)$ is the set of all letters which are true at a w.r.t. V.

Let $\mathcal{M}(0,n,V) := \langle F(0,n), V\rangle$. The model $\mathcal{M}(1,n,V)$ is generated by $\mathcal{M}(0,n,V)$ as follows. For any $a_j \in \bigcirc_n$ we join to the model $\langle F(0,n), V\rangle$ the set of new elements $Pr(a_j)$, where

$$Pr(a_j) := \{s_{X,1,a_j} \mid X \subseteq \{p_1, \ldots, p_k\}, X \neq Var_V(Next^{-1}(a_j))\},$$

(where $Next(Next^{-1}(a_j)) = a_j$). We extend the valuation V on these new elements by setting $Var_V(s_{X,1,a_j}) := X$. Thus if $Next(a_\xi) = a_j$ holds then

$$Var_V(a_\xi) \notin \{Var_V(x) \mid x \in Pr(a_j)\}.$$

For any $x \in Pr(a_j)$, we set $Next(x) = a_j$ and xRy for any $y \in F(0,n)$ or $y = x$. This completes the definition of $\mathcal{M}(1,n,V)$. To comment this definition, the elements of $Pr(a_j)$ can be viewed as new immediate time-predecessors for a_j: for any $x \in Pr(a_j)$, $Next(x) = a_j$. We set

$$Slp_1(\mathcal{M}(1,n,V)) := [\mathcal{M}(1,n,V) - \mathcal{M}(0,n,V)].$$

To comment this notation, the set $Slp_1(\mathcal{M}(1,n,V))$ consists of all new immediate time-predecessors for elements of $\mathcal{M}(0,n,V)$, they can be viewed as the past-slice of new elements gathered round the time loop $F(0,n)$. So, viewing $Next$ and R as time-flow, $Slp_1(\mathcal{M}(1,n,V))$ are all initial time-points of $\mathcal{M}(1,n,V)$ (i.e., for all $y \in Slp_1(\mathcal{M}(1,n,V))$ there are no elements x where $Next(x) = y$).

Assume that the model $\mathcal{M}(m,n,V)$ is already constructed, $m \geq 1$, and $Slp_m(\mathcal{M}(1,n,V))$ (as it has been above for $Slp_1(\mathcal{M}(1,n,V))$ in $\mathcal{M}(1,n,V)$) are all initial time-points of $\mathcal{M}(m,n,V)$, i.e. there are no elements x where $Next(x) = y$ for $y \in Slp_m(\mathcal{M}(1,n,V))$.

To build the model $\mathcal{M}(m+1,n,V)$ based on $\mathcal{M}(m,n,V)$, we do the construction similarly to our first step. First, take $Slp_m(\mathcal{M}(m,n,V)$. For any $x_j \in Slp_m(\mathcal{M}(m,n,V))$ we join to the model $\mathcal{M}(m,n,V)$ the set of new elements $Pr(x_j)$, where $Pr(x_j) := \{s_{X,m+1,x_j} \mid X \subseteq \{p_1,\ldots,p_k\}\}$. The valuation V is extended on these new elements by setting $Var_V(s_{X,1,a_j}) := X$. For any $x \in Pr(x_j)$, we set $Next(x) = x_j$ and xRy for any y, where x_jRy or $y = x$. This complete the definition of the model $\mathcal{M}(m+1,n,V)$ and we set

$$Slp_{m+1}(\mathcal{M}(m+1,n,V)) := [\mathcal{M}(m+1,n,V) - \mathcal{M}(m,n,V)].$$

To illustrate this construction, we depict a fragment of the frame $\mathcal{M}(m,n,V)$ in Fig. 2 below.

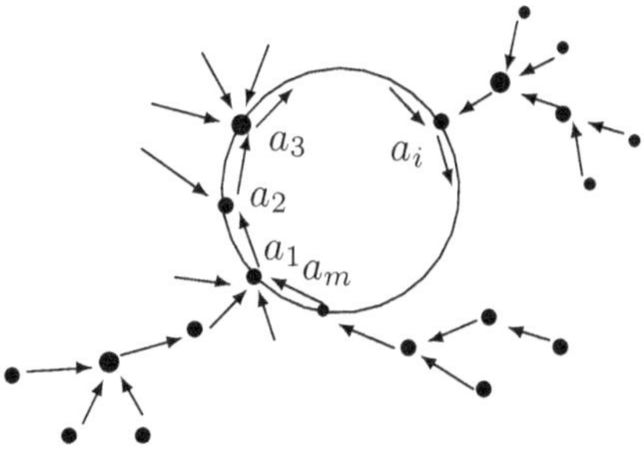

Figure 2. A fragment of the frame $\mathcal{M}(m,n,V)$

The model $\mathcal{M}(\omega,n,V)$ is the result of all our inductive steps:

$$\mathcal{M}(\omega,n,V) := \bigcup_{m\geq 1} \mathcal{M}(m,n,V).$$

We call the model (the frame) $\mathcal{M}(0,n,V)$ *kernel model (kernel frame)* of the model (frame) $\mathcal{M}(\omega,n,V)$. Recall that the *disjoint union* $\mathcal{M}_1 \sqcup \mathcal{M}_2$ of some models $\mathcal{M}_1$ and $\mathcal{M}_2$ is the model whose base set is the disjoint union of the base sets of $\mathcal{M}_1$ and $\mathcal{M}_2$, where the relations R and $Next$, as binary relations, are disjoint unions of corresponding relations from $\mathcal{M}_1$ and $\mathcal{M}_2$, and where the valuation of propositional letters remains the same on the components $\mathcal{M}_1$ and $\mathcal{M}_2$. The model $\mathcal{M}(k,\mathbf{NXT})$ is the disjoint union of all models $\mathcal{M}(\omega, n, V)$ for all n and all possible perfect valuations V of the propositional letters $p_1, \ldots, p_k$ in the kernel models $\mathcal{M}(0, n, V)$:

$$\mathcal{M}(k, \mathbf{NXT}) := \bigsqcup_{n,V} \mathcal{M}(\omega, n, V).$$

DEFINITION 13. *A model $\mathcal{M}$ is said to be* definable *iff any element of $\mathcal{M}$ is definable in $\mathcal{M}$, i.e., for any element $a \in \mathcal{M}$, there is a formula which is true only at a.*

LEMMA 14. *The model $\mathcal{M}(k, \mathbf{NXT})$ is definable.*

Proof. Indeed, all elements of the kernels $\mathcal{M}(0, n, V)$ are definable in their own models $\mathcal{M}(0, n, V)$ because V is perfect and because of Lemma 11. For distinct kernels $\mathcal{M}(0, n, V)$ with the same number n, the elements from distinct kernels are distinguished by formulas because these kernels are not isomorphic. For kernels $\mathcal{M}(0, n_1, V)$ and $\mathcal{M}(0, n_2, V)$ with $n_1 \neq n_2$, the elements of $\mathcal{M}(0, n_1, V)$ are distinguished from all elements of $\mathcal{M}(0, n_2, V)$ by formulas due to items (ii) and (iii) of Lemma 12, and because each valuation V is perfect. So any element of any kernel is definable in the set of all kernels.

All elements of $Slp_1(\mathcal{M}(1, n, V))$ are distinguished by formulas from the set of all elements of all kernels, again by (ii)–(iv), Lemma 12. So, any element of any $\mathcal{M}(1, n, V)$ is definable in the set of all elements of all $\mathcal{M}(1, n, V)$. Assume that any element of any $\mathcal{M}(m, n, V)$ for $m \leq t \in N$ is definable in the set of all elements of all $\mathcal{M}(m_1, n, V)$ for $m_1 \leq t \in N$. Consider any element

$$x \in Slp_{t+1}(\mathcal{M}(t+1, n, V)) = [\mathcal{M}(t+1, n, V) - \mathcal{M}(t, n, V)].$$

Using (iii) and (iv) from Lemma 12 we can conclude that x is distinguished by formulas from all other elements of $\mathcal{M}(t+1, n, V)$ and, also, from all elements of all $\mathcal{M}(m_1, n, V)$ for $m_1 \leq t+1 \in N$ because all elements of all kernels $\mathcal{M}(0, n, V)$ are definable as we showed above. Hence, any element of any $\mathcal{M}(m, n, V)$ for $m \leq t+1$ is definable in the set of all elements of all $\mathcal{M}(m_2, n, V)$ for $m_2 \leq t+1$. So, continuing with this inductive procedure we show that all elements are definable. ∎

DEFINITION 15. *Let $\mathcal{M}$ be a model with a valuation V, and let V_1 be an arbitrary new valuation V_1 of some set of propositional letters q_j in the frame of $\mathcal{M}$. The valuation V_1 is* definable in the model *$\langle \mathcal{M}, V \rangle$ if, for any letter q_j, there is a formula A_{q_j} whose propositional letters are all from the domain of V, such that $V_1(q_j) = V(A_{q_j}) := \{a \mid a \Vdash_V A_{q_j}\}$.*

DEFINITION 16. *Given a logic L and a model $\mathcal{M}$ with a valuation of a finite set of propositional letters $p_1, \ldots, p_k$, $\mathcal{M}$ is said to be k*-characterizing *for L iff the following holds. For any formula $A(p_1, \ldots, p_k)$, whose propositional letters occur in the list $p_1, \ldots, p_k$, $A(p_1, \ldots, p_k) \in L$ iff $\mathcal{M} \Vdash A(p_1, \ldots, p_k)$.*

THEOREM 17. *The model $\mathcal{M}(k, \mathbf{NXT})$ is k-characterizing for* **NXT**.

Proof. All formulas occurring in **NXT** are valid in $\mathcal{M}(k, \mathbf{NXT})$ because this model is composed of frames $F(m, n)$ which are **NXT**-frames. Assume that A is a formula with k propositional letters $p_1, \ldots, p_k$ and $A \notin \mathbf{NXT}$. By Lemma 3, there is $F(m, n) := [1, m] \Rightarrow \bigcirc_n$ which refutes A by a valuation S: $1 \nVdash_S A$.

At the first stage, consider the sets of variables valid w.r.t. S in m and a_n: $Vars_S(m) := \{p_j \mid p_j \in \{p_1, \ldots, p_k\},\ m \Vdash_S p_j\}$ and $Vars_S(a_n) := \{p_j \mid p_j \in \{p_1, \ldots, p_k\}, a_n \Vdash_S p_j\}$. If $Vars_S(m) = Vars_S(a_n)$ we can delete m from $F(m, n)$ and set $Next(m-1) = a_n$. This transformation will not alter the truth-values of formulas at the remaining elements, so the resulting model will refute A again. So, we may assume $Vars_S(m) \neq Vars_S(a_n)$.

If S is perfect in $\bigcirc_m$ then $F(m, n) := [1, m] \Rightarrow \bigcirc_n$ has been included in $\mathcal{M}(k, \mathbf{NXT})$ as an open submodel at step $m+1$ of our construction of $\mathcal{M}(k, \mathbf{NXT})$. Recall that for two models $\mathcal{M}_1$ and $\mathcal{M}_2$, $\mathcal{M}_1$ is a submodel of $\mathcal{M}_2$ iff $\mathcal{M}_1$ (i) is based on some subset of the base set of $\mathcal{M}_2$, (ii) has the relations R and $Next$ transferred from $\mathcal{M}_2$ and (iii) the valuation in $\mathcal{M}_1$ is transferred from $\mathcal{M}_2$. $\mathcal{M}_1$ is an open submodel of $\mathcal{M}_2$ iff it is a submodel and

$$\forall x \in \mathcal{M}_1, \forall y \in \mathcal{M}_2\ (xRy \Rightarrow y \in \mathcal{M}_1) \& (Next(x) = y \Rightarrow y \in \mathcal{M}_1).$$

So, $\langle F(m, n), S \rangle$ is an open submodel of $\mathcal{M}(k, \mathbf{NXT})$. The last, in particular, means that the truth values for formulas in $\mathcal{M}(k, \mathbf{NXT})$ and in $F(m, n)$ are the same, and, consequently, A is refuted in $\mathcal{M}(k, \mathbf{NXT})$. Hence, to complete our theorem it is sufficient to show that we could restrict the case of perfect valuations on $\bigcirc_n$ only.

Assume S is imperfect on $\bigcirc_n$. If $n = 2$ we merely contract elements of $\bigcirc_2$ into one element. Otherwise, for some two distinct elements a_i and a_j from $\bigcirc_n$, sequences

$$Seq(a_i) := \{Vars_S(a_i), Vars_S(Next(a_i)), Vars_S(Next^2(a_i)), \ldots, \tag{1}$$

$$Vars_S(Next^{m-1}(a_i))\}$$

$$Seq(a_j) := \{Vars_S(a_j), Vars_S(Next(a_j)), Vars_S(Next^2(a_j)), \ldots, \tag{2}$$

$$Vars(Next^{m-1}(a_j))\}$$

coincide. Then $a_j = Next^r(a_i)$ where $1 \leq r \leq n-1$. Choose a_i and a_j among all possible pairs making the valuation imperfect, i.e. when sequences (1) and (2) coincide, such that the number r is minimal. Clearly, $r \leq || \bigcirc_n ||/2$. Therefore

$$||\{a_i, Next(a_i), \ldots, Next^{r-1}(a_i)\}|| \leq ||\{Next(a_i)^r, \ldots, \qquad (3)$$
$$Next^{r+(m-r)-1}(a_i)\}||.$$

We set a mapping f of the base set of $\bigcirc_n$ into itself as follows:

$$\forall p(0 \leq u \leq r-1)[f(Next^u(a_i)) := Next^{r+u}(a_i)];$$

$$\forall u(r \leq u \leq m-1)[f(Next^u(a_i)) := Next^u(a_i)].$$

Because of (3), f is a one-to-one mapping on $\{Next^u(a_i) \mid 0 \leq u \leq r-1\}$, and evidently f preserves the truth-values of letters $p_1, \ldots, p_k$ w.r.t. S. Consider the model $f(\bigcirc_n)$ based on the f-image of $\bigcirc_n$ with the transferred valuation S, which is a loop w.r.t. R and where

$$Next(f(Next^{m-1}(a_i)) := Next^r(a_i),$$

and where $Next$ remains the same for other elements. From (3), the model $f(\bigcirc_n)$ is based on the frame $\bigcirc_{n-r}$. f is not a p-morphism, but by conventional induction on the length of formulas C (with propositional letters from the domain of S) it could be easy verified that, for any $x \in \bigcirc_n$,

$$(\bigcirc_n, x) \Vdash_S C \Leftrightarrow (f(\bigcirc_n), f(x)) \Vdash_S C.$$

Thus, if we replace $\bigcirc_n$ in $F(m,n)$ by its image $f(\bigcirc_n)$, the resulting model will again refute the formula A. By this transformation we reduced at least by one the number of pairs of elements in $\bigcirc_n$ making S imperfect. Carrying on with this procedure, we transform our original frame to one having a perfect valuation S. If our original frame $F(m,n)$ is $F(0,n)$ it is sufficient to repeat only the final part of our reasoning above. ■

Using the above k-characterizing models for **NXT**, we can describe admissible inferences of **NXT** in terms of inferences valid in these models.

DEFINITION 18. *For a given inference* **inf** $:= A_1, \ldots, A_m/B$ *and a model* $\mathcal{M}$ *with a valuation* V *of all letters (variables) of* **inf**, *we say that* **inf** *is valid in* $\mathcal{M}$ *(notation* $\mathcal{M} \Vdash_V$**inf***) iff the following holds:*

$$[\mathcal{M} \Vdash_V A_1 \& \ldots \& \mathcal{M} \Vdash_V A_m] \Rightarrow [\mathcal{M} \Vdash_V B].$$

We say that **inf** *is valid in a frame* $\mathcal{F}$ *if it is valid in any model based on* $\mathcal{F}$. *If* **inf** *is invalid in* $\mathcal{M}$ $(\mathcal{F})$, *we say that* $\mathcal{M}$ $(\mathcal{F})$ *refutes* **inf**.

An attempt to describe admissible inferences of **NXT** as those which are valid in any class of frames generating **NXT** would fail. Indeed, $\bigcirc x/\Box x$ is admissible for **NXT**, but $\bigcirc x/\Box x$ is invalid in any finite frame sound for **NXT** with non-trivial linear part. So, finite frames must be excluded from consideration. But using Theorem 17 and the definition of admissible inferences we immediately obtain

THEOREM 19. *An inference* **inf** *is admissible in* **NXT** *iff, for any* k, **inf** *is valid in the frame of* $\mathcal{M}(k, \mathbf{NXT})$ *w.r.t. any definable valuation.*

4 Main Results, Decidability w.r.t. Admissible Inferences

Theorem 19 gives us a description of all admissible inferences, and we can use it in order to show that a particular inference **inf** is admissible. Say, if it turns out that **inf** is valid in all frames pointed in Theorem 19 for *all* valuations then **inf** should be admissible. But a problem arising from this approach is as follows. First, we have to verify whether **inf** is valid in infinite models. It is sufficient to verify only definable valuations, but it is a problem to recognize whether a valuation is definable. Therefore, we need more advanced technique.

DEFINITION 20. *An inference* **inf** *is said to have the* normal reduced form *if*

$$\mathbf{inf} = \frac{\bigvee_{1\leq j\leq m}(\bigwedge_{1\leq i\leq n}[x_i^{k(j,i,0)} \wedge (\Diamond x_i)^{k(j,i,1)} \wedge (\bigcirc x_i)^{k(j,i,2)}])}{x_1},$$

where all x_s *are letters (variables),* $k(i,j,z) \in \{0,1\}$ *and for any wff* t, $t^0 := f, t^1 := \neg f$.

Given an inference in the normal reduced form nrf(inf) it is said to be *a normal reduced form for an inference* **inf** iff, for any logic L, **inf** is admissible in L iff nrf(**inf**) is. Using the same ideas and the structure of the proofs as for Lemma 3.1.3 and Theorem 3.1.11 in [Rybakov, 1997] we obtain

THEOREM 21. *There exists an algorithm which, for any given inference* **inf**, *constructs its normal reduced form nrf(***inf***).*

Let **nrf**(**inf**) be an inference in normal reduced form. $Pr_{nrf}(inf) := \{\varphi_i \mid i \in I\}$ is the set of all disjunctive members of the premise of **nrf**(**inf**), and $Sub(nrf(inf))$ is the set of all subformulas of formulas from the inference **nrf**(**inf**).

LEMMA 22. *If* **nrf(inf)** *is not admissible in* **NXT** *then there is a valuation S of its variables in the one-element frame $F_1 := \{c\}$ (i.e. $Next(c) = c, cRc$) where $c \Vdash_S \bigvee Pr_{nrf}(inf)$.*

Proof. If **nrf(inf)** fails to be admissible in **NXT** then there is a substitution ε such that

$$[\bigvee Pr_{nrf}(inf)]^{\varepsilon} \in \mathbf{NXT}, \text{ but } x_1^{\varepsilon} \notin \mathbf{NXT}.$$

Then $[\bigvee Pr_{nrf}(inf)]^{\varepsilon}$ must be valid in the frame F_1 which is **NXT**-frame. ■

LEMMA 23. *If* **nrf(inf)** *is not admissible in* **NXT** *then there is a valuation S for variables of* **nrf(inf)** *in some frame $\mathcal{M}(n,k,V)$, for some $n \geq 1$ and k, refuting* **nrf(inf)** *where*

(1) $||\mathcal{M}(n,k,V)|| \leq g(||Sub(nrf(inf))||)$, and g is a computable function;
(2) For any $x \in Slp_n(\mathcal{M}(n,k,V))$, $\exists \varphi_i \in Pr_{nrf}(inf)$ where

$$x \Vdash_S \varphi_i \text{ and } Next(x) \Vdash_S \varphi_i.$$

Proof. Assume **nrf(inf)** is not admissible in **NXT**. From Theorem 19 **nrf(inf)** is invalidated in an $\mathcal{M}(k,\mathbf{NXT})$ by a definable valuation S. In particular, for some $a \in \mathcal{M}(k,\mathbf{NXT})$, the conclusion x_1 of **nrf(inf)** is refuted at a: $a \nVdash_S x_1$. Assume a has depth $d \geq 2$ w.r.t. R in

$$a^R := \{b \mid b \in \mathcal{M}(k,\mathbf{NXT}), aRb\} = F(n_1,k_1)$$

for some n_1 and k_1, such that $n_1 \geq 1$.

Consider any element $x \in Slp_{n_1}(\mathcal{M}(\omega,k_1,V))$. It is clear that the frame x^R coincides with the frame $F(n_1,k_1)$ and holds the definable valuation S transferred from $\mathcal{M}(k,\mathbf{NXT})$. Fix this $F(n_1,k_1)$ with given S. First we will satisfy part (2) of our lemma: we have to guarantee that the same premise disjunct φ_i is valid w.r.t S at the initial element 1 and in $Next(1)$ in some extension $F(n_{1,1},k_1)$ of the frame $F(n_1,k_1)$. For any frames (models) F_1 and F_2, we write $F_1 \sqsubseteq F_2$ if F_1 is an open subframe (submodel) of F_2, i.e. $|F_1| \subseteq |F_2|$ and if $a \in F_1$ and $b \in F_2$ then $aRb \Rightarrow b \in \mathcal{F}_1$ and $Next(a) = b \Rightarrow b \in F_1$.

Step 1. Validating ***(2).***

LEMMA 24. *The frame $F(n_1,k_1)$ is an open subframe of some model $F(n_2,k_1)$ (based on the same loop-model $F(0,k_1)$), which is an open submodel of $\mathcal{M}(\omega,k_1,V)$ w.r.t. S (i.e. $n_1 \leq n_2$), where*

$$\exists \varphi_i \in Pr_{nrf}(inf)((F(n_2,k_1),1) \Vdash_S \varphi_i \text{ and } (F(n_2,k_1),Next(1)) \Vdash_S \varphi_i).$$

Proof. For any element $c \in \mathcal{M}(\omega, k_1, V)$, $Vars_S(c)$ denotes the set of all variables from the domain of S which are valid in c w.r.t. S. Take the R-minimal element a of $F(n_1, k_1)$ (i.e. $a := 1$), and take the following infinite decreasing sequence of elements of $\mathcal{M}(\omega, k_1, V)$ dropping down from a:

$$St(a) := \{x \mid x \in \mathcal{M}(\omega, k_1, V), \exists n \in N(Next^n(x) = a), \quad (4)$$
$$\forall i\ ((i \leq n\&0 \leq i) \Rightarrow Vars_S(a) = Vars_S(Next^i(x))\}.$$

The valuation S is definable, therefore, for some formulas α_j, $S(x_j) := V(\alpha_j)$ for all letters (variables) x_j of the rule nrf(inf). Below, $Form(S)$ is the set of all formulas α_j and all their subformulas. Take an element a_m from $St(a)$ where the set $\{\beta \mid \beta \in Form(S), a_m \Vdash_S \Diamond\beta\}$ is the greatest. Let $Next^k(a_m) = a$. Let, as above in (4),

$$St(a_m) := \{x \mid x \in \mathcal{M}(\omega, k_1, V), \exists n \in N(Next^n(x) = a_m),$$
$$\forall i(\ (i \leq n\&0 \leq i) \Rightarrow Vars_S(a_m) = Vars_S(Next^i(x))\}$$

Recall that, for any formula γ, $\deg_O(\gamma)$ is the modal degree of γ w.r.t. $\bigcirc$. For any formula γ from $Form(S)$ with $deg_O(\gamma) \leq 1$, if $x \in St(a_m)$ then

$$Next^2(x) = a_m \Rightarrow (x \Vdash_S \gamma \Leftrightarrow Next(x) \Vdash_S \gamma).$$

This immediately follows by simple induction on the length of γ using the choice of $St(a_m)$. Assume we showed already that, for any formula γ from $Form(S)$ with $deg_O(\gamma) \leq t, 1 \leq t$, if $x \in St(a_m)$ then

$$Next^{t+1}(x) = a_m \Rightarrow (Next^{t-1}(x) \Vdash_S \gamma \Leftrightarrow Next^t(x) \Vdash_S \gamma). \quad (5)$$

Using this assumption and again, as above, simple induction on the length of formulas γ, it immediately follows that for any $\gamma \in Form(S)$ with $deg_O(\gamma) \leq t+1, 1 \leq t$, if $x \in St(a_m)$ then

$$Next^{t+2}(x) = a_m \Rightarrow (Next^t(x) \Vdash_S \gamma \Leftrightarrow Next^{t+1}(x) \Vdash_S \gamma).$$

Thus, the inductive procedure described above works and (5) holds for any $t \geq 1$. We take the element $b \in St(a_m)$ where $Next^h(b) = a_m$ and $h := max\{deg_O(\gamma) \mid \gamma \in Form(S)\} + 3$. Then, using (5), for any formula $\gamma \in Form(S)$, $Next(b) \Vdash_S \gamma \Leftrightarrow b \Vdash_S \gamma$. The premise of our rule **nrf(inf)** is valid at b and $Next(b)$, and what is proven above implies that the disjuncts of $Pr_{nrf}(r)$, which are valid in b and $Next(b)$ must coincide. So $b^R = F(n_2, k_1)$ has all necessary properties. ■

Step 2. Reducing n_2 in $F(n_2,k_1)$.

To continue the proof of our lemma, consider any frame $F(n_2,k_1)$ produced by Lemma 24 from the original $F(n_1,k_1)$, where $F(n_1,k_1)=x^R$ for $x \in Slp_{n_1}(\mathcal{M}(\omega,k_1,V))$. Then, as we proved, in the model $F(n_2,k_1)$, the following holds $\exists\varphi_i \in Pr_{nrf}(inf)$ $1 \Vdash_S \varphi_i$ and $Next(1) \Vdash_S \varphi_i$.

We need to bound effectively the size of each such frame $F(n_2,k_1)$. So, $F(n_2,k_1)=[1,n_2]\Rightarrow\bigcirc_{k_1}$. First we will reduce the number of elements in the interval $[1,n_2]$. For any $\varphi_i \in Pr_{nrf}(inf)$, we denote by $max(\varphi_i)$ the maximal w.r.t. R element in the set

$$Tr(\varphi_i):=\{m \mid (m\in[1,n_2])\&(m \Vdash_S \varphi_i)\}.$$

For any $x \in F(n_2,k_1)$, $\varphi_i(x)$ is the unique formula $\varphi \in Pr_{nrf}(inf)$, where $x \Vdash_S \varphi$ in $\mathcal{M}(\omega,k_1,V)$. Let $F_0 := ([1,n_2]\Rightarrow\bigcirc_{k_1})$. The frame F_1 is the following rarefied version of the frame F_0:

$$F_1 := \{[1,1]\cup[max(\varphi_i(Next(1))),n_2]\Rightarrow\bigcirc_{k_1},$$

and F_1 has the accessibility relation R and the valuation S transferred from the model $F(n_1,k_1)$, and $Next(1):=max(\varphi_i(Next(1)))$, and $Next$ remains the same for other elements. It is straightforward to derive that $(F_1,1)\Vdash_S\varphi_i(1)$. This is because (i) all letters x_i, for which $\bigcirc x_i$ were valid w.r.t S at 1 in $F(n_2,k_1)$, are valid w.r.t S at the element $max(\varphi_i(Next(1))$ also, and (ii) for any x_i,

$$(F(n_2,k_1),Next(1))\Vdash_S\Diamond x_i\Leftrightarrow(F(n_2,k_1),max(\varphi_i(Next(1))))\Vdash_S\Diamond x_i.$$

Note that the set $[max(\varphi_i(Next(1))+1,n_2]$ contains already no elements at which $\varphi_i(Next(1))$ is valid by S in $F(n_2,k_1)$. Assume that we already constructed a submodel F_m of $F(n_2,k_1)$, where

$$F_m := [a_1,a_2,\ldots,a_m,a_m+1,a_m+2\ldots,n_2]\Rightarrow\bigcirc_{k_1},$$

all elements a_k are certain elements of $[1,n_2]$ (shown in increasing order) with the same valuation S and the accessibility relation R transferred from $F(n_2,k)$, where $a_1=1$, and where

$$\forall j\in[1,m](F_m,a_j)\Vdash_S\varphi_i(a_j), \tag{6}$$

$$\forall a_j\in[a_1,a_m+1]\forall x\in[a_m+2,n_2],((F_m,x)\nVdash_S\varphi_i(a_j)), \tag{7}$$

and, for any a_k,

$$k \leq m \Rightarrow (F_m, a_k) \Vdash_S \varphi_i(a_k). \tag{8}$$

We will construct a rarefied version F_{m+1} of F_m as follows. Let

$$F_{m+1} := [a_1, a_2, \ldots, a_m, a_m + 1] \cup [max(\varphi_i(Next(a_m + 1))), n_2] \Rightarrow \bigcirc_{k_1},$$

where F_{m+1} has the accessibility relation R and the valuation S transferred from the model F_m, and where $Next(a_{m+1}) := max(\varphi_i(Next(a_m+1)))$, and $Next$ remains as before for all other elements. Again, as in the initiation of our inductive procedure, we can conclude that

$$(F_{m+1}, a_m + 1) \Vdash_S \varphi_i(a_m + 1).$$

This is because (i) $max(\varphi_i(Next(a_m+1)))$ holds w.r.t. S and all letters x_i, for which $\bigcirc x_i$ were valid w.r.t. S at $a_m + 1$ in $F(n_2, k_1)$, and because (ii) for any x_i,

$$(F(n_2, k_1), Next(a_m + 1)) \Vdash_S \Diamond x_i \Leftrightarrow (F_m, max(\varphi_i(Next(1)))) \Vdash_S \Diamond x_i.$$

Note that the set $[max(\varphi_i(Next(a_m+1)))+1, n_2]$ contains now no elements at which $\varphi_i(Next(a_m + 1))$ is valid w.r.t. S in $F(n_2, k_1)$. So, the inductive step is completed and (6), (7) and (8) hold for any m. If it will happen that $max(\varphi_i(Next(a_m + 1)))$ coincides with n_2, we terminate the procedure.

Note that this procedure terminates before $2 \times ||Pr(inf)|| + 1$-st step because in each step the disjunct $\varphi_i(Next(a_{m+1}))$ should be distinct from all disjuncts $\varphi_i(Next(a_m))$, $\varphi_i(Next(a_{m-1})), \ldots$ used in previous steps. So, the rarefication procedure described at the termination step gives us frames $F(n_3, k_1)$ that are subframes of frames $F(n_2, k_1)$ with the same $F(0, k_1)$ and (cf. (6), (7) and (8)) such that

$$(F(n_3, k_1), 1) \Vdash_S \varphi_i(1), \; (F(n_3, k_1), Next(1)) \Vdash_S \varphi_i(1),$$

$$\forall j \in F(n_3, k_1), \; \exists \varphi_i(u) \in Pr_{nrf}(inf)[(F(n_3, k_1), j) \Vdash_S \varphi_i(u)]$$

where the number of elements n_3 in the interval $[1, n_3]$ is effectively computed from the size of **nrf(inf)**.

Step 3. Refining the loop $\bigcirc_{k_1}$

Now we will refine the loop $\bigcirc_{k_1}$ in the frames $F(n_3, k_1)$ in order to bound effectively the number of elements in $\bigcirc_{k_1}$. Consider the loop $\bigcirc_{k_1}$ as the list $[a_1, \ldots, a_{k_1}, a_1]$, assuming that $\forall j \in [1, k_1 - 1](Next(a_j) := a_{j+1})$ and $Next(a_{k_1}) := a_1$. Let

$$R(\bigcirc_{k_1}) := \{a_1\} \cup \{a_j \mid a_j \in \bigcirc_{k_1}, j = max(m \mid a_m \Vdash_S (\varphi_i)),$$

$$\varphi_i \in Pr_{nrf}(inf), \varphi_i \neq \varphi(a_1)\}.$$

We transfer directly the valuation S and R from $\bigcirc_{k_1}$ onto $R(\bigcirc_{k_1})$. Note that $||R(\bigcirc_{k_1})|| \leq ||Pr_{nrf}(inf)|| + 1$. Suppose $R(\bigcirc_{k_1})$ has exactly m_l elements. Now we do not assume that $R(\bigcirc_{k_1})$ inherits $Next$ from $\bigcirc_{k_1}$, we define a new operation $Next_1$ on $R(\bigcirc_{k_1})$ as follows:

$$Next_1(a_j) := a \in R(\bigcirc_{k_1}), \text{ where } (\bigcirc_{k_1}, a) \Vdash_S \varphi_i(Next(a_j)). \tag{9}$$

By our choice of $R(\bigcirc)_{k_1}$ the operation $Next_1$ is one-to-one mapping. However it is not clear immediately that $R(\bigcirc_{k_1})$ is a loop w.r.t. $Next_1$, but it is the case:

LEMMA 25. *$R(\bigcirc_{k_1})$ forms a loop w.r.t. $Next_1$.*

Proof. Assume the opposite. Consider the following sequence

$$a_1, Next_1(a_1), \ldots, Next_1^{m_l - 1}(a_1).$$

Because it is not a loop terminated by $Next_1^{m_l-1}(a_1) = a_1$, a looping should happen before reaching a_1, i.e.

$$Next_1^m(a_1) = Next_1(Next_1^{m-1}(a_1)) = Next_1^t(a_1), \tag{10}$$

where $1 < t < m - 1 < m_l - 1$. We can assume that t is a minimal number among all possible ones for all possible m. So, $Next_1^t(a_1)$ is the first entry point from our hypothetical defective loop. But $Next_1^t(a_1)$ is also the entry point from its immediate predecessor $Next_1^{t-1}(a_1)$:

$$Next_1(Next_1^{t-1}(a_1)) = Next_1^t(a_1).$$

So, $t - 1 < t < m - 1$, consequently, $Next_1^{t-1}(a_1) \neq Next_1^{m-1}(a_1)$ by our assumption about minimality of t in (10). But then two distinct elements have the same $Next_1$-successor which is impossible by definitions of $Next_1$ in (9) and the definition of $R(\bigcirc_{k_1})$, a contradiction. ■

To continue the proof of our initial lemma note that $\forall a_j \in R(\bigcirc_{k_1})$,

$$\forall \varphi_i \in Pr_{nrf}(inf)[(\bigcirc_{k_1}, a_j) \Vdash_S \varphi_i \Leftrightarrow (R(\bigcirc_{k_1}), a_j) \Vdash_S \varphi_i].$$

So, we can replace the loop $\bigcirc_{k_1}$ in any frame $F(n_3, k_1)$ by $R(\bigcirc_{k_1})$. Then we can contract all the intervals $[1, n_3]$ of frames $F(n_3, k_1)$ if they have the same entry point in $R(\bigcirc_{k_1})$ and are isomorphic as models w.r.t. S. The resulting model $\mathcal{M}_r$ will again refute **nrf(inf)** by S (as it has been in the initial model). Now, in order to modify $\mathcal{M}_r$ in accordance with all conditions of our lemma, we can stretch the intervals $[1, n_3]$ till a maximal n_3 is accounted for. This is possible by Lemma 24. Then we can insert in the obtained model some missing (comparing with the structure of models $\mathcal{M}(n, k, V)$) intervals $[1, n_3]$ making proper duplication of existing ones. The resulting model $\mathcal{M}_f$ will have all required properties. ■

LEMMA 26. *If an inference* **nrf(inf)** *satisfies all conclusions of Lemmas 22 and 23 then* **nrf(inf)** *is not admissible in* **NXT**.

Proof. From Lemma 23 there is a valuation S of letters (variables) from **nrf(inf)** in the frame $\mathcal{M}(n, k, V)$ to the open subframe of a disjoint component $\mathcal{M}(\omega, k, V)$ of $\mathcal{M}(t, \mathbf{NXT})$ for some n and k, with the following properties: V refutes **nrf(inf)**, $||\mathcal{M}(n, k, V)|| \leq g(||Sub(nrf(inf))||$ where g is a computable function, and for any $x \in Slp_n(\mathcal{M}(n, k, V))$, $\exists \varphi_i \in Pr_{nrf}(inf)$ such that $x \Vdash_S \varphi_i$ and $Next(x) \Vdash_S \varphi_i$.

Take the model $\mathcal{M}(\omega, k, V)$. By Lemma 14, each element x of the model $\mathcal{M}(\omega, k, V)$ is definable in $\mathcal{M}(t, \mathbf{NXT})$ by a formula $\gamma(x)$. Choose the following valuation for letters x_i from **nrf(inf)** in $\mathcal{M}(\omega, k, V)$:

$$V_1(x_i) := V(\bigvee\{\gamma(x) \mid (x \Vdash_S x_i)\&(x \in \mathcal{M}(n, k, V))\})\cup \tag{11}$$

$$\{b \mid b \in [\mathcal{M}(\omega, k, V) - \mathcal{M}(n, k, V)],$$

$$Next(b)^R \cap Slp_n(\mathcal{M}(\omega, k, V)) = b(n), b(n) \Vdash_S x_i\}.$$

It is clear that this valuation is definable in the model $\mathcal{M}(\omega, k, V)$ and in $\mathcal{M}(t, \mathbf{NXT})$ also. V_1 evidently coincides with S on the frame $\mathcal{M}(n, k, V)$. Because of (2) from Lemma 23, the premise of **nrf(inf)** will be valid w.r.t. V_1 in $\mathcal{M}(\omega, k, V)$, so **nrf(inf)** is refuted on $\mathcal{M}(\omega, k, V)$ by the valuation definable in $\mathcal{M}(t, \mathbf{NXT})$.

Using Lemma 22, a valuation S_1 makes the premise of **nrf(inf)** valid on the one-element frame F_1. Define

$$V_2(x_i) := \{b \mid b \in \mathcal{M}(t, \mathbf{NXT}) - \mathcal{M}(\omega, k, V)\} \text{ if } F_1 \Vdash_{S_1} x_i, \text{ else} \tag{12}$$

$$V_2(x_i) := \emptyset, \text{ otherwise.}$$

Again V_2 is definable in $\mathcal{M}(t, \mathbf{NXT})$ and the valuation $V_3(x_i) := V_1(x_i) \cup V_2(x_i)$ (where V_1 and V_2 are given by (11) and (12)) is a definable valuation refuting $\mathbf{nrf}(\mathbf{inf})$ in the whole frame $\mathcal{M}(t, \mathbf{NXT})$. Then, by Theorem 19, the inference $\mathbf{nrf}(\mathbf{inf})$ is not admissible in **NXT**. ■

Using Lemma 26, Lemma 23 and Theorem 19 we immediately derive:

THEOREM 27. *The logic* **NXT** *is decidable w.r.t. admissible inferences, i.e. there is an algorithm which, for any given inference* **inf** *determines whether* **inf** *is admissible in* **NXT**.

BIBLIOGRAPHY

[Barringer *et al.*, 1999] H. Barringer, M. Fisher, D. Gabbay and G. Gough. *Advances in Temporal Logic*, Vol. 16 of Applied logic series, Kluwer Academic Publishers, Dordrecht, 1999.

[Bruns and Godefroid, 2001] G. Bruns and P. Godefroid. Temporal Logic Query-Checking. In *Proceedings of 16th Annual IEEE Symposium on Logic in Computer Science (LICS'01)*, pp. 409–417, Boston, MA, 2001.

[Cyrluk and Narendran, 1994] D. Cyrluk and P. Narendran. *Ground Temporal Logic: A Logic for Hardware Verification.* Lecture Notes in computer Science, V. 818. *From Computer-aided Verification (CaV'94)*, David Dill, ed., pp. 247–259. Springer-Verlag, Stanford, CA, 1994.

[Demri and Goré, 1999] S. Demri and R. Goré. Cut-Free Display Calculi for Nominal Tense Logics, *TABLEAUX*, pp. 155–170, 1999.

[Gabbay, 1981] D. M. Gabbay. An Irreflevivity Lemma with Applications to Axiomatizations of Conditions of Linear Frames. In *Aspects of Phoilosophical Logic*, U. Monnich, ed., p. 67–89. Reidel, Dordrecht, 1981.

[Gabbay *et al.*, 2003] D. M. Gabbay, A. Kurucz, F. Wolter and M. Zakharyaschev. *Many-Dimensional Modal Logics: Theory and Applications.* Elsevier, 2003.

[Emerson, 1990] E. A. Emerson. Temporal and and Modal Logics. In *Handbook of Theoretical Computer Science*, J. van Leeuwen, ed., pp. 996–1072. Elsevier, the Netherlands, 1990.

[van der Hoek and Wooldridge, 2002] W. van der Hoek and M. Wooldridge. Model Checking Knowledge and Time. In *SPIN 2002, Proc. of the Ninth International SPIN Workshop on Model Checking of Software*, Grenoble, pp. 95–111, 2002.

[Iemhoff, 2001] R. Iemhoff. On the admissible rules of Intuitionistic Propositional Logic. *Journal of Symbolic Logic*, **66**, 281–294, 2001.

[Ghilardi, 1999] S. Ghilardi. Unification in intuitionistic logic. *Journal of Symbolic Logic*, **64**, 859–880, 1999.

[Friedman, 1975] H. Friedman. One Hundred and Two Problems in Mathematical Logic. *Journal of Symbolic Logic*, **40**, 113–130, 1975.

[Kröger, 1987] F. Kröger. Temporal Logic of Programs. *EATCS Monographs on Theoretical Computer Science*, Volume 8. Springer Verlag, 1987.

[Laroussinie *et al.*, 2002] F. Laroussinie, N. Markey, and P. Schnoebelen. Temporal Logic with Forgettable Past. *IEEE Symp. Logic in Computer Science*, LICS, pp. 383–392, 2002.

[Laureys, 1999] T. Laureys. *From Event-Based Semantics to Linear Temporal Logic.* The Logical and Computational Aspects of a Natural Language Interface for Hardware Verification, School of Cognitive Science, University of Edinburgh, 1999.

[McMillan, 1993] K. L. McMillan. *Symbolic Model Checking.* Kluwer, Boston, MA, 1993.

[Manna and Pnueli, 1992] Z. Manna and A. Pnueli. *The Temporal Logic of Reactive and Concurrent Systems.* Springer-Verlag, 1992.

[Mints, 1976] G. E. Mints. Derivability of Admissible Rules. *Journal of Soviet Mathematics*, **6**, 417–421, 1976.

[Lorenzen, 1955] P. Lorenzen. *Einführung in die operative Logik und Mathematik.* Berlin-Göttingen, Heidelberg, Springer-Verlag, 1955.

[Pnueli, 1977] A. Pnueli. The Temporal Logic of Programs. In *Proceedings of 18th Symposium on the Foundations of Computer Science*, Providence, 1977.

[Prior, 1967] A. Prior. *Past, Present and Future.* Oxford University Press, 1967.

[Rybakov, 2003] V. V. Rybakov. Refined Common Knowledge Logics or Logics of Common Information, *Archive for Mathematical Logic*, **42**, 179–200, 2003.

[Rybakov, 2001] V. V. Rybakov. Construction of an Explicit Basis for Rules Admissible in Modal System S4, *Mathematical Logic Quarterly*, **47**, 441–451, 2001.

[Rybakov *et al.*, 2000a] V. V. Rybakov, M. Terziler, and V. Remazki. Basis in Semi-Reduced Form for the Admissible Rules of the Intuitionistic Logic IPC, *Mathematical Logic Quarterly*, **46**, 207–218, 2000.

[Rybakov and Fedorishin, 2000] V. V. Rybakov and B. Fedorishin. Faces of Monotonicity and Wisdom Formulas problem, *Bulletin of the Section Logic*, **29**, 181–192, 2000.

[Rybakov *et al.*, 2000b] V. V. Rybakov, M. Terziler, and C. Gencer. Unification and Passive Inference Rules for Modal Logics, *Journal of Applied Non-Classical Logic*, **10**, 79–91, 2000.

[Rybakov, 1997] V. V. Rybakov. *Admissible Logical Inference Rules,* Studies in Logic and the Foundations of Mathematics, Vol. 136, Elsevier, North-Holland, New-York-Amsterdam, 1997.

[Segerberg, 1970] K. Segerberg. Modal Logics with Linear Alternative Relations. *Theoria*, **36**, 301–322, 1970.

[van Benthem, 1991] J. van Benthem. *The Logic of Time — A Model-Theoretic Investigation into the Varieties of Temporal Ontology and Temporal Discourse.*- Kluwer, 1991.

[Venema, 1993] Y. Venema. Derivation Rules as Anti-Axioms in Modal Logic, *Journal of Symbolic Logic*, **58**, 1003–1034, 1993.

e-Learning Logic and Mathematics: What We Have and What We Still Need

ERICA MELIS AND JÖRG SIEKMANN

1 Introduction

Intelligent tutoring systems provide a promising application area for techniques from many subfields of Artificial Intelligence (AI), including knowledge representation, user modelling, rule-based systems, automated diagnosis, automated reasoning, adaptive hypermedia, natural language processing and automated reasoning. But as it turned out this is not just a one way road, but a give and take in both ways as these educational applications led to new research problems as well, among others, open student modelling, tutorial dialogues, and adaptive hypermedia, see, e.g., [Aleven and Koedinger, 2000; Brusilovski, 1996]. This also applies to automated reasoning as we shall see in the sequel.

With the widespread availability of the Web, there is the great opportunity that educational tools developed at one place can be used anywhere as long as they are encapsulated in Web-applications that are interoperable and compliant with standard input languages.

While e-learning tools are now widely used in life-long learning applications such as industrial training courses for specialist knowhow and skills, these systems slowly but surely enter into school teaching [Koedinger et al, 1997; Matsuda and vanLehn, 2005], academic teaching, and professional training as well. Some countries even embrace these new opportunities on a grand scale. China has currently about 90 million Internet users with a yearly growth rate of more than 10%. Provided this trend is not disturbed by external events, China will have more Internet users than the US by 2007 and it has been predicted that within the next ten years, China will have more officially registered Internet users than the rest of the world [China, 2005]. This is the fastest growing market in the world right now and the Chinese government intends to use this as a backbone of its next five year development plan on a grand scale — in particular to develop and educate the western parts of the country.

In this article, we shall focus on e-learning for mathematics (and logic) with special emphasis on its relationship to automated theorem proving and deduction. As first publications show (e.g. [Melis, 2000]) the field of automated theorem proving begins to recognise the potential of educational systems and some research groups of the deduction community have started to work on deductive components inside of educational systems [Melis *et al.*, 2001b; Baumgartner and Furbach, 2003]. Others developed theorem proving systems as interactive tools for education. Historically first was Patrick Suppes' use of a deduction system at Stanford [Suppes, 1981] and Peter Andrews' TPS system to teach higher order logic at Carnegie Mellon University [TPS, 2004]. Other recent work includes [Lowe and Duncan, 1997; Buchberger *et al.*, 1997; Bornat, ; Sieg and Byrnes, 1998; Sommer *et al.*, 2000; Melis *et al.*, 2001b]. Most of these systems, however are based primarily on their developers' experience and insight rather than on objective cognitive psychological or pedagogical results and — even more importantly — have generally not been empirically validated for their educational impact and cognitive adequacy.

As experienced in other areas before, it is not the development of the technology per se that has an impact on education but only a technology responding to the actual needs of the learner. In order to build *useful and usable* educational applications, empirical investigation and observations have to set the stage for technologies effectively supporting the learner. As we will see in Section 2, there are a number of cognitive and pedagogical results that we can rely upon and we should pay attention to.

Most of this article focuses on mathematics (including logic) education.[1] It distinguishes the usage of theorem proving/deduction within modules the student is not using directly and the usage of systems as interactive problem solving tools.

The paper is organised as follows. In Section 2 we shall give a brief account of some relevant (empirical) psychological and pedagogical results. Then we analyse in Section 3 how current techniques can or cannot realise these needs and provide some examples. Finally, we pose some challenges and propose future work in Section 4.

2 What e-Learning Systems Need

Learning resembles research and discovery in many ways [White and Shimoda, 1999]. However, a student is not like an expert mathematician or a developer of a theorem proving system. A system that is useful for learning

[1]However, most needs and some techniques apply more generally to e-learning systems per se.

drastically differs from systems that are valuable for a mathematician or other expert users.

This statement may sound like a platitude, but it points to a potentially rich and new research programme. A student's goals in learning (mathematics) may be just to understand existing problem solutions and proofs or else to learn how to find a solution or to prove a theorem on her own (with or without system support). The goal might be to train solution procedures or to learn how to gather information and search for solutions in the literature.

Cognitive psychology sheds light on human learning: current paradigms and instructional theories assume learning requires that a student *constructs* knowledge, dependencies and procedural skills in her mind [Piaget, 1977; Vygotsky, 1978]. As many empirical results show (see e.g. [Mandl *et al.*, 1997]), this should be promoted inter alia by:

- contextual real life learning experience
- a personalised learning context
- active, explorative learning opportunities
- feedback on the learner's activities
- coaching and stimulating meta-reasoning
- an appropriate level of abstraction and detailed presentation of the solution
- an appropriate user interface.

Let us elaborate these aspects in turn:

Context. If a learner can match her learning experience with a real-life context it will be more likely that she learns more rapidly and that she can transfer this knowledge better to problem solving in reality [Andriessen and Sandberg, 1999]. A 'real-life context' could be introduced by realistic visual impressions from video clips, pictures or diagrams or simply by a verbal description of a concrete situation representing a problem from the experience of the targeted learner group.

Another relevant feature concerns the complexity of real world problems and the various phases their solution requires. Such a problem solving cycle may require very different activities and skills from a pure textbook or academic problem: recognising the problem in the real life setting, developing the mathematical model, then mathematical problem solving and possibly revision and, finally, translating the solution back from the mathematical result into the real life setting.

Personalisation. Personalised learning experience is crucial not just for the motivation but also for efficiency and effectiveness of learning [Andriessen and Sandberg, 1999]. There are several ways of adapting an e-learning tool to the student. The most obvious one is to tailor the learning material (in particular the examples and exercises) to the capabilities of the student and to her learning goals. Furthermore, a course on statistics for example should be tailored to the special interest of the student: a student of physics expects different statistics examples than, say, a chemist or an electrical engineering student.

Active Learning. One of the most important ingredients of effective learning is *active problem solving* and exploration of alternatives and the discovery of faulty steps and assumptions [VanLehn *et al.*, 2001]. In particular learning from errors and failures is an important ingredient of active learning and should be maximally exploited.

Feedback. Empirical investigations corroborate that an appropriate feedback during problem solving improves subsequent performance [Jacobs, 2001]. An intelligent learning assistant should therefore diagnose mistakes, adaptively scaffold the individual process of problem solving and generate appropriate feedback and hints, see e.g., [Narciss, 2001; Tsovaltzi, 2005].

Meta-reasoning. Meta-reasoning plays a crucial role in successful problem solving [Polya, 1945; Schoenfeld, 1985; Melis and Ullrich, 2003] which includes planning, monitoring, self-regulation and self-explanation [Chi *et al.*, 1989]. Take, for example, Polya's 'How to Solve It' [1945]: it has the form of a 'how-to manual', i.e. a formulation of a set of heuristics cast in the form of brief commands within a frame of the four following problem solving stages

1. Understand the problem
2. Devise a plan
3. Carry out the plan
4. Look back at the solution.

These stages should be followed and actively monitored by the student, and exactly how this can be done is the subject of this very influential book. So far, however, this kind of reasoning is rarely taught, let alone implemented in current e-learning systems.

Cognitively Adequate Presentation. The presentation of a solution and in particular, the presentation of proofs have to be appropriately structured [Catrambone and Holyoak, 1990] and to hide irrelevant details in order to be comprehensive and transferable. The presentation should exploit multiple modalities such as text, formulas, diagrams, but also speech and animations if appropriate.

User Interface. The user interface of an e-learning tool has to be designed such that the student can focus on the essentials of the particular learning task, undisturbed much as possible by system peculiarities. Anderson [Anderson and Pelletier, 1991] claims that the design of the user interface substantially influences what a student will learn using the LISP tutor [Anderson *et al.*, 1995] as a case in point. Depending on the user interface, in particular, depending on the input language, a learner will either just learn a particular programming syntax or else be able to focus on general programming strategies. Moreover, just clicking buttons is unlikely to stimulate serious learning. This applies to learning mathematics just as well. Since e-learning tools have to be usable rather than just useful, a user-centred design of the interface built according to current cognitive requirements is key.

3 What we Have: ActiveMath

Learning environments have to meet realistic and complex needs, both technically as well as psychologically. So let us look at the the pedagogical and technical goals of our research for the ActiveMath system.

Pedagogical Goals

ActiveMath aims at an interactive and exploratory learning process and assumes the student to be responsible for the actual learning session. Therefore, the system supports relative freedom for navigating through a course and the user defines choices. The system supports a dynamic student model and by default, the student model is scrutable, i.e., inspectable and modifiable. Moreover, dependencies between the mathematical concepts can be inspected in a dictionary in order to help the student to learn the overall structure of a domain (e.g., analysis, algebra or number theory).

ActiveMath can adapt a course to the learner's goals, prerequisites and learning scenarios. In colleges and universities, the same subject is taught differently for different groups of users in different contexts, e.g., statistics has to be taught differently for students of mathematics, for students of economics, or in medicine. Therefore, the adaptive choice of content to be presented as well as examples and exercises is pivotal. In addition, an adaptation of examples and exercises to the student's *actual* capabilities is highly desirable in order to keep the learner in the zone of proximal development [Vygotsky, 1978] rather than overtax or undertax her.

Moreover, web-based systems can be used in several learning contexts, e.g., long-distance learning, homework, or teacher-assisted learning. Personalization is required in all of them because even for teacher-assisted learning in a classroom with, say, 30 students and one teacher, the teacher cannot really respond to all the individual needs. ActiveMath's current version provides adaptive content and adaptive presentation features.

Technical Goals

Building hyper-media content with assured quality is a time-consuming and costly process, hence the content should be *reusable* in different contexts. As most of today's interactive textbooks consist of a collection of predefined documents, typically canned `HTML` pages and multimedia animations, it is difficult to reuse them in another context. A re-combination of the encoded learning objects or a new adaptation of the course presentation and content of the course to other users is impossible most of the time.
ActiveMath's generic and semantically annotated knowledge representation supports re-usability and interoperability. In particular, it is compliant with the emerging mathematical knowledge representation and communication standards such as Dublin Core, `OpenMath`, `MathML`, and LOM.[2] Some of the buzzwords here are metadata, ontological XML, and standardized content packaging. Such features of the knowledge representation ensure a longer life cycle even with new and changing technologies in browsers and other devices.

In order to use the potential power of existing web-based technology e-learning systems need an open architecture to integrate and connect to new components including student management systems such as Ilias, Moodle, Sakay, WebCT as well as assessment tools, collaboration tools, and problem solving tools.

3.1 Architecture

The architecture of ActiveMath, as sketched in Figure 1, strictly realizes the principle of separation of (declarative) knowledge from functionalities as well as the separation of different kinds of knowledge. For instance, pedagogical knowledge is stored in a pedagogical rule base, the educational content is stored in MBase, and the knowledge about the user is stored in the student model. This principle has proved valuable in many AI-applications and eases modifications as well as configurability and reuse of the system.

ActiveMath has a client-server architecture whose client is a browser (or several browsers in case of multi-user mode). This architecture serves not only its openness but also its *platform independence*, and a browser such as Netscape, Mozilla, or IE with MathPlayer is sufficient to work with

[2] `http://ltsc.ieee.org/wg12/`

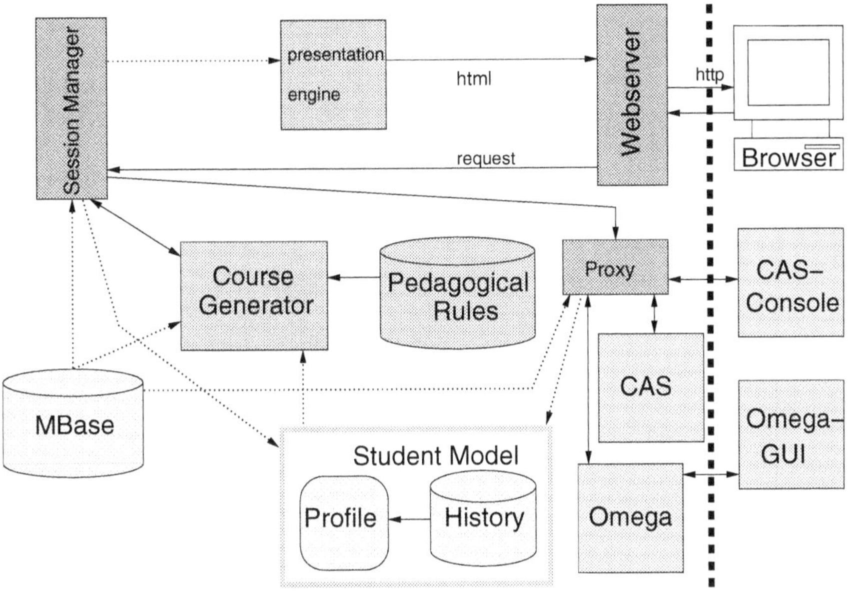

Figure 1. Architecture of ActiveMath

ActiveMath. The components of ActiveMath have been designed in a *modular* way in order to guarantee exchangeability and robustness.

The actual flow of control (and data) in the above diagram is as follows: when the student has chosen her goal concepts and learning scenario, the session manager sends a request to the *course generator*. The course generator is responsible for choosing and arranging the content to be learned. The course generator contacts the *mathematical knowledge base* in order to fetch the identifiers (IDs) of the mathematical concepts that are required for learning the goal concepts, queries the student model in order to find out about the student's prior knowledge and preferences, and uses *pedagogical rules* to select, annotate, and arrange the content — including examples and exercises — in a way that is suitable for this particular learner in this particular session. The resulting instructional graph of IDs is sent to the *presentation engine* which retrieves the actual mathematical content corresponding to the IDs and transforms the `XML`-data to output-pages which are then finally presented via the student's browser.

The *course* generator and the suggestion mechanism [Melis and Andres, 2004] previously worked with the rule-based system Jess [Friedman-Hill, 1997] that evaluates the (pedagogical) rules in order to decide which particular adaptation and content to select and which actions to suggest. Jess uses the Rete algorithm [Forgy, 1982] for optimization.

External systems such as the computer algebra systems Maple [Maple, 1986] and MuPad [Sorgatz and Hillebrand, 1995] and the proof planner MULTI [Melis and Meier, 2000] communicate with the ACTIVEMATH system. They serve as cognitive tools [Lajoie and Derry, 1993] and support the learner in complex interactive exercises. They also assist in generating feedback by evaluating the learner's input. Finally, a diagnosis is passed to the student model in order to update the student model.

In exercises ACTIVEMATH does not necessarily guide the user strictly along a predefined expert solution. It only evaluates whether the student's input is mathematically equivalent to an admissible subgoal, i.e., maybe it is irrelevant, but not outside the solution space (see [Buedenbender *et al.*, 2002]). Moreover, the external systems can support the user with automated problem solving, i.e., they may take over some parts in the human problem solving process and thereby help the user to focus on important learning tasks and to delegate routine tasks.

Actually, the diagnoses of a student's performance is well known to be an 'AI-complete' problem and hence several context-dependent equivalence checkings have been implemented so far. Moreover, most tutor systems encode the possible problem solving steps and the most typical misconceptions into their solution space or into systems that execute them. From this encoding, the system diagnoses the misconception of a student. This is, however, infeasible in realistic applications with large solution spaces as it is in general impossible to represent all potential misconceptions of a student [VanLehn *et al.*, 2002].

The *presentation engine* generates personalized web pages based on two frameworks: Maverick and Velocity. Maverick[3] is a minimalist model view controller (MVC) framework for web publishing using Java and J2EE, focusing solely on MVC logic. It provides a wiring between URLs, Java controller classes and view templates.

The presentation engine is a reusable component that takes a structure of `OMDocs` and transforms them into a presentation output that can be PDF (print format) or `HTML` with different maths-presentations such as Unicode or `MathML` (screen format) [Ullrich *et al.*, 2004]. Basically, the presentation pipeline comprises two stages: stage 1 encompasses Fetching, Pre-Processing and Transformation, while stage 2 consists of Assembly, Personalization and optional Compilation. Stage 1 deals with individual content fragments or items, which are written in `OMDoc` and stored in a knowledge base. Content items in the knowledge base do not depend on the user who is to view

[3]Maverick: `http://mav.sourceforge.net/`

them, they have unique identifiers and can be handled separately. It is only in stage 2 that items are composed to user-specific pages.

3.2 Adaptivity

ActiveMath adapts the course generation and presentation to the student's

- technical equipment (*customization*)
- environment variables, e.g., curriculum, native language, and the field of study (*contextualization*) and
- her cognitive and educational needs and preferences such as learning goals, and prerequisite knowledge (*personalization*).

As for personalization, individual preferences (such as the style of presentation), goal-competencies, and mastery-level are taken into account by the course generator. The goal-competencies are characterized by concepts that are to be learned and by the competency-level to be achieved: knowledge (k), comprehension (c), or application (a).

The learner can initialize her student model by self-assessment of her mastery-level of concepts and choose her learning goals and learning scenario, for instance, the preparation for an exam or learning from scratch for k-competency level. The course generator processes this information and updates the student model and generates pages/sessions as depicted in the sreenshots of Figures 2 and 3. These two screenshots differ in the underlying scenarios as the captions indicate.

The adaptation to the capabilities of the learner is carried out by the course generator and later, during the actual session, by the suggestion mechanism. The course generator checks whether the mastery-level of prerequisite concepts is sufficient for the goal competency. If not, it presents the missing concepts and/or explanations, examples and exercises for these concepts to the learner when a new session is requested. The suggestion mechanism acts dynamically in response to the student's activities. Essentially, this mechanism works with two blackboards, a diagnosis blackboard and a suggestion blackboard on which particular knowledge sources operate.

We also investigated special scenarios that support a student's metacognitive activities, such as those proposed in Polya's book [1945]. A Polya-scenario structures the solution space with headlines such as "understand the problem", "make a plan", "execute the plan", and "look back at the solution". It augments and structures exercises with additional prompts similar to the above headlines [Melis and Ullrich, 2003].

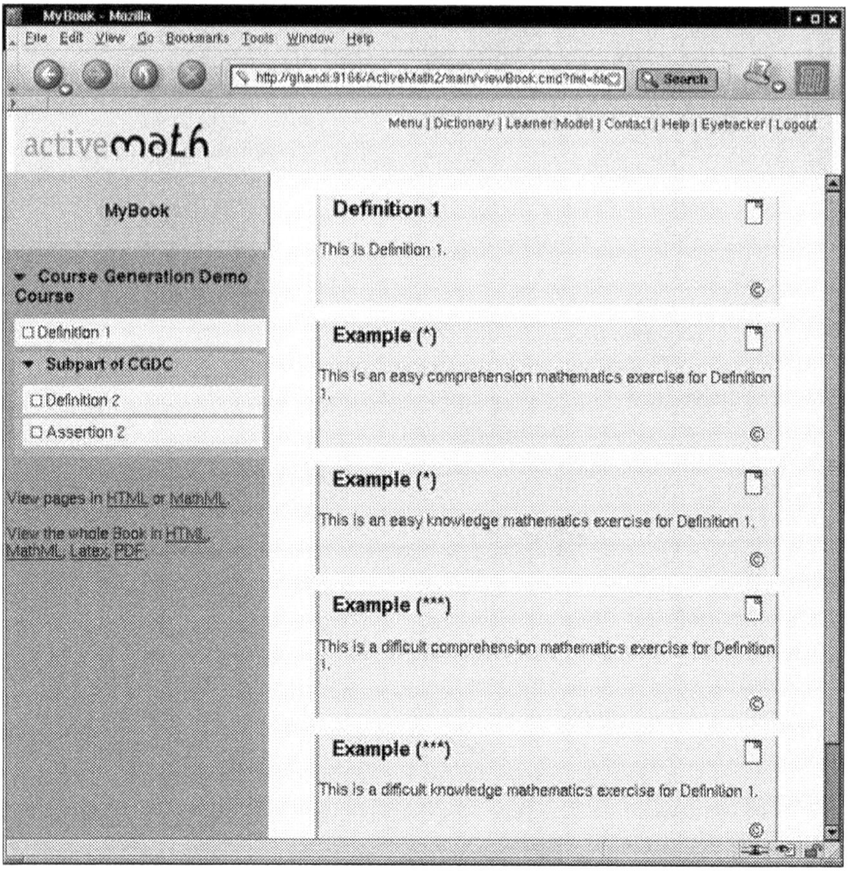

Figure 2. A screen shot of an ACTIVEMATH session for exam preparation

3.3 The Student Model

User modeling has been a research area in AI for some time. It actually started by developing techniques for student modeling and still continues with the investigation of representational issues as well as diagnostic and updating techniques.

As ACTIVEMATH's presentation is user-adaptive, we need to incorporate persistent information about the student as well as a representation of the student's learning progress. Therefore, *static* (wrt. the current session) properties such as field, scenario, goal concepts, and preferences as well as *dynamic* properties such as the mastery values for concepts and the student's

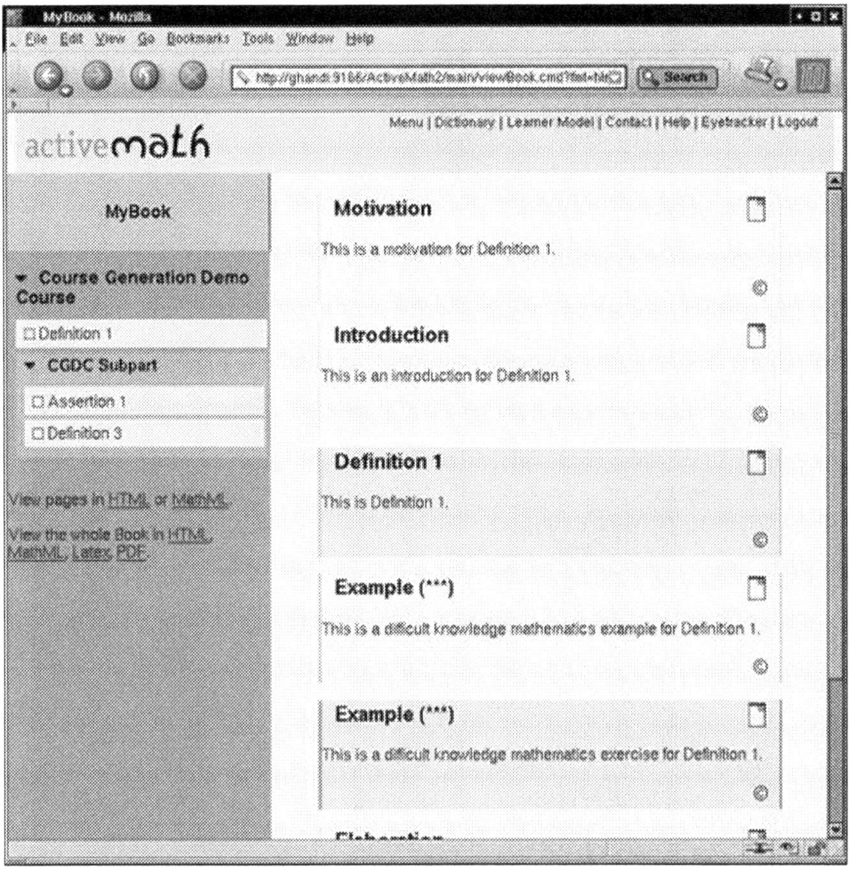

Figure 3. k-level session of ACTIVEMATH

actual behaviour, are stored in the current student model.

The profile is initialized with the learner's entries entered into ACTIVE-MATH's registration page which holds the preferences (static), scenario, goals (static for the current session), and self-assessment values for knowledge, comprehension, and application of concepts (dynamic).

The history component stores the information about the learner's actions. Its elements contain information such as the IDs of the content of a read page or the ID of an exercise, the reading time, and the success rate of the exercise. We also developed a "poor man's eye-tracker" which allows to trace the student's attention and reading time [Ullrich and Melis, 2002].

To represent the concept of mastery, the current (dynamic) profile contains values for a subset of the competences of Bloom's mastery taxonomy [Bloom, 1956]:

- Knowledge (K)
- Comprehension (C)
- Application (A)

Finishing an exercise or going to another page triggers an update of the student model. Different types of learner actions may exhibit different competencies, hence reading a concept mainly updates 'knowledge' values, reading examples mainly updates 'comprehension', and solving exercises mainly updates 'application'. When the student model receives the notification that a student has finished reading a page, an evaluator fetches the list of its items and their types (concept, example, ...) and delivers an update of the values of those items. When the learner finishes an exercise, an evaluator delivers an update of the values of the involved concepts that depend on the difficulty and on the rating of how successful the solution was.

The student model is inspectable and modifiable by the student as shown in Figure 4. Our experience is that students like to inspect their student model in order to plan what to learn next.

3.4 Knowledge Representation

As opposed to the purely syntactic representation formats for mathematical knowledge such as LaTex or `HTML`, the knowledge representation used by ActiveMath is the *semantic* `XML`-language `OMDoc` [Kohlhase, 2000] which is an extension of `OpenMath` [Caprotti and Cohen, 1998]. `OpenMath` provides a collection of `OpenMath` objects together with a grammar for the representation of mathematical objects and sets of standardized symbols (the content-dictionaries). That is, `OpenMath` talks about objects rather than syntax.

`OpenMath` does not have the means to represent the content of a mathematical *document* nor its structure, whereas `OMDoc` defines logical units such as "definition", "theorem", and "proof" with semantical annotations. In addition, the purely mathematical `OMDoc` representation is augmented by educational metadata such as the difficulty of a learning object or the type of an exercise.

This representation has several advantages, among them

- it is human and machine understandable
- the presentation can automatically and dynamically be linked to concepts and learning objects and thus,

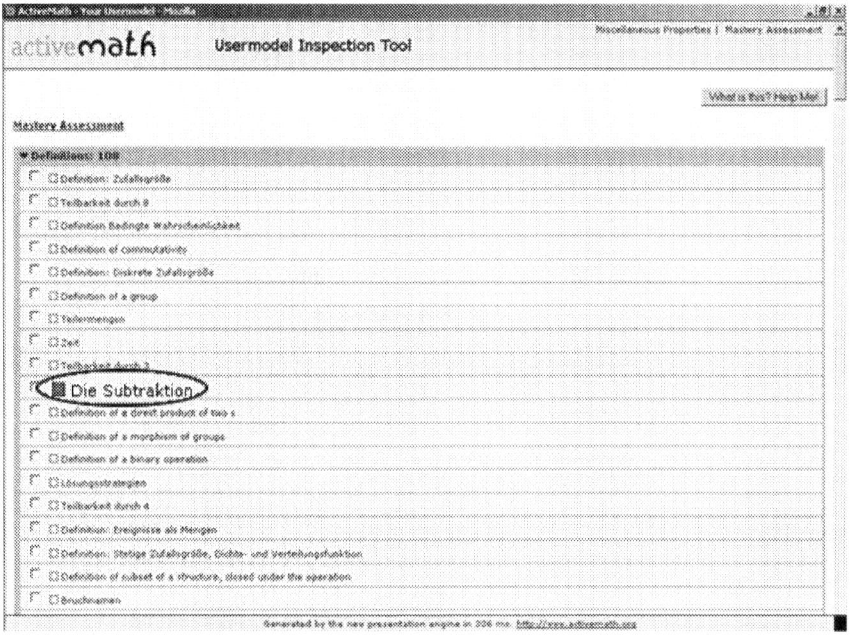

Figure 4. Inspection of the student model (mastery-level)

- concepts can easily be fetched from ACTIVEMATH's dictionary when clicking on a concept or formula during the course
- and mathematical objects can in principle be copied and pasted.

For more details on the representation of mathematical knowledge in ACTIVEMATH, see [Melis *et al.*, 2003].

The web-based ACTIVEMATH system has been under development now for several years at the CCeL (Competence Centre for e-Learning) of the DFKI and at the University of Saarland. Its research and development is supported by several projects of the BMBF (the German Ministry for Research and Development) and the European Union. A demo (and demo guide) is available at `http://www.activemath.org`.

3.5 Automated Reasoning Tools for System Functionalities

Deduction systems have been used as internal modules in learning systems for user modeling [Kobsa and Pohl, 1995], diagnosis [Hoppe, 1994], and course generation [Baumgartner and Furbach, 2003; Melis, 2005]. We discuss the last usage here.

Personalization In order to present the learning material according to the user's needs and abilities, the characteristics of the individual user have to be stored and updated and personalization actions can then be inferred.

So far, ActiveMath uses relatively simple deductive techniques to infer the personalization of content and its presentation from the information in the user model and from pedagogical knowledge in a rule base [Ullrich, 2003]. The course generator of ActiveMath dynamically generates mathematical courses by

1. retrieving the appropriate content from a knowledge base and

2. by applying pedagogical knowledge that is formalized in rules.

In the second stage, when the content is already assembled from the collection of concepts the user has to learn in order to meet the learning goals, the pedagogical rules are applied to select instructional items that are related to these concepts. They also determine the order of items in the learning material. The rules have a *condition* and an *action* part. The condition part of a rule specifies the conditions that have to be fulfilled for the rule to be applicable, the action part specifies the actions to be taken when the rule is applied. The course generator uses the pedagogical rules in order to decide: (i) which information should be presented on a page; (ii) in which order this information should appear on a single page; (iii) how many exercises and examples should be presented and how difficult they should be; (iv) whether or not to include exercises and examples that make use of a particular service system. Since the work with service systems requires a certain minimal familiarity with these systems, ActiveMath presents these exercises only, if the capability to use them is confirmed in the user model. The following are examples of pedagogical rules for two different types of decisions. The rule

```
(defrule PatternForExamPrep
(scenario ExamPrep) =>
          (assert (definitions assertions methods exercises)))
```

determines the kind of items and the order in which they will appear on the course pages. In this example, the learner selected *preparation* for an exam (indicated by the fact `(scenario ExamPrep)`). When this rule fires, the facts `(definition, assertions, methods, exercises)` are asserted, i.e., added to the course. This implies that these items will appear on a page in the specified order.

In turn, these facts may cause other rules to fire, e.g., those choosing exercises with appropriate levels of difficulty:

```
(defrule RequireAppropriateExercise
  (exercises)
  (definition (name ?definition)
              (userKnowledge ?user-knowledge))
  (test (< ?user-knowledge 0.3))
  =>(assert  (choose-exercise-for  ?definition (0.3 0.5 0.7))))
```

This rules determines that if exercises should be presented at all (indicated by `(exercises)`) and if there exists a definition d, then d's name is bound to the variable `?definition` and the learner's knowledge of d is bound to `?user-knowledge`. Then the `test` is evaluated and its value determines whether the rule fires or not. The above rule fires, if the learner's knowledge is less than 0.3. Then the fact `(choose-exercise-for ?definition (0.3 0.5 0.7))` is inserted into the assembled set of items and this triggers the selection of examples for d with difficulty levels 0.3, 0.5, and 0.7 respectively.

3.6 Automated Deduction Systems as Tools

Several systems have been developed for teaching mathematics, e.g., the interactive CMU proof tutor [Sieg and Byrnes, 1996], the EPGY theorem proving environment [Sommer *et al.*, 2000] where Otter [McCune, 1990] checks the correctness of the student's input, living Book [Baumgartner and Furbach, 2003] in which a theorem prover checks the correctness of truth values and normal forms, and Jape [Aczel *et al.*, 1999]. These systems have a deductive component, however, since the (deductive) service systems are fixed and do not adapt to the student, the students have to adapt to the system.

Adaptation to the Student. It may not always be the best idea to make only correct suggestions, as the students might then just click on these suggestions rather than learn anything. So sometimes it is better to include faulty suggestions. The decision of when it is most appropriate to present a faulty suggestion, depends on the learning goal, the learning context, the learning history, and the competency of the student, as well as on the pedagogical strategy.

The interaction console of the proof planner MULTI which is used as a mathematical service in ActiveMath, offers adapted suggestions by configurable suggestion agents [Pollet *et al.*, 2003]. A configuration is a set of agents and an agent encodes pedagogical knowledge. For instance, in order to understand why a method is applicable it may be useful for a student to encounter a situation in which a method is not. Sometimes alternative proof strategies should be learned, where these strategies represent different ways to prove a theorem. For instance, proofs of properties of residue classes in group theory can be tackled by three different proof planning strategies.

The first strategy tries to apply some related theorems, the second strategy reduces the problem to an equation for which a general solution must be found, and the third strategy introduces a case split over the (finitely many) elements of a residue class. The decision for the suggestion of a strategy depends on the knowledge of a student (whether she knows the theorems that are the prerequisites for the first strategy) but it could also be the result of her performance in previous exercises in which, e.g., the other strategies have been trained already.

Active Explorative Learning. An interesting application of proof planning in a system that teaches how to prove theorems is as a 'domain reasoner'. As there can be more ways to prove a given theorem than a tutoring system anticipates and stores, a student may come up with a proof idea which does not match any expert solution the system stored. Now, an interactive proof planner is a more (albeit not always) helpful assistant under these circumstances as it will just generate a new proof using the student's input as islands in this particular proof planning mode. If the system can successfully complete the proof it is accepted – and otherwise not.

Active problem solving and exploration play an essential role in apprenticeship learning and interactive service systems such as a proof planner or a computer algebra system can be used as cognitive tools here. [4]

As its name suggests, ACTIVEMATH emphasizes the active role of the student and this feature is supported inter alia by the integration of some computer algebra systems, MUPAD and the proof planner MULTI. They provide the backbone for interactive problem solving and for dynamically producing feedback to the user's actions. For university students as users, the philosophy of ACTIVEMATH suggests that the user controls her exercise activities herself and no single pre-determined solution needs to be followed as long as any correct solution results eventually [Buedenbender *et al.*, 2002].

Standard theorem proving systems could, in principle, be employed for the exploration of mathematical proofs as well, however essential obstacles that prevent most of the systems from being used for mathematics (except for logic) learning are their low-level logical input language, the small-grained logic-level inference steps, and the poor user interface. Some user interfaces such as PCoq [PCoq, ongoing] are useful for experts who want to prove a theorem, but not for students who want to learn and understand how theorems from an undergraduate textbook, say, should be proved. Similarly, the size and nature of the inference steps (resolution or natural deduction) are too fine grained to be used in a classroom exercise.

[4] The term *cognitive tool* was coined in [Lajoie and Derry, 1993] and generally denotes instruments supporting cognitive processes by extending the limits of the human cognitive capacities, e.g., the working memory.

In learning mathematical proof we need a higher level of abstraction including the particular mathematical vernacular of the area to be taught. This is where proof planning comes into play. For proof planning, there is empirical evidence that instruction with proof planning examples, which work at a higher level of abstraction, can be effective for learning [Melis *et al.*, 2001c].

Moreover, in order to support the student's discovery of failed proof attempts and their repair we made explicit some information that was implicitly available in the proof planning process and we designed messages about failures that can be useful for a learner. Failures include failing meta-level application conditions of methods (e.g., inconsistency of collected constraints), failing object-level conditions of proof planning methods (e.g., missing precondition), as well as erroneous input for the construction of mathematical objects (instantiations of meta-variables). Moreover, it includes meta-reasoning about ways to backtrack, subsequent introduction of a case split into the proof plan, proof by analogy, or not yet sufficiently determined meta-variables.

Feedback. Theorem proving systems and computer algebra systems can be used to automatically check the student's input and return a 'correct' ('incorrect') as, e.g., in the systems MathDox [Cohen *et al.*, 1999] ACTIVEMATH [Melis *et al.*, 2001b] and EPGY [Sommer *et al.*, 2000]. These automated problem solvers can also serve as back-engines for generating example proofs or example computations.

Although classical theorem provers based on a machine oriented logic are not very useful for teaching mathematical proofs, they are used to advantage however in courses on logic as for example eTPS [TPS, 2004].

Meta-Reasoning. Although still a far cry from effective meta-reasoning, first attempts to present or to employ Polya's problem solving heuristics have been made. Cairns [Cairns and Gow, 2001] gives an interpretation of Polya's stages in his presentation of a proof that includes pointers to prerequisites and to applications of some theorem. Similarly, ACTIVEMATH uses Polya's stages in a presentation-scenario that provides structure and links to related topics. In the Polya-presentation scenario, the stages *Understand the Problem*, *Devise a Plan*, *Carry out the Plan*, and *Look Back at the Solution* are realised by assembling certain types of learning objects by their metadata. This way, proof examples and exercises can be presented at the appropriate stage of searching for a solution to the given problem.

Cognitively Adequate Presentations. An adequate presentation of a proof for education would have to address at least the following issues:

- hierarchically structuring a (complex) proof to make it more comprehensible

- emphasize the proof *process* in order to support the student's proof activities and derivational transfer.

- cognitive overload

As for hierarchical structuring, the presentation of proof trees in a box style is a good way to make the structure visible. However, the box structure is logic-oriented and does not help much for mathematical proofs for which a hierarchy helps to separate and express (and remember) proof ideas from details.

Most current approaches to natural language presentation of proofs [Dahn, 1998; Fiedler, 1999; PCoq, ongoing] target a presentation similar to proofs in a book. However, these presentations are result-oriented rather than process-oriented. More abstraction is introduced by presentations based on tactics using proof plans such as in [Holland-Minkley *et al.*, 1999] and [Melis and Leron, 1999].

As for emphasizing the proof process, presenting the search for a *partial proof plan* is often useful, because it makes the proof situation explicit. For instance, the collection of constraints of a meta-variable could be presented in order to provide a clue on how to (interactively) construct a mathematical object [Zimmer and Melis, 2004]. This can help the student to *construct an object* which is the most difficult task in many proofs.

4 Challenges and Future Work

Tools for learning mathematics (and logic) are leaving the lab to be used in practice,[5] however there is still plenty of room for improvement. The following open or only partially solved problems represent a spectrum from the user-centered design of user interfaces and presentations up to the formalization of pedagogical strategies. Most of these challenges require interdisciplinary research.

Proof Presentation and User Interfaces. Learning mathematics involves people, situations, and goals that are very different from those presupposed for an automated theorem proving system. Apart from other things, learning may have the goal to understand a given proof or find a proof. In order to help the student to *understand* solutions and proofs, we need a comprehensible proof presentation at various levels of abstraction, detail and explanation upon request. In order to support *learning by doing*,

[5]For example the geometry tutor from Carnegie Mellon is used now in more than 2000 schools in the US.

however, a better proof presentation alone is not sufficient as it requires more advanced user-adaptive user interfaces specially designed for learning.

Currently the design of a user interface for a theorem proving system typically starts with the special technical features and capabilities of the underlying system rather than with a user-centered approach. This is not just our own unfortunate experience with the user interface of the ΩMEGA system [Siekmann *et al.*, 1999], which is inadequate for an average student. A student concentrates on learning and problem solving and any focus on the tool produces a cognitive overload that gets in the way of learning and mathematical skills. The structured (and foldable) hypermedia presentation of worked-out problem solutions as well as the presentation of information relevant to the problem solving *process* is in fact much better than a traditional textbook proof, which is in the tradition of minimalistic proof presentation developed over the centuries in mathematics.

As part of the user interface, a simple to use input editor is needed. Some work in this direction was done at RISC [Nakagawa and Buchberger, 2001] with the Mathematica functionality, in Nice [Dirat *et al.*, 2000], in Grenoble [Nicaud *et al.*, 2002], and for ACTIVEMATH which by now features a full-fledged palette-based input editor that generates OpenMath.

Feedback in Problem Solving. A system should reason about the student's input and not just use pre-computed solutions and proofs in order to guide the student's problem solving. Moreover, it is certainly not sufficient to just respond 'correct' or 'incorrect'. Interesting feedback has to include the provision of counter-examples, similar proofs, explanations and hints.

Meta-Reasoning. The integration of heuristics and meta-reasoning into a learning tool for learning mathematics is still a challenge. An advanced tool support might offer means far beyond Polya's ideas. For instance, the student could use an online search tool for the Internet to find similar problems, analogous solutions, or the concepts which are prerequisites for her proof. Semantic search techniques, managing (little) theories, browsing theories, and maintaining and managing mathematical ontologies are currently research topics of the MKM Conference series and as partially implemented in ACTIVEMATH search/dictionary tool.

Proof Planning In order to use proof planning in an educational setting, several research directions are promising, among them

- more advanced support for reasoning about failed proof attempts and appropriate support for revising a proof plan,
- support for checking whether a path is heuristically promising or dead-ended altogether,

- support for the construction of mathematical objects.
- Island planning as suggested in [Melis, 1996] is a good starting point in order to develop a proof idea first and to leave the rest of the details to the proof system.

5 Conclusion

The student is not like me.[6] It is insufficient to rely on our own intuition on what may be useful or not. It is also not enough to use standard psychological results about human learning when designing a learning tool. The actual proof of usefulness and usability comes from observations on how students actually use a system and from empirical evidence in controlled experiments that measure the effect of a system.

BIBLIOGRAPHY

[PCoq, ongoing] http://www-sop.inria.fr/lemme/pcoq/.

[Aczel *et al.*, 1999] J.C. Aczel, P. Fung, R. Bornat, M. Oliver, T. OShea, and B. Sufrin. Using computers to learn logic: Undergraduates experiences. In G. Cumming, T. Okamote, and L. Gomez, editors, *Advanced Research in Computers and Communications in Education. Proccedings of the 7th International Conference on Computers in Education*, Amsterdam, 1999. IOS Press.

[Koedinger et al, 1997] K.R. Koedinger, J.R. Anderson, W.H. Hadley and M.A. Mark. Intelligent Tutoring Goes to School in the big City. *International Journal of Artificial Intelligence in Education*, vol 8: 30–43, 1997.

[Aleven and Koedinger, 2000] V. Aleven and K.R. Koedinger. The need for tutorial dialog to support self-explanation. In C. P. Rose and R. Freedman, editors, *Building Dialogue Systems for Tutorial Applications*, AAAI Fall Symposium, pages 65–73. AAAI Press, 2000.

[Anderson *et al.*, 1995] J. Anderson, A. Corbett, K. Koedinger, and R. Pelletier. Cognitive tutors: Lessons learned. *The Journal of the Learning Sciences*, 4(2):167–207, 1995.

[Anderson and Pelletier, 1991] J.R. Anderson and R. Pelletier. A development system for model-tracing tutors. In L. Birnbaum, editor, *The International Conference on the Learning Sciences*, 1991.

[Andriessen and Sandberg, 1999] J. Andriessen and J. Sandberg. Where is education heading and how about AI. *International Journal of Artificial Intelligence in Education*, 10:130–150, 1999.

[Baumert *et al.*, 1997] J. Baumert, R. Lehmann, M. Lehrke, B. Schmitz, M. Clausen, I. Hosenfeld, O. Köller, and J. Neubrand. *Mathematisch-naturwissenschaftlicher Unterricht im internationalen Vergleich.* Leske und Budrich, 1997.

[Baumgartner, 2003] P. Baumgartner, editor. *IJCAI workshop on Knowledge Representation and Deduction for Education.* AAAI, 2003.

[Baumgartner and Furbach, 2003] P. Baumgartner and U. Furbach. Automated deduction techniques for the management of personalized documents. *Annals of Mathematics and Artificial Intelligence, Special Issue*, 38(1-3):211–228, 2003.

[Bloom, 1956] B.S. Bloom, editor. *Taxonomy of educational objectives: The classification of educational goals: Handbook I, cognitive domain.* Longmans, Green, New York, Toronto, 1956.

[6] Mantra from a personal communication with Ken Koedinger.

[Bornat,] R. Bornat. The JAPE web site. http://www.jape.org.uk.

[Brusilovski, 1996] P. Brusilovski. Methods and techniques of adaptive hypermedia. *User MOdeling and User Adapted Interaction*, 6(2-3):87–129, 1996.

[Buchberger *et al.*, 1997] B. Buchberger, T. Jebelean, F. Kriftner, M. Marin, E. Tomuta and D. Vasaru. An Overview of the Theorema Project. Proceedings if ISSAC, 1997.

[Buedenbender *et al.*, 2002] J. Buedenbender, E. Andres, A. Frischauf, G. Goguadze, P. Libbrecht, E. Melis, and C. Ullrich. Using computer algebra systems as cognitive tools. In S.A. Cerri, G. Gouarderes, and F. Paraguacu, editors, *6th International Conference on Intelligent Tutor Systems (ITS-2002)*, number 2363 in Lecture Notes in Computer Science, pages 802–810. Springer-Verlag, 2002.

[Cairns and Gow, 2001] P. Cairns and J. Gow. On dynamically presenting a topology course. In *First International Workshop on Mathematical Knowledge Management (MKM 2001)*, September 2001.

[Caprotti and Cohen, 1998] O. Caprotti and A. M. Cohen. Draft of the OpenMath standard. OpenMath Consortium, http://www.nag.co.uk/projects/OpenMath/omstd/, 1998.

[Catrambone and Holyoak, 1990] R. Catrambone and K.J. Holyoak. Learning subgoals and methods for solving probability problems. *Memory and Cognition*, 18(6):593–603, 1990.

[Chi *et al.*, 1989] M.T. Chi, M. Bassok, M.W. Lewis, P. Reimann, and R. Glaser. Self-explanations: How students study and use examples in learning to solve problems. *Cognitive Science*, 13:145–182, 1989.

[China, 2005] China Network Information Centre. Statistic Report on the Development and Status of Networks in China, 2005.

[Cohen *et al.*, 1999] A. Cohen, H. Cuypers and H. Sterk. *Algebra Interactive*, Springer-Verlag, 1999.

[Conati *et al.*, 1997] C. Conati, A.S. Gertner, K. VanLehn, and M. Druzdzel. On-line student modeling for coached problem solving using Baysian networks. In A. Jameson, C. Paris, and C. Tasso, editors, *User Modeling: Proceedings of the Sixth International Conference, UM97*, pages 231–242, 1997.

[Dahn, 1998] B.I. Dahn. Using ILF as a user interface for many theorem provers. In R.C. Backhouse, editor, *Proceedings of the Workshop on User Interfaces for Theorem Provers*, volume 98-08 of *Eindhoven University of Technology*, pages 75–86. Department of Mathematics and Computing Science, 1998.

[Dirat *et al.*, 2000] L. Dirat, M. Buffa, J.-M. Fedou, and P. Sander. Jome, un composant logiciel pour le tele-enseignement des mathematiques via le web. In *Actes du Colloque Internationalal sur les Technologies de l'Information et de la Communication dans les Enseignements d'Ingenieurs et dans l'Industrie (TICE*, pages 7–16, 2000.

[Fiedler, 1999] A. Fiedler. Using a cognitive architecture to plan dialogs for the adaptive explanation of proofs. In *Proceedings of the 16th International Joint Conference on Artificial Intelligence (IJCAI)*. Morgan Kaufman, 1999.

[Forgy, 1982] C.L. Forgy. Rete: a fast algorithm for the many pattern/many object pattern match problem. *Artificial Intelligence*, pages 17–37, 1982.

[Franke and Kohlhase, 2000] A. Franke and M. Kohlhase. MBase: Representing mathematical knowledge in a relational data base. In F. Pfenning, editor, *Proc. 17th International Conference on Automated Deduction (CADE)*, Lecture Notes on Artificial Intelligence. Springer-Verlag, 2000.

[Friedman-Hill, 1997] E. Friedman-Hill. Jess, the java expert system shell. Technical Report SAND98-8206, Sandia National Laboratories, 1997.

[Holland-Minkley *et al.*, 1999] A.M. Holland-Minkley, R. Barzilay, and R.L. Constable. Verbalization of high-level formal proofs. In *National Conference on Artificial Intelligence (AAAI-99)*, 1999.

[Hoppe, 1994] U. Hoppe. Deductive error diagnosis and inductive error generalization for intelligent tutoring systems. *Journal of Artificial Intelligence in Education*, 5(1):27–49, 1994.

[Jacobs, 2001] B. Jacobs. Aufgaben stellen und Feedback geben. Technical report, Medienzentrum der Philosophischen Fakultät der Universität des Saarlandes, 2001.

[Kobsa and Pohl, 1995] A. Kobsa and W. Pohl. The user modeling shell system bgp-ms. *User Modeling and User Adapted Interaction*, 4:59–106, 1995. http://www.ics.uci.edu/ kobsa/papers/1995-UMUAI-kobsa.ps.gz.

[Kohlhase, 2000] M. Kohlhase. OMDoc: Towards an internet standard for the administration, distribution and teaching of mathematical knowledge. In *Proceedings Artificial Intelligence and Symbolic Computation AISC'2000*, 2000.

[Lajoie and Derry, 1993] S. Lajoie and S. Derry, editors. *Computers as Cognitive Tools.* Erlbaum, Hillsdale, NJ, 1993.

[Leron, 1983] U. Leron. Structuring mathematical proofs. *The American Mathematical Monthly*, 90:174–185, 1983.

[Lowe and Duncan, 1997] H. Lowe and D. Duncan. XBarnacle: Making theorem provers more accessible. In W. McCune, editor, *Proceedings of the Fourteenth Conference on Automated Deduction (CADE)*, volume 1249 of *Lecture Notes in Artificial Intelligence*, pages 404–408. Springer, 1997.

[Maple, 1986] B.W. Char, G.J. Fee, K.O. Geddes, G.H. Gonnet and M.B. Monagan. A Tutorial Introduction to MAPLE. *Journal of Symbolic Computation*, 2(2):179–200, 1986.

[Matsuda and vanLehn, 2005] N. Matsuda and K. vanLehn Advanced Geometry Tutor: An intelligent Tutor that Teaches Proof-Writing with Construction. *Artificial Intelligence in Education*, Ch-K. Looi, G. McCalla, B. Bredewig and J.Breuker (editors) pages 443–450, IOS Press, 2005.

[McCune, 1990] W. W. McCune. Otter 2.0 Users' Guide. Argonne Natioanl Laboratory 1990. ANL-90/9, Maths and CS Division, Argonne, Illinois.

[Mandl *et al.*, 1997] H. Mandl, H. Gruber, and A. Renkl. *Enzyklopädie der Psychologie*, volume 4, chapter Lernen und Lehren mit dem Computer, pages 436–467. Hogrefe, 1997.

[Meier and Melis, 2005] A. Meier and E. Melis. Failure reasoning in multiple-strategy proof planning. In *Electronic Notes in Theoretical Computer Science*, M. Bonacina and T. Boy de la Tour, eds. pp. 67–90, Elsevier, 2005.

[Melis, 1996] E. Melis. Island planning and refinement. Seki Report SR-96-10, Universität des Saarlandes, FB Informatik, 1996.

[Melis, 2000] E. Melis, editor. *Proceedings of CADE-17 Workshop on Deduction in Education*, 2000.

[Melis, 2005] E. Melis. Why Proof Planning for Maths Education and How? In D. Hutter and W. Stephan, editors, *Mechanizing Mathematical Reasoning: Essays in Honor of Jörg Siekmann*, Lecture Notes in Artficial Intelligence, no 2605, pp. 364–376. Springer-Verlag, 2005.

[Melis and Andres, 2004] E. Melis and E. Andres. Global Feedback in ACTIVEMATH. *Internatioal Journal of Computers in Mathematics and Science Teaching*, 24:197–220, 2005.

[Melis and Leron, 1999] E. Melis and U. Leron. A proof presentation suitable for teaching proofs. In S.P. Lajoie and M. Vivet, editors, *9th International Conference on Artificial Intelligence in Education*, pages 483–490, Le Mans, 1999. IOS Press.

[Melis and Meier, 2000] E. Melis and A. Meier. Proof planning with multiple strategies. In *First Internatioanl Conference on Computational Logic.* J. Lloyd, V. Dahl and U. Furbach, eds. LNAI vol 1861. pp. 644–659, 2000.

[Melis *et al.*, 2001a] E. Melis, E. Andres, A. Franke, G. Goguadse, P. Libbrecht, M. Pollet, and C. Ullrich. ACTIVEMATH system description. In *Artificial Intelligence and Education*, pages 580–582, 2001.

[Melis *et al.*, 2001b] E. Melis, J. Buedenbender, E. Andres, A. Frischauf, G. Goguadse, P. Libbrecht, M. Pollet, and C. Ullrich. ACTIVEMATH: A generic and adaptive web-based learning environment. *Artificial Intelligence and Education*, 12(4):385–407, winter 2001.

[Melis *et al.*, 2001c] E. Melis, Ch. Glasmacher, C. Ullrich, and P. Gerjets. Automated proof planning for instructional design. In *Annual Conference of the Cognitive Science Society*, pages 633–638, 2001.

[Melis *et al.*, 2003] E. Melis, J. Buedenbender, E. Andres, A. Frischauf, G. Goguadse, P. Libbrecht, M. Pollet, and C. Ullrich. Knowledge representation and management in ACTIVEMATH. *International Journal on Artificial Intelligence and Mathematics, Special Issue on Management of Mathematical Knowledge*, 38(1-3):47–64, 2003.

[Melis and Ullrich, 2003] E. Melis and C. Ullrich. Polya-scenarios in ACTIVEMATH. In *AI in Education, AIED-2003.* U. Hoppe, F. Verdejo, and J. Kay, editors, pp. 141–147. IOS Press, 2003.

[Murray *et al.*, 1999] T. Murray, C. Condit, T. Shen, J. Piemonte, and S. Khan. Metalinks - a framework and authoring tool for adaptive hypermedia. In S.P. Lajoie and M. Vivet, editors, *Proceedings of the International Conference on Artificial Intelligence and Education*, pages 744–746. IOS Press, 1999.

[Nakagawa and Buchberger, 2001] K. Nakagawa and B. Buchberger. Presenting proofs using logicographic symbols. In A. Fiedler and H. Horacek, editors, *IJAR-2001 Workshop on Proof Transformation and Presentation*, 2001.

[Narciss, 2001] S. Narciss. Informatives Tutorielles Feedback. Habilschrift,Technische Universität Dresden, Fak. Mathematik und Naturwissenschaften, 2004.

[Nicaud *et al.*, 2002] J-F. Nicaud, D. Bouhineau, and T. Huguet. Aplusix-editor: A new kind of software for the learning of algebra. In S.A. Cerri, G. Gouarderes, and F. Paraguacu, editors, *Intelligent Tutoring Systems, 6th International Conference, ITS2002*, volume 2363 of *LNCS*, pages 178–187, 2002.

[Piaget, 1977] J. Piaget. *Equilibration of Cognitive Structures.* Viking, New York, 1977.

[Pollet *et al.*, 2003] M. Pollet, E. Melis, and A. Meier. User interface for adaptive suggestions for interactive proof. In *Proceedings of the International Workshop on User Interfaces for Theorem Provers (UITP)*, pp. 133–142, 2003.

[Polya, 1945] G. Polya. *How to Solve it.* Princeton University Press, Princeton, 1945.

[Schoenfeld, 1985] A.H. Schoenfeld. *Mathematical Problem Solving.* Academic Press, New York, 1985.

[Schoenfeld, 1990] A.H. Schoenfeld, editor. *A Source Book for College Mathematics Teaching.* Mathematical Association of America, Washington, DC, 1990.

[Sieg and Byrnes, 1996] W. Sieg and J. Byrnes. Normal natural deduction proofs. Technical Report CMU-PHIL-74, Department of Philosophy, Carnegie Mellon University, Pittsburgh, 1996.

[Sieg and Byrnes, 1998] W. Sieg and J. Byrnes. Normal natural deduction proofs (in classical logic). *Studia Logica*, 60:67–106, 1998.

[Siekmann *et al.*, 1999] J. Siekmann, S. Hess, Ch. Benzmüller, L. Cheikhrouhou, A. Fiedler, M. Kohlhase H. Horacek, K. Konrad, A. Meier, E. Melis, and V. Sorge. LΩUI: Lovely ΩMEGA User Interface. *Formal Aspects of Computing*, 11(3):326–342, 1999.

[Sommer *et al.*, 2000] R. Sommer, M. Rozenfeld, and R. Ravaglia. A proof environment for teaching mathematics. In E. Melis, editor, *Deduction in Education, CADE-17 workshop*, pages 35–43, 2000.

[Sorgatz and Hillebrand, 1995] A. Sorgatz and R. Hillebrand. MuPAD – Ein Computeralgebra System I, *Linux Magazin*, 12/95, 1995. http://www.geo.unibonne.de/software/wwpad/BIB

[Suppes, 1981] P. Suppes, ed. University-level Computer-assisted Instruction at Stanford: 1968–1980. Stanford. Institute for Mathematical Studies in teh Social Sciences. 1981.

[Thery *et al.*, 1992] L. Thery, Y. Bertot, and G. Kahn. Real theorem provers deserve real user-interfaces. Rapports de Recherche 1684, Institute National de Recherche en Informatique et en Automatique, Sophia Antipolis, 1992.

[TPS, 2004] P.B. Andrews, C.E. Brown, F. Pfenning, M. Bishop, S. Issar and H. Xi. ETPS: A System to Help Students Write Formal Proofs. *Journal of Automated Reasoning*, 32: 75–92, 2004

[Tsovaltzi, 2005] A. Fiedler and D. Tsovaltzi. Domain-Knowledge Manipulation for Dialogue-Adaptive Hinting. *Artificial Intelligence in Education*, C.-K. Looi et al. (editors), IOS Press, pages: 801–803, 2005.

[Ullrich, 2003] C. Ullrich. Pedagogical rules in ActiveMath and their pedagogical foundations. Seki Report SR-03-03, Universität des Saarlandes, FB Informatik, 2003.

[Ullrich and Melis, 2002] C. Ullrich and E. Melis. The Poor Man's Eyetracker in ACTIVEMATH. *Proceedings of the World Conference on E-Learning in Corporate, Government, Healthcare, and Higher Education (eLearn-2002)*, vol.4, pages:2313–2316, 2002.

[Ullrich *et al.*, 2004] C. Ullrich, P. Libbrecht, S. Winterstein and M. Muehlenbrock. A flexible and efficient presentation architecture for adaptive hypermedia: description and technical evaluation. *Proceedigns fo the 4th IEEE Internatioanl Conference on Advanced Learning Technologies (ICALT 2004)*. pp. 21–25, Kinshuk, C. Looi, E. Sutinen, D. Sampson, eds., 2004

[VanLehn *et al.*, 2001] K. VanLehn, S. Siler, C. Murray, T. Yamauchi, and W.B. Baggett. Human tutoring: Why do only some events cause learning? *Cognition and Instruction*, 2001.

[VanLehn *et al.*, 2002] K. VanLehn, C. Lynch, L. Taylor, A. Weinstein, R. Shelby, K. Schulze, D. Treacy, and M. Wintersgill. Minimally invasive tutoring of complex physics problem solving. In S.A. Cerri, G. Gouarderes, and F. Paraguacu, editors, *Intelligent Tutoring Systems, 6th International Conference, ITS 2002*, number 2363 in LNCS, pages 367–376. Springer-Verlag, 2002.

[Vygotsky, 1978] L. Vygotsky. *The Development of Higher Psychological Processes*. Harvard University Press, Cambridge, 1978.

[Weber and Brusilovsky, 2001] G. Weber and P. Brusilovsky. ELM-ART an adaptive versatile system for web-based instruction. *Artificial Intelligence and Education*, 2001.

[White and Shimoda, 1999] B.Y. White and T.A. Shimoda. Enabling students to construct theories of collaborative inquiry and reflective learning: Computer support for metacognitive development. *International Journal of Artificial Intelligence in Education*, 10:151–182, 1999.

[Zimmer and Melis, 2004] J. Zimmer and E. Melis. Constraint solving for proof planning. *Journal of Automated Reasoning*, 33:51–88, 2004.

On Neighbourhood Semantics Thirty Years Later

VALENTIN SHEHTMAN

1 Introduction

The research of Dov Gabbay includes different parts of logic, and in many cases it essentially influenced further development. Problems considered here are motivated by two his papers, [Gabbay, 1975] and [Gabbay and de Jongh, 1974]. These papers appeared at the beginning of the 1970s, at the time of remarkable events in modal logic, when all of a sudden the whole area was found full of difficult problems and nice theorems, like flowers growing in high mountains, and young researchers came for new interesting discoveries.

General problems in neighbourhood semantics were first addressed by D. Gabbay and M. Gerson in [Gabbay, 1975; Gerson, 1975a; Gerson, 1975b]. At that time the interest in neighbourhood semantics was rather moderate, but now it is clear that neighbourhood approach may be quite useful in different kinds of modal logic: spatial, epistemic, conditional etc. [van Benthem and Sarenac, 2004; Aiello, 2002].

This paper studies neighbourhood completeness and compactness for modal and intermediate logics. We show that completeness and compactness are closely related: noncompact logics (in Thomason's sense [Thomason, 1972]) may be helpful for distinguishing Kripke and neighbourhood completeness, which is the problem studied in [Gabbay, 1975]. Three technical details are crucial here: K. Fine's frame [Fine, 1974], the axiomatisation of binary finite trees [Gabbay and de Jongh, 1974], and the ultrabouqet construction [Shehtman, 1998].

Some of the results presented here were published, but only in Russian, in a hardly available paper [Shehtman, 1980] and later in the author's Thesis [Shehtman, 2000] (even less available). So we give a slightly modified exposition of these results.

The plan of the paper is as follows. Section 2 contains very basic material on Kripke and neigbourhood semantics; but some notions (such as different kinds of compactness in modal logic) are not widely known. In Section 3 we recall properties of ultrabouqets of topological spaces proved in [Shehtman, 1998]. Ultrabouqets of neighbourhood **K4**-frames are considered in Section 4. The latter construction is new, and it is used only in Section 9, so the readers can skip it if they are interested in other parts of the paper. Section 5 presents an example of a TKN-noncompact extension of **Grz** from [Shehtman, 1980]. Basing on it, we construct a new rather simple example of a relatively incomplete modal logic in Section 6. The same is done for intermediate logics: Section 7 contains an example of TK-noncompactness from [Shehtman, 1980], and Section 8 — a new example of relative incompleteness. Note that the earlier example of a relatively incomplete intermediate logic is quite complicated. This example is recalled in Section 10, but without the laborious proof (given in full detail in [Shehtman, 2000]). Section 9 also proves new results: N-compactness for all extensions of **GL** and **Grz**. Section 10 contains some hints for further results and some questions.

2 Preliminaries

The material of this Section is rather standard, most of it can be found in the first chapters of [Chagrov and Zakharyaschev, 1997], but our notation is slightly different.

In this paper we consider monomodal and intermediate propositional logics. So *modal formulas* are constructed from the countable set of propositional variables $PV = \{p, q, \dots\}$, the constant $\bot$, and the connectives $\rightarrow$ $\wedge$, $\vee$, $\Box$; the derived connectives are: $\neg$, $\top$, $\leftrightarrow$, $\Diamond$. *Intuitionistic formulas* are modal formulas without occurrences of $\Box$.

A *modal logic* is a set of modal formulas containing all classical tautologies, the axiom $\Box(p \rightarrow q) \rightarrow (\Box p \rightarrow \Box q)$ and closed under Modus Ponens, $\Box$-introduction, and Substitution; we consider only consistent logics (i.e. not containing $\bot$).

An *intermediate logic* is a consistent set of intuitionistic formulas closed under (intuitionistic) Substitutions and Modus Ponens and containing the standard axioms of Heyting Calculus [Chagrov and Zakharyaschev, 1997].

The minimal modal logic is denoted as usual by **K**. Intuitionistic logic (denoted by **H**) is the smallest intermediate logic.

For a logic $\mathbf{\Lambda}$ and a formula A, the notation $\mathbf{\Lambda} \vdash A$ is used as an equivalent to $A \in \mathbf{\Lambda}$. For a set of formulas Γ and a modal (or intermediate) logic $\mathbf{\Lambda}$, the minimal modal (resp., intermediate) logic containing $\mathbf{\Lambda} \cup \Gamma$ is denoted by $\mathbf{\Lambda} + \Gamma$.

Some particular modal logics used in this paper are

$$\mathbf{K4} = \mathbf{K} + \Box p \to \Box\Box p,$$
$$\mathbf{S4} = \mathbf{K4} + \Box p \to p,$$
$$\mathbf{Grz} = \mathbf{S4} + AG \text{ (Grzegorczyk logic)},$$
$$\mathbf{GL} = \mathbf{K4} + AL \text{ (Löb logic)},$$

where

$$AG := \neg(p \wedge \Box(p \to \Diamond(\neg p \wedge \Diamond p))),$$
$$AL := \Box(\Box p \to p) \to \Box p.$$

Note that the AG is usually written in a different (equivalent) form: as $\Box(\Box(p \to \Box p) \to p) \to p$.

The intermediate logic of our special interest is Gabbay – De Jongh's $\mathbf{H} + Br_2$, where

$$Br_2 := (\bigwedge_{i=0}^{n}(P_i \to \bigvee_{j\neq i} P_j) \to \bigvee_{j\neq i} P_j) \to \bigvee_{i=0}^{n} P_i,$$

and $P_0 := p,\ P_1 := q,\ P_2 := (p \leftrightarrow q)$.

This logic was introduced in [Gabbay and de Jongh, 1974], with the axiom Br_2 in an equivalent form, where P_i are just propositional variables. The above form (proposed by S.K. Sobolev) uses only two variables.

A *neighbourhood frame* is a pair $\mathcal{X} = (X, \Box)$ consisting of a non-empty set with an operation on its subsets $\Box : 2^X \longrightarrow 2^X$. In this paper we consider only **K4**-*frames*, i.e. those, in which $\Box(V_1 \cap V_2) = \Box V_1 \cap \Box V_2$, $\Box X = X$, $\Box V \subseteq \Box\Box V$ and **S4**-*frames* (also satisfying $\Box V \subseteq V$). The latter are nothing but topological spaces ($\Box$ is the interior operator). $\Diamond V = -\Box(-V)$[1] is the closure operator in topological spaces: A *(neighbourhood) model* over $\mathcal{X}$ is a pair $M = (\mathcal{X}, \theta)$, where $\theta : PV \longrightarrow 2^X$ is a *valuation* in $\mathcal{X}$.

The map θ extends to all formulas in the well-known way:
$\theta(\bot) = \varnothing$, $\theta(A \to B) = -\theta(A) \cup \theta(B)$, $\theta(A \wedge B) = \theta(A) \cap \theta(B)$,
$\theta(A \vee B) = \theta(A) \cup \theta(B)$, $\theta(\Box A) = \Box\theta(A)$.
The notation $M, x \vDash A$ means $x \in \theta(A)$, which is also read as "a modal formula A is *true at world* x of M". A is called

- *true in* M (notation: $M \vDash A$) if A is true at all worlds of M;
- *valid at* $\mathcal{X}, x$ (notation: $F, x \vDash A$) if A is true at world x under all valuations;
- *valid in* $\mathcal{X}$ (notation: $\mathcal{X} \vDash A$) if A is valid at all worlds of $\mathcal{X}$.

[1]$(-V)$ denotes the compement of V.

$\mathbf{ML}(\mathcal{X})$ denotes the *modal logic of* $\mathcal{X}$, i.e., the set of all modal formulas valid in $\mathcal{X}$. If $\mathbf{\Lambda} \subseteq \mathbf{ML}(\mathcal{X})$, then $\mathcal{X}$ is called a $\mathbf{\Lambda}$*-frame*. For a class of frames $\mathcal{C}$, we denote $\mathbf{ML}(\mathcal{C}) := \bigcap\{\mathbf{L}(\mathcal{X}) \mid \mathcal{X} \in \mathcal{C}\}$ (the *modal logic of* $\mathcal{C}$, or the *modal logic determined by* $\mathcal{C}$). A modal logic determined by some class of neighbourhood frames is called *neighbourhood complete*, or *N-complete.*

A *Kripke frame* is a pair $F = (W, R)$ consisting of a non-empty set with a binary relation. F is associated with the neighbourhood frame $N(F) = (W, \Box)$ such that $\Box V = \{x \in W \mid R(x) \subseteq V\}$. In this paper all Kripke frames are transitive.

A *(Kripke) model* over F is a pair $M = (F, \theta)$, where $\theta : PV \longrightarrow 2^W$ is a valuation; so we can consider M as a neghbourhood model over $N(F)$. For Kripke frames we also use the notations $M, x \vDash A$, $M \vDash A$, $F \vDash A$, $F, x \vDash A$, $\mathbf{ML}(F)$ and the corresponding terminology, as explained above. So transitive Kripke frames are exactly $\mathbf{K4}$-frames; reflexive transitive Kripke frames (quasi-ordered sets) are exactly $\mathbf{S4}$-frames. Logics determined by classes of Kripke frames are called *Kripke complete* (or *K-complete*).

For semantics of intuitionistic formulas we need topological spaces rather than arbitrary neighbourhood frames. Every intuitionistic formula A translates as a modal formula A^T (by putting $\Box$ in front of every its subformula). A valuation θ in a space $\mathcal{X}$ is called *intuitionistic* if $\theta(s)$ is open for any propositional variable s; then $(\mathcal{X}, \theta)$ is an *intuitionistic topological model.* For any intuitionistic formula A we put:

$$\theta^I(A) := \theta(A^T).$$

Thus we obtain the *intuitionistic extension* θ^I of θ.

We shall also use some special notation. For intutuionistic formulas A, B we put

$$\theta^\bullet(A \to B) := \theta(A, B) := \theta^I(A) - \theta^I(B).$$

Thus $x \notin \theta^I(A \to B)$ iff $x \in \Diamond\theta^\bullet(A \to B)$, and $\theta^\bullet(\neg A) = \theta^I(A)$.

The intuitionistic notions of truth and validity are defined similarly to the modal case:

$M, x \Vdash A$ means $x \in \theta^I(A)$ and is also read as "A is *intuitionistically true at world* x of M", or "M, x *forces* A". A is called

- *intuitionistically true in* M (notation: $M \Vdash A$) if A is intuitionistically true at all worlds of M;
- *intuitionistically valid at* $\mathcal{X}, x$ (notation: $\mathcal{X}, x \Vdash A$) if A is intuitionistically true at world x under all intuitionistic valuations;
- *intuitionistically valid in* $\mathcal{X}$ (notation: $\mathcal{X} \Vdash A$) if A is intuitionistically valid at all worlds of $\mathcal{X}$.

The set of all intuitionistic formulas valid in $\mathcal{X}$ is denoted by $\mathbf{IL}(\mathcal{X})$ (the *intermediate logic of* $\mathcal{X}$)

Let us recall sufficient conditions for validity of AG and Br_2.

LEMMA 1. *For a Kripke* **S4***-frame* $F = (W, R)$

(1) $F \vDash AG$ *iff* F *is Nötherian, i.e. it does not contain infinite ascending chains:* $x_0 R x_1 R \ldots$

(2) If R *is a partial Nötherian order, then* $F \Vdash Br_2$ *if every world has at most two immediate successors:* $\forall x \exists y, z\ (xRy\ \&\ xRz\ \&\ R(x) = \{x\} \cup R(y) \cup R(z))$.

Let us also recall two standard facts:

LEMMA 2. *Let* θ, ψ *be a modal and an intuitionistic valuation in the same topological space such that for any* $s \in PV$

$$\psi(s) = \Box\theta(s).$$

Then for any intuitionistic A

$$\psi^I(A) = \theta(A^T).$$

COROLLARY 3. *Let* $\mathcal{X}$ *be a topological space,* A *an intuitionistic formula. Then*

(1) for any $x \in \mathcal{X}$ $\quad \mathcal{X}, x \Vdash A$ *iff* $\mathcal{X}, x \vDash A^T$,

(2) $\mathcal{X} \Vdash A$ *iff* $\mathcal{X} \vDash A^T$.

A **K4**-frame $\mathcal{X} = (X, \Box)$ is associated with a topological space $\mathcal{X}^+ = (X, \Box^+)$ such that $\Box^+ Y = \Box Y \cap Y$. The topological terminology referring to $\mathcal{X}^+$ will be also used for $\mathcal{X}$; so for example, we say that a subset Y is *closed in* $\mathcal{X}$ if it is closed in $\mathcal{X}^+$ (which is equivalent to $\Diamond Y \subseteq Y$); Y is *open in* $\mathcal{X}$ iff $Y \subseteq \Box Y$. So a closed point x may be of two kinds: *reflexive*, with $\Diamond\{x\} = \{x\}$, and *irreflexive*, with $\Diamond\{x\} = \varnothing$.

If $\mathcal{X}$ is corresponds to a Kripke frame (W, R), then $\Diamond Y = R^{-1}(Y)$, so x is closed iff x is R-minimal, and the above reflexivity notion corresponds to the standard one.

LEMMA 4. *Let* x *be a reflexive closed point in a* **K4***-frame. Then* $x \in \Box V$ *implies* $x \in V$.

Proof. Suppose $x \in \Box V$, but $x \notin V$. Then $V \subseteq -\{x\}$, and so $\Box V \subseteq \Box(-\{x\})$. Thus $x \in \Box(-\{x\})$, i.e. $x \notin \Diamond\{x\}$ — a contradiction. ∎

DEFINITION 5. Let $\mathcal{X} = (X, \Box)$ be a neighbourhood frame, $X_1 \subseteq X$. The *restriction of* $\mathcal{X}$ *to* X_1 (or the *subframe obtained by restriction to* X_1, notation: $\mathcal{X} \upharpoonright X_1$) is $\mathcal{X}_1 = (X_1, \Box_1)$, where $\Box_1 V := \Box V \cap X_1$ for $V \subseteq X_1$. X_1 (and $\mathcal{X}_1$) is called *open* if X_1 is open.

LEMMA 6. *Let* $\mathcal{X}_1$ *be an open subframe of a* **K4**-*frame* $\mathcal{X}$. *Let* ψ *be a valuation in* $\mathcal{X}$, ψ_1 *a valuation in* $\mathcal{X}_1$ *such that* $\psi_1(s) = \psi(s) \cap X_1$ *for any* $s \in PV$. *Then for any modal formula* A, $\psi_1(A) = \psi(A) \cap X_1$.

Proof. Easy, by induction on A. Here is the induction step for $A = \Box B$: suppose $\psi_1(B) = \psi(B) \cap X_1$; then
$\psi_1(A) = \Box_1 \psi_1(B) = \Box_1(\psi(B) \cap X_1) = \Box(\psi(B) \cap X_1) \cap X_1 =$
$\Box\psi(B) \cap \Box X_1 \cap X_1 = \Box\psi(B) \cap X_1 = \psi(A) \cap X_1$. ■

LEMMA 7. *If* $\mathcal{X}_1 \subseteq \mathcal{X}$ *is open, then* $\mathbf{ML}(\mathcal{X}) \subseteq \mathbf{ML}(\mathcal{X}_1)$.

Proof. By Lemma 6, if $(\mathcal{X}_1, \psi_1) \not\models A$, then $(\mathcal{X}, \psi) \not\models A$, where $\psi(s) = \psi_1(s)$ for any $s \in PV$. ■

DEFINITION 8. For a Kripke frame $F = (W, R)$ and a set $V \subseteq W$ we define the *subframe* $F \upharpoonright V := (V, R \cap (V \times V))$. A *subframe of (transitive)* F *generated by a world* x is $F^x := F \upharpoonright \overline{R}(x)$, where $\overline{R}(x) := R(x) \cup \{x\}$.

A frame F is called *rooted* with the root x if $F = F^x$, i.e. if $W = \overline{R}(x)$.

A *p-morphism* from a Kripke frame $F = (W, R)$ onto a Kripke frame $F' = (W', R')$ is a surjective map $f : W \longrightarrow W'$ such that for any $x \in W$

$$f(R(x)) = R'(f(x)).$$

$f : F \twoheadrightarrow F'$ denotes that f is a p-morphism from F onto F'. The following two lemmas are well-known.

LEMMA 9. (Generation Lemma)
$\mathbf{L}(F) = \bigcap\{\mathbf{L}(F^x) \mid x \in W\}$.

LEMMA 10. (P-morphism Lemma)
$f : F \twoheadrightarrow F'$ *implies* $\mathbf{ML}(F) \subseteq \mathbf{ML}(F')$ *(and* $\mathbf{IL}(F) \subseteq \mathbf{IL}(F')$ *if* F, F' *are* **S4**-*frames). More precisely, if* $f : F \twoheadrightarrow F'$ *and for every* $s \in PV$, $\varphi(s) = f^{-1}(\varphi'(s))$, *then*

$$(F, \varphi), x \models A \textit{ iff } (F', \varphi'), f(x) \models A$$

for any world x *and modal formula* A, *and similarly for the intutuionistic case.*

DEFINITION 11. A formula A (modal or intuitionistic) is a *logical consequence of a logic* $\mathbf{\Lambda}$ *(respectively, modal or intermediate) in neighbourhood semantics* (notation: $\mathbf{\Lambda} \models_N A$) if A is valid in all neighbourhood $\mathbf{\Lambda}$-frames.

Similarly, A is a *logical consequence of* $\mathbf{\Lambda}$ *in Kripke semantics* (notation: $\mathbf{\Lambda} \vDash_K A$) if A is valid in all Kripke $\mathbf{\Lambda}$-frames.

One can easily check the following

LEMMA 12.

(1) $C_K(\mathbf{\Lambda}) := \{A \mid \mathbf{\Lambda} \vDash_K A\}$ *is the smallest K-complete logic containing* $\mathbf{\Lambda}$.

(2) $C_N(\mathbf{\Lambda}) := \{A \mid \mathbf{\Lambda} \vDash_N A\}$ *is the smallest N-complete logic containing* $\mathbf{\Lambda}$.

So we have

$$\mathbf{\Lambda} \subseteq C_N(\mathbf{\Lambda}) \subseteq C_K(\mathbf{\Lambda}).$$

DEFINITION 13. A modal or intermediate logic $\mathbf{\Lambda}$ is called *relatively complete* if $C_N(\mathbf{\Lambda}) = C_K(\mathbf{\Lambda})$.

We can also consider *finitary logical consequence.*

DEFINITION 14. $\mathbf{\Lambda} \vDash_N^0 A$ if $\mathbf{\Lambda}_1 \vDash_N A$ for some finitely axiomatisable $\mathbf{\Lambda}_1 \subseteq \mathbf{\Lambda}$. The relation $\mathbf{\Lambda} \vDash_K^0 A$ is defined analogously.

Let
$C_K^0(\mathbf{\Lambda}) := \{A \mid \mathbf{\Lambda} \vDash_K^0 A\}$,
$C_N^0(\mathbf{\Lambda}) := \{A \mid \mathbf{\Lambda} \vDash_N^0 A\}$.
The following diagram is clear:

$$\begin{array}{ccc} \mathbf{\Lambda} \subseteq C_N^0(\mathbf{\Lambda}) & \subseteq & C_N(\mathbf{\Lambda}) \\ \cap\!| & & \cap\!| \\ C_K^0(\mathbf{\Lambda}) & \subseteq & C_K(\mathbf{\Lambda}) \end{array}$$

DEFINITION 15. A logic $\mathbf{\Lambda}$ is called

- *TK-compact* if $C_K^0(\mathbf{\Lambda}) = C_K(\mathbf{\Lambda})$,
- *TN-compact* if $C_N^0(\mathbf{\Lambda}) = C_N(\mathbf{\Lambda})$,
- *TKN-compact* if $C_N(\mathbf{\Lambda}) \subseteq C_K^0(\mathbf{\Lambda})$.

Obviously, every finitely axiomatisable logic is both TK-compact and TN-compact. There is also the following diagram of properties:

$$\begin{array}{ccccc} \text{K-completeness} & \Rightarrow & \text{TK-compactness} & \Rightarrow & \text{TKN-compactness} \\ \Downarrow & & & & \\ \text{N-completeness} & \Rightarrow & \text{TN-compactness} & \Rightarrow & \text{TKN-compactness} \end{array}$$

DEFINITION 16. A set of modal formulas Γ is called *satisfiable* in a frame F if there exists a model M over F and a world x such that $M, x \models A$ for every $A \in \Gamma$.

DEFINITION 17. Let $\mathbf{\Lambda}$ be a modal logic. A set of modal formulas Γ is called $\mathbf{\Lambda}$-*N-satisfiable* (respectively, $\mathbf{\Lambda}$-*K-satisfiable*) if it is satisfiable in some neighbourhood (respectively, Kripke) $\mathbf{\Lambda}$-frame. Γ is called *finitely* $\mathbf{\Lambda}$-*N-satisfiable* if every its finite subset is $\mathbf{\Lambda}$-N-satisfiable; the definition of finite $\mathbf{\Lambda}$-K-satisfiability is analogous.

DEFINITION 18. A modal logic $\mathbf{\Lambda}$ is called

- *N-compact* if every finitely $\mathbf{\Lambda}$-N-satisfiable set is $\mathbf{\Lambda}$-N-satisfiable,
- *strongly neighbourhood (SN-) complete* if it is both N-complete and N-compact.

K-compactness and SK-completeness are defined in a similar way.

An equivalent definition of SN-completeness is the following: every $\mathbf{\Lambda}$-consistent set of formulas is $\mathbf{\Lambda}$-N-satisfiable.

3 Ultrabouqets of topological spaces

The notion of an ultrabouqet exists in several versions, cf. [Shehtman, 1998; Shehtman, 1999]. Let us begin with the case of topological spaces.

DEFINITION 19. Let (X_n, x_n), $n \in \omega$ be sets with designated points. Their *bouqet* $\bigvee\limits_{n\in\omega} (X_n, x_n)$ is obtained from the disjoint union $\bigsqcup\limits_{n\in\omega} X_n$ by identifying all points x_n.

We denote the designated point of $\bigvee\limits_{n\in\omega} (X_n, x_n)$ by x_*.

DEFINITION 20. Let $\mathcal{X}_n = (X_n, \Box_n)$ be topological spaces, and for every n, let x_n be a closed point in $\mathcal{X}_n$. Let $\mathcal{U}$ an ultrafilter in ω. Then we define the *ultrabouqet* $\bigvee\limits_{\mathcal{U}}(\mathcal{X}_n, x_n)$ as the bouqet $\bigvee\limits_{n\in\omega} (X_n, x_n)$ with the topology, in which a subset V is open iff the following conditions hold:

(1) every part $V \cap (X_n - \{x_n\})$ is open;

(2) if $x_* \in V$, then $\{n \mid x_n \in \Box_n(V \cap X_n)\} \in \mathcal{U}$.

A particular case of this construction is when every $\mathcal{X}_n$ corresponds to a Kripke frame $F_n = (W_n, R_n)$ with root x_n and $R^{-1}(x_n) = \{x_n\}$. Then (1) and (2) can be written as follows:

(1) $R_n(V \cap (X_n - \{x_n\}) \subseteq V$;

(2) if $x_* \in V$, then $\{n \mid W_n \subseteq V\} \in \mathcal{U}$.

DEFINITION 21. Let $\mathcal{X}_n$, x_n be the same as in Definition 20, ψ_n a valuation in $\mathcal{X}_n$. Then we define the valuation $\psi = \bigvee_{\mathcal{U}} \psi_n$ in $\bigvee_{\mathcal{U}}(\mathcal{X}_n, x_n)$ as follows:
for any propositional variable s,
$x \in \psi(s)$ iff $x \in \psi_n(s)$ (whenever $x \in X_n - \{x_n\}$),
$x_* \in \psi(s)$ iff $\forall^\infty n\, x_n \in \psi_n(s)$,
where for a predicate $\mathcal{P}$, $\forall^\infty n\, \mathcal{P}(n)$ means[2] $\{n \mid \mathcal{P}(n)\} \in \mathcal{U}$.

LEMMA 22. *Let $\mathcal{X}_n$, x_n, ψ_n be the same as in Definition 21. Then for any modal formula A,*

(1) $x \in \psi(A)$ iff $x \in \psi_n(A)$ (for $x \in X_n - \{x_n\}$),

(2) $x_* \in \psi(A)$ iff $\forall^\infty n\, x_n \in \psi_n(A)$.

Proof. (1) Follows easily by induction on A; note that $(X - \{x_n\})$ is open both in $\mathcal{X}$ and $\mathcal{X}_n$.

(2) Also by induction, cf. [Shehtman, 1998, Lemma 5.5]. ∎

LEMMA 23. *Let $\mathcal{X}_n$, x_n be the same as in Definition 20, $(\mathcal{X}, x_*) = \bigvee_{\mathcal{U}}(\mathcal{X}_n, x_n)$. Then for any modal formula A,*

(1) $\mathcal{X}, x \vDash A$ iff $\mathcal{X}_n, x \vDash A$ (whenever $x \in X_n - \{x_n\}$),

(2) $\mathcal{X}, x_* \vDash A$ iff $\forall^\infty n\, \mathcal{X}_n, x_n \vDash A$.

Proof. (1) (Only if.) Assume $\mathcal{X}, x \vDash A$ and consider an arbitrary valuation ψ_n in $\mathcal{X}_n$. Let ψ be "the same" valuation in $\mathcal{X}$, i.e. $\psi(s) = \psi_n(s)$ for every $s \in PV$. By our assumption, $x \in \psi(A)$; hence $x \in \psi_n(A)$ by Lemma 22 (1).

(If.) Assume $\mathcal{X}_n, x \vDash A$ and consider an arbitrary valuation ψ in $\mathcal{X}$. Let ψ_n be its "restriction" to $\mathcal{X}_n$, i.e. $\psi_n(s) = \psi(s) \cap X_n$ for every $s \in PV$[3]. By assumption, $x \in \psi(A)$; hence $x \in \psi_n(A)$ by Lemma 22 (1).

(2) (Only if.) Assume $\mathcal{X}, x_* \vDash A$ and suppose $\forall^\infty n\, \mathcal{X}_n, x_n \vDash A$ does not hold. Since $\mathcal{U}$ is an ultrafilter, this implies $\forall^\infty n\, \mathcal{X}_n, x_n \nvDash A$. Then consider valuations ψ_n in $\mathcal{X}_n$ such that

- $x_n \notin \psi_n(A)$ if $\mathcal{X}_n, x_n \nvDash A$;

[2] $\forall^\infty$ is read as "for almost all".

[3] More precisely, this means: $x \in \psi(s)$ iff $x \in \psi_n(s)$ (for $x \in X - \{x_n\}$) and $x_* \in \psi(s)$ iff $x_n \in \psi_n(s)$.

- ψ_n is arbitrary otherwise.

Let $\psi = \bigvee_{\mathcal{U}} \psi_n$ (Definition 21); then $\forall^\infty n\, x_n \in \psi_n(\neg A)$ implies $x_* \in \psi(\neg A)$ by Lemma 22. This contradicts our assumption.

(2) (If.) Assume $\forall^\infty n\, \mathcal{X}_n, x_n \vDash A$ and consider an arbitrary valuation ψ in $\mathcal{X}$. For each n, take a valuation ψ_n in $\mathcal{X}_n$ such that $\psi_n(s) = \psi(s) \cap X_n$ for any $s \in PV$. Then $\psi = \bigvee_{\mathcal{U}} \psi_n$; in fact, $x_* \in \psi(s)$ iff $\forall n\, x_n \in \psi_n(s)$ iff $\exists n\, x_n \in \psi_n(s)$. So $x_* \in \psi(s)$ implies $\forall^\infty n\, x_n \in \psi_n(s)$, and $x_* \notin \psi(s)$ implies $\neg\forall^\infty n\, x_n \in \psi_n(s)$. By assumption, $\forall^\infty n\, x_n \in \psi_n(A)$; hence $x_* \in \psi(A)$ by Lemma 22 (2). ■

LEMMA 24. *Let $\mathcal{X}_n$, x_n, $\mathcal{X}$ be the same as in Lemma 23. Then for any intuitionistic formula A,*

(1) $\mathcal{X}, x \vDash A$ iff $\mathcal{X}_n, x \vDash A$ (whenever $x \in X_n - \{x_n\}$),

(2) $\mathcal{X}, x_* \vDash A$ iff $\forall^\infty n\, \mathcal{X}_n, x_n \vDash A$.

Proof. Follows readily from Lemma 23 and Corollary 3. ■

4 Ultrabouqets of K4-frames

Now let us extend the notion of an ultrabouqet to neighbourhood **K**4-frames.

DEFINITION 25. Let $\mathcal{X}_n = (X_n, \Box_n)$, $n \in \omega$ be a family of **K**4-frames, with designated closed points x_n which are all reflexive or all irreflexive. Let $(X, x_*) = \bigvee_{n\in\omega} (X_n, x_n)$ be the corresponding bouqet, $\mathcal{U}$ an ultrafilter in ω. For $V \subseteq X$ we put

$$V_n := V \cap X_n{}^4,$$
$$\Box V := V^1 \cup V^0,$$

where

$$V^1 := \bigcup_n (\Box_n V_n - \{x_n\}),$$

and

$$V^0 := \begin{cases} \{x_*\} & \text{if } \forall^\infty n\, x_n \in \Box_n V_n; \\ \varnothing & \textit{otherwise.} \end{cases}$$

The frame $(X, \Box)$ is called the *ultrabouqet* of the family $(\mathcal{X}_n, x_n)_{n\in\omega}$ w.r.t $\mathcal{U}$ and denoted by $\bigvee_{\mathcal{U}} (\mathcal{X}_n, x_n)$.

[4] More precisely, $y \in V_n$ iff ($y \neq x_n$ & $y \in V \cap X_n$ or $y = x_n$ & $x_* \in V$).

Note that in the reflexive case $x_* \in \Box V$ implies $x_* \in V$ (and thus x_* is reflexive). In fact, if $\forall^\infty n\, x_n \in \Box_n V_n$, then for some n, $x_n \in \Box_n V_n$; hence $x_n \in V_n$ by Lemma 4, i.e. $x_* \in V$.

DEFINITION 26. Let $\mathcal{X}_n$, x_n be the same as in Definition 25, and let ψ_n be valuations in $\mathcal{X}_n$. Consider the valuation $\psi = \bigvee_{\mathcal{U}} \psi_n$ in $\bigvee_{\mathcal{U}}(\mathcal{X}_n, x_n)$ such that for any $s \in PV$,
$x \in \psi(s)$ iff $x \in \psi_n(s)$ (whenever $x \in X_n - \{x_n\}$),
$x_* \in \psi(s)$ iff $\forall^\infty n\, x_n \in \psi_n(s)$.

LEMMA 27. *Let $\mathcal{X}_n$, x_n, ψ_n, ψ be the same as in Definition 26. Then for any modal formula A,*

(1) $x \in \psi(A)$ iff $x \in \psi_n(A)$ (for $x \in X_n - \{x_n\}$),

(2) $x_* \in \psi(A)$ iff $\forall^\infty n\, x_n \in \psi_n(A)$.

Proof. Both statements are proved by induction on the length of A. Let us consider the only nontrivial case: $A = \Box B$.

(1) We have:

$$x \in \psi(A) = \Box\psi(B) \text{ iff } x \in \psi(B)^1 \text{ iff } x \in \Box_n \psi(B)_n,$$

$$x \in \psi_n(A) \text{ iff } x \in \Box_n \psi_n(B).$$

Since x_n is closed, we also have $x \in \Box_n(X_n - \{x_n\})$, and thus

$$x \in \Box_n \psi(B)_n \text{ iff } x \in \Box_n(\psi(B)_n - \{x_n\}),$$

$$x \in \Box_n \psi_n(B) \text{ iff } x \in \Box_n(\psi_n(B) - \{x_n\}).$$

By induction hypothesis,

$$\psi(B)_n - \{x_n\} = \psi_n(B) - \{x_n\},$$

hence

$$x \in \psi(A) \text{ iff } x \in \psi_n(A).$$

(2) `Reflexive case.` By Definition 25 we have:

$$(\sharp) \quad x_* \in \psi(A) = \Box\psi(B) \text{ iff } \forall^\infty n\, x_n \in \Box_n \psi(B)_n.$$

Now assume $x_* \in \psi(A)$. By the remark after Definition 25 it follows that $x_* \in \psi(B)$, and thus for any n, $x_n \in \psi(B)_n$. Hence by induction hypothesis (1),

($\sharp\sharp$) $\psi(B)_n = \{x_n\} \cup (\psi(B)_n - \{x_n\}) = \{x_n\} \cup (\psi_n(B) - \{x_n\}) = \psi_n(B) \cup \{x_n\}$.

By induction hypothesis (2), $x_* \in \psi(B)$ implies $\forall^\infty n\ x_n \in \psi_n(B)$, and thus from ($\sharp\sharp$) we have

$$(\sharp\sharp\sharp) \quad \forall^\infty n\ \psi_n(B) = \psi(B)_n.$$

Eventually from ($\sharp\sharp$) and ($\sharp\sharp\sharp$) we obtain

($\natural$) $\forall^\infty n\, x_n \in \Box_n \psi_n(B) = \psi_n(A)$.

Conversely, assume ($\natural$). Then by Lemma 4, $\forall^\infty n\ x_n \in \psi_n(B)$, and thus $x_* \in \psi(B)$ by induction hypothesis (2). Hence by the same argument as above we obtain ($\sharp\sharp\sharp$). Now it follows that $\forall^\infty n\, x_n \in \Box_n \psi(B)_n$, which implies $x_* \in \psi(A)$ by ($\sharp$).

`Irreflexive case.`

By Definition 4,

$$x_* \in \psi(A) \text{ iff } \forall^\infty n\, x_n \in \Box_n \psi(B)_n.$$

By induction hypothesis,

$$\psi_n(B) - \{x_n\} = \psi(B)_n - \{x_n\},$$

hence

$$x_n \in \Box_n(\psi_n(B) - \{x_n\}) \text{ iff } x_n \in \Box_n(\psi(B)_n - \{x_n\}).$$

Since $\Diamond\{x_n\} = \varnothing$, we also have $x_n \in \Box_n(-\{x_n\})$, and thus

$$x_n \in \Box_n \psi_n(B) \text{ iff } x_n \in \Box_n \psi(B)_n.$$

This eventually implies

$$x_* \in \psi(A) \text{ iff } \forall^\infty n\, x_n \in \Box_n \psi_n(B) = \psi_n(A).$$

■

Hence similarly to Lemma 23, we obtain

LEMMA 28. *Let $\mathcal{X}_n$, x_n be the same as in Definition 25, $(\mathcal{X}, x_*) = \bigvee_{\mathcal{U}}(\mathcal{X}_n, x_n)$ Then for any modal formula A,*

(1) $\mathcal{X}, x \vDash A$ iff $\mathcal{X}_n, x \vDash A$ (whenever $x \in X_n - \{x_n\}$),

(2) $\mathcal{X}, x_* \vDash A$ iff $\forall^\infty n\, \mathcal{X}_n, x_n \vDash A$.

In particular, it follows that an ultrabouqet of **K4**-frames is a **K4**-frame.

5 TKN-noncompactness above Grz

DEFINITION 29. Let us define modal formulas β_n, γ_n by induction.

$$\begin{array}{ll} \beta_0 = \Box p, & \gamma_0 = \Box\neg p, \\ \beta_1 = \neg p \wedge \Diamond\beta_0 \wedge \neg\Diamond\gamma_0, & \gamma_1 = p \wedge \Diamond\gamma_0 \wedge \neg\Diamond\beta_0, \\ \beta_{n+1} = \Diamond\beta_n \wedge \Diamond\gamma_{n-1} \wedge \neg\Diamond\gamma_n, & \gamma_{n+1} = \Diamond\gamma_n \wedge \Diamond\beta_{n-1} \wedge \neg\Diamond\beta_n. \end{array}$$

Also let

$$\alpha_n = \Diamond\beta_{n+1} \wedge \Diamond\gamma_{n+1} \wedge \neg\Diamond\beta_{n+2} \wedge \neg\Diamond\gamma_{n+2},$$

$$\varepsilon_n = \Diamond\alpha_n \wedge \Diamond\beta_{n+2},\ \theta_n = \varepsilon_{n+1} \wedge \neg\Diamond\alpha_n,$$

$\delta_n = \varepsilon_n \to \Diamond\theta_n$, $\mathbf{\Lambda}_1 = \mathbf{Grz}+\{\delta_n \mid n \geq 0\}$, $\mathbf{\Lambda}_1^{(n)} = \mathbf{Grz}+\{\delta_m \mid n \geq m \geq 0\}$.

LEMMA 30. *The following formulas are* **S4***-theorems:*

(1) $\beta_n \to \Diamond\beta_m$ *for* $n \geq m \geq 0$,

(2) $\beta_n \to \Diamond\gamma_0$ *for* $n \geq 2$,

(3) $\varepsilon_n \to \Diamond\beta_1$ *for* $n \geq 0$.

Proof. By definition, $\mathbf{S4} \vdash \beta_n \to \Diamond\beta_{n-1}$. Hence by induction it follows that $\mathbf{S4} \vdash \beta_n \to \Diamond\beta_m$ for $n \geq m$. Since $\mathbf{S4} \vdash \beta_2 \to \Diamond\gamma_1$, $\gamma_1 \to \Diamond\gamma_0$ by definition, it follows that $\mathbf{S4} \vdash \beta_2 \to \Diamond\gamma_0$, and thus we obtain $\mathbf{S4} \vdash \beta_n \to \Diamond\gamma_0$ for $n \geq 2$.

For the proof of (3), note that $\mathbf{S4} \vdash \varepsilon_n \to \Diamond\alpha_n$, $\alpha_n \to \Diamond\beta_{n+1}$ by definition and $\mathbf{S4} \vdash \beta_{n+1} \to \Diamond\beta_1$ by (1). ∎

LEMMA 31. *A topological space* $\mathcal{X}$ *refutes AG iff there exist sets* $X_0, X_1 \subseteq X$ *such that*

$$X_0 \neq \varnothing,\ X_0 \cap X_1 = \varnothing,\ X_0 \subseteq \Diamond X_1,\ X_1 \subseteq \Diamond X_0.$$

Proof. (If.) Take a valuation φ such that $\varphi(p) = X_0$ and consider the model $(\mathcal{X}, \varphi)$. Then we have: $X_1 \subseteq \varphi(\neg p)$, $X_1 \subseteq \varphi(\Diamond p)$, and thus

$$X_0 \subseteq \varphi(\Diamond(\neg p \wedge \Diamond p)).$$

Hence $(\mathcal{X}, \varphi) \vDash p \to \Diamond(\Diamond p \wedge \neg p)$, and therefore

$$X_0 \subseteq \varphi(p \wedge \Box(p \to \Diamond(\neg p \wedge \Diamond p))).$$

Since $X_0 \neq \varnothing$, this implies that $\mathcal{X}$ refutes AG.

(Only if.) Consider a model $(\mathcal{X}, \varphi)$ and a point u such that

$$u \in \varphi(p \wedge \Box(p \to \Diamond(\neg p \wedge \Diamond p))).$$

Take a neighbourhood V of x such that $V \subseteq \varphi(p \to \Diamond(\neg p \wedge \Diamond p))$ and put $X_0 := \varphi(p) \cap V$, $X_1 := \varphi(\neg p \wedge \Diamond p) \cap V$. It follows that X_0, X_1 are the sets required. ■

Remark. This lemma means that AG behaves like a subframe formula for topological spaces: $\mathcal{X} \not\models AG$ iff there exists a subreduction from $\mathcal{X}$ onto a two-element cluster, i.e. an interior map from a subspace of $\mathcal{X}$ onto the two-element space with the weakest topology.

LEMMA 32. *A topological space $\mathcal{X}$ refutes AG iff there exist sets $Y_n \subseteq X$, $n \geq 0$ such that $Y_0 \neq \varnothing$ and for any n,*

$$Y_n \cap Y_{n+1} = \varnothing, \; Y_n \subseteq \Diamond Y_{n+1}.$$

Proof. (If.) Let $Y = \bigcup\limits_{n \in \omega} Y_n$. For $y \in Y$ put

$$m(y) := \min\{n \mid y \in Y_n\},$$

and let

$$X_0 := \{y \in Y \mid m(y) \text{ is even}\},$$
$$X_1 := \{y \in Y \mid m(y) \text{ is odd}\}.$$

Let us show that X_0, X_1 satisfy the conditions from the previous lemma. In fact, $X_0 \cap X_1 = \varnothing$ is trivial, and obviously, $X_0 \supseteq Y_0 \neq \varnothing$. It remains to prove by induction that

$$(*) \quad \forall n \forall y \in Y \; (m(y) = n \Rightarrow y \in \Diamond X_0 \cap \Diamond X_1),$$

i.e. that X_0, X_1 are dense in Y. In fact, assume that (*) holds for any $k < n$. Let $m(y) = n$, then $y \in Y_n \subseteq \Diamond Y_{n+1}$. If n is even, then obviously, $y \in X_0 \subseteq \Diamond X_0$; so we have to show that $y \in \Diamond X_1$.

Take an arbitrary neighbourhood V of y; then it contains a point $z \in Y_{n+1}$. Since $Y_n \cap Y_{n+1} = \varnothing$, we have either $m(z) = n+1$ (in which case $z \in X_1$, by definition), or $m(z) < n$. In the latter case $z \in \Diamond X_1$ by induction hypothesis, and thus $V \cap X_1 \neq \varnothing$. Hence $y \in \Diamond X_1$.

If n is odd, the argument is similar.

(Only if.) Take the sets X_0, X_1 from Lemma 31 and put

$$Y_n := \begin{cases} X_0 & \text{if } n \text{ is even;} \\ X_1 & \text{if } n \text{ is odd.} \end{cases}$$

■

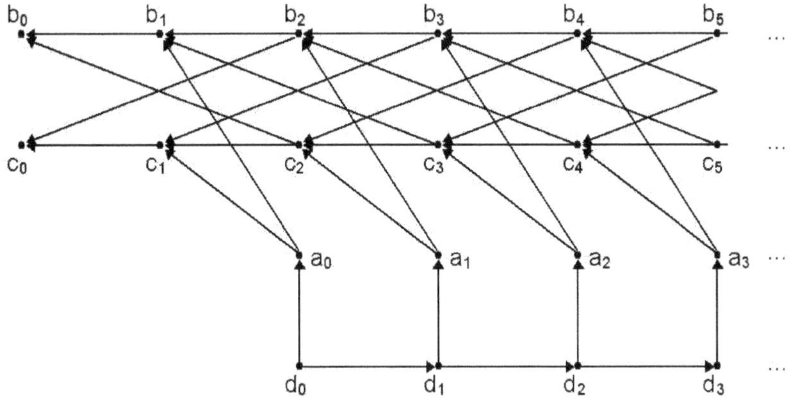

Figure 1.

LEMMA 33. $\neg\varepsilon_0 \in C_N(\mathbf{\Lambda}_1)$.

Proof. For a topological space $\mathcal{X}$, suppose $\mathcal{X} \vDash \mathbf{\Lambda}_1$, but $\varphi(\varepsilon_0) \neq \varnothing$ for some valuation φ in $\mathcal{X}$. Let $Y_n := \varphi(\theta_n)$. Then by definition, $Y_n \subseteq \varphi(\varepsilon_{n+1})$, and $\varphi(\varepsilon_{n+1}) \subseteq \Diamond Y_{n+1}$, since $\mathcal{X} \vDash \delta_{n+1}$. Hence $Y_n \subseteq \Diamond Y_{n+1}$.

On the other hand, $Y_n \cap Y_{n+1} = \varnothing$, since

$$Y_n \subseteq \varphi(\varepsilon_{n+1}) \subseteq \varphi(\Diamond\alpha_{n+1}),\ Y_{n+1} \subseteq \varphi(\neg\Diamond\alpha_{n+1}).$$

Finally, $\varnothing \neq \varphi(\varepsilon_0) \subseteq \Diamond Y_0$, since $\mathcal{X} \vDash \delta_0$, and thus $Y_0 \neq \varnothing$.

So by Lemma 32, $\mathcal{X} \nvDash AG$, which contradicts our assumption. ■

DEFINITION 34. $\Phi = (W, \leq)$ (Fine's frame) is the partially order set shown in Fig. 1.

DEFINITION 35. Let $\Phi_n = (W_n, \leq)$ be the restriction of Φ to the set $W_n := W - \{d_m \mid m \geq n+2\}$ (the n-truncated Fine's frame)[5].

LEMMA 36. *$\Phi_n \vDash \delta_m$ for any $m \leq n$.*

Proof. For an arbitrary model (Φ_n, φ), we have to prove that $\varphi(\varepsilon_m) \subseteq \varphi(\Diamond\theta_m)$. So let us assume $x \vDash \varepsilon_m$ and show that $x \vDash \Diamond\theta_m$ (in this model). First note that either (1) or (2) holds:

[5] To simplify notation, we use the same symbol $\leq$ for the relation in Φ_n.

(1) $b_0 \models p,\ c_0 \models \neg p$,
(2) $c_0 \models p,\ b_0 \models \neg p$.

In fact, if $b_0, c_0 \models p$, then $\varphi(\Box\neg p) = \varnothing$, since either b_0 or c_0 is accessible from any world of Φ_n. Thus $\varphi(\gamma_0) = \varnothing$, which implies $\varphi(\beta_k) = \varnothing$ for any $k \geq 2$) (remember that $\mathbf{S4} \vdash \beta_k \to \Diamond\gamma_0$ by Lemma 30). This contradicts $x \models \varepsilon_m$.

A similar argument shows, that p cannot be false at both $b_0,\ c_0$.

Next, if (1) holds, by induction we obtain that for any k

(3) $\varphi(\beta_k) = \{b_k\},\ \varphi(\gamma_k) = \{c_k\}$.

In fact, we obviously have

$$b_0 \in \varphi(\beta_0),\ c_0 \in \varphi(\gamma_0)$$

and thus $\varphi(\neg\Diamond\gamma_0) \subseteq \{b_0, b_1\}$. Since $b_0 \models p$, it follows that $\varphi(\beta_1) \subseteq \{b_1\}$.

On the other hand, $\Phi_n \models \varepsilon_m \to \Diamond\beta_1$ by Lemma 30 and soundness; thus $x \models \Diamond\beta_1$, and so $\varphi(\beta_1) \neq \varnothing$, and the only remaining option is $\varphi(\beta_1) = \{b_1\}$.

A similar argument shows that $\varphi(\gamma_1) = \{c_1\}$.

Now we can apply induction for the proof of (3); for the induction step note that for any y

$$y \models \beta_{k+1} \text{ iff } y \leq b_k \ \&\ y \leq c_{k-1} \ \&\ y \not\leq c_k \text{ iff } y = b_{k+1},$$

and similarly for $y \models \gamma_{k+1}$.

Next, if (2) holds, in the same way we obtain

(4) $\varphi(\beta_k) = \{c_k\},\ \varphi(\gamma_k) = \{b_k\}$.

Now since

$$y = a_k \text{ iff } y \leq b_{k+1} \ \&\ y \leq c_{k+1} \ \&\ y \not\leq b_{k+2} \ \&\ y \not\leq c_{k+2},$$

it follows that (in any case)

(5) $\varphi(\alpha_k) = \{a_k\}$.

By assumption, $x \models \varepsilon_m$, so we have

$$x \leq a_m \text{ and } (x \leq b_{m+2} \text{ or } x \leq c_{m+2}).$$

Therefore $x \leq d_m$.

But $m \leq n$, and thus $d_{m+1} \in \Phi_n$. It remains to note that $d_{m+1} \models \theta_m$. In fact, we can again apply (3), (4), (5): $d_{m+1} \models \varepsilon_{m+1}$, since $d_{m+1} \leq a_{m+1}, b_{m+3}, c_{m+3}$; at the same time $d_{m+1} \not\models \Diamond\alpha_m$, since $d_{m+1} \not\leq a_m$.

So we obtain $x \models \Diamond\theta_m$, as required. ■

LEMMA 37. $\Phi_n \models \mathbf{\Lambda}_1^{(n)}$

Proof. Since Φ_n is Nötherian, by Lemma 1 it follows that $\Phi_n \vDash AG$. The remaining axioms of $\mathbf{\Lambda}_1^{(n)}$ are valid, by Lemma 36. ■

LEMMA 38. $\Phi_n \nvDash \neg\varepsilon_0$

Proof. Take a valuation φ in Φ_n such that $\varphi(p) = \{b_0,\ c_1\}$. The same induction as in the proof of Lemma 36 shows that

$$\varphi(\beta_k) = \{b_k\},\ \varphi(\gamma_k) = \{c_k\}.$$

This implies $a_0 \in \varphi(\alpha_0)$, and thus $d_0 \in \varphi(\varepsilon_0)$. ■

THEOREM 39. *The logic $\mathbf{\Lambda}_1$ is TKN-noncompact.*

Proof. $\neg\varepsilon_0 \in C_N(\mathbf{\Lambda}_1)$, by Lemma 33.

On the other hand, $\neg\varepsilon_0 \notin C_K^0(\mathbf{\Lambda}_1)$. In fact, suppose $\mathbf{S4} + A \vDash_K \neg\varepsilon_0$ for some $A \in \mathbf{\Lambda}_1$. Then A is provable in **Grz** with a finite set of extra axioms, i.e. $A \in \mathbf{\Lambda}_1^{(n)}$ for some finite n. It follows that $\mathbf{\Lambda}_1^{(n)} \vDash_K \neg\varepsilon_0$, which contradicts Lemmas 37 and 38. ■

6 Relative incompleteness above Grz

Now let us slightly modify the counterexample from the previous Section to obtain another counterexample. We use the same special formulas as in Definition 29.

DEFINITION 40. Let

$$\delta'_n = \varepsilon_0 \to \Box\delta_n,\ \mathbf{\Lambda}_2 = \mathbf{Grz} + \{\delta'_n \mid n \geq 0\}.$$

LEMMA 41.

(1) $\Phi_n \vDash \delta'_m$ for any $m \leq n$.

(2) $\Phi_n, x \vDash \delta'_m$ for any $x \neq d_0$ and for any m.

Proof. Consider an arbitrary model $M = (\Phi_n, \varphi)$, and assume $M, x \vDash \varepsilon_0$. The proof of Lemma 36 shows that for any k, we have either

(3) $\varphi(\beta_k) = \{b_k\},\ \varphi(\gamma_k) = \{c_k\}$

or

(4) $\varphi(\beta_k) = \{c_k\},\ \varphi(\gamma_k) = \{b_k\}$,

and also

(5) $\varphi(\alpha_k) = \{a_k\}$.

Thus $x \leq a_0$ and either $x \leq b_2$ or $x \leq c_2$, which is possible only if $x = d_0$. So (2) follows readily.

The claim (1) is a trivial consequence of Lemma 36: $\Phi_n \vDash \delta_m$ for any $m \leq n$. ■

LEMMA 42. $\neg\varepsilon_0 \in C_K(\mathbf{\Lambda}_2)$.

Proof. Suppose there is a Kripke frame $F = (V, R)$ such that $F \vDash \mathbf{\Lambda}_2$, but $x_0 \vDash \varepsilon_0$ for some $x_0 \in V$ in some model over F. Then there exists an infinite ascending chain starting from x_0 such that $x_n R x_{n+1}$ and $x_n \vDash \varepsilon_n$. In fact, if $x_n \vDash \varepsilon_n$, we also have $x_n \vDash \delta_n$ $(= \varepsilon_n \to \Diamond\theta_n)$, and thus there exists $x_{n+1} \in R(x_n)$ such that $x_{n+1} \vDash \theta_n$ (and so $x_{n+1} \vDash \varepsilon_{n+1}$).

But then F is not Nötherian, which contradicts $F \vDash \mathbf{Grz}$. ■

LEMMA 43. *Let $\mathcal{U}$ be a non-principal ultrafilter on ω, $\mathcal{X} = \bigvee\limits_{\mathcal{U}}(\Phi_n, d_0)$. Then $\mathcal{X} \vDash \mathbf{\Lambda}_2$, but $\mathcal{X} \nvDash \neg\varepsilon_0$.*

Proof. Let x_0 be the root of $\mathcal{X}$. By Lemmas 23 and 41, we have: $\mathcal{X}, x \vDash \delta'_m$ for any $x \neq x_0$, for any m. We also have $\mathcal{X}, x_0 \vDash \delta'_m$ since $\{n \mid n \geq m\} \subseteq \{n \mid \Phi_n, d_0 \vDash \delta'_m\}$ and $\mathcal{U}$ is non-principal.

Thus $\mathcal{X} \vDash \delta'_m$.

As we know, every Φ_n validates $\mathbf{Grz}$ (Lemma 37), so $\mathcal{X} \vDash \mathbf{Grz}$, by Lemma 23.

On the other hand, ε_0 is satisfiable at Φ_n, d_0 for any n, so it is satisfiable at $\mathcal{X}, x_0$, by Lemma 22. Hence $\mathcal{X} \nvDash \neg\varepsilon_0$. ■

THEOREM 44. $\mathbf{\Lambda}_2$ *is relatively incomplete.*

Proof. In fact, $\neg\varepsilon_0 \in (C_K(\mathbf{\Lambda}_2) - C_N(\mathbf{\Lambda}_2))$ by Lemmas 42, 43. ■

We also have

THEOREM 45. $\mathbf{\Lambda}_2$ *is TK-noncompact.*

Proof. Almost the same as for Theorem 39. We already know that $\neg\varepsilon_0 \in C_K(\mathbf{\Lambda}_2)$. To show that $\neg\varepsilon_0 \notin C^0_K(\mathbf{\Lambda}_2)$, it suffices to note that $\mathbf{\Lambda}_2^{(n)} \nvDash_K \neg\varepsilon_0$, where

$$\mathbf{\Lambda}_2^{(n)} = \mathbf{Grz} + \{\delta'_m \mid 0 \leq m \leq n\}.$$

This follows from Lemmas 41, 38. ■

7 TK-noncompactness for intermediate logics

This Section is an intuitionistic analogue of Section 5. Let us first define some intuitionistic formulas.

DEFINITION 46.

$$\begin{array}{ll} B'_0 = \neg(p \wedge q), & C'_0 = \neg(\neg p \wedge q), \\ B'_1 = C'_0 \to B'_0 \vee q, & C'_1 = B'_0 \to C'_0 \vee p, \\ B'_{n+1} = C'_n \to B'_n \vee C'_{n-1}, & C'_{n+1} = B'_n \to C'_n \vee B'_{n-1} \end{array}$$

for $n \geq 0$.

Also let

$$A'_n = B'_{n+2} \wedge C'_{n+2} \to B'_{n+1} \vee C'_{n+1},$$

$$E'_n = A'_n \vee B'_{n+2},\ D'_n = (A'_n \to E'_{n+1}) \to E'_n,$$

$$\mathbf{\Lambda}_3 = \mathbf{H} + \{D'_n \mid n \geq 0\} + Br_2,$$

$$\mathbf{\Lambda}_3^{(n)} = \mathbf{H} + \{D'_m \mid n \geq m \geq 0\} + Br_2.$$

LEMMA 47. *If $m \leq n$, then*

$$\mathbf{H} \vdash B'_m \to B'_n,\ C'_m \to C'_n,\ B'_m \to C'_{n+2},\ C'_m \to B'_{n+2}.$$

Proof. By induction; note that $\mathbf{H} \vdash B'_m \to B'_{m+1},\ C'_m \to C'_{m+1}, B'_m \to C'_{m+2}, C'_m \to B'_{m+2}$, by definition. ■

DEFINITION 48. The frame $\Phi^- := \Phi \upharpoonright W^-$, where Φ is Fine's frame (Definition 34), $W^- := \{a_n \mid n \geq 0\} \cup \{d_n \mid n \geq 0\}$ is called the *willow*.

The willow is a tree shown in the picture.

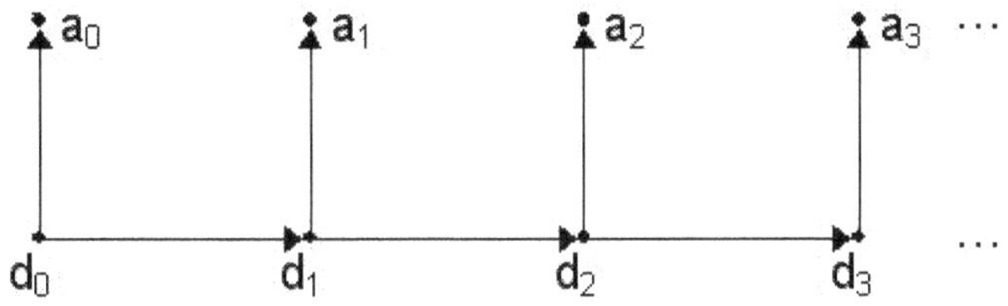

LEMMA 49. $E'_0 \in C_K(\mathbf{\Lambda}_3)$.

Proof. Analogous to the proof of Lemma 33. For a Kripke frame $F = (V, R)$ suppose $F \vDash \mathbf{\Lambda}_3$, while $\varphi(E'_0) \neq V$ for some model (F, φ). Let $e_0 \notin \varphi(E'_0)$. Let us construct a p-morphism from a subframe of F^{e_0} onto the willow Φ^-. This will give us a refutation of Br_2 in F.

Let

$$\mathcal{D}_n := \varphi(\bigwedge_{m<n} A'_m, E'_n),$$

$$\mathcal{A}_n := \varphi(\bigwedge_{m<n} A'_m \wedge B'_{n+2} \wedge C'_{n+2}, B'_{n+1} \wedge C'_{n+1}).$$

Next, for $x \in V$ let

$$N(x) := \{n \mid x \in R^{-1}(\mathcal{A}_n)\},$$

and also

$$\mathcal{A}_n^+ := \{x \in V \mid N(x) = \{n\}\}.$$

Now let us prove some auxiliary facts.

(1) $\mathcal{D}_n \subseteq R^{-1}(\mathcal{D}_{n+1})$.

In fact, for $x \in \mathcal{D}_n$ we have $x \not\Vdash E'_n$, and thus $x \not\Vdash A'_n \to E'_{n+1}$, since $F \Vdash D'_n$, by assumption. So for some $y \in R(x)$

$$y \Vdash A'_n \ \& \ y \not\Vdash E'_{n+1}.$$

Since $x \in \mathcal{A}_n$, we have $y \Vdash \bigwedge\limits_{m<n} A'_m$, and so $y \in \varphi(\bigwedge\limits_{m<n+1} A'_n, E'_{n+1}) = \mathcal{D}_{n+1}$.

(2) $k \leq n \Rightarrow \mathcal{D}_k \subseteq R^{-1}(\mathcal{D}_n)$.

This follows easily from (1) by induction.

(3) $\mathcal{D}_n \subseteq R^{-1}(\mathcal{A}_n)$.

In fact, $x \in \mathcal{D}_n$ implies $x \not\Vdash E'_n$, and thus $x \not\Vdash A'_n$. But then for some $y \in R(x)$

$$y \Vdash B'_{n+2} \wedge C'_{n+2},\ y \not\Vdash B'_{n+1} \vee C'_{n+1}.$$

Since $x \Vdash \bigwedge\limits_{m<n} A'_m$, it follows that $y \in \mathcal{A}_n$.

(4) $k \leq n \Rightarrow \mathcal{D}_k \subseteq R^{-1}(\mathcal{A}_n)$.

This follows from (2) and (3).

(5) If $x \in R^{-1}(\mathcal{A}_m) \cap R^{-1}(\mathcal{A}_k)$, $m > k$, then $x \in \mathcal{D}_n$ for some $n \leq k$.

In fact, assume $x \in R^{-1}(\mathcal{A}_m) \cap R^{-1}(\mathcal{A}_k)$. Since $y \not\Vdash A'_k$ for any $y \in \mathcal{A}_k$, we also have $x \not\Vdash A'_k$, $x \not\Vdash A'_m$, and so the set $S = \{l \mid x \not\Vdash A'_l\}$ is non-empty.

Let $n = \min \mathrm{S}$. Then obviously, $x \Vdash \bigwedge\limits_{i<n} A'_i$. On the other hand, $x \in R^{-1}(\mathcal{A}_m)$ implies $x \not\Vdash B'_{m+1}$, and thus $x \not\Vdash B'_{n+2}$ (since $n+2 \leq k+2 \leq m+1$ implies $\mathbf{H} \vdash B'_{n+2} \to B'_{m+1}$, by Lemma 47).

Since $n \in S$, we also have $x \not\Vdash A'_n$, and thus $x \in \mathcal{D}_n$.

(6) If $k \leq n$, then $\mathcal{D}_k \cap \mathcal{A}_n = \varnothing$.

In fact, assume $k \leq n$. Then $\mathcal{D}_k \subseteq R^{-1}(\mathcal{D}_{n+1})$, by (2). By definition,

$$\mathcal{D}_{n+1} \subseteq -\varphi(E'_{n+1}) \subseteq -\varphi(B'_{n+3}),$$

hence

$$\mathcal{D}_k \subseteq R^{-1}(\mathcal{D}_{n+1}) \subseteq -\varphi(B'_{n+3}).$$

On the other hand,

$$\mathcal{A}_n \subseteq \varphi(B'_{n+2}) \subseteq \varphi(B'_{n+3}),$$

by definition and since $\mathbf{H} \vdash B'_{n+2} \to B'_{n+3}$ (Lemma 47). This yields (6).

(7) $\mathcal{A}_n \subseteq \mathcal{A}_n^+$.

In fact, if $x \in \mathcal{A}_n$, then $n \in N(x)$, and thus by (5), either $N(x) = \{n\}$ or $x \in \mathcal{D}_k$ for some $k \leq n$. The latter contradicts (6).

(8) $R^{-1}(\mathcal{A}_n) = R^{-1}(\mathcal{A}_n^+)$.

This follows from the inclusions

$$\mathcal{A}_n \subseteq \mathcal{A}_n^+ \subseteq R^{-1}(\mathcal{A}_n).$$

(9) $R^{-1}(\mathcal{A}_n) \cap \mathcal{A}_m^+ = \varnothing$ if $m \neq n$.

This is obvious by definition.

(10) $\mathcal{D}_m \cap \mathcal{A}_n^+ = \varnothing$ for any m, n.

In fact, $\mathcal{D}_m \subseteq R^{-1}(\mathcal{A}_m)$ by (3), and also

$$\mathcal{D}_m \subseteq R^{-1}(\mathcal{D}_{m+1}) \subseteq R^{-1}(\mathcal{A}_{m+1})$$

by (1), (3). Now (10) follows from (9), since either $n \neq m$ or $n \neq m+1$.

(11) $\mathcal{D}_n \cap \mathcal{D}_m = \varnothing$ for $m \neq n$.

In fact, we may assume $m < n$. By definition, we have $\mathcal{D}_n \subseteq \varphi(A_m)$ and $\mathcal{D}_m \cap \varphi(A_m) = \varnothing$, whence (11) follows.

Now due to (5), (9), (10), (11), we obtain the following partition of the set $V_0 := \{x \mid N(x) \neq \varnothing\}$:

$$V_0 = \bigcup_{n \geq 0} \mathcal{A}_n^+ \cup \bigcup_{n \geq 0} \mathcal{D}_n$$

Then let us define a map $f : V_0 \longrightarrow W^-$ by putting

$$f(x) := \begin{cases} a_n & \text{if } x \in \mathcal{A}_n^+; \\ d_n & \text{if } x \in \mathcal{D}_n. \end{cases}$$

We claim that f is p-morphism from $F_0 := F \restriction V_0$ onto Φ^-.

In fact, if $f(x) = a_n$, then $x \in \mathcal{A}_n^+$, and so obviously, $R(x) \cap \mathcal{A}_m^+ = \varnothing$ for $m \neq n$; and also $R(x) \cap \mathcal{D}_m = \varnothing$ for any m, by (9). This means $f(R(x)) = \{a_n\} = \leq (f(x))$.

If $f(x) = d_n$, then $x \in \mathcal{D}_n$, and so $R(x)$ intersects every $\mathcal{D}_m$ for $m \geq n$, by (2), and thus every $\mathcal{A}_m^+$ for $m \geq n$, by (3) and (6). On the other hand, $x \notin R^{-1}(\mathcal{A}_m^+) = \varnothing$ for $m < n$.

In fact, $x \Vdash A'_m$, by the definition of $\mathcal{D}_m$, while $\mathcal{A}_m^+ \subseteq R^{-1}(\mathcal{A}_m) \subseteq -\varphi(A'_m)$, also by definitions. Thus $f(R(x)) = \leq (f(x))$, i.e. $f : F_0 \twoheadrightarrow \Phi^-$.

Since the willow itself is p-morphically mapped onto the tree $T_{3,2}$ (see the picture),

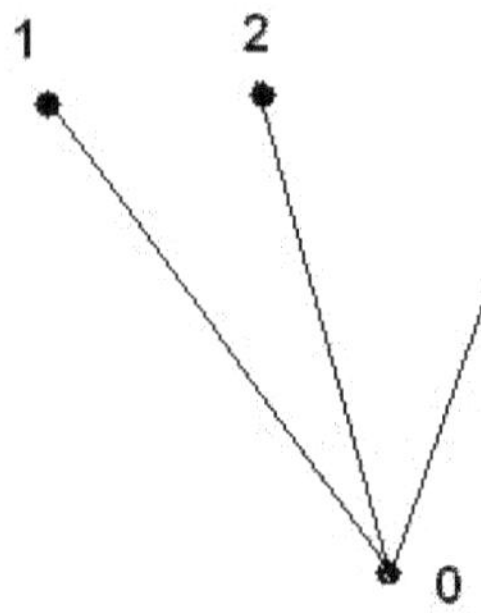

we have a combined p-morphism $g : F_0 \twoheadrightarrow T_{3,2}$, and from Lemma 10 it follows that Br_2 is refuted in F_0 under the valuation θ such that

$$\theta(p) = g^{-1}(1), \ \theta(q) = g^{-1}(2).$$

But we also obtain a refutation of Br_2 in (V, R) under the valuation θ' such that

$$\theta'(p) = \theta(p) \cup Z \text{ and } \theta'(q) = \theta(q) \cup Z,$$

where $Z := \{x \in V \mid N(x) = \varnothing\}$ (note that θ' is intutuionistic since $R(Z) \subseteq Z$). In fact, $\bigwedge\limits_{0 \leq i \leq 2} (P_i \to \bigvee\limits_{j \neq i} P_j)$ remains true at x, since p, q are true at all points of Z. ■

LEMMA 50. $\Phi_n \Vdash D'_m$ *for any* $m \leq n$.

Proof. Similar to Lemma 36. Assuming that $M = (\Phi_n, \varphi)$ is an intuitionistic Kripke model and $M, x \nVdash E'_m$, let us show that $x \nVdash A'_m \to E'_{m+1}$.

Our first claim is that either (1) or (2) below holds in M:

(1) $b_0 \Vdash p \wedge q, \ c_0 \Vdash \neg p \wedge q$,

(2) $c_0 \Vdash p \wedge q, \ b_0 \Vdash \neg p \wedge q$.

In fact, suppose $b_0 \nVdash p \wedge q$, $c_0 \nVdash p \wedge q$. Since either b_0 or c_0 is accessible from every world of Φ_n, we obtain $x \Vdash \neg(p \wedge q)$, i.e., $x \Vdash B'_0$. But $\mathbf{H} \vdash B'_0 \to B'_{m+2}$, by Lemma 47, hence $x \Vdash B'_{m+2}$, which contradicts $x \nVdash E'_m$. Thus $b_0 \Vdash p \wedge q$ or $c_0 \Vdash p \wedge q$.

In the same way from $\mathbf{H} \vdash C'_0 \to B'_{m+1}$ (Lemma 47) it follows that $b_0 \Vdash \neg p \wedge q$ or $c_0 \Vdash \neg p \wedge q$. Thus only two options, (1) or (2), are possible.

Now assume that (1) holds. Then we have:

(3) $\varphi^\bullet(B'_1) = \{b_1\}$.

Let us first show that $\varphi^\bullet(B'_1) \subseteq \{b_1\}$. In fact, if $y \in \varphi^\bullet(B'_1)$, then $y \Vdash C'_0$, and thus $y \not\leq c_0$ (since $c_0 \nVdash C'_0$), i.e., $y \in \{b_0, b_1\}$. But $b_0 \Vdash q$, thus $b_0 \Vdash B'_1$, and so $y = b_1$.

On the other hand, by our assumption, $x \not\Vdash E'_m$, so we have $x \not\Vdash B'_{m+2}$, and thus $x \not\Vdash B'_1$ (since $\mathbf{H} \vdash B'_1 \to B'_{m+2}$). Therefore $\varphi^\bullet(B'_1) \neq \varnothing$, and (3) follows.

(4) $\varphi^\bullet(B'_0) = \{b_0\}$.

In fact, by (1), $b_0 \Vdash p \wedge q$. On the other hand, $c_0 \not\Vdash p$ by (1) and $b_1 \not\Vdash p$ by (3); thus $y \not\Vdash p$ for any $y \neq b_0$. Hence (4) follows.

(5) $\varphi^\bullet(C'_1) = \{c_1\}$.

The proof is analogous to (3).

(6) $\varphi^\bullet(C'_0) = \{c_0\}$.

The proof is analogous to (4).

(7) for any k, $\varphi^\bullet(B'_k) = \{b_k\}$, $\varphi^\bullet(C'_k) = \{c_k\}$.

This follows by induction from (3)–(6) using the equivalence

$$y = b_{k+1} \text{ iff } y \leq b_k \;\&\; y \leq c_{k-1} \;\&\; y \not\leq c_k.$$

Next, if (2) holds, in the same way we obtain

(8) for any k, $\varphi^\bullet(B'_k) = \{c_k\}$, $\varphi^\bullet(C'_k) = \{b_k\}$.

Now by the same argument as in the proof of Lemma 36, we obtain (in both cases, (1) or (2)):

(9) for any k, $\varphi^\bullet(A'_k) = \{a_k\}$.

By our assumption, $M, x \not\Vdash E'_m$, so $x \leq a_m$, and also $x \leq b_{m+2}$ or $x \leq c_{m+2}$. So it follows that $x \leq d_m$. Next, since $m \leq n$, we have $d_{m+1} \in \Phi_n$ (Definition 35). By (9), (7), (8), $d_{m+1} \in \varphi(A'_m, E'_{m+1})$ — since $d_{m+1} \not\leq a_m$ and $d_{m+1} \leq a_{m+1}$, $d_{m+1} \leq b_{m+3}$, $d_{m+1} \leq c_{m+3}$. Therefore, $x \not\Vdash A'_m \to E'_{m+1}$. ■

LEMMA 51. $\Phi_n \Vdash \mathbf{\Lambda}_3^{(n)}$

Proof. Since Φ_n is of branching 2, by Lemma 1 it follows that $\Phi_n \Vdash Br_2$. The axioms D'_m are valid, by Lemma 50. ■

LEMMA 52. $\Phi_n, d_0 \not\Vdash E'_0$

Proof. Consider a valuation φ on Φ_n such that

$$\varphi(p) = \{b_0\},\ \varphi(q) = \{c_0\}.$$

By the same argument as in the proof of Lemma 50 it follows that for any k

$$\varphi^\bullet(B'_k) = \{b_k\},\ \varphi^\bullet(C'_k) = \{c_k\},\ \varphi^\bullet(A'_k) = \{a_k\}.$$

Since $d_0 \leq a_0$, $d_0 \leq b_2$, we obtain $d_0 \notin \varphi(E'_0)$. ■

THEOREM 53. $\mathbf{\Lambda}_3$ *is TK-noncompact.*

Proof. $E'_0 \in C_K(\mathbf{\Lambda}_3)$ by Lemma 49, and for any n $E'_0 \notin C_K(\mathbf{\Lambda}_3^{(n)})$, by Lemmas 51, 52. Then $E'_0 \notin C_K^0(\mathbf{\Lambda}_3)$, cf. the proof of Theorem 39. ■

8 Relative incompleteness for intermediate logics

This Section is an intuitionistic analogue of Section 5. Now we modify $\mathbf{\Lambda}_3$ to obtain a relatively incomplete logic.

DEFINITION 54. Let

$$D''_n := E'_0 \vee D'_n,\ \mathbf{\Lambda}_4 := \mathbf{H} + \{D''_n \mid n \geq 0\}.$$

LEMMA 55.

(1) $\Phi_n \Vdash D''_m$ *for any* $m \leq n$.

(2) $\Phi_n, x \Vdash D''_m$ *for any* $x \neq d_0$ *and for any* m.

Proof. (1) follows readily from Lemma 50.

To prove (2), let us show that $\Phi_n, x \Vdash E'_0$ for any $x \neq d_0$.

In fact, consider a model $M = (\Phi_n, \varphi)$ and suppose $M, x \not\Vdash E'_0$. Then according to the proof of Lemma 50, we obtain that either

$$\varphi^\bullet(B'_k) = \{b_k\},\ \varphi^\bullet(C'_k) = \{c_k\}$$

or

$$\varphi^\bullet(B'_k) = \{c_k\},\ \varphi^\bullet(C'_k) = \{b_k\},$$

and also

$$\varphi^\bullet(A'_k) = \{a_k\}.$$

Hence $x \leq a_0$ and either $x \leq b_2$ or $x \leq c_2$, which eventually implies $x = d_0$. This is a contradiction. ■

LEMMA 56. $E'_0 \in C_K(\mathbf{\Lambda}_4)$.

Proof. Similar to Lemma 49. Suppose $F = (V, R) \Vdash \mathbf{\Lambda}_4$ and $e_0 \notin \varphi(E'_0)$, then $(F, \varphi), e_0 \Vdash D'_n$ for any n. Next, note that the proof of Lemma 49 does not fully use the validity of D'_n; it actually yields that if $\forall n\, (F, \varphi) \Vdash D'_n$ and $\varphi(E'_0) \neq V$, then $F \not\Vdash Br_2$. This implies our assertion. ■

LEMMA 57. *Let* $(\mathcal{X}, x_0) = \bigvee_{\mathcal{U}} (\Phi_n, d_0)$ *be the same as in Lemma 43. Then* $\mathcal{X} \Vdash \mathbf{\Lambda}_4$, *but* $\mathcal{X} \not\Vdash E'_0$.

Proof. By Lemmas 24 and 55, we obtain that every D''_m is valid at any $x \neq x_0$. Since $\Phi_n, d_0 \Vdash D''_m$ for $n \geq m$, from Lemma 24 it also follows that $\mathcal{X}, x_0 \Vdash D''_m$. Since $\Phi_n, d_0 \not\Vdash E'_0$ by Lemma 52, we obtain $\mathcal{X}, x_0 \not\Vdash E'_0$, again by Lemma 24. ■

THEOREM 58. $\mathbf{\Lambda}_4$ *is relatively incomplete.*

Proof. By Lemmas 56, 57, we have $E'_0 \in (C_K(\mathbf{\Lambda}_4) - C_N(\mathbf{\Lambda}_4))$. ■

Remark $\mathbf{\Lambda}_4$ is also TK-noncompact, but $\mathbf{\Lambda}_3$ is slightly simpler.

9 N-compactness for transitive modal logics

DEFINITION 59. A neighbourhood **K4**-frame $\mathcal{X}$ is called *local* T_1 if the corresponding topological space $\mathcal{X}^+$ is local T_1 (in the sense of [Shehtman, 1998]), i.e. if every point is closed in some its neighbourhood.

LEMMA 60. *If $\mathcal{X}$ is a topological space and $\mathcal{X} \vDash AG$, then $\mathcal{X}$ is local T_1.*

Proof. Suppose the contrary, and let $x \in \mathcal{X}$ be a point such that $\Diamond\{x\} \cap U \neq \{x\}$ for any open $U \ni x$, i.e.

$$(\Diamond\{x\} - \{x\}) \cap U \neq \varnothing.$$

Hence

($\sharp$) $x \in \Diamond(\Diamond\{x\} - \{x\})$.

Now Lemma 42 show that $\mathcal{X} \nvDash AG$. In fact, take $X_0 = \{x\}$, $X_1 = (\Diamond\{x\} - \{x\})$; then $X_0 \subseteq \Diamond X_1$ by ($\sharp$), and obviously, $X_0 \neq \varnothing$, $X_0 \cap X_1 = \varnothing$, $X_1 \subseteq X_0$. ■

LEMMA 61.

Every **GL**-*frame is local* T_1.

Proof. By Lemma 60, since $\mathcal{X} \vDash AL$ implies $\mathcal{X}^+ \vDash AG$. The latter is rather well-known: the modality $\Box^+ A = \Box A \wedge A$ satisfies Grzecorczyk axiom if $\Box$ satisfies Löb axiom; the proof is either syntactical or by applying Kripke models. ■

THEOREM 62. *Let $\mathbf{\Lambda} \supseteq$ **K4** be a modal logic, S a set of modal formulas. If S is finitely $\mathbf{\Lambda}$-satisfiable in local T_1-frames, then S is $\mathbf{\Lambda}$-satisfiable.*

Proof. Similar to [Shehtman, 1999, Theorem 3.1]. Suppose $S = \{A_n \mid n \in \omega\}$, $B_n = \bigwedge_{i=0}^{n} A_i$. By assumption, there exists a local T_1 $\mathbf{\Lambda}$-frame $\mathcal{X}_n$, a valuation θ_n and a point x_n such that $(\mathcal{X}_n, \theta_n), x_n \models B_n$.

Let $\mathcal{Y}_n$ be an open subspace of $\mathcal{X}_n$, in which x_n is closed. By Lemma 7, $\mathcal{Y}_n \vDash \mathbf{\Lambda}$, and by Lemma 6, $\mathcal{Y}_n, \psi_n, x_n \vDash B_n$ for some valuation ψ_n. Now there are two cases.

`Case 1`. The set $\{n \mid x_n$ is reflexive in $\mathcal{Y}_n\}$ is infinite.

Let $\{n_1, n_1, \cdots,\}$ be the increasing enumeration of this set; then $n_k \geq k$, and obviously, $\mathcal{Y}_{n_k}, \psi_{n_k}, x_{n_k} \vDash B_k$. To simplify the notation, let $\mathcal{Z}_k = \mathcal{Y}_{n_k}$, $\varphi_k = \psi_{n_k}$, $z_k = x_{n_k}$; thus $z_k \in \varphi_k(B_k)$.

Take a non-principal ultrafilter $\mathcal{U}$ in ω, and consider the ultrabouqet $(\mathcal{Z}, z_*) = \bigvee_{\mathcal{U}}(\mathcal{Z}_n, z_n)$. Then $\mathcal{Z} \models \mathbf{\Lambda}$ by Lemma 28.

On the other hand, $z_k \in \phi_k(B_k)$ implies

$$\forall n \geq k\ z_n \in \varphi_n(A_k),$$

and thus

$$\forall^\infty n\ z_n \in \varphi_n(A_k),$$

since $\mathcal{U}$ is non-principal. Now take the valuation $\varphi = \bigvee_{\mathcal{U}} \varphi_n$. Then by Lemma 27, $z_* \in \varphi(A_k)$, and therefore $(\mathcal{Z}, \varphi), z_* \models S$.

Case 2. The set $\{n \mid x_n \text{ is reflexive}\}$ is finite. Then the set $\{n \mid x_n \text{ is irreflexive}\}$ is infinite, and we can repeat the same argument as in Case 1. ■

THEOREM 63.

(1) Every extension of **GL** *is N-compact.*

(2) Every extension of **Grz** *is N-compact.*

Proof. Follows readily from Theorem 62 and Lemmas 61, 60. ■

10 Final remarks

General theory of neighbourhood semantics and other modifications in Kripke semantics in modal logic is far beyond our understanding. However very interesting results on various kinds of semantics were recently obtained by T. Litak [2005], and this gives a hope for further perspectives.

Let us briefly discuss some topics and open problems related to this paper.

10.1 More counterexamples

Our logics $\mathbf{\Lambda}_1$, $\mathbf{\Lambda}_2$ are extensions of **Grz**; it is very likely that similar counterexamples can be constructed above **GL** and between **S4** and **Grz** (cf. [Rybakov, 1977] studying the same properties in Kripke semantics). Moreover, the methods from [Rybakov, 1977; Litak, 2002] allow us to construct a continuum of logics of this kind. However the following question seems more difficult:

Is it true that for any proper extension $\mathbf{\Lambda}$ *of* **S4** *the interval* $[\mathbf{S4}, \mathbf{\Lambda}]$ *contains uncountably many K-incomplete (N-incomplete, etc.) logics?*

Let us also recall another open problem (Kuznetsov, 1974):

Is every intermediate logic N-complete?

and two other related problems:

Is every intermediate logic N-compact?

Is every extension of **S4** *N-compact?*

As for the latter, one can slightly improve Theorem 63, because ultrabouqets can be defined for a larger class of spaces. In fact, let us call a point x in a topological space *weakly closed* if $\Diamond\{x\}$ is a "cluster", i.e. $y \in \Diamond\{x\}$ iff $x \in \Diamond\{y\}$. A space is *weakly local* T_1 if every its point is locally weakly closed. This class of spaces also allows for a certain ultrabouqet construction, and therefore Theorem 63 transfers to extensions of $\mathbf{Grz}_n$, the logic of all Kripke frames with clusters of cardinality $\leq n$. But this argument is not sufficient to cover all logics above **S4**.

10.2 Finitely axiomatisable incomplete logics

In this paper incomplete logics are TK-noncompact, and thus not finitely axiomatisable. But examples of incomplete finitely axiomatisable (f.a.) logics are also known.

N-incomplete f.a. extension of **S4** was first constructed in [Gerson, 1975b], a somewhat simpler N-incomplete f.a. logic above **Grz** can be found in [Shehtman, 1980]; it is obtained as $\mathbf{Grz} + D_0^T$, for D_0 defined below.

A K-incomplete intermediate logic is constructed in [Shehtman, 1977]; see also [Chagrov and Zakharyaschev, 1997, Ch. 6]. The same logic happens to be relatively incomplete, but the proof is quite complicated [Shehtman, 1980]. For the reader's conveneience, let us recall some details of this construction. The basic formulas are almost the same as in Definition 46:

$$\begin{array}{ll} B_0 = q \to p, & C_0 = p \to q, \\ B_1 = C_0 \to B_0 \vee q, & C_1 = B_0 \to C_0 \vee p, \\ B_{n+1} = C_n \to B_n \vee C_{n-1}, & C_{n+1} = B_n \to C_n \vee B_{n-1}. \end{array}$$

Also let

$$A_n = B_{n+2} \wedge C_{n+2} \to B_{n+1} \vee C_{n+1},$$

$$E_n = A_n \vee B_{n+2}, \ D_n = (A_n \to E_{n+1}) \to E_n,$$

$$\mathbf{\Lambda}_5 = \mathbf{H} + D_0 + Br_2.$$

Then the logic $\mathbf{\Lambda}_5$ is relatively incomplete; namely,

$$E_0 \in C_K(\mathbf{\Lambda}_5) - C_N(\mathbf{\Lambda}_5).$$

The first part $\mathbf{\Lambda}_5 \models_K E_0$ is proved similarly to Lemma 43. To prove $\mathbf{\Lambda}_5 \not\models_N E_0$, we have to construct a $\mathbf{\Lambda}_5$-space $\mathcal{Y}$ such that $\mathcal{Y} \not\Vdash E_0$.

This construction is nontrivial. The space $\mathcal{Y}$ is obtained from Fine's frame Φ by adding a continuum of extra points, in order to make Br_2 valid. Namely, consider partitions $e = (S_1, S_2, S_3)$ of ω with infinite members (we call them just 'partitions'). A filter $\mathcal{F}$ is called *subordinate* to e if

- $\mathcal{F}$ contains all cofinite subsets of ω,
- $-S_1, -S_2 \notin \mathcal{F}$,
- $-S_3 \in \mathcal{F}$.

One can show that the set of filters subordinate to e is non-empty and satisfies the conditions of Zorn Lemma. So let $\mathcal{F}(e)$ be a maximal element of this set. Let $Y = W \cup \mathcal{E} \cup \mathcal{E}'$, where $\mathcal{E}$ is the set of all partitions, $\mathcal{E}'$ is a copy of $\mathcal{E}$ (more precisely, $\mathcal{E}' = \{e' \mid e \in \mathcal{E}\}$, where $e' = e \times \{\varnothing\}$). Let $\preceq$ be a certain well-ordering of $\mathcal{E}$. Then the space $\mathcal{Y}$ is Y with the topology, where a set V is open iff

- $\leq (V \cap W) \subseteq V$,
- $(\forall e \in \mathcal{E} \cap V)\ \{n \mid a_n \in V\} \in \mathcal{F}(e)$,
- $\forall e, f \in \mathcal{E}\ (e' \in V\ \&\ f \preceq e \Rightarrow f \in V\ \&\ f' \in V)$.

($\leq$ denotes the original relation in Φ).

The question, whether there exists a simpler (say, countable) counterexample of this kind, remains open. Here is another question:

Do there exist f.a. logics that are N-complete, but K-incomplete?

For instance, one can try to axiomatise the ultrabouqets from Sections 4, 6 or the above defined space $\mathcal{Y}$.

10.3 Löwenheim – Skolem property

Classical first order logical consequence does not distinguish between infinite cardinalities: a theory with an infinite model always has a countable model. Unlike this, the relation $\models_K$ in modal or intuitionistic logic is quite sensible to cardinality, as the results by S.K. Thomason, A. Chagrov and M. Kracht show, see [Thomason, 1975]; [Chagrov and Zakharyaschev, 1997, Theorem 6.35]; [Kracht, 1999]. What happens in neighbourhood semantics in this respect, is still unclear:

Does there exist an N-complete modal logic that is not determined by any countable neighbourhood frame?

Acknowledgements

The work on this paper was supported by the Russian Foundation for Basic Research, Projects No. 02-01-22003, 02-01-01041, by ECO-NET 2004 project No. 08111TL, and by "Jumellage" (Twinship) program CNRS-IUM.

BIBLIOGRAPHY

[Aiello, 2002] M. Aiello. Spatial reasoning: theory and practice, ILLC Dissertation series, no. 2002-02. University of Amsterdam, 2002.

[Chagrov and Zakharyaschev, 1997] A. Chagrov and M. Zakharyaschev. Modal logic. Oxford University Press, 1997.

[Fine, 1974] K. Fine. An incomplete logic containing **S4**. Theoria, v.40 (1974), No.1, 23-29.

[Gabbay, 1975] D. Gabbay. A modal logic that is complete for neighbourhood frames but not in Kripke frames. Theoria , v.41 (1975), No. 3, 148-153.

[Gabbay and de Jongh, 1974] D. Gabbay, D. de Jongh. Sequence of decidable finitely axiomatizable intermediate logics with the disjunction property. Journal of Symbolic Logic, v.39 (1974), 67-78.

[Gerson, 1975a] M. Gerson. An extension of **S4** complete for the neighbourhood semantics but incomplete for the relational semantics. Studia Logica, 34:333–342, 1975.

[Gerson, 1975b] M. Gerson. The inadequacy of neighbourhood semantics for modal logics. Journal of Symbolic Logic, 40:141–147, 1975.

[Kracht, 1999] M. Kracht. Modal logics that need very large frames. Notre Dame Journal of Formal Logic, v. 40(1999), 141 - 173.

[Litak, 2002] T. Litak. A continuum of incomplete intewrmediate logics. Reports on Mathematical Logic v. 36 (2002), 131–141.

[Litak, 2005] T. Litak. An algebraic approach to incompleteness in modal logic. PhD Thesis. Japan Advanced Institute of Science and Technology, 2005. http://www.jaist.ac.jp/ litak/papers/myphd.pdf

[Rybakov, 1977] V. Rybakov. Non-compact extensions of the logic **S4**. Algebra and Logic, v. 18 (1977), 472-490. (In Russian)

[Shehtman, 1977] V. Shehtman. On incomplete propositional logics. Doklady AN SSSR, v. 235(1977), No. 3, 542-545 (In Russian).

[Shehtman, 1980] V. Shehtman. Topological models of propositional logics. In: Semiotika i informatika, No. 15 (1980), 74-98. (In Russian)

[Shehtman, 1998] V. Shehtman. On strong neighbourhood completeness of modal and intermediate propositional logics (Part I). In: M. Kracht et al. (eds). Advances in Modal Logic, v. 1. CSLI Publications, Stanford, 1998, pp. 209-222.

[Shehtman, 1999] V. Shehtman. On strong neighbourhood completeness of modal and intermediate propositional logics, II. In: JFAK. Essays dedicated to Johan Van Benthem on the occasion of his 50th birthday. Ed. by J. Gerbrandy et al. Vossiuspers, Amsterdam University Press, 1999 (CD).

[Shehtman, 2000] V. Shehtman. Modal logics of topological spaces. Habilitation Thesis, Moscow, 2000.

[Thomason, 1972] S.K. Thomason. Noncompactness in propositional modal logic. Journal of Symb. Logic, v.37(1972), 716-720.

[Thomason, 1975] S.K. Thomason. The logical consequence relation of propositional tense logic. Zeitschrift für Math. Logik und Grundlagen der Math., Bd.21(1975), 29-40.

[van Benthem and Sarenac, 2004] J. van Benthem and D. Sarenac. The geometry of knowledge. ILLC Prepublication Series, 2004, PP 2004/20. http://www.illc.uva.nl/Publications

Psychological Nature of Verification of Informal Mathematical Proofs

Patrick Suppes

It is a pleasure to dedicate this article to Dov Gabbay on the occasion of his sixtieth birthday. I still have fond memories of him as a very young person, when I met him for the first time four decades ago at Stanford.

That discovering or finding proofs is essentially a psychological process is widely recognized. Distinguished mathematicians such as Poincare and Hadamard have written very personal statements about their experience of discovery. There is, in fact, no serious body of opinion disputing this matter.

The question of verification is very different. On the one side we have the important concept of a formal proof, well developed and used in logic and foundational studies of mathematics. The essential character of such proofs is their being verifiable by an algorithm, many of which have now been implemented on digital computers and used in teaching to check student proofs. There have also been important research applications. A well-known example is the computer analysis of a large number of cases in the proof of the four-color theorem. The complete proof was originally not, and I believe, there is still not, a published formal proof of the theorem itself. But there is already a smattering of research articles focused on theorems that have formal proofs.

My focus here is on the informal proofs that still dominate the research literature and undoubtedly will do so for the foreseeable future. What is the basis for saying that an informal proof is valid? It cannot be that it has been checked by some familiar algorithm of formal verification or computation. Certainly some parts will often have this character, but all those informal proofs that have not been formalized, but are judged correct, must have a different basis. What is it?

The familiar and almost standard answer is an appeal to understanding, a concept notable for its psychological vagueness. In saying this, I do not mean to suggest that when a mathematician says that he understands a proof, and on a another occasion, in reference to a another proof that he does not understand it, that on both occasions he is just talking nonsense. In both cases I accept that usually there is something with content, and indeed, correctness about his or her state of mind. Moreover, the reference to state of mind does not mean a purely subjective claim is being made. It has the same status as all kinds of other empirical claims we make to each other and to ourselves all the time. "This floor is dirty", "That door is cracked", "That legal brief is not well written", "That point seems irrelevant", and so on endlessly. What is important in the present case is that in traditional discussions of proofs, before Hilbert's formal theory of them, mathematicians liked to talk about understanding. An *apriori* concept of understanding was made a central part of Kantian philosophy, but even though Kant's views were influential, they are certainly not part of the mainstream thinking about proofs in recent decades. Useful philosophical discussion of the desirability of a stable, even if not, *apriori*, concept of understanding can be found in Friedman [1974] and Kitcher [1976, 1982]. An excellent very recent analysis of proofs and understanding is to be found in Tappenden [2005]. He ends up with a view with which I am sympathetic. It is that there is no definite and sharp concept of understanding feasible, and perhaps not possible. There are too many ways of approaching the broad ill-defined concept as used in casual remarks by mathematicians to hope to glean from them a satisfactory exact concept, one suitable in itself for mathematical analysis. In general terms the situation is the same for the concept of an informal proof. Not a subject for direct, detailed formal analysis. Only the Hilbert-style formal proofs of mathematical logic, not the working informal proofs standard in all parts of mathematics, have an appropriate mathematical representation.

The need for definiteness and concreteness is succinctly expressed by Hilbert in the second paragraph of his well-known article "The Foundations of Mathematics":

> No more than any other science can mathematics be found by logic alone; rather, as a condition for the use of logical and inferences and the performance of logical operations, something must already be given to us in our faculty of representation [[in der Vorstellung]], certain extralogical concrete objects that are intuitively [[anschaulich]] present as immediate experience prior to all thought. If logical inference is to be reliable, it must be possible to survey these objects completely in all their parts, and the fact that they occur, that they differ from one another, and that they follow each other, or are concatenated, is immediately

> given intuitively, together with the objects, as something that neither can be reduced to anything else nor requires reduction. This is the basic philosophical position that I regard as requisite for mathematics and, in general, for all scientific thinking, understanding, and communication. [Hilbert, 1927, pp. 464–465]

Starting from physically printed symbols, or other familiar physical displays, as what is, in Hilbert's phrase, "given to us in our faculty of representation," we can ask what representation we form of that given—here I am taking the given to be the printed symbols—where symbols are printed letters of the alphabet as well as proper mathematical symbols. How these physical objects are represented takes me to my first psychological concept.

Mental representations. Let me begin, not at the beginning, but near it, with a quotation from Aristotle's *De Anima*, the best systematic work on psychology in ancient times:

> We must understand as true generally of every sense (1) that sense is that which is repetitive of the form of sensible objects without the matter, just as the wax receives the impression of the signet-ring without the iron or the gold, and receives the impression of the gold or bronze, but not as gold or bronze; so in every case sense is affected by that which has colour, or flavour, or sound, but by it, not *qua* having a particular identity, but *qua* having a certain quality, and in virtue of its formula; (2) the sense organ in its primary meaning is that in which this potentiality lies. [*De Anima*, 424a17–424a25]

To paraphrase the Aristotelian view, when we perceive a candle we receive exactly the form of the candle, but not the matter. We have then a mental representation that is approximately isomorphic to the physical candle. In many ways Aristotle's concept of form is close to the modern definition of isomorphism, abstracting as it always does from the particular physical realization, or, put another way, the non-relevant properties or relations of the structures that are isomorphic.

Introducing now, for the immediate context, abstract concepts that, in the usual parlance, do not have physical representation, Aristotle also has a place for them as intelligible forms of the intellect or mind, but I shall not pursue the details here, which are at some points obscure in Aristotle's text, but then often clarified, at least in part, in Thomas Aquinas' *Commentary on the De Anima.*

To mention only one other philosopher for whom similar concepts were important, even if there are many other differences in their theories of perception and intellect, I quote a passage from David Hume's *A Treatise of Human Nature*:

> When we have been accustom'd to observe a constancy in certain impressions, and have found, that the perception of the sun or ocean, for instance, returns upon us after an absence or annihilation with like parts and in like order, as at its first appearance, we are not apt to regard these interrupted perceptions as different, (which they really are) but on the contrary consider them as individually the same, upon account of their resemblance. But as this interruption of their existence is contrary to their perfect identity, and makes us regard the first impression as annihilated, and the second as newly created, we find ourselves somewhat at a loss, and are involv'd in a kind of contradiction. In order to free ourselves from this difficulty, we disguise, as much as possible, the interruption, or rather remove it entirely, by supposing that these interrupted perceptions are connected by a real existence, of which we are insensible. This supposition, or idea of continu'd existence, acquires a force and vivacity from the memory of these broken impressions, and from that propensity, which they give us, to suppose them the same; and according to the precedent reasoning, the very essence of belief consists in the force and vivacity of the conception. [*Treatise, p. 199*]

Here the concept of resemblance, which is Hume's term for similarity, or in more mathematical terms, isomorphism, is used to construct the concept of an individual from many resembling perceptions, as he puts the matter a few pages later:

> Thus the principle of individuation is nothing but the *invariableness* and *uninterruptedness* of any object, thro' a suppos'd variation of time, by which the mind can trace it in the different periods of its existence, without any break of the view, and without being oblig'd to form the idea of multiplicity or number. [*Treatise*, p. 201]

From a psychological standpoint, Hume's "construction" of an individual is much more complicated, and also more realistic than Aristotle's, as an account of how the perception of physical objects or processes is taking place. Hume's fundamental principle remains sound in all essential respects. Spelling out the details has usefully occupied several generations of psychologists.

Here is Hume's summarizing passage on how we come to have the mental concept of an individual back of the perceptions that are immediate:

> Our memory presents us with a vast number of instances of perceptions perfectly resembling each other, that return at different distances of time, and after considerable interruptions. This resemblance gives us a propension to consider these interrupted perceptions as the same; and also a propension to connect them by a continu'd existence, in order to justify this identity, and avoid the contradiction, in which the interrupted appearance of these perceptions seems necessarily to involve us. Here then we have a propensity to feign the continu'd existence of all sensible objects; and as this propensity arises from some lively impressions of the memory, it bestows a vivacity on that fiction; or in other words, makes us believe the continu'd existence of body. If sometimes we ascribe a continu'd existence to objects, which are perfectly new to us, and of whose constancy and coherence we have no experience, 'tis because the manner, in which they present themselves to our senses, resembles that of constant and coherent objects; and this resemblance is a source of reasoning and analogy, and leads us to attribute the same qualities to the similar objects. [*Treatise*, pp. 208–209]

I move now from Hume to the modern psychological literature on mental representation. The work I turn to will not say anything about the mental representation of informal proofs, the topic, I want to reach after a few more paragraphs. The strongest claims in the psychological literature about representation are, not surprisingly for mental images. Here is a working definition from *Principles of Mental Imagery* [1999] by Ronald A. Finke, an excellent book that reviews and summarizes many of the experiments and some of the philosophical literature on mental imagery.

> Mental imagery is defined as *the mental invention or recreation of an experience that in at least some respects resembles the experience of actually perceiving an object or an event, either in conjunction with, or in the absence of, direct sensory stimulation.* [Finke, 1989, p. 2]

Finke formulates several principles that hold for the relation between physical objects and their mental images, principles that he supports by many citations from the experimental work on perception. The three principles most pertinent to the present analysis are the principle of perceptual equivalence, the principle of spatial equivalence, and the principle of transformational equivalence. What they amount to is a strong thesis of structural isomorphism between physical objects and their mental images, or, to put it more generally, as Finke does, between experiences and their mental images. I don't agree with all parts of Finke's argument, but do with the major aspects, already anticipated in the texts of Aristotle and Hume cited above. Readers skeptical of this claim of structural isomorphism should read Finke's presentation of a large number of supporting experimental results.

In some respects my thesis is stronger than Finke's, for I also claim such a structural isomorphism for mental representations of language, but I will present this a little later more directly in terms of brain images.

Now broadly what I have to say about informal proofs at the psychological level is this. First, the psychological aspect is essential, because it is a mark of an informal proof that the steps in it have formal gaps. No algorithm of proof verification has been explicitly used to cover these gaps. So the basis of accepting the gaps is the implicit psychological claim that the mathematician offering the proof as a correct one has a good "understanding" of what would be needed to fill in the gaps. The usual remark would be, "Filling in those gaps is, of course, trivial. Everyone who is competent in this area of mathematics recognizes the correctness of the steps taken in the informal proof."

This summarizing remark states an important fact about standard informal proofs that I do not in any way want to challenge. An excellent defense of their importance, even in discussions of foundations, is to be found in a well-known article of Kreisel [1967], entitled "Informal Rigour and Completeness Proofs". Two other useful articles are Kreisel [1981a/1981b]. Kreisel stresses the experience and developed intuition of mathematicians specializing in any area of mathematics are what makes their informal proofs reliable and valid. His exposition of these matters is clear and useful, and I have nothing directly to add to what he says.

But Kreisel does not move at all in the direction of examining the psychological processes by which experience trains the mind, so to speak. This is not the place to attempt a serious presentation of the psychological theories that have been developed for learning from experience. But the past experience of expert mathematicians is just one of the great examples, of which many can be given for every significant human activity, of such learning. There is particular empirical work in psychology especially relevant to my focus here. I have in mind the learning of concepts. Mathematicians and philosophers talk often and with ease about learning a great variety of concepts, and how some are almost always more difficult to learn than others.

However, it is not part of their thinking and intellectual culture to ask for detailed theories of how concepts are learned, in the sense of having systematic, testable theories about such learning, and even less, experiments, which they would generally regard as outside their domain. Many questions about concepts arise that are of interest to everybody, but do not have simple answers, certainly not ones to be found within pure mathematics itself.

There are currently three different theoretical approaches in psychology to the mental representation of concepts. One is the classical mathematical and philosophical one of definability of a concept in terms of others already given. Certainly one of the great triumphs of foundational studies of mathematics is the unintuitive and surprising result that most standard mathematical concepts can be defined within axiomatic set theory, which has little more than the concept of set membership as its primitive concept. But the concepts of a less formal nature are not susceptible to such a radical reduction. So the second approach is the use of prototypes. How do I know a cow I see across the road is a cow. I compare the perceptual image I have as I look at it to one or more prototype images and by that internal comparison decide, for example, it is a cow, not a horse. Of course, my description of the use of prototypes in the process of recognition is much too sketchy, for any skeptic to take it seriously. I mean here only to convey the general idea. For a good survey of prototype theory, as well as the other two, see Murphy [2002].

The third approach is the exemplar one, made famous in the history of philosophy by Berkeley [1710], in his telling but humorous criticisms of John Locke's theory [1690] of concepts. Hume [1739] agrees with Berkeley and nicely summarizes the exemplar theory in the following passage:

> A great philosopher [Berkeley] has disputed the receiv'd opinion in this particular, and has asserted, that all general ideas are nothing but particular ones, annexed to a certain term, which gives them a more extensive signification, and makes them recall upon occasion other individuals, which are similar to them. [Hume, p .17]

Some direct support for this exemplar theory is found in a study of mine with my colleagues of brain representations in terms of the electromagnetic waves generated by presentation of simple visual images to subjects, for example, simple shapes such as a triangle or a square, and full screens of a single color such as red or green. For English speakers, evidence supporting Hume's "annexing to a term" was supported; for example, the brain response to a line drawing of a triangle was similar to that for the stimulus word *triangle*. Six of the ten stimuli of shapes and colors, as prototypes, and the corresponding stimuli of auditory words, as test samples, are shown in Figure 1. The experiment is reported in detail and the figure is reproduced from Suppes, P., Han, Epelboim, and Lu [1999]. (An unpublished experiment replicated these results for Mandarin speakers.) Use of the word *prototype* just above does not mean this is a prototype theory; in this experimental context a prototype is an average of many brain waves activated on many different trials by a particular stimulus word such as *triangle*.

This brief excursion to a brain experiment relevant to the exemplar theory of concepts should not be misunderstood. It does not, and is not meant, to support the claim that the concept of a triangle is represented just by the word *triangle*. It is meant to help establish that there is no accessible definition or prototype for such concepts as that of a triangle. Rather, the common man has associated a great variety of instances, i.e., exemplars, and these are attached by association to the word *triangle*, which, in one plausible theory, provides a content-address in memory for them.

This is not the place to review the complicated competing analyses in philosophy for the correct characterization of the mental representation of a concept like that of triangle or group (in the mathematical sense). Suffice it to say, what is on standard offer is not able to give any serious psychological account of the thinking relevant to verifying the correctness of informal proofs. Moreover, I do not think I can take the subject very much farther either, in any detail. But I do want to sketch an application of associative networks that has some promise, and has some ingredients that are reasonable conjectures about how the brain is producing the underlying computations that lead to affirming as correct the successive steps in an informal proof. I concentrate on some simpler examples, namely, the psychological (and underlying brain) associative processes of judging as true or false ordinary empirical sentences. (Remember, this is the sketch of a conjectured theory, but one I think promising.)

The theory of the computation of such truths, i.e., the truths of ordinary empirical statements, is an important aspect of how the mind works. It is by no means anything like the whole story of the computations in which the mind, or the brain, is involved. Very much more is involved in even the simplest computations of perception. Just think of the necessary computations to decide that the image from a certain perspective of a person 200 meters away is indeed an image of your oldest child. In many ways, what I shall have to say about computation will be much simpler than that of organizing such perceptual input to form beliefs about what I am seeing.

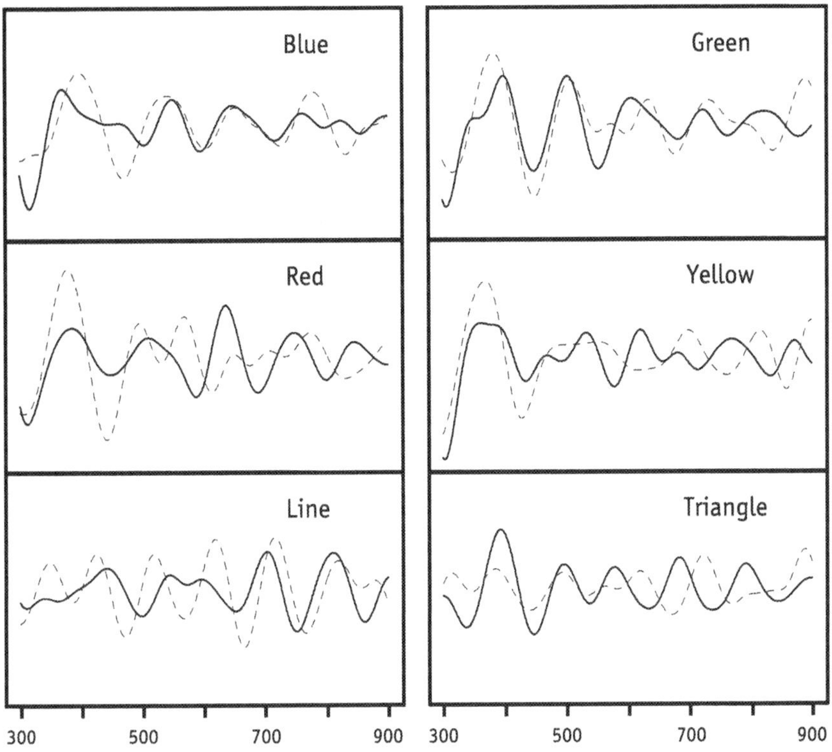

Figure 1. Comparison of averaged and filtered brain waves generated by visual images and auditory words. The six brain-wave prototypes (solid curved lines) were generated by the four color stimuli, named in the top four panels, and by the two visual shapes line and triangle in the bottom two panels. The test samples (dotted lines) were generated by the spoken names of the six visual stimuli averaged over both speakers. No translations along the x axis to improve the fit were made. The x axis is measured in ms after the onset of the visual image or its spoken name. The y axis is measured in microvolts, but no scale is shown, because after averaging and filtering the scale is only of relative magnitude.

Put another way, because of the great importance of perception in all our activities, we are continually forced to make a dazzling array of computations in processing stimulus input that actually reaches the cortex of our brains as electromagnetic signals, reflecting a marked degree of abstraction from the vivid language of ordinary talk about processes and things. It is mind boggling when first encountered. Indeed, the electromagnetic signals that the cortex processes seem inherently more difficult to understand than the sensible forms of Aristotle's theory, as set forth in the *De Anima*.

In spite of my references to electromagnetic signals, discussion here will be at a still more abstract psychological level most of the time, but I will turn back on several occasions to the brain rather than the mind, because, historically, the literature on the mind is almost entirely absent any serious theory of computation. So, to continue this general point about computation and how the mind works, all the neurophysiological processes of perception that reduce observed features of things and processes to electromagnetic signals sent to the cortex are ignored in ordinary or philosophical talk about experience. And so modern philosophy of mind is not concerned with the details of how we learn about the ways in which phenomena in the world are connected. I choose the last word deliberately. The approach to computation about such things and processes, characteristic of our minds, was well recognized by Hume, the godfather of the central mechanism of association, and already foreshadowed by Aristotle. What I shall insist on here is the universal role of association as the main method of computation in the brain, (and in the mind, if you will) in dealing with ordinary experience.

The point is an important one, even if there is not space here to muster all the arguments that I think are relevant. I will make the following general point about computation. It is sometimes felt that a very clear criticism of behaviorism and connectionism in relation to much cognitive behavior is easily made. Rules play too central a role, it is claimed, to believe an associationist account could be correct. This, however, rests upon a deep mathematical misunderstanding of what can be done with quite simple methods of association or conditioning. It is evident from the many proofs that any computable function can be computed by a universal Turing machine, by a universal register machine, or by any of six or seven other devices. Very elementary primitive ideas are quite sufficient, once there is any method of recursion available, to prove that the basic device to do the computing can be quite simple in conception.

All of these remarks are a kind of prolegomena to what I have to say about the computation of the truth of ordinary empirical statements. One point I want to make is that I shall not, in this discussion of truth, distinguish between belief and truth. It is possible to be too zealous, from a philosophical standpoint, and not accept a discussion of the truth of ordinary statements, as opposed to 'Which ones do you believe to be true?' In fact, in much ordinary discourse, claims about belief are used to express doubts about truth, not as a separate point of positive emphasis.

Associative networks. Anyway, I want to give a sketch of the theory of how such ordinary computations of truth are made. In doing so, I draw on a recent article of mine with Jean-Yves Beziau [2005]. The basic idea is that the computations are made by an associative network with brain representations of words being the nodes and the links between being the associations. More generally, auditory, visual, and other kinds of brain images can also be nodes.

In the initial state, not all nodes are linked, and there are, in this simple formulation, just two states, *quiescent* and *active*. No learning or forgetting is considered. It is assumed, without being formulated here, that, after a given sentence is responded to as being either true or false, all the activated states return to quiescent. The axioms, which are not stated here, are formulated just for the evaluation of a single sentence, not for giving an account of how the process works over a longer period of time. The way to think about the networks introduced is that a person is asked to say whether a sentence about familiar phenomena is true or false. It is very natural to ask, and not to have a quibble about 'Do you believe this, even though you don't know whether it is true?' We take things that are so obvious everyone accepts them as true. An example, often used in experiments on these matters, are simple geography sentences. Here is one: *Warsaw is not the capital of Austria.* This sentence input comes from outside the associative network in the brain. I will consider only spoken words forming a sentence, although what is said also applies to visual presentation of words and sentences, as well. So, as the sentence is spoken, the sound pressure image of each word that comes to the ear is drastically transformed by a sequence of auditory computations leading to the auditory nerve fibers which conduct electromagnetic signals to the cortex. Such signals are examples of those mentioned earlier. In earlier work, as already discussed above, I have been much concerned with seeing if we can identify such brain signals as brain representations of words. Some additional references are Suppes, Lu, and Han [1997] and Suppes, Han, Epelboim, and Lu [1999a, 1999b].

We activate quiescent states, that is, we move them from the quiescent to the active state by using the energy for this activation from that brought into the cortex by the brain representation/indexbrain representation of the verbal stimulus input. With the activation of the brain representation of words by external stimuli, we next assume that the associations between activated brain representations are also activated by using this same source of energy. It seems too complicated to try to apply any principle of energy conservation as part of the analysis.

However, it is assumed in the theory that energy can be passed along from one node to another by a phenomenon characterized some decades ago in psychological research as *spreading activation.* For example, in a sentence about a city like Rome or Paris, some familiar properties are closely associated with these cities and the brain representation of these properties may well be activated shortly after the activation of the brain representations of the words *Rome* or *Paris*, even though the names of these properties, or verbal descriptions of them, did not occur in any current utterance. This is what goes under the heading of *spreading activation.* Some form of it is essential to activate the nodes and links needed in judging truth, for, often, we must depend upon a search for properties, which means, in terms of processing, a search for brain representations of properties, to settle a question of truth or falsity. A good instance of this, to be seen in the one example considered here, is the 1–1 property, characteristic of such a word as *capital:* x *is capital of* y. Where here, x is ordinarily a city and y a country. There are some exceptions to this being 1–1, but they are quite rare and, in ordinary discourse, the 1–1 property is automatically assumed. But this is only one of many other examples, easily given, that arise in ordinary conversation.

One other notion introduced in the axioms for computing truth to be given shortly is the notion of the *associative core* of a sentence, in our notation, $c(S)$ of a sentence S. For example, in the kinds of geography sentences given in the experiments referenced above, where similar syntactic forms are given and the sentences are given about every four seconds, persons apparently quickly learn to consider only the key reference words, which vary in an otherwise fixed sentential context, or occur in a small number of such contexts. So, for example, the associative core of the sentence *Berlin is the capital of Germany* is a string of brain representations of the three words *Berlin*, *capital* and *Germany*, for which I use the notation BERLIN/CAPITAL/GERMANY, with, obviously, the words in caps being used to denote the brain representations.

A more complicated concept is obviously needed for more general use. In the initial state of the network associations are all quiescent, e.g., PARIS $\sim$ CAPITAL, and after activation we use the notation PARIS $\approx$ CAPITAL. In the example itself we show only the activated associations and the activated nodes of the network, which are brain representations of words, visual or auditory images, and so forth. The steps of the associative computation are numbered in temporal steps t_1, etc.

EXAMPLE. *Rome is the capital of France.*

t_1.	ROME, CAPITAL, FRANCE	Activation
t_2.	PARIS, 1–1 Property	Spreading activation
t_3.	ROME $\approx$ CAPITAL, CAPITAL $\approx$ 1–1 Property CAPITAL $\approx$ FRANCE, PARIS $\approx$ CAPITAL PARIS $\approx$ FRANCE	Activation
t_4.	ITALY	Spreading activation
t_5.	PARIS/CAPITAL/FRANCE ROME/CAPITAL/ITALY	Activation
t_6.	TRUE $\approx$ PARIS/CAPITAL/FRANCE TRUE $\approx$ ROME/CAPITAL/ITALY	Spreading activation
t_7.	FALSE $\approx$ ROME/CAPITAL/FRANCE	Spreading activation

This sketch of an example, without stating the axioms and providing other technical details, is meant only to provide a limited intuitive sense of how the theory can be developed for simple empirical sentences. Most important, there is for example, no account of learning associations. Only an idealized performance setup is used.

Even less can be easily said about the intuitive steps, without explicit formal verification, in informal proofs. But I would defend the proposition that in such proofs we continually use patterns of associations that are more complicated and subtle than those needed in my truth examples. Yet I suggest, it is a feasible psychological project to survey the main features of such patterns in the informal proofs that occur in a given area of mathematics. Memory of many such patterns is undoubtedly a mark of being an expert in a given domain. Perhaps even more important, is having a feel of how to judge correctly the similarity of a prior pattern, known to be valid, to a new one being evaluated. Such experienced judgments of similarity are not at all special to proofs, but occur in every area of experience from case studies of the law to athletic skills of every variety. The content is special to the domain, but the general empirical character is not. (For an introduction to the formal aspects of the large psychological literature on similarity, often an intransitive relation due to thresholds, see Chapter 14 on proximity spaces and Chapter 16 on representations and thresholds of Suppes, Krantz, Luce, and Tversky [1989], which also contains extensive references.)

What I have said is too general, but can quickly be extended to more specific considerations by examining some examples of informal proofs. For reasons of space, I restrict myself to two.

EXAMPLE 1. Proof of an Archimedean axiom, taken from Royden [1963]. The axiom C referred to in the proof is the standard completeness axiom:

every nonempty set S of real numbers which has an upper bound has a least upper bound. Here is the theorem and informal proof, as given by Royden:

> **Axiom of Archimedes:** *Given any real number x, there is an integer n such that $x < n$.*
>
> *Proof:* Let S be the set of integers k such that $k \leq x$. Since S has the upper bound x, it has a least upper bound y by axiom C. Since y is the least upper bound for S, $y - \frac{1}{2}$ cannot be an upper bound for S, and so there is a $k \varepsilon S$ such that $k > y - \frac{1}{2}$. But $k + 1 > y + \frac{1}{2} > y$, and so $(k + 1) \not\varepsilon S$. Since $k + 1$ is an integer not is S, we must have $k + 1$ greater than x by the definition of S. [Royden, p. 25]

As expected, there is no filling out of obvious simple arguments. For example, "$y - \frac{1}{2}$ cannot be an upper bound for S." The formal expansion is obvious but tedious. What is important here, and critical for informal proofs, is the power of ordinary language along with a minimum of notation to describe the argument that could easily be written as an algorithm. In saying this I am not claiming that we know how to write general algorithms for any such gaps. In general form they may not exist, because of well-known undecidability results, or high lower bounds on such decision procedures as Tarski's for the first-order theory of real closed fields. The last sentence of the proof exhibits a similar use of ordinary language to summarize informally the argument.

EXAMPLE 2. This one concerns equivalents of the axiom of choice. I take the example from my own book on axiomatic set theory Suppes [1960/1972]. A useful maximal principle, due independently to Teichmüller [1939] and Tukey [1940] is characterized by defining when a set is of *finite character*, which is true of a set A if and only if

(i) *A is a nonempty set of sets,*

(ii) *every finite subset of a member of A is also a member of A.*

The intuitive idea behind this formulation is that a property is of finite character if a set has the property when and only when all of its finite subsets have the property.

> **Teichmüller-Tukey Lemma**: T. *Any set of finite character has a maximal element.*
>
> The theorem of interest is:
>
> *Theorem: The Teichmüller-Tukey Lemma* T *is equivalent to Zorn's Lemma Z.*

Recall that Zorn's Lemma states that if $A \neq 0$ and if the sum of each non-empty chain which is a subset of A is in A, then A has a maximal element, where A is a chain if and only if A is a set of sets and for any two sets B and C in A either $B \subseteq C$ or $C \subseteq B$.

Proof. We prove only the first half, namely, that Zorn's Lemma (Z) implies the Teichmüller-Tukey Lemma.

> Let A be a set of finite character, and let C be any chain which is a subset of A. To apply Z we need to prove that $\cup C \in A$. Let F be a finite subset of $\cup C$. Then F is a subset of the union of a finite collection D of members of C, for each element of F must belong to some member of C and there are only a finite number of elements in F. Now since D is finite and is a subset of the chain C, it has a largest member, say E; and F must be a subset of E, for otherwise C would not be a chain. $E \in A$, whence since A is a set of finite character, $F \in A$; but then also $\cup C \in A$. The hypothesis of Z is thus satisfied by A and by virtue of Z, A has a maximal element. [Suppes, 1960/1972, p. 249]

■

A first rough comparison to the length of this informal proof to formal ones that assumed the same background of prior theorems may be made using empirical data in Suppes and Sheehan [1981, p. 79] on the length of nine formal proofs made by students in a course I taught for many years on set theory, for which a computer-based proof checker was developed and then regularly used. (Details are in the article just cited.) The mean length, is terms of number of lines with explicit inference rules, was 41.8, with the min = 25 and the max = 61. As expected, the informal proof given above is much shorter.

The central characteristic of this informal proof, like the previous one, is the use of ordinary English sentences with some embedded mathematical symbols to summarize intuitively individual arguments, each of which correspond approximately to a number of steps in a formal proof.

A psychological point about this linguistic feature of many informal written proofs is the implication for understanding such sentences. Undoubtedly the problem of being satisfied with a personal verification of an informal proof is quite dependent on the intuitive mathematical clarity of the written form of the informal proof. So, often it is not the overall structure of the proof, but the difficulty of comprehending individual sentences. To comprehend such sentences the reader needs to be able to build a mathematical model satisfying the sentence, and often some other sentences as well as visual graphs and the like, at least in sufficient detail to feel the model is enough. Mathematicians are good at this. It is an essential part of what they have learned. Intellectually, this differs from algorithmic checking in a way that is parallel to the difference between model theory and proof theory.

Finally, I want to emphasize that the model sketches, as I am calling them, used to check informal proofs, are special to mathematicians only in part, not in their general psychological features, which are surely shared by architects, builders, and designers of all kinds who rely on a variety of images, externalized on paper or on a computer screen, but also images of the imagination, to control thinking about whatever problem is current.

Brain representations. On the basis of the sound methodological principle that properties of the mind are really properties of the brain, it is useful to see what can be said about informal proofs from the standpoint of brain activity, even though it is obvious we have at present many mental concepts we cannot characterize in terms of what we know about the brain.

The first observation is one that brings us back to formal proofs. It is widely recognized by almost everyone working on the subject that all computations are physical computations. In other words, any actual computations require a physical embodiment. This does not mean that digital computers are in any sense the universal model of how computations are made. Natural computations in the biological world have an endless variety of physical embodiments. The computational nature of DNA as the genetic code is probably the greatest single scientific discovery of the second half of the twentieth century. But the problems of understanding the physical computations of the seemingly simple motions of the thousands of insect species are overwhelming in their complexity and diversity. How, for example, does an ordinary house fly compute its escape route of flight from a detected predator?

Formal proofs are certainly recognized as having relatively easy implementation as physical computations on a digital computer. Whatever abstract talk there is about the meaning and implications of a formal proof, the verification of the proof is a recursive physical process, painfully explicit in its details.

Something similar has to be true of informal proofs, with brain computations replacing digital ones. A much too simple model of such brain computations was given above in the analysis of how the truth is computed of ordinary empirical statements about highly familiar matters, such as the most obvious geographic or demographic facts about Europe and the United States. At the psychological level, the method of computation proposed is that of association, recognized since Hume as fundamental, but also recognized for many years as fundamental in the brain's activities. We know from the large literature on Turing machines and register machines, how simple a basic mechanism of computing any computable function can be, and the same is true of the mechanism of association. Informal proofs are a triumphant application of associations. At least until recently, many psychologists unfamiliar with the detailed analysis of mechanisms of computations were inclined to be skeptical of such a claim. But the general complexity analogy with digital computers is too obvious to tolerate any wholesale rejection of association as the primary basis of the brain's computations. It is a long way from Minsky's simple universal Turing machine [1967] with four symbols and seven internal states to the complexity of a digital computer with programs able to defeat the best human chess players, for example, or make a trillion computations to predict the weather. So it is with the brain, from the simple associations of little worms like *Aplysia* to the most intricate mathematical proofs.

I have introduced, in these last paragraphs, many inadequately developed ideas about how the brain works. Full details are available nowhere, but will undoubtedly be a subject of intense research for some years to come. There is not space to try to say in a careful way what I think we do, at the present, know. So I will end with two remarks, one speculative and one empirical.

The first remark concerns meaning. It is a standard complaint of many years that Hilbert's formal systems for the foundations of mathematics turn mathematics into a meaningless game. The arithmetic of numerals, as opposed to numbers, is nothing like the rich content of genuine number theory or geometry, a source of endless intuitions and meaningful relations. The new home of Hilbert-style formalisms is, of course, in computer science, and much more broadly, the programming efforts throughout the world to use formal languages to write computer programs, which implement solutions to a vast array of tasks and problems. There is no additional sense or meaning given to the computer as part of these programs. It is formalism all the way down. Moreover, it is hard to think of its being any other way.

Detailed thinking about the brain moves in the same direction. There is no mysterious Fregean sense lurking somewhere in the cortex, ready to supply meaning as needed. The meaning of a word, a phrase or a sentence, like the meaning of a perceptual image, is to be found in a welter of associations, or, to put it more soberly, in associative networks that are, in humans, if not in *Aplysia*, of great complexity. Of course, to put it this way is too bald and simple, as if reference to complexity were sufficient to explain how a predictive model of the weather computes an estimate of what the weather will be like the day after tomorrow. It helps not at all, in concrete terms to say the program uses a terabyte of memory and computes at the rate of two terabits a second. Sustained research will be required for the indefinite future to untangle just how the brain is computing any important task, but the associative nature of the computations is, on present evidence, a reasonable conjecture.

The second and final remark is more down to earth and empirical. An early question that arises in thinking about how the brain processes ordinary language, including that used in informal proofs, is, how does the brain process linguistic input? (The question of speech production seems even more complicated.) Here I have in mind the relatively simple question of just what can we say about the initial processing of the words and sentences heard. (I consider here only spoken language, but most of what I have to say applies just as well to the temporal process of reading words and sentences.) Now the analysis of sentences as sequences of words that occur in them, and the analysis of words as sequences of syllables, and, at the more detailed level, phonemes, is widely accepted as being approximately correct. In addition, for many reasons, it is a sensible hypothesis to expect that the methods of brain computation are likely to preserve approximately the structured order of words in sentences, syllables in words, and so forth.

To test this idea in brain data is to test a hypothesis of structural isomorphism between spoken sentences and their brain representations. I summarize some unpublished work with my younger colleagues Marcos Perreau Guimaraes, Timothy Uy, and Dik K. Wong. Sentences are presented visually, one word at a time on a computer screen, at the temporal pace of ordinary speech. Electric waves in the cortex, time-locked to the presentation of each sentence, are recorded for each subject using standard electroencephalography (EEG) techniques. Various linear models, such as those based on Fourier transforms and filters, or one-layer neural networks, are used to eliminate noise and find an approximately invariant signal [Wong, Guimaraes, Uy, and Suppes, 2004]. The measure of success at the first level is being able to classify correctly a significant number of test trials not used in estimating the parameters of the model being evaluated. The sentences were of the geographic type mentioned earlier, and the subjects were asked to judge each one as true or false, and so indicate by typing 1 for true and 2 for false. Excellent, but far from perfect, recognition results were obtained by sets of sentences of size 24, 48 and 100. We also isolated in the temporal sequence of presentation, individual words to which the same models were applied.

Let f be a one-one function mapping each sentence s to its brain representation $f(s)$. Let g be a corresponding function for words. Then our test of structural isomorphism is whether or not we could find empirical support for the structural equation, where $s = w_1 w_2 \cdots w_n$,

$$f(w_1 w_2 \cdots w_n) = g(w_1) g(w_2) \cdots g(w_n).$$

It perhaps seems too obvious that this equation should hold, just by asking how else could the brain process sentences. But already at the level of speech such precise identification is not always easy, so it is a serious problem whether or not such results can be substantiated in the brain. We have good support, but not as good as our recognition of sentences, which is not surprising, since this same relation holds for speech.

Finding support for such a natural isomorphism seems necessary to get started, but it is clearly a long journey of further results to get to such questions as how informal proofs are processed in the brain. In this case, much more than the initial isomorphism of recognition is needed. Semantic computation in the spirit of the model sketches mentioned earlier are essential. Still, at a certain level the task is well-defined, and I see, at the present, no alternative conception that is more promising. Whatever empirical route does prove successful, I find it unimaginable that there can be a fully satisfactory theory of the verification of informal proofs that is *apriori* and devoid of psychological, and ultimately neural, concepts and data.

BIBLIOGRAPHY

[Aristotle, 1975] Aristotle. *De Anima (On the soul).* Cambridge, MA: Harvard University Press, 4th edn. English translation by W. S. Hett. First published in 1936, 1975.

[Berkeley, 1710] G. Berkeley. *Principles of human knowledge.* (Jeremy Pepyat, Dublin), 1710.

[Finke, 1989] R. A. Finke. *Principles of mental imagery.* Cambridge, MA: MIT Press, 1989.

[Friedman, 1974] M. Friedman. Explanation and scientific understanding. *Journal of Philosophy* **73**: 250–261, 1974.

[Hilbert, 1927] D. Hilbert. The foundations of mathematics. In J. van Heijenoort, ed., *From Frege to Gödel: A source book in mathematical logic, 1879–1931.* Cambridge, MA: Harvard University Press, 1927.

[Hume, 1951] D. Hume. (1739/1951). *A treatise of human nature.* London: John Noon. Quotations taken from L. A. Selby-Bigge's edition, Oxford University Press, London, 1951.

[Kitcher, 1976] P. Kitcher. Explanation, conjunction and unification. *Journal of Philosophy* **73**: 207–212, 1976.

[Kitcher, 1982] P. Kitcher. Explanatory unification. *Philosophy of Science* **48**: 507–531, 1982.

[Kreisel, 1967] G. Kreisel. Informal rigour and completeness proofs. In I. Lakatos, ed., *Studies in logic and the foundations of mathematics*, pp. 138–171. Amsterdam: North-Holland Publishing Company, 1967.

[Kreisel, 1981a] G. Kreisel. Neglected possibilities of processing assertions and proofs mechanically: Choice of problems and data. In P. Suppes, ed., *University-level computer-assisted instruction at Stanford: 1968–1980.* Stanford, CA: Stanford University, Institute for Mathematical Studies in the Social Sciences, pp. 131–148, 1981.

[Kreisel, 1981b] G. Kreisel. Extraction of bounds: Interpreting some tricks of the trade. In P. Suppes, ed., *University-level computer-assisted instruction at Stanford: 1968–1980.* Stanford, CA: Stanford University, Institute for Mathematical Studies in the Social Sciences, pp. 149–164, 1981.

[Locke, 1690] J. Locke. *An essay concerning human understanding.* (Thomas Basset, London), 1690.

[Minsky, 1967] M. L. Minsky. *Computation: Finite and infinite machines.* Englewood Cliffs, N.J.: Prentice-Hall, 1967.

[Murphy, 2002] Murphy, G. L. (2002). *The big book of concepts.* Cambridge, MA: MIT Press.

[Royden, 1963] H. L. Royden. *Real analysis.* New York: The Macmillan Company, 284 pp, 1963.

[Suppes and Beziau, 2003] P. Suppes, and J.-Y. Beziau. Semantic computations of truth, based on associations already learned. *Journal of Applied Logic* **2**: 457–467, 2003.

[Suppes *et al.*, 1999a] P. Suppes, B. Han, J. Epelboim, and Z.-L. Lu. Invariance between subjects of brain wave representations of language. *Proceedings National Academy of Sciences* **96**: 12953–12958, 1999.

[Suppes *et al.*, 1999b] P. Suppes, B. Han, J. Epelboim, and Z.-L. Lu. Invariance of brain-wave representations of simple visual images and their names. *Proceedings of the National Academy of Sciences* **96**: 14658–14663, 1999.

[Suppes *et al.*, 1997] P. Suppes, Z.-L. Lu, and B. Han. Brain wave recognition of words. *Proceedings National Academy of Sciences* **94**: 14965–14969, 1997.

[Suppes *et al.*, 1989] P. Suppes, D. H. Krantz, R. D. Luce, and A. Tversky. *Foundations of Measurement, Vol. II. Geometrical, Threshold, and Probabilistic Representations.* New York: Academic Press, 493 pp, 1989.

[Suppes and Sheehan, 1981] P. Suppes and J. Sheehan. CAI course in axiomatic set theory. In P. Suppes (ed.), *University-level Computer-assisted Instruction at Stanford: 1968-1980.* Stanford, CA: Stanford University, Institute for Mathematical Studies in the Social Sciences, pp. 3–80, 1981.

[Suppes, 1960] P. Suppes. *Axiomatic Set Theory.* New York: Van Nostrand, 265 pp. Slightly revised edition published by Dover, New York, 1972, 267 pp, 1960.

[Tappenden, 2005] J. Tappenden. Proof style and understanding in mathematics I: Visualization, unification and axiom choice. In P. Mancosu, K. F. Jorgensen, and S. A. Pedersen, eds., *Visualization, explanation and reasoning styles in mathematics*, pp. 147–206. The Netherlands: Springer, 2005.

[Teichmüller, 1939] O. Teichmüller. Braucht der Algebraiker das Auswahlaxiom? *Deutsche Math*, Vol. 4, pp. 567–577, 1939.

[Tukey, 1940] J. W. Tukey. *Convergence and uniformity in Topology.* Annals of Math. Studies, No. 2, Princeton, 1940.

[Wong *et al.*, 2004] D. K. Wong, M. P. Guimaraes, E. T. Uy, and P. Suppes. Classification of individual trials based on the best independent component of EEG-recorded sentences. *Neurocomputing*, **61**: 479–484, 2004.

Objective Bayesian Nets

JON WILLIAMSON

1 Introduction

Any theory of rationality must at some stage address the following key question:

Belief Representation What is the best way to represent an agent's rational belief state?

It is the aim of this paper to sketch a solution to the belief representation problem.

The proposed solution has two facets. First, *objective Bayesianism* tells us which degrees of belief an agent should adopt: she should adopt as her belief function a probability function, from all those that satisfy constraints imposed by her background knowledge, that is maximally non-committal, i.e. that maximises entropy (§2). Second, recent developments in probabilistic expert systems tell us how best to represent a probability function: a *Bayesian net* offers an efficient, clear and informative representation (§3).

Combining these two facets in §4, we use a Bayesian net to represent the agent's optimal belief function—such a Bayesian net will be called an *objective Bayesian net*.

The method for constructing an objective Bayesian net given in §4 requires that the agent's background knowledge be formulated as a set of quantitative constraints on her degrees of belief. However knowledge is often qualitative; the question arises as to how objective Bayesian nets can be constructed in the presence of such knowledge. In §5 we shall see that qualitative knowledge of influence relationships (e.g. causal influence) can be transformed into quantitative constraints on degrees of belief.

An objective Bayesian net is derived from background knowledge. Thus to understand how to perform an operation on an objective Bayesian net, one should perform the corresponding operation on background knowledge and derive the associated objective Bayesian net. For instance, when an objective Bayesian net needs to be updated, the updated net should be the same as the net generated by updated background knowledge (§6). The combination of two Bayesian nets should be the same as the net generated by the combination of their associated knowledge bases (§7).

Having presented the theory of objective Bayesian nets in Part I, we turn briefly to applications in Part II. We shall see that apart from their use in a general theory of rationality, objective Bayesian nets also shed light on a number of specific modes of reasoning. They can be used to perform inference in a probabilistic logic (§8), to justify the assumptions behind causal models (§9), to guide logical (§10) and semantic (§11) reasoning, and to develop a framework for argumentation (§12) and recursive modelling (§13).

Part I: Theory

2 Objective Bayesianism

Suppose a patient has a high fever, a dry cough and appears confused—to what extent should one believe that he has Legionnaire's disease?

Bayesians hold that an agent's degrees of belief ought to satisfy the axioms of probability. Thus the above degree of belief has the form of a conditional probability statement, $p(l|fdc)$ where l signifies that the patient has Legionnaire's disease, f that he has a high fever, d that he has a dry cough and c that he appears confused. Subjective Bayesians stop there and consider an agent to be rational whatever probability function he adopts as her initial belief function. But objective Bayesians go further, insisting (i) that an agent's degrees of belief should also respect background knowledge—they should for example be calibrated with known frequencies (if she knows only the incidence rate of Legionnaire's disease in the population then $p(l)$ should match that rate)—and (ii) that the agent should commit to outcomes only to the extent warranted by background knowledge (e.g. if she knows nothing concerning l then she should not commit to l; instead she should equivocate between l and $\neg l$, i.e. set $p(l) = p(\neg l) = 1/2$).

More precisely, objective Bayesians suppose that an agent's background knowledge β delimits a set $\mathbb{P}_\beta$ of probability functions that are compatible with that knowledge, and that the agent should choose a function $p \in \mathbb{P}_\beta$ that maximises entropy as her belief function. The entropy of a probability function p is

$$(1) \quad H(p) = -\sum_{v\in\Omega} p(v) \log p(v),$$

where Ω is the space of all possible indivisible outcomes, e.g. $\Omega = \{\pm l \pm f \pm d \pm e\}$.[1]

[1] The notation $\pm l$ refers to *either* l or $\neg l$; thus there are 2^4 indivisible outcomes, $\neg l \neg f \neg d \neg e, \neg l \neg d \neg de, \ldots, lfde$. We shall assume throughout that Ω is finite. The extension to the infinite is discussed in [Williamson, 2005c, §19].

Entropy is interpreted as a measure of the uncertainty or lack of commitment of a probability function: the more middling the probabilities, the higher the entropy and the higher the uncertainty; the nearer the probabilities are to the extremes of 0 or 1, the lower the entropy and the more the probability function commits to certain outcomes.[2] A probability function in $\mathbb{P}_\beta$ that has maximum entropy is compatible with background knowledge but is maximally non-committal in other respects.[3] Such a probability function is to be desired as a representation of one's degrees of belief because it is guided by empirical information yet is on average maximally cautious when it comes to risky decisions, which tend to be embarked upon when one has more extreme degrees of belief.[4]

For a set $\mathbb{Q}$ of probability functions we shall write $p \uparrow \mathbb{Q}$ as shorthand for $p \in \{q \in \mathbb{Q} : H(q) \text{ is maximised}\}$. Objective Bayesians maintain then that one should take $p \uparrow \mathbb{P}_\beta$ as one's belief function, given background knowledge β. This principle is often called the *maximum entropy principle*; it considerably narrows down the values one can ascribe to $p(l|fdc)$.[5]

Two questions remain: How should one best represent the probability function $p \uparrow \mathbb{P}_\beta$? How should one calculate probabilities like $p(l|fdc)$? One can appeal to Bayesian nets to address these questions: a Bayesian net offers an efficient and perspicuous representation of a probability function and offers an efficient way to calculate conditional probabilities.

3 Bayesian Nets

Consider a domain $V = \{A_1, A_2, \ldots, A_n\}$ of finitely many variables, each of which has finitely many possible values. Let $a_i@A_i$ signify that a_i is an assignment of a value to variable A_i. Associated with V there is set $\Omega_V = \{a_1 a_2 \cdots a_n : a_i@A_i, 1 \leqslant i \leqslant n\}$ of indivisible outcomes.

A *Bayesian net* $\mathfrak{p}_V = (G, S)$ contains

- a directed acyclic graph G whose nodes are variables in V (e.g. Fig. 1),
- a probability specification S which contains the probability distribution of each variable in V conditional on its parents in G (Table 1 contains an example distribution—B, C, and D each take two possible values, superscripted by 0 and 1).

A Bayesian net is also subject to an assumption, the *Markov Condition*, which holds that each variable A_i is probabilistically independent of its

[2] [Shannon, 1948].
[3] [Jaynes, 1957].
[4] [Williamson, 2005b].
[5] Plausibly $\mathbb{P}_\beta$ will be a closed convex set of probability functions, in which case $p \uparrow \mathbb{P}_\beta$ is uniquely determined—see [Williamson, 2005a, §5.3].

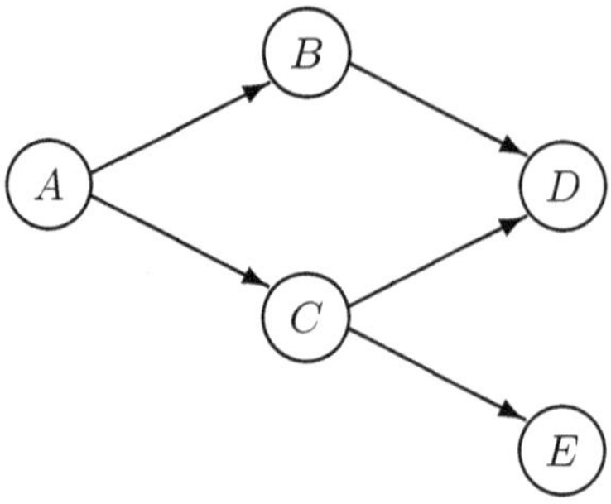

Figure 1. A directed acyclic graph.

Table 1. The probability distribution of D conditional on B and C.

$p(d^0\|b^0c^0) = 0.7$	$p(d^1\|b^0c^0) = 0.3$
$p(d^0\|b^0c^1) = 0.9$	$p(d^1\|b^0c^1) = 0.1$
$p(d^0\|b^1c^0) = 0.2$	$p(d^1\|b^1c^0) = 0.8$
$p(d^0\|b^1c^1) = 0.4$	$p(d^1\|b^1c^1) = 0.6$

non-descendants ND_i in G conditional on its parents Par_i in G, written $A_i \perp\!\!\!\perp ND_i \mid Par_i$.

A Bayesian net determines a unique probability function p over Ω_V since the Markov Condition implies that

$$p(a_1a_2\cdots a_n) = \prod_{i=1}^{n} p(a_i|par_i),$$

where $par_i@Par_i$ is determined by $a_1a_2\cdots a_n$, and since the probabilities in this product are all contained in S.

A Bayesian net $\mathfrak{p}$ offers an attractive representation of a probability function p for a number of reasons. First, $\mathfrak{p}$ perspicuously represents probabilistic independencies satisfied by p in the sense that one can simply read independencies off the graph: for $X, Y, Z \subseteq V$, $X \perp\!\!\!\perp Y \mid Z$ if Z *blocks* each path between X and Y, i.e., for each path between $A_i \in X$ and $A_j \in Y$, there is some node on the path in Z whose adjacent arrows meet head-to-tail or tail-to-tail, or there is a node on the path whose adjacent arrows meet head-to-head and Z contains neither that node nor any of its descendants. Second, $\mathfrak{p}$ is an efficient representation in the sense that relatively few probability specifiers $p(a_i|par_i)$ determine a large number of probabilities $p(a_1a_2\cdots a_n)$ (this depends on the structure of G: roughly speaking the sparser the graph G, the smaller the specification S). Third, $\mathfrak{p}$ admits efficient probabilistic inference: there are algorithms for quickly determining conditional probabilities from the Bayesian net (again, the efficiency of these algorithms depends on the structure of the graph).[6]

Bayesian nets are typically constructed in one of two ways. One is to employ a machine learning methodology to construct a net that represents the frequency distribution of a database of past observations of assignments to variables in V. The other is to elicit a graph and probability specifiers from an expert to construct a net that represents the expert's (subjective Bayesian) belief function. Here we are interested in objective Bayesian probability rather than frequency or subjective Bayesian probability—clearly neither of these two approaches are appropriate for representing an objective Bayesian belief function. We thus need a technique for constructing a Bayesian net that represents a probability function, from all those that satisfies constraints imposed by background knowledge, that maximises entropy.

4 Objective Bayesian Nets

An *objective Bayesian net*, or *obnet* for short, is a Bayesian net that represents an objective Bayesian probability function p, i.e. a probability function that maximises entropy subject to constraints imposed by background knowledge β.

An objective Bayesian net can be constructed using the following strategy:

Step 1 determine conditional independencies that $p \uparrow \mathbb{P}_\beta$ must satisfy,

Step 2 represent these by a directed acyclic graph G that satisfies the Markov Condition with respect to p,

Step 3 maximise entropy to calculate the numerical parameters $p(a_i|par_i)$ in the probability specification S.

We shall briefly run through each of these steps in turn—this procedure for constructing an objective Bayesian net is presented in more detail in [Williamson, 2005a, §§5.6–5.7].

Step 1: Determine Conditional Independencies

We shall suppose β can be construed as probabilistic constraints $\pi_1, \ldots, \pi_m$ on the probability function p. For example, π_1 might be $p(a_1|a_2) \geqslant 0.7$, and π_2 might be $p(a_2a_4) = p(a_3)^2p(a_2)$.

6 [Neapolitan, 1990].

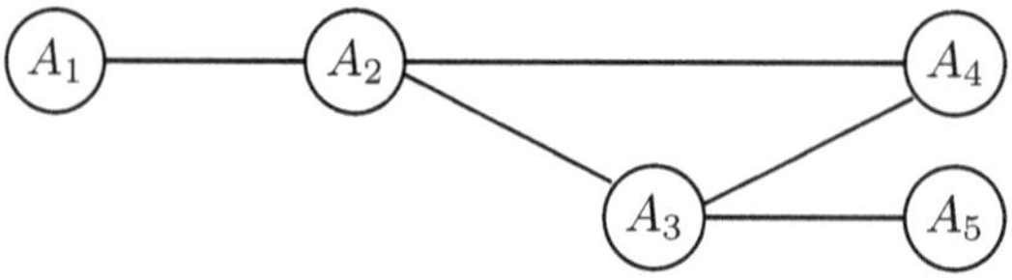

Figure 2. Constraint graph.

Construct an undirected graph, the *constraint graph*, by taking the variables in V as nodes and by connecting two nodes with an edge if they occur in the same constraint.

Suppose, for example, that $V = \{A_1, \ldots, A_5\}$, π_1 is a constraint involving A_1 and A_2, π_2 involves A_2, A_3, A_4, π_3 involves A_3, A_5, and π_4 involves A_4. Then the constraint graph is depicted in Fig. 2.

The constraint graph tells us about probabilistic independencies that the maximum entropy function will satisfy, since the following key property holds:[7] if Z separates X from Y in the constraint graph then $X \perp\!\!\!\perp Y \mid Z$ for $p \upharpoonright \mathbb{P}_\beta$.

In Fig. 2, for example, A_2 separates A_1 from A_3, A_4 and A_5, so we know that a maximum entropy function renders A_1 probabilistically independent of A_3, A_4 and A_5 conditional on A_2.

Step 2: Construct a Graph Satisfying the Markov Condition

One can transform the constraint graph into a directed acyclic graph G that satisfies the Markov Condition via the following algorithm:[8]

- triangulate the constraint graph,
- re-order V according to maximum cardinality search,
- let $D_1, \ldots, D_l$ be the cliques of the triangulated constraint graph ordered according to highest labelled node,
- set $E_j = D_j \cap (\bigcup_{i=1}^{j-1} D_i)$ for $j = 1, \ldots, l$,
- set $F_j = D_j \backslash E_j$ for $j = 1, \ldots, l$,
- take variables in V as the nodes of G,
- add an arrow from each vertex in E_j to each vertex in F_j $(j = 1, \ldots, l)$,
- ensure that there is an arrow between each pair of vertices in D_j $(j = 1, \ldots, l)$.

The resulting directed graph often looks much like the undirected constraint graph—in our example G is depicted in Fig. 3.

[7] [Williamson, 2005a, Theorem 5.3].

[8] See [Williamson, 2005a, §5.7] for an explanation of the graph-theoretic terminology.

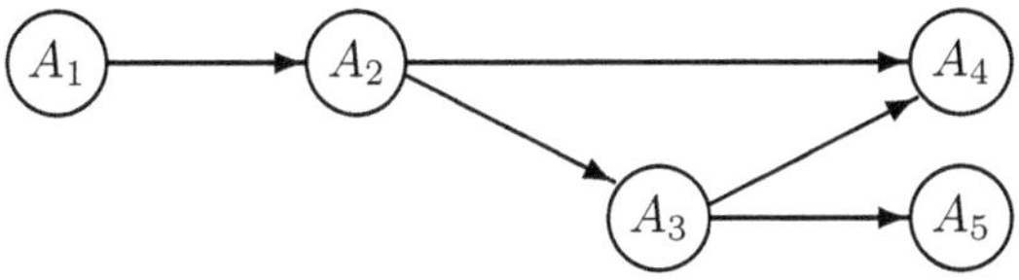

Figure 3. Graph satisfying the Markov Condition.

Step 3: Determine Probability Specification

Having found a graph G that satisfies the Markov Condition, to construct an objective Bayesian net it only remains to determine the probability distribution of each variable conditional on its parents.

Here it helps to rewrite the entropy equation as $H = \sum_{i=1}^{n} H_i$ where

$$H_i = -\sum \left(\prod_{A_j \in Anc_i} p(a_j|par_j) \right) \log p(a_i|par_i),$$

(Anc_i being the set of ancestors of A_i in G.)

One can then use numerical optimisation techniques or Lagrange multiplier methods to find the parameters $p(a_i|par_i)$ that maximise entropy. This entropy maximisation problem will in practice be a smaller problem than the original problem of maximising entropy over the whole domain (Equation 1) since there will in practice be far fewer parameters of the form $p(a_i|par_i)$ than there were of the form $p(v) = p(a_1 a_2, \ldots, a_n)$ (this is because in practice while one may know of many observations or constraints, each constraint tends to involve relatively few variables in comparison with n, as n becomes large, so G tends to be sparse).

We see then that by pursuing this three-step procedure it is quite straightforward to construct an obnet, given a set of probabilistic constraints.

Quantitative probabilistic constraints are clearly required in order to apply the maximum entropy principle. However background knowledge does not always take the form of a set of quantitative constraints on degrees of belief—an agent may know of qualitative causal relationships, for instance. The task of converting qualitative constraints into quantitative constraints is a significant challenge for objective Bayesianism.[9] We shall see next how qualitative knowledge of influence relationships (e.g. causal influences) can be converted into quantitative constraints on an agent's belief function p.

[9] [Williamson, 2005c, §18].

5 Influence Relations

We turn now to the question of how one can construct an objective Bayesian net when background knowledge includes qualitative knowledge of influence relationships.

We shall take the following to be the defining feature of the notion of *influence*: learning of the existence of new variables that are not influences of the other variables should not change degrees of belief concerning those other variables.[10]

The causal relation, for example, is an influence relation. If an agent learns of a new variable that is known not to be a cause of any the variables she already knew about, then this new information provides no reason for the agent to change her degrees of belief concerning those other variables. In the absence of any reason for change, her degrees of belief should stay the same. (In contrast, learning of new causes may motivate a change in degrees of belief: at first glance the flooding of glacial valleys in Kyrgyzstan and the insect population of southern England seem quite unrelated, but the knowledge that global warming affects both these variables may warrant an increase in the degree to which one believes insect populations will rise given that glacial flooding is increasing.) Causality and other examples of influence relationships will be discussed in Part II.

Given the above implicit definition of influence, it is straightforward to see that qualitative knowledge concerning influences can be transferred into quantitative constraints on degrees of belief. Suppose $V \supseteq U$ is a set of variables containing variables in U together with other variables that are known not to be influences of variables in U. As long as any other knowledge concerning variables in $V \backslash U$ does not itself warrant a change in degrees of belief on U, then $p^V_{\beta \restriction U} = p^U_{\beta_U}$, i.e. one's belief function on the whole domain V formed on the basis of all one's background knowledge β, when restricted to U, should match the belief function one would have adopted on domain U given just the part β_U of one's knowledge involving U. Thus knowledge of influences is transferred into equality constraints on degrees of belief.

Once qualitative knowledge has been transferred into quantitative constraints on degrees of belief, the three-step procedure of §4 for constructing an objective Bayesian net can be directly applied. However, the fact that the new constraints are equality constraints leads to a simplification: these new constraints can be ignored in Step 1 of the process.[11] We thus have a slightly modified three-step procedure:

Step 1 determine conditional independencies that $p \uparrow \mathbb{P}_\beta$ must satisfy from

[10] [Williamson, 2005a, §11.4].

[11] [Williamson, 2005a, Theorem 5.6].

the constraint graph, *ignoring constraints yielded by knowledge of influences,*

Step 2 represent these independencies by a directed acyclic graph G that satisfies the Markov Condition with respect to p,

Step 3 maximise entropy to calculate the numerical parameters $p(a_i|par_i)$ in the probability specification S *(remembering to take equality constraints yielded by knowledge of influences into account).*

Thus knowledge of influences does not add to the complexity of an objective Bayesian net, in the sense that the graph in the net is just as sparse as it would have been if there were no such knowledge.

A further simplification is possible in the case in which the agent knows *all* the influence relationships amongst the variables and has no quantitative knowledge that overrides the equality constraints generated by these influence relationships (n.b. quantitative information regarding the strengths of the influence relationships will not override the equality constraints).[12] As long as the *influence graph*—i.e. the directed graph in which there is an arrow from variable A to variable B if and only if A directly influences B—is acyclic, we can go straight to Step 2: the influence graph itself satisfies the Markov Condition.[13] Step 3 is also simpler in this case: we can maximise entropy by maximising each component H_i of the modified entropy equation sequentially (rather than maximising their sum).[14] This breaks down the entropy maximisation problem into n smaller problems. In this case, then, the objective Bayesian net is just the influence graph plus sequentially-determined conditional probability distributions.

Having discussed the construction of obnets, we now turn to how they might be updated (§6) and combined (§7).

6 Updating

An objective Bayesian net represents the degrees of belief that an agent should adopt and these rational degrees of belief are determined by the agent's background knowledge. So when her background knowledge changes, so too should the obnet. The extent to which the net changes will depend on the extent to which background knowledge changes.

If the new knowledge consists of an observation o of the values of some of the variables, then the new probability function $p' \uparrow \mathbb{P}_{\beta \cup \{o\}}$ is just the old function conditional on the observation, i.e. $p' = p(\cdot|o)$ where $p \uparrow \mathbb{P}_\beta$.[15]

[12] [Williamson, 2005a, pp. 99–100].
[13] [Williamson, 2005a, Theorem 5.7].
[14] [Williamson, 2005a, Theorem 5.8].
[15] [Williams, 1980, pp. 134–135].

This type of update is known as *Bayesian conditionalisation*. It is simple to modify a Bayesian net to represent its Bayesian conditionalisation update: the graph in the net remains the same but the probability specification gets updated using standard propagation algorithms.[16]

More generally, when the new knowledge consists of new constraints on the agent's degrees of belief that are consistent with the old constraints, the new probability function $p' \upharpoonright \mathbb{P}_{\beta'}$ is the probability function satisfying β' that is *closest* to the old function $p \upharpoonright \mathbb{P}_{\beta}$ in the sense that it minimises the *cross entropy distance* to p, $d(p',p) = \sum_v p'(v) \log p'(v)/p(v)$.[17] Thus we need to modify the objective Bayesian net $\mathfrak{p}$ representing $p \upharpoonright \mathbb{P}_{\beta}$ to form its *cross entropy update* $\mathfrak{p}'$ that represents the $p' \in \mathbb{P}_{\beta'}$ which minimises $d(p',p)$. This involves reconstructing the part of the graph of $\mathfrak{p}$ that involves variables in the new constraints and their ancestors in the graph and updating the associated conditional probability distributions; the rest of the net stays the same.[18]

In other cases, the whole net may need be reconstructed. If the new constraints are inconsistent with the old, background knowledge can not be simply augmented, it must change: some element of background knowledge must be repealed to eradicate the inconsistency. In this case a new objective Bayesian net must be constructed around the changed constraints, via the three-step procedure of §4 and §5. Similarly if the new knowledge consists of knowledge of new variables as well as new constraints, a reconstruction of the net will be required, unless the new knowledge does not warrant a change of degrees of belief involving the old variables (in particular if the new variables are known not to be influences of the old). In this latter case one can just augment the old net by adding the new variables to the graph, adding arrows to the new variables from old variables that occur in the same constraints (and amongst new variables that occur in the same constraints) and adding the probability distributions of new variables conditional on their parents.

We see then that the updating of an objective Bayesian net hinges on the updating of background knowledge. This yields a *foundational* approach to updating—the warrant for degrees of belief, background knowledge, is the crucial determinant of those degrees of belief; one does not update by cohering with past degrees of belief but by satisfying constraints imposed by this knowledge.

[16][Neapolitan, 1990, Chapters 6–7].

[17]N.b. $\beta \subseteq \beta'$. Here $p' \upharpoonright \mathbb{P}_{\beta'}$ is the function minimising $d(p',c)$, where c is the *central function* that gives the same probability to each elementary outcome [Paris, 1994, p. 120]. As long as constraints are all *affine*, this as the same function as that found by minimising $d(p,c)$ first and then minimising $d(p',p)$—see [Williams, 1980, pp. 139–140].

[18][Williamson, 2005a, §12.11].

7 Combining

In certain circumstances it is useful to consider the combination $\mathfrak{p}$ of two objective Bayesian nets $\mathfrak{q}$ and $\mathfrak{r}$, written $\mathfrak{p} = \mathfrak{q} \star \mathfrak{r}$. (More generally, given a set $\mathfrak{Q}$ of obnets we can denote their combination by $\mathfrak{p} = \star\mathfrak{Q}$.) For example, two or more agents may need to come to some consensus and act as one agent, and the question arises as to which belief function this group agent should adopt.[19]

From the foundational point of view, the combination of a set of obnets should be determined from the combination of the set of background knowledge bases that underpin the respective obnets: $\mathfrak{p} = \mathfrak{q} \star \mathfrak{r}$ should represent $p \uparrow \mathbb{P}_\beta$ where $\beta = \gamma \star \delta$, the combination of the knowledge base γ that determines $\mathfrak{q}$ and the knowledge base δ that determines $\mathfrak{r}$.

So the combination of obnets boils down to the combination of knowledge bases. How should knowledge bases be combined? This is a rather subtle question that turns on the origins of the constraints in the knowledge bases. Consider an example. Suppose Quentin's background knowledge γ contains the constraint $q(a) = 0.7$, while Ronette's background knowledge δ contains $r(a) = 0.8$. Clearly these are incompatible assignments of probability if reinterpreted as constraints on a single function p. But the way this inconsistency is resolved depends on the origins of these constraints. Suppose that both constraints originated from observed frequencies: for Quentin a falls under a reference class which has observed frequency 0.7 of a-type outcomes, while for Ronette a falls under a reference class which has observed frequency 0.8 of a-type outcomes. If Ronette's reference class is *narrower* than Quentin's, then her constraint should override Quentin's, and only the constraint $p(a) = 0.8$ should appear in the combined knowledge base $\beta = \gamma \star \delta$. On the other hand, if neither reference class is narrower than the other then neither constraint is defeated by the other and the best one can do is include the constraint $p(a) \in [0.7, 0.8]$ in β.[20] In general, we can say that $\mathbb{P}_\beta$ is the smallest closed convex set of probability functions generated by *undefeated* constraints in $\gamma \cup \delta$.

In sum then, a combination of objective Bayesian nets will depend on defeasibility relationships amongst constraints in the associated knowledge bases. If one agent's knowledge is better than all the others' then the group obnet should match that agent's obnet. Typically though the combined obnet will need to be constructed afresh from the combined background knowledge.

---

19 [Gillies, 1991].

20 [Williamson, 2005a, §5.3].

Part II: Applications

We shall now quickly run through some applications of the theory developed above. A more detailed treatment is given in [Williamson, 2005a].

8 Probability Logic

The simplest probability logics are concerned with questions of the form: given premiss sentences and their probabilities, what probability should attach to a conclusion sentence?

For instance,

$$a_1 \wedge \neg a_2,^{0.9} (\neg a_4 \vee a_3) \rightarrow a_2,^{0.2} a_5 \vee a_3,^{0.3} a_4.^{0.7} \models a_5 \rightarrow a_1.^{?}$$

is short for the following question: given that $a_1 \wedge \neg a_2$ has probability 0.9, $(\neg a_4 \vee a_3) \rightarrow a_2$ has probability 0.2, $a_5 \vee a_3$ has probability 0.3 and a_4 has probability 0.7, what probability should $a_5 \rightarrow a_1$ have?

Such questions can be given an objective Bayesian interpretation: supposing background knowledge consists of the constraints $p(a_1 \wedge \neg a_2) = 0.9, p((\neg a_4 \vee a_3) \rightarrow a_2) = 0.2, p(a_5 \vee a_3) = 0.3, p(a_4) = 0.7$, what degree of belief should be awarded to $a_5 \rightarrow a_1$?

An objective Bayesian net can be constructed to answer this question. The first step is to determine conditional independencies that must be satisfied by the probability function, out of all those that satisfy these constraints, that maximises entropy. To do this we link variables that occur in the same constraint, as in Fig. 2; separation in this graph determines conditional independencies. The second step is to transform this graph into a directed acyclic graph satisfying the Markov Condition, such as Fig. 3. The third step is to maximise entropy to determine the probability distribution of each variable conditional on its parents in the directed graph. This yields a Bayesian net. Finally we use the net to calculate the probability of the conclusion

$$\begin{aligned} p(a_5 \rightarrow a_1) &= p(\neg a_5 \wedge a_1) + p(a_5 \wedge a_1) + p(\neg a_5 \wedge \neg a_1) \\ &= p(a_1) + p(\neg a_5 | \neg a_1)(1 - p(a_1)) \end{aligned}$$

Thus we must calculate $p(a_1)$ and $p(\neg a_5|\neg a_1)$ from the net, which can be done using standard algorithms.

This application of obnets to probability logic is quite straightforward because background knowledge is quantitative. Other applications use the apparatus of §5 to exploit qualitative knowledge, as we shall now see.

9 Causal Modelling

Many types of causal model (e.g. structural equation models) consist of information about the qualitative causal relationships amongst a set of variables together with the quantitative strengths of these causal relationships.

In order to easily infer a causal model from data, a number of fundamental assumptions are made about connections between causal relationships and empirical phenomena. Perhaps the key assumption is the following:

Causal Markov Condition (CMC) each variable is probabilistically independent of its non-effects conditional on its direct causes.

A fundamental problem facing proponents of causal modelling is the question of the justification of the Causal Markov Condition. One approach—taken by [Pearl, 2000] for example—is to make a number of other assumptions that are collectively stronger than the CMC and which together imply CMC. For example Pearl assumes universal determinism, that variables are functions of just their direct causes and error terms that are not in the variable set, and that error terms are probabilistically independent.

Objective Bayesian nets offer a less drastic solution to this conundrum. The components of the causal model can be thought of as an agent's background knowledge β. As we saw in §5, causality is an influence relation, and if β contains just causal relationships and their strengths then the graph in the obnet generated by β is just the causal graph. By construction, the Markov Condition is guaranteed to hold for this graph. But the Markov Condition for the causal graph is just the Causal Markov Condition. Thus the Causal Markov Condition must hold, where the probabilities that CMC talks about are interpreted as the degrees of belief that an agent ought to adopt if all she knows is the causal model.

Thus objective Bayesian nets offer a framework for causal reasoning. But obnets can also be applied to other influence relations. We shall turn to other examples of influence relations now.

10 Logical Reasoning

A sentence a is a *logical influence* of sentence b if either a or $\neg a$ is a necessary component of some set of sentences that logically imply either b or $\neg b$, i.e. $\pm a, d \models \pm b$ for some sentence d, and $d \not\models \pm b$.

By analogy with causal influence, logical influence is plausibly an influence relation: learning of variables that are not logical influences of the others provides no reason to change one's degrees of belief concerning those other variables. Hence objective Bayesian nets can be used to represent an agent's degrees of belief in sentences given qualitative knowledge of logical influence relationships.

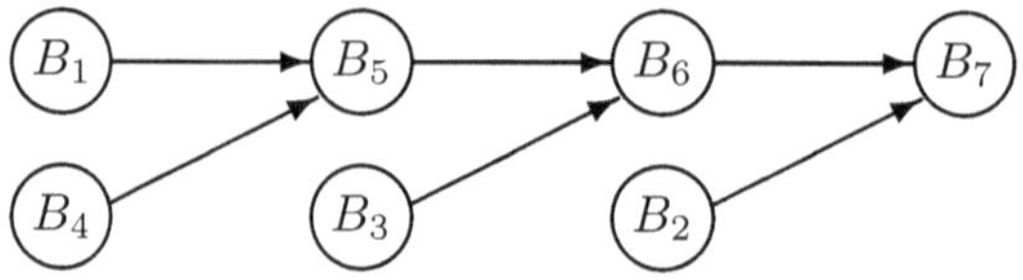

Figure 4. A logical influence graph.

For example, suppose β consists of the following proof:
1: $\phi \to \psi$ [hypothesis]
2: $\theta \to \phi$ [hypothesis]
3: $(\theta \to (\phi \to \psi)) \to ((\theta \to \phi) \to (\theta \to \psi))$ [axiom]
4: $(\phi \to \psi) \to (\theta \to (\phi \to \psi))$ [axiom]
5: $\theta \to (\phi \to \psi)$ [by 1, 4]
6: $(\theta \to \phi) \to (\theta \to \psi)$ [3, 5]
7: $\theta \to \psi$ [2, 6]

This proof yields not only qualitative knowledge of logical influences but also quantitative constraints, namely $p(b_5|b_1b_4) = 1, p(b_6|b_3b_5) = 1, p(b_7|b_2b_6) = 1$, where variable B_i takes assignment b_i (respectively $\neg b_i$) just when the sentence on line i of the proof is true (respectively false). Then the graph in the obnet generated by β maps the structure of the proof, as in Fig. 4. The probability specification in the obnet contains the probabilities yielded by the quantitative constraints $p(b_5|b_1b_4) = 1, p(b_6|b_3b_5) = 1, p(b_7|b_2b_6) = 1, p(\neg b_5|b_1b_4) = 0, p(\neg b_6|b_3b_5) = 0, p(\neg b_7|b_2b_6) = 0$; all other probabilities in the specification, e.g. $p(b_6|\neg b_3b_5)$, will be set to $\frac{1}{2}$ by maximising entropy. This net can then be used to calculate arbitrary probabilities, e.g. $p(b_1|\neg b_7)$.

11 Semantic Reasoning

A concept a is a *semantic influence* of concept b if a (or its complement) is a b (or its complement). For example, 'flu is a semantic influence of virus, because 'flu is a virus.

Plausibly, semantic influence is an influence relation. Learning that 'flu and herpes are both viruses provides no reason to change degrees of belief involving 'flu and herpes: one's degree of belief that a patient has herpes given that he has 'flu and that they are both viruses should be the same as it would be in the absence of the knowledge that they are both viruses. Thus learning of non-semantic-influences should not change degrees of belief over other variables. (On the other hand, learning of semantic influences may warrant a change in degrees of belief: learning of 'flu and that 'flu is a short-term illness and a virus may increase one's degree of belief that a patient has virus given that he has a short-term illness.)

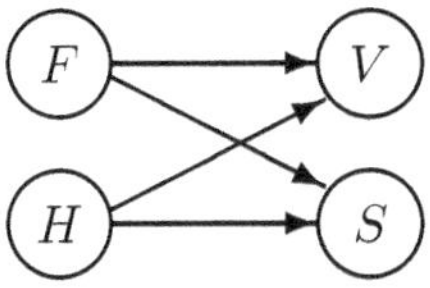

Figure 5. A semantic influence graph.

Since semantic influence is an influence relation, objective Bayesian nets can be used to represent an agent's degrees of belief given qualitative semantic knowledge. Suppose the agent's background knowledge β consists of the following semantic knowledge:

- 'flu is a virus,
- herpes is a virus,
- 'flu is a short-term illness,
- herpes is not a short-term illness.

This consists of qualitative semantic knowledge, but also imposes the constraints $p(v|fx) = 1, p(v|hx) = 1, p(s|fx) = 1, p(\neg s|hx) = 1$, where v signifies virus, f 'flu, h herpes, s short-term illness and x is an arbitrary assignment. The resulting obnet will consist of the semantic graph Fig. 5 (a semantic graph is sometimes called a *semantic network* in AI) together with the entropy maximising probability specifiers e.g. $p(v|fh) = 1, p(v|f\neg h) = 1, p(v|\neg fh) = 1, p(v|\neg f\neg h) = 1/2$. One can use the obnet to calculate probabilities such as $p(h|vs)$.

12 Argumentation

A proposition a is an *argumentative influence* of proposition b if a, or its negation, is an argument in favour of, or against, b. Plausibly, this is another example of an influence relationship: learning of propositions that are not argumentative influences of other propositions does not warrant a change in one's degrees of belief involving the other propositions.

If so, an objective Bayesian net can be used to represent an agent's degrees of belief given knowledge of argumentation structure. As before, given full knowledge of argumentation structure, the obnet will consist of an argument graph together with probability specifiers that maximise entropy.

13 Recursive Modelling

In a *recursive model* the values that variables take may themselves be structured, containing further variables. Such models can be used to represent nested relationships. For example, the fact that smoking causes cancer causes governments to restrict tobacco advertising. This can be represented by a recursive model of the form $SC \longrightarrow A$ where SC is a variable taking value $S \longrightarrow C$ or value $S \not\longrightarrow C$, S represents smoking, C cancer and A advertising, the latter three variables just take the value true or false, and the arrow represents causal connection. Another example: Fig. 4 can be thought of as a recursive model if each variable B_i takes as one value the sentence on line i of the proof used to generate the graph.

A variable A is *superior* to variable B if B occurs at a lower level to A. In the above causal model, SC is superior to S and C, but not to A. Arguably, superiority is an influence relation: learning of more structure at lower levels does not warrant a change in degrees of belief concerning higher levels. Full knowledge of superiority relationships leads to an obnet which contains arrows from superiors to their direct inferiors.

In fact a recursive model soon looks quite complicated if all these superiority arrows are included in the model. But one can eliminate them from the model if one imposes a new Markov Condition, called the *Recursive Markov Condition*, which holds that each variable is probabilistically independent of those other variables that are neither its inferiors nor at the same level, conditional on its direct superiors. This yields a *recursive Bayesian net*, a formalism that is explored in some detail in [Williamson and Gabbay, 2005].

14 Concluding Remarks

We have explored a new, third way of constructing a Bayesian net: like a subjectively elicited Bayesian net, an objective Bayesian net represents an agent's degrees of belief; like a Bayesian net learned from a frequency distribution, an obnet is objectively determined from data. Objective Bayesian nets combine the best aspects of the other two methods: an obnet can make use of frequency information where available, but can also incorporate qualitative knowledge that is not reflected in frequencies.

A theory of rationality must tell us about knowledge (how it should be gleaned, updated, combined, and so on), about belief, and about decision-making, and must also offer a practical framework for their integration. Objective Bayesian nets provide the belief module: given knowledge, an obnet can be constructed to represent the agent's degrees of belief; given an obnet, a decision theory can advise the agent as to which decisions to make on the basis of her beliefs. Objective Bayesian nets are thus a crucial component of our normative toolkit.

BIBLIOGRAPHY

[Gillies, 1991] Gillies, D. (1991). Intersubjective probability and confirmation theory. *British Journal for the Philosophy of Science*, 42:513–533.

[Jaynes, 1957] Jaynes, E. T. (1957). Information theory and statistical mechanics. *The Physical Review*, 106(4):620–630.

[Neapolitan, 1990] Neapolitan, R. E. (1990). *Probabilistic reasoning in expert systems: theory and algorithms.* Wiley, New York.

[Paris, 1994] Paris, J. B. (1994). *The uncertain reasoner's companion.* Cambridge University Press, Cambridge.

[Pearl, 2000] Pearl, J. (2000). *Causality: models, reasoning, and inference.* Cambridge University Press, Cambridge.

[Shannon, 1948] Shannon, C. (1948). A mathematical theory of communication. *The Bell System Technical Journal*, 27:379–423 and 623–656.

[Williams, 1980] Williams, P. M. (1980). Bayesian conditionalisation and the principle of minimum information. *British Journal for the Philosophy of Science*, 31:131–144.

[Williamson, 2005a] Williamson, J. (2005a). *Bayesian nets and causality: philosophical and computational foundations.* Oxford University Press, Oxford.

[Williamson, 2005b] Williamson, J. (2005b). Motivating objective Bayesianism: from empirical constraints to objective probabilities. In Harper, W. L. and Wheeler, G. R., editors, *Probability and Inference: Essays in Honour of Henry E. Kyburg Jr.* King's College Publications, London.

[Williamson, 2005c] Williamson, J. (2005c). Philosophies of probability: objective Bayesianism and its challenges. In Irvine, A., editor, *Handbook of the philosophy of mathematics.* Elsevier. Handbook of the Philosophy of Science volume 9.

[Williamson and Gabbay, 2005] Williamson, J. and Gabbay, D. (2005). Recursive causality in Bayesian networks and self-fibring networks. In Gillies, D., editor, *Laws and models in the sciences*, pages 173–221. King's College Publications, London. With comments pp. 223–245.

Epistemic Bubbles

JOHN WOODS

> The action of thought is excited by the initiation of doubt and ceases when belief is attained; so that the production of belief is the sole function of thought.
>
> Charles S. Peirce

> But since one does not think about knowledge except when he is *thinking*, except, that is, when the intellectual or cognitive interest is dominant, the professional philosopher is only too prone to think of all experiences as if they were of a type he is specially engaged in, and hence unconsciously or intentionally to project its traits into experiences to which they are alien ... But a discussion of knowledge perverted at the outset by such a misconception is not likely to proceed prosperously.
>
> John Dewey

1 The New Logic and Knowledge

The issues I shall discuss here arise from an approach to logic that Dov Gabbay and I have been developing in recent years. The essential assumptions of "the new logic" are easy to state. (Here, as in most things, the devil is in the details). We find this an extremely interesting time to be doing logic, especially if one takes the position that logic's deeply important turn towards the mathematical is by now largely a completed development. As the programme of mathematical logic settles into the maturely congenial complacency that often attends occupancy of the mainstream, logic is showing signs of recovering its historical mission as a science of reasoning. (In a sense, the new logic is a variation of the old logic.) The etiology of this recovery is multi-faceted. Important contributions have been made in computer science, cognitive and social psychology, economics, linguistics, argumentation theory and logic itself.[1] Along with a number of like-minded researchers, the approach we take to logic is *agent-centred*, *task-oriented* and *resource-bound*. As conceived of here, agents are reasoners, tasks are cognitive, and resources are the cognitive wherewithal for an agent to discharge his tasks.

Since reasoning is a natural concomitant of cognition, the logician's historic interest in reasoning motivates his attention to the nature of cognitive agency, a matter to which Gabbay and I have given some consideration in the first two volumes of *A Practical Logic of Cognitive Systems* [Gabbay & Woods, 2003a; Gabbay & Woods, 2005b]. The logician's requirement to attend to the nature and function of cognitive systems commits him to say something in turn about the epistemological presuppositions that underlie his conception of cognitive agency. Except for their interest in epistemic interpretations of modal operators or in such things as decision theory, logicians don't usually have much to say about epistemology. In our view, this is a mistake. The following pages are a modest first-step towards repairing the logician's epistemological omissions.

So, then, a principal function of reasoning is to facilitate cognition. This means the reasoning agent is also a cognitive agent. If logic is to press forward as a renewed science of reasoning, it would do well to reflect on what cognitive agency is like, on what it is like to be a knower. The revival of logic's interest in reasoning on the ground confronts the investigator with

---

[1]Within the logic community these developments include modal logics and their epistemic and deontic variations [Von Wright, 1951; Hintikka:1962; Kripke, 1963; Gabbay, 1976; Lenzen, 1978; Chellas, 1980; Hilpinen, 1981; Gochet & Gribomont, 2005], probabilistic and abductive logics [Magnani, 2001; Williamson, 2002; Gabbay & Woods, 2005b], dynamic logics [Harel, 1979; Van Benthem, 1996; Gochet, 2002], situation logics [Barwise & Perry, 1983] game-theoretic logics [Hintikka & Sandu, 1997], temporal and tense logics [Prior, 1967; Van Benthem, 1983], time and action logics [Gabbay *et al.*, 1994], systems of belief dynamics [Alchourron *et al.*, 1985; Gabbay *et al.*, 2004b], practical logics [Gabbay & Woods, 2003a; Gabbay & Woods, 2005a], and various attempts to float the programme of informal logic. The informal logic movement comprises three over-lapping orientations. One is argumentation theory [Johnson, 1996; Johnson, 2000; Freeman, 1991; Woods, 2003]. Another is fallacy theory [Hamblin, 1970; Woods & Walton, 1989; Hansen & Pinto, 1995; Walton, 1995; Woods, 2004]. Completing the trio is dialogue-logic [Hamblin, 1970; Barth & Krabbe, 1982; Hintikka, 1981; MacKenzie, 1990; Walton & Krabbe, 1995; Gabbay & Woods, 2001a; Gabbay & Woods, 2001b]. Work in computer science, AI and cognitive psychology communities includes important developments in defeasible, non-monotonic and autoepistemic reasoning, and logic programming [Sandewall, 1972; Kowalski, 1979; McCarthy, 1980; Reiter, 1980; Moore, 1985; Pereira, 2002; Schlecta, 2004]. Work from economics includes [Simon, 1973; Gigerenzer & Selton, 2001]. An especially valuable contribution from linguistics is [Carlson & Pelletier, 1995]. Thus what we are calling the new logic is best conceived of as a confederacy of very wide-ranging approaches (and not everywhere pairwise compatible) to agent-based reasoning "on the hoof".

issues that are more complex (and more obscure) than those that set the research programmes in the four main domains of mathematical logic: set theory, model theory, proof theory and recursion theory. In these mainstream venues the logician's preoccupation is with properties of linguistic entities or of linguistic entities in connection with set theoretic structures. He pursues his objectives without regard to the role of agents, and independently of other contextual features. This presents the new logician with one of his primary tasks. He must try to determine the extent to which conditions on the presence of these mainstream properties are susceptible to adaptation as conditions on reasoning. For example, the new logician must have something to say about the consequence relation. At least equally important is the account he gives of drawing consequences. Whereupon, the investigator is faced with an important distinction between the consequences a set of sentences *has* and the consequences that an agent is sanctioned *to draw* from them.

Logicians must take care to orient their account of reasoning to what it is like to reason in quest of the satisfaction of cognitive targets. And to do this they must have some idea of the constraints under which cognizers operate, and of the resources available to them. In making this preliminary exploration, I am mindful of the tug of an important question. Given that the outline of cognitive agency to be presented here is, as far as it goes, a tenable one, does it provide any reason for re-directing the present flow of the new logic, and, if so, why? To take just one example, would standard rules of systems of belief dynamics have to be greatly reconfigured in the light of such investigations?

A logician's account of cognitive agency involves (we say) exposure of and reflection on the epistemological presuppositions of cognitive practice. Another way of saying this is that the human cognitive agent instantiates a certain epistemological model. In its descriptive phase, the model is a record of actual cognitive practice, including the agent's own conception of his cognitivity (his cognitive auto-psychology, so to speak). It is normative phase, the model attempts to capture the conditions under which cognitive practice is sound. There is widespread agreement that the descriptive component is the proper business of the cognitive sciences, most especially the requisite branches of psychology, and that the normative component is the natural object of the logician's attention. Gabbay and I demur from this view. I shall not take the time here to explain our reservations in any detail.[2] But briefly, as we see it, the sheer difficulty of *establishing* the normative legitimacy of a model's regulae is greatly underestimated by the

[2]Interested readers may find it useful to consult [Woods, 2003, ch. 8] and [Gabbay & Woods, 2003a; Gabbay & Woods, 2005b].

received opinion, and, in any case, unsuccessfully handled in standard theories [Gabbay & Woods, 2003b]. Our own view is that the norms of rational practice are not specifiable *à priori*, but are imminent in actual practice. For this reason we are drawn to the

> *Actually Happens Rule.* To see what agents should do, look first to what they actually do. Then repair the account if there is particular reason to do so.

Needless to say, Actually Happens is a highly defeasible rule. But it has two particular virtues that make it an attractive assumption to make about human cognitive behaviour on the ground. One is that beings like us make errors, lots of them. The other is that, by and large, cognition is something that we are actually very good at. Accordingly, it is fundamental to the underlying epistemological enterprise to give due and judicious weight to this pair of facts.

2 Cognitive Abundance

As the Actually Happens Rule clearly suggests, what stands out about the cognitive practices of the human tribe is how incessantly and systematically successful they appear to be. We survive, we prosper, and from time to time we build great civilizations. Whether in running our own individual lives, or in the coordinations that underly its massive and complex collective ventures the human animal discloses a huge talent for knowledge.[3] It is rather singular, therefore, that the starting point of traditional epistemology should be the tenuousness, if not outright scarcity, of knowledge, and the toughness of the conditions under which it grudgingly makes itself available to us. There is something about this philosophical fastidiousness that is right-headed; but something about it is dead wrong. What's wrong about it is its tendency to downplay the sheer luxuriance of the human cognitive record. What is right about it is the notice it takes (albeit clumsily and often tacitly) of a problematic obscurity that attends this impressive record, and which, as we shall see, is wholly impossible to remove. These two traits repose in the coils of the above pair of tenacious and commonplace facts. One is that we are awfully good at knowing things. (The Abundance Thesis) The other — also captured by the defeasibility of the Actually Happens Rule — is that we sometimes make mistakes (and know it). (Fallibilism)

We should take care not to overstate the begrudgments of the epistemological tradition. Notwithstanding the tough exclusions of Plato, and of all

[3]Such a view is suggested by contextualist approaches such as [Lewis, 1983; Lewis, 1996] and counterfactual approaches in the manner of [Dretske, 1983; Nozick, 1993].

the standard scepticisms, there have been notable attempts by epistemologists to preserve the intuition that most (anyhow a great deal) of what we think we know we do know. Descartes attempted to demonstrate that this is so, and failed. Other philosophers have simply taken it as given. Even so, the weight overall of the tradition heavily qualifies the intuition of cognitive abundance. It does so for a reason. The reason is that the tradition's dominant paradigm stands in an awkward tension with presumptions of abundance.

In these days of causal theories of knowledge and other forms of naturalized epistemology, it is easy to overlook the grip on contemporary thinking exercised by the original insights of the Greek founders of philosophical epistemics. Given that a true belief can be the appearance of knowledge rather than the actuality of it, genuine knowledge involves the satisfaction of some third condition, *logos*. *Logos* would close the gap between the appearance and the reality of knowledge. Given its even earlier role in the birth of theoretical physics, it is not implausible to suppose that the earliest epistemologists were minded to think of a proposition endorsed by *logos* as the output of theory. By these lights, knowledge is theoretically validated true belief. Thus it is an important legacy of the early Greek epistemologists, and an enduring emphasis ever since, that knowledge — genuine knowledge — is a form of intellection, and that the intellectual work that generates and validates knowledge is a kind of *case-making*. One selects a proposition; one reflects upon it; one builds a case for it; and then one embraces it.

In Plato's hands, *logos* had already taken on the somewhat broader character of reasons for believing; and the idea that knowledge must satisfy some third condition, after belief and truth, has been with us ever since with scarcely an interruption. Whether taken in the strict sense of theoretical validation or the more general sense of reason-giving, the imprint of *logos* on western epistemology has been paradigmatically deep. Again, it presents us with a conception in which

PROPOSITION 1 (The Heritage of Logos). *Knowledge is a kind of case-making. One knows that P only if one has one's disposal a case of requisite strength to make for P.*

COROLLARY (a) *Seen this way, cognition takes on an irreducibly argumentative (or dialectical) character.*[4]

The case-making model (CM-model) of knowledge is at variance with its apparent abundance and with our apparent ease in attaining it. For what is very often missing from our actual cognitive practice is precisely this factor

---

[4] It is hardly surprising therefore that the Greek word *syllogismos* is ambiguous as between "(deductive) argument" and "(deductive) reasoning".

of case-making.

There are two ways in which the CM-model calls into question the hypothesis of cognitive abundance. In the one instance, *logos* (or whatever else is given the role of knowledge's third condition) is interpreted too strictly. So, if a knowledge of P is the true belief that P plus, as the third condition, provability in some or other theory of P-hood, most of the things we think we know we don't. Similarly, if the third condition is taken as conditional probability high enough to confer what logicians call inductive strength, a great deal of what we think we know we don't. Apart from unrealistically austere interpretations of the third condition, there is a second kind of case. A great deal of what we think we know is unsupported by anything that realistically qualifies as case-making. In the first instance, the cases that have been presumed to have been made fail to hit an unrealistically high standard of sufficiency. In the second instance, no case is made.

Here, too, we should be at pains not to overstate the dominance of the CM-model. It is perfectly true that there are ways of giving reasons in which there isn't the slightest pretense of case-making. There appear to be lots of situations in which the belief that P is supported by considerations as commonplace as that one remembers seeing that P or that one became aware of P on the sayso of one's mother. Even so, the rich history of our Attic inheritance does show a dominant preference for more robust notions of reasons than these. Arising from this resistance are two possibilities, each of which has had quite long runs. One is to reject the putative knowledge that P. The other is to take the stated reasons enthymematically, as but part of a more elaborately argued but tacitly engaged argument for P. (Thus the citation of one's memory is presumed to be embedded in surrounding generalizations about the reliability of memory, and one's citation of the sayso of another is presumed to be lodged in quite a complex set of principles about epistemic cooperation).

Anyone drawn to the Abundance Hypothesis is bound to resist these alternatives, the first because it contradicts the hypothesis outright and the second because there is nothing in our actual cognitive practice to suggest that we typically *make* these quite complex cases that are thought to be tacitly in play. Of course, it is not to be denied that theories about the general reliability of memory or about the overall efficiency of epistemic cooperation might well be *true*. If so, it is perfectly open to someone — an epistemologist or a philosophically minded cognitive psychologist, for example — to make a case for my knowledge that P on precisely those grounds. So case-making of this sort is possible. But it is not commonplace. It is rare. Most would-be knowers haven't a clue about how to construct them.

It is easy to see why the CM-model would be so attractive to logicians. Logicians are case-makers *par excellence*. But if our present reservations have merit, logicians must discipline their enthusiasm. In their capacity as investigators of how actual individuals employ reasoning to assist in the transaction of their cognitive agendas on the ground, logicians should think twice about imposing case-making standards of unattainable strictness. They should also show forbearance in not making case-making of an even realistic degrees of strictness an invariable condition of knowledge on the ground. In other works, logicians should lighten up. They should turn their attention to the hard question of how reasoning works in contexts of cognition in which case-making is absent.

Of course, all of this pivots on a decision to take the Abundance Hypothesis seriously. The epistemological tradition reveals a readiness to give up on it that may well be premature (and I think is premature). For the purposes of this chapter, I intend to show the hypothesis a certain steadfastness. But steadfastness is not dogmatism. It is possible that the Abundance Hypothesis will have to be abandoned. Suppose that this chances to be so. This would provoke an interesting and difficult question about the value of knowledge, a question to which one of the answers is that

PROPOSITION 2 (Knowledge debased). *Since we survive and prosper and sometimes build great civilizations in the absence of knowledge, that is to say, under conditions of widespread ignorance, knowledge is of no essential value to these achievements.*

Needless to say, the CM-model could be wrong and the Abundance Hypothesis correct. Anyone who finds Proposition (2) to be sufficiently unappealing may also see it as motivating serious consideration of this possibility.

3 An Asymmetry

Here is how we are presently positioned. The abundance hypothesis is taken as a fixed starting point. The CM-model is rejected as inadequate. We have it from this pair that

PROPOSITION 3 (Recognizing knowledge). *Given their general success as cognitive agents, individual humans owe this achievement either to their facility at recognizing the presence of knowledge or in default of such facility. Epistemologists should give their attention to this. For example, if knowledge is recognizable how can fallibilism be true. If knowledge is not recognizable, how can it be a guide in life?*

Each of these forwards an interesting and philosophically problematic possibility. In each case, analytical headway will depend in part on what a cognitive agent is like, on how he is "put together" and on what he is "good for".

So, then, what is a cognitive agent like? What is it like to be a knower? Here is part of an answer to that question. Legions of epistemologists have called attention to the fact that in one of its most common meanings the word "believe" admits of a crucially important first-third person asymmetry. In this usage, when Y says of X that X *believes* that P, X would say of *himself* that P. In other words, in this present use, belief is first-person knowledge-ascription. Believing that P is, from the first-person perspective, knowing that P. Philosophers have been right to observe that *self*-ascription of belief constitute a kind of attenuated or qualified subscription to the proposition at hand. But in the present meaning of the term, the *other*-ascription of a belief that P leaves it entirely open that the person to whom the belief is attributed might hold P assertively and without qualification or hesitation (and would be right to). Accordingly, for the sense of "believes" in question,

PROPOSITION 4 (Belief as knowledge-ascription). *Whenever it is true for Y to say of X that X believes that P, it is also true that X takes himself as knowing that P.*[5]

This is part of what a cognitive agent is like. Cognitive agents satisfy the Asymmetry Thesis. The Asymmetry Thesis has a noticeably Gricean cast to it. While widely-held, not everyone is a Gricean. I won't debate the pros and cons of this issue here. Suffice it to say that those who disbelieve the Asymmetry Thesis are at liberty to judge the adequacy of what follows with requisite relativity.

4 Irritation and Relief

Consider now an agent X's cognitive target K. Suppose that K is attainable only when X is in the requisite epistemic state k. We could think of K as wanting to know who shot Cock Robin. Let k be the state in which it is true to say that X knows that so-and-so shot Cock Robin. X's target K is occasion of a kind of *cognitive irritation.*[6] X is so constituted and so related to K that he aspires, or is disposed, to be in a state in which the irritation is relieved.[7] We have known at least since the presocratics that although

[5]Not conversely, however. When Y asserts that X believes that P, it does not follow that Y would say of himself that he knows that P.

[6]We must take care with the metaphor of irritation. Not every irritation of the human system that is put right by the requisite causal adjustments is something the human agent is either conscious of or openly desirous of remedying. Given that cognition can be so deeply implicit, we require the same latitude be extended to the idea of cognitive irritants.

[7]Such aspirations flow from what St. Augustine calls "the *eros* of the mind". In [Woods, 2005a] it is called "cognitive yearning".

being in k is the state that X is required to be in for K to be attained, it is *not* required for X's cognitive irritation to be relieved. Irritation-relief is one thing. Cognitive attainment is another. From the third-person perspective, it would appear that this is not a difficult contrast to command. But, from the first-person perspective, it is a contrast that collapses, and is recoverable, if at all, only in the person's own reflective aftermath. When the perspective of that reflective aftermath is at hand, the first-person can now say what the third-person could have said all along: X only believed that so-and-so killed Cock Robin, P, rather than knowing it. When X is in a state of belief that relieves the cognitive irritation occasioned by K, he is in a condition which he takes to constitute attainment of K. Not only is that state, b, not the same as k, but X's being in k carries no *phenomenological* markers over and above those carried by b. Accordingly,

PROPOSITION 5 (Phenomenologically structured inapparency). *By the psychological structure of an individual's cognitive agency, the difference between being in b and being in k is phenomenologically inapparent. So when one indeed is not in k, being in b disguises that fact.*

So we must allow that

PROPOSITION 6 (The downside of belief). *Belief is both a condition of knowledge and an impediment to its attainment.*

In so saying, we see that the traditional approach to knowledge is defective. It rightly insists on the indispensability of belief for knowledge, but it ignores, or downplays, its impedimental role. If this is right, then the capacity for, indeed the likelihood of, false apparency is structured by the phenomenology of cognitive states and reinforced by one's auto-psychology. It seems to be a rather *quite general* factor intrinsic to the possession of any b-state in relation to any K that calls for attainment by way of a k-state.

In our disposition to confuse relief with attainment there need be not the slightest hint of fatigue, intoxication, or brain damage. So our present confusion seems not to arise from what psychologists call "performance errors". Given that such confusions appear to be intrinsic to the phenomenological structure of cognitive states, it lies more in the ambit of the "competence error", hence reflective of an objective fact about how individual cognitive agents are constituted.

Some readers may find Proposition 6 to be excessively paradoxical. Others may see it as trivial. Granted, to believe that P is from the first-person point of view to experience it as known. But whereas belief is in this way an impediment, it is also a way to advance knowledge, since we also know (*de dicto*) that some of our beliefs are false. Knowing this adds an element of conditionality to our beliefs and a disposition to change our beliefs and to set out mechanisms of self-correction [Stalnaker, 1984]. By these lights, Proposition 6 is both needlessly paradoxical and rather small beer. For the broader fact is that if one cannot in the here and now recognize the falsity of what he experiences as true, this is something he can do later under press of new information or upon further reflection.

These are the standard objections. They are also driven by rather deep intuitions. They require to be answered. This I shall do in the section after this one. But I want first to acquaint the reader with the notion of *epistemic bubbles.*

5 Satisfaction and Embubblement

Relief of cognitive irritation carries with it an element of satisfaction. In this context, "satisfaction" carries an engaging ambiguity. One's belief that P produces the satisfaction constituted by the relief of the irritation occasioned by not knowing what one wanted to know. In this sense, belief is a kind of balm, calamine lotion for excited minds. But belief also constitutes satisfaction in an epistemic sense, according to which when one is satisfied that P one takes oneself as knowing that P (or knowing that it is reasonable to accept P). Accordingly,

PROPOSITION 7 (Satisfaction: "by" and "that"). *Whenever an agent X actualizes a third-person attribution of belief with respect to some proposition P, X is in a state b such that X is satisfied by b and is satisfied that P.*

COROLLARY 7 (a) *To the extent that b-states constitute satisfaction in this second sense, a b-state with respect to P is a natural inhibitor of further enquiry into P.*

A useful way of summing up our present reflections on the phenomenologically structured collapse of the distinction between belief and knowledge in the first-person case is to say that

PROPOSITION 8 (Epistemic bubbles). *A cognitive agent X occupies an epistemic bubble precisely when he is unable to command the distinction between his thinking that he knows P and his knowing P.*[8]

COROLLARY 8 (a) *When in an epistemic bubble, cognitive agents always resolve the tension between their thinking that they know P and their knowing P in favour of knowing that P.*

6 Third-person Corrigibility

Philosophers have long been aware of first-and third-person asymmetries. But it is an awareness that has not prompted a proper appreciation of the

[8]One sees stirrings of the bubble idea in [Rozeboom, 1967].

problems to which asymmetries give rise. A case in point is the uncritical assumption of what we shall call "third-person corrigibility". On this view, the first-person's phenomenologically guaranteed exposure to the risk of false apparency is open to correction from the third person perspective. Accordingly, whereas I myself cannot (here and now) command the distinction between thinking I know that P and knowing that P, I am open to instruction from a third-party, who may set me straight with a counterexample, or challenge me to supply supporting reasons. If I am suitably critical and reflective, this is a role that I might sometimes play for myself. Or so the story goes.

But consider this. Each of us is involved in his or her own epistemic bubble. Third-person corrections take the form of what the correcting agent takes himself and others to know. If Y is such an agent and if Y is directing his corrections to something that X claims to know (and which, from his perspective, he cannot help but think he knows), Y operates from Y's own first-person perspective. On the face of it, this may appear no impediment to correction, since the empirical record is replete with instances of corrective success. But the record confuses *change* with *correction*. It does so because the change in question is precisely a change to what the agent can only think of as target-attaining. Y may indeed prevail upon X to abandon P for Q. In this he might succeed. But he cannot succeed unless X now ceases taking himself as knowing that P. What cannot be denied is that, owing to critical intervention by others, or by himself after the fact, or, as we might say, by Nature herself, the contents of X's epistemic bubble *change*. Whether the changes count as corrections of mistakes or cognitive improvements of other kinds depends entirely on what X and Y now know. But since X and Y are enclosed in their respective bubbles, the distinction between taking themselves as knowing and actually knowing collapses. Accordingly,

PROPOSITION 9 (Apparent corrigibility). *Since each of us is in his own epistemic bubble, the distinction between merely apparent correction and genuinely successful correction exceeds the agent's command.*

COROLLARY 9 (a) *As before, the cognitive agent from his own first-person perspective favours the option of genuinely sound correction.*

COROLLARY 9 (b) *Within an epistemic bubble the distinction between belief-change and belief-correction is also "resolved" in favour of the latter.*

Whereupon we have it that

PROPOSITION 10 (The bubble thesis). *Epistemic bubbles cannot be popped by cognitive resources available to beings like us.*

Traditional epistemology has a large stake in discouraging if not the very idea of epistemic bubbles then certainly the sheer extent of its applicability. It leads, these epistemologists say, to the worst kind of scepticism — an instance of what Ralph Barton Perry dubbed the "egocentric predicament". It lands us in the disagreeable position of having to concede that for beings like us it is impossible to determine whether we or any of our fellows is ever in a state of knowledge. What, pray, could be a more intellectually destructive thing to concede? Surely this is a defeatist throwing up of our collective hands. Surely this is radically odious and perverse philosophy, making all attempts at recovery subject to the same corrosiveness that flows from the intractability and pervasiveness of epistemic bubbles. Discouraging as prospects for recovery would appear to be, philosophers have routinely mounted two types of attack against the Bubble Thesis.

1. *Attack one* on Proposition 10 attempts to attenuate the idea of epistemic bubbles. While conceding that, for a great many propositions, the Bubble Thesis is correct, there are *kinds* of propositions to which it does not apply. Hence: "analytic", "self-evident", "indubitable", "axiomatic", and so on.
2. *Attack two* is aimed at the ubiquity factor, i.e., the claim that there is no third-person perspective that escapes the enclosure of an epistemic bubble. Perhaps the card most often played by these critics is that of the deductive closure of the properties that fall within the ambit of attack number one. Thus, given that P's analyticity (etc.) breaks it free from a holder's epistemic bubble, so too for any Q that is in the deductive closure of P.

It is not difficult to see that these attacks are doomed to fail. One may think that P is analytic without its being so, and one might think that Q follows from an analytic truth without its doing so.

When together the present pair of objections constitute one of the more dominant conceptions of *foundationalism*, albeit with the details left out. It is an interesting question whether foundationalism has a future in epistemology. This, too, is not the place to engage that question. In any case, the issue before us is not foundationalism but rather *incorrigibilism*. Let us say that

PROPOSITION 11 (Incorrigibilism). *Incorrigibilism is true for any proposition P such that being in a b-state with respect to P is impossible unless one is also in k-state toward it.*

The epistemological tradition is chock-a-block with attempts to give to P the widest possible interpretation. I have no doubt that there are values of P for which incorrigibilism is true. To be sure, such Ps are bubble-resistant. But they offer no relief to the legions of other propositions P which incorrigibilism fails. It is these propositions that are my present concern.

In our routine cognitive activities we are not aware of our containment within epistemic bubbles. We are precluded from such recognition by the threat of blindspot [Sorensen, 1988]. Given the phenomenological structure of cognition, in satisfying the third-person sense of "believe" with respect to P, I know that P (hence know that P is true). If *concurrently* I were to satisfy the third-person sense of "believe" with respect to the proposition "I cannot know whether P is true", I would have constructed a blindspot *for myself.* I will have lost all purchases on what my position actually is with respect to P. I would not know what I was talking about. Accordingly,

PROPOSITION 12 (Immunization). *Although a cognitive agent may well be aware of the Bubble Thesis and may accept it as true, the phenomenological structure of cognitive states precludes such awareness as a concomitant feature of our general cognitive awareness.*

COROLLARY 12 (a) *Even when an agent X is in a cognitive state S in which he takes himself as knowing the Bubble Thesis to be true, S is immune to for as long as it is operative for X.*

If the Bubble Thesis is scepticism, it is well to note that the functioning cognitive agent pays it no heed, and is not, operationally-speaking, in the slightest disturbed by it. This, of course, is a Humean point. Scepticism is true, Hume observed, but there can be no question of the possibility (or the value) of actually bending our cognitive practices to it. This happy behavioural discompliance Hume took to be a matter of habit. It is an understatement, to say the least. But perhaps "habit" commended itself to Hume as alternative to "cause", which he also had thought that he had effectively disabled for serious philosophical service. Were it available to him, "cause" would have carried a heavier explanatory load. But better still is "belief", to which we shall shortly return.

7 Degrees of Belief

It might be objected that the beliefs that satisfy the Asymmetry Thesis constitute only a small subclass of the beliefs an agent is subject to, and that for those latter the thesis clearly fails. Beliefs in this larger subclass may be called, correspondingly, qualified beliefs, and those in the smaller subclass we could think of an unqualified.[9] Such beliefs may well include

[9]Another way of labeling the same constrast is with the distinction between partial and full belief [Adler, 2002].

most propositions about the distant future and past, propositions about the mental states of others and, on some tellings, most of the propositions of theoretical science. A qualified or partial belief that P is a state of mind in which the agent experiences P as other than known. Accordingly, it might also be said that existence of qualified belief affords us away to escape our respective embubblements. Why not simply resolve to substitute for our unqualified beliefs beliefs of the highest strength compatible with their successful resistance of the asymmetry thesis? This is not an escape that holds out much promise.

PROPOSITION 13 (The unwilledness of belief). *The problem is that the substitution of partial for full belief is not something that lies within our psychological control. It is not in general a performable escape.*

This brings us to a second objection. If it were true that most of a human agent's beliefs are qualified or partial beliefs, if it were the case that most beliefs come in degrees sufficiently below the bar of fullness to render the asymmetry thesis untrue, the Bubble Thesis would indeed be pretty small beer. For only a slight percentage of our beliefs would then comport with it. Doesn't this trivialize the Bubble Thesis? The answer is that it does not. The suggestion that it does confuses minimality with triviality. What counts here is not the extent to which the Bubble Thesis is true of all senses of "belief", but whether it is true of those states experienced in the first-person as states of *knowledge*? The answer is that it fits such states in every case.

For what I have just said to be true, something else cannot be true. This is the idea that although belief is a condition on knowledge, it need not be full belief. In one variation, one might be said to know that P if P is true and one's belief that P is and one's reasons for believing P. It is not a possibility that I am inclined to avail myself of. Here is why. As long as we are operating within the true-belief-plus-reasons parameters, it is implausible that when reasons for P are insufficient to override the partiality of the belief that P, they are likewise insufficient to close the presumed gap between belief and knowledge. A stronger way of saying this is that belief is like pregnancy. No pregnancy is a partial pregnancy. No belief is a partial belief. Of course, this is not to say that there *are* no partial beliefs. It is just that partial belief is not belief (as white chocolate is not chocolate). I may suspect that P, I may be inclined to believe that P, I may be pretty sure that P, I may believe that it is practically certain that P, I might strongly believe that P might be true, I might find P to be highly plausible, I might be prepared to bet that P, and so on. These are cognitively important states which no epistemology can afford to be indifferent to.[10] But nothing is gained by taking them all as kinds of *belief*. Conceiving them thus adds not a jot of clarity as to their nature and function, and it runs roughshod over a centrally important distinction between a state of mind in which an agent takes himself to know that P and a state of mind in which he takes himself not to know P.

8 The Fugitivity of Truth and Reasons

Most of the going philosophical theories of truth emphasize its "disquotational" character, as conveyed by Convention T and like biconditionals:

- P is true if and only if P.

In its right-to-left implication, Convention T has it that if P then P is true. This bears closely on the epistemic bubble phenomenon. It is an implication that holds up in opaque contexts. So if an agent believes that P, he believes that P is true. Since "believes" is here a correct ascription only from the third-person perspective (or the first-person *ex post facto* perspective), it follows that, when in that condition, a cognitive agent takes himself to know that P *is true*.

Here, then, another staple of traditional epistemology collapses in the first-person case.

PROPOSITION 14 (Inapparent falsity). *The putative distinction between a merely apparent truth and a genuine truth collapses from the perspective of the first-person awareness, i.e., it collapses within epistemic bubbles.*

COROLLARY 14 (a) *As before, when in an epistemic bubble, cognitive agents always resolve the tension between only the apparently true and the genuinely true in favour of the genuinely true.*

Corollary (14a) reminds us of the remarkable perceptiveness of Peirce, who points out in the epigraph under the title of this essay "that the production of belief is the *sole* function of thought" (emphasis added). What Peirce is calling attention to is that enquiry *stops* when belief is attained. It is wholly natural that this should be so. But, as Peirce was also aware, the propensity of belief to suspend thinking constitutes a significant economic advantage. It discourages the over-use of (often) scarce resources.

Accordingly,

PROPOSITION 15 (Fugitivity of truth). *Within epistemic bubbles, truth is a fugitive property. That is, one can never attain it without thinking that one has done so; but thinking that one has attained it is not attaining it.*

[10]For one thing, people are disposed to take their beliefs and partial beliefs as guides to action.

COROLLARY 15 (a) *Within epistemic bubbles, cognitive agents lack the means to apprehend these fugitives in a principled way. That is to say, the distinction between merely apparent and genuine truth-apprehension also collapses.*

COROLLARY 15 (b) *It is easy to see that reasons for believing P are subject to this same fugitivity. One cannot have a reason for believing P without thinking one does. But thinking that one has a reason to believe P is not having a reason to believe P.*

Of course, all of this is so from the point of view of epistemic bubbles, imposed by the phenomenological structure of cognitive agency. As we saw, the received wisdom is that epistemic bubbles can be effectively lanced by third-person intervention (whether exercised by another agent or the original agent in *ex post facto* critical reflection upon his own situation). Embedded in this supposition is the powerful intuition that objective measures exist for the discovery and repair of cognitive error, an intuition that every epistemology yet developed has honoured, save only for the most paradoxically sceptical of them. It is also a view well-entrenched in the cognitive psychology literature. But, as we have seen, it fails. From the first-person point of view, which is the point of view of every person, one cannot command the distinction between belief-change and belief-correction.

9 The *CM*-model Redux

As we have it now, two of the three parameters of the standard tripartite conception of knowledge suffer the same fate as knowledge itself. Within epistemic bubbles they are all fugitives. Since the CM-model is arguably the most prominent examplar of this tripartitism, this is devilish news for it. But further bad news also lies in wait.

The CM-model sees knowledge as a kind of case-making. So conceived of, your knowing that P pivots in large part on what you have to say for yourself. If you lack the means to put your case for P, you are in a condition of cognitive inadequacy. You may think you know it, but you don't. Knowledge is not true belief. It is *defendedly* true belief, and here your defence has failed.

If we are to judge from our actual cognitive practice, the CM-model is a considerable liability for knowledge. It is a distortion of the cognitive record. This comes about in two ways, neither of which is given adequate weight in the epistemological tradition. In the first way, reasons are frequently — indeed routinely — beyond the knower's power to articulate then and there, if ever [Woods, 2000]. Nor are these inarticulacies performance errors. They are not experienced by knowers as errors of *any* kind. Even so, possibly they are errors; perhaps we should conceive of them as imperfections of competence. But, short of an affection for Proposition (2), there would be a cost. Masses of what we take for knowledge would be wiped out.

The second way in which the *CM*-model is inimical to knowledge flows from the dialectical structure of case-making. Of the three conditions on knowledge,[11] only belief (in the third person case) is not a fugitive property. In the sense in which we are using the word here, there is no contrast to draw between believing that *P* and thinking that you believe that *P*. Truth and reasonedness are different. They are fugitives. In each case, the contrast between appearance and reality collapses phenomenologically. Thanks to the structure of his *b*-states, the cognitive agent "opts" in each case for the right-hand member of the relevant pair.[12] But, given the fugitivity of these properties, all that the would-be knower can be guaranteed of is that it is the *left-hand* members of these pairs that actually obtain. The best that he can do in defending his knowledge-claim that *P* with reasons for holding *P* is to proffer what he takes for reasons. This triggers a nasty little problem for *CM*-knowledge.

PROPOSITION 16 (A problem for the *CM*-model). *If the reasons requirement is a condition on knowledge, then knowledge itself lies obscured by the fugitivity of reasons (and truth). If, on the other hand, it is not a condition as knowledge, the CM-model is not the right model.*

10 Causation

Given the centrality of *b*-state capability in a human agent's orientation towards and participation in the world, the relation of *cause* has a correspondingly deep function. Here, too, it is striking to see the extent to which philosophers have been drawn to a certain picture of belief-acquisition. We might call it the "propositional select-reflect-choose" model (the *P*-model, for short), according to which a belief is acquired when a proposition is selected from a space of propositions (or otherwise presents itself), and then is reflected upon, and is accepted, rejected or deferred in light of considerations having to do with considerations for and against.

Certainly the *P*-model is often clearly in play in human life. (It is a standard variation of the *CM*-model.) But it carries suggestions that cannot be true of belief-formation as such. One is that belief-acquisition is voluntary. (See again Proposition (2)). The other is that belief-formation is, case by case, discretionary.[13] With regard to these assumptions, a fur-

[11] Or, with the Gettier-problem in mind, three+ conditions.

[12] the scare-quotes are a useful reminder that in no literal sense is one able to choose the character of how things present themselves to one. See the section to follow.

[13] A belief is voluntary when it can be willed into existence. A belief is discretionary

ther distinction lends some support to the P-model. For in the manner understood by Cohen and others [Cohen, 1992; Woods, 2001; Pinto, 2003; Woods, 2003] the *acceptance* of a proposition may be said to be both voluntary and discretionary, whereas a state of belief towards it may not. While, these assumptions appear to work for acceptance, they don't work for belief. What is more, given the time and effort required to submit a proposition to scrutiny prior to accepting, rejecting or deferring commitment to it, the b-states that human agents routinely are in from one moment to the next are not reflected upon and not subjected to the scrutiny required by consideration of evidence or presentation of reasons.

This can be explained by factors other than that the individual has an interest in making his cognitive way through life without needlessly overspending his limited cognitive resources. The more fundamental explanation is that given the dire necessity to be in b-states as conditions of survival, prosperity (and mental health), the sheer drive of the necessity is reciprocated by an abundant and low-cost accommodation of it, namely, a *causal* accommodation.

Against the P-model *or* intellection-model, one finds the *causal responsiveness* (CR) model, which sees knowledge in reliabilist terms. While the Bubble Thesis is logically independent of epistemic causalism, it is especially congenial to the CR-model. It is a congeniality secured by the deep centrality of belief to the bubble account of knowledge and to the causal nexi in which belief is necessarily embedded.

Cohen has performed a valuable service to epistemology in emphasizing the contrast between belief and acceptance. It is a contrast that helps us see how deeply wrong it is to speak of beliefs as if it were in some direct way *up to us* whether to have them or not. A habit so bad should excite some interest in learning how it came to pass. An obvious part of the answer is that "belief" 's ambiguity sometimes leads us to attribute to belief characteristics of acceptance. In other words, ambiguity sometimes gives rise to equivocation. Another perhaps less obvious part of the explanation is that the causal thralldom constituted by our belief that P is disguised not only by the sense of welcome given to P by believing it, but also by the plain fact of our ability, and often our readiness, to reflect on whether P is actually true or properly grounded. These actions do indeed lie within our discretion and are matters of the will. They are "up to us". It may strike us that if it is up to us whether to submit a believed proposition to scrutiny, it

when to have it or not is subject to the agent's approval; it is "up to" the agent. In some respects, voluntariness and discretionariness are opposite sides of the same coin. On the one hand, I may acquire a belief by willing that I have it. On the other, I may preclude my having a belief by refusing to have it (and I may rid myself of a belief by expelling it).

must also be up to us whether to continue to believe it. But it is precisely this that is not up to us.

11 The Realist Stance

Given what we know of the empirical record, human beings are rigged to *aim* for knowledge and to be *satisfied* with belief. "Rig" here is a large expository convenience, leaving undealt with the important and difficult question of how much of the rigging is nature and how much nurture. This is not our question here. It suffices to note that this is how the human cognitive agent is structured. Bearing on this is a related matter which also raises questions about hardwiring and learning. Here, too, what concerns us is that this is also how the agent is (somehow) structured. The "it" in question is our propensity to experience and understand the world as if realism were true, to take, so to speak, the *realist stance* [Woods, 2003]. Doing so is as natural, and as urgent, as breathing. Part of what makes this so may well have something to do with its survival-value. Realism is an especially efficient way of paying attention.

The realism to which we are so drawn is a realism of externality and independence. Aside from our inner states, the world is out there and is, in its basics, the way it is without regard to what we may chance to think about it. Correspondingly, a realist proto-epistemology commends itself. There is little that our thinking makes so, and when we get something right, is usually the case that it was right all along and would have been right had we not been part of the scene. So viewed, truth is an objective property of truths, and judging something to be true involves an engagement, both highly complex and difficult to describe, with the outer world.

Realism is a siren. It is irresistible. Even when we are openly and transparently irrealist in some of our theoretical ventures, realism re-imposes itself sooner or later. Sextus' insistence that it cannot be known whether there is an external world is an insistence that *reality is such that* it cannot be known whether there is an external world, that, in the light of how things actually are, such knowledge cannot be attained. We are entrenched recidivizers under press of our realist urges. Post-paradox set theory is a modern case in point. Once Russell's contradiction was seen as striking intuitive set theory dead, Russell was clear about how the rehabilitation of sets would have to go. Given that sets could not now respond to a philosophical analysis, there remained only the *ad hoc* option of mathematical (i.e., nominal) definition [Russell,1903, pp. 15, 27, 112]. Thus stipulated, these new "sets" would be offered to the research community for the only validation open to them, namely, absorption into settled usage. The originators of the new set theories, those who were philosophically serious, entertained no notion of realism with regard to their inventions. In the end, however, ZF won the competition for settled usage, and within a generation the realist impulse simply reinstated itself without fanfare or even commentary. ZF came to be regarded as an intuitive and authoritative description of what sets really are.

Of course, there are some attractive and powerfully argued defences of anti-realism, e.g., *constructive empiricism* [VanFraassen, 1980]. It is well to observe that can't-help-it-realism is nothing like the view that everything is real or even is such that we can't help thinking it real. Can't-help-it realism is wholly compatible with local judgements of unreality, for example, with the judgement that phlogiston is unreal. However, trouble lies in wait when judgements of unreality attain ever greater degrees of globality. One can certainly experience the world aphlogistonistically. But one cannot experience the world acausally.

The link that binds can't-help-it realism to the Bubble Thesis is truth. We are naturally disposed to be realists about truth; and since belief is disquotational, believing that P is believing that P is true. When beings like us satisfy a third-person ascription of belief with respect to a proposition P, we take our b-states as a k-states with respect to P's truth. We find ourselves in a position in which P's truth aggressively presents itself. Thus does the realism of truth reinforce and thicken the satisfaction of belief.

In our somewhat Kantian moments we might say that realism is less a deliverance of the intellect than an irreducible condition of there being any experience at all. Beings like us cannot be in the world as knowing beings except that the world is experienced as external and independent. It is as pervasive and embedded an aspect of experience as the imperative that we deal with the world from the inside out.

If we are interested in deliverances of the intellect, we should pause to reflect on a point of importance. Human agents possess a remarkable duality. We are capable of discerning what our experience imposes upon us, and we are (sometimes) capable of figuring things out for ourselves, or so it appears. Sometimes the results of these figurings refine our appreciation of what experience intrudes upon us. Sometimes such figurings will correct the details of experience. In some cases, we may even reason our way to outcomes radically hostile to what our experiential natures require us to believe. In such cases, the deliverances of our intellects are at variance with our experience in ways interesting enough to command the attention of philosophers. Scepticism is the most ancient of such deliverances. It may upon reflection be accepted as true. It may sustain entire intellectual schools. But it cannot be experienced. It cannot be comported with behaviourally. It cannot be acted on (except professionally — people publish books about it).

There is a distinction that cognitive scientists have found it useful to draw. It is the distinction between Strong and Weak AI, initially proposed by John Searle. According to Strong AI, computers (or computer programs) have minds in the same sense that we ourselves do. According to Weak AI, while there is no sufficient reason to think that Strong AI is actually true, it is a methodological aid to our theoretical probes of cognition to assume that it is. A similar distinction might be drawn between Strong and Weak Realism. Let us understand by Strong Realism that realism is true and is a constraint on knowledge. Accordingly, Weak Realism would be agnostic about Strong Realism but would assert that we owe our epistemic successes (such as they may be) to operating in the world as if Strong Realism were true. So we see that can't-help-it-realism is a variant of the Weak member of the pair.

It should be repeated that neither can't-help-it-realism nor the Bubble Thesis calls into question the cognitive flourishing of the human race. There is nothing in these views to disturb the conviction that by and large, and given our interests, beings like us have the *right* beliefs, and have them under conditions of intersubjectivity that underwrite all the requisite co-ordinations. So, again, we survive, we prosper and occasionally we build great civilizations, for all of which massive cognitive coordination is indispensable and delivered on. Yet nothing in our cognitive flourishing upsets either the fact of error or the pervasiveness of embubblement. The beliefs that produce this flourishing are driven by Weak Realism and are taken for k-states. But, again, one cannot be in a k-state without being in a b-state, and being in a b-state, when it is not a k-state, is being in a state that conceals the fact.

12 Error

Before moving on, however, we should say a word about the spectre of scepticism. In its most general form scepticism asserts that it is compatible with everything we know that some class of believed propositions P is untrue. Whatever punch attends scepticism is largely determined by the values of 'P'.[14] According to the Bubble Hypothesis, the values of 'P' is virtually unconstrained. On the face of it, then, ours is a wrenchingly radical scepticism, hardly worth the time of effort to dismiss it. In fact, this is much too fast. This can be seen by noticing a distinction between

1. *error-susceptibility scepticism*

[14] Unconstrained, except possibly for such old chestnuts as one's knowledge of one's own pain and the like.

2. *defective-reasons scepticism.*

In epistemology's vast 2,500 year-old literature, sceptism is dominantly one or other variation on (2). The sceptic argues that in our purported knowledge of the external world or other minds or the the past, there is something about the *subject-matters* of these classes of epistemic claims that makes our beliefs about them, in some way or other, *unreasonable.*

Scepticism in sense (1) is significantly different. It pivots entirely on three commonplace assumptions: first, that for beings like us error is to some extent unavoidable; second, that by its very nature error's efficacy lies embedded in its concealment; and, third, that since every cognitive agent operates from within a bubble, any act of error-detection and error-correction is subject in its own right to the concealedness of error. Thus scepticism_1 is a thesis about bubbles, and makes no complaint against particular *classes* of knowledge-claims. Since it conceives of reasons for as a fugitive property, itself subject to the distorting potential of embubblement, scepticism_1 has no stake in showing that knowledge-claims about the external world or other minds or the past (or whatever other standard target of scepticism_2) are unreasonable, or that, in holding such beliefs, human agents are unreasonable. Scepticism_1 hardly deserves the name of scepticism. It is a common sense thesis about error and embubblement.

Of all the more or less established approaches to epistemology, perhaps it is *fallibilism* that is least at home with scepticism_1. Fallibilism is the doctrine that

PROPOSITION 17 (Fallibilism). *Not only are the cognitive strategies deployed by beings like us liable to error, and not only is this known to us, but because error is corrigible, our use of such strategies is a reasonable way for us to manage our cognitive behaviour.*

Fallibilism is an *error-tolerant* epistemology, and stands apart from the weight of the tradition in which *error-elimination* is a central part of the project of knowledge. Granted that for beings like us error is to some degree unavoidable, why should we think it reasonable to deploy strategies that we are certain will on occasion lead us astray? Why is it not better simply to opt for strategies that won't let us down in this way? The answer pivots on the nature of error and the nature of *us.* The two natures fall into an interesting kind of alignment. It lies in the nature of error to conceal itself. It lies in the nature of us to be attracted to the error-free. Since the distinction between apparent and genuine freedom from error is not phenomenologically available to us, it is easy enough to understand why we would be error-susceptible. It is also easy to see why it can't be considered unreasonable to proceed with our cognitive lives even at the cost of occasional error, since the alternative is incomparably worse — paralysis. (Downing tools is not a cognitive option.)

Fallibilism has a downside and an upside. The downside is that individuals make errors, lots of them. The upside is that we have rather good feedback mechanisms, which gives us an aptitude for *post ex facto* error-recognition, and both the inclination and the wherewithal for error-correction. As standardly conceived of, fallibilism's upside more than compensates for its downside and, this being so, it can be allowed that pursuing error-prone strategies is not an unreasonable thing for beings like us to do. But if we are right to say that the phenomenological structure of human cognitive agency precludes the popping of epistemic bubbles, fallibilism itself is shot through with unexpungeable fugitivity. Wherewith a pressing question: Has the fallibilist assurance of the reasonableness of adopting error-prone strategies now lost its footing?

13 Cognitive Economies

There is a distinction that economists and rational choice theorists find it useful to draw. It can be drawn for agents or for the action-strategies employed by agents. An agent *satisfices* when he settles for an outcome which is "good enough". (Correlatively, a satisficing strategy is one that halts upon the attainment of the good-enough). An agent *maximizes* when he settles only for the largest gain. (Similarly for strategies.) An agent *optimizes* when he settles only for the best outcome. (Similarly for strategies.) All this bears on our discussion in an especially curious way. When an agent is seized with the desire to know whether P is true, we may regard him as having set himself on the course of cognitive optimization. However, given that, phenomenologically speaking, attainment is indistinguishable from relief, belief often plays the role of *faux* optimization, and makes of the agent an inadvertent satisficer. It is worth emphasizing that this is so whenever the agent has set himself an epistemic target, i.e., a target that is attainable only by way of the (requisite) state of knowledge.

What this means in particular is that the epistemic character of an agent's target is *not* set by the epistemic *bona fides* of its propositional *object.* If someone wants to know whether P is *true*, he has an epistemic target. But the same must be said of an agent who wants to know whether P is *false.* The epistemicity of the target flows from the fact that it takes a state of knowledge to attain it, whatever its propositional content.

This is evident if we examine the following list of what an agent X might want to know:

1. whether P *is true*
2. whether P *is false*

3. whether P *is probable*
4. whether P *is improbable*
5. whether P *is plausible*
6. whether P *is implausible*
7. whether P *has explanatory force*
8. whether P *is explanatorily incoherent*
9. whether P *is justified*
10. whether P *is indefencible*
11. whether P *is supported by reasons*
12. whether P *is groundless*
13. why P *is true*
14. what makes P *probable*
15. what confers *plausibility* on P
16. what constitutes the *explanatory force* of P
17. *the way* to Central Station
18. how to construct an indirect *proof*
19. how to construct an indirect proof *of* P

In each case, the content of X's cognitive target changes. In some cases the properties associated with P are cognitive virtues ("true", "justified", "explanative", "reasonable", and so on). In other cases, the properties associated with P are cognitive vices ("false", "indefensible", "explanatorily incoherent", "groundless", and so on). But throughout what the agent's target aims at is cognitive virtue, namely, knowledge. He wants to *know* whether P has these virtues or vices.

14 Curiosity

Philosophers (Aristotle is one) have from time to time made fleeting reference to the role of curiosity in human life. Its function has been greatly underestimated. When, on those comparatively rare occasions it is given any recognition, the emphasis falls on curiosity as a goal of theoretical enquiry. Clearly it does play that role, but attending to it to the exclusion of other functions it also performs leaves the inference that in both its exercise and its fulfillment curiosity is a *discretionary* and *postponable* urge in human agents. Rightly, too, since our theoretical targets are frequently met with failure, and even their successes often follow after long periods of fallowness.

In a more basic sense, curiosity is neither discretionary nor postponable. In this sense, curiosity is the drive to be in b-states; or more carefully, a drive that only a b-state can still. Stilling this drive in its incessant and multivarious manifestations is as fundamental to human well-being as an adequately functioning liver. Again, it is as urgent a necessity as breathing. B-states give us the stability, coherence and intelligibility with which to negotiate the passage form one event to the next. Belief-incapability would deprive life of all leverage, of all its junctures of negotiation and command, and would afflict its sufferers with radical paralysis and uttermost impotence and disorientation; and, of course, panic.

As we saw, being in a b-state with regard to some matter is, from the first-person point of view, being in a k-state with regard to it. Given the indispensability to human life of b-states in rich abundance, that which is thus indispensable (and routinely and freely available) to human agents is, from the first person perspective, knowledge. By these lights, knowledge is richly, abundantly and freely available; it is something that, like breathing, we are rigged for and good at. Accordingly,

PROPOSITION 18 (Epistemic indispensability). *Beings like us owe survival, prosperity — and the occasional great civilizations that arise — to the capacious and largely unfettered capacity for b-states which, from the first-person point of view, are k-states.*

PROPOSITION 19 (Doxastic irrisistibility). *Whether full or partial, belief states are not chosen. They befall us like the measles.*

15 The Duality of Reasons

In the epistemological tradition it is customary to ascribe two functions to reasons for belief. One is a case-making function in regard to the belief's propositional object. The other is a normative function, aimed at the *rightness* of the agent's b-state. Part of what makes for the distinctiveness of the CM-model is its habit of running these two traits together. Thus what makes for the rightness of my b-state with respect to P is the case that I manage to make in support of P. The rightness of my state is indissolubly linked to the correctness of my advocacy. But, again, if the CM-model fails, might we not expect a reconfiguration of the case-making and normative dimensions of reasons? Yes. The normative (or "rightness") dimension of reasons is a virtually invariable component of being in a b-state at all; whereas having a case to make for the propositional object of that state is certainly not an invariable component of our cognitive successes and, in fact, very often is unlamentedly absent from them.

Perhaps there are cases in which although someone has the full belief that

P, i.e., experiences P as known, he somehow finds that he is astonished that he believes P. Of course, he might be astonished that P *is the case*. But, thinking that P is the case, the present astonishment is that he would *think* it at all. If such states are possible, then one is satisfied that P and astonished that one is satisfied that P. One would be a stranger to one's beliefs. One would be alienated from one's beliefs. It is an extremely odd, and rare, turn of events. If it scored wide swaths across the sweep of one's total beliefs, it would be an understatement that such is the way of madness.

By these lights, the normativity that is traditionally associated with protocols for case-making is relocated to the phenomenology of belief itself. In believing that P, it is routinely the case that its constituent satisfactions are expressions of the rightness the agent enjoys in being in that state. Although not a matter of strict inconsistency, it is worth repeating that it is deeply anomalous to be in a state of belief with respect to P in the absence of any trace of the doxastic comfort that betokens its rightness. If a human being has a belief on a given occasion, the default position is that his is not an alien belief. From this it follows that alien belief is indeed anomalous. In so saying, it would be decidedly wrong to leave the impression that our cognitive successes leave no room for *doubt*. The present point has nothing to do with doubt. It is a point about non-doubt, about full-bore belief. Similarly, nothing about the present point is the slightest discouragement of the commonplace that what I take as known may well occasion dissatisfaction, apprehensiveness or great unhappiness. In knowing that his government will fall later today, the Prime Minister might have been made more miserable than ever before in life, but his dissatisfaction is not epistemic. One could almost say that the more he has been shattered by what he has learned, the more his knowledge of it is *epistemically* satisfying, the more it is the *right* b-state for him to be in.

The disposition that we have to suppose our b-states to be attended by reasons (albeit tacitly much of the time) proceeds from the phenomenological structure of the satisfaction-component of those states. As we saw, a b-state with respect to P satisfies the irritation occasioned by not knowing whether P and it provides that the agent is satisfied that P is the case. Although there are exceptions, it is easy to see that in the general case the satisfaction that flows from an agent's b-state incorporates the presumption that that is a reasonable b-state for him to be in. Here again blind-spots loom. If I am in a state which constitutes me as satisfied that P, I cannot without anomaly concurrently avow that I have no inclination to suppose it reasonable for me to be in that relation to it. In default of further information, I could hardly know what my own position is.

Of course, there are exceptions, for which there is a sense of the word "just" that does the needed ancillary work. Wild hunches come to mind here. Sometimes we are wholly convinced that P yet not in the slightest degree inclined to see our embrace of it as a reasonable state in which to be. So we explain ourselves: "I just believe that P". Note well: we explain *ourselves*. We do not explain that P; still less do we "prove" that P. If you have a hunch that P you might even say that you "just know" that P. Period. The work done by "just" is to call attention to the anomaly. As with other cases, anomaly-recognition is a means of anomaly-relief. That is to say, calling attention to the anomalousness of my situation is sometimes sufficient to mitigate it, or even at times to eliminate it altogether. So

PROPOSITION 20 (Reasonableness of *b*-states). *While it is not a logically necessary condition on being in a b-state that the agent be disposed to experience it as a reasonable state for him to be in, it is a default condition for such states; it is a generic feature of satisfaction-that.*

Proposition 20 is the conduit through which reasons enter the general picture, as follows;

PROPOSITION 21 (From reasonableness to reasons). *Though here, too, the connection is not a logically strict one, being disposed to presume to take his b-state as a reasonable one for him to be in, is itself disposition to presume that reasons exist for the agent to be satisfied that P.*

There is a common meaning of "belief" according to which it is beliefs are the outputs of causal processes. Since belief is fundamental to knowledge, knowledge is fundamentally a causal matter. Knowledge is widely held to be a state of mind, and so it is. It is a state of mind concurring with the satisfaction of further conditions. As traditionally conceived of, these further conditions are not themselves states of mind. So the only state of mind that one's knowledge of P could be is a *b*-state. Thus, as far as its mental state component is concerned, knowledge too is the causal output of cognitive devices. As we have already seen, the further conditions of truth and reasonableness embed fugitive properties. Since the only state of mind that a *k*-state with regard to P is the corresponding *b*-state with respect to P, and since the phenomenology of *b*-states conceals whether it is a k-state, the concurrence that exists within the traditional triple belief, truth, reasons is that a belief is caused to exist in such a way that its object is apprehended as true and reasonable. While these latter two may be objective conditions on knowledge, from the first-person perspective they are phenomenologically present in the *b*-state that takes itself as a *k*-state.

It is well to emphasize the dispositional and presumptive character of a believer's link to both reasonability and reasons. It helps us understand how uncommon it is for us to give voice on demand to the reasons we presume to exist for a believing that P. Large numbers of philosophers are minded to think of this elusiveness as an epistemic spoiler. The truth is that, given the phenomenological structure of human cognitive agency, it could not possibly be an epistemic spoiler in the general case. What it is, however, is a *dialectical* spoiler.

16 Epistemic and Dialectical Entanglements

Proposition (21) is rather important. When a person is in a b-state with respect to P, (21) endorses a certain presumption. It is the agent's presumption that reasons exist for his being in that *b*-state. The existence of the presumption is correct, but it needs to be handled with care. The necessity is occasioned by a further presumption to which we must now attend, namely, that the rightness/case-making duality of reasons is exhaustive of what reasons are. If one asserts or presupposes that when one is in a b-state there are reasons for it, it cannot reasonably be ruled out in advance that such reasons might be asked for. It is widely assumed that when such reasons *are* asked for, the request takes the form of an *epistemic challenge*, that it arises from the challenger's conviction that his addressee's knowledge-claim is defective. Let there be no doubting it; often this is precisely what a request for reasons signifies. When this is so, it is usually attended by a further assumption, namely, that a call for reasons for believing P can only be met by making a case for P. Thus we find ourselves oddly placed. We want to accept Proposition (21), but we do not want to accept the CM-model of knowledge. So how are we to proceed? Perhaps it is best to say it outright:

PROPOSITION 22 (Giving reasons causally). *It is frequently the case that in giving reasons for believing that P it suffices to make mention of features of the belief's causal circumstances.*

Some people will dislike Proposition (22). Some of them may think it confuses reasons and causes. Others may see it as committing the genetic fallacy. Perhaps they might like to reconsider. [Davidson, 1963; Pietroski, 2000] and [O'Connor, 2001] bear on the reasons and causes question, and [Honderich, 2005, p. 331] on the genetic fallacy charge. I shall have nothing more to say about these matters here.

It will prove an efficient *entré* to the business of this section to attend what human agents often actually say when pressed for holding a belief. X claims to know that Thanksgiving falls on the second Monday of October. If asked he might say, "I remember it that way." X asserts that Charles and Camilla are soon to marry. If asked he might respond, "I heard it somewhere". The list lengthens: "I probably was told it by my wife", "I saw it with my own eyes" and so on. It takes little reflection to be made aware that each of these responses is *question-begging*. It is a fact of crucial importance. It drives us to acknowledge a distinction long ago adumbrated by Aristotle.

Everyone knows that Aristotle was the founder of syllogistic logic. Syllogistic logic is purpose-built for antagonistic *arguments*. Aristotle was also the founder of the logic of immediate inference, of which the best-known part is the Square of Opposition. This is not the place to explore these accounts in any detail. But it is necessary to sound a caution that Aristotle himself was careful to make. It is that a sound principle of *reasoning* (i.e., of immediate inference) can be a bad principle of *argument* (i.e., of syllogistic manoeuvering). Aristotle illustrates this point with a brief remark about begging the question. Consider the immediate inference:

1. All men are mortal
2. Therefore, some men are mortal.

Aristotle thinks that this begs the question.[15] But that is not an indictment of it just as it stands. What matters is that this reasoning would be wholly unacceptable if used in a dispute with another party, but acceptable if used as a part of one's own solo reasoning. What makes this so is that in an argument about whether some men are Greeks, that all men are Greeks cannot possibly convince someone whose position is that not even some are. Aristotle's view is that question-begging is not inherently a defective practice, but that it is a dialectically-unavailing move to make against a contrary-minded *opponent*.

Bearing on this in a central way is an ambiguity that attaches to the idea of "giving reasons for". In one sense — we might call it *giving* $reasons_a$ — one gives reasons for P in response to an argumentative challenge from a party who denies P or doubts it, or denies or doubts that reasons for it exist. (It is easy to see the affinity of reason-$giving_a$ and case-making.) If we now re-visit the list of reasons that an agent might actually give, we see that their question-begging character guarantees their failure as answers to this *kind* of challenge. Even so, we have already made the point that, often enough, reasons such as these are all, then and there, that an agent has to say for himself. To this we now add the point that depending on the reasons for citing them, these reasons need not be at all defective. This we can see by introducing the other half of the contrast between giving

[15]But is Aristotle right? Couldn't I get you to accept Q by appeal to an entailing P when you accept P and do not (antecedently) appreciate the entailment? No. Aristotle's case assumes that it is Q that is in dispute. If you dent Q and accept P, you cannot consistently concede the entailment.

reasons$_a$. We label this half "giving reasons$_{non-a}$". In so doing we pick up on the point that often reasons are requested, and offered, for one's believing P in which there isn't the slightest degree of adversarial challenge. Reasons$_a$ are responses to the counterclaim that P isn't worthy of belief. They are responses to an argumentative challenge which cannot abide a question-begging answer. Reasons$_{non-a}$ are non-argumentative. They can be responses to mere expressions of curiosity. (Question: "How did you come to know that?" Answer: "I think I may have heard it on the news.")[16]

It is easy to see that, quite generally, reasons$_a$ are harder to come up with than reasons$_{non-a}$. Part of what explains this is that giving a reason$_{non-a}$ takes little more effort than taking the way out offered by "just". "I remember it" is hardly more reassuring than "Somehow I just know it", and not all more so than "I've forgotten how I know it". Giving reasons$_a$ is subject to a number of constraints, among which we should note the following.

- *time*: by and large it takes a good deal longer to make a case for P than to identify whatever it is that makes one's belief that P a reasonable state to be in (e.g., "I heard it somewhere").
- *dialectical effort*: case-making in adversarial *milieux* is subject to dialectical conventions, some of which are rather subtle and difficult to pick up on and deploy effectively.
- *search costs*: case-making is often inhibited by simple ignorance, by not commanding, then and there, the requisites of an effective response to adversarial challenge.
- *systematic inarticulacy*: most of what we know is subject to the vagaries of what is called *common knowledge*. Accordingly, to take just one example, nearly everyone in the requisite linguistic community knows that the difference between a sentence and a non-sentence, but hardly anyone can supply the applicable reason.

These and other considerations make it plain that

PROPOSITION 23 (Costs). *Giving reasons$_a$ for a belief that P carries costs to a degree that disqualifies it as a general condition on the reasonability of belief.*

[16] We note in passing that even challenges are not always met with reasons$_a$. Sometimes they give rise to *assurances*. A case in point: A challenger says "Surely it can't be true that Johan is an intuitionist "*Reply*: "Oh, I know very well that he is, I assure you". *Response*: "No kidding? Well, well!"

What then are we to make of these responses to the call for reasons? If they fail as answers to *adversarial* challenges, does this dispossess them of all case-making import? It would appear not. For in each case, what the respondent draws attention to is some or other aspect of what got him to believe that P in the first place. What the list has in common is the causalist's presumption that often it is in the manner in which one's b-state was induced that the rightness of being in it is constituted. So we may say that

PROPOSITION 24 (Causes as reasons). *For large classes of cases, citing the causal circumstances of a b-state with respect to P is a way of giving reasons for the reasonableness of being in that state.*

Why would this be so? Again, it arises from the dominant role played in our cognitive lives by *belief*. Perhaps it was Peirce who understood more than most philosophers, the extent to which we are in thrall to our beliefs. A belief, Peirce thought, is a kind of *force majeure*; it is something "*plus fort que moi*". In places, Peirce also speaks of the "Insistence of an Idea" [Peirce, 1992, pp. 121–122]. Beliefs are states of minds forced on by the operations of our cognitive devices. There is nothing discretionary about them. Our b-states are subject to neither freedom of the will nor democratic consensus. Beliefs are encumbrances which we take on passively. When in a b-state with respect to P, a cognitive agent is made to think he knows that P. Being in such states is not only something that he is causally responsive to, these are also states that bring him satisfaction. Although forced on him by his cognitive devices, the agent does not experience his b-states as intrusive or unwanted. It is a welcome servitude. Accordingly, the sense of rightness that attends the mere fact of belief is doubly occasioned; the agent is satisfied by his satisfaction *that* P. In having a b-state with respect to P, the agent experiences this as the right state to be in.

All of this is wholly compatible with the flux of belief, with the amazing dynamism that drives our doxastic lives. Most beliefs are short-lived, and anyone who is the slightest bit reflective knows it. Yet knowing it in no way attenuates the authority — or the sheer grip — of a belief for as long as it is held, however briefly. One of the extraordinary facts about the human believer is this bottomless facility for falling in and out of short-lived thralldoms. In this he might call to mind the fickle lover. But there is nothing facile about his doxastic transformations. By and large one's new b-state is the right state now to be in and one's old b-state is the right state no longer to be in. For do we not survive, prosper and from time to time build great civilizations?

Proposition (24) conceives of the human agent as a (largely tacit) *reifier* of the sense of reasonableness that attends his b-states. It ascribes an inference to the existence of reasons on the basis of those aspectual feelings of reasonableness involved in what the agent takes himself as knowing. It is important to see that Proposition (24) doesn't endorse this inference; it attributes it. All the same, it would be helpful if we were able to determine the extent to which this is a warranted thing for a cognitive agent to suppose. No one doubts that human knowers often are quite right to think that what they take as their k-states are supported by reasons. What is interesting about Proposition (24) is its scope. It attributes to the cognitive agent the quite *general* disposition to suppose that reasons exist for his k-states, and that this existence is something that can be plausibly inferred from the phenomenological reasonableness that attends the experience of (what he takes for) knowledge.

Pages ago we suggested that there might well be a non-case-making sense of "reasons" which is canonical for all knowledge. Proposition (24) records an inference to the existence of reasons attending b-states in a quite general way. What would justify such a suggestion? What would such reasons be like? There are two approaches we might take in answering these questions. The first is "the causal-normative approach", the other "the cognitive flourishing approach".

On the causal-normative approach, every b-state has a cause (or causes), and since b-states are experienced as the right states to be in, as states that are satisfying in the two senses we've already taken note of, then what causes b-states also causes these satisfactions. Accordingly,

PROPOSITION 25 (Reasons as normative causes). *It may be postulated that since such normative causes are indeed general concomitants of what agents take as their k-states, they may be designated as phenomenologically grounded approximations of the traditional idea of reasons as a general condition on knowledge.*

Of course, Proposition (25) gives a subjective interpretation of reasons. Subjectivity can't be helped. It is a byproduct of embubblement. Even so, something rather more objective-seeming arises from the cognitive flourishing approach. If we take cognitive flourishing as a given, as close-to-objective a fact about knowledge as can be got, then

PROPOSITION 26 (Reasons as causes of cognitive flourishing). *(Reasons as causes of cognitive flourishing) It may be postulated that in as much as our cognitive flourishing is a function of our collective and evolving web of belief, and in as much as every component of the web has a b-cause, such causes may be designated as intersubjectively grounded approximations of the traditional idea of reasons as a general condition on knowledge*

Proposition (26) embeds an abductive inference. It takes the form of an inference to the best explanation. Given the sheer extent of humanity's cognitive flourishing, given that this cognitive flourishing is at any time the net resultant of the *b*-states that have obtained until then, it may be inferred that by and large given our cognitive interests, those have been the right b-states to have been in. Accordingly, it may be taken as a default that my present *b*-state is a right state to be in. So might we not now say that

PROPOSITION 27 (B-states as reasons). *Just being in a b-state with respect to P is (defeasible) reason to think it a right state to be in.*

17 Knowledge of Knowledge

Like anyone else, an epistemologist might wish to know where he left his umbrella or how she is going to pay her taxes this year. In their professional or theoretical moments, philosophers seek a knowledge of knowledge. Given the way in which philosophers conceive of their theoretical tasks, the quest for a knowledge of knowledge is far from an ordinary thing. Philosophers have always had considerable difficulty in setting forth the attainment standards, and the methods for meeting them, that are embedded in this theoretical quest. But indisputably the dominant general view has been that the attainment standards for a philosopher's knowledge of knowledge are high in ways that might recall (myriad differences aside) the bar that the sciences have set for themselves. Suppose that we grant the dominant presumption that a philosophically successful quest for a knowledge of knowledge must meet the standards, whatever they are in fine, that qualify the result as a philosophical *theory*. In other words, epistemology holds *itself* to the *P*-model, which is a variant of the *CM*-model of knowledge. It will assist us if we label the knowledge that epistemologists seek a knowledge of as "knowledge_1" and the knowledge which, once acquired, achieves the epistemologist's desired result as "knowledge_2". We now have the means of making a simple point.

PROPOSITION 28 (False canonicity). *It is an epistemological mistake — and a bad one — to make the standards for the attainment of* $\textit{knowledge}_2$ *canonical for* $\textit{knowledge}_1$*, i.e., for knowledge in general. (See, again, the quotation from Dewey in an epigraph under this chapter's title.)*

COROLLARY 28 (a) *Accordingly, the P-model cannot be the right model for knowledge.*

Proposition (28) gives us a certain insight into the "third condition" on knowledge has been seriously misconceived by philosophers. There is no better example of this claim than the original presocratic founders of epistemology, for whom *episteme* = *endoxon* + *aletheia* + *logos*, i.e., true belief plus a third condition. If logos is the third condition, and if it is given anything like the presocratic meaning of "endorsed-by-theory", then *either* most of knowledge is not knowledge in fact (since most of knowledge is knowledge_1), *or* most of knowledge is indeed knowledge and the tripartite definition of knowledge fails for any such interpretation of the third condition (since knowledge_1 cannot be held to the standards of knowledge_2). Given the resolute determination of the cognitive agent to persist in his embrace of knowledge whenever he is, in the third-person sense, in the requisite state of belief, and given that satisfaction of the third condition on the present strict interpretations of it are nowhere close to being conditions on or causal requirements of an agent's being in a state of such belief, it can hardly be surprising that strict versions of the third conditions secure no footing in the phenomenology of cognition — that is to say, within epistemic bubbles.

18 Lessons For The New Logic

One of the questions that the logic of cognitive systems should keep its eye on is whether its findings necessitate a change in how the standard accounts deal with cognition on the ground. It is time to face that question. We begin with the Abundance Hypothesis. Although I have been assuming its truth in these pages, some consideration must be given to its possible falsity. One possibility, therefore, is that we know a lot less than we think we *do*. If this were so, we would be required to consider whether the devices of belief-fixation sanctioned in the going systems of belief-dynamics and various other sectors of the new logic would, as they now stand, have to be given up on, and with them the idea that cognitive rationality is implicitly defined by them. It is a good question but a bad idea. To see why, it is well to note that the Abundance Hypothesis is the conclusion of a backwards-chaining inference, as follows:

1. We survive and prosper, and sometimes build great civilizations.
2. This couldn't be so in the absence of cognitive abundance, that is, in conditions of massive ignorance.
3. So the Abundance Hypothesis is true.

The inference is defective. What counts for our survival, prosperity and so on is not what we know but what we do. It is true that much of what we do is rooted in what we think we know. It is also true that in the absence of those thinkings lots of doings would not be done. It may therefore be granted that necessary for our flourishing is that we be in states of mind not greatly unlike the states of mind that flourishing humans have always been in, and that conditions on belief-change and belief-update not deviate greatly in the general case from those that are currently in force.

Action is driven by thought before it is driven by knowledge. What I am here suggesting is that the actions required for survival, prosperity and so on owe nothing intrinsically to knowledge and everything to thought. What counts is the attachment of the behaviour that makes us flourish to what we think we know and to the conditions under which those states of mind arise. If this is right, then it is business as usual for the new logic.

PROPOSITION 29 (The unimportance of scant knowledge). *Let C be the contributions to the logic of cognitive systems currently in play under the assumption that the Abundance Hypothesis is true. Let it now be conceded that the hypothesis is not true. The integrity of C would be largely unaffected by this state of affairs.*

Proposition (2) considered this possibility right at the beginning. It raised the question of the value of knowledge to human flourishing. What we are now saying is that if our flourishing is abundant and our knowledge scant, then the value of knowledge is likewise scant. We should not think of our present Proposition (29) as occasion for complacency. The large sprawl of research programmes loosely collected under the name of the new logic has not yet produced much in the way of consensus about the detailed role of reasoning in the cognitive agendas of beings like us. Much of the new logic remains over-committed to the old logic, hardly a surprising happenstance given the depth and solidity of the achievements of mainstream logic. Another way of saying this is the presumed set C of theoretical insights into the logical structure of cognitive agency is the object of much vexation and dissensus. Proposition (29) is no vindication of C, but neither is it a condemnation of it. What Proposition (29) tells us is that even if the Abundance Hypothesis were false, C would not be significantly more vexed and dissensus-riven than it is at present.

This is far from saying that the falsity of the Abundance Hypothesis would leave C wholly unaffected. A case in point is the normativity of C. Let S be any principle in C whose normative force lies in its role in the facilitation of cognitive abundance. Its normativity would now need to be re-expressed in some such terms as:

COROLLARY 29 (a) *Any such S owes its normative force to the role it plays in fostering human flourishing.*

We may think that it is all fairly straightforward; that the Bubble Thesis upsets the Abundance Hypothesis, and does so without damage to our flourishing. This is too fast. The Bubble Thesis is rooted in the Asymmetry Thesis. The Asymmetry Thesis, if true, lays down some exceedingly important constraints on the acquisition of the states of mind necessary for our flourishing. A condition of their being had at all is that they be *experienced* as cases of knowing, hence as states of mind that satisfy the conditions of truth and reasonedness. There is something Kantian about this set-up. Notwithstanding that truth and reasonedness are fugitives — and knowledge too — it is not possible for us to be in the states of mind necessary for human flourishing unless the ways in which we proceed collectively presuppose that the Abundance Hypothesis is, in effect, *not* false. This bears a striking similarity to the fate of the Causal Abundance Hypothesis, as we might call it. Famously Hume drew our attention to the fact that, if true, its truth is not discernible to observation or reason. We might even suppose the worse, that it is totally false. Even so, our experience would not change. We would continue to experience the world causally. Doing so would be a condition of our having experiences at all.

If this Kantian point holds for *b*-states, if it is a condition on there being *b*-states that their objects be experienced as known, we have reason to suppose that it is a condition of there being *b*-states that the world be experienced as cognitively abundant, as if the Abundance Hypothesis were true. It would not be prudent for any research programme that sought to expose the links between reasoning and the possession of *b*-states to overlook the point.

We seek knowledge and settle for belief. We cannot do otherwise. Partly this is because we are can't-help-it realists. Partly it is because of the Asymmetry Thesis. Partly it is because we do not experience ourselves as embubbled. If we ceased being seekers after knowledge, if we ceased thinking that what is sought after is routinely attained, there would be no reason to suppose that we would attain those *b*-states with the timeliness and the frequency that is necessary to influence our actions flourishingly. If the Abundance Hypothesis is false, then thinking it true is an error. Even so, we have it as a further corollary of Proposition (29) that

COROLLARY 29 (b) *If it is an error to take the Abundance Hypothesis as true, it is a benign error, since it is an error indispensable to human flourishing.*

We saw at the beginning the logic of cognitive systems has a natural interest in two modes of enquiry. One we may think of as descriptive and the other as normative epistemology. Descriptive epistemology would detail the conditions under which human agents transact their cognitive tasks.

Normative epistemology would disclose the conditions under which cognitive practice on the hoof is rational. Descriptive psychology subsumes what I have called an agent's cognitive psychology. Cognitive psychology is a matter of how we see the mental states of cognitive agents operating within cognitive practice. But it also involves our taking note of how the cognizing agent conceives *of himself* in the transaction of his own cognitive — with his auto-psychology. Here, too, we meet with a fateful asymmetry. We ourselves may see the agent's success as a matter of acquiring the requisite beliefs independently of whether they are ever also knowledge. But the agent himself cannot see himself in such ways. This gives us reason to reinstate the Abundance Hypothesis, if not in the epistemology of cognitive agency, then in its cognitive auto-psychology. Needless to say, the Abundance Hypothesis presupposes an epistemology all of its own. As he sees himself, we may suppose that the agent on the ground implements that epistemology. To the extent to which it remains important to the logician of cognitive systems to take account of the cognitive agent's own psychological appreciation of himself, it is also important that he give attention to the details of the epistemology the agent sees himself as implementing. An obvious question would be the kind of epistemology an agent can *consistently* suppose himself to be running, both in the circumstances that we take him to be in and the circumstances that he takes himself to be in. Given that the agent on the ground is systematically deceived about the epistemic status of his own *b*-states in the here and now, and given the utter benigness of this self-deception, another hugely interesting requirement for the logician of cognitive systems is — beginning with the standard fallacies — to produce a general theory of *error* on which this fact is preserved and elucidated.[17]

[17]This is the task that Gabbay and I have set ourselves in [Gabbay & Woods, 2007]. Here is an indication of how that work is shaping up. It is widely held that a pattern of reasoning is fallacious when four conditions are met. (1) The reasoning is erroneous. (2) The reasoning is attractive; i.e., its erroneousness is unapparent. (3) The reasoning has universal appeal; i.e., it is widely employed. (4) The reasoning is incorrigible; i.e., levels of post-diagnostic recidivism are high. Let us call this the EAUI conception of fallacy (which may be pronounced "Yowee"). The EAUI conception has had a long history, originating with Aristotle. Fallacy theorists in this tradition have concentrated their attention on the first condition, and have tended to regard the others as more or less well-understood just as they stand. This is a regrettable turn of events. No account of fallacies can pretend to completeness as long as it leaves these three conditions in their present largely unexamined state. Worse still, an account of error in reasoning cannot proceed in a principled way without taking into account what the human reasoner's target is and what resources are available for its attainment. More particularly, the relevant account of error must disarm the objection that satisfaction of the last three conditions is reason to believe that the first condition is *not* met. For a piece of reasoning that is attractive, universal and incorrigible suggests that it is not an error.

If the Abundance Hypothesis is false, it is nevertheless remains possible that what is lost of abundance is compensated for by strategic success. In particular, there is reason to suppose that the following would be true:

PROPOSITION 30 (The enough principle). *Given that we survive, flourish and sometimes guild great civilizations, we are, in matters of what we think we know, right enough about enough of the right things enough of the time.*

But it must also be said that our prosperity and so on lies open to a less sunny supposition, in which there is a distinction between harmful and harmless error. Roughly speaking, a harmful error is one that impedes human flourishing. A harmless error is one that does not harm human flourishing. Accordingly, we might have it that

PROPOSITION 31 (The harmless principle). *Given that we survive, prosper and sometimes build great civilizations, we avoid harmful errors enough of the time about enough of the right things.*

Perhaps we can agree that the extent to which the Abundance Hypothesis fails is the extent to which something like the Harmless Principle is the minimal explanation of human flourishing. And, once again, the call goes out for a theory of error. Of course, the Abundance Hypothesis might not be false, as I have been assuming from the outset. What I have been saying of late is that, if false, it hardly matters that it is. What we must now reflect upon is, if true, whether *that* matters.

If the Abundance Hypothesis is true and the Bubble Thesis is also true, we achieve our cognitive luxuriance in the absence of the ability to recognize the presence of the knowledge on which it rests. How, then, as we asked in Proposition (3), can knowledge be a guide to life? It may have occurred to readers that here, too, we have an interesting question but a bad idea. Although fallibilism effectively guarantees that we make mistakes, it gives no support to the proposition that we err widely. The Abundance Hypothesis is wholly compatible with fallibilism. This matters. Knowledge is the error-free possession of *b*-states. If there is a great deal of knowledge, there is correspondingly not much error. The scantness of error is doubly

In the case of the more or less traditional list of fallacies — the Gang of Eighteen — the above pair of observations fall into an attractive kind of alignment. Arising from it is a certain model in which all four conditions manage to be satisfied. Thus a piece of reasoning is erroneous in relation to a target that embeds a standard that it fails to meet. It is attractive, universal and incorrigible in relation to a more modest target, embedding a lesser standard which the reasoning does meet. The reason *is* erroneous (in relation to the higher standard) and *looks* good because, in relation to the lower standard, it *is* good. Accordingly, the identity, and appropriateness, of a reasoner's target and its embedded standard precede any assessment of fallaciousness. And we must take seriously the possibility that the usual run of fallacies are errors only in relation to targets that reasoner's don't usually set for themselves.

aspected, if true. Not only would we make comparatively few errors, but a great many of those we do make would be corrected in a timely way. If all this were true, it would be occasion to call into question one of the presuppositions of Proposition (3). It is the idea that the Bubble Thesis makes knowledge unrecognizable. It may well be true that embubblement precludes the *perception* of the presence of knowledge in the here and now, but perhaps it doesn't preclude an *inference* to the same effect. Given that b-states are more often than not k-states, and given that the errors that attend the exceptions to it are more often than not detected and corrected, then that I am in a b-state with regard to P, concerning which no error has yet been discerned, there is defeasible reason to suppose that my b-state is a k-state with respect to P. If this inference holds up, then it serves as a recognition procedure for knowledge. It comes to this:

PROPOSITION 32 (Inferring the presence of knowledge). *If one thinks that one knows that P then, in the absence of particular indications that thinking so is an error, one may defeasibly conclude that P is an object of one's knowledge, and recognizable as such.*

It won't work. It is a condition of my being in a b-state with regard to P is that there is nothing in particular to indicate that I am in error in thinking that I am in a state of knowledge with regard to it. That being so, every b-state would count defeasibly as a state of knowledge. Of course this is precisely the case from the agent's first-person perspective. From that perspective, the agent's standing position is to take any b-state for a k-state and then wait to see what happens. What he is not able to do on the spot is distinguish those b-state that are from the b-states that aren't k-states. This is the sense in which knowledge is unrecognizable. What is more, if the Bubble Thesis is also true, our future corrections of these inapparent errors (if the redundancy might be forgiven) is by way of subsequent b-states subject to the same indeterminacy.

This in no way discredits the rule licensing the claim that in the general case it is reasonable to take what we think we know as knowledge. It is one thing to infer the presence of knowledge. It is another thing entirely to discern its presence. If the rule has any legitimacy, it delivers the goods in the first instance, and does nothing of the kind in the second.

The Bubble Thesis likewise makes error unrecognizable, never mind, that given the Abundance Hypothesis, we might well be justified in taking the defeasible position that for the most part my identification of b-states with k-states is not in error. It is one thing to infer the absence of error. It is another thing entirely to discern its absence (or its presence). If this is right, it is indeed terrible news for fallibilism. Fallabilism postulates — indeed it pivots on — the detectability and corrigibility of error. But you cannot detect what you cannot recognize, and you cannot correct what you cannot detect.

This is a deeply consequential development. If the actual empirical record is anything to go on, with regard to most of his b-states an agent's cognitive behaviour embeds the presumption that fallibilism is true. It is part of the cognitive auto-psychology of beings like us that, in regard to most things, error is a theoretical possibility, and in some much slighter degree an actuality. It is also part of this psychology that it lies in the nature of error, and of us, that error is undetectable when committed, but both detectable and corrigible after the fact. It may also be said that these are assumptions that leave a large footprint on cognitive practice, and that in their absence we would operate our cognitive lives very differently, differently enough to put in jeopardy our human flourishing. In their absence what earthly purpose would be served by our feedback mechanisms?

Perhaps it is not altogether surprising that we now find ourselves faced with an earlier asymmetry. From our theoretical perch, we see the cognitive agent operating with presumptions that aren't available to him. From the agent's first-person point of view he is implementing fallibilist epistemology. But the epistemology presupposed by the agent's auto-cognitive psychology, cannot be validly attributed to him by the third-person theorist. From the first-person perspective, the agent has no option but to proceed fallibistically. From the theorist's third-person perspective, the agent is kidding himself.

As before, the present situation imposes on the logician of cognitive systems. As before, he has occasion to flesh out the epistemology that deceptively underlies the agent's conception of what he is up to on the ground, as well as the quite different epistemology that the other-observer has occasion to describe. Both are necessary, since it is fundamental to the discharge of the second task that agents on the ground take themselves as implementing the imperatives of the first task. Important as these projects are, nothing matches the importance of accounting for error if error is undetectable. Two and a half-millenis ago Aristotle made the crushing discovery that the mechanism for the detection of error (viz. syllogisms) was itself subject to the same virus. Not everything that looks like a syllogism is a syllogism. Left untreated, this leaves judgements of syllogisity open to error. One of the great achievements in the history of logic was Aristotle's almost successful proof of the perfectability of syllogisms, by which it could be determined finitely and in quite determinate ways whether an argument was in fact a syllogism. No logician since has shown such sensitivity to the importance of error in logic. The modern logician would do well to show it a like sensitivity. If errors are undetectable, what would a coherent theory of error look like?

Well, which is it? Is the Abundance Hypothesis true or false? Perhaps we may now say that

PROPOSITION 33 (Useless or unusable). *If the Bubble Thesis is true, nothing that is so far known about the structure of cognition answers that question. If it is false, it doesn't matter; we continue to flourish. If it is true, it is not a usable truth, since knowledge is unrecognizable.*

I said at the beginning that I would stand by the Abundance Hypothesis. Nothing we have unearthed in these pages requires that it be given up on. Even so, if, as I believe, it doesn't much matter what its truth value is, this is something for the logician of cognitive agency to take principled and systematic note of.[18]

BIBLIOGRAPHY

[Adler, 2002] Jonathan Adler. *Belief's Own Ethics*. MIT Press, Cambridge MA, 2002.

[Alchourron *et al.*, 1985] C.A. Alchourron, P.G. Gärdenfors and D. Makinson. "On the logic of theory change: partial meet, contraction and revision functions". *The Journal of Symbolic Logic*, 50:510–530, 1985.

[Barwise & Perry, 1983] Jon Barwise and John Perry. *Situations and Attitudes*. MIT Press, Cambridge MA, 1983.

[Barth & Krabbe, 1982] E.M. Barth and E.C.W. Krabbe. *From Axiom to Dialogue*. de Gruyter, Berlin and New York, 1982.

[Carlson & Pelletier, 1995] Gregory N. Carlson and Francis Jeffry Pelletier, eds. *The Generic Book*. Chicago University Press, Chicago IL, 1995.

[Carnap, 1950] Rudolf Carnap. *The Logical Foundations of Probability*. University of Chicago Press, Chicago IL, 1950.

[Chellas, 1980] Brian Chellas. *Modal Logic: An Introduction*. Cambridge University Press, Cambridge, 1980.

[Cohen, 1992] L. Jonathan Cohen. *An Essay on Belief and Acceptance*. Clarendon Press, Oxford, 1992.

[Davidson, 1963] Donald Davidson. "Actions, reasons and causes". *Journal of Philosophy*, 60:685–700, 1963.

[Dretske, 1983] Fred Dretske. *Knowledge and the Flow of Information*. MIT Press, Cambridge MA, 1983.

[Freeman, 1991] James B. Freeman. *Dialectics and the Microstructure of Argument*. Foris, Dordrecht, 1991.

[Gabbay, 1976] Dov M. Gabbay. *Investigations in Modal and Tense Logics with Applications*. Reidel, Dordrecht and Boston, 1976.

[18] For trenchant criticisms of and helpful suggestions regarding an earlier draft, I am indebted to Jonathan Adler, Robert Pinto and Harvey Siegel. I also thank Peter Bruza, David Hitchcock, Theo Kuipers, Luis Lamb, Lorenzo Magnani, Kent Peacock and Bas van Fraassen for stimulating and provocative conversations about various issues discussed here, and Carol Woods and Dawn Collins for technical assistance. I would also like to salute my friend and collaborator Dov Gabbay on his 60th birthday, to thank him for his richly important contributions to logic, and to express keen anticipation of the splendid work yet to come.

[Gabbay *et al.*, 1994] Dov M. Gabbay, I. Hodkinson and M. Reynolds. *Temporal Logic: Mathematical Foundation and Computational Aspects*, vol. 1. Oxford University Press, Oxford, 1994.

[Gabbay *et al.*, 2002] Dov M. Gabbay, Odinaldo Roderigues and John Woods. "Belief contraction, anti-formulae and resource-overdraft: Part I Deletion in resource bounded logics". *Logic Journal of the IGPL*, 10:601–652, 2002.

[Gabbay & Woods, 2003a] Dov M. Gabbay and John Woods. *Agenda Relevance: A Study in Formal Pragmatics*. Elsevier/North-Holland, Amsterdam, 2003.

[Gabbay & Woods, 2003b] Dov M. Gabbay and John Woods. "Normative models of rationality: The disutility of some approaches". *Logic Journal of IGPL*, 11:597–613, 2003.

[Gabbay *et al.*, 2004a] Dov M. Gabbay, Odinaldo Roderigues and John Woods. "Belief contraction, anti-formulae and resource-overdraft: Part II Deletion in resource unbounded logics". In Shahid Rahman, John Symons, Dov M. Gabbay and Jean Paul van Bendegem (eds.), *Logic Epistemology and the Unity of Science*, pp. 291–326. Kluwer, Dordrecht and Boston, 2004.

[Gabbay *et al.*, 2004b] Dov M. Gabbay, Gabriella Pigozzi and John Woods. "Controlled revision: A preliminary account". *Journal of Logic and Computation*, 13:5–27, 2004.

[Gabbay & Woods, 2001a] Dov M. Gabbay and John Woods. "Non-cooperation in dialogue logic: Getting beyond the goody two-shoes model of argument". *Synthese*, 127:161–186, 2001.

[Gabbay & Woods, 2001b] Dov M. Gabbay and John Woods. "More on non-cooperation in dialogue logic". *Logic Journal of the IGPL*, 9:321–339, 2001.

[Gabbay & Woods, 2005a] Dov M. Gabbay and John Woods. "Fallacies as cognitive virtues". In Ahti-Veikko Pietarinen (ed.), *Logic, Games and Philosophy: Foundational Perspectives*. Springer, Amsterdam, 2005. To appear.

[Gabbay & Woods, 2005b] Dov M. Gabbay and John Woods. *The Reach of Abduction: Insight and Trial*. Vol. 2 of *A Practical Logic of Cognitive Systems*. Elsevier/North-Holland, Amsterdam, 2005.

[Gabbay & Woods, 2005c] Dov M. Gabbay and John Woods. *The Practical Turn in Logic*. In Dov M. Gabbay and F. Guenthner (eds.), *Handbook of Philosophical Logic*. Kluwer, Dordrecht and Boston, 2005. 2nd edition.

[Gabbay & Woods, 2007] Dov M. Gabbay and John Woods. *Seductions and Shortcuts: Fallacies in the Cognitive Economy*. Vol. 3 of *A Practical Logic of Cognitive Systems*. Elsevier, Amsterdam, 2007. To appear.

[Gigerenzer & Selton, 2001] G. Gigerenzer and R. Selten. "Rethinking rationality". In G. Gigerenzer and R. Selten (eds.), *Bounded Rationality: The Adaptive Toolbox*, pp. 1–12. MIT Press, Cambridge MA, 2001.

[Gochet, 2002] P. Gochet. "The dynamic turn in twentieth century logic". *Synthese*, 130:175–184, 2002.

[Gochet & Gribomont, 2005] P. Gochet and Pascal Gribomont. "Epistemic logic". In Dov M. Gabbay and John Woods (eds.), *Logic and Modalities in the Twentieth Century*, vol. 6 of the *Handbook of the History of Logic*, Elsevier/North-Holland, Amsterdam, 2005. To appear.

[Hamblin, 1970] C.L. Hamblin. *Fallacies*. Methuen, London, 1970.

[Hansen & Pinto, 1995] Hans V. Hansen and Robert C. Pinto, eds. *Fallacies: Classical and Contemporary Readings*, Pennsylvania State University Press, University Park PA, 1995.

[Harel, 1979] D. Harel. *First-Order Dynamic Logic*. Springer-Verlag, Berlin, 1979.

[Hilpinen, 1981] Risto Hilpinen, ed. *Deontic Logic: Introductory and Systematic Readings*. Reidel, Dordrecht, 1981.

[Hintikka:1962] Jaakko Hintikka. *Knowledge and Belief*. Cornell University Press, Ithaca NY, 1962.

[Hintikka, 1981] Jaakko Hintikka. *Modern Logic — A Survey*. Reidel, Boston, 1981.

[Hintikka & Sandu, 1997] Jaakko Hintikka and Gabriel Sandu. "Gametheoretical semantics". In Johan van Benthem and Alice ter Meulen (eds.), *Handbook of Logic and Language*, pp. 361–410. Elsevier and MIT Press, Amsterdam, New York and Cambridge MA, 1997. New edition.

[Horgan & Tienson, 1999] T. Horgan and J. Tienson. "Authors' replies". *Acta Analytica*, 22:275–287, 1999.

[Honderich, 2005] Ted Honderich, ed. *The Oxford Companion to Philosophy*. Oxford University Press, Oxford, 2005. New edition.

[Johnson, 1996] Ralph H. Johnson. *The Rise of Informal Logic*. Vale Press, Newport News VA, 1996

[Johnson, 2000] Ralph H. Johnson. *Manifest Rationality: A Pragmatic Theory of Argument*. Lawrence Erlbaum Associates, London, 2000.

[Kowalski, 1979] R.A. Kowalski. *Logic for Problem Solving*. Elsevier, New York, 1979.

[Kripke, 1963] Saul A. Kripke. "Semantical considerations on modal logic". *Acta Philosophica Fennica*, 16:83–94, 1963.

[Lenzen, 1978] Wolfgang Lenzen. "Recent work in epistemic logic". *Acta Philosophica Fennica*, 30:1–137, 1978.

[Lewis, 1983] David Lewis. *Philosophical Papers, Vol. 1*, Oxford University Press, Oxford, 1983.

[Lewis, 1996] David Lewis. "Elusive knowledge". *Australasian Journal of Philosophy*, 74:549–567, 1996.

[MacKenzie, 1990] J. MacKenzie. "Four dialogue systems". *Studia Logica*, XLIX:567–583, 1990.

[McCarthy, 1980] J. McCarthy. "Circumscription — a form of non-monotonic reasoning". *Artificial Intelligence*, 13:27–39, 1980.

[Magnani, 2001] Lorenzo Magnani. *Abduction, Reason and Science: Processes of Discovery and Explanation*. Kluwer and Plenum, New York, 2001.

[Moore, 1985] R. Moore. "Semantical considerations on non-monotonic logics". *Artificial Intelligence*, 25:75–94, 1985.

[Nozick, 1993] Robert Nozick. *The Nature of Rationality*. Princeton University Press, Princeton NJ, 1993.

[O'Connor, 2001] T. O'Connor. *Persons and Causes*. Oxford University Press, Oxford, 2001.

[Peirce, 1992] C.S. Peirce. *Reasoning and the Logic of Things: The 1898 Cambridge Conference's Lectures by Charles Sanders Peirce*. Harvard University Press, Cambridge MA, 1992. Edited by Kenneth Laine Ketner.

[Pereira, 2002] Louis Moniz Pereira. "Philosophical incidence of logic programming". In Dov M. Gabbay, Ralph H. Johnson, Hans Jürgen Ohlbach and John Woods, (eds.), *Handbook of the Logic of Argument and Inference: The Turn Towards the Practical*, vol. 1, pp. 421–444. North-Holland, Amsterdam, 2002.

[Pietroski, 2000] P.M. Pietroski. *Causing Actions*. Oxford University Press, Oxford, 2000.

[Pinto, 2003] Robert Pinto. "The uses of Argument in Communicative Contexts". In Blair, Farr, Hansen, Johnson and Tindal (eds.), *Informal Logic at 25: Proceedings of the Windsor Conference*. CD-ROM. Ontario Society for the Study of Argumentation, 2003.

[Prior, 1967] A.N. Prior. *Past, Present and Future*. Oxford University Press, Oxford, 1967.

[Reiter, 1980] Raymond Reiter. "A Logic for default reasoning". *Artificial Intelligence*, 12:81–132, 1980.

[Rozeboom, 1967] William Rozeboom. "Why I know so much more than you do". *American Philosophical Quarterly*, 4:281–290, 1967.

[Russell,1903] Bertrand Russell. *The Principles of Mathematics*. George Allen and Unwin, London, 1903. Second Edition 1937.

[Sandewall, 1972] E. Sandewall. "An approach to the frame problem and its implementation". *Machine Intelligence.* Vol. 7, Edinburgh University Press, Edinburgh, 1972.
[Schlecta, 2004] Karl Schlecta. *Coherent Systems.* Elsevier, Amsterdam, 2004.
[Simon, 1973] Herbert Simon. "The structure of ill-structured problems". *Artificial Intelligence*, 4(1973):181–202.
[Sorensen, 1988] Roy A. Sorensen, *Blindspots*, Clarendon Press, Oxford, 1988.
[Stalnaker, 1984] Robert Stalnaker. *Inquiry.* MIT Press, Cambridge, MA, 1984.
[Van Benthem, 1983] Johan van Benthem. *The Logic of Time.* Reidel, Dordrecht, 1983.
[Van Benthem, 1996] Johan van Benthem. *Exploring Logical Dynamics.* CSLI Publications, Stanford CA, 1996.
[VanFraassen, 1980] Bas C. van Fraassen. *The Scientific Image.* Clarendon Press, Oxford, 1980.
[Von Wright, 1951] Georg von Wright. *An Essay in Modal Logic.* North-Holland, Amsterdam, 1951.
[Walton & Krabbe, 1995] Douglas Walton and E.C.W. Krabbe. *Commitment in Dialogue.* SUNY Press, Albany NY, 1995.
[Walton, 1995] Douglas Walton. *A Pragmatic Theory of Fallacy.* University of Alabama Press, Tuscaloosa AL, 1995.
[Williamson, 2002] Jon Williamson. "Probability logic". In Dov M. Gabbay, Ralph Johnson, Hans J. Ohlbach and John Woods (eds.), *Handbook of the Logic of Argument and Inference: The Turn Towards the Practical*, pp. 397–424. North-Holland, Amsterdam, 2002.
[Woods & Walton, 1989] John Woods and Douglas Walton. *Fallacies: Selected Papers 1972–1982.* Foris de Gruyter, Berlin and New York, 1989.
[Woods, 2000] John Woods. "Speaking your mind: Inarticulacies constitutional and circumstantial". *Argumentation*, 16:59–78, 2000.
[Woods, 2001] John Woods. *Aristotle's Earlier Logic.* Hermes Science Publications, Oxford, 2001.
[Woods, 2003] John Woods. *Paradox and Paraconsistency: Conflict Resolution in the Abstract Sciences.* Cambridge University Press, Cambridge and New York, 2003.
[Woods, 2004] John Woods. *The Death of Argument: Fallacies in Agent-Based Reasoning.* Kluwer, Dordrecht and Boston, 2004.
[Woods, 2005a] John Woods. "Cognitive yearning and fugitive truth". In Kent Peacock and Andrew Irvine, (eds.), *Mistakes of Reason: Essays in Honour of John Woods.* University of Toronto Press, Toronto, 2005. To appear.
[Woods, 2005b] John Woods. "Eight theses reflecting on Stephen Toulmin". In David Hitchcock (ed.), *The Uses of Argument*, pp. 495–504. Ontario Society for the Study of Argumentation, St. Catharines ON, 2005.
[Woods, 2006] John Woods, . "Logical error". Forthcoming in 2006.

INDEX